INTRODUCTION TO PROBABILITY AND MATHEMATICAL STATISTICS

SECOND EDITION

Lee J. Bain
University of Missouri—Rolla

Max Engelhardt
University of Idaho

BROOKS/COLE
CENGAGE Learning™

Australia • Brazil • Japan • Korea • Mexico • Singapore • Spain • United Kingdom • United States

BROOKS/COLE
CENGAGE Learning™

Introduction to Probability and Mathematical Statistics, Second Edition
Lee J. Bain and Max Engelhardt

Library of Congress Control Number: 91-25923

ISBN-13: 978-0-534-38020-5

ISBN-10: 0-534-38020-4

Brooks/Cole
10 Davis Drive
Belmont, CA 94002-3098
USA

Cengage Learning is a leading provider of customized learning solutions with office locations around the globe, including Singapore, the United Kingdom, Australia, Mexico, Brazil, and Japan. Locate your local office at: **www.cengage.com/global**

Cengage Learning products are represented in Canada by Nelson Education, Ltd.

To learn more about Brooks/Cole, visit **www.cengage.com/brookscole**

Purchase any of our products at your local college store or at our preferred online store **www.cengagebrain.com**

Printed in the United States of America
12 13 14 15 16 16 15 14 13 12
FD303

Special Continuous Distributions

Notation and Parameters	Continuous pdf $f(x)$	Mean	Variance	MGF $M_X(t)$
Weibull				
$X \sim \text{WEI}(\theta, \beta)$	$\frac{\beta}{\theta^\beta} x^{\beta - 1} e^{-(x/\theta)^\beta}$	$\theta\Gamma\left(1 + \frac{1}{\beta}\right)$	$\theta^2\left[\Gamma\left(1 + \frac{2}{\beta}\right) - \Gamma^2\left(1 + \frac{1}{\beta}\right)\right]$	*
$0 < \theta$ $0 < \beta$	$0 < x$			
Extreme Value				
$X \sim \text{EV}(\theta, \eta)$	$\frac{1}{\theta}\exp\{[(x-\eta)/\theta] - \exp[(x-\eta)/\theta]\}$	$\eta - \gamma\theta$	$\frac{\pi^2\theta^2}{6}$	$e^{\eta t}\Gamma(1 + \theta t)$
$0 < \theta$		$\gamma \doteq 0.5772$ (Euler's const.)		
Cauchy				
$X \sim \text{CAU}(\theta, \eta)$	$\frac{1}{\theta\pi\{1 + [(x - \eta)/\theta]^2\}}$	**	**	**
$0 < \theta$				
Pareto				
$X \sim \text{PAR}(\theta, \kappa)$	$\frac{\kappa}{\theta(1 + x/\theta)^{\kappa + 1}}$	$\frac{\theta}{\kappa - 1}$	$\frac{\theta^2\kappa}{(\kappa - 2)(\kappa - 1)^2}$	**
$0 < \theta$ $0 < \kappa$	$0 < x$	$1 < \kappa$	$2 < \kappa$	
Chi-Square				
$X \sim \chi^2(\nu)$	$\frac{1}{2^{\nu/2}\Gamma(\nu/2)} x^{\nu/2 - 1} e^{-x/2}$	ν	2ν	$\left(\frac{1}{1 - 2t}\right)^{\nu/2}$
$\nu = 1, 2, \ldots$	$0 < x$			

Special Continuous Distributions

Notation and Parameters	Continuous pdf $f(x)$	Mean	Variance	MGF $M_X(t)$
Student's t				
$X \sim t(\nu)$	$\dfrac{\Gamma\left(\frac{\nu+1}{2}\right)}{\Gamma\left(\frac{\nu}{2}\right)} \dfrac{1}{\sqrt{\nu\pi}} \left(1 + \dfrac{x^2}{\nu}\right)^{-\frac{\nu+1}{2}}$	0	$\dfrac{\nu}{\nu - 2}$	**
$\nu = 1, 2, \ldots$		$1 < \nu$	$2 < \nu$	
Snedecor's F				
$X \sim F(\nu_1, \nu_2)$	$\dfrac{\Gamma\left(\frac{\nu_1+\nu_2}{2}\right)}{\Gamma\left(\frac{\nu_1}{2}\right)\Gamma\left(\frac{\nu_2}{2}\right)} \left(\dfrac{\nu_1}{\nu_2}\right)^{\frac{\nu_1}{2}} x^{\frac{\nu_1}{2} - 1}$	$\dfrac{\nu_2}{\nu_2 - 2}$	$\dfrac{2\nu_2^2(\nu_1 + \nu_2 - 2)}{\nu_1(\nu_2 - 2)^2(\nu_2 - 4)}$	**
$\nu_1 = 1, 2, \ldots$ $\nu_2 = 1, 2, \ldots$	$\times \left(1 + \dfrac{\nu_1}{\nu_2} x\right)^{-\frac{\nu_1+\nu_2}{2}}$	$2 < \nu_2$	$4 < \nu_2$	
Beta				
$X \sim \text{BETA}(a, b)$	$\dfrac{\Gamma(a+b)}{\Gamma(a)\Gamma(b)} x^{a-1}(1-x)^{b-1}$	$\dfrac{a}{a+b}$	$\dfrac{ab}{(a+b+1)(a+b)^2}$	*
$0 < a$ $0 < b$	$0 < x < 1$			

*Not tractable.
**Does not exist.

CONTENTS

CHAPTER 3

SPECIAL PROBABILITY DISTRIBUTIONS 90

CHAPTER 4

JOINT DISTRIBUTIONS 136

CHAPTER 5

PROPERTIES OF RANDOM VARIABLES 171

CHAPTER 6

FUNCTIONS OF RANDOM VARIABLES 193

CHAPTER 7

LIMITING DISTRIBUTIONS 231

CHAPTER 8

STATISTICS AND SAMPLING DISTRIBUTIONS 263

* Advanced (or optional) topics

CHAPTER 9

POINT ESTIMATION 288

CHAPTER 10

SUFFICIENCY AND COMPLETENESS 335

CHAPTER 11

INTERVAL ESTIMATION 358

CHAPTER 12

TESTS OF HYPOTHESES 389

CHAPTER 13

CONTINGENCY TABLES AND GOODNESS-OF-FIT 442

CHAPTER 14

NONPARAMETRIC METHODS 468

* Advanced (or optional) topics

PREFACE

This book provides an introduction to probability and mathematical statistics. Although the primary focus of the book is on a mathematical development of the subject, we also have included numerous examples and exercises that are oriented toward applications. We have attempted to achieve a level of presentation that is appropriate for senior-level undergraduates and beginning graduate students.

The second edition involves several major changes, many of which were suggested by reviewers and users of the first edition. Chapter 2 now is devoted to general properties of random variables and their distributions. The chapter now includes moments and moment generating functions, which occurred somewhat later in the first edition. Special distributions have been placed in Chapter 3. Chapter 8 is completely changed. It now considers sampling distributions and some basic properties of statistics. Chapter 15 is also new. It deals with regression and related aspects of linear models.

As with the first edition, the only prerequisite for covering the basic material is calculus, with the lone exception of the material on general linear models in Section 15.4; this assumes some familiarity with matrices. This material can be omitted if so desired.

Our intent was to produce a book that could be used as a textbook for a two-semester sequence in which the first semester is devoted to probability concepts and the second covers mathematical statistics. Chapters 1 through 7 include topics that usually are covered in a one-semester introductory course in probability, while Chapters 8 through 12 contain standard topics in mathematical statistics. Chapters 13 and 14 deal with goodness-of-fit and nonparametric statistics. These chapters tend to be more methods-oriented. Chapters 15 and 16 cover material in regression and reliability, and these would be considered as optional or special topics. In any event, judgment undoubtedly will be required in the choice of topics covered or the amount of time allotted to topics if the desired material is to be completed in a two-semester course.

It is our hope that those who use the book will find it both interesting and informative.

ACKNOWLEDGMENTS

We gratefully acknowledge the numerous suggestions provided by the following reviewers:

Dean H. Fearn
California State University–Hayward

Joseph Glaz
University of Connecticut

Victor Goodman
Rensselaer Polytechnic Institute

Shu-ping C. Hodgson
Central Michigan University

Robert A. Hultquist
Pennsylvania State University

Alan M. Johnson
University of Arkansas, Little Rock

Benny P. Lo
Ohlone College

D. Ramachandran
California State University–Sacramento

Douglas A. Wolfe
Ohio State University

Linda J. Young
Oklahoma State University

Thanks also are due to the following users of the first edition who were kind enough to relate their experiences to the authors: H. A. David, Iowa State University; Peter Griffin, California State University–Sacramento.

Finally, special thanks are due for the moral support of our wives, Harriet Bain and Linda Engelhardt.

Lee J. Bain
Max Engelhardt

CHAPTER 1

PROBABILITY

1.1 INTRODUCTION

In any scientific study of a physical phenomenon, it is desirable to have a mathematical model that makes it possible to describe or predict the observed value of some characteristic of interest. As an example, consider the velocity of a falling body after a certain length of time, t. The formula $v = gt$, where $g \doteq 32.17$ feet per second per second, provides a useful mathematical model for the velocity, in feet per second, of a body falling from rest in a vacuum. This is an example of a **deterministic model**. For such a model, carrying out repeated experiments under ideal conditions would result in essentially the same velocity each time, and this would be predicted by the model. On the other hand, such a model may not be adequate when the experiments are carried out under less than ideal conditions. There may be unknown or uncontrolled variables, such as air temperature or humidity, that might affect the outcome, as well as measurement error or other factors that might cause the results to vary on different performances of the experiment. Furthermore, we may not have sufficient knowledge to derive a more complicated model that could account for all causes of variation.

There are also other types of phenomena in which different results may naturally occur by chance, and for which a deterministic model would not be appropriate. For example, an experiment may consist of observing the number of particles emitted by a radioactive source, the time until failure of a manufactured component, or the outcome of a game of chance.

The motivation for the study of probability is to provide mathematical models for such nondeterministic situations; the corresponding mathematical models will be called **probability models** (or **probabilistic models**). The term **stochastic**, which is derived from the Greek word *stochos*, meaning "guess," is sometimes used instead of the term *probabilistic*.

A careful study of probability models requires some familiarity with the notation and terminology of set theory. We will assume that the reader has some knowledge of sets, but for convenience we have included a review of the basic ideas of set theory in Appendix A.

1.2

NOTATION AND TERMINOLOGY

The term **experiment** refers to the process of obtaining an observed result of some phenomenon. A performance of an experiment is called a **trial** of the experiment, and an observed result is called an **outcome**. This terminology is rather general, and it could pertain to such diverse activities as scientific experiments or games of chance. Our primary interest will be in situations where there is uncertainty about which outcome will occur when the experiment is performed. We will assume that an experiment is repeatable under essentially the same conditions, and that the set of all possible outcomes can be completely specified before experimentation.

Definition 1.2.1

The set of all possible outcomes of an experiment is called the **sample space**, denoted by S.

Note that one and only one of the possible outcomes will occur on any given trial of the experiments.

Example 1.2.1 An experiment consists of tossing two coins, and the observed face of each coin is of interest. The set of possible outcomes may be represented by the sample space

$$S = \{\text{HH, HT, TH, TT}\}$$

which simply lists all possible pairings of the symbols H (heads) and T (tails). An alternate way of representing such a sample space is to list all possible ordered pairs of the numbers 1 and 0, $S = \{(1, 1), (1, 0), (0, 1), (0, 0)\}$, where, for example, (1, 0) indicates that the first coin landed heads up and the second coin landed tails up.

Example 1.2.2 Suppose that in Example 1.2.1 we were not interested in the individual outcomes of the coins, but only in the total number of heads obtained from the two coins. An appropriate sample space could then be written as $S^* = \{0, 1, 2\}$. Thus, different sample spaces may be appropriate for the same experiment, depending on the characteristic of interest.

Example 1.2.3 If a coin is tossed repeatedly until a head occurs, then the natural sample space is $S = \{\text{H}, \text{TH}, \text{TTH}, \ldots\}$. If one is interested in the number of tosses required to obtain a head, then a possible sample space for this experiment would be the set of all positive integers, $S^* = \{1, 2, 3, \ldots\}$, and the outcomes would correspond directly to the number of tosses required to obtain the first head. We will show in the next chapter that an outcome corresponding to a sequence of tosses in which a head is never obtained need not be included in the sample space.

Example 1.2.4 A light bulb is placed in service and the time of operation until it burns out is measured. At least conceptually, the sample space for this experiment can be taken to be the set of nonnegative real numbers, $S = \{t \mid 0 \leqslant t < \infty\}$.

Note that if the actual failure time could be measured only to the nearest hour, then the sample space for the actual observed failure time would be the set of nonnegative integers, $S^* = \{0, 1, 2, 3, \ldots\}$. Even though S^* may be the observable sample space, one might prefer to describe the properties and behavior of light bulbs in terms of the conceptual sample space S. In cases of this type, the discreteness imposed by measurement limitations is sufficiently negligible that it can be ignored, and both the measured response and the conceptual response can be discussed relative to the conceptual sample space S.

A sample space S is said to be **finite** if it consists of a finite number of outcomes, say $S = \{e_1, e_2, \ldots, e_N\}$, and it is said to be **countably infinite** if its outcomes can be put into a one-to-one correspondence with the positive integers, say $S = \{e_1, e_2, \ldots\}$.

Definition 1.2.2

If a sample space S is either finite or countably infinite, then it is called a **discrete sample space**.

A set that is either finite or countably infinite also is said to be **countable**. This is the case in the first three examples. It is also true for the last example when failure times are recorded to the nearest hour, but not for the conceptual sample space. Because the conceptual space involves outcomes that may assume any value in some interval of real numbers (i.e., the set of nonnegative real numbers), it could be termed a **continuous sample space**, and it provides an example where a discrete sample space is not an appropriate model. Other, more complicated experiments exist, the sample spaces of which also could be characterized as continuous, such as experiments involving two or more continuous responses.

Example 1.2.5 Suppose a heat lamp is tested and X, the amount of light produced (in lumens), and Y, the amount of heat energy (in joules), are measured. An appropriate sample space would be the Cartesian product of the set of all nonnegative real numbers with itself,

$$S = [0, \infty) \times [0, \infty) = \{(x, y) \mid 0 \leqslant x < \infty \text{ and } 0 \leqslant y < \infty\}$$

Each variable would be capable of assuming any value in some subinterval of $[0, \infty)$.

Sometimes it is possible to determine bounds on such physical variables, but often it is more convenient to consider a conceptual model in which the variables are not bounded. If the likelihood of the variables in the conceptual model exceeding such bounds is negligible, then there is no practical difficulty in using the conceptual model.

Example 1.2.6 A thermograph is a machine that records temperature continuously by tracing a graph on a roll of paper as it moves through the machine. A thermographic recording is made during a 24-hour period. The observed result is the graph of a continuous real-valued function $f(t)$ defined on the time interval $[0, 24] = \{t \mid 0 \leqslant t \leqslant 24\}$, and an appropriate sample space would be a collection of such functions.

Definition 1.2.3

An **event** is a subset of the sample space S. If A is an event, then A has **occurred** if it contains the outcome that occurred.

To illustrate this concept, consider Example 1.2.1. The subset

$$A = \{\text{HH, HT, TH}\}$$

contains the outcomes that correspond to the event of obtaining "at least one head." As mentioned earlier, if one of the outcomes in A occurs, then we say that the event A has occurred. Similarly, if one of the outcomes in $B = \{\text{HT, TH, TT}\}$ occurs, then we say that the event "at least one tail" has occurred.

Set notation and terminology provide a useful framework for describing the possible outcomes and related physical events that may be of interest in an experiment. As suggested above, a subset of outcomes corresponds to a physical event, and the event or the subset is said to occur if any outcome in the subset occurs. The usual set operations of union, intersection, and complement provide a way of expressing new events in terms of events that already have been defined. For example, the event C of obtaining "at least one head and at least one tail" can be expressed as the intersection of A and B, $C = A \cap B = \{\text{HT, TH}\}$. Similarly, the event "at least one head or at least one tail" can be expressed as the union $A \cup B = \{\text{HH, HT, TH, TT}\}$, and the event "no heads" can be expressed as the complement of A relative to S, $A' = \{\text{TT}\}$.

A review of set notation and terminology is given in Appendix A.

In general, suppose S is the sample space for some experiments, and that A and B are events. The intersection $A \cap B$ represents the outcomes of the event "A and B," while the union $A \cup B$ represents the event "A or B." The complement A' corresponds to the event "not A." Other events also can be represented in terms of intersections, unions, and complements. For example, the event "A but not B" is said to occur if the outcome of the experiment belongs to $A \cap B'$, which sometimes is written as $A - B$. The event "exactly one of A or B" is said to occur if the outcome belongs to $(A \cap B') \cup (A' \cap B)$. The set $A' \cap B'$ corresponds to the event "neither A nor B." The set identity $A' \cap B' = (A \cup B)'$ is another way to represent this event. This is one of the set properties that usually are referred to as De Morgan's laws. The other such property is $A' \cup B' = (A \cap B)'$.

More generally, if $A_1, \ldots, A_k$ is a finite collection of events, occurrence of an outcome in the intersection $A_1 \cap \cdots \cap A_k$ (or $\bigcap_{i=1}^{k} A_i$) corresponds to the occurrence of the event "every A_i; $i = 1, \ldots, k$." The occurrence of an outcome in the union $A_1 \cup \cdots \cup A_k$ (or $\bigcup_{i=1}^{k} A_i$) corresponds to the occurrence of the event "at least one A_i; $i = 1, \ldots, k$." Similar remarks apply in the case of a countably infinite collection $A_1, A_2, \ldots$, with the notations $A_1 \cap A_2 \cap \cdots$ (or $\bigcap_{i=1}^{\infty} A_i$) for the intersection and $A_1 \cup A_2 \cup \cdots$ (or $\bigcup_{i=1}^{\infty} A_i$) for the union.

The intersection (or union) of a finite or countably infinite collection of events is called a **countable intersection** (or **union**).

We will consider the whole sample space S as a special type of event, called the **sure event**, and we also will include the empty set $\varnothing$ as an event, called the **null event**. Certainly, any set consisting of only a single outcome may be considered as an event.

Definition 1.2.4

An event is called an **elementary event** if it contains exactly one outcome of the experiment.

In a discrete sample space, any subset can be written as a countable union of elementary events, and we have no difficulty in associating every subset with an event in the discrete case.

In Example 1.2.1, the elementary events are {HH}, {HT}, {TH}, and {TT}, and any other event can be written as a finite union of these elementary events. Similarly, in Example 1.2.3, the elementary events are {H}, {TH}, {TTH}, ..., and any event can be represented as a countable union of these elementary events.

It is not as easy to represent events for the continuous examples. Rather than attempting to characterize these events rigorously, we will discuss some examples.

In Example 1.2.4, the light bulbs could fail during any time interval, and any interval of nonnegative real numbers would correspond to an interesting event for that experiment. Specifically, suppose the time until failure is measured in hours. The event that the light bulb "survives at most 10 hours" corresponds to the interval $A = [0, 10] = \{t \mid 0 \leqslant t \leqslant 10\}$. The event that the light bulb "survives more than 10 hours" is $A' = (10, \infty) = \{t \mid 10 < t < \infty\}$. If $B = [0, 15)$, then $C = B \cap A' = (10, 15)$ is the event of "failure between 10 and 15 hours."

In Example 1.2.5, any Cartesian product based on intervals of nonnegative real numbers would correspond to an event of interest. For example, the event

$$(10, 20) \times [5, \infty) = \{(x, y) \mid 10 < x < 20 \text{ and } 5 \leqslant y < \infty\}$$

corresponds to "the amount of light is between 10 and 20 lumens and the amount of energy is at least 5 joules." Such an event can be represented graphically as a rectangle in the xy plane with sides parallel to the coordinate axes.

In general, any physical event can be associated with a reasonable subset of S, and often a subset of S can be associated with some meaningful event. For mathematical reasons, though, when defining probability it is desirable to restrict the types of subsets that we will consider as events in some cases. Given a collection of events, we will want any countable union of these events to be an event. We also will want complements of events and countable intersections of events to be included in the collection of subsets that are defined to be events. We will assume that the collection of possible events includes all such subsets, but we will not attempt to describe all subsets that might be called events.

An important situation arises in the following developments when two events correspond to disjoint subsets.

Definition 1.2.5

Two events A and B are called **mutually exclusive** if $A \cap B = \varnothing$.

If events are mutually exclusive, then they have no outcomes in common. Thus, the occurrence of one event precludes the possibility of the other occurring. In Example 1.2.1, if A is the event "at least one head" and if we let B be the event "both tails," then A and B are mutually exclusive. Actually, in this example $B = A'$ (the complement of A). In general, complementary events are mutually exclusive, but the converse is not true. For example, if C is the event "both heads," then B and C are mutually exclusive, but not complementary.

The notion of mutually exclusive events can be extended easily to more than two events.

Definition 1.2.6

Events $A_1, A_2, A_3, \ldots,$ are said to be **mutually exclusive** if they are pairwise mutually exclusive. That is, if $A_i \cap A_j = \varnothing$ whenever $i \neq j$.

One possible approach to assigning probabilities to events involves the notion of relative frequency.

RELATIVE FREQUENCY

For the experiment of tossing a coin, we may declare that the probability of obtaining a head is 1/2. This could be interpreted in terms of the relative frequency with which a head is obtained on repeated tosses. Even though the coin may be tossed only once, conceivably it could be tossed many times, and experience leads us to expect a head on approximately one-half of the tosses. At least conceptually, as the number of tosses approaches infinity, the proportion of times a head occurs is expected to converge to some constant p. One then might define the probability of obtaining a head to be this conceptual limiting value. For a balanced coin, one would expect $p = 1/2$, but if the coin is unbalanced, or if the experiment is conducted under unusual conditions that tend to bias the outcomes in favor of either heads or tails, then this assignment would not be appropriate.

More generally, if $m(A)$ represents the number of times that the event A occurs among M trials of a given experiment, then $f_A = m(A)/M$ represents the **relative frequency** of occurrence of A on these trials of the experiment.

Example 1.2.7 An experiment consists of rolling an ordinary six-sided die. A natural sample space is the set of the first six positive integers, $S = \{1, 2, 3, 4, 5, 6\}$. A simulated die-rolling experiment is performed, using a "random number generator" on a computer. In Figure 1.1, the relative frequencies of the elementary events $A_1 = \{1\}$, $A_2 = \{2\}$, and so on are represented as the heights of vertical lines. The first graph shows the relative frequencies for the first $M = 30$ rolls, and the second graph gives the results for $M = 600$ rolls. By inspection of these graphs, obviously the relative frequencies tend to "stabilize" near some fixed value as M increases. Also included in the figure is a dotted line of height 1/6, which is the value that experience would suggest as the long-term relative frequency of the outcomes of rolling a die. Of course, in this example, the results are more relevant to the properties of the random number generator used to simulate the experiment than to those of actual dice.

FIGURE 1.1 Relative frequencies of elementary events for die-rolling experiment

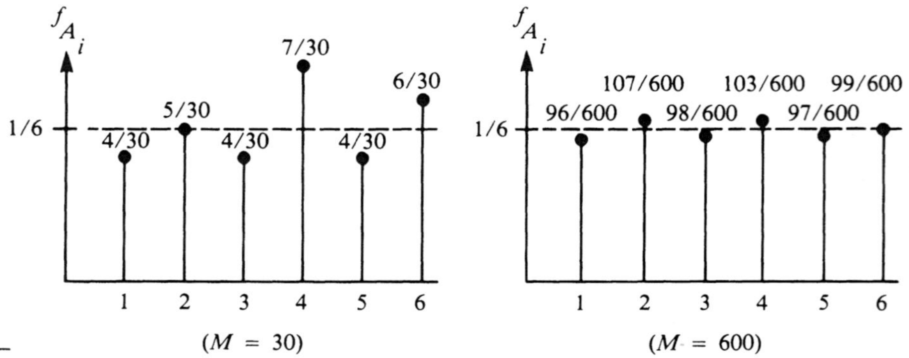

If, for an event A, the limit of f_A as M approaches infinity exists, then one could assign probability to A by

$$P(A) = \lim_{M \to \infty} f_A \tag{1.2.1}$$

This expresses a property known as **statistical regularity.** Certain technical questions about this property require further discussion. For example, it is not clear under what conditions the limit in equation (1.2.1) will exist, or in what sense, or whether it will necessarily be the same for every sequence of trials. Our approach to this problem will be to define probability in terms of a set of axioms and eventually show that the desired limiting behavior follows.

To motivate the defining axioms of probability, consider the following properties of relative frequencies. If S is the sample space for an experiment and A is an event, then clearly $0 \leqslant m(A)$ and $m(S) = M$, because $m(A)$ counts the number of occurrences of A, and S occurs on each trial. Furthermore, if A and B are mutually exclusive events, then outcomes in A are distinct from outcomes in B, and consequently $m(A \cup B) = m(A) + m(B)$. More generally, if $A_1, A_2, \ldots$ are pairwise mutually exclusive, then $m(A_1 \cup A_2 \cup \cdots) = m(A_1) + m(A_2) + \cdots$. Thus, the following properties hold for relative frequencies:

$$0 \leqslant f_A \tag{1.2.2}$$

$$f_S = 1 \tag{1.2.3}$$

$$f_{A_1 \cup A_2 \cup \cdots} = f_{A_1} + f_{A_2} + \cdots \tag{1.2.4}$$

if $A_1, A_2, \ldots$ are pairwise mutually exclusive events.

Although the relative frequency approach may not always be adequate as a practical method of assigning probabilities, it is the way that probability usually is interpreted. However, many people consider this interpretation too restrictive. By regarding probability as a **subjective** measure of belief that an event will occur, they are willing to assign probability in any situation involving uncertainty without assuming properties such as repeatability or statistical regularity. Statistical methods based on both the relative frequency approach and the subjective approach will be discussed in later chapters.

1.3

DEFINITION OF PROBABILITY

Given an experiment with an associated sample space S, the primary objective of probability modeling is to assign to each event A a real number $P(A)$, called the probability of A, that will provide a measure of the likelihood that A will occur when the experiment is performed.

Mathematically, we can think of $P(A)$ as a **set function**. In other words, it is a function whose domain is a collection of sets (events), and the range of which is a subset of the real numbers.

Some set functions are not suitable for assigning probabilities to events. The properties given in the following definition are motivated by similar properties that hold for relative frequencies.

Definition 1.3.1

For a given experiment, S denotes the sample space and $A, A_1, A_2, \ldots$ represent possible events. A set function that associates a real value $P(A)$ with each event A is called a **probability set function**, and $P(A)$ is called the **probability** of A, if the following properties are satisfied:

$$0 \leq P(A) \quad \text{for every } A \tag{1.3.1}$$

$$P(S) = 1 \tag{1.3.2}$$

$$P\left(\bigcup_{i=1}^{\infty} A_i\right) = \sum_{i=1}^{\infty} P(A_i) \tag{1.3.3}$$

if $A_1, A_2, \ldots$ are pairwise mutually exclusive events.

These properties all seem to agree with our intuitive concept of probability, and these few properties are sufficient to allow a mathematical structure to be developed.

One consequence of the properties is that the null event (empty set) has probability zero, $P(\varnothing) = 0$ (see Exercise 11). Also, if A and B are two mutually exclusive events, then

$$P(A \cup B) = P(A) + P(B) \tag{1.3.4}$$

Similarly, if $A_1, A_2, \ldots, A_k$ is a finite collection of pairwise mutually exclusive events, then

$$P(A_1 \cup A_2 \cup \cdots \cup A_k) = P(A_1) + P(A_2) + \cdots + P(A_k) \tag{1.3.5}$$

(See Exercise 12.)

In the case of a finite sample space, notice that there is at most a finite number of nonempty mutually exclusive events. Thus, in this case it would suffice to verify equation (1.3.4) or (1.3.5) instead of (1.3.3).

Example 1.3.1 The successful completion of a construction project requires that a piece of equipment works properly. Assume that either the "project succeeds" (A_1) or it fails because of one and only one of the following: "mechanical failure" (A_2) or "electrical failure" (A_3). Suppose that mechanical failure is three times as likely as electrical failure, and successful completion is twice as likely as mechanical failure. The resulting assignment of probability is determined by the equations $P(A_2) = 3P(A_3)$ and $P(A_1) = 2P(A_2)$. Because one and only one of these events will occur, we also have from (1.3.2) and (1.3.5) that $P(A_1) + P(A_2) + P(A_3) = 1$. These equations provide a system that can be solved simultaneously to obtain $P(A_1) = 0.6$, $P(A_2) = 0.3$, and $P(A_3) = 0.1$. The event "failure" is represented by the union $A_2 \cup A_3$, and because A_2 and A_3 are assumed to be mutually exclusive, we have from equation (1.3.5) that the probability of failure is $P(A_2 \cup A_3) = 0.3 + 0.1 = 0.4$.

PROBABILITY IN DISCRETE SPACES

The assignment of probability in the case of a discrete sample space can be reduced to assigning probabilities to the elementary events. Suppose that to each elementary event $\{e_i\}$ we assign a real number p_i, so that $P(\{e_i\}) = p_i$. To satisfy the conditions of Definition 1.3.1, it is necessary that

$$p_i \geqslant 0 \quad \text{for all } i \tag{1.3.6}$$

$$\sum_i p_i = 1 \tag{1.3.7}$$

Because each term in the sum (1.3.7) corresponds to an outcome in S, it is an ordinary summation when S is finite, and an infinite series when S is countably infinite. The probability of any other event then can be determined from the above assignment by representing the event as a union of mutually exclusive elementary events, and summing the corresponding values of p_i. A concise notation for this is given by

$$P(A) = \sum_{e_i \in A} P(\{e_i\}) \tag{1.3.8}$$

With this notation, we understand that the summation is taken over all indices i such that e_i is an outcome in A. This approach works equally well for both finite and countably infinite sample spaces, but if A is a countably infinite set the summation in (1.3.8) is actually an infinite series.

Example 1.3.2 If two coins are tossed as in Example 1.2.1, then $S = \{\text{HH, HT, TH, TT}\}$; if the coins are balanced, it is reasonable to assume that each of the four outcomes is equally likely. Because $P(S) = 1$, the probability assigned to each elementary event must be 1/4. Any event in a finite sample space can be written as a finite union of distinct elementary events, so the probability of any event is a sum including the constant term 1/4 for each elementary event in the union. For example, if $C = \{\text{HT, TH}\}$ represents the event "exactly one head," then

$$P(C) = P(\{\text{HT}\}) + P(\{\text{TH}\}) = 1/4 + 1/4 = 1/2$$

Note that the "equally likely" assumption cannot be applied indiscriminately. For example, in Example 1.2.2 the number of heads is of interest, and the sample space is $S^* = \{0, 1, 2\}$. The elementary event $\{1\}$ corresponds to the event $C = \{\text{HT, TH}\}$ in S. Rather than assigning the probability 1/3 to the outcomes in S^*, we should assign $P(\{1\}) = 1/2$ and $P(\{0\}) = P(\{2\}) = 1/4$.

In many problems, including those involving games of chance, the nature of the outcomes dictates the assignment of equal probability to each elementary event. This type of model sometimes is referred to as the **classical probability model**.

CLASSICAL PROBABILITY

Suppose that a finite number of possible outcomes may occur in an experiment, and that it is reasonable to assume that each outcome is equally likely to occur. Typical problems involving games of chance—such as tossing a coin, rolling a die, drawing cards from a deck, and picking the winning number in a lottery—fit this description. Note that the "equally likely" assumption requires the experiment to be carried out in such a way that the assumption is realistic. That is, the coin should be balanced, the die should not be loaded, the deck should be shuffled, the lottery tickets should be well mixed, and so forth.

This imposes a very special requirement on the assignment of probabilities to the elementary outcomes. In particular, let the sample space consist of N distinct outcomes,

$$S = \{e_1, e_2, \ldots, e_N\} \tag{1.3.9}$$

The "equally likely" assumption requires of the values p_i that

$$p_1 = p_2 = \cdots = p_N \tag{1.3.10}$$

and, to satisfy equations (1.3.6) and (1.3.7), necessarily

$$p_i = P(\{e_i\}) = \frac{1}{N} \tag{1.3.11}$$

In this case, because all terms in the sum (1.3.8) are the same, $p_i = 1/N$, it follows that

$$P(A) = \frac{n(A)}{N} \tag{1.3.12}$$

where $n(A)$ represents the number of outcomes in A. In other words, if the outcomes of an experiment are equally likely, then the problem of assigning probabilities to events is reduced to counting how many outcomes are favorable to the occurrence of the event as well as how many are in the sample space, and then finding the ratio. Some techniques that will be useful in solving some of the more complicated counting problems will be presented in Section 1.6.

The formula presented in (1.3.12) sometimes is referred to as **classical probability**. For problems in which this method of assignment is appropriate, it is fairly easy to show that our general definition of probability is satisfied. Specifically, for any event A,

$$P(A) = \frac{n(A)}{N} \geqslant 0$$

$$P(S) = \frac{n(S)}{N} = \frac{N}{N} = 1$$

$$P(A \cup B) = \frac{n(A \cup B)}{N} = \frac{n(A) + n(B)}{N} = P(A) + P(B)$$

if A and B are mutually exclusive.

RANDOM SELECTION

A major application of classical probability arises in connection with choosing an object or a set of objects "at random" from a collection of objects.

Definition 1.3.2

If an object is chosen from a finite collection of distinct objects in such a manner that each object has the same probability of being chosen, then we say that the object was chosen **at random**.

Similarly, if a subset of the objects is chosen so that each subset of the same size has the same probability of being chosen, then we say that the subset was chosen **at random**. Usually, no distinction is made when the elements of the subset are listed in a different order, but occasionally it will be useful to make this distinction.

Example 1.3.3 A game of chance involves drawing a card from an ordinary deck of 52 playing cards. It should not matter whether the card comes from the top or some other part of the deck if the cards are well shuffled. Each card would have the same probability, 1/52, of being selected. Similarly, if a game involves drawing five cards, then it should not matter whether the top five cards or any other five cards are drawn. The probability assigned to each possible set of five cards would be the reciprocal of the total number of subsets of size 5 from a set of size 52. In Section 1.6 we will develop, among other things, a method for counting the number of subsets of a given size.

1.4 SOME PROPERTIES OF PROBABILITY

From general properties of sets and the properties of Definition 1.3.1 we can derive other useful properties of probability. Each of the following theorems pertains to one or more events relative to the same experiment.

Theorem 1.4.1 If A is an event and A' is its complement, then

$$P(A) = 1 - P(A') \tag{1.4.1}$$

Proof

Because A' is the complement of A relative to S, $S = A \cup A'$. Because $A \cap A' = \varnothing$, A and A' are mutually exclusive, so it follows from equations (1.3.2) and (1.3.4) that

$$1 = P(S) = P(A \cup A') = P(A) + P(A')$$

which established the theorem. ■

This theorem is particularly useful when an event A is relatively complicated, but its complement A' is easier to analyze.

Example 1.4.1 An experiment consists of tossing a coin four times, and the event A of interest is "at least one head." The event A contains most of the possible outcomes, but the complement, "no heads," contains only one, $A' = \{TTTT\}$, so $n(A') = 1$. It can be shown by listing all of the possible outcomes that $n(S) = 16$, so that $P(A') = n(A')/n(S) = 1/16$. Thus, $P(A) = 1 - P(A') = 1 - 1/16 = 15/16$.

Theorem 1.4.2 For any event A, $P(A) \leqslant 1$.

Proof

From Theorem (1.4.1), $P(A) = 1 - P(A')$. Also, from Definition (1.3.1), we know that $P(A') \geqslant 0$. Therefore, $P(A) \leqslant 1$. ■

Note that this theorem combined with Definition (1.3.1) implies that

$$0 \leqslant P(A) \leqslant 1 \qquad \textbf{(1.4.2)}$$

Equations (1.3.3), (1.3.4), and (1.3.5) provide formulas for the probability of a union in the case of mutually exclusive events. The following theorems provide formulas that apply more generally.

Theorem 1.4.3 For any two events A and B,

$$P(A \cup B) = P(A) + P(B) - P(A \cap B) \qquad \textbf{(1.4.3)}$$

Proof

The approach will be to express the events $A \cup B$ and A as unions of mutually exclusive events. From set properties we can show that

$$A \cup B = (A \cap B') \cup B$$

and

$$A = (A \cap B) \cup (A \cap B')$$

See Figure 1.2 for an illustration of these identities.

FIGURE 1.2 Partitioning of events

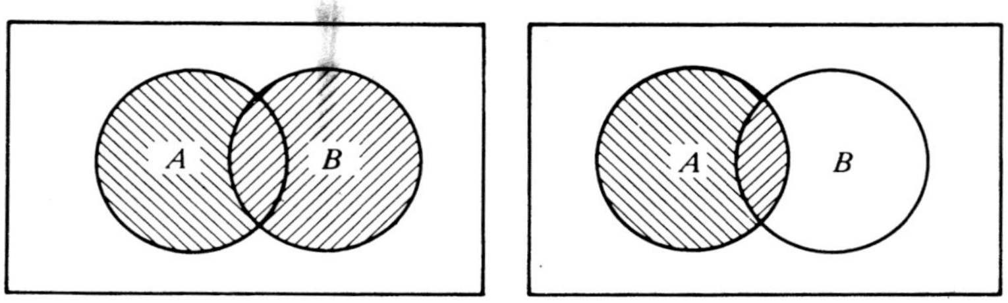

$A \cup B = (A \cap B') \cup B$ $\qquad$ $A = (A \cap B) \cup (A \cap B')$

It also follows that the events $A \cap B'$ and B are mutually exclusive because $(A \cap B') \cap B = \varnothing$, so that equation (1.3.4) implies

$$P(A \cup B) = P(A \cap B') + P(B)$$

Similarly, $A \cap B$ and $A \cap B'$ are mutually exclusive, so that

$$P(A) = P(A \cap B) + P(A \cap B')$$

The theorem follows from these equations:

$$\begin{aligned} P(A \cup B) &= P(A \cap B') + P(B) \\ &= [P(A) - P(A \cap B)] + P(B) \\ &= P(A) + P(B) - P(A \cap B) \end{aligned}$$

■

Example 1.4.2 Suppose one card is drawn at random from an ordinary deck of 52 playing cards. As noted in Example 1.3.3, this means that each card has the same probability, 1/52, of being chosen.

Let A be the event of obtaining "a red ace" and let B be the event "a heart." Then $P(A) = 2/52$, $P(B) = 13/52$, and $P(A \cap B) = 1/52$. From Theorem (1.4.3) we have $P(A \cup B) = 2/52 + 13/52 - 1/52 = 14/52 = 7/26$.

Theorem 1.4.3 can be extended easily to three events.

Theorem 1.4.4 For any three events A, B, and C,

$$\begin{aligned} P(A \cup B \cup C) = P(A) &+ P(B) + P(C) \\ &- P(A \cap B) - P(A \cap C) - P(B \cap C) \\ &+ P(A \cap B \cap C) \end{aligned} \tag{1.4.4}$$

Proof

See Exercise 16. ■

It is intuitively clear that if every outcome of A is also an outcome of B, then A is no more likely to occur than B. The next theorem formalizes this notion.

Theorem 1.4.5 If $A \subset B$, then $P(A) \leqslant P(B)$.

Proof

See Exercise 17. ■

Property (1.3.3) provides a formula for the probability of a countably infinite union when the events are mutually exclusive. If the events are not mutually exclusive, then the right side of property (1.3.3) still provides an upper bound for this probability, as shown in the following theorem.

Theorem 1.4.6 **Boole's Inequality** If $A_1, A_2, \ldots$ is a sequence of events, then

$$P\left(\bigcup_{i=1}^{\infty} A_i\right) \leqslant \sum_{i=1}^{\infty} P(A_i) \tag{1.4.5}$$

Proof

Let $B_1 = A_1$, $B_2 = A_2 \cap A_1'$, and in general $B_i = A_i \cap \left(\bigcup_{j=1}^{i-1} A_j\right)'$. It follows that $\bigcup_{i=1}^{\infty} A_i = \bigcup_{i=1}^{\infty} B_i$ and $B_1, B_2, \ldots$ are mutually exclusive. Because $B_i \subset A_i$, it follows from Theorem 1.4.5 that $P(B_i) \leqslant P(A_i)$, and thus

$$P\left(\bigcup_{i=1}^{\infty} A_i\right) = P\left(\bigcup_{i=1}^{\infty} B_i\right) = \sum_{i=1}^{\infty} P(B_i) \leqslant \sum_{i=1}^{\infty} P(A_i)$$ ■

A similar result holds for finite unions. In particular,

$$P(A_1 \cup A_2 \cup \cdots \cup A_k) \leqslant P(A_1) + P(A_2) + \cdots + P(A_k) \tag{1.4.6}$$

which can be shown by a proof similar to that of Theorem 1.4.6.

Theorem 1.4.7 **Bonferroni's Inequality** If $A_1, A_2, \ldots, A_k$ are events, then

$$P\left(\bigcap_{i=1}^{k} A_i\right) \geqslant 1 - \sum_{i=1}^{k} P(A_i') \tag{1.4.7}$$

Proof

This follows from Theorem 1.4.1 applied to $\bigcap_{i=1}^{k} A_i = \left(\bigcup_{i=1}^{k} A_i'\right)'$, together with inequality (1.4.6). ■

1.5

CONDITIONAL PROBABILITY

A major objective of probability modeling is to determine how likely it is that an event A will occur when a certain experiment is performed. However, in numerous cases the probability assigned to A will be affected by knowledge of the occurrence or nonoccurrence of another event B. In such an example we will use the terminology "conditional probability of A given B," and the notation $P(A \mid B)$ will be used to distinguish between this new concept and ordinary probability $P(A)$.

Example 1.5.1 A box contains 100 microchips, some of which were produced by factory 1 and the rest by factory 2. Some of the microchips are defective and some are good (nondefective). An experiment consists of choosing one microchip at random from the box and testing whether it is good or defective. Let A be the event "obtaining a defective microchip"; consequently, A' is the event "obtaining a good microchip." Let B be the event "the microchip was produced by factory 1" and B' the event "the microchip was produced by factory 2." Table 1.1 gives the number of microchips in each category.

TABLE 1.1 **Numbers of defective and nondefective microchips from two factories**

	B	B'	Totals
A	15	5	20
A'	45	35	80
Totals	60	40	100

The probability of obtaining a defective microchip is

$$P(A) = \frac{n(A)}{n(S)} = \frac{20}{100} = 0.20$$

Now suppose that each microchip has a number stamped on it that identifies which factory produced it. Thus, before testing whether it is defective, it can be determined whether B has occurred (produced by factory 1) or B' has occurred (produced by factory 2). Knowledge of which factory produced the microchip affects the likelihood that a defective microchip is selected, and the use of conditional probability is appropriate. For example, if the event B has occurred, then the only microchips we should consider are those in the first column of Table 1.1, and the total number is $n(B) = 60$. Furthermore, the only defective chips to consider are those in both the first column and the first row, and the total number is $n(A \cap B) = 15$. Thus, the conditional probability of A given B is

$$P(A \mid B) = \frac{n(A \cap B)}{n(B)} = \frac{15}{60} = 0.25$$

Notice that if we divide both the numerator and denominator by $n(S) = 100$, we can express conditional probability in terms of some ordinary unconditional probabilities,

$$P(A \mid B) = \frac{n(A \cap B)/n(S)}{n(B)/n(S)} = \frac{P(A \cap B)}{P(B)}$$

This last result can be derived under more general circumstances as follows.

Suppose we conduct an experiment with a sample space S, and suppose we are given that the event B has occurred. We wish to know the probability that an event A has occurred given that B has occurred, written $P(A \mid B)$. That is, we want the probability of A relative to the reduced sample space B. We know that B can be partitioned into two subsets,

$$B = (A \cap B) \cup (A' \cap B)$$

$A \cap B$ is the subset of B for which A is true, so the probability of A given B should be proportional to $P(A \cap B)$, say $P(A \mid B) = kP(A \cap B)$. Similarly, $P(A' \mid B) = kP(A' \cap B)$. Together these should represent the total probability relative to B, so

$$\begin{aligned} P(A \mid B) + P(A' \mid B) &= k[P(A \cap B) + P(A' \cap B)] \\ &= kP[(A \cap B) \cup (A' \cap B)] \\ &= kP(B) \\ &= 1 \end{aligned}$$

and $k = 1/P(B)$. That is,

$$P(A \mid B) = \frac{P(A \cap B)}{P(A \cap B) + P(A' \cap B)} = \frac{P(A \cap B)}{P(B)}$$

and $1/P(B)$ is the proportionality constant that makes the probabilities on the reduced sample space add to 1.

Definition 1.5.1

The **conditional probability** of an event A, given the event B, is defined by

$$P(A \mid B) = \frac{P(A \cap B)}{P(B)} \tag{1.5.1}$$

if $P(B) \neq 0$.

Relative to the sample space B, conditional probabilities defined by (1.5.1) satisfy the original definition of probability, and thus conditional probabilities enjoy all the usual properties of probability on the reduced sample space. For example, if two events A_1 and A_2 are mutually exclusive, then

$$P(A_1 \cup A_2 \mid B) = \frac{P[(A_1 \cup A_2) \cap B]}{P(B)}$$

$$= \frac{P[(A_1 \cap B) \cup (A_2 \cap B)]}{P(B)}$$

$$= \frac{P(A_1 \cap B) + P(A_2 \cap B)}{P(B)}$$

$$= P(A_1 \mid B) + P(A_2 \mid B)$$

This result generalizes to more than two events. Similarly, $P(A \mid B) \geqslant 0$ and $P(S \mid B) = P(B \mid B) = 1$, so the conditions of a probability set function are satisfied.

Thus, the properties derived in Section 1.4 hold conditionally. In particular,

$$P(A \mid B) = 1 - P(A' \mid B) \tag{1.5.2}$$

$$0 \leqslant P(A \mid B) \leqslant 1 \tag{1.5.3}$$

$$P(A_1 \cup A_1 \mid B) = P(A_1 \mid B) + P(A_2 \mid B) - P(A_1 \cap A_2 \mid B) \tag{1.5.4}$$

The following theorem results immediately from equation (1.5.1):

Theorem 1.5.1 For any events A and B,

$$P(A \cap B) = P(B)P(A \mid B) = P(A)P(B \mid A) \tag{1.5.5}$$

■

This sometimes is referred to as the **Multiplication Theorem** of probability. It provides a way to compute the probability of the joint occurrence of A and B by multiplying the probability of one event and the conditional probability of the other event. In terms of Example 1.5.1, we can compute directly $P(A \cap B) = 15/100 = 0.15$, or we can compute it as $P(B)P(A \mid B) = (60/100)(15/60) = 0.15$ or $P(A)P(B \mid A) = (20/100)(15/20) = 0.15$.

Formula (1.5.5) also is quite useful in dealing with problems involving **sampling without replacement**. Such experiments consist of choosing objects one at a time from a finite collection, without replacing chosen objects before the next choice. Perhaps the most common example of this is dealing cards from a deck.

Example 1.5.2 Two cards are drawn without replacement from a deck of cards. Let A_1 denote the event of getting "an ace on the first draw" and A_2 denote the event of getting "an ace on the second draw."

The number of ways in which different outcomes can occur can be enumerated, and the results are given in Table 1.2. The enumeration of possible outcomes can be a tedious problem, and useful techniques that are helpful in such counting problems are discussed in Section 1.6. The values in this example are based on the so-called multiplication principle, which says that if there are n_1 ways of doing one thing and n_2 ways of doing another, then there are $n_1 \cdot n_2$ ways of doing both. Thus, for example, the total number of ordered two-card hands that can be formed from 52 cards (without replacement) is $52 \cdot 51 = 2652$. Similarly, the number of ordered two-card hands in which both cards are aces is $4 \cdot 3$, the number in which the first card is an ace and the second is *not* an ace is $4 \cdot 48$, and so forth. The appropriate products for all cases are provided in Table 1.2.

TABLE 1.2 **Partitioning the numbers of ways to draw two cards**

	A_1	A'_1	
A_2	$4 \cdot 3$	$48 \cdot 4$	$4 \cdot 51$
A'_2	$4 \cdot 48$	$48 \cdot 47$	$48 \cdot 51$
	$4 \cdot 51$	$48 \cdot 51$	$52 \cdot 51$

For example, the probability of getting "an ace on the first draw and an ace on the second draw" is given by

$$P(A_1 \cap A_2) = \frac{4 \cdot 3}{52 \cdot 51}$$

Suppose one is interested in $P(A_1)$ without regard to what happens on the second draw. First note that A_1 may be partitioned as

$$A_1 = (A_1 \cap A_2) \cup (A_1 \cap A'_2)$$

so that

$$\begin{aligned} P(A_1) &= P(A_1 \cap A_2) + P(A_1 \cap A'_2) \\ &= \frac{4 \cdot 3}{52 \cdot 51} + \frac{4 \cdot 48}{52 \cdot 51} \\ &= \frac{4 \cdot 51}{51 \cdot 52} = \frac{4}{52} \end{aligned}$$

This same result would have occurred if A_1 had been partitioned by another event, say B, which deals only with the face value of the second card. This follows because $n(B \cup B') = 51$, and relative to the $52 \cdot 51$ ordered pairs of cards,

$$n(A_1) = 4 \cdot n(B) + 4 \cdot n(B') = 4 \cdot n(B \cup B') = 4 \cdot 51$$

The numerators of probabilities such as $P(A_1)$, $P(A'_1)$, $P(A_2)$, and $P(A'_2)$, which deal with only one of the draws, appear in the margins of Table 1.2. These probabilities may be referred to as **marginal probabilities.** Note that the marginal probabilities in fact can be computed directly from the original 52-card sample space, and it is not necessary to consider the sample space of ordered pairs at all. For example, $P(A_1) = 4 \cdot 51/52 \cdot 51 = 4/52$, which is the probability that would be obtained for one draw from the original 52-card sample space. Clearly, this result would apply to sampling-without-replacement problems in general. What may be less intuitive is that these results also apply to marginal probabilities such as $P(A_2)$, and not just to the outcomes on the first draw. That is, if the outcome of the first draw is not known, then $P(A_2)$ also can be computed from the original sample space and is given by $P(A_2) = 4/52$. This can be verified in this example because

$$A_2 = (A_2 \cap A_1) \cup (A_2 \cap A_1')$$

and

$$P(A_2) = \frac{4 \cdot 3}{52 \cdot 51} + \frac{48 \cdot 4}{52 \cdot 51}$$
$$= \frac{4}{52}$$

Indeed, if the result of the first draw is not known, then the second draw could just as well be considered as the first draw.

The conditional probability that an ace is drawn on the second draw given that an ace was obtained on the first draw is

$$P(A_2 \mid A_1) = \frac{P(A_1 \cap A_2)}{P(A_1)}$$
$$= \frac{(4 \cdot 3)/(52 \cdot 51)}{(4 \cdot 51)/(52 \cdot 51)}$$
$$= \frac{3}{51}$$

That is, given that A_1 is true, we are restricted to the first column of Table 1.2, and the relative proportion of the time that A_2 is true on the reduced sample space is $(4 \cdot 3)/[(4 \cdot 3 + (4 \cdot 48)]$. Again, it may be less obvious, but it is possible to carry this problem one step further and compute $P(A_2 \mid A_1)$ directly in terms of the 51-card conditional sample space, and obtain the much simpler solution that $P(A_2 \mid A_1) = 3/51$, there being three aces remaining in the 51 remaining cards in the conditional sample space. Thus, it is common practice in this type of problem to compute the conditional probabilities and marginal probabilities directly from the one-dimensional sample spaces (one marginal and one conditional space), rather than obtain the joint probabilities from the joint sample space of ordered pairs. For example,

$$P(A_1 \cap A_2) = P(A_1)P(A_2 \mid A_1)$$
$$= \frac{4}{52} \cdot \frac{3}{51}$$

This procedure would extend to three or more draws (without replacement) where, for example, if A_3 denotes obtaining "an ace on the third draw," then

$$P(A_1 \cap A_2 \cap A_3) = P(A_1)P(A_2 \mid A_1)P(A_3 \mid A_1 \cap A_2)$$
$$= \frac{4}{52} \cdot \frac{3}{51} \cdot \frac{2}{50}$$

An indication of the general validity of this approach for computing conditional probabilities is obtained by considering $P(A_2 \mid A_1)$ in the example. Relative to the joint sample space of ordered pairs, $204 = 4 \cdot 51$, where 4 represents the number of ways the given event A_1 can occur on the first draw and 51 is the total number of possible outcomes in the conditional sample space for the second draw; also, $12 = 4 \cdot 3$ represents the number of ways the given event A_1 can occur times the number of ways a success, A_2, can occur in the conditional sample space. Because the number of ways A_1 can occur is a common multiplier in the numerator and denominator when counting ordered pairs, one may equivalently count directly in the one-dimensional conditional space associated with the second draw.

The computational advantage of this approach is obvious, because it allows the computation of the probability of an event in a complicated higher-dimensional product space as a product of probabilities, one marginal and the others conditional, of events in simpler one-dimensional sample spaces.

The above discussion is somewhat tedious, but it may provide insight into the physical meaning of conditional probability and marginal probability, and also into the topic of sampling without replacement, which will come up again in the following sections.

TOTAL PROBABILITY AND BAYES' RULE

As noted in Example 1.5.2, it sometimes is useful to partition an event, say A, into the union of two or more mutually exclusive events. For example, if B and B' are events that pertain to the first draw from a deck, and if A is an event that pertains to the second draw, then it is worthwhile to consider the partition $A = (A \cap B) \cup (A \cap B')$ to compute $P(A)$, because this separates A into two events that involve information about both draws. More generally, if $B_1, B_2, \ldots, B_k$

are mutually exclusive and **exhaustive**, in the sense that $B_1 \cup B_2 \cup \cdots \cup B_k = S$, then

$$A = (A \cap B_1) \cup (A \cap B_2) \cup \cdots \cup (A \cap B_k)$$

This is useful in the following theorem.

Theorem 1.5.2 **Total Probability** If $B_1, B_2, \ldots, B_k$ is a collection of mutually exclusive and exhaustive events, then for any event A,

$$P(A) = \sum_{i=1}^{k} P(B_i)P(A \mid B_i) \tag{1.5.6}$$

Proof

The events $A \cap B_1, A \cap B_2, \ldots, A \cap B_k$ are mutually exclusive, so it follows that

$$P(A) = \sum_{i=1}^{k} P(A \cap B_i) \tag{1.5.7}$$

and the theorem results from applying Theorem 1.5.1 to each term in this summation. ■

Theorem 1.5.2 sometimes is known as the **Law of Total Probability**, because it corresponds to mutually exclusive ways in which A can occur relative to a partition of the total sample space S.

Sometimes it is helpful to illustrate this result with a **tree diagram**. One such diagram for the case of three events B_1, B_2, and B_3 is given in Figure 1.3.

FIGURE 1.3 Tree diagram showing the Law of Total Probability

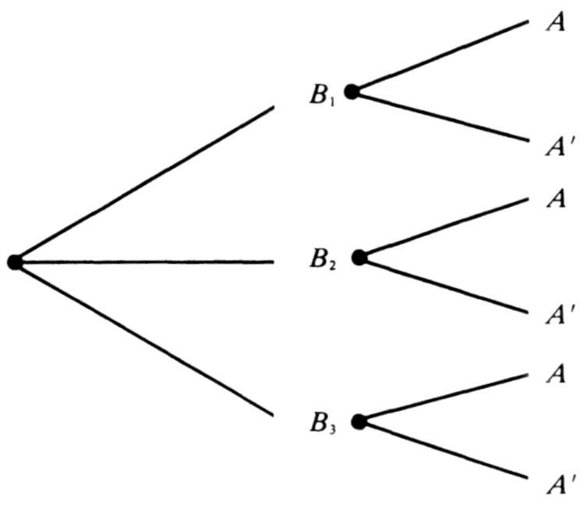

The probability associated with branch B_i is $P(B_i)$, and the probability associated with each branch labeled A is a conditional probability $P(A \mid B_i)$, which may be different depending on which branch, B_i, it follows. For A to occur, it must occur jointly with one and only one of the events B_i. Thus, only $A \cap B_1$, $A \cap B_2$, or $A \cap B_3$ must occur, and the probability of A is the sum of the probabilities of these joint events, $P(B_i)P(A \mid B_i)$.

Example 1.5.3 Factory 1 in Example 1.5.1 has two shifts, and the microchips from factory 1 can be categorized according to which shift produced them. As before, the experiment consists of choosing a microchip at random from the box and testing to see whether it is defective. Let B_1 be the event "produced by shift 1" (factory 1), B_2 the event "produced by shift 2" (factory 1), and B_3 the event "produced by factory 2." As before, let A be the event "obtaining a defective microchip." The categories are given by Table 1.3.

TABLE 1.3 **Numbers of defective and nondefective microchips from a common lot**

	B_1	B_2	B_3	Totals
A	5	10	5	20
A'	20	25	35	80
Totals	25	35	40	100

Various probabilities can be computed directly from the table. For example, $P(B_1) = 25/100$, $P(B_2) = 35/100$, $P(B_3) = 40/100$, $P(A \mid B_1) = 5/25$, $P(A \mid B_2) = 10/35$, and $P(A \mid B_3) = 5/40$. It is possible to compute $P(A)$ either directly from the table, $P(A) = 20/100 = 0.20$, or by using the Law of Total Probability:

$$\begin{aligned} P(A) &= P(B_1)P(A \mid B_1) + P(B_2)P(A \mid B_2) + P(B_3)P(A \mid B_3) \\ &= \left(\frac{25}{100}\right)\left(\frac{5}{25}\right) + \left(\frac{35}{100}\right)\left(\frac{10}{35}\right) + \left(\frac{40}{100}\right)\left(\frac{5}{40}\right) \\ &= 0.05 + 0.10 + 0.05 = 0.20 \end{aligned}$$

This problem is illustrated by the tree diagram in Figure 1.4.

FIGURE 1.4 Tree diagram for selection of microchips from combined lot

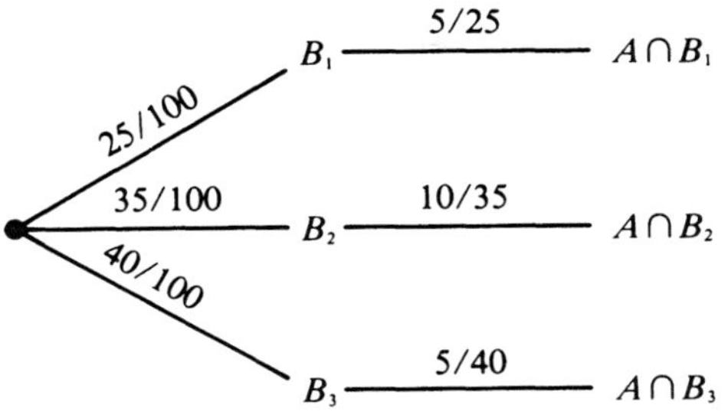

Example 1.5.4 Consider the following variation on Example 1.5.3. The microchips are sorted into three separate boxes. Box 1 contains the 25 microchips from shift 1, box 2 contains the 35 microchips from shift 2, and box 3 contains the remaining 40 microchips from factory 2. The new experiment consists of choosing a box at random, then selecting a microchip from the box. This experiment is illustrated in Figure 1.5.

FIGURE 1.5 Selection of microchips from three different sources

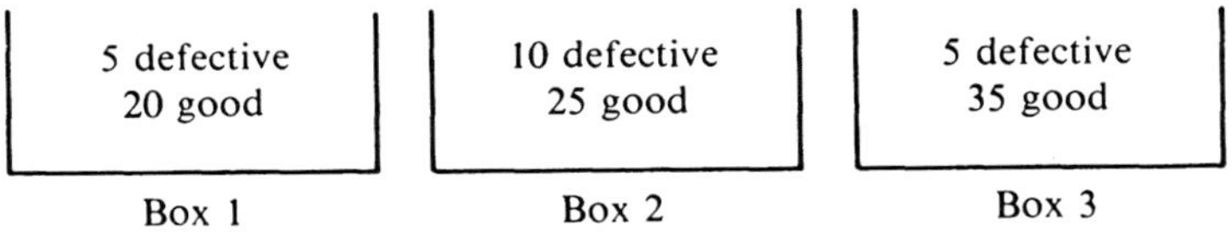

In this case, it is not possible to compute $P(A)$ directly from Table 1.3, but it still is possible to use equation (1.5.6) by redefining the events B_1, B_2, and B_3 to be respectively choosing "box 1," "box 2," and "box 3." Thus, the new assignment of probability to B_1, B_2, and B_3 is $P(B_1) = P(B_2) = P(B_3) = 1/3$, and

$$P(A) = \left(\frac{1}{3}\right)\left(\frac{5}{25}\right) + \left(\frac{1}{3}\right)\left(\frac{10}{35}\right) + \left(\frac{1}{3}\right)\left(\frac{5}{40}\right)$$

$$= \frac{57}{280}$$

As a result of this new experiment, suppose that the component obtained is defective, but it is not known which box it came from. It is possible to compute the probability that it came from a particular box given that it was defective, although a special formula is required.

Theorem 1.5.3 **Bayes' Rule** If we assume the conditions of Theorem 1.5.2, then for each $j = 1, 2, \ldots, k$,

$$P(B_j \mid A) = \frac{P(B_j)P(A \mid B_j)}{\sum_{i=1}^{k} P(B_i)P(A \mid B_i)} \tag{1.5.8}$$

Proof

From Definition 1.5.1 and Multiplication Theorem 1.5.5 we have

$$P(B_j \mid A) = \frac{P(A \cap B_j)}{P(A)} = \frac{P(B_j)P(A \mid B_j)}{P(A)}$$

The theorem follows by replacing the denominator with the right side of (1.5.6). ■

For the data of Example 1.5.4, the conditional probability that the microchip came from box 1, given that it is defective, is

$$P(B_1 \mid A) = \frac{(1/3)(5/25)}{(1/3)(5/25) + (1/3)(10/35) + (1/3)(5/40)}$$

$$= \frac{56}{171} = 0.327$$

Similarly, $P(B_2 \mid A) = 80/171 = 0.468$ and $P(B_3 \mid A) = 35/171 = 0.205$.

Notice that these differ from the unconditional probabilities, $P(B_i) = 1/3 = 0.333$. This reflects the different proportions of defective items in the boxes. In other words, because box 2 has a higher proportion of defectives, choosing a defective item effectively increases the likelihood that it was chosen from box 2.

For another illustration, consider the following example.

Example 1.5.5 A man starts at the point O on the map shown in Figure 1.6. He first chooses a path at random and follows it to point B_1, B_2, or B_3. From that point, he chooses a new path at random and follows it to one of the points A_i, $i = 1, 2, \ldots, 7$.

FIGURE 1.6 Map of possible paths

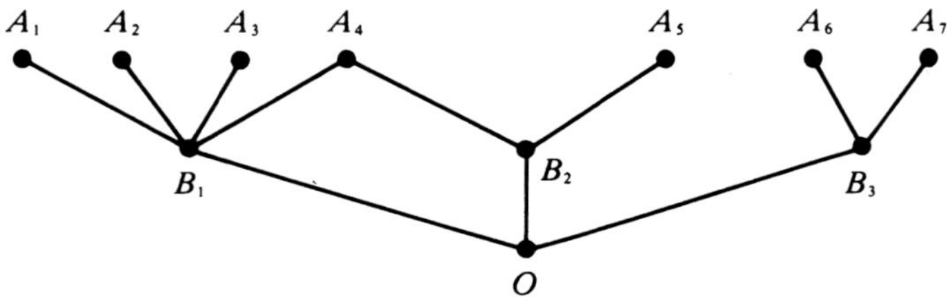

It might be of interest to know the probability that the man arrives at point A_4. This can be computed from the Law of Total Probability:

$$P(A_4) = P(B_1)P(A_4 \mid B_1) + P(B_2)P(A_4 \mid B_2) + P(B_3)P(A_4 \mid B_3)$$
$$= \left(\frac{1}{3}\right)\left(\frac{1}{4}\right) + \left(\frac{1}{3}\right)\left(\frac{1}{2}\right) + \left(\frac{1}{3}\right)(0) = \frac{1}{4}$$

Suppose the man arrives at point A_4, but it is not known which route he took. The probability that he passed through a particular point, B_1, B_2, or B_3, can be computed from Bayes' Rule. For example,

$$P(B_1 \mid A_4) = \frac{(1/3)(1/4)}{(1/3)(1/4) + (1/3)(1/2) + (1/3)(0)} = \frac{1}{3}$$

which agrees with the unconditional probability, $P(B_1) = 1/3$.

This is an example of a very special situation called "independence," which we will pursue in the next section. However, this does not occur in every case. For example, an application of Bayes' Rule also leads to $P(B_2 \mid A_4) = 2/3$, which does not agree with $P(B_2) = 1/3$. Thus, if he arrived at point A_4, it is twice as likely that he passed through point B_2 as it is that he passed through B_1. Of course, the most striking result concerns point B_3, because $P(B_3 \mid A_4) = 0$, while $P(B_3) = 1/3$. This reflects the obvious fact that he cannot arrive at point A_4 by passing through point B_3. The practical value of conditioning is obvious when considering some action such as betting on whether the man passed through point B_3.

INDEPENDENT EVENTS

In some situations, knowledge that an event A has occurred will not affect the probability that an event B will occur. In other words, $P(B \mid A) = P(B)$. We saw this happen in Example 1.5.5, because the probability of passing through point B_1 was 1/3 whether the knowledge that the man arrived at point A_4 was taken into account. As a result of the Multiplication Theorem (1.5.5), an equivalent formulation of this situation is $P(A \cap B) = P(A)P(B \mid A) = P(A)P(B)$. In general, when this happens the two events are said to be independent or stochastically independent.

Definition 1.5.2

Two events A and B are called **independent events** if

$$P(A \cap B) = P(A)P(B) \qquad \textbf{(1.5.9)}$$

Otherwise, A and B are called **dependent events.**

As already noted, an equivalent formulation can be given in terms of conditional probability.

Theorem 1.5.4 If A and B are events such that $P(A) > 0$ and $P(B) > 0$, then A and B are independent if and only if either of the following holds:

$$P(A \mid B) = P(A) \qquad P(B \mid A) = P(B)$$ ■

We saw examples of both independent and dependent events in Example 1.5.5. There was also an example of mutually exclusive events, because $P(B_3 \mid A_4) = 0$, which implies $P(B_3 \cap A_4) = 0$. There is often confusion between the concepts of independent events and mutually exclusive events. Actually, these are quite different notions, and perhaps this is seen best by comparisons involving conditional probabilities. Specifically, if A and B are mutually exclusive, then $P(A \mid B) = P(B \mid A) = 0$, whereas for independent nonnull events the conditional probabilities are nonzero as noted by Theorem 1.5.4. In other words, the property of being mutually exclusive involves a very strong form of dependence, because, for nonnull events, the occurrence of one event precludes the occurrence of the other event.

There are many applications in which events are assumed to be independent.

Example 1.5.6 A "system" consists of several components that are hooked up in some particular configuration. It is often assumed in applications that the failure of one component does not affect the likelihood that another component will fail. Thus, the failure of one component is assumed to be independent of the failure of another component.

A **series system** of two components, C_1 and C_2, is illustrated by Figure 1.7. It is easy to think of such a system in terms of two electrical components (for example, batteries in a flashlight) where current must pass through both components for the system to function. If A_1 is the event "C_1 fails" and A_2 is the event "C_2 fails," then the event "the system fails" is $A_1 \cup A_2$. Suppose that $P(A_1) = 0.1$ and $P(A_2) = 0.2$. If we assume that A_1 and A_2 are independent, then the probability that the system fails is

$$\begin{aligned} P(A_1 \cup A_2) &= P(A_1) + P(A_2) - P(A_1 \cap A_2) \\ &= P(A_1) + P(A_2) - P(A_1)P(A_2) \\ &= 0.1 + 0.2 - (0.1)(0.2) = 0.28 \end{aligned}$$

The probability that the system works properly is $1 - 0.28 = 0.72$.

FIGURE 1.7 Series system of two components

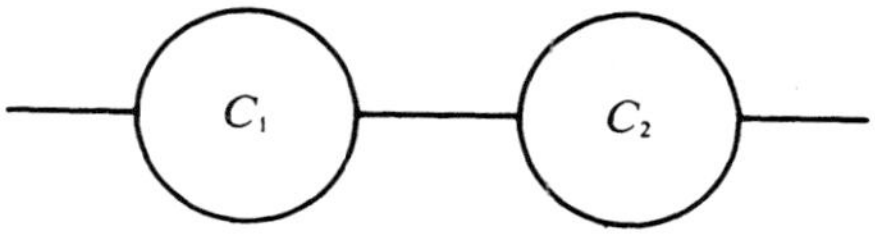

Notice that the assumption of independence permits us to factor the probability of the joint event, $P(A_1 \cap A_2)$, into the product of the marginal probabilities, $P(A_1)P(A_2)$.

Another common example involves the notion of a **parallel system**, as illustrated in Figure 1.8. For a parallel system to fail, it is necessary that both components fail, so the event "the system fails" is $A_1 \cap A_2$. The probability that this system fails is $P(A_1 \cap A_2) = P(A_1)P(A_2) = (0.1)(0.2) = 0.02$, again assuming the components fail independently.

Note that the probability of failure for a series system is greater than the probability of failure of either component, whereas for a parallel system it is less. This is because both components must function for a series system to function, and consequently the system is more likely to fail than an individual component. On the other hand, a parallel system is a **redundant system**: One component can fail, but the system will continue to function provided the other component functions. Such redundancy is common in aerospace systems, where the failure of the system may be catastrophic.

A common example of dependent events occurs in connection with repeated **sampling without replacement** from a finite collection. In Example 1.5.2 we considered the results of drawing two cards in succession from a deck. It turns out that the events A_1 (ace on the first draw) and A_2 (ace on the second draw) are dependent because $P(A_2) = 4/52$, while $P(A_2 \mid A_1) = 3/51$.

Suppose instead that the outcome of the first card is recorded and then the card is replaced in the deck and the deck is shuffled before the second draw is made. This type of sampling is referred to as **sampling with replacement**, and it would be reasonable to assume that the draws are independent trials. In this case $P(A_1 \cap A_2) = P(A_1)P(A_2)$.

There are many other problems in which it is reasonable to assume that repeated trials of an experiment are independent, such as tossing a coin or rolling a die repeatedly.

It is possible to show that independence of two events also implies the independence of some related events.

FIGURE 1.8 Parallel system of two components

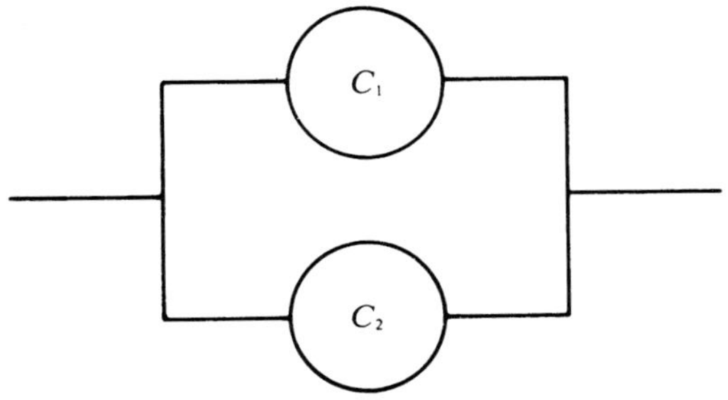

Theorem 1.5.5 Two events A and B are independent if and only if the following pairs of events are also independent:

1. A and B'.
2. A' and B.
3. A' and B'.

Proof

See Exercise 38. ■

It is also possible to extend the notion of independence to more than two events.

> ***Definition 1.5.3***
>
> The k events $A_1, A_2, \ldots, A_k$ are said to be **independent** or **mutually independent** if for every $j = 2, 3, \ldots, k$ and every subset of distinct indices $i_1, i_2, \ldots, i_j$,
>
> $$P(A_{i_1} \cap A_{i_2} \cap \cdots \cap A_{i_j}) = P(A_{i_1})P(A_{i_2}) \cdots P(A_{i_j}) \tag{1.5.10}$$

Suppose A, B, and C are three mutually independent events. According to the definition of mutually independent events, it is not sufficient simply to verify pairwise independence. It would be necessary to verify $P(A \cap B) = P(A)P(B)$, $P(A \cap C) = P(A)P(C)$, $P(B \cap C) = P(B)P(C)$, and also $P(A \cap B \cap C) = P(A)P(B)P(C)$. The following examples show that pairwise independence does not imply this last three-way factorization and vice versa.

Example 1.5.7 A box contains eight tickets, each labeled with a binary number. Two are labeled 111, two are labeled 100, two 010, and two 001. An experiment consists of drawing one ticket at random from the box. Let A be the event "the first digit is 1," B the event "the second digit is 1," and C the event "the third digit is 1." This is illustrated by Figure 1.9. It follows that $P(A) = P(B) = P(C) = 4/8 = 1/2$ and that $P(A \cap B) = P(A \cap C) = P(B \cap C) = 2/8 = 1/4$; thus A, B, and C are pairwise independent. However, they are *not* mutually independent, because

$$P(A \cap B \cap C) = \frac{2}{8} = \frac{1}{4} \neq \frac{1}{8} = P(A)P(B)P(C)$$

FIGURE 1.9 Selection of numbered tickets

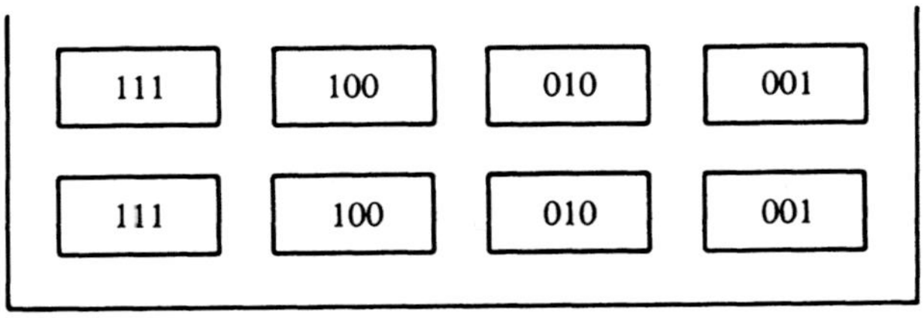

Example 1.5.8 In Figure 1.9, let us change the number on one ticket in the first column from 111 to 110, and the number of one ticket in the second column from 100 to 101. We still have

$$P(A) = P(B) = P(C) = \frac{1}{2}$$

but

$$P(B \cap C) = \frac{1}{8} \neq \frac{1}{4} = P(B)P(C)$$

and

$$P(A \cap B \cap C) = \frac{1}{8} = P(A)P(B)P(C)$$

In this case we have three-way factorization, but not independence of all pairs.

1.6

COUNTING TECHNIQUES

In many experiments with finite sample spaces, such as games of chance, it may be reasonable to assume that all possible outcomes are equally likely. In that case, a realistic probability model should result by following the classical approach and taking the probability of any event A to be $P(A) = n(A)/N$, where N is the total number of possible outcomes and $n(A)$ is the number of these outcomes that correspond to occurrence of the event A. Counting the number of ways in which an event may occur can be a tedious problem in complicated experiments. A few helpful counting techniques will be discussed.

MULTIPLICATION PRINCIPLE

First note that if one operation can be performed in n_1 ways and a second operation can be performed in n_2 ways, then there are $n_1 \cdot n_2$ ways in which both operations can be carried out.

Example 1.6.1 Suppose a coin is tossed and then a marble is selected at random from a box containing one black (B), one red (R), and one green (G) marble. The possible outcomes are HB, HR, HG, TB, TR, and TG. For each of the two possible outcomes of the coin there are three marbles that may be selected for a total of $2 \cdot 3 = 6$ possible outcomes. The situation also is easily illustrated by a tree diagram, as in Figure 1.10.

FIGURE 1.10 Tree diagram of two-stage experiment

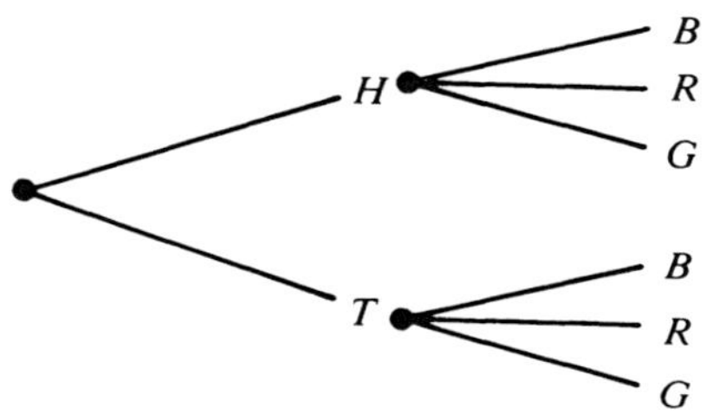

Another application of the multiplication principle was discussed in Example 1.5.2 in connection with counting the number of ordered two-card hands.

Note that the multiplication principle can be extended to more than two operations. In particular, if the ith of r successive operations can be performed in n_i ways, then the total number of ways to carry out all r operations is the product

$$\prod_{i=1}^{r} n_i = n_1 n_2 \cdots n_r \tag{1.6.1}$$

One standard type of counting problem is covered by the following theorem.

Theorem 1.6.1 If there are N possible outcomes of each of r trials of an experiment, then there are N^r possible outcomes in the sample space. ■

Example 1.6.2 How many ways can a 20-question true–false test be answered? The answer is 2^{20}.

Example 1.6.3 How many subsets are there from a set of m elements? In forming a subset, one must decide for each element whether to include that element in the subset. Thus for each of m elements there are two choices, which give a total of 2^m possible subsets. This includes the null set, which corresponds to the case of not including any element in the subset.

As suggested earlier, the way an experiment is carried out or the method of sampling may affect the sample space and the probability assignment over the sample space. In particular, sampling items from a finite population with and without replacement are two common schemes. Sampling without replacement was illustrated in Example 1.5.2. Sampling with replacement is covered by Theorem 1.6.1.

Example 1.6.4 If five cards are drawn from a deck of 52 cards with replacement, then there are $(52)^5$ possible hands. If the five cards are drawn without replacement, then the more general multiplication principle may be applied to determine that there are $52 \cdot 51 \cdot 50 \cdot 49 \cdot 48$ possible hands. In the first case, the same card may occur more than once in the same hand. In the second case, however, a card may not be repeated.

Note that in both cases in the above example, **order** is considered important. That is, two five-card hands may eventually end up with the same five cards, but they are counted as different hands in the example if the cards were obtained in a different order. For example, let all five cards be spades. The outcome (ace, king, queen, jack, ten) is different from the outcome (king, ace, queen, jack, ten). If order had not been considered important, both of these outcomes would be considered the same; indeed, there would be several different ordered outcomes corresponding to this same (unordered) outcome. On the other hand, only one outcome corresponds to all five cards being the ace of spades (in the sampling-with-replacement case), whether the cards are ordered or unordered.

This introduces the concept of **distinguishable** and **indistinguishable** elements. Even though order may be important, a new result or arrangement will not be obtained if two indistinguishable elements are interchanged. Thus, fewer ordered arrangements are possible if some of the items are indistinguishable. We also noted earlier that there are fewer distinct results if order is not taken into account, but the probability of any one of these unordered results occurring then would be greater. Note also that it is common practice to assume that order is not important when drawing without replacement, unless otherwise specified, although we did consider order important in Example 1.6.4.

PERMUTATIONS AND COMBINATIONS

Some particular formulas that are helpful in counting the number of possible arrangements for some of the cases mentioned will be given. An ordered arrangement of a set of objects is known as a **permutation**.

Theorem 1.6.2 The number of permutations of n distinguishable objects is $n!$.

Proof

This follows by applying the multiplication principle. To fill n positions with n distinct objects, the first position may be filled n ways using any one of the n objects, the second position may be filled $n - 1$ ways using any of the remaining $n - 1$ objects, and so on until the last object is placed in the last position. Thus, by the multiplication principle, this operation may be carried out in $n \cdot (n-1) \cdot \cdots \cdot 1 = n!$ ways. ■

For example, the number of arrangements of five distinct cards is $5! = 120$.

One also may be interested in the number of ways of selecting r objects from n distinct objects and then ordering these r objects.

Theorem 1.6.3 The number of permutations of n distinct objects taken r at a time is

$$_nP_r = \frac{n!}{(n-r)!} \tag{1.6.2}$$

Proof

To fill r positions from n objects, the first position may be filled in n ways using any one of the n objects, the second position may be filled in $n - 1$ ways, and so on until $n - (r - 1)$ objects are left to fill in the rth position. Thus, the total number of ways of carrying out this operation is

$$n \cdot (n-1) \cdot (n-2) \cdot \cdots \cdot (n-(r-1)) = \frac{n!}{(n-r)!}$$ ■

Example 1.6.5 The number of permutations of the four letters a, b, c, d taken two at a time is $4!/2! = 12$. These are displayed in Figure 1.11. In picking two out of the four letters, there are six unordered ways to choose two letters from the four, as given by the top row. Each combination of two letters then can be permuted $2!$ ways to get the total of 12 ordered arrangements.

FIGURE 1.11 Permutations of four objects taken two at a time

ab	*ac*	*ad*	*bc*	*bd*	*cd*
ba	*ca*	*da*	*cb*	*db*	*dc*

Example 1.6.6 A box contains n tickets, each marked with a different integer, 1, 2, 3, ..., n. If three tickets are selected at random *without* replacement, what is the probability of obtaining tickets with consecutive integers? One possible solution would be to let the sample space consist of all ordered triples (i, j, k), where i, j, and k are different integers in the range 1 to n. The number of such triples is ${}_nP_3 = n!/(n-3)! = n(n-1)(n-2)$. The triples that consist of consecutive integers would be (1, 2, 3), (2, 3, 4), ..., $(n-2, n-1, n)$ or any of the triples formed by permuting the entries in these. There would be $3! \cdot (n-2) = 6 \cdot (n-2)$ such triples. The desired probability is

$$\frac{6 \cdot (n-2)}{{}_nP_3} = \frac{6 \cdot (n-2)}{n(n-1)(n-2)} = \frac{6}{n(n-1)}$$

If the order of the objects is not important, then one may simply be interested in the number of **combinations** that are possible when selecting r objects from n distinct objects. The symbol $\binom{n}{r}$ usually is used to denote this number.

Theorem 1.6.4 The number of combinations of n distinct objects chosen r at a time is

$$\binom{n}{r} = \frac{n!}{r!(n-r)!} \tag{1.6.3}$$

Proof

As suggested in the preceding example, ${}_nP_r$ may be interpreted as the number of ways of choosing r objects from n objects and then permuting the r objects $r!$ ways, giving

$${}_nP_r = \binom{n}{r} r! = \frac{n!}{(n-r)!}$$

Dividing by $r!$ gives the desired expression for $\binom{n}{r}$. ■

Thus, the number of combinations of four letters taken two at a time is $\binom{4}{2} = \frac{4!}{2!2!} = 6$, as noted above. If order is considered, then the number of arrangements becomes $6 \cdot 2! = 12$ as before. Thus, $\binom{4}{2}$ counts the number of paired symbols in either the first or second row, but not both, in Figure 1.11.

It also is possible to solve the probability problem in Example 1.6.6 using combinations. The sample space would consist of all combinations of the n integers 1, 2, ..., n taken three at a time. Equivalently, this would be the collection of all subsets of size 3 from the set $\{1, 2, 3, \ldots, n\}$, of which there are

$$\binom{n}{3} = \frac{n!}{3!(n-3)!} = \frac{n(n-1)(n-2)}{6}$$

The $n - 2$ combinations or subsets of consecutive integers would be $\{1, 2, 3\}$, $\{2, 3, 4\}, \ldots, \{n-2, n-1, n\}$. As usual, no distinction should be made of subsets that list the elements in a different order. The resulting probability is

$$\frac{(n-2)}{[n(n-1)(n-2)/6]} = \frac{6}{n(n-1)}$$

as before.

This shows that some problems can be solved using either combinations or permutations. Usually, if there is a choice, the combination approach is simpler because the sample space is smaller. However, combinations are not appropriate in some problems.

Example 1.6.7 In Example 1.6.6, suppose that the sampling is done *with* replacement. Now, the same number can be repeated in the triples (i, j, k), so that the sample space has n^3 outcomes. There are still only $6(n - 2)$ triples of consecutive integers, because repeated integers cannot be consecutive. The probability of consecutive integers in this case is $6(n - 2)/n^3$. Integers can be repeated in this case, so the combination approach is not appropriate.

Example 1.6.8 A familiar use of the combination notation is in expressing the binomial expansion

$$(a + b)^n = \sum_{k=0}^{n} \binom{n}{k} a^k b^{n-k} \tag{1.6.4}$$

In this case, $\binom{n}{k}$ is the coefficient of $a^k b^{n-k}$, and it represents the number of ways of choosing k of the n factors $(a + b)$ from which to use the a term, with the b term being used from the remaining $n - k$ factors.

Example 1.6.9 The combination concept can be used to determine the number of subsets of a set of m elements. There are $\binom{m}{j}$ ways of choosing j elements from the m elements, so there are $\binom{m}{j}$ subsets of j elements for $j = 0, 1, \ldots, m$. The case $j = 0$ corresponds to the null set and is represented by $\binom{m}{0} = 1$, because $0!$ is defined to be equal to 1, for notational convenience. Thus the total number of subsets including the null set is given by

$$\sum_{j=0}^{m} \binom{m}{j} = (1+1)^m = 2^m \tag{1.6.5}$$

Example 1.6.10 If five cards are drawn from a deck of cards without replacement, the number of five-card hands is

$$\binom{52}{5} = \frac{52!}{5!47!}$$

If order is taken into account as in Example 1.6.4, then the number of ordered five-card hands is

$${}_{52}P_5 = \binom{52}{5} 5! = \frac{52!}{47!}$$

Similarly, in Example 1.5.2 the number of ordered two-card hands was given to be

$$\binom{52}{2} \cdot 2! = 52 \cdot 51$$

INDISTINGUISHABLE OBJECTS

The discussion to this point has dealt with arrangements of n *distinguishable* objects. There are also many applications involving objects that are not all distinguishable.

Example 1.6.11 You have five marbles, two black and three white, but otherwise indistinguishable. In Figure 1.12, we represent all the distinguishable arrangements of two black (B) and three white (W) marbles.

FIGURE 1.12 Distinguishable arrangements of five objects, two of one type and three of another

$BBWWW$	$BWBWW$	$WBBWW$	$BWWBW$	$WBWBW$
$WWBBW$	$WWBWB$	$WWWBB$	$WBWWB$	$BWWWB$

Notice that arrangements are distinguishable if they differ by exchanging marbles of different colors, but not if the exchange involves the same color. We will refer to these 10 different arrangements as *permutations* of the five objects even though the objects are not all distinguishable.

A more general way to count such permutations first would be to introduce labels for the objects, say $B_1 B_2 W_1 W_2 W_3$. There are 5! permutations of these distinguishable objects, but within each color there are permutations that we don't want to count. We can compensate by dividing by the number of permutations of black objects (2!) and of white objects (3!). Thus, the number of permutations of nondistinguishable objects is

$$\frac{5!}{2!3!} = 10$$

This is a special case of the following theorem.

Theorem 1.6.5 The number of distinguishable permutations of n objects of which r are of one kind and $n - r$ are of another kind is

$$\binom{n}{r} = \frac{n!}{r!(n-r)!} \tag{1.6.6}$$

■

Clearly, this concept can be generalized to the case of permuting k types of objects.

Theorem 1.6.6 The number of permutations of n objects of which r_1 are of one kind, r_2 of a second kind, ..., r_k of a kth kind is

$$\frac{n!}{r_1!r_2! \cdots r_k!} \tag{1.6.7}$$

Proof

This follows from the argument of Example 1.6.11, except with k different colors of balls. ■

Example 1.6.12 You have 10 marbles—two black, three white, and five red, but otherwise not distinguishable. The number of different permutations is

$$\frac{10!}{2!3!5!} = 2520$$

The notion of permutations of n objects, not all of which are distinguishable, is related to yet another type of operation with *n distinct* objects.

PARTITIONING

Let us select r objects from n distinct objects and place them in a box or "cell," and then place the remaining $n - r$ objects in a second cell. Clearly, there are $\binom{n}{r}$ ways of doing this (because permuting the objects within a cell will not produce a new result), and this is referred to as the number of ways of **partitioning** n objects into two cells with r objects in one cell and $n - r$ in the other. The concept generalizes readily to partitioning n distinct objects into more than two cells.

Theorem 1.6.7 The number of ways of partitioning a set of n objects into k cells with r_1 objects in the first cell, r_2 in the second cell, and so forth is

$$\frac{n!}{r_1!r_2! \cdots r_k!}$$

where $\sum_{i=1}^{k} r_i = n$. ■

Note that partitioning assumes that the number of objects to be placed in each cell is fixed, and that the order in which the objects are placed into cells is not considered.

By successively selecting the objects, the number of partitions also may be expressed as

$$\binom{n}{r_1}\binom{n-r_1}{r_2}\cdots\binom{n-r_1-\cdots-r_{k-1}}{r_k} = \frac{n!}{r_1!r_2!\cdots r_k!}$$

Example 1.6.13 How many ways can you distribute 12 different popsicles equally among four children? By Theorem 1.6.7 this is

$$\frac{12!}{3!3!3!3!} = 369{,}600$$

This is also the number of ways of arranging 12 popsicles, of which three are red, three are green, three are orange, and three are yellow, if popsicles of the same color are otherwise indistinguishable.

PROBABILITY COMPUTATIONS

As mentioned earlier, if it can be assumed that all possible outcomes are equally likely to occur, then the classical probability concept is useful for assigning probabilities to events, and the counting techniques reviewed in this section may be helpful in computing the number of ways an event may occur.

Recall that the method of sampling, and assumptions concerning order, whether the items are indistinguishable, and so on, may have an effect on the number of possible outcomes.

Example 1.6.14 A student answers 20 true–false questions at random. The probability of getting 100% on the test is $P(100\%) = 1/2^{20} = 0.00000095$. We wish to know the probability of getting 80% right, that is, answering 16 questions correctly. We do not care which 16 questions are answered correctly, so there are $\binom{20}{16}$ ways of choosing exactly 16 correct answers, and $P(80\%) = \binom{20}{16}\Big/2^{20} = 0.0046$.

Example 1.6.15 **Sampling Without Replacement** A box contains 10 black marbles and 20 white marbles, and five marbles are selected without replacement. The probability of getting exactly two black marbles is

$$P(\text{exactly 2 black}) = \frac{\binom{10}{2}\binom{20}{3}}{\binom{30}{5}} = 0.360 \tag{1.6.8}$$

There are $\binom{30}{5}$ total possible outcomes. Also there are $\binom{10}{2}$ ways of choosing the two black marbles from the 10 black marbles, and $\binom{20}{3}$ ways of choosing the remaining three white marbles from the 20 white marbles. By the multiplication principle, there are $\binom{10}{2}\binom{20}{3}$ ways of achieving the event of getting two black marbles. Note that order was not considered important in this problem, although all 30 marbles are considered distinct in this computation, both in considering the total number of outcomes in the sample space and in considering how many outcomes correspond to the desired event occurring. Even though the question does not distinguish between the order of outcomes, it is possible to consider the question relative to the larger sample space of equally likely ordered outcomes. In that case one would have ${}_{30}P_5 = \binom{30}{5}\cdot 5!$ possible outcomes and

$$P(\text{exactly 2 black}) = \frac{\binom{10}{2}\binom{20}{3}\cdot 5!}{\binom{30}{5}\cdot 5!} \tag{1.6.9}$$

which gives the same answer as before.

It also is possible to attack this problem by the conditional probability approach discussed in Section 1.5. First consider the probability of getting the outcome BBWWW in the specified order. Here we choose to use the distinction between B and W but not the distinction within the B's or within the W's. By the conditional probability approach, this joint probability may be expressed as

$$P(\text{BBWWW}) = \frac{10}{30}\,\frac{9}{29}\,\frac{20}{28}\,\frac{19}{27}\,\frac{18}{26}$$

Similarly,

$$P(\text{BWBWW}) = \frac{10}{30}\,\frac{20}{29}\,\frac{9}{28}\,\frac{19}{27}\,\frac{18}{26}$$

and so on. Thus, each particular ordering has the same probability. If we do not wish to distinguish between the ordering of the black and white marbles, then

$$P(\text{exactly 2 black}) = \binom{5}{2} \cdot \frac{10}{30}\,\frac{9}{29}\,\frac{20}{28}\,\frac{19}{27}\,\frac{18}{26} \qquad \textbf{(1.6.10)}$$

which again is the same as equation (1.6.8). That is, there are $\binom{5}{2} = 10$ different particular orderings that have two black and three white marbles (see Figure 1.12). One could consider $\binom{5}{2}$ as the number of ways of choosing two positions out of the five positions in which to place two black marbles. If a particular order is not required, the probability of a successful outcome is greater.

We could continue to consider all 30 marbles distinct in this framework, but because only the order between black and white was considered in computing a particular sequence, it follows that there are only $\binom{5}{2}$ unordered sequences rather than 5! sequences. Thus, although two black marbles may be distinct, permuting them does not produce a different result: The order of the black marbles within themselves was not considered important when defining the ordered sequences; only the order between black and white was considered. Thus the coefficient $\binom{5}{2}$ could also be interpreted as the number of permutations of five things of which two were alike and three were alike (see Figure 1.12).

Thus, we have seen that it is possible to think of the black and white marbles as being indistinguishable within themselves in this problem, and the same value for $P(\text{exactly 2 black})$ is obtained; however, the computation is no longer carried out over an original basic sample space of equally likely outcomes. For example, on the first draw one would just have the two possible outcomes, B and W, although these two outcomes obviously would not be equally likely, but rather $P(B) = 10/30$ and $P(W) = 20/30$. Indeed, the assumption that the black marbles and white marbles are indistinguishable within themselves appears more natural in the conditional probability approach. Nevertheless, the distinctness assumption is a convenient aid in the first approach to obtain the more basic equally likely sample space, even though the question itself does not require distinguishing within a color.

Example 1.6.16 **Sampling with Replacement** If the five marbles are drawn *with replacement* in Example 1.6.15, then the conditional probability approach seems most natural and analogous to (1.6.10),

$$P(\text{exactly 2 black}) = \binom{5}{2}\left(\frac{10}{30}\right)^2\left(\frac{20}{30}\right)^3 \tag{1.6.11}$$

Of course, in this case the outcomes on each draw are independent.

If one chooses to use the classical approach in this case, it is more convenient to consider the sample space of 30^5 equally likely *ordered* outcomes; in Example 1.6.15 it is more convenient just to consider the sample space of $\binom{30}{5}$ unordered outcomes as in equation (1.6.8), rather than the ordered outcomes as in equation (1.6.9). For event A, one then has "exactly 2 black,"

$$P(A) = \frac{n(A)}{N} = \frac{\binom{5}{2}10^2 30^3}{30^5}$$

The form in this case remains quite similar to equation (1.6.11), although the argument would be somewhat different. There are $\binom{5}{2}$ different patterns in which the ordered arrangements may contain two black and three white marbles, and for each pattern there are $10^2 30^3$ distinct arrangements that can be formed in this sample space.

Because many diverse types of probability problems can be stated, a unique approach often may be needed to identify the mutually exclusive ways that an event can occur in such a manner that these ways can be readily counted. However, certain classical problems (such as those illustrated in Examples 1.6.15 and 1.6.16) can be recognized easily and general probability distribution functions can be determined for them. For these problems, the individual counting problems need not be analyzed so carefully each time.

SUMMARY

The purpose of this chapter was to develop the concept of probability in order to model phenomena where the observed result is uncertain before experimentation. The basic approach involves defining the sample space as the set of all possible outcomes of the experiment, and defining an event mathematically as the set of outcomes associated with occurrence of the event. The primary motivation for assigning probability to an event involves the long-term relative frequency interpretation. However, the approach of defining probability in terms of a simple set of axioms is more general, and it allows the possibility of other methods of assignment and other interpretations of probability. This approach also makes it possible to derive general properties of probability.

The notion of conditional probability allows the introduction of additional information concerning the occurrence of one event when assigning probability to another. If the probability assigned to one event is not affected by the information that another event has occurred, then the events are considered independent. Care should be taken not to confuse the concepts of independent and mutually exclusive events. Specifically, mutually exclusive events are dependent, because the occurrence of one precludes the occurrence of the other. In other words, the conditional probability of one given the other is zero.

One of the primary methods of assigning probability, which applies in the case of a finite sample space, is based on the assumption that all outcomes are equally likely to occur. To implement this method, it is useful to have techniques for counting the number of outcomes in an event. The primary techniques include formulas for counting ordered arrangements of objects (permutations) and unordered sets of objects (combinations).

To express probability models by general formulas, it is convenient first to introduce the concept of a "random variable" and a function that describes the probability distribution. These concepts will be discussed in the next chapter, and general solutions then can be provided for some of the basic counting problems most often encountered.

EXERCISES

1. A gum-ball machine gives out a red, a black, or a green gum ball.

(a) Describe an appropriate sample space.

(b) List all possible events.

(c) If R is the event "red," then list the outcomes in R'.

(d) If G is the event "green," then what is $R \cap G$?

2. Two gum balls are obtained from the machine in Exercise 1 from two trials. The order of the outcomes is important. Assume that at least two balls of each color are in the machine.

(a) What is an appropriate sample space?

(b) How many total possible events are there that contain eight outcomes?

(c) Express the following events as unions of elementary events. C_1 = getting a red ball on the first trial, C_2 = getting at least one red ball, $C_1 \cap C_2$, $C_1' \cap C_2$.

3. There are four basic blood groups: O, A, B, and AB. Ordinarily, anyone can receive the blood of a donor from their own group. Also, anyone can receive the blood of a donor from the O group, and any of the four types can be used by a recipient from the AB group.

All other possibilities are undesirable. An experiment consists of drawing a pint of blood and determining its type for each of the next two donors who enter a blood bank.

(a) List the possible (ordered) outcomes of this experiment.

(b) List the outcomes corresponding to the event that the second donor can receive the blood of the first donor.

(c) List the outcomes corresponding to the event that each donor can receive the blood of the other.

4. An experiment consists of drawing gum balls from a gum-ball machine until a red ball is obtained. Describe a sample space for this experiment.

5. The number of alpha particles emitted by a radioactive sample in a fixed time interval is counted.

(a) Give a sample space for this experiment.

(b) The elapsed time is measured until the first alpha particle is emitted. Give a sample space for this experiment.

6. An experiment is conducted to determine what fraction of a piece of metal is gold. Give a sample space for this experiment.

7. A randomly selected car battery is tested and the time of failure is recorded. Give an appropriate sample space for this experiment.

8. We obtain 100 gum balls from a machine, and we get 20 red (R), 30 black (B), and 50 green (G) gum balls.

(a) Can we use, as a probability model for the color of a gum ball from the machine, one given by $p_1 = P(\text{R}) = 0.2$, $p_2 = P(\text{B}) = 0.3$, and $p_3 = P(\text{G}) = 0.5$?

(b) Suppose we later notice that some yellow (Y) gum balls are also in the machine. Could we use as a model $p_1 = 0.2$, $p_2 = 0.3$, $p_3 = 0.5$, and $p_4 = P(\text{Y}) = 0.1$?

9. In Exercise 2, suppose that each of the nine possible outcomes in the sample space is equally likely to occur. Compute each of the following:

(a) $P(\text{both red})$.

(b) $P(C_1)$.

(c) $P(C_2)$.

(d) $P(C_1 \cap C_2)$.

(e) $P(C_1' \cap C_2)$.

(f) $P(C_1 \cup C_2)$.

10. Consider Exercise 3. Suppose, for a particular racial group, the four blood types are equally likely to occur.

(a) Compute the probability that the second donor can receive blood from the first donor.

(b) Compute the probability that each donor can receive blood from the other.

(c) Compute the probability that neither can receive blood from the other.

11. Prove that $P(\varnothing) = 0$. *Hint*: Let $A_i = \varnothing$ for all i in equation (1.3.3).

12. Prove equation (1.3.5). *Hint:* Let $A_i = \varnothing$ for all $i > k$ in equation (1.3.3).

13. When an experiment is performed, one and only one of the events A_1, A_2, or A_3 will occur. Find $P(A_1)$, $P(A_2)$, and $P(A_3)$ under each of the following assumptions:

(a) $P(A_1) = P(A_2) = P(A_3)$.

(b) $P(A_1) = P(A_2)$ and $P(A_3) = 1/2$.

(c) $P(A_1) = 2P(A_2) = 3P(A_3)$.

14. A balanced coin is tossed four times. List the possible outcomes and compute the probability of each of the following events:

(a) exactly three heads.

(b) at least one head.

(c) the number of heads equals the number of tails.

(d) the number of heads exceeds the number of tails.

15. Two part-time teachers are hired by the mathematics department and each is assigned at random to teach a single course, in trigonometry, algebra, or calculus. List the outcomes in the sample space and find the probability that they will teach different courses. Assume that more than one section of each course is offered.

16. Prove Theorem 1.4.4. *Hint*: Write $A \cup B \cup C = (A \cup B) \cup C$ and apply Theorem 1.4.3.

17. Prove Theorem 1.4.5. *Hint:* If $A \subset B$, then we can write $B = A \cup (B \cap A')$, a disjoint union.

18. If A and B are events, show that:

(a) $P(A \cap B') = P(A) - P(A \cap B)$.

(b) $P(A \cup B) = 1 - P(A' \cap B')$.

19. Let $P(A) = P(B) = 1/3$ and $P(A \cap B) = 1/10$. Find the following:

(a) $P(B')$.

(b) $P(A \cup B')$.

(c) $P(B \cap A')$.

(d) $P(A' \cup B')$.

20. Let $P(A) = 1/2$, $P(B) = 1/8$, and $P(C) = 1/4$, where A, B, and C are mutually exclusive. Find the following:

(a) $P(A \cup B \cup C)$.

(b) $P(A' \cap B' \cap C')$.

21. The event that *exactly one* of the events A or B occurs can be represented as $(A \cap B') \cup (A' \cap B)$. Show that

$$P[(A \cap B') \cup (A' \cap B)] = P(A) + P(B) - 2P(A \cap B)$$

22. A track star runs two races on a certain day. The probability that he wins the first race is 0.7, the probability that he wins the second race is 0.6, and the probability that he wins both races is 0.5. Find the probability that:

(a) he wins at least one race.

(b) he wins exactly one race.

(c) he wins neither race.

23. A certain family owns two television sets, one color and one black-and-white set. Let A be the event the color set is on and B the event the black-and-white set is on. If $P(A) = 0.4$, $P(B) = 0.3$, and $P(A \cup B) = 0.5$, find the probability of each event:

(a) both are on.

(b) the color set is on and the other is off.

(c) exactly one set is on.

(d) neither set is on.

24. Suppose $P(A_i) = 1/(3 + i)$ for $i = 1, 2, 3, 4$. Find an upper bound for $P(A_1 \cup A_2 \cup A_3 \cup A_4)$.

25. A box contains three good cards and two bad (penalty) cards. Player A chooses a card and then player B chooses a card. Compute the following probabilities:

(a) $P(A \text{ good})$.

(b) $P(B \text{ good} \mid A \text{ good})$.

(c) $P(B \text{ good} \mid A \text{ bad})$.

(d) $P(B \text{ good} \cap A \text{ good})$ using (1.5.5).

(e) Write out the sample space of ordered pairs and compute $P(B \text{ good} \cap A \text{ good})$ and $P(B \text{ good} \mid A \text{ good})$ directly from definitions. (*Note:* Assume that the cards are distinct.)

(f) $P(B \text{ good})$.

(g) $P(A \text{ good} \mid B \text{ good})$.

26. Repeat Exercise 25, but assume that player A looks at his card, replaces it in the box, and remixes the cards before player B draws.

27. A bag contains five blue balls and three red balls. A boy draws a ball, and then draws another without replacement. Compute the following probabilities:

(a) $P(2 \text{ blue balls})$.

(b) $P(1 \text{ blue and } 1 \text{ red})$.

(c) $P(\text{at least } 1 \text{ blue})$.

(d) $P(2 \text{ red balls})$.

28. In Exercise 27, suppose a third ball is drawn without replacement. Find:

(a) $P(\text{no red balls left after third draw})$.

(b) $P(1 \text{ red ball left})$.

(c) $P(\text{first red ball on last draw})$.

(d) $P(\text{a red ball on last draw})$.

29. A family has two children. It is known that at least one is a boy. What is the probability that the family has two boys, given at least one boy? Assume $P(\text{boy}) = 1/2$.

30. Two cards are drawn from a deck of cards without replacement.

(a) What is the probability that the second card is a heart, given that the first card is a heart?

(b) What is the probability that both cards are hearts, given that at least one is a heart?

31. A box contains five green balls, three black balls, and seven red balls. Two balls are selected at random without replacement from the box. What is the probability that:

(a) both balls are red?

(b) both balls are the same color?

32. A softball team has three pitchers, A, B, and C, with winning percentages of 0.4, 0.6, and 0.8, respectively. These pitchers pitch with frequency 2, 3, and 5 out of every 10 games, respectively. In other words, for a randomly selected game, $P(A) = 0.2$, $P(B) = 0.3$, and $P(C) = 0.5$. Find:

(a) $P(\text{team wins game}) = P(W)$.

(b) $P(A \text{ pitched game} \mid \text{team won}) = P(A \mid W)$.

33. One card is selected from a deck of 52 cards and placed in a second deck. A card then is selected from the second deck.

(a) What is the probability the second card is an ace?

(b) If the first card is placed into a deck of 54 cards containing two jokers, then what is the probability that a card drawn from the second deck is an ace?

(c) Given that an ace was drawn from the second deck in (b), what is the conditional probability that an ace was transferred?

34. A pocket contains three coins, one of which had a head on both sides, while the other two coins are normal. A coin is chosen at random from the pocket and tossed three times.

(a) Find the probability of obtaining three heads.

(b) If a head turns up all three times, what is the probability that this is the two-headed coin?

35. In a bolt factory, machines 1, 2, and 3 respectively produce 20%, 30%, and 50% of the total output. Of their respective outputs, 5%, 3%, and 2% are defective. A bolt is selected at random.

(a) What is the probability that it is defective?

(b) Given that it is defective, what is the probability that it was made by machine 1?

36. Drawer A contains five pennies and three dimes, while drawer B contains three pennies and seven dimes. A drawer is selected at random, and a coin is selected at random from that drawer.

(a) Find the probability of selecting a dime.

(b) Suppose a dime is obtained. What is the probability that it came from drawer B?

37. Let $P(A) = 0.4$ and $P(A \cup B) = 0.6$.

(a) For what value of $P(B)$ are A and B mutually exclusive?

(b) For what value of $P(B)$ are A and B independent?

38. Prove Theorem 1.5.5. *Hint:* Use Exercise 18.

39. Three independent components are hooked in series. Each component fails with probability p. What is the probability that the system does not fail?

40. Three independent components are hooked in parallel. Each component fails with probability p. What is the probability that the system does not fail?

41. Consider the following system with assigned probabilities of malfunction for the five components. Assume that malfunctions occur independently.

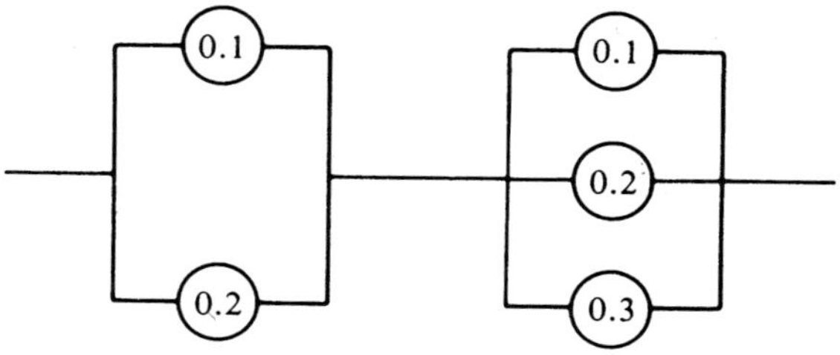

What is the probability the system does not malfunction?

42. The probability that a marksman hits a target is 0.9 on any given shot, and repeated shots are independent. He has two pistols; one contains two bullets and the other contains only one bullet. He selects a pistol at random and shoots at the target until the pistol is empty. What is the probability of hitting the target exactly one time?

43. Rework Exercise 27, assuming that the balls are chosen *with replacement*.

44. In a marble game a shooter may (A) miss, (B) hit one marble out and stick in the ring, or (C) hit one marble out and leave the ring. If B occurs, the shooter shoots again.

(a) If $P(A) = p_1$, $P(B) = p_2$, and $P(C) = p_3$, and these probabilities do not change from shot to shot, then express the probability of getting out exactly three marbles on one turn.

(b) What is the probability of getting out exactly x marbles in one turn?

(c) Show that the probability of getting one marble is greater than the probability of getting zero marbles if

$$p_1 < \frac{1 - p_2}{2 - p_2}$$

45. In the marble game in Exercise 44, suppose the probabilities depend on the number of marbles left in the ring, N. Let

$$P(A) = \frac{1}{N+1} \qquad P(B) = \frac{0.2N}{N+1} \qquad P(C) = \frac{0.8N}{N+1}$$

Rework Exercise 44 under this assumption.

46. A, B, and C are events such that $P(A) = 1/3$, $P(B) = 1/4$, and $P(C) = 1/5$. Find $P(A \cup B \cup C)$ under each of the following assumptions:

(a) If A, B, and C are mutually exclusive.

(b) If A, B, and C are independent.

47. A bowl contains four lottery tickets with the numbers 111, 221, 212, and 122. One ticket is drawn at random from the bowl, and A_i is the event "2 in the ith place"; $i = 1, 2, 3$. Determine whether A_1, A_2, and A_3 are independent.

48. Code words are formed from the letters A through Z.

(a) How many 26-letter words can be formed without repeating any letters?

(b) How many five-letter words can be formed without repeating any letters?

(c) How many five-letter words can be formed if letters can be repeated?

49. License plate numbers consist of two letters followed by a four-digit number, such as SB7904 or AY1637.

(a) How many different plates are possible if letters and digits can be repeated?

(b) Answer (a) if letters can be repeated but digits cannot.

(c) How many of the plates in (b) have a four-digit number that is greater than 5500?

50. In how many ways can three boys and three girls sit in a row if boys and girls must alternate?

51. How many *odd* three-digit numbers can be formed from the digits 0, 1, 2, 3, 4 if digits can be repeated, but the first digit cannot be zero?

52. Suppose that from 10 distinct objects, four are chosen at random with replacement.

(a) What is the probability that no object is chosen more than once?

(b) What is the probability that at least one object is chosen more than once?

53. A restaurant advertises 256 types of nachos. How many topping ingredients must be available to meet this claim if plain corn chips count as one type?

54. A club consists of 17 men and 13 women, and a committee of five members must be chosen.

(a) How many committees are possible?

(b) How many committees are possible with three men and two women?

(c) Answer (b) if a particular man must be included.

55. A football coach has 49 players available for duty on a special kick-receiving team.

(a) If 11 must be chosen to play on this special team, how many different teams are possible?

(b) If the 49 include 24 offensive and 25 defensive players, what is the probability that a randomly selected team has five offensive and six defensive players?

56. For positive integers $n > r$, show the following:

(a) $$\binom{n}{r} = \binom{n}{n-r}$$

(b) $$\binom{n}{r} = \binom{n-1}{r-1} + \binom{n-1}{r}.$$

57. Provide solutions for the following sums:

(a) $$\binom{4}{0} + \binom{4}{2} + \binom{4}{4}.$$

(b) $$\binom{6}{0} + \binom{6}{2} + \binom{6}{4} + \binom{6}{6}.$$

(c) $$\sum_{i=0}^{n} \binom{2n}{2i}.$$ *Hint:* Use Exercise 56(b).

58. Seven people show up to apply for jobs as cashiers at a discount store.

(a) If only three jobs are available, in how many ways can three be selected from the seven applicants?

(b) Suppose there are three male and four female applicants, and all seven are equally qualified, so the three jobs are filled at random. What is the probability that the three hired are all of the same sex?

(c) In how many different ways could the seven applicants be lined up while waiting for an interview?

(d) If there are four females and three males, in how many ways can the applicants be lined up if the first three are female?

59. The club in Exercise 54 must elect three officers: president, vice-president, and secretary. How many different ways can this turn out?

60. How many ways can 10 students be lined up to get on a bus if a particular pair of students refuse to follow each other in line?

61. Each student in a class of size n was born in a year with 365 days, and each reports his or her birth date (month and day, but not year).

(a) How many ways can this happen?

(b) How many ways can this happen with no repeated birth dates?

(c) What is the probability of no matching birth dates?

(d) In a class of 23 students, what is the probability of at least one repeated birth date?

62. A kindergarten student has 12 crayons.

(a) How many ways can three blue, four red, and five green crayons be arranged in a row?

(b) How many ways can 12 distinct crayons be placed in three boxes containing 3, 4, and 5 crayons, respectively?

63. How many ways can you partition 26 letters into three boxes containing 9, 11, and 6 letters?

64. How many ways can you permute 9 a's, 11 b's, and 6 c's?

65. A contest consists of finding all of the code words that can be formed from the letters in the name "ATARI." Assume that the letter A can be used twice, but the others at most once.

(a) How many five-letter words can be formed?

(b) How many two-letter words can be formed?

(c) How many words can be formed?

66. Three buses are available to transport 60 students on a field trip. The buses seat 15, 20, and 25 passengers, respectively. How many different ways can the students be loaded on the buses?

67. A certain machine has nine switches mounted in a row. Each switch has three positions, a, b, and c.

(a) How many different settings are possible?

(b) Answer (a) if each position is used three times.

68. Suppose 14 students have tickets for a concert.

(a) Three students (Bob, Jim, and Tom) own cars and will provide transportation to the concert. Bob's car has room for three passengers (nondrivers), while the cars owned by Jim and Tom each has room for four passengers. In how many different ways can the 11 passengers be loaded into the cars?

(b) At the concert hall the students are seated together in a row. If they take their seats in random order, find the probability that the three students who drove their cars have adjoining seats.

69. Suppose the winning number in a lottery is a four-digit number determined by drawing four slips of paper (without replacement) from a box that contains nine slips numbered consecutively 1 through 9 and then recording the digits in order from smallest to largest.

(a) How many different lottery numbers are possible?

(b) Find the probability that the winning number has only odd digits.

(c) How many different lottery numbers are possible if the digits are recorded in the order they were drawn?

70. Consider four dice A, B, C, and D numbered as follows: A has 4 on four faces and 0 on two faces; B has 3 on all six faces; C has 2 on four faces and 6 on two faces; and D has 5 on three faces and 1 on the other three faces. Suppose the statement $A > B$ means that the face showing on A is greater than on B, and so forth. Show that $P[A > B] = P[B > C] = P[C > D] = P[D > A] = 2/3$. In other words, if an opponent chooses a die, you can always select one that will defeat him with probability 2/3.

71. A laboratory test for steroid use in professional athletes has detection rates given in the following table:

	Test Result	
Steroid Use	+	−
Yes	.90	.10
No	.01	.99

If the rate of steroid use among professional athletes is 1 in 50:

(a) What is the probability that a professional athlete chosen at random will have a negative test result for steroid use?

(b) If the athlete tests positive, what is the probability that he has actually been using steroids?

72. A box contains four disks that have different colors on each side. Disk 1 is red and green, disk 2 is red and white, disk 3 is red and black, and disk 4 is green and white. One disk is selected at random from the box. Define events as follows: A = one side is red, B = one side is green, C = one side is white, and D = one side is black.

(a) Are A and B independent events? Why or why not?

(b) Are B and C independent events? Why or why not?

(c) Are any pairs of events mutually exclusive? Which ones?

CHAPTER 2

RANDOM VARIABLES AND THEIR DISTRIBUTIONS

2.1 INTRODUCTION

Our purpose is to develop mathematical models for describing the probabilities of outcomes or events occurring in a sample space. Because mathematical equations are expressed in terms of numerical values rather than as heads, colors, or other properties, it is convenient to define a function, known as a random variable, that associates each outcome in the experiment with a real number. We then can express the probability model for the experiment in terms of this associated random variable. Of course, in many experiments the results of interest already are numerical quantities, and in that case the natural function to use as the random variable would be the identity function.

Definition 2.1.1

Random Variable A **random variable**, say X, is a function defined over a sample space, S, that associates a real number, $X(e) = x$, with each possible outcome e in S.

Capital letters, such as X, Y, and Z will be used to denote random variables. The lower case letters $x, y, z, \ldots$ will be used to denote possible values that the corresponding random variables can attain. For mathematical reasons, it will be necessary to restrict the types of functions that are considered to be random variables. We will discuss this point after the following example.

Example 2.1.1 A four-sided (tetrahedral) die has a different number—1, 2, 3, or 4—affixed to each side. On any given roll, each of the four numbers is equally likely to occur. A game consists of rolling the die twice, and the score is the *maximum* of the two numbers that occur. Although the score cannot be predicted, we can determine the set of possible values and define a random variable. In particular, if $e = (i, j)$, where $i, j \in \{1, 2, 3, 4\}$, then $X(e) = \max(i, j)$. The sample space, S, and X are illustrated in Figure 2.1.

FIGURE 2.1 Sample space for two rolls of a four-sided die

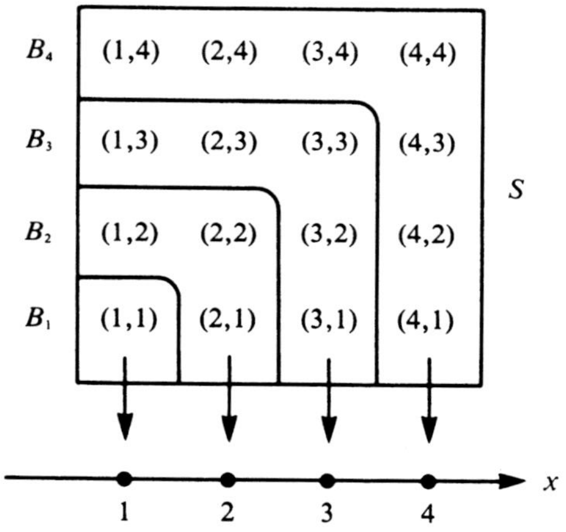

Each of the events B_1, B_2, B_3, and B_4 of S contains the pairs (i, j) that have a common maximum. In other words, X has value $x = 1$ over B_1, $x = 2$ over B_2, $x = 3$ over B_3, and $x = 4$ over B_4.

Other random variables also could be considered. For example, the random variable $Y(e) = i + j$ represents the total on the two rolls.

The concept of a random variable permits us to associate with any sample space, S, a sample space that is a set of real numbers, and in which the events of interest are subsets of real numbers. If such a real-valued event is denoted by A, then we would want the associated set

$$B = \{e \mid e \in S \text{ and } X(e) \in A\} \tag{2.1.1}$$

to be an event in the underlying sample space S. Even though A and B are subsets of different spaces, they usually are referred to as **equivalent events**, and we write

$$P[X \in A] = P(B) \tag{2.1.2}$$

The notation $P_X(A)$ sometimes is used instead of $P[X \in A]$ in equation (2.1.2). This defines a set function on the collection of real-valued events, and it can be shown to satisfy the three basic conditions of a probability set function, as given by Definition 1.3.1.

Although the random variable X is defined as a function of e, it usually is possible to express the events of interest only in terms of the real values that X assumes. Thus, our notation usually will suppress the dependence on the outcomes in S, such as we have done in equation (2.1.2).

For instance, in Example 2.1.1, if we were interested in the event of obtaining a score of "at most 3," this would correspond to $X = 1$, 2, or 3, or $X \in \{1, 2, 3\}$. Another possibility would be to represent the event in terms of some interval that contains the values 1, 2, and 3 but not 4, such as $A = (-\infty, 3]$. The associated equivalent event in S is $B = B_1 \cup B_2 \cup B_3$, and the probability is $P[X \in A] = P(B) = 1/16 + 3/16 + 5/16 = 9/16$. A convenient notation for $P[X \in A]$, in this example, is $P[X \leqslant 3]$. Actually, any other real event containing 1, 2, and 3 but not 4 could be used in this way, but intervals, and especially those of the form $(-\infty, x]$, will be of special importance in developing the properties of random variables.

As mentioned in Section 1.3, if the probabilities can be determined for each elementary event in a discrete sample space, then the probability of any event can be calculated from these by expressing the event as a union of mutually exclusive elementary events, and summing over their probabilities.

A more general approach for assigning probabilities to events in a real sample space can be based on assigning probabilities to intervals of the form $(-\infty, x]$ for all real numbers x. Thus, we will consider as random variables only functions X that satisfy the requirements that, for all real x, sets of the form

$$B = [X \leqslant x] = \{e \mid e \in S \text{ and } X(e) \in (-\infty, x]\} \tag{2.1.3}$$

are events in the sample space S. The probabilities of other real events can be evaluated in terms of the probabilities assigned to such intervals. For example, for the game of Example 2.1.1, we have determined that $P[X \leqslant 3] = 9/16$, and it also follows, by a similar argument, that $P[X \leqslant 2] = 1/4$. Because $(-\infty, 2]$ contains 1 and 2 but not 3, and $(-\infty, 3] = (-\infty, 2] \cup (2, 3]$, it follows that $P[X = 3] = P[X \leqslant 3] - P[X \leqslant 2] = 9/16 - 1/4 = 5/16$.

Other examples of random variables can be based on the sampling problems of Section 1.6.

Example 2.1.2 In Example 1.6.15, we discussed several alternative approaches for computing the probability of obtaining "exactly two black" marbles, when selecting five (without replacement) from a collection of 10 black and 20 white marbles. Suppose we are concerned with the general problem of obtaining x black marbles, for arbitrary x. Our approach will be to define a random variable X as the number of black marbles in the sample, and to determine the probability $P[X = x]$ for every possible value x. This is easily accomplished with the approach given by equation (1.6.8), and the result is

$$P[X = x] = \frac{\binom{10}{x}\binom{20}{5-x}}{\binom{30}{5}} \qquad x = 0, 1, 2, 3, 4, 5 \tag{2.1.4}$$

Random variables that arise from counting operations, such as the random variables in Examples 2. 1.1. and 2.1.2, are integer-valued. Integer-valued random variables are examples of an important special type known as discrete random variables.

2.2 DISCRETE RANDOM VARIABLES

Definition 2.2.1

If the set of all possible values of a random variable, X, is a countable set, $x_1, x_2, \ldots, x_n$, or $x_1, x_2, \ldots$, then X is called a **discrete random variable**. The function

$$f(x) = P[X = x] \qquad x = x_1, x_2, \ldots \tag{2.2.1}$$

that assigns the probability to each possible value x will be called the **discrete probability density function** (discrete pdf).

If it is clear from the context that X is discrete, then we simply will say pdf. Another common terminology is **probability mass function** (pmf), and the possible values, x_i, are called **mass points** of X. Sometimes a subscripted notation, $f_X(x)$, is used.

The following theorem gives general properties that any discrete pdf must satisfy.

Theorem 2.2.1 A function $f(x)$ is a discrete pdf if and only if it satisfies both of the following properties for at most a countably infinite set of reals $x_1, x_2, \ldots$:

$$f(x_i) \geqslant 0 \tag{2.2.2}$$

for all x_i, and

$$\sum_{\text{all } x_i} f(x_i) = 1 \tag{2.2.3}$$

Proof

Property (2.2.2) follows from the fact that the value of a discrete pdf is a probability and must be nonnegative. Because $x_1, x_2, \ldots$ represent all possible values of X, the events $[X = x_1]$, $[X = x_2]$, $\ldots$ constitute an exhaustive partition of the sample space. Thus,

$$\sum_{\text{all } x_i} f(x_i) = \sum_{\text{all } x_i} P[X = x_i] = 1$$

Consequently, any pdf must satisfy properties (2.2.2) and (2.2.3) and any function that satisfies properties (2.2.2) and (2.2.3) will assign probabilities consistent with Definition 1.3.1. ■

In some problems, it is possible to express the pdf by means of an equation, such as equation (2.1.4). However, it is sometimes more convenient to express it in tabular form. For example, one way to specify the pdf of X for the random variable X in Example 2.1.1 is given in Table 2.1.

TABLE 2.1 **Values of the discrete pdf of the maximum of two rolls of a four-sided die**

x	1	2	3	4
$f(x)$	1/16	3/16	5/16	7/16

Of course, these are the probabilities, respectively, of the events B_1, B_2, B_3, and B_4 in S.

A graphic representation of $f(x)$ is also of some interest. It would be possible to leave $f(x)$ undefined at points that are not possible values of X, but it is convenient to define $f(x)$ as zero at such points. The graph of the pdf in Table 2.1 is shown in Figure 2.2.

FIGURE 2.2 Discrete pdf of the maximum of two rolls of a four-sided die

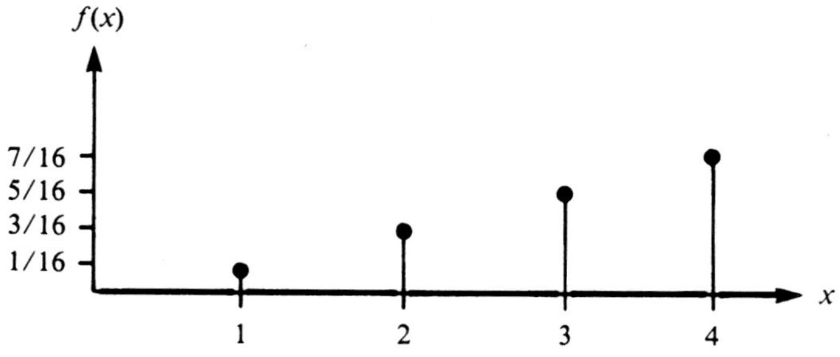

Example 2.2.1 Example 2.1.1 involves two rolls of a four-sided die. Now we will roll a 12-sided (dodecahedral) die twice. If each face is marked with an integer, 1 through 12, then each value is equally likely to occur on a single roll of the die. As before, we define a random variable X to be the maximum obtained on the two rolls. It is not hard to see that for each value x there are an odd number, $2x - 1$, of ways for that value to occur. Thus, the pdf of X must have the form

$$f(x) = c(2x - 1) \quad \text{for } x = 1, 2, \ldots, 12 \tag{2.2.4}$$

One way to determine c would be to do a more complete analysis of the counting problem, but another way would be to use equation (2.2.3). In particular,

$$1 = \sum_{x=1}^{12} f(x) = c \sum_{x=1}^{12} (2x - 1) = c\left[2 \sum_{x=1}^{12} x - 12\right]$$

$$= c\left[\frac{2(12)(13)}{2} - 12\right] = c(12)^2$$

So $c = 1/(12)^2 = 1/144$.

As mentioned in the last section, another way to specify the distribution of probability is to assign probabilities to intervals of the form $(-\infty, x]$, for all real x. The probability assigned to such an event is given by a function called the cumulative distribution function.

Definition 2.2.2

The **cumulative distribution function** (CDF) of a random variable X is defined for any real x by

$$F(x) = P[X \leqslant x] \tag{2.2.5}$$

FIGURE 2.3 The CDF of the maximum of two rolls of a four-sided die

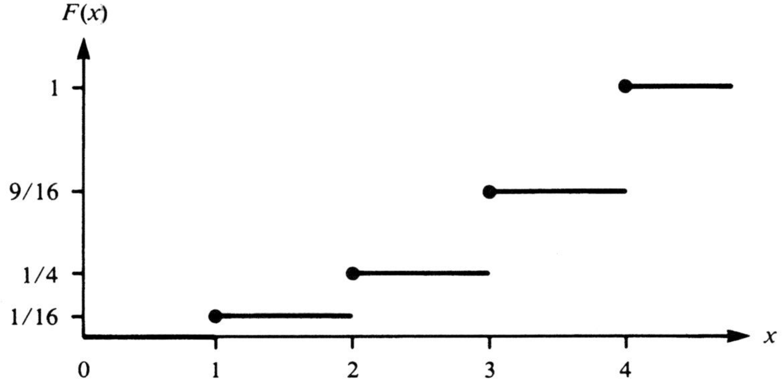

The function $F(x)$ often is referred to simply as the **distribution function** of X, and the subscripted notation, $F_X(x)$, sometimes is used.

For brevity, we often will use a short notation to indicate that a distribution of a particular form is appropriate. If we write $X \sim f(x)$ or $X \sim F(x)$, this will mean that the random variable X has pdf $f(x)$ and CDF $F(x)$.

As seen in Figure 2.3, the CDF of the distribution given in Table 2.1 is a nondecreasing step function. The step-function form of $F(x)$ is common to all discrete distributions, and the sizes of the steps or jumps in the graph of $F(x)$ correspond to the values of $f(x)$ at those points. This is easily seen by comparing Figures 2.2 and 2.3.

The general relationship between $F(x)$ and $f(x)$ for a discrete distribution is given by the following theorem.

Theorem 2.2.2 Let X be a discrete random variable with pdf $f(x)$ and CDF $F(x)$. If the possible values of X are indexed in increasing order, $x_1 < x_2 < x_3 < \cdots$, then $f(x_1) = F(x_1)$, and for any $i > 1$,

$$f(x_i) = F(x_i) - F(x_{i-1}) \tag{2.2.6}$$

Furthermore, if $x < x_1$ then $F(x) = 0$, and for any other real x

$$F(x) = \sum_{x_i \leq x} f(x_i) \tag{2.2.7}$$

where the summation is taken over all indices i such that $x_i \leq x$. ∎

The CDF of any random variable must satisfy the properties of the following theorem.

Theorem 2.2.3 A function $F(x)$ is a CDF for some random variable X if and only if it satisfies the following properties:

$$\lim_{x \to -\infty} F(x) = 0 \tag{2.2.8}$$

$$\lim_{x \to \infty} F(x) = 1 \tag{2.2.9}$$

$$\lim_{h \to 0^+} F(x + h) = F(x) \tag{2.2.10}$$

$$a < b \text{ implies } F(a) \leqslant F(b) \tag{2.2.11}$$

■

The first two properties say that $F(x)$ can be made arbitrarily close to 0 or 1 by taking x arbitrarily large, and negative or positive, respectively. In the examples considered so far, it turns out that $F(x)$ actually assumes these limiting values. Property (2.2.10) says that $F(x)$ is *continuous from the right.* Notice that in Figure 2.3 the only discontinuities are at the values 1, 2, 3, and 4, and the limit as x approaches these values from the right is the value of $F(x)$ at these values. On the other hand, as x approaches these values from the left, the limit of $F(x)$ is the value of $F(x)$ on the lower step, so $F(x)$ is not (in general) continuous from the left. Property (2.2.11) says that $F(x)$ is *nondecreasing*, which is easily seen to be the case in Figure 2.3. In general, this property follows from the fact that an interval of the form $(-\infty, b]$ can be represented as the union of two disjoint intervals

$$(-\infty, b] = (-\infty, a] \cup (a, b] \tag{2.2.12}$$

for any $a < b$. It follows that $F(b) = F(a) + P[a < x \leqslant b] \geqslant F(a)$, because $P[a < x \leqslant b] \geqslant 0$, and thus equation (2.2.11) is obtained.

Actually, by this argument we have obtained another very useful result, namely.

$$P[a < X \leqslant b] = F(b) - F(a) \tag{2.2.13}$$

This reduces the problem of computing probabilities for events defined in terms of intervals of the form $(a, b]$ to taking differences with $F(x)$.

Generally, it is somewhat easier to understand the nature of a random variable and its probability distribution by considering the pdf directly, rather than the CDF, although the CDF will provide a good basis for defining continuous probability distributions. This will be considered in the next section.

Some important properties of probability distributions involve numerical quantities called expected values.

Definition 2.2.3

If X is a discrete random vaιiable with pdf $f(x)$, then the **expected value** of X is defined by

$$E(X) = \sum_{x} x f(x) \tag{2.2.14}$$

The sum (2.2.14) is understood to be over all possible values of X. Furthermore, it is an ordinary sum if the range of X is finite, and an infinite series if the range of X is infinite. In the latter case, if the infinite series is not absolutely convergent, then we will say that $E(X)$ does not exist. Other common notations for $E(X)$ include μ, possibly with a subscript, μ_X. The terms **mean** and **expectation** also are often used.

The mean or expected value of a random variable is a "weighted average," and it can be considered as a measure of the "center" of the associated probability distribution.

Example 2.2.2 A box contains four chips. Two are labeled with the number 2, one is labeled with a 4, and the other with an 8. The average of the numbers on the four chips is $(2 + 2 + 4 + 8)/4 = 4$. The experiment of choosing a chip at random and recording its number can be associated with a discrete random variable X having distinct values $x = 2$, 4, or 8, with $f(2) = 1/2$ and $f(4) = f(8) = 1/4$. The corresponding expected value or mean is

$$\mu = E(X) = 2\left(\frac{1}{2}\right) + 4\left(\frac{1}{4}\right) + 8\left(\frac{1}{4}\right) = 4$$

as before. Notice that this also could model selection from a larger collection, as long as the possible observed values of X and the respective proportions in the collection, $f(x)$, remain the same as in the present example.

There is an analogy between the distribution of probability to values, x, and the distribution of mass to points in a physical system. For example, if masses of 0.5, 0.25, and 0.25 grams are placed at the respective points $x = 2$, 4, and 8 cm on the horizontal axis, then the value $2(0.5) + 4(0.25) + 8(0.25) = 4$ is the "center of mass" or balance point of the corresponding system. This is illustrated in Figure 2.4.

FIGURE 2.4 The center-of-mass interpretation of the mean

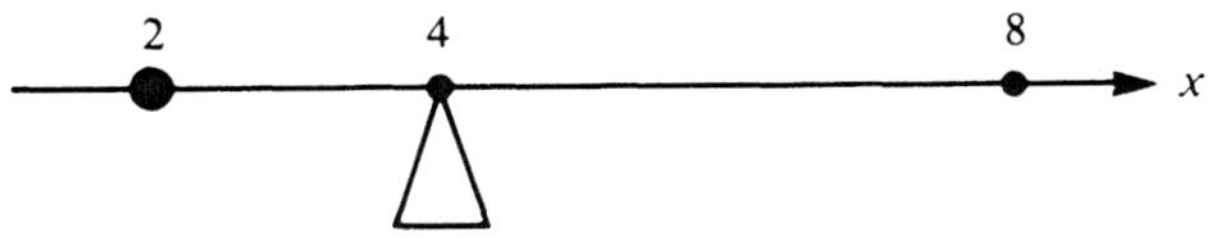

In the previous example $E(X)$ coincides with one of the possible values of X, but this is not always the case, as illustrated by the following example.

Example 2.2.3 A game of chance is based on drawing two chips at random without replacement from the box considered in Example 2.2.2. If the numbers on the two chips match, then the player wins \$2; otherwise, she loses \$1. Let X be the amount won by the player on a single play of the game. There are only two possible values, $X = 2$ if both chips bear the number 2, and $X = -1$ otherwise. Furthermore, there are $\binom{4}{2} = 6$ ways to draw two chips, and only one of these outcomes correspond to a match. The distribution of X is $f(2) = 1/6$ and $f(-1) = 5/6$, and consequently the expected amount won is $E(X) = (-1)(5/6) + (2)(1/6) = -1/2$. Thus, the expected amount "won" by the player is actually an expected loss of one-half dollar.

The connection with long-term relative frequency also is well illustrated by this example. Suppose the game is played M times in succession, and denote the relative frequencies of winning and losing by f_W and f_L, respectively. The average amount the player wins is $(-1)f_L + (2)f_W$. Because of statistical regularity, we have that f_L and f_W approach $f(-1)$ and $f(2)$, respectively, and thus the player's average winnings approach $E(X)$ as M approaches infinity.

Notice also that the game will be more equitable if the payoff to the player is changed to \$5 rather than \$2, because the resulting expected amount won then will be $(-1)(5/6) + (5)(1/6) = 0$. In general, for a game of chance, if the net amount won by a player is X, then the game is said to be a **fair game** if $E(X) = 0$.

2.3 CONTINUOUS RANDOM VARIABLES

The notion of a discrete random variable provides an adequate means of probability modeling for a large class of problems, including those that arise from the operation of counting. However, a discrete random variable is not an adequate model in many situations, and we must consider the notion of a continuous random variable. The CDF defined earlier remains meaningful for continuous random variables, but it also is useful to extend the concept of a pdf to continuous random variables.

Example 2.3.1 Each work day a man rides a bus to his place of business. Although a new bus arrives promptly every five minutes, the man generally arrives at the bus stop at a random time between bus arrivals. Thus, we might take his waiting time on any given morning to be a random variable X.

Although in practice we usually measure time only to the nearest unit (seconds, minutes, etc.), in theory we could measure time to within some arbitrarily small unit. Thus, even though in practice it might be possible to regard X as a discrete random variable with possible values determined by the smallest appropriate time unit, it usually is more convenient to consider the idealized situation in which X is assumed capable of attaining any value in some interval, and not just discrete points.

Returning to the man waiting for his bus, suppose that he is very observant and noticed over the years that the frequency of days when he waits no more than x minutes for the bus is proportional to x for all x. This suggests a CDF of the form $F(x) = P[X \leqslant x] = cx$, for some constant $c > 0$. Because the buses arrive at regular five-minute intervals, the range of possible values of X is the time interval $[0, 5]$. In other words, $P[0 \leqslant X \leqslant 5] = 1$, and it follows that $1 = F(5) = c \cdot 5$, and thus $c = 1/5$, and $F(x) = x/5$ if $0 \leqslant x \leqslant 5$. It also follows that $F(x) = 0$ if $x < 0$ and $F(x) = 1$ if $x > 5$.

Another way to study this distribution would be to observe the relative frequency of bus arrivals during short time intervals of the same length, but distributed throughout the waiting-time interval $[0, 5]$. It may be that the frequency of bus arrivals during intervals of the form $(x, x + \Delta x]$ for small Δx was proportional to the length of the interval, Δx, regardless of the value of x. The corresponding condition this imposes on the distribution of X is

$$P[x < X \leqslant x + \Delta x] = F(x + \Delta x) - F(x) = c\,\Delta x$$

for all $0 \leqslant x < x + \Delta x \leqslant 5$ and some $c > 0$. Of course, this implies that if $F(x)$ is differentiable at x, its derivative is constant, $F'(x) = c > 0$. Note also that for $x < 0$ or $x > 5$, the derivative also exists, but $F'(x) = 0$ because $P[x < X \leqslant x + \Delta x] = 0$ when x and $x + \Delta x$ are not possible values of X, and the derivative does not exist at all at $x = 0$ or 5.

In general, if $F(x)$ is the CDF of a continuous random variable X, then we will denote its derivative (where it exists) by $f(x)$, and under certain conditions, which will be specified shortly, we will call $f(x)$ the probability density function of X. In our example, $F(x)$ can be represented for values of x in the interval $[0, 5]$ as the integral of its derivative:

$$F(x) = \int_{-\infty}^{x} f(t)\,dt = \int_{0}^{x} \frac{1}{5}\,dt = \frac{x}{5}$$

The graphs of $F(x)$ and $f(x)$ are shown in Figure 2.5.

FIGURE 2.5 CDF and pdf of waiting time for a bus

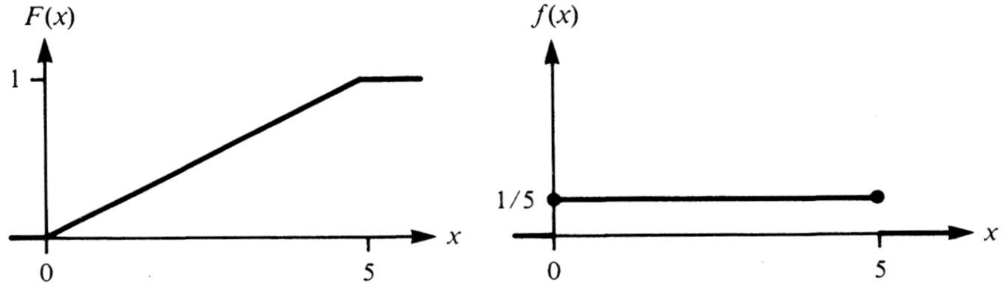

This provides a general approach to defining the distribution of a continuous random variable X.

Definition 2.3.1

A random variable X is called a **continuous random variable** if there is a function $f(x)$, called the **probability density function** (pdf) of X, such that the CDF can be represented as

$$F(x) = \int_{-\infty}^{x} f(t)\,dt \tag{2.3.1}$$

In more advanced treatments of probability, such distributions sometimes are called "absolutely continuous" distributions. The reason for such a distinction is that CDFs exist that are continuous (in the usual sense), but which cannot be represented as the integral of the derivative. We will apply the terminology **continuous distribution** only to probability distributions that satisfy property (2.3.1).

Sometimes it is convenient to use a subscripted notation, $F_X(x)$ and $f_X(x)$, for the CDF and pdf, respectively.

The defining property (2.3.1) provides a way to derive the CDF when the pdf is given, and it follows by the Fundamental Theorem of Calculus that the pdf can be obtained from the CDF by differentiation. Specifically,

$$f(x) = \frac{d}{dx} F(x) = F'(x) \tag{2.3.2}$$

wherever the derivative exists. Recall from Example 2.3.1 that there were two values of x where the derivative of $F(x)$ did not exist. In general, there may be many values of x where $F(x)$ is not differentiable, and these will occur at discontinuity points of the pdf, $f(x)$. Inspection of the graphs of $f(x)$ and $F(x)$ in Figure 2.5 shows that this situation occurs in the example at $x = 0$ and $x = 5$. However, this will not usually create a problem if the set of such values is finite, because an integrand can be redefined arbitrarily at a finite number of values x without affecting the value of the integral. Thus, the function $F(x)$, as represented in property (2.3.1), is unaffected regardless of how we treat such values. It also follows by similar considerations that events such as $[X = c]$, where c is a constant, will have probability zero when X is a continuous random variable. Consequently, events of the form $[X \in I]$, where I is an interval, are assigned the same probability whether I includes the endpoints or not. In other words, for a continuous random variable X, if $a < b$,

$$\begin{aligned} P[a < X \leqslant b] &= P[a \leqslant X < b] = P[a < X < b] \\ &= P[a \leqslant X \leqslant b] \end{aligned} \tag{2.3.3}$$

and each of these has the value $F(b) - F(a)$.

Thus, the CDF, $F(x)$, assigns probabilities to events of the form $(-\infty, x]$, and equation (2.3.3) shows how the probability assignment can be extended to any interval.

Any function $f(x)$ may be considered as a possible candidate for a pdf if it produces a legitimate CDF when integrated as in property (2.3.1). The following theorem provides conditions that will guarantee this.

Theorem 2.3.1 A function $f(x)$ is a pdf for some continuous random variable X if and only if it satisfies the properties

$$f(x) \geqslant 0 \tag{2.3.4}$$

for all real x, and

$$\int_{-\infty}^{\infty} f(x)\, dx = 1 \tag{2.3.5}$$

Proof

Properties (2.2.9) and (2.2.11) of a CDF follow from properties (2.3.5) and (2.3.4), respectively. The other properties follow from general results about integrals. ■

Example 2.3.2 A machine produces copper wire, and occasionally there is a flaw at some point along the wire. The length of wire (in meters) produced between successive flaws is a continuous random variable X with pdf of the form

$$f(x) = \begin{cases} c(1 + x)^{-3} & x > 0 \\ 0 & x \leqslant 0 \end{cases} \tag{2.3.6}$$

where c is a constant. The value of c can be determined by means of property (2.3.5). Specifically, set

$$1 = \int_{-\infty}^{\infty} f(x)\, dx = \int_{0}^{\infty} c(1 + x)^{-3}\, dx = c\left(\frac{1}{2}\right)$$

which is obtained following the substitution $u = 1 + x$ and an application of the power rule for integrals. This implies that the constant is $c = 2$.

Clearly property (2.3.4) also is satisfied in this case.

The CDF for this random variable is given by

$$F(x) = P[X \leqslant x] = \int_{-\infty}^{x} f(t)\, dt$$

$$= \begin{cases} \int_{-\infty}^{0} 0\, dt + \int_{0}^{x} 2(1 + t)^{-3}\, dt & x > 0 \\ \int_{-\infty}^{x} 0\, dt & x \leqslant 0 \end{cases}$$

$$= \begin{cases} 1 - (1 + x)^{-2} & x > 0 \\ 0 & x \leqslant 0 \end{cases}$$

Probabilities of intervals, such as $P[a \leqslant X \leqslant b]$, can be expressed directly in terms of the CDF or as integrals of the pdf. For example, the probability that a flaw occurs between 0.40 and 0.45 meters is given by

$$P[0.40 \leqslant X \leqslant 0.45] = \int_{0.40}^{0.45} f(x)\, dx = F(0.45) - F(0.40) = 0.035$$

Consideration of the frequency of occurrences over short intervals was suggested as a possible way to study a continuous distribution in Example 2.3.1. This approach provides some insight into the general nature of continuous distributions. For example, it may be observed that the frequency of occurrences over short intervals of length Δx, say $[x, x + \Delta x]$, is at least approximately proportional to the length of the interval, Δx, where the proportionality factor depends on x, say $f(x)$. The condition this imposes on the distribution of X is

$$\begin{aligned} P[x \leqslant X \leqslant x + \Delta x] &= F(x + \Delta x) - F(x) \\ &\doteq f(x)\, \Delta x \end{aligned} \tag{2.3.7}$$

where the error in the approximation is negligible relative to the length of the interval, Δx. This is illustrated in Figure 2.6. for the copper wire example.

The exact probability in equation (2.3.7) is represented by the area of the shaded region under the graph of $f(x)$, while the approximation is the area of the corresponding rectangle with height $f(x)$ and width Δx.

The smaller the value of Δx, the closer this approximation becomes. In this sense, it might be reasonable to think of $f(x)$ as assigning "probability density" for the distribution of X, and the term probability density function seems appropriate for $f(x)$. In other words, for a continuous random variable X, $f(x)$ is *not* a probability, although it does determine the probability assigned to arbitrarily small intervals. The area between the x-axis and the graph of $f(x)$ assigns probability to intervals, so that for $a < b$,

$$P[a \leqslant X \leqslant b] = \int_{a}^{b} f(x)\, dx \tag{2.3.8}$$

FIGURE 2.6 Continuous assignment of probability by pdf

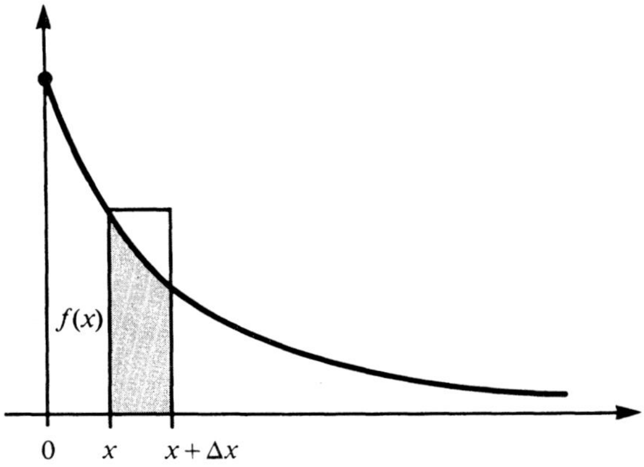

In Example 2.3.2, we could take the probability that the length between successive flaws between 0.40 and 0.45 meters to be approximately $f(0.40)(0.05) = 2(1.4)^{-3}(0.05) = 0.036$, or we could integrate the pdf between the limits 0.40 and 0.45 to obtain the exact answer, 0.035. For longer intervals, integrating $f(x)$ as in equation (2.3.8) would be more reasonable.

Note that in Section 2.2 we referred to a probability density function or density function for a discrete random variable, but the interpretation there is different, because probability is assigned at discrete points in that case rather than in a continuous manner. However, it will be convenient to refer to the "density function" or pdf in both continuous and discrete cases, and to use the same notation, $f(x)$ or $f_X(x)$, in the later chapters of the book. This will avoid the necessity of separate statements of general results that apply to both cases.

The notion of expected value can be extended to continuous random variables.

Definition 2.3.2

If X is a continuous random variable with pdf $f(x)$, then the **expected value** of X is defined by

$$E(X) = \int_{-\infty}^{\infty} x f(x)\, dx \tag{2.3.9}$$

if the integral in equation (2.3.9) is absolutely convergent. Otherwise we say that $E(X)$ does not exist.

As in the discrete case, other notations for $E(X)$ are μ or μ_X, and the terms **mean** or **expectation** of X also are commonly used. The center-of-mass analogy is still valid in this case, where mass is assigned to the x-axis in a continuous manner and in accordance with $f(x)$. Thus, μ can also be regarded as a central measure for a continuous distribution.

In Example 2.3.2, the mean length between flaws in a piece of wire is

$$\mu = \int_{-\infty}^{0} x \cdot 0\, dx + \int_{0}^{\infty} x \cdot 2(1 + x)^{-3}\, dx$$

If we make the substitution $t = 1 + x$, then

$$\mu = 2 \int_{1}^{\infty} (t - 1)t^{-3}\, dt = 2\left(1 - \frac{1}{2}\right) = 1$$

Other properties of probability distributions can be described in terms of quantities called percentiles.

Definition 2.3.3

If $0 < p < 1$, then a $100 \times p$th **percentile** of the distribution of a continuous random variable X is a solution x_p to the equation

$$F(x_p) = p \tag{2.3.10}$$

In general, a distribution may not be continuous, and if it has a discontinuity, then there will be some values of p for which equation (2.3.10) has no solution. Although we emphasize the continuous case in this book, it is possible to state a general definition of percentile by defining a pth percentile of the distribution of X to be a value x_p such that $P[X \leqslant x_p] \geqslant p$ and $P[X \geqslant x_p] \geqslant 1 - p$.

In essence, x_p is a value such that $100 \times p$ percent of the population values are at most x_p and $100 \times (1 - p)$ percent of the population values are at least x_p. This is illustrated for a continuous distribution in Figure 2.7. We also can think in terms of a proportion p rather than a percentage $100 \times p$ of the population, and in this context x_p is called a pth **quantile** of the distribution.

FIGURE 2.7 A $100 \times p$th percentile

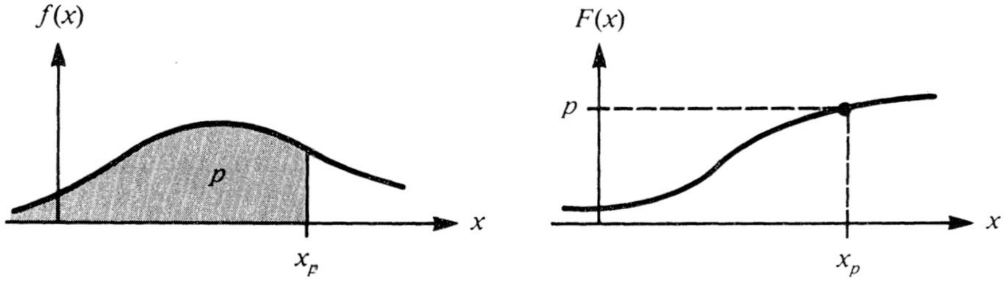

A **median** of the distribution of X is a 50th percentile, denoted by $x_{0.5}$ or m. This is an important special case of the percentile such that half of the population values are above it and half are below it. The median is used in some applications instead of the mean as a central measure.

Example 2.3.3 Consider the distribution of lifetimes, X (in months), of a particular type of component. We will assume that the CDF has the form

$$F(x) = 1 - e^{-(x/3)^2} \qquad x > 0$$

and zero otherwise. The median lifetime is

$$m = 3[-\ln(1 - 0.5)]^{1/2} = 3\sqrt{\ln 2} = 2.498 \text{ months}$$

It is desired to find the time t such that 10% of the components fail before t. This is the 10th percentile:

$$x_{0.10} = 3[-\ln(1 - 0.1)]^{1/2} = 3\sqrt{-\ln(0.9)} = 0.974 \text{ months}$$

Thus, if the components are guaranteed for one month, slightly more than 10% will need to be replaced.

Another measure of central tendency, which is sometimes considered, is the mode.

Definition 2.3.4

If the pdf has a unique maximum at $x = m_0$, say max $f(x) = f(m_0)$, then m_0 is called the **mode** of X.

In the previous example, the pdf of the distribution of lifetimes is

$$f(x) = \left(\frac{2}{9}\right)xe^{-(x/3)^2} \qquad x > 0$$

The solution to $f'(x) = 0$ is the unique maximum of $f(x)$, $x = m_0 = 3\sqrt{2}/2 = 2.121$ months.

In general, the mean, median, and mode may be all different, but there are cases in which they all agree.

Definition 2.3.5

A distribution with pdf $f(x)$ is said to be **symmetric** about c if $f(c - x) = f(c + x)$ for all x.

FIGURE 2.8 The pdf of a symmetric distribution

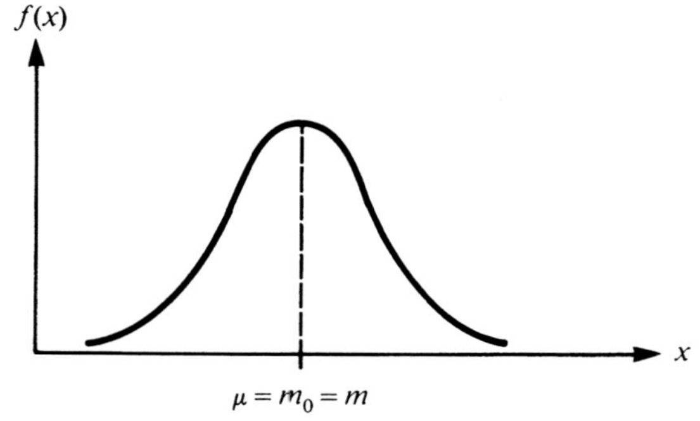

In other words, the "centered" pdf $g(x) = f(c - x)$ is an even function, in the usual sense that $g(x) = g(-x)$. The graph of $y = f(x)$ is a "mirror image" about the vertical line $x = c$. Asymmetric distributions, such as the one in Example 2.3.2, are called **skewed** distributions.

If $f(x)$ is symmetric about c and the mean μ exists, then $c = \mu$. If additionally, $f(x)$ has a unique maximum at m_0 and a unique median m, then $\mu = m_0 = m$. This is illustrated in Figure 2.8.

MIXED DISTRIBUTIONS

It is possible to have a random variable whose distribution is neither purely discrete nor continuous. A probability distribution for a random variable X is of **mixed type** if the CDF has the form

$$F(x) = aF_d(x) + (1 - a)F_c(x)$$

where $F_d(x)$ and $F_c(x)$ are CDFs of discrete and continuous type, respectively, and $0 < a < 1$.

Example 2.3.4 Suppose that a driver encounters a stop sign and either waits for a random period of time before proceeding or proceeds immediately. An appropriate model would allow the waiting time to be either zero or positive, both with nonzero probability. Let the CDF of the waiting time X be

$$F(x) = 0.4F_d(x) + 0.6F_c(x)$$
$$= 0.4 + 0.6(1 - e^{-x})$$

where $F_d(x) = 1$ and $F_c(x) = 1 - e^{-x}$ if $x \geqslant 0$, and both are zero if $x < 0$. The graph of $F(x)$ is shown in Figure 2.9. Thus, the probability of proceeding immediately is $P[X = 0] = 0.4$. The probability that the waiting time is less than 0.5

FIGURE 2.9 The CDF of a mixed distribution

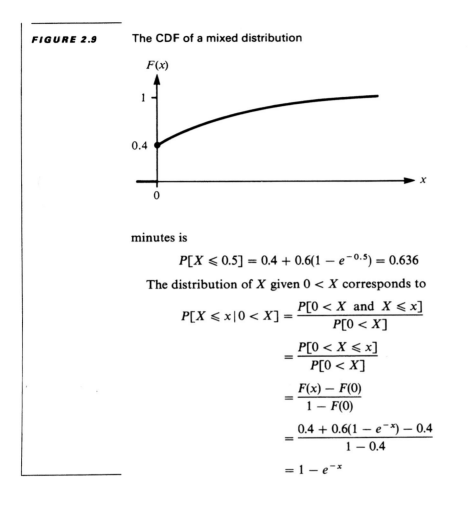

minutes is

$$P[X \leqslant 0.5] = 0.4 + 0.6(1 - e^{-0.5}) = 0.636$$

The distribution of X given $0 < X$ corresponds to

$$\begin{aligned} P[X \leqslant x \mid 0 < X] &= \frac{P[0 < X \text{ and } X \leqslant x]}{P[0 < X]} \\ &= \frac{P[0 < X \leqslant x]}{P[0 < X]} \\ &= \frac{F(x) - F(0)}{1 - F(0)} \\ &= \frac{0.4 + 0.6(1 - e^{-x}) - 0.4}{1 - 0.4} \\ &= 1 - e^{-x} \end{aligned}$$

2.4

SOME PROPERTIES OF EXPECTED VALUES

It is useful to consider, more generally, the expected value of a function of X. For example, if the radius of a disc is a random variable X, then the area of the disc, say $Y = \pi X^2$, is a function of X. In general, let X be a random variable with pdf $f(x)$, and denote by $u(x)$ a real-valued function whose domain includes the possible values of X. If we let $Y = u(X)$, then Y is a random variable with its own pdf, say $g(y)$. Suppose, for example, that X is a discrete random variable with pdf $f(x)$. Then $Y = u(X)$ is also a discrete random variable with pdf $g(y)$ and expected value defined in accordance with Definition 2.2.3, namely $E(Y) = \sum_y yg(y)$. Of course, evaluation of $E(Y)$ directly from the definition requires knowing the pdf $g(y)$. The following theorem provides another way to evaluate this expected value. The proof, which requires advanced methods, will be discussed in Chapter 6.

Theorem 2.4.1 If X is a random variable with pdf $f(x)$ and $u(x)$ is a real-valued function whose domain includes the possible values of X, then

$$E[u(X)] = \sum_x u(x)f(x) \qquad \text{if } X \text{ is discrete} \tag{2.4.1}$$

$$E[u(X)] = \int_{-\infty}^{\infty} u(x)f(x)\,dx \qquad \text{if } X \text{ is continuous} \tag{2.4.2}$$

■

It is clear that the expected value will have the "linearity" properties associated with integrals and sums.

Theorem 2.4.2 If X is a random variable with pdf $f(x)$, a and b are constants, and $g(x)$ and $h(x)$ are real-valued functions whose domains include the possible values of X, then

$$E[ag(X) + bh(X)] = aE[g(X)] + bE[h(X)] \tag{2.4.3}$$

Proof

Let X be continuous. It follows that

$$\begin{aligned} E[ag(X) + bh(X)] &= \int_{-\infty}^{\infty} [ag(x) + bh(x)]f(x)\,dx \\ &= a\int_{-\infty}^{\infty} g(x)f(x)\,dx + b\int_{-\infty}^{\infty} h(x)f(x)\,dx \\ &= aE[g(X)] + bE[h(X)] \end{aligned}$$

The discrete case is similar. ■

An obvious result of this theorem is that

$$E(aX + b) = aE(X) + b \tag{2.4.4}$$

An important special expected value is obtained if we consider the function $u(x) = (x - \mu)^2$

Definition 2.4.1

The **variance** of a random variable X is given by

$$\text{Var}(X) = E[(X - \mu)^2] \tag{2.4.5}$$

Other common notations for the variance are σ^2, σ_X^2, or $V(X)$, and a related quantity, called the **standard deviation** of X, is the positive square root of the variance, $\sigma = \sigma_X = \sqrt{\text{Var}(X)}$.

The variance provides a measure of the variability or amount of "spread" in the distribution of a random variable.

Example 2.4.1 In the experiment of Example 2.2.2, $E(X^2) = 2^2(1/2) + 4^2(1/4) + 8^2(1/4) = 22$, and thus $\text{Var}(X) = 22 - 4^2 = 6$ and $\sigma_X = \sqrt{6} = 2.45$. For comparison, consider a slightly different experiment where two chips are labeled with zeros, one with a 4, and one with a 12. If one chip is selected at random, and Y is its number, then $E(Y) = 4$ as in the original example. However, $\text{Var}(Y) = 24$ and $\sigma_Y = 2\sqrt{6} > \sigma_X$, which reflects the fact that the probability distribution of Y has more spread than that of X.

Certain special expected values, called moments, are useful in characterizing some features of the distribution.

Definition 2.4.2

The ***k*th moment about the origin** of a random variable X is

$$\mu'_k = E(X^k) \tag{2.4.6}$$

and the ***k*th moment about the mean** is

$$\mu_k = E[X - E(X)]^k = E(X - \mu)^k \tag{2.4.7}$$

Thus $E(X^k)$ may be considered as the kth moment of X or as the first moment of X^k. The first moment is the mean, and the simpler notation μ, rather than μ'_1, generally is preferred. The first moment about the mean is zero,

$$\mu_1 = E[X - E(X)] = E(X) - E(X) = 0$$

The second moment about the mean is the variance,

$$\mu_2 = E[(X - \mu)^2] = \sigma^2$$

and the second moment about the origin is involved in the following theorem about the variance.

Theorem 2.4.3 If X is a random variable, then

$$\mathrm{Var}(X) = E(X^2) - \mu^2 \tag{2.4.8}$$

Proof

$$\begin{aligned}\mathrm{Var}(X) &= E(X^2 - 2\mu X + \mu^2)\\ &= E(X^2) - 2\mu E(X) + \mu^2\\ &= E(X^2) - 2\mu^2 + \mu^2\end{aligned}$$

which yields the theorem. ■

It also follows immediately that

$$E(X^2) = \sigma^2 + \mu^2 \tag{2.4.9}$$

As noted previously, the variance provides a measure of the amount of spread in a distribution or the variability among members of a population. A rather extreme example of this occurs when X assumes only one value, say $P[X = c] = 1$. In this case $E(X) = c$ and $\mathrm{Var}(X) = 0$.

The remark following Theorem 2.4.2 dealt with the expected value of a linear function of a random variable. The following theorem deals with the variance.

Theorem 2.4.4 If X is a random variable and a and b are constants, then

$$\mathrm{Var}(aX + b) = a^2\,\mathrm{Var}(X) \tag{2.4.10}$$

Proof

$$\begin{aligned}\mathrm{Var}(aX + b) &= E[(aX + b - a\mu_X - b)^2]\\ &= E[a^2(X - \mu_X)^2]\\ &= a^2\,\mathrm{Var}(X)\end{aligned}$$

■

This means that the variance is affected by a change of scale, but not by a translation.

Another natural measure of variability would be the **mean absolute deviation**, $E|X - \mu|$, but the variance is generally a more convenient quantity with which to work.

The mean and variance provide a good deal of information about a population distribution, but higher moments and other quantities also may be useful. For example, the third moment about the mean, μ_3, is a measure of asymmetry or "skewness" of a distribution.

Theorem 2.4.5 If the distribution of X is symmetric about the mean $\mu = E(X)$, then the third moment about μ is zero, $\mu_3 = 0$.

Proof

See Exercise 28. ■

We can conclude that if $\mu_3 \neq 0$, then the distribution is not symmetric, but not conversely, because distributions exist that are not symmetric but which do have $\mu_3 = 0$ (see Exercise 29).

BOUNDS ON PROBABILITY

It is possible, in some cases, to find bounds on probabilities based on moments.

Theorem 2.4.6 If X is a random variable and $u(x)$ is a nonnegative real-valued function, then for any positive constant $c > 0$,

$$P[u(X) \geqslant c] \leqslant \frac{E[u(X)]}{c} \tag{2.4.11}$$

Proof

If $A = \{x \mid u(x) \geqslant c\}$, then for a continuous random variable,

$$\begin{aligned}
E[u(X)] &= \int_{-\infty}^{\infty} u(x) f(x)\, dx \\
&= \int_A u(x) f(x)\, dx + \int_{A^c} u(x) f(x)\, dx \\
&\geqslant \int_A u(x) f(x)\, dx \\
&\geqslant \int_A c f(x)\, dx \\
&= cP[X \in A] \\
&= cP[u(X) \geqslant c]
\end{aligned}$$

A similar proof holds for discrete variables. ■

A special case, known as the **Markov inequality**, is obtained if $u(x) = |x|^r$ for $r > 0$, namely

$$P[|X| \geqslant c] \leqslant \frac{E(|X|^r)}{c^r} \tag{2.4.12}$$

Another well-known result, the **Chebychev inequality**, is given by the following theorem.

Theorem 2.4.7 If X is a random variable with mean μ and variance σ^2, then for any $k > 0$,

$$P[|X - \mu| \geqslant k\sigma] \leqslant \frac{1}{k^2} \tag{2.4.13}$$

Proof

If $u(X) = (X - \mu)^2$, $c = k^2\sigma^2$, then using equation (2.4.11),

$$P[(X - \mu)^2 \geqslant k^2\sigma^2] \leqslant \frac{E(X - \mu)^2}{k^2\sigma^2} \leqslant \frac{1}{k^2}$$

and the result follows. ■

An alternative form is

$$P[|X - \mu| < k\sigma] \geqslant 1 - \frac{1}{k^2} \tag{2.4.14}$$

and if we let $\varepsilon = k\sigma$, then

$$P[|X - \mu| < \varepsilon] \geqslant 1 - \frac{\sigma^2}{\varepsilon^2} \tag{2.4.15}$$

and

$$P[|X - \mu| \geqslant \varepsilon] \leqslant \frac{\sigma^2}{\varepsilon^2} \tag{2.4.16}$$

Letting $k = 2$, we see that a random variable will be within two standard deviations of its mean with probability at least 0.75. Although this may not be a tight bound in all cases, it is surprising that such a bound can be found to hold for all possible discrete and continuous distributions. A tighter bound, in general, cannot be obtained, as shown in the following example.

Example 2.4.2 Suppose that X takes on the values -1, 0, and 1 with probabilities 1/8, 6/8, and 1/8, respectively. Then $\mu = 0$ and $\sigma^2 = 1/4$. For $k = 2$,

$$\begin{aligned} P[-2(0.5) < X - 0 < 2(0.5)] &= P[-1 < X < 1] \\ &= P[X = 0] \\ &= \frac{3}{4} = 1 - \frac{1}{k^2} \end{aligned}$$

It also is possible to show that if the variance is zero, the distribution is concentrated at a single value. Such a distribution is called a **degenerate distribution**.

Theorem 2.4.8 Let $\mu = E(X)$ and $\sigma^2 = \text{Var}(X)$. If $\sigma^2 = 0$, then $P[X = \mu] = 1$.

Proof

If $x \neq \mu$ for some observed value x, then $|x - \mu| \geqslant 1/i$ for some integer $i \geqslant 1$, and conversely. Thus,

$$[X \neq \mu] = \bigcup_{i=1}^{\infty} \left[|X - \mu| \geqslant \frac{1}{i} \right]$$

and using Boole's inequality, equation (1.4.5), we have

$$P[X \neq \mu] \leqslant \sum_{i=1}^{\infty} P\left[|X - \mu| \geqslant \frac{1}{i} \right]$$

and using equation (2.4.16) we obtain

$$P[X \neq \mu] \leqslant \sum_{i=1}^{\infty} i^2 \sigma^2 = 0$$

which implies that $P[X = \mu] = 1$. ■

APPROXIMATE MEAN AND VARIANCE

If a function of a random variable, say $H(X)$, can be expanded in a Taylor series, then an expression for the approximate mean and variance of $H(X)$ can be obtained in terms of the mean and variance of X.

For example, suppose that $H(x)$ has derivatives $H'(x)$, $H''(x)$, ... in an open interval containing $\mu = E(X)$. The function $H(x)$ has a Taylor approximation about μ,

$$H(x) \doteq H(\mu) + H'(\mu)(x - \mu) + \tfrac{1}{2}H''(\mu)(x - \mu)^2 \qquad \textbf{(2.4.17)}$$

which suggests the approximation

$$E[H(X)] \doteq H(\mu) + \tfrac{1}{2}H''(\mu)\sigma^2 \tag{2.4.18}$$

and, using the first two terms,

$$\text{Var}[H(X)] \doteq [H'(\mu)]^2\sigma^2 \tag{2.4.19}$$

where $\sigma^2 = \text{Var}(X)$.

The accuracy of these approximations depends primarily on the nature of the function $H(x)$ as well as on the amount of variability in the distribution of X.

Example 2.4.3 Let X be a positive-valued random variable, and let $H(x) = \ln x$, so that $H'(x) = 1/x$ and $H''(x) = -1/x^2$. It follows that

$$E[\ln X] \doteq \ln \mu + \left(\frac{1}{2}\right)\left(-\frac{1}{\mu^2}\right)\sigma^2$$

$$= \ln \mu - \frac{\sigma^2}{2\mu^2}$$

and

$$\text{Var}[\ln X] \doteq \left(\frac{1}{\mu}\right)^2\sigma^2 = \frac{\sigma^2}{\mu^2}$$

2.5 MOMENT GENERATING FUNCTIONS

A special expected value that is quite useful is known as the moment generating function.

Definition 2.5.1

If X is a random variable, then the expected value

$$M_X(t) = E(e^{tX}) \tag{2.5.1}$$

is called the **moment generating function** (MGF) of X if this expected value exists for all values of t in some interval of the form $-h < t < h$ for some $h > 0$.

In some situations it is desirable to suppress the subscript and use the simpler notation $M(t)$.

Example 2.5.1 Assume that X is a discrete finite-valued random variable with possible values $x_1, \dots, x_m$. The MGF is

$$M_X(t) = \sum_{i=1}^{m} e^{tx_i} f_X(x_i)$$

which is a differentiable function of t, with derivative

$$M'_X(t) = \sum_{i=1}^{m} x_i e^{tx_i} f_X(x_i)$$

and, in general, for any positive integer r,

$$M_X^{(r)}(t) = \sum_{i=1}^{m} x_i^r e^{tx_i} f_X(x_i)$$

Notice that if we evaluate $M_X^{(r)}(t)$ at $t = 0$ we obtain

$$M_X^{(r)}(0) = \sum_{i=1}^{m} x_i^r f_X(x_i) = E(X^r)$$

the rth moment about the origin. This also suggests the possibility of expanding in a power series about $t = 0$, $M_X(t) = c_0 + c_1 t + c_2 t^2 + \cdots$, where $c_r = E(X^r)/r!$.

These properties hold for any random variable for which an MGF exists, although a general proof is somewhat harder.

Theorem 2.5.1 If the MGF of X exists, then

$$E(X^r) = M_X^{(r)}(0) \quad \text{for all } r = 1, 2, \dots \tag{2.5.2}$$

and

$$M_X(t) = 1 + \sum_{r=1}^{\infty} \frac{E(X^r)t^r}{r!} \tag{2.5.3}$$

Proof

We will consider the case of a continuous random variable X. The MGF for a continuous random variable is

$$M_X(t) = \int_{-\infty}^{\infty} e^{tx} f_X(x)\, dx$$

When the MGF exists, it can be shown that the rth derivative exists, and it can be obtained by differentiating under the integral sign,

$$M_X^{(r)}(t) = \int_{-\infty}^{\infty} x^r e^{tx} f_X(x)\, dx$$

from which it follows that for all $r = 1, 2, \ldots$

$$E(X^r) = \int_{-\infty}^{\infty} x^r f_X(x)\,dx = \int_{-\infty}^{\infty} x^r e^{0x} f_X(x)\,dx = M_X^{(r)}(0)$$

When the MGF exists, it also can be shown that a power series expansion about zero is possible, and from standard results about power series, the coefficients have the form $M_X^{(r)}(0)/r!$. We combine this with the above result to obtain

$$M_X(t) = 1 + \sum_{r=1}^{\infty} \frac{M_X^{(r)}(0)t^r}{r!} = 1 + \sum_{r=1}^{\infty} \frac{E(X^r)t^r}{r!}$$

The discrete case is similar. ■

Example 2.5.2 Consider a continuous random variable X with pdf $f(x) = e^{-x}$ if $x > 0$, and zero otherwise. The MGF is

$$\begin{aligned} M_X(t) &= \int_0^{\infty} e^{tx} e^{-x}\,dx \\ &= \int_0^{\infty} e^{-(1-t)x}\,dx \\ &= \frac{1}{1-t} e^{-(1-t)x}\Big|_0^{\infty} \\ &= \frac{1}{1-t} \qquad t < 1 \end{aligned}$$

The rth derivative is $M_X^{(r)}(t) = r!(1-t)^{-r-1}$, and thus the rth moment is $E(X^r) = M_X^{(r)}(0) = r!$. The mean is $\mu = E(X) = 1! = 1$, and the variance is $\mathrm{Var}(X) = E(X^2) - \mu^2 = 2 - 1 = 1$.

Example 2.5.3 A discrete random variable X has pdf $f(x) = (1/2)^{x+1}$ if $x = 0, 1, 2, \ldots$, and zero otherwise. The MGF of X is

$$\begin{aligned} M_X(t) &= \sum_{x=0}^{\infty} e^{tx}(1/2)^{x+1} \\ &= (1/2) \sum_{x=0}^{\infty} (e^t/2)^x \end{aligned}$$

We make use of the well-known identity for the geometric series,

$$1 + s + s^2 + s^3 + \cdots = \frac{1}{1-s} \qquad -1 < s < 1$$

with $s = e^t/2$. The resulting MGF is

$$M_X(t) = \frac{1}{2 - e^t} \qquad t < \ln 2$$

The first derivative is $M'_X(t) = e^t(2 - e^t)^{-2}$, and thus $E(X) = M'_X(0) = e^0(2 - e^0)^{-2} = 1$. It is possible to obtain higher derivatives, but the complexity increases with the order of the derivative.

PROPERTIES OF MOMENT GENERATING FUNCTIONS

Theorem 2.5.2 If $Y = aX + b$, then $M_Y(t) = e^{bt}M_X(at)$.

Proof

$$\begin{aligned} M_Y(t) &= E(e^{tY}) \\ &= E(e^{t(aX+b)}) \\ &= E(e^{atX}e^{bt}) \\ &= e^{bt}E(e^{atX}) \\ &= e^{bt}M_X(at) \end{aligned}$$

■

One possible application is in computing the rth moment about the mean, $E[(X - \mu)^r]$. Because $M_{X-\mu}(t) = e^{-\mu t}M_X(t)$,

$$E[(X - \mu)^r] = \frac{d^r}{dt^r}\left[e^{-\mu t}M_X(t)\right]\Big|_{t=0} \tag{2.5.4}$$

It can be shown that MGFs uniquely determine a distribution.

Example 2.5.4 Suppose, for example, that X and Y are both integer-valued with the same set of possible values—say 0, 1, and 2—and that X and Y have the same MGF,

$$M(t) = \sum_{x=0}^{2} e^{tx}f_X(x) = \sum_{y=0}^{2} e^{ty}f_Y(y)$$

if we let $s = e^t$ and $c_i = f_Y(i) - f_X(i)$ for $i = 0, 1, 2$, then we have $c_0 + c_1 s + c_2 s^2 = 0$ for all $s > 0$. The only possible coefficients are $c_0 = c_1 = c_2 = 0$, which implies that $f_X(i) = f_Y(i)$ for $i = 0, 1, 2$, and consequently X and Y have the same distribution.

In other words, X and Y cannot have the same MGF but different pdf's. Thus, the form of the MGF determines the form of the pdf.

This is true in general, although harder to prove in general.

Theorem 2.5.3 **Uniqueness** If X_1 and X_2 have respective CDFs $F_1(x)$ and $F_2(x)$, and MGFs $M_1(t)$ and $M_2(t)$, then $F_1(x) = F_2(x)$ for all real x if and only if $M_1(t) = M_2(t)$ for all t in some interval $-h < t < h$ for some $h > 0$. ■

For nonnegative integer-valued random variables, the derivation of moments often is made more tractable by first considering another type of expectation known as a factorial moment.

FACTORIAL MOMENTS

Definition 2.5.2

The rth **factorial moment** of X is

$$E[X(X-1)\cdots(X-r+1)] \tag{2.5.5}$$

and the **factorial moment generating function** (FMGF) of X is

$$G_X(t) = E(t^X) \tag{2.5.6}$$

if this expectation exists for all t in some interval of the form $1 - h < t < 1 + h$.

The FMGF is more tractable than the MGF in some problems.

Also note that the FMGF sometimes is called the **probability generating function**. This is because for nonnegative integer-valued random variables X, $P[X = r] = G_X^{(r)}(0)/r!$, which means that the FMGF uniquely determines the distribution. Also note the following relationship between the FMGF and MGF:

$$G_X(t) = E(t^X) = E(e^{X \ln t}) = M_X(\ln t)$$

Theorem 2.5.4 If X has a FMGF, $G_X(t)$, then

$$G'_X(1) = E(X) \tag{2.5.7}$$

$$G''_X(1) = E[X(X-1)] \tag{2.5.8}$$

$$G_X^{(r)}(1) = E[X(X-1)\cdots(X-r+1)] \tag{2.5.9}$$

Proof

See Exercise 35. ■

It is possible to compute regular moments from factorial moments. For example, notice that $E[X(X-1)] = E(X^2 - X) = E(X^2) - E(X)$, so that

$$E(X^2) = E(X) + E[X(X-1)] \tag{2.5.10}$$

Example 2.5.5 We consider the discrete distribution of Example 2.5.3. The FMGF of X is

$$G_X(t) = M_X(\ln t)$$

$$= \frac{1}{2 - t} \qquad t < 2$$

Notice that higher derivatives are easily obtained for the FMGF, which was not the case for the MGF. In particular, the rth derivative is

$$G_X^{(r)}(t) = r!(2 - t)^{-r-1}$$

Consequently, $E(X) = G'_X(1) = 1!(2 - 1)^{-2} = 1$, and $E[X(X - 1)] = G''_X(1) = 2!(2 - 1)^{-3} = 2$. It follows that $E(X^2) = E(X) + 2 = 3$, and thus, $\text{Var}(X) = 3 - 1^2 = 2$.

SUMMARY

The purpose of this chapter was to develop a mathematical structure for expressing a probability model for the possible outcomes of an experiment when these outcomes cannot be predicted deterministically. A random variable, which is a real-valued function defined on a sample space, and the associated probability density function (pdf) provide a reasonable approach to assigning probabilities when the outcomes of an experiment can be quantified. Random variables often can be classified as either discrete or continuous, and the method of assigning probability to a real event A involves summing the pdf over values of A in the discrete case, and integrating the pdf over the set A in the continuous case. The cumulative distribution function (CDF) provides a unified approach for expressing the distribution of probability to the possible values of the random variable.

The moments are special expected values, which include the mean and variance as particular cases, and also provide descriptive measures for other characteristics such as skewness of a distribution.

Bounds for the probabilities of certain types of events can be expressed in terms of expected values. An important bound of this sort is given by the Chebychev inequality.

EXERCISES

1. Let $e = (i, j)$ represent an arbitrary outcome resulting from two rolls of the four-sided die of Example 2.1.1. Tabulate the discrete pdf and sketch the graph of the CDF for the following random variables:

(a) $Y(e) = i + j$.

(b) $Z(e) = i - j$.

(c) $W(e) = (i - j)^2$.

2. A game consists of first rolling an ordinary six-sided die once and then tossing an unbiased coin once. The score, which consists of adding the number of spots showing on the die to the number of heads showing on the coin (0 or 1), is a random variable, say X. List the possible values of X and tabulate the values of:

(a) the discrete pdf.

(b) the CDF at its points of discontinuity.

(c) Sketch the graph of the CDF.

(d) Find $P[X > 3]$.

(e) Find the probability that the score is an odd integer.

3. A bag contains three coins, one of which has a head on both sides while the other two coins are normal. A coin is chosen at random from the bag and tossed three times. The number of heads is a random variable, say X.

(a) Find the discrete pdf of X. (*Hint:* Use the Law of Total Probability with B_1 = a normal coin and B_2 = two-headed coin.)

(b) Sketch the discrete pdf and the CDF of X.

4. A box contains five colored balls, two black and three white. Balls are drawn successively without replacement. If X is the number of draws until the last black ball is obtained, find the discrete pdf $f(x)$.

5. A discrete random variable has pdf $f(x)$.

(a) If $f(x) = k(1/2)^x$ for $x = 1, 2, 3$, and zero otherwise, find k.

(b) Is a function of the form $f(x) = k[(1/2)^x - 1/2]$ for $x = 0, 1, 2$ a pdf for any k?

6. Denote by $[x]$ the greatest integer not exceeding x. For the pdf in Example 2.2.1, show that the CDF can be represented as $F(x) = ([x]/12)^2$ for $0 < x < 13$, zero if $x \leqslant 0$, and one if $x \geqslant 13$.

7. A discrete random variable X has a pdf of the form $f(x) = c(8 - x)$ for $x = 0, 1, 2, 3, 4, 5$, and zero otherwise.

(a) Find the constant c.

(b) Find the CDF, $F(x)$.

(c) Find $P[X > 2]$.

(d) Find $E(X)$.

8. A nonnegative integer-valued random variable X has a CDF of the form $F(x) = 1 - (1/2)^{x+1}$ for $x = 0, 1, 2, \ldots$ and zero if $x < 0$.

(a) Find the pdf of X.

(b) Find $P[10 < X \leqslant 20]$.

(c) Find $P[X \text{ is even}]$.

9. Sometimes it is desirable to assign numerical "code" values to experimental responses that are not basically of numerical type. For example, in testing the color preferences of experimental subjects, suppose that the colors blue, green, and red occur with probabilities

1/4, 1/4, and 1/2, respectively. A different integer value is assigned to each color, and this corresponds to a random variable X that can take on one of these three integer values.

(a) Can $f(x) = (1/4)^{|x|}(1/2)^{1-|x|}$ for $x = -1, 1, 0$ be used as a pdf for this experiment?

(b) Can $f(x) = \binom{2}{x}(1/2)^2$ for $x = 0, 1, 2$ be used?

(c) Can $f(x) = (1 - x)/4$ for $x = -1, 0, 2$ be used?

10. Let X be a discrete random variable such that $P[X = x] > 0$ if $x = 1, 2, 3$, or 4, and $P[X = x] = 0$ otherwise. Suppose the CDF is $F(x) = .05x(1 + x)$ at the values $x = 1, 2, 3$, or 4.

(a) Sketch the graph of the CDF.

(b) Sketch the graph of the discrete pdf $f(x)$.

(c) Find $E(X)$.

11. A player rolls a six-sided die and receives a number of dollars corresponding to the number of dots on the face that turns up. What amount should the player pay for rolling to make this a "fair" game?

12. A continuous random variable X has pdf given by $f(x) = c(1 - x)x^2$ if $0 < x < 1$ and zero otherwise.

(a) Find the constant c.

(b) Find $E(X)$.

13. A function $f(x)$ has the following form:

$$f(x) = kx^{-(k+1)} \qquad 1 < x < \infty$$

and zero otherwise.

(a) For what values of k is $f(x)$ a pdf?

(b) Find the CDF based on (a).

(c) For what values of k does $E(X)$ exist?

14. Determine whether each of the following functions could be a CDF over the indicated part of the domain:

(a) $F(x) = e^{-x}$; $0 \leqslant x < \infty$.

(b) $F(x) = e^{x}$; $-\infty < x \leqslant 0$.

(c) $F(x) = 1 - e^{-x}$; $-1 \leqslant x < \infty$.

15. Find the pdf corresponding to each of the following CDFs:

(a) $F(x) = (x^2 + 2x + 1)/16$; $-1 \leqslant x \leqslant 3$.

(b) $F(x) = 1 - e^{-\lambda x} - \lambda x e^{-\lambda x}$; $0 \leqslant x < \infty$; $\lambda > 0$.

16. If $f_i(x)$, $i = 1, 2, \ldots, n$, are pdf's, show that

$$\sum_{i=1}^{n} p_i f_i(x) \text{ is a pdf where } p_i \geqslant 0 \text{ and } \sum_{i=1}^{n} p_i = 1$$

17. A random variable X has a CDF such that

$$F(x) = \begin{cases} x/2 & 0 < x \leqslant 1 \\ x - 1/2 & 1 < x \leqslant 3/2 \end{cases}$$

(a) Graph $F(x)$.
(b) Graph the pdf $f(x)$.
(c) Find $P[X \leqslant 1/2]$.
(d) Find $P[X \geqslant 1/2]$.
(e) Find $P[X \leqslant 1.25]$.
(f) What is $P[X = 1.25]$?

18. A continuous random variable X has a pdf of the form $f(x) = 2x/9$ for $0 < x < 3$, and zero otherwise.
(a) Find the CDF of X.
(b) Find $P[X \leqslant 2]$.
(c) Find $P[-1 < X < 1.5]$.
(d) Find a number m such that $P[X \leqslant m] = P[X \geqslant m]$.
(e) Find $E(X)$.

19. A random variable X has the pdf

$$f(x) = \begin{cases} x^2 & \text{if } 0 < x \leqslant 1 \\ 2/3 & \text{if } 1 < x \leqslant 2 \\ 0 & \text{otherwise} \end{cases}$$

(a) Find the median of X.
(b) Sketch the graph of the CDF and show the position of the median on the graph.

20. A continuous random variable X has CDF given by

$$F(x) = \begin{cases} 0 & \text{if } x < 1 \\ 2(x - 2 + 1/x) & \text{if } 1 \leqslant x \leqslant 2 \\ 1 & \text{if } 2 < x \end{cases}$$

(a) Find the $100 \times p$th percentile of the distribution with $p = 1/3$.
(b) Find the pdf of X.

21. Verify that the following function has the four properties of Theorem 2.2.3, and find the points of discontinuity, if any:

$$F(x) = \begin{cases} 0.25e^{x} & \text{if } -\infty < x < 0 \\ 0.5 & \text{if } 0 \leqslant x < 1 \\ 1 - e^{-x} & \text{if } 1 \leqslant x < \infty \end{cases}$$

22. For the CDF, $F(x)$, of Exercise 21, find a CDF of discrete type, $F_d(x)$, and a CDF of continuous type, $F_c(x)$, and a number $0 < a < 1$ such that

$$F(x) = aF_d(x) + (1 - a)F_c(x)$$

23. Let X be a random variable with discrete pdf $f(x) = x/8$ if $x = 1, 2, 5$, and zero otherwise. Find:

(a) $E(X)$.

(b) $\text{Var}(X)$.

(c) $E(2X + 3)$.

24. Let X be continuous with pdf $f(x) = 3x^2$ if $0 < x < 1$, and zero otherwise. Find:

(a) $E(X)$.

(b) $\text{Var}(X)$.

(c) $E(X^r)$.

(d) Find $E(3X - 5X^2 + 1)$.

25. Let X be continuous with pdf $f(x) = 1/x^2$ if $1 < x < \infty$, and zero otherwise.

(a) Does $E(X)$ exist?

(b) Does $E(1/X)$ exist?

(c) For what values of k does $E(X^k)$ exist?

26. At a computer store, the annual demand for a particular software package is a discrete random variable X. The store owner orders four copies of the package at \$10 per copy and charges customers \$35 per copy. At the end of the year the package is obsolete and the owner loses the investment on unsold copies. The pdf of X is given by the following table:

x	0	1	2	3	4
$f(x)$	.1	.3	.3	.2	.1

(a) Find $E(X)$.

(b) Find $\text{Var}(X)$.

(c) Express the owner's net profit Y as a linear function of X, and find $E(Y)$ and $\text{Var}(Y)$.

27. The measured radius of a circle, R, has pdf $f(r) = 6r(1 - r)$, $0 < r < 1$. Find:

(a) the expected value of the radius.

(b) the expected circumference.

(c) the expected area.

28. Prove Theorem 2.4.5 for the continuous case. *Hint:* Use the transformation $y = x - \mu$ in the integral and note that $g(y) = yf(\mu + y)$ is an odd function of y.

29. Consider the discrete random variable X with pdf given by the following table:

x	-3	-1	0	2	$2\sqrt{2}$
$f(x)$	1/4	1/4	$(6-3\sqrt{2})/16$	1/8	$3\sqrt{2}/16$

The distribution of X is not symmetric. Why? Show that $\mu_3 = 0$.

30. Let X be a nonnegative continuous random variable with CDF $F(x)$ and $E(X) < \infty$. Use integration by parts to show that

$$E(X) = \int_0^\infty [1 - F(x)]\,dx$$

Note: For any continuous random variable with $E(|X|) < \infty$, this result extends to

$$E(X) = -\int_{-\infty}^0 F(x)\,dx + \int_0^\infty [1 - F(x)]\,dx$$

31. (a) Use Chebychev's inequality to obtain a lower bound on $P[5/8 < X < 7/8]$ in Exercise 24. Is this a useful bound?

(b) Rework (a) for the probability $P[1/2 < X < 1]$.

(c) Compare this bound to the exact probability.

32. Consider the random variable X of Example 2.1.1, which represents the largest of two numbers that occur on two rolls of a four-sided die.

(a) Find the expected value of X.

(b) Find the variance of X.

33. Suppose $E(X) = \mu$ and $\text{Var}(X) = \sigma^2$. Find the approximate mean and variance of:

(a) e^X.

(b) $1/X$ (assuming $\mu \neq 0$).

(c) $\ln(X)$ (assuming $X > 0$).

34. Suppose that X is a random variable with MGF $M_X(t) = (1/8)e^t + (1/4)e^{2t} + (5/8)e^{5t}$.

(a) What is the distribution of X?

(b) What is $P[X = 2]$?

35. Prove Theorem 2.5.4 for a nonnegative integer-valued random variable X.

36. Assume that X is a continuous random variable with pdf

$$f(x) = \exp[-(x + 2)] \text{ if } -2 < x < \infty \text{ and zero otherwise.}$$

(a) Find the moment generating function of X.

(b) Use the MGF of (a) to find $E(X)$ and $E(X^2)$.

37. Use the FMGF of Example 2.5.5 to find $E[X(X-1)(X-2)]$, and then find $E(X^3)$.

38. In Exercise 26, suppose instead of ordering four copies of the software package, the store owner orders c copies ($0 \leqslant c \leqslant 4$). Then the number sold, say S, is the smaller of c or X.

(a) Express the net profit Y as a linear function of S.

(b) Find $E(Y)$ for each value of c and indicate the solution c that maximizes the expected profit.

39. Show that $\sigma^2 = E[X(X-1)] - \mu(\mu - 1)$.

40. Let $\psi_X(t) = \ln [M_X(t)]$, where $M_X(t)$ is a MGF. The function $\psi_X(t)$ is called the **cumulant generating function** of X, and the value of the rth derivative evaluated at $t = 0$, $\kappa_r = \psi_X^{(r)}(0)$, is called the rth cumulant of X.

(a) Show that $\mu = \psi_X'(0)$.

(b) Show that $\sigma^2 = \psi_X''(0)$.

(c) Use $\psi_X(t)$ to find μ and σ^2 for the random variable of Exercise 36.

(d) Use $\psi_X(t)$ to find μ and σ^2 for the random variable of Example 2.5.5.

CHAPTER 3

SPECIAL PROBABILITY DISTRIBUTIONS

3.1 INTRODUCTION

Our purpose in this chapter is to develop some special probability distributions. In many applications by recognizing certain characteristics, it is possible to determine that the distribution has a known special form. Typically, a special distribution will depend on one or more parameters, and once the numerical value of each parameter has been ascertained, the distribution is completely determined.

Special discrete distributions will be derived using the counting techniques of Chapter 1. Special continuous distributions also will be presented, and relationships between various special distributions will be discussed.

3.2

SPECIAL DISCRETE DISTRIBUTIONS

We will use the counting techniques of Chapter 1 to derive special discrete distributions.

BERNOULLI DISTRIBUTION

On a single trial of an experiment, suppose that there are only two events of interest, say E and its complement E'. For example, E and E' could represent the occurrence of a "head" or a "tail" on a single coin toss, obtaining a "defective" or a "good" item when drawing a single item from a manufactured lot, or, in general, "success" or "failure" on a particular trial of an experiment. Suppose that E occurs with probability $p = P(E)$, and consequently E' occurs with probability $q = P(E') = 1 - p$.

A random variable, X, that assumes only the values 0 or 1 is known as a **Bernoulli variable**, and a performance of an experiment with only two types of outcomes is called a **Bernoulli trial**. In particular, if an experiment can result only in "success" (E) or "failure" (E'), then the corresponding Bernoulli variable is

$$X(e) = \begin{cases} 1 & \text{if } e \in E \\ 0 & \text{if } e \in E' \end{cases} \tag{3.2.1}$$

The pdf of X is given by $f(0) = q$ and $f(1) = p$. The corresponding distribution is known as a **Bernoulli distribution**, and its pdf can be expressed as

$$f(x) = p^x q^{1-x} \qquad x = 0, 1 \tag{3.2.2}$$

Example 3.2.1 In Example 2.1.1, we considered rolls of a four-sided die. A bet is placed that a 1 will occur on a single roll of the die. Thus, $E = \{1\}$, $E' = \{2, 3, 4\}$, and $p = 1/4$.

In an earlier example, we considered drawing marbles at random from a collection of 10 black and 20 white marbles. In such a problem, we might regard "black" as success and "white" as failure, or vice versa, in a single draw. If obtaining a black marble is regarded as a success, then $p = 10/30 = 1/3$ and $q = 20/30 = 2/3$.

Notice that $E(X) = 0 \cdot q + 1 \cdot p = p$ and $E(X^2) = 0^2 \cdot q + 1^2 \cdot p = p$, so that $\text{Var}(X) = p - p^2 = p(1 - p) = pq$.

An important distribution arises from counting the number of successes on a fixed number of independent Bernoulli trials.

BINOMIAL DISTRIBUTION

Often it is possible to structure a more complicated experiment as a sequence of **independent Bernoulli trials**, where the quantity of interest is the number of successes on a certain number of trials.

Example 3.2.2 **Sampling with Replacement** In Example 1.6.16, we considered the problem of drawing five marbles from a collection of 10 black and 20 white marbles, where the marbles are drawn one at a time, and each time replaced before the next draw. We shall let X be the number of black marbles drawn, and consider $f(2) = P[X = 2]$. To draw exactly two black (B), and consequently three white (W), it would be necessary to obtain some permutation of two B's and three W's, BBWWW, BWBWW, and so on (see Figure 1.12 for a complete listing). There are $\binom{5}{2} = 10$ possible permutations of this type, and each one has the same probability of occurrence, namely $(10/30)^2(20/30)^3$, which is the product of two values of $P(B) = 10/30$ and three values of $P(W) = 20/30$, multiplied together in some order. The probability can be obtained this way because draws made with replacement can be regarded as independent Bernoulli trials. Thus,

$$f(2) = \binom{5}{2}\left(\frac{10}{30}\right)^2\left(\frac{20}{30}\right)^3$$

which agrees with the solution (1.6.11).

This approach can be used to derive the more general binomial distribution. In a sequence of n independent Bernoulli trials with probability of success p on each trial, let X represent the number of successes. The discrete pdf of X is given by

$$b(x;\, n,\, p) = \binom{n}{x}p^x q^{n-x} \qquad x = 0, 1, \ldots, n \tag{3.2.3}$$

For the event $[X = x]$ to occur, it is necessary to have some permutation of x successes (E) and $n - x$ failures (E'). There are $\binom{n}{x}$ such permutations, and each occurs with probability $p^x q^{n-x}$, which is the product of x values of $p = P(E)$ and $n - x$ values of $q = P(E')$. Of course, the order of multiplication is unimportant, and formula (3.2.3), which is known as the pdf of the **binomial distribution**, is established. The notation $b(x; n, p)$, which we have used instead of $f(x)$, reflects the dependence on the parameters n and p.

The general properties (2.2.2) and (2.2.3) are satisfied by equation (3.2.3), because $0 \leqslant p \leqslant 1$ and

$$\sum_{x=0}^{n} b(x;\, n,\, p) = \sum_{x=0}^{n} \binom{n}{x}p^x q^{n-x} = (p + q)^n = 1^n = 1 \tag{3.2.4}$$

The CDF of a binomial distribution is given at integer values by

$$B(x;\, n,\, p) = \sum_{k=0}^{x} b(k;\, n,\, p) \qquad x = 0, 1, \ldots, n \tag{3.2.5}$$

Some values of $B(x; n, p)$ are provided in Table 1 in Appendix C for various values of n and p. The following identity is easily verified:

$$B(x;\, n,\, p) = 1 - B(n - x - 1;\, n,\, 1 - p) \tag{3.2.6}$$

Values of the pdf can be obtained easily from Table 1 because

$$b(x;\, n,\, p) = B(x;\, n,\, p) - B(x - 1;\, n,\, p) \tag{3.2.7}$$

A short notation to designate that X has the binomial distribution with parameters n and p is $X \sim B(x; n, p)$ or an alternative notation

$$X \sim \text{BIN}(n,\, p) \tag{3.2.8}$$

The binomial distribution arises in connection with many games of chance, such as rolling dice or tossing coins.

Example 3.2.3 A coin is tossed independently n times. Denote by $p = P(H)$ the probability of obtaining a head on a single toss. If $p = 1/2$ we say that the coin is fair or **unbiased**; otherwise it is said to be **biased**. For example, if X is the number of heads obtained by tossing an unbiased coin 20 times, then $X \sim \text{BIN}(20, 1/2)$. There is a connection between this example and Example 1.6.14, which dealt with randomly choosing the answers to a 20-question true–false test. If the questions were answered according to the results of 20 coin tosses, then the distribution of the number of correct answers also would be BIN(20, 1/2). Thus, the probability of exactly 80% of the answers being correct would be

$$b\left(16;\, 20,\, \frac{1}{2}\right) = \binom{20}{16}\left(\frac{1}{2}\right)^{16}\left(\frac{1}{2}\right)^{4} = 0.0046$$

which also was obtained in Example 1.6.14 by using counting techniques and the classical approach. In applications where $p \neq 1/2$, the classical approach no longer works because the permutations of successes and failures are not all equally likely, although the binomial distribution still applies.

For example, suppose that, instead of a true–false test, a 20-question multiple-choice test with four choices per question is answered at random. This could be carried out by rolling the four-sided die of Example 3.2.1 20 times, and the distribution of the number of correct answers would be the same as the distribution of occurrences of any particular one of the values 1, 2, 3, or 4. The appropriate distribution of either variable would be BIN(20, 1/4). In this case, the probability of exactly 80% correct is

$$b\left(16;\, 20,\, \frac{1}{4}\right) = \binom{20}{16}\left(\frac{1}{4}\right)^{16}\left(\frac{3}{4}\right)^{4} = 0.00000036$$

Actually, even the probability of 50% correct is rather small, namely

$$b\left(10; 20, \frac{1}{4}\right) = \binom{20}{10}\left(\frac{1}{4}\right)^{10}\left(\frac{3}{4}\right)^{10} = 0.0099$$

The probability of at most 50% correct is $B(10; 20, 1/4) = 0.9960$.

The binomial distribution also is useful in evaluating certain games of chance.

Example 3.2.4 A game of chance consists of rolling three ordinary six-sided dice. The player bets \$1 per game, and wins \$1 for each occurrence of the number 6 on any of the dice, retaining the original bet in that case. Thus, the net amount won would be a discrete random variable, say Y, with possible values 1, 2, 3, or -1, where the latter value corresponds to the dollar bet, which would be a net loss if no die shows a 6.

One possible approach would be to work out the distribution of Y and then compute $E(Y)$ directly. Instead we will use the fact that Y is a function of a binomial random variable, X, which is the number of 6's on the three dice. In particular, $X \sim \text{BIN}(3, 1/6)$ and $Y = u(X)$, where $u(x)$ is given by

x:	0	1	2	3
$u(x)$:	-1	1	2	3

It follows that

$$\begin{aligned} E(Y) &= E[u(X)] \\ &= \sum_{x=0}^{3} u(x)\binom{3}{x}\left(\frac{1}{6}\right)^{x}\left(\frac{5}{6}\right)^{3-x} \\ &= -1\left(\frac{125}{216}\right) + 1\left(\frac{75}{216}\right) + 2\left(\frac{15}{216}\right) + 3\left(\frac{1}{216}\right) \\ &= -\frac{17}{216} \doteq -0.08 \end{aligned}$$

Thus, the expected amount won is actually an expected loss. In other words, if the player bets \$1 on each play, in the long run he or she should expect to lose roughly \$8 for every 100 plays.

Now we will derive some general properties of the binomial distribution.

If $X \sim \text{BIN}(n, p)$, then

$$\begin{aligned} M_X(t) &= \sum_{x=0}^{n} e^{tx}\binom{n}{x}p^x q^{n-x} \\ &= \sum_{x=0}^{n} \binom{n}{x}(pe^t)^x q^{n-x} \\ &= (pe^t + q)^n \qquad -\infty < t < \infty \end{aligned} \tag{3.2.9}$$

from the binomial expansion.

We note that $M'_X(t) = n(pe^t + q)^{n-1}pe^t$, and so $M'_X(0) = np$.

It is also possible to derive the variance by first evaluating $E(X^2) = M''_X(0)$. However, we will use this opportunity to illustrate the use of the factorial moment generating function, or FMGF, $G_X(t)$. Specifically, if $X \sim \text{BIN}(n, p)$, then

$$\begin{aligned} G_X(t) = E(t^X) &= \sum_{x=0}^{n} t^x \binom{n}{x} p^x q^{n-x} \\ &= \sum_{x=0}^{n} \binom{n}{x} (pt)^x q^{n-x} \\ &= (pt + q)^n \end{aligned}$$

$$G'_X(t) = n(pt + q)^{n-1}p$$

$$G''_X(t) = (n - 1)n(pt + q)^{n-2}p^2$$

Thus, $E(X) = G'_X(1) = np$ and $E[X(X - 1)] = G''_X(1) = (n - 1)np^2$, so that $E(X^2) = np + (n - 1)np^2 = np + (np)^2 - np^2$ and $\text{Var}(X) = np + (np)^2 - np^2 - (np)^2 = np(1 - p) = npq$.

The results on the mean, variance, and MGF of the binomial distribution are summarized in Appendix B.

HYPERGEOMETRIC DISTRIBUTION

In Example 1.6.15, we found the probability of obtaining exactly two black marbles out of five selected at random without replacement from a collection of 10 black and 20 white marbles. This type of problem can be generalized to obtain an important special discrete distribution known as the hypergeometric distribution.

Suppose a population or collection consists of a finite number of items, say N, and there are M items of type 1 and the remaining $N - M$ items are of type 2. Suppose n items are drawn at random *without replacement*, and denote by X the number of items of type 1 that are drawn. The discrete pdf of X is given by

$$h(x; n, M, N) = \frac{\binom{M}{x}\binom{N-M}{n-x}}{\binom{N}{n}} \tag{3.2.10}$$

The underlying sample space is taken to be the collection of all subsets of size n, of which there are $\binom{N}{n}$, and there are $\binom{M}{x}\binom{N-M}{n-x}$ outcomes that correspond to the event $[X = x]$. Equation (3.2.10), which is the pdf of the **hypergeometric distribution**, follows by the classical method of assigning probabilities. The notation $h(x; n, M, N)$, which is used here instead of $f(x)$, reflects the dependence on the parameters n, M, and N. The required properties (2.2.2) and (2.2.3) are clearly satisfied because $\binom{M}{x}\binom{N-M}{n-x}$ counts the number of subsets of size n with exactly x items of type 1, and thus the total number of subsets of size n can be represented either by $\binom{N}{n}$ or by $\sum_{x=0}^{n}\binom{M}{x}\binom{N-M}{n-x}$. It follows that

$$\sum_{x=0}^{n} h(x;\, n,\, M,\, N) = 1$$

In Example 1.6.15, the parameter values are $n = 5$, $N = 30$, and $M = 10$, where black marbles are regarded as type 1. Suppose the number of marbles selected is increased to $n = 25$, and the probability of obtaining exactly eight black ones is desired. This is

$$h(8;\, 25,\, 10,\, 30) = \frac{\binom{10}{8}\binom{20}{17}}{\binom{30}{25}}$$

Notice that the possible values of X in this case are $x = 5, 6, 7, 8, 9$, or 10. This is because the selected subset cannot have more than $M = 10$ black marbles or more than $N - M = 20$ white marbles. In general, the possible values of X in (3.2.10) are

$$\max\,(0,\, n - N + M) \leqslant x \leqslant \min\,(n,\, M) \tag{3.2.11}$$

and $h(x;\, n,\, M,\, N)$ is zero otherwise.

The hypergeometric distribution is important in applications such as deciding whether to accept a lot of manufactured items.

Example 3.2.5 Recall Example 1.5.1, in which a box contained 100 microchips, 80 good and 20 defective. The number of defectives in the box is unknown to a purchaser, who decides to select 10 microchips at random without replacement and to consider the microchips in the box acceptable if the 10 items selected include no more than three defectives. The number of defectives selected, X, has the hypergeometric distribution with $n = 10$, $N = 100$, and $M = 20$ (according to Table 1.1), and the probability of the lot being acceptable is

$$P[X \leqslant 3] = \sum_{x=0}^{3} \frac{\binom{20}{x}\binom{80}{10-x}}{\binom{100}{10}} = 0.890$$

Thus, the probability of accepting a lot with 20% defective items is fairly high. Suppose, on the other hand, that the box contained 50 good and 50 defective items. The same acceptance criterion would yield

$$P[X \leqslant 3] = \sum_{x=0}^{3} \frac{\binom{50}{x}\binom{50}{10-x}}{\binom{100}{10}} = 0.159$$

which means, as we might expect, that a lot with a higher percentage of defective items is less likely to be accepted.

In the preceding example, probabilities of the form $P[X \leqslant x]$ were important. This is the CDF of X, and we will adopt a special notation for the CDF of a hypergeometric distribution, namely

$$H(x; n, M, N) = \sum_{i=0}^{x} h(i; n, M, N) \tag{3.2.12}$$

Any hypergeometric probability of interest can be expressed in terms of equation (3.2.12). For example, consider the sampling problem of Example 1.6.15, where X is the number of black marbles in a sample of size five. It follows that

$$P[X \leqslant 2] = H(2; 5, 10, 30)$$

$$P[X = 2] = P[X \leqslant 2] - P[X \leqslant 1] = H(2; 5, 10, 30) - H(1; 5, 10, 30)$$

$$P[X > 3] = 1 - P[X \leqslant 3] = 1 - H(3; 5, 10, 30)$$

$$P[X \geqslant 3] = 1 - P[X \leqslant 2] = 1 - H(2; 5, 10, 30)$$

and so on.

A short notation to designate that X has the hypergeometric distribution with parameters n, M, and N is

$$X \sim \text{HYP}(n, M, N) \tag{3.2.13}$$

It can be shown by a straightforward but rather tedious derivation that $E(X) = nM/N$ and $\text{Var}(X) = n(M/N)(1 - M/N)(N - n)/(N - 1)$, but we will postpone further discussion of this point until Chapter 5, where these quantities will be obtained.

The main properties of the hypergeometric distribution are summarized in Appendix B.

Under certain conditions, the binomial distribution can be used to approximate the hypergeometric distribution.

Theorem 3.2.1 If $X \sim \text{HYP}(n, M, N)$, then for each value $x = 0, 1, \ldots, n$, and as $N \to \infty$ and $M \to \infty$ with $M/N \to p$, a positive constant,

$$\lim_{N\to\infty} \frac{\binom{M}{x}\binom{N-M}{n-x}}{\binom{N}{n}} = \binom{n}{x} p^x (1-p)^{n-x} \tag{3.2.14}$$

Proof

The proof is based on rearranging the factorials in equation (3.2.10) to obtain an expression of the form $\binom{n}{x}$ times a product of ratios that converges to $p^x(1-p)^{n-x}$ as $M \to \infty$. ■

This provides an approximation when the number selected, n, is small relative to the size of the collection, N, and the number of items of type 1, M. This is intuitively reasonable because the binomial distribution is applicable when we sample *with* replacement, while the hypergeometric distribution is applicable when we sample *without* replacement. If the size of the collection sampled from is large, then it should not make a great deal of difference whether a particular item is returned to the collection before the next one is selected.

Example 3.2.6 Ten seeds are selected from a bin that contains 1000 flower seeds, of which 400 are red flowering seeds, and the rest are of other colors. How likely is it to obtain exactly five red flowering seeds? Strictly speaking, this is hypergeometric and $h(5; 10, 100, 1000) = 0.2013$. The binomial approximation is $b(5; 10, 0.4) = 0.2007$.

If we work this directly, with the method of conditional probability, it provides some additional insight into Theorem 3.2.1. Suppose we draw the 10 seeds, one at a time, without replacement from the bin. To obtain five red flowering seeds, it is necessary to obtain some permutation of five red and five other, of which there are $\binom{10}{5}$ possible. Each one would have the same probability of occurrence, namely

$$\frac{400}{1000} \cdot \frac{399}{999} \cdot \cdots \cdot \frac{396}{996} \cdot \frac{600}{995} \cdot \frac{599}{994} \cdot \cdots \cdot \frac{596}{991}$$

which is close to $(0.4)^5(0.6)^5$.

There are other special discrete distributions that are based on independent Bernoulli trials. For example, suppose a certain satellite launch has a 0.7 probability of success. The probability remains constant at 0.7 for repeated launches, and the success or failure of one launch does not depend on the outcome of other launches. It is feared that funding for the satellite program will be cut off if success is not achieved within three trials. What is the probability that the first success will occur within three trials? The probability of success occurring on the first trial is $p_1 = 0.7$, the probability of the first success occurring on the second trial is $p_2 = 0.3(0.7)$, and the probability of the first success occurring on the third trial is $(0.3)^2 0.7$. The probability of first success within three trials is

$0.7 + 0.3(0.7) + (0.3)^2 0.7 = 0.973$. General solutions to this type of problem lead to the geometric and negative binomial distributions.

GEOMETRIC AND NEGATIVE BINOMIAL DISTRIBUTIONS

We again consider a sequence of independent Bernoulli trials with probability of success $p = P(E)$. In the case of the binomial distribution, the number of trials was a fixed number n, and the variable of interest was the number of successes. Now we consider the number of trials required to achieve a specified number of successes.

If we denote the number of trials required to obtain the *first* success by X, then the discrete pdf of X is given by

$$g(x; p) = pq^{x-1} \qquad x = 1, 2, 3, \ldots \tag{3.2.15}$$

For the event $[X = x]$ to occur, it is necessary to have a particular permutation consisting of $x - 1$ failures followed by a success. Because the trials are independent, this probability is the product of p with $x - 1$ factors of $q = 1 - p$, as given by equation (3.2.15).

The general properties (2.2.2) and (2.2.3) are satisfied by equation (3.2.15), because $0 < p < 1$ and

$$\begin{aligned}\sum_{x=1}^{\infty} g(x; p) &= p \sum_{x=1}^{\infty} q^{x-1} = p(1 + q + q^2 + \cdots) \\ &= p\left(\frac{1}{1-q}\right) = \frac{p}{p} = 1\end{aligned} \tag{3.2.16}$$

The distribution of X is known as the **geometric distribution**, which gets its name from its relationship with the geometric series that was used to evaluate equation (3.2.16). This also sometimes is known as the **Pascal distribution**. We will use special notation that designates that X has pdf (3.2.15):

$$X \sim \text{GEO}(p) \tag{3.2.17}$$

It also follows from the properties of the geometric series that the CDF of X is

$$G(x; p) = \sum_{i=1}^{x} pq^{i-1} = 1 - q^x \qquad x = 1, 2, 3, \ldots \tag{3.2.18}$$

Example 3.2.7 The probability a certain baseball player gets a hit is 0.3, and we assume that times at bat are independent. The probability that he will require five times at bat to get his first hit is $g(5; 0.3) = 0.7^4(0.3)$. Given that he has been at bat 10 times without a hit, the probability is still $0.7^4(0.3)$ that it will require five more times at bat for him to get his first hit. Also, the probability that five or fewer at bats are required to obtain the first hit is given by

$$G(5; 0.3) = 1 - (0.7)^5 = 0.83193$$

The geometric distribution is the only discrete probability distribution that has a so-called no-memory property.

Theorem 3.2.2 **No-Memory Property** If $X \sim \text{GEO}(p)$, then

$$P[X > j + k \mid X > j] = P[X > k]$$

Proof

$$\begin{aligned} P[X > j + k \mid X > j] &= \frac{P[X > j + k]}{P[X > j]} \\ &= \frac{(1-p)^{j+k}}{(1-p)^j} \\ &= (1-p)^k \\ &= P[X > k] \end{aligned}$$

■

Thus, knowing that j trials have passed without a success does not affect the probability of k more trials being required to obtain a success. That is, having several failures in a row does not mean that you are more "due" for a success.

Example 1.2.3 involved tossing a coin until the first head occurs. If X is the number of tosses, and if $p = P(H)$, then $X \sim \text{GEO}(p)$. It was noted that an outcome corresponding to never obtaining a head was unnecessary. Of course, this is because the probability of never obtaining a head is zero. Specifically, if A represents the event "a head is never obtained," then A' is the event "at least one head," and

$$P(A) = 1 - P(A') = 1 - \sum_{x=1}^{\infty} g(x; p) = 1 - 1 = 0$$

The mean of $X \sim \text{GEO}(p)$ is obtained as follows:

$$\begin{aligned} E(X) &= \sum_{x=1}^{\infty} xpq^{x-1} \\ &= \sum_{x=1}^{\infty} p \frac{d}{dq} q^x \\ &= p \frac{d}{dq} \sum_{k=0}^{\infty} q^k \\ &= p \frac{d}{dq} (1-q)^{-1} \\ &= p(1-q)^{-2} \\ &= \frac{1}{p} \end{aligned}$$

By a similar argument, $E(X^2) = (1 + q)/p^2$ and, consequently, $\text{Var}(X) = q/p^2$. The properties of a geometric distribution are summarized in Appendix B.

It should be noted that some authors consider a slightly different variable, Y, defined as the number of failures that occur *before* the first success. Thus, $Y = X - 1$, and

$$P[Y = y] = (1 - p)^y p \qquad y = 0, 1, 2, \ldots \tag{3.2.19}$$

This probability distribution also sometimes is referred to as a geometric distribution.

NEGATIVE BINOMIAL DISTRIBUTION

In repeated independent Bernoulli trials, let X denote the number of trials required to obtain r successes. Then the probability distribution of X is the **negative binomial distribution** with discrete pdf given by

$$f(x; r, p) = \binom{x-1}{r-1} p^r q^{x-r} \qquad x = r, r + 1, \ldots \tag{3.2.20}$$

For the event $[X = x]$ to occur, one must obtain the rth success on the xth trial by obtaining "$r - 1$ successes in the first $x - 1$ trials" in any order, and then obtaining a "success on the xth trial." Thus, the probability of the first event may be expressed as

$$\binom{x-1}{r-1} p^{r-1} (1 - p)^{(x-1)-(r-1)}$$

and multiplying this by p, the probability of the second event, produces equation (3.2.20).

A special notation, which designates that X has the negative binomial distribution (3.2.20), is

$$X \sim \text{NB}(r, p) \tag{3.2.21}$$

Example 3.2.8 Team A plays team B in a seven-game world series. That is, the series is over when either team wins four games. For each game, $P(A \text{ wins}) = 0.6$, and the games are assumed independent. What is the probability that the series will end in exactly six games? We have $x = 6$, $r = 4$, and $p = 0.6$ in equation (3.2.20), and

$$\begin{aligned} P(A \text{ wins series in } 6) &= f(6; 4, 0.6) \\ &= \binom{5}{3}(0.6)^4(0.4)^2 \\ &= 0.20736 \end{aligned}$$

$$P(B \text{ wins series in } 6) = f(6;\, 4,\, 0.4)$$
$$= \binom{5}{3}(0.4)^4(0.6)^2$$
$$= 0.09216$$
$$P(\text{series goes 6 games}) = 0.20736 + 0.09216$$
$$= 0.29952$$

The general properties (2.2.2) and (2.2.3) are satisfied by equation (3.2.20), because $0 < p < 1$ and

$$\sum_{x=r}^{\infty} \binom{x-1}{r-1} p^r q^{x-r} = p^r \sum_{i=0}^{\infty} \binom{i+r-1}{r-1} q^i$$
$$= p^r(1-q)^{-r} = 1 \tag{3.2.22}$$

Note that $\sum_{i=0}^{\infty} \binom{i+r-1}{r-1} q^i$ is the series expansion of $(1-q)^{-r}$ as given in any standard mathematical handbook. The name "negative binomial" distribution results from its relationship to this binomial series expansion with negative exponent, $-r$, which was used to establish equation (3.2.22).

Again, some authors consider the alternate variable Y, which is defined to be the number of failures that occur prior to obtaining the rth success. That is, $X = Y + r$ and

$$f_Y(y;\, r,\, p) = p[Y = y] = \binom{y+r-1}{r-1} p^r(1-p)^y$$
$$= f_X(y+r;\, r,\, p) \qquad y = 0, 1, 2, \ldots \tag{3.2.23}$$

In Example 3.2.8, the terminology now would become

$$P(A \text{ wins series in } 6) = P(2 \text{ losses occur before 4 wins})$$
$$= P[Y = 2]$$
$$= \binom{5}{3}(0.6)^4(0.4)^2$$

as before.

It can be shown that if $X \sim \text{NB}(r, p)$, then $E(X) = r/p$, $\text{Var}(X) = rq/p^2$, and $M_X(t) = [pe^t/(1 - qe^t)]^r$. These properties are left as an exercise (see Exercise 19).

A summary of the properties of the geometric and negative binomial distributions is given in Appendix B.

BINOMIAL RELATIONSHIP TO NEGATIVE BINOMIAL

The negative binomial problem sometimes is referred to as **inverse binomial sampling**. Suppose $X \sim \text{NB}(r, p)$ and $W \sim \text{BIN}(n, p)$. It follows that

$$P[X \leqslant n] = P[W \geqslant r] \tag{3.2.24}$$

That is, $W \geqslant r$ corresponds to the event of having r or more successes in n trials, and that means n or fewer trials will be needed to obtain the first r successes. Clearly, the negative binomial distribution can be expressed in terms of the binomial CDF by the relationship

$$\begin{aligned} F(x; r, p) = P[X \leqslant x] &= 1 - B(r - 1; x, p) \\ &= B(x - r; x, q) \end{aligned} \tag{3.2.25}$$

If in Example 3.2.8 we are interested in the probability that team A wins the world series in six or fewer games, then $x = 6$, $r = 4$, $p = 0.6$, and

$$\begin{aligned} P(A \text{ wins series in 6 or less}) &= P[X \leqslant 6] \\ &= F(6; 4, 0.6) \\ &= \sum_{x=4}^{6} \binom{x-1}{3}(0.6)^4(0.4)^{x-4} \\ &= B(2; 6, 0.4) \\ &= \sum_{w=0}^{2} \binom{6}{w}(0.4)^w(0.6)^{6-w} \\ &= 0.5443 \end{aligned}$$

That is, the probability of winning the series in six *or fewer* games is equivalent to the probability of suffering two or fewer losses in six games. The main advantage of writing the negative binomial CDF in terms of a binomial CDF is that the binomial CDF is tabulated much more extensively in the literature. Also, known approximations to the binomial can be applied.

We will consider one more discrete distribution, which cannot be obtained directly by counting arguments, but which can be derived as a limiting form of the binomial distribution.

POISSON DISTRIBUTION

A discrete random variable X is said to have the **Poisson distribution** with parameter $\mu > 0$ if it has discrete pdf of the form

$$f(x; \mu) = \frac{e^{-\mu}\mu^x}{x!} \qquad x = 0, 1, 2, \ldots \tag{3.2.26}$$

A special notation that designates that a random variable X has the Poisson distribution with parameter μ is

$$X \sim \text{POI}(\mu) \tag{3.2.27}$$

Properties (2.2.2) and (2.2.3) clearly are satisfied, because $\mu > 0$ implies $f(x; \mu) \geqslant 0$ and

$$\sum_{x=0}^{\infty} f(x; \mu) = e^{-\mu} \sum_{x=0}^{\infty} \frac{\mu^x}{x!} = e^{-\mu} e^{\mu} = 1 \tag{3.2.28}$$

The CDF of $X \sim \text{POI}(\mu)$, denoted by

$$F(x; \mu) = \sum_{k=0}^{x} f(k; \mu) \tag{3.2.29}$$

cannot be expressed in a simpler functional form, but it can be tabulated. Values of $F(x; \mu)$ are provided in Table 2 (Appendix C) for various values of x and μ.

If $X \sim \text{POI}(\mu)$, then

$$\begin{aligned} M_X(t) &= \sum_{x=0}^{\infty} e^{tx} e^{-\mu} \frac{\mu^x}{x!} \\ &= e^{-\mu} \sum_{x=0}^{\infty} \frac{(\mu e^t)^x}{x!} \\ &= e^{-\mu} e^{\mu e^t} \end{aligned}$$

Thus,

$$M_X(t) = e^{\mu(e^t - 1)} \qquad -\infty < t < \infty \tag{3.2.30}$$

It follows that $M'_X(t) = e^{\mu(e^t - 1)}\mu e^t = M_X(t)\mu e^t$, and thus $M'_X(0) = M_X(0)\mu e^0 = \mu$. Similarly, $M''_X(t) = [M_X(t) + M'_X(t)]\mu e^t$, so that $M''_X(0) = (1 + \mu)\mu$ and, thus $\text{Var}(X) = E(X^2) - \mu^2 = \mu(1 + \mu) - \mu^2 = \mu$.

A summary of properties of the Poisson distribution is given in Appendix B.

It is possible to derive equation (3.2.26) as a limiting form of the binomial pdf if $n \to \infty$ and $p \to 0$ with $np = \mu$ constant.

Theorem 3.2.3 If $X \sim \text{BIN}(n, p)$, then for each value $x = 0, 1, 2, \ldots$, and as $p \to 0$ with $np = \mu$ constant,

$$\lim_{n \to \infty} \binom{n}{x} p^x (1 - p)^{n-x} = \frac{e^{-\mu}\mu^x}{x!} \tag{3.2.31}$$

Proof

$$\begin{aligned} \binom{n}{x} p^x (1 - p)^{n-x} &= \frac{n!}{x!(n - x)!} \left(\frac{\mu}{n}\right)^x \left(1 - \frac{\mu}{n}\right)^{n-x} \\ &= \frac{\mu^x}{x!} \left(\frac{n}{n}\right)\left(\frac{n - 1}{n}\right) \cdots \left(\frac{n - x + 1}{n}\right)\left(1 - \frac{\mu}{n}\right)^n \left(1 - \frac{\mu}{n}\right)^{-x} \end{aligned}$$

The result follows by taking limits on both sides of the equation, and recalling from calculus that

$$\lim_{n\to\infty}\left(1-\frac{\mu}{n}\right)^n = e^{-\mu}$$

and, for fixed x,

$$\lim_{n\to\infty}\left(1-\frac{\mu}{n}\right)^{-x} = 1$$

■

One consequence of Theorem 3.2.3 is that $f(x; np)$ provides an approximation to the more complicated $b(x; n, p)$ when n is large and p is small.

Example 3.2.9 Suppose that 1% of all transistors produced by a certain company are defective. A new model of computer requires 100 of these transistors, and 100 are selected at random from the company's assembly line. The exact probability of obtaining a specified number of defectives, say 3, is $b(3;\ 100, 0.01) = 0.0610$, whereas the Poisson approximation is $f(3; 1) = 0.0613$.

As a general rule, the approximation gives reasonable results provided $n \geqslant 100$ and $p \leqslant 0.01$, and when x is close to np.

Theorem 3.2.3 also gives some insight into the types of problems for which the Poisson distribution provides an adequate probability model: When the variable of interest results, at least approximately, from a large number of independent Bernoulli-type experiments, each with a small probability of success. For example, the number of accidents in a year at a particular intersection may be approximately a Poisson variable, because a large number of vehicles may pass the intersection in a year and the probability of any one vehicle having an accident is small. In this case, the parameter μ would be directly affected by the number of vehicles per year and the degree of risk at the intersection.

POISSON PROCESSES

Consider a physical situation in which a certain type of event is recurring, such as telephone calls or defects in a long piece of wire. Let $X(t)$ denote the number of such events that occur in a given interval $[0, t]$, and suppose that the following assumptions hold. First, the probability that an event will occur in a given short interval $[t, t + \Delta t]$ is approximately proportional to the length of the interval, Δt, and does not depend on the position of the interval. Furthermore, suppose that the occurrences of events in nonoverlapping intervals are independent, and the probability of two or more events in a short interval $[t, t + \Delta t]$ is negligible. If these assumptions become valid as $\Delta t \to 0$, then the distribution of $X(t)$ will be Poisson. The assumptions and conclusions are stated mathematically in the following theorem. Note that $o(\Delta t)$ denotes a function of Δt such that $\lim_{\Delta t \to 0} o(\Delta t)/\Delta t = 0$. In other words, $o(\Delta t)$ is negligible relative to Δt.

Theorem 3.2.4 **Homogeneous Poisson Process** Let $X(t)$ denote the number of occurrences in the interval $[0, t]$, and $P_n(t) = P[n$ occurrences in an interval $[0, t]]$. Consider the following properties:

1. $X(0) = 0$,
2. $P[X(t + h) - X(t) = n \mid X(s) = m] = P[X(t + h) - X(t) = n]$ for all $0 \leqslant s \leqslant t$ and $0 < h$.
3. $P[X(t + \Delta t) - X(t) = 1] = \lambda \Delta t + o(\Delta t)$ for some constant $\lambda > 0$, and
4. $P[X(t + \Delta t) - X(t) \geqslant 2] = o(\Delta t)$.

If properties 1 through 4 hold, then for all $t > 0$,

$$P_n(t) = P[X(t) = n] = e^{-\lambda t}(\lambda t)^n/n! \tag{3.2.32}$$

Proof

Now n events may occur in the interval $[0, t + \Delta t]$ by having 0 events in $[t, t + \Delta t]$ and n events in $[0, t]$, or one event in $[t, t + \Delta t]$ and $n - 1$ events in $[0, t]$, or two or more events in $[t, t + \Delta t]$; thus for $n > 0$;

$$\begin{aligned} P_n(t + \Delta t) &= P_{n-1}(t)P_1(\Delta t) + P_n(t)P_0(\Delta t) + o(\Delta t) \\ &= P_{n-1}(t)[\lambda \Delta t + o(\Delta t)] + P_n(t)[1 - \lambda \Delta t - o(\Delta t)] + o(\Delta t) \end{aligned}$$

but

$$\begin{aligned} \frac{dP_n(t)}{dt} &= \lim_{\Delta t \to 0} \frac{P_n(t + \Delta t) - P_n(t)}{\Delta t} \\ &= \lim_{\Delta t \to 0} \frac{P_{n-1}(t)\lambda \Delta t + P_n(t) - P_n(t)\lambda \Delta t - P_n(t)}{\Delta t} \\ &= \lambda[P_{n-1}(t) - P_n(t)] \end{aligned}$$

For $n = 0$,

$$\begin{aligned} P_0(t + \Delta t) &= P_0(t)P_0(\Delta t) \\ &= P_0(t)[1 - \lambda \Delta t - o(\Delta t)] \end{aligned}$$

$$\begin{aligned} \frac{dP_0(t)}{dt} &= \lim_{\Delta t \to 0} \frac{P_0(t + \Delta t) - P_0(t)}{\Delta t} \\ &= \lim_{\Delta t \to 0} \frac{-\lambda \Delta t P_0(t) - o(\Delta t)P_0(t)}{\Delta t} \\ &= -\lambda P_0(t) \end{aligned}$$

Assuming the initial condition $P_0(0) = 1$, the solution to the above differential equation is verified easily to be

$$P_0(t) = e^{-\lambda t}$$

Similarly, letting $n = 1$,

$$\frac{dP_1(t)}{dt} = \lambda[P_0(t) - P_1(t)]$$

$$= \lambda[e^{-\lambda t} - P_1(t)]$$

which gives

$$P_1(t) = \lambda t e^{-\lambda t}$$

It can be shown by induction that

$$P_n(t) = e^{-\lambda t}(\lambda t)^n/n! \qquad n = 0, 1, 2, \ldots$$ ■

Thus, $X(t) \sim \text{POI}(\lambda t)$, where $\mu = E[X(t)] = \lambda t$. The proportionality constant λ reflects the **rate of occurrence** or **intensity** of the Poisson process. Because λ is assumed constant over t, the process is referred to as a **homogeneous Poisson process** (HPP). Because λ is constant and the increments are independent, it turns out that one does not need to be concerned about the location of the interval under question, and the model $X \sim \text{POI}(\mu)$ is applicable for any interval of length t, $[s, s + t]$, with $\mu = \lambda t$. The constant λ is the rate of occurrence per unit length, and the interval is t units long.

Finally, we will consider a rather simple type of discrete distribution known as the discrete uniform distribution.

DISCRETE UNIFORM DISTRIBUTION

Many problems, essentially those involving classical assignment of probability, can be modeled by a discrete random variable that assumes all of its values with the same probability. It usually is possible to relate such problems to a set of consecutive integers $1, 2, \ldots, N$.

A discrete random variable X has the **discrete uniform distribution** on the integers $1, 2, \ldots, N$ if it has a pdf of the form

$$f(x) = \frac{1}{N} \qquad x = 1, 2, \ldots, N \tag{3.2.33}$$

A special notation for this situation is

$$X \sim \text{DU}(N) \tag{3.2.34}$$

Example 3.2.10 Games of chance such as lotteries or rolling unbiased dice obviously are modeled by equation (3.2.33). For example, the number obtained by rolling an ordinary six-sided die would correspond to DU(6). Of course, this assumes that the die is not loaded or biased in some way in favor of certain numbers.

Another example that we have considered is the multiple-choice test of Example 3.2.3. If, on any question, we associate the four choices with the integers 1, 2, 3, and 4, then the response, X, on any given question that is answered at random is DU(4). This also models the outcome, X, of rolling a four-sided die. The discrete pdf and CDF of $X \sim$ DU(4) are given in Figure 3.1.

FIGURE 3.1 Discrete pdf and CDF of number obtained on a single roll of a four-sided die

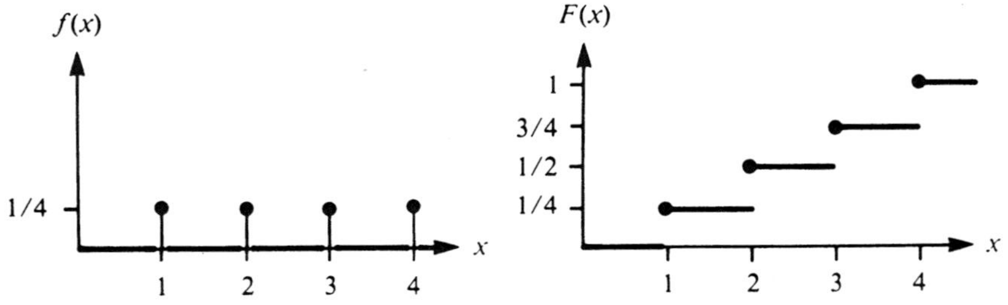

Comparing Figure 3.1 with Figures 2.2 and 2.3, the distribution of the maximum over two rolls favors the larger values, whereas the result of a single roll does not. This should not be a surprise because there are more ways to achieve the larger values based on the maximum.

The mean is obtained as follows:

$$E(X) = \sum_{x=1}^{N} \frac{x}{N} = \left(\frac{1}{N}\right)(1 + 2 + \cdots + N)$$

$$= \frac{(1/N)N(N+1)}{2}$$

$$= \frac{N+1}{2}$$

Similarly, $E(X^2) = (N + 1)(2N + 1)/6$ and $\text{Var}(X) = (N^2 - 1)/12$.

3.3

SPECIAL CONTINUOUS DISTRIBUTIONS

We will now discuss several special continuous distributions.

UNIFORM DISTRIBUTION

Suppose that a continuous random variable X can assume values only in a bounded interval, say the open interval (a, b), and suppose that the pdf is constant, say $f(x) = c$ over the interval. Property (2.3.5) implies $c = 1/(b - a)$, because $1 = \int_a^b c\,dx = c(b - a)$. If we define $f(x) = 0$ outside the interval, then property (2.3.4) also is satisfied.

This special distribution is known as the **uniform distribution** on the interval (a, b). The pdf is

$$f(x; a, b) = \frac{1}{b - a} \qquad a < x < b \tag{3.3.1}$$

and zero otherwise. A notation that designates that X has pdf of the form (3.3.1) is

$$X \sim \text{UNIF}(a, b) \tag{3.3.2}$$

This is the continuous counterpart of the discrete uniform distribution, which was discussed in Section 2.2. It provides a probability model for selecting a point "at random" from an interval (a, b). A more specific example is given by the random waiting times for the bus passenger in Example 2.3.1. As noted earlier, it does not matter whether we include the endpoints $a = 0$ and $b = 5$.

Perhaps a more important application occurs in the case of computer simulation, which relies on the generation of "random numbers." Random number generators are functions in the computer language, or in some cases subroutines in programs, which are designed to produce numbers that behave as if they were data from UNIF(0, 1).

The CDF of $X \sim \text{UNIF}(a, b)$ has the form

$$F(x; a, b) = \begin{cases} 0 & x \leqslant a \\ \dfrac{x - a}{b - a} & a < x < b \\ 1 & b \leqslant x \end{cases} \tag{3.3.3}$$

The general form of the graphs of $f(x; a, b)$ and $F(x; a, b)$ can be seen in Figure 2.5, where, in general, the endpoints would be a and b, rather than 0 and 5.

If $X \sim \text{UNIF}(a, b)$, then

$$\begin{aligned}
E(X) &= \int_a^b x\left(\frac{1}{b-a}\right) dx \\
&= \frac{b^2 - a^2}{2(b-a)} \\
&= \frac{(b+a)(b-a)}{2(b-a)} \\
&= \frac{a+b}{2}
\end{aligned}$$

Furthermore,

$$\begin{aligned}
E(X^2) &= \int_a^b x^2\left(\frac{1}{b-a}\right) dx \\
&= \frac{b^3 - a^3}{3(b-a)} \\
&= \frac{(b^2 + ab + a^2)(b-a)}{3(b-a)} \\
&= \frac{b^2 + ab + a^2}{3}
\end{aligned}$$

Thus,

$$\begin{aligned}
\text{Var}(X) &= \frac{b^2 + ab + a^2}{3} - \frac{(a+b)^2}{4} \\
&= \frac{(b-a)^2}{12}
\end{aligned}$$

In this case, we can conclude that the mean of the distribution is the midpoint, and the variance is proportional to the square of the length of the interval (a, b). This is consistent with our interpretations of mean and variance as respective measures of the "center" and "variability" in a population.

For example, the temperature reading (in Fahrenheit degrees) at a randomly selected time at some location is a random variable $X \sim \text{UNIF}(50, 90)$, and the reading at a second location is a random variable $Y \sim \text{UNIF}(30, 110)$. The means are the same, $\mu_X = \mu_Y = 70$, but the variances are different, $\sigma_X^2 = 400/3 < \sigma_Y^2 = 1600/3$.

The $100 \times p$th percentile, which is obtained by equating the right side of equation (3.3.3) to p and solving for x, is $x_p = a + (b - a)p$.

The main properties of a uniform distribution are summarized in Appendix B.

GAMMA DISTRIBUTION

A continuous distribution that occurs frequently in applications is called the gamma distribution. The name results from its relationship to a function called the gamma function.

Definition 3.3.1

The **gamma function**, denoted by $\Gamma(\kappa)$ for all $\kappa > 0$, is given by

$$\Gamma(\kappa) = \int_0^\infty t^{\kappa-1} e^{-t}\, dt \tag{3.3.4}$$

For example, if $\kappa = 1$, then $\Gamma(1) = \int_0^\infty e^{-t}\, dt = 1$. The gamma function has several useful properties, as stated in the following theorem.

Theorem 3.3.1 The gamma function satisfies the following properties:

$$\Gamma(\kappa) = (\kappa - 1)\Gamma(\kappa - 1) \qquad \kappa > 1 \tag{3.3.5}$$

$$\Gamma(n) = (n - 1)! \qquad n = 1, 2, \ldots \tag{3.3.6}$$

$$\Gamma\left(\frac{1}{2}\right) = \sqrt{\pi} \tag{3.3.7}$$

Proof

See Exercise 35. ■

A continuous random variable X is said to have the **gamma distribution** with parameters $\kappa > 0$ and $\theta > 0$ if it has pdf of the form

$$f(x; \theta, \kappa) = \frac{1}{\theta^\kappa \Gamma(\kappa)} x^{\kappa-1} e^{-x/\theta} \qquad x > 0 \tag{3.3.8}$$

and zero otherwise. The function given by equation (3.3.8) satisfies the general properties (2.3.4) and (2.3.5), with the latter resulting from the substitution $t = x/\theta$ in the integral $\int_0^\infty f(x;\theta,\kappa)\, dx$, resulting in $\Gamma(\kappa)/\Gamma(\kappa) = 1$.

A special notation, which designates that X has pdf given by equation (3.3.8), is

$$X \sim \text{GAM}(\theta, \kappa) \tag{3.3.9}$$

The parameter κ is called a **shape parameter** because it determines the basic shape of the graph of the pdf. Specifically, there are three basic shapes, depending on whether $\kappa < 1$, $\kappa = 1$, or $\kappa > 1$. This is illustrated in Figure 3.2, which shows the graphs of equation (3.3.8) for $\kappa = 0.5$, 1, and 2.

FIGURE 3.2 The pdf's of gamma distributions

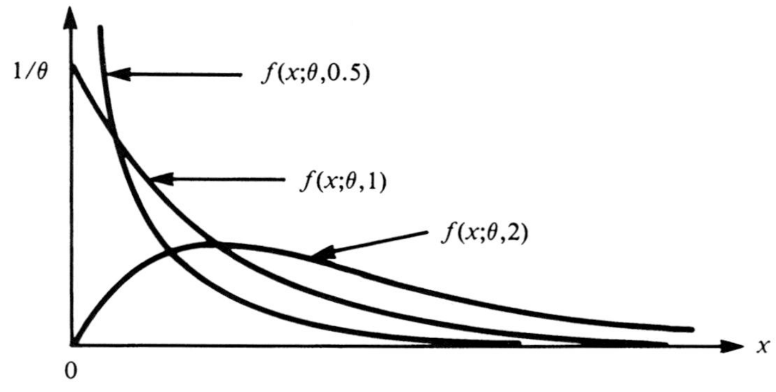

Note that the y-axis is an asymptote of $y = f(x;\ \theta,\ \kappa)$ if $\kappa < 1$, while $f(0;\ \theta,\ 1) = 1/\theta$; and if $\kappa > 1, f(0;\ \theta,\ \kappa) = 0$.

The CDF of $X \sim \text{GAM}(\theta, \kappa)$ is

$$F(x;\ \theta,\ \kappa) = \int_0^x \frac{1}{\theta^\kappa \Gamma(\kappa)}\, t^{\kappa - 1} e^{-t/\theta}\, dt \tag{3.3.10}$$

The substitution $u = t/\theta$ in this integral yields

$$F(x;\ \theta,\ \kappa) = F(x/\theta;\ 1,\ \kappa) \tag{3.3.11}$$

which depends on θ only through the variable x/θ. Such a parameter usually is called a **scale parameter**.

Usually, it is important to have a scale parameter in a model so that the results will not depend on which scale of measurement is used. For example, if X represents time in months and it is assumed that $X \sim \text{GAM}(\theta, \kappa)$ with $\theta = 12$, then

$$P[X \leqslant 24 \text{ months}] = F(24/12;\ 1,\ \kappa) = F(2;\ 1,\ \kappa)$$

If one considers the time Y to be measured in weeks, then an equivalent model still can be achieved by considering Y to be a gamma variable with $\theta = 4 \cdot 12 = 48$. For example,

$$\begin{aligned} P[X \leqslant 24 \text{ months}] = P[Y \leqslant 96 \text{ weeks}] &= F(96/48;\ 1,\ \kappa) \\ &= F(2;\ 1,\ \kappa) \end{aligned}$$

as before. Thus, different scales of measurement can be accommodated by changing the value of the scale parameter in this case without changing to a different general form.

The CDF obtained in equation (3.3.10) generally cannot be solved explicitly, but if κ is a positive integer, say $\kappa = n$, then the integral can be expressed as a sum.

Theorem 3.3.2 If $X \sim \text{GAM}(\theta, n)$, where n is a positive integer, then the CDF can be written

$$F(x; \theta, n) = 1 - \sum_{i=0}^{n-1} \frac{(x/\theta)^i}{i!} e^{-x/\theta} \tag{3.3.12}$$

Proof

This follows by repeated integration by parts on integral (3.3.10). ■

Notice that the terms in the sum in equation (3.3.12) resemble the terms of a Poisson sum with μ replaced by x/θ.

Example 3.3.1 The daily amount (in inches) of measurable precipitation in a river valley is a random variable $X \sim \text{GAM}(0.2, 6)$. It might be of interest to know the probability that the amount of precipitation will exceed some level, say 2 inches. This would be

$$\begin{aligned} P[X > 2] &= \int_2^\infty \frac{1}{(0.2)^6 \Gamma(6)} x^{6-1} e^{-(x/0.2)}\, dx \\ &= 1 - F(2; 0.2, 6) \\ &= \sum_{i=0}^{5} \frac{10^i}{i!} e^{-10} = 0.067 \end{aligned}$$

which can be found in Table 2 in Appendix C with $\mu = 10$.

The mean of $X \sim \text{GAM}(\theta, \kappa)$ is obtained as follows:

$$\begin{aligned} E(X) &= \int_0^\infty x \frac{1}{\theta^\kappa \Gamma(\kappa)} x^{\kappa-1} e^{-x/\theta}\, dx \\ &= \frac{1}{\theta^\kappa \Gamma(\kappa)} \int_0^\infty x^{(1+\kappa)-1} e^{-x/\theta}\, dx \\ &= \frac{\theta^{1+\kappa}\Gamma(1+\kappa)}{\theta^\kappa \Gamma(\kappa)} \int_0^\infty \frac{1}{\theta^{1+\kappa}\Gamma(1+\kappa)} x^{(1+\kappa)-1} e^{-x/\theta}\, dx \\ &= \frac{\theta^{1+\kappa}\Gamma(1+\kappa)}{\theta^\kappa \Gamma(\kappa)} \\ &= \theta \frac{\kappa\Gamma(\kappa)}{\Gamma(\kappa)} \\ &= \kappa\theta \end{aligned}$$

Similarly, $E(X^2) = \theta^2\kappa(1 + \kappa)$, and thus

$$\text{Var}(X) = \theta^2\kappa(1 + \kappa) - (\kappa\theta)^2 = \kappa\theta^2$$

In the previous example, the daily amount of precipitation is gamma distributed with mean of 1.2 inches and variance of 0.24.

Of course, the moments also can be obtained using the MGF,

$$M_X(t) = \int_0^\infty e^{tx} \frac{x^{\kappa-1}e^{-x/\theta}}{\theta^\kappa\Gamma(\kappa)}\,dx$$
$$= \frac{1}{\theta^\kappa\Gamma(\kappa)}\int_0^\infty x^{\kappa-1}e^{(t-1/\theta)x}\,dx$$

After the substitution $u = -(t - 1/\theta)x$, we obtain

$$M_X(t) = \left(\frac{1}{\theta} - t\right)^{-\kappa} \frac{1}{\theta^\kappa\Gamma(\kappa)}\int_0^\infty u^{\kappa-1}e^{-u}\,du$$

$$M_X(t) = (1 - \theta t)^{-\kappa} \qquad t < 1/\theta \tag{3.3.13}$$

The rth derivative, in this case, is

$$M_X^{(r)}(t) = (\kappa + r - 1)\cdots(\kappa + 1)\kappa\theta^r(1 - \theta t)^{-\kappa-r}$$
$$= \frac{\Gamma(\kappa + r)}{\Gamma(\kappa)}\theta^r(1 - \theta t)^{-\kappa-r}$$

and $M_X^{(r)}(0)$ yields the rth moment of X,

$$E(X^r) = \frac{\Gamma(\kappa + r)}{\Gamma(\kappa)}\theta^r \tag{3.3.14}$$

Strictly speaking, this derivation is valid only if r is a positive integer, but it is possible to show by a direct argument that (3.3.14) is valid for any real $r > -\kappa$.

The power series has the form

$$M_X(t) = 1 + \sum_{r=1}^{\infty} \frac{\Gamma(\kappa + r)}{\Gamma(\kappa)} \frac{\theta^r}{r!} t^r \tag{3.3.15}$$

A special case of the gamma distribution with $\theta = 2$ and $\kappa = \nu/2$ is referred to as a chi-square distribution with ν degrees of freedom; this distribution is discussed in more detail in Chapter 8. It will be seen that cumulative chi-square tables can be used to evaluate gamma cumulative probabilities.

When $\kappa = 1$, we obtain a special case known as the exponential distribution.

EXPONENTIAL DISTRIBUTION

A continuous random variable X has the **exponential distribution** with parameter $\theta > 0$ if it has a pdf of the form

$$f(x; \theta) = \frac{1}{\theta} e^{-x/\theta} \qquad x > 0 \tag{3.3.16}$$

and zero otherwise. The CDF of X is

$$F(x; \theta) = 1 - e^{-x/\theta} \qquad x > 0 \tag{3.3.17}$$

so that θ is a scale parameter.

The notation $X \sim \text{GAM}(\theta, 1)$ could be used to designate that X has pdf (3.3.16), but a more common notation is

$$X \sim \text{EXP}(\theta) \tag{3.3.18}$$

The exponential distribution, which is an important probability model for lifetimes, sometimes is characterized by a property that is given in the following theorem.

Theorem 3.3.3 For a continuous random variable X, $X \sim \text{EXP}(\theta)$ if and only if

$$P[X > a + t \mid X > a] = P[X > t] \tag{3.3.19}$$

for all $a > 0$ and $t > 0$.

Proof (only if)

$$\begin{aligned} P[X > a + t \mid X > a] &= \frac{P[X > a + t \text{ and } X > a]}{P[X > a]} \\ &= \frac{P[X > a + t]}{P[X > a]} \\ &= \frac{e^{-(a+t)/\theta}}{e^{-a/\theta}} \\ &= P[X > t] \end{aligned}$$

■

This shows that the exponential distribution satisfies property (3.3.19), which is known as the **no-memory property**. We will not attempt to show that the exponential distribution is the only such continuous distribution.

If X is the lifetime of a component, then property (3.3.19) asserts that the probability that the component will last more than $a + t$ time units given that it has already lasted more than a units is the same as that of a new component lasting more than t units. In other words, an old component that still works is just as reliable as a new component. Failure of such a component is not the result of fatigue or wearout.

Example 3.3.2 Suppose a certain solid-state component has a lifetime or failure time (in hours) $X \sim \text{EXP}(100)$. The probability that the component will last at least 50 hours is

$$P[X \geqslant 50] = 1 - F(50; 100) = e^{-0.5} = 0.6065$$

It follows from the relationship to the gamma distribution that $E(X) = 1 \cdot \theta = \theta$ and $\text{Var}(X) = 1 \cdot \theta^2 = \theta^2$. Thus, in the previous example, the mean lifetime of a component is $\mu = 100$ hours, and the standard deviation, σ, is also 100 hours.

The exponential distribution is also a special case of another important continuous distribution called the Weibull distribution.

WEIBULL DISTRIBUTION

A widely used continuous distribution is named after the physicist W. Weibull, who suggested its use for numerous applications, including fatigue and breaking strength of materials. It is also a very popular choice as a failure-time distribution.

A continuous random variable X is said to have the **Weibull distribution** with parameters $\beta > 0$ and $\theta > 0$ if it has a pdf of the form

$$f(x; \theta, \beta) = \frac{\beta}{\theta^\beta} x^{\beta - 1} e^{-(x/\theta)^\beta} \qquad x > 0 \tag{3.3.20}$$

and zero otherwise. A notation that designates that X has pdf (3.3.20) is

$$X \sim \text{WEI}(\theta, \beta) \tag{3.3.21}$$

The parameter β is called a **shape parameter**. This is similar to the situation we encountered with the gamma distribution, because there are three basic shapes, depending on whether $\beta < 1$, $\beta = 1$, or $\beta > 1$. This is illustrated in Figure 3.3, which shows the graphs of pdf (3.3.20) for $\beta = 0.5$, 1, and 2.

Notice that the y-axis is an asymptote of $y = f(x; \theta, \beta)$ if $\beta < 1$, while $f(0; \theta, 1) = 1/\theta$; and if $\beta > 1$, then $f(0; \theta, \beta) = 0$.

One advantage of the Weibull distribution is that the CDF can be obtained explicitly by integrating pdf (3.3.20):

$$F(x; \theta, \beta) = 1 - e^{-(x/\theta)^\beta} \qquad x > 0 \tag{3.3.22}$$

It is also clear that (3.3.22) can be written as $F(x/\theta; 1, \beta)$, which means that θ is a scale parameter, as discussed earlier in the chapter.

The special case with $\beta = 2$ is known as the **Rayleigh distribution**.

FIGURE 3.3 The pdf's of Weibull distributions

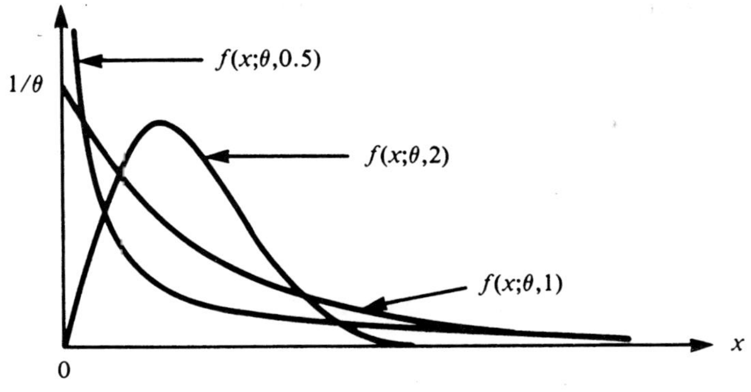

Example 3.3.3 The distance (in inches) that a dart hits from the center of a target may be modeled as a random variable $X \sim \text{WEI}(10, 2)$. The probability of hitting within five inches of the center is

$$P[X \leqslant 5] = F(5;\, 10, 2) = 1 - e^{-(5/10)^2} = 0.221$$

The mean of $X \sim \text{WEI}(\theta, \beta)$ is obtained as follows:

$$E(X) = \int_0^\infty x \frac{\beta}{\theta^\beta} x^{\beta - 1} e^{-(x/\theta)^\beta}\, dx$$

$$= \frac{\beta}{\theta^\beta} \int_0^\infty x^{(1+\beta)-1} e^{-(x/\theta)^\beta}\, dx$$

Following the substitution $t = (x/\theta)^\beta$, and some simplification,

$$E(X) = \theta \int_0^\infty t^{(1+1/\beta)-1} e^{-t}\, dt = \theta\Gamma\left(1 + \frac{1}{\beta}\right)$$

Similarly, $E(X^2) = \theta^2\Gamma(1 + 2/\beta)$, and thus

$$\text{Var}(X) = \theta^2\left[\Gamma\left(1 + \frac{2}{\beta}\right) - \Gamma^2\left(1 + \frac{1}{\beta}\right)\right]$$

It follows from equation (3.3.22) that the $100 \times p$th percentile has the form $x_p = \theta[-\ln(1 - p)]^{1/\beta}$.

PARETO DISTRIBUTION

A continuous random variable X is said to have the **Pareto distribution** with parameters $\theta > 0$ and $\kappa > 0$ if it has a pdf of the form

$$f(x; \theta, \kappa) = \frac{\kappa}{\theta}\left(1 + \frac{x}{\theta}\right)^{-(\kappa+1)} \qquad x > 0 \tag{3.3.23}$$

and zero otherwise. A notation to designate that X has pdf (3.3.23) is

$$X \sim \text{PAR}(\theta, \kappa) \tag{3.3.24}$$

The parameter κ is a **shape parameter** for this model, although there is not as much variety in the possible basic shapes as we found with the gamma and Weibull models. The CDF is given by

$$F(x; \theta, \kappa) = 1 - \left(1 + \frac{x}{\theta}\right)^{-\kappa} \qquad x > 0 \tag{3.3.25}$$

Because equation (3.3.25) also can be expressed as $F(x/\theta; 1, \kappa)$, θ is a scale parameter for the Pareto distribution.

The length of wire between flaws, X, discussed in Example 2.3.2 is an example of a Pareto distribution, namely $X \sim \text{PAR}(1, 2)$, and the graph of $f(x; 1, 2)$ is shown in Figure 2.6. This model also has been used to model biomedical problems, such as survival time following a heart transplant.

Another related distribution, which is also sometimes referred to as the Pareto distribution, has a pdf of the form

$$f(y) = \left(\frac{\kappa}{a}\right)\left(\frac{y}{a}\right)^{-(\kappa+1)} \qquad y > a \tag{3.3.26}$$

and zero otherwise, where $a > 0$ and $\kappa > 0$.

It is straightforward to show that $E(X) = \theta/(\kappa - 1)$ and $\text{Var}(X) = \theta^2\kappa/[(\kappa - 2)(\kappa - 1)^2]$, and that the $100 \times p$th percentile is $x_p = \theta[(1 - p)^{-1/\kappa} - 1]$.

NORMAL DISTRIBUTION

The normal distribution was first published by Abraham de Moivre in 1733 as an approximation for the distribution of the sum of binomial random variables. It is the single most important distribution in probability and statistics.

A random variable X follows the **normal distribution** with mean μ and variance σ^2 if it has the pdf

$$f(x; \mu, \sigma) = \frac{1}{\sqrt{2\pi}\,\sigma}\, e^{-[(x-\mu)/\sigma]^2/2} \tag{3.3.27}$$

for $-\infty < x < \infty$, where $-\infty < \mu < \infty$ and $0 < \sigma < \infty$. This is denoted by

$$X \sim \text{N}(\mu, \sigma^2) \tag{3.3.28}$$

The normal distribution also is referred to frequently as the **Gaussian distribution**.

The normal distribution arises frequently in physical problems; there is a theoretical reason for this, which will be developed in Chapter 7.

First, we will verify that the normal pdf integrates to 1, and then we will verify that the values of the parameters μ and σ^2 are indeed the mean and variance of X.

Making a change of variable, $z = (x - \mu)/\sigma$ with $dx = \sigma\, dz$, gives

$$I = \int_{-\infty}^{\infty} f(x; \mu, \sigma)\, dx = \int_{-\infty}^{\infty} \frac{1}{\sqrt{2\pi}} e^{-z^2/2}\, dz$$

$$= 2 \int_{0}^{\infty} \frac{1}{\sqrt{2\pi}} e^{-z^2/2}\, dz$$

If we let $w = z^2/2$, then $z = \sqrt{2w}$ and $dz = (w^{-1/2}/\sqrt{2})\, dw$, so

$$I = \int_{0}^{\infty} \frac{w^{-1/2}}{\sqrt{\pi}} e^{-w}\, dw = \frac{\Gamma(1/2)}{\sqrt{\pi}} = 1$$

which follows from equation (3.3.7).

The integrand obtained following the substitution $z = (x - \mu)/\sigma$ is an important special case known as the **standard normal pdf**. We will adopt a special notation for this pdf, namely

$$\phi(z) = \frac{1}{\sqrt{2\pi}} e^{-z^2/2} \qquad -\infty < z < \infty \tag{3.3.29}$$

If Z had pdf (3.3.29), then $Z \sim \mathrm{N}(0, 1)$, and the standard normal CDF is given by

$$\Phi(z) = \int_{-\infty}^{z} \phi(t)\, dt \tag{3.3.30}$$

Some basic geometric properties of the standard normal pdf can be obtained by the methods of calculus. Notice that

$$\phi(z) = \phi(-z) \tag{3.3.31}$$

for all real z, so $\phi(z)$ is an even function of z. In other words, the standard normal distribution is symmetric about $z = 0$. Furthermore, because of the special form of $\phi(z)$, we have

$$\phi'(z) = -z\phi(z) \tag{3.3.32}$$

and

$$\phi''(z) = (z^2 - 1)\phi(z) \tag{3.3.33}$$

Consequently, $\phi(z)$ has a unique maximum at $z = 0$ and inflection points at $z = \pm 1$. Note also that $\phi(z) \to 0$ and $\phi'(z) = -z/[\sqrt{2\pi} \exp(z^2/2)] \to 0$ as $z \to \pm\infty$.

It is also possible, using equations (3.3.32) and (3.3.33), to find $E(Z)$ and $E(Z^2)$. Specifically,

$$\begin{aligned}
E(Z) &= \int_{-\infty}^{\infty} z\phi(z)\,dz \\
&= -\int_{-\infty}^{\infty} \phi'(z)\,dz \\
&= -\phi(z)\,|_{-\infty}^{\infty} \\
&= 0
\end{aligned}$$

and

$$\begin{aligned}
E(Z^2) &= \int_{-\infty}^{\infty} z^2\phi(z)\,dz \\
&= \int_{-\infty}^{\infty} [\phi''(z) + \phi(z)]\,dz \\
&= \phi'(z)\,|_{-\infty}^{\infty} + \int_{-\infty}^{\infty} \phi(z)\,dz \\
&= 0 + 1 \\
&= 1
\end{aligned}$$

Similar results follow for the more general case $X \sim \mathrm{N}(\mu, \sigma^2)$. Based on the substitution $z = (x - \mu)/\sigma$, we have

$$\begin{aligned}
E(X) &= \int_{-\infty}^{\infty} x\,\frac{1}{\sqrt{2\pi}\,\sigma}\exp\left[-\frac{1}{2}\left(\frac{x-\mu}{\sigma}\right)^2\right] dx \\
&= \int_{-\infty}^{\infty} (\mu + \sigma z)\phi(z)\,dz \\
&= \mu\int_{-\infty}^{\infty} \phi(z)\,dz + \sigma\int_{-\infty}^{\infty} z\phi(z)\,dz \\
&= \mu
\end{aligned}$$

and

$$\begin{aligned}
E(X^2) &= \int_{-\infty}^{\infty} x^2\,\frac{1}{\sqrt{2\pi}\,\sigma}\exp\left[-\frac{1}{2}\left(\frac{x-\mu}{\sigma}\right)^2\right] dx \\
&= \int_{-\infty}^{\infty} (\mu + \sigma z)^2\phi(z)\,dz \\
&= \mu^2\int_{-\infty}^{\infty} \phi(z)\,dz + 2\mu\sigma\int_{-\infty}^{\infty} z\phi(z)\,dz + \sigma^2\int_{-\infty}^{\infty} z^2\phi(z)\,dz \\
&= \mu^2 + \sigma^2
\end{aligned}$$

It follows that $\mathrm{Var}(X) = E(X^2) - \mu^2 = (\mu^2 + \sigma^2) - \mu^2 = \sigma^2$.

FIGURE 3.4 A normal pdf

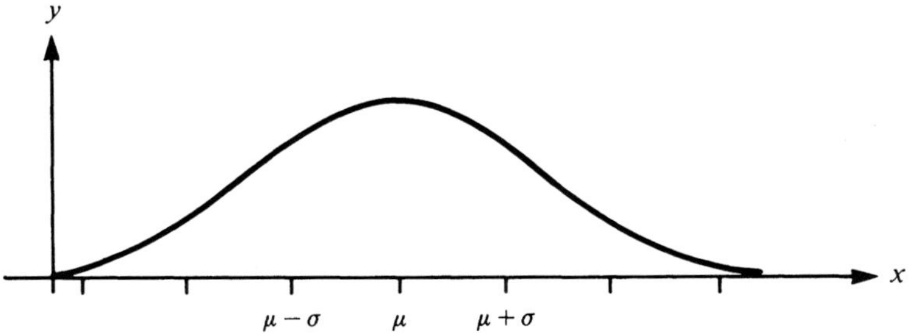

The graph of $y = f(x; \mu, \sigma)$ is shown in Figure 3.4.

The general topic of transformation of variables will be discussed later, but it is convenient to consider the following theorem at this point.

Theorem 3.3.4 If $X \sim N(\mu, \sigma^2)$, then

1. $Z = \dfrac{X - \mu}{\sigma} \sim N(0, 1)$

2. $F_X(x) = \Phi\left(\dfrac{x - \mu}{\sigma}\right)$ **(3.3.34)**

Proof

$$\begin{aligned} F_Z(z) &= P[Z \leqslant z] \\ &= P\left[\frac{X - \mu}{\sigma} \leqslant z\right] \\ &= P[X \leqslant \mu + z\sigma] \\ &= \int_{-\infty}^{\mu + z\sigma} \frac{1}{\sqrt{2\pi}\,\sigma} \exp\left[-\frac{1}{2}\left(\frac{x - \mu}{\sigma}\right)^2\right] dx \end{aligned}$$

After the substitution $w = (x - \mu)/\sigma$, we have

$$\begin{aligned} F_Z(z) &= \int_{-\infty}^{z} \frac{1}{\sqrt{2\pi}} e^{-w^2/2}\, dw \\ &= \Phi(z) \end{aligned}$$

Part 1 follows by differentiation, $f_Z(z) = F'_Z(z) = \dfrac{1}{\sqrt{2\pi}} e^{-z^2/2}$.

We obtain Part 2 as follows:

$$F_X(x) = P[X \leqslant x]$$
$$= P\left[\frac{X - \mu}{\sigma} \leqslant \frac{x - \mu}{\sigma}\right]$$
$$= \Phi\left(\frac{x - \mu}{\sigma}\right) \qquad \textbf{(3.3.35)}$$

■

Standard normal cumulative probabilities, $\Phi(z)$, are provided in Table 3 in Appendix C for positive values of z. Because of the symmetry of the normal density, cumulative probabilities may be determined for negative values of z by the relationship

$$\Phi(-z) = 1 - \Phi(z) \qquad \textbf{(3.3.36)}$$

We will let z_γ denote the γth percentile of the standard normal distribution, $\Phi(z_\gamma) = \gamma$. For example, for $\gamma = 0.95$, $z_{0.95} = 1.645$ from Table 3 (Appendix C). By symmetry, we know that $\Phi(-1.645) = 1 - 0.95 = 0.05$. That is, $z_{0.05} = -z_{1-0.05}$. It follows that

$$P[z_{0.05} < Z < z_{0.95}] = 0.95 - 0.05 = 0.90$$

or

$$P[-z_{1-0.05} < Z < z_{1-0.05}] = 0.90 \qquad \textbf{(3.3.37)}$$

Some authors find it more convenient to use z_α to denote the value that has α area to the right; however, we will use the notation above, which is more consistent with our notation for percentiles. Thus, in general,

$$P[-z_{1-\alpha/2} < Z < z_{1-\alpha/2}] = 1 - \alpha \qquad \textbf{(3.3.38)}$$

which corresponds to equation (3.3.37) with $\alpha = 0.10$ and where $z_{1-\alpha/2} = z_{0.95} = 1.645$. Similarly, if $\alpha = 0.05$, then $z_{1-\alpha/2} = z_{0.975} = 1.96$, and

$$P[-1.96 < Z < 1.96] = 0.95$$

It often is useful to consider normal probabilities in terms of standard deviations from the mean. For example, if $X \sim \mathrm{N}(\mu, \sigma^2)$, then

$$P[\mu - 1.96\sigma < X < \mu + 1.96\sigma] = F_X(\mu + 1.96\sigma) - F_X(\mu - 1.96\sigma)$$
$$= \Phi\left(\frac{\mu + 1.96\sigma - \mu}{\sigma}\right) - \Phi\left(\frac{\mu - 1.96\sigma - \mu}{\sigma}\right)$$
$$= \Phi(1.96) - \Phi(-1.96)$$
$$= 0.95$$

That is, 95% of the area under a normal pdf is within 1.96 standard deviations of the mean, 90% is within 1.645 standard deviations, and so on.

Example 3.3.4 Let X represent the lifetime in months of a battery, and assume that approximately $X \sim \mathrm{N}(60, 36)$. The fraction of batteries that will fail within a four-year warranty period is given by

$$P[X \leqslant 48] = \Phi\left(\frac{48 - 60}{6}\right) = \Phi(-2) = 0.0228$$

If one wished to know what warranty period would correspond to 5% failures, then

$$P[X \leqslant x_{0.05}] = \Phi\left(\frac{x_{0.05} - 60}{6}\right) = 0.05$$

which means that $(x_{0.05} - 60)/6 = -1.645$, and $x_{0.05} = -1.645(6) + 60 = 50.13$ months.

In general, we see that the $100 \times p$th percentile is

$$x_p = \mu + z_p \sigma \tag{3.3.39}$$

In the previous example, note also that

$$P[X \leqslant 0] = \Phi\left(\frac{0 - 60}{6}\right) = \Phi(-10) \doteq 0$$

Thus, although the normal random variable theoretically takes on values over the whole real line, it still may provide a reasonable model for a variable that takes on only positive values, if very little probability is associated with the negative values. Another possibility is to consider a truncated normal model when the variable must be positive, although we need not bother with that here; the probability assigned to the negative values is so small that the truncated model would be essentially the same as the untruncated model. Of course, there is still no guarantee that the normal model is a good choice for this variable. In particular, the normal distribution is symmetric, and it is not uncommon for lifetimes to follow skewed distributions. The question of model selection is a statistical topic to be considered later, but we will not exclude the possibility of using a normal model for positive variables in examples, with the understanding that it may be approximating the more theoretically correct truncated normal model.

Some additional properties are given by the following theorem.

Theorem 3.3.5 If $X \sim \mathrm{N}(\mu, \sigma^2)$, then

$$M_X(t) = e^{\mu t + \sigma^2 t^2/2} \tag{3.3.40}$$

$$E(X - \mu)^{2r} = \frac{(2r)!\,\sigma^{2r}}{r!\,2^r} \qquad r = 1, 2, \ldots \tag{3.3.41}$$

$$E(X - \mu)^{2r-1} = 0 \qquad r = 1, 2, \ldots \tag{3.3.42}$$

Proof

To show equation (3.3.40), we note that the MGF for a standard normal random variable is given by

$$M_Z(t) = \int_{-\infty}^{\infty} \frac{1}{\sqrt{2\pi}} e^{tz} e^{-z^2/2}\, dz$$

$$= \int_{-\infty}^{\infty} \frac{1}{\sqrt{2\pi}} e^{-(z-t)^2/2 + t^2/2}\, dz = e^{t^2/2}$$

The integral of the first factor in the second integral is 1, because it is the integral of a normal pdf with mean t and variance 1. Because $X = Z\sigma + \mu$,

$$M_X(t) = M_{\sigma Z + \mu}(t) = e^{\mu t} M_Z(\sigma t) = e^{\mu t + \sigma^2 t^2/2}$$

Equations (3.3.41) and (3.3.42) follow from a series expansion:

$$M_{X-\mu}(t) = e^{\sigma^2 t^2/2}$$

$$= \sum_{r=0}^{\infty} \frac{\sigma^{2r} t^{2r}}{2^r r!}$$

$$= \sum_{r=0}^{\infty} \frac{\sigma^{2r} (2r)!\, t^{2r}}{2^r r!\, (2r)!}$$

This expansion contains only even integer powers, and the coefficient of $t^{2r}/(2r)!$ is the $2r$th moment of $(X - \mu)$. ■

The mean of a normal distribution is an example of a special type of parameter known as a location parameter, and the standard deviation is a scale parameter.

3.4 LOCATION AND SCALE PARAMETERS

In each of the following definitions, $F_0(z)$ represents a completely specified CDF, and $f_0(z)$ is the pdf.

Definition 3.4.1

Location Parameters A quantity η is a **location parameter** for the distribution of X if the CDF has the form

$$F(x; \eta) = F_0(x - \eta) \quad \textbf{(3.4.1)}$$

In other words, the pdf has the form

$$f(x; \eta) = f_0(x - \eta) \quad \textbf{(3.4.2)}$$

Example 3.4.1 A distribution that often is encountered in life-testing applications has pdf

$$f(x;\, \eta) = e^{-(x-\eta)} \qquad x > \eta$$

and zero otherwise. The location parameter, η, in this application usually is called a **threshold** parameter because the probability of a failure before η is zero. This is illustrated in Figure 3.5.

FIGURE 3.5 An exponential pdf with a threshold parameter

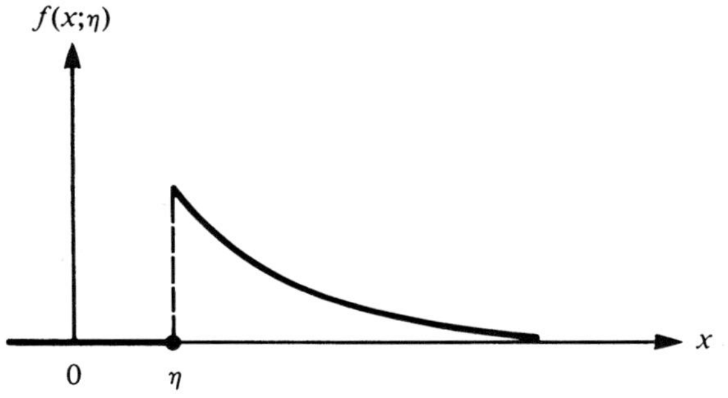

It is more common for a location parameter to be a measure of central tendency of X, such as a mean or a median.

Example 3.4.2 Consider the pdf

$$f_0(z) = \frac{1}{2}\, e^{-|z|} \qquad -\infty < z < \infty$$

If X has pdf of the form

$$f(x;\, \eta) = \frac{1}{2}\, e^{-|x-\eta|} \qquad -\infty < x < \infty$$

FIGURE 3.6 A double-exponential pdf with a location parameter

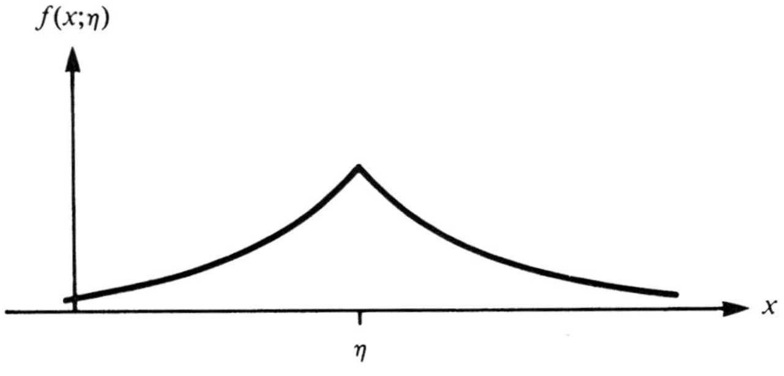

then the location parameter, η, is the mean of the distribution. Because $f(x;\, \eta)$ is symmetric about η, and has a unique maximum, η is also the median and the mode in this example. This is illustrated in Figure 3.6.

The notion of a scale parameter was mentioned earlier in this chapter. A more precise definition now will be given.

Definition 3.4.2

Scale Parameter A positive quantity θ is a **scale parameter** for the distribution of X if the CDF has the form

$$F(x;\, \theta) = F_0\left(\frac{x}{\theta}\right) \tag{3.4.3}$$

In other words, the pdf has the form

$$f(x;\, \theta) = \frac{1}{\theta}\, f_0\left(\frac{x}{\theta}\right) \tag{3.4.4}$$

A frequently encountered example of a random variable, the distribution of which has a scale parameter, is $X \sim \text{EXP}(\theta)$.

The standard deviation, σ, often turns out to be a scale parameter, but sometimes it is more convenient to use something else. For example, if $X \sim \text{WEI}(\theta, 2)$, then θ is a scale parameter, but it is not the standard deviation of X.

Often, both types of parameters are required.

Definition 3.4.3

Location-Scale Parameter Quantities η and $\theta > 0$ are called **location-scale parameters** for the distribution of X if the CDF has the form

$$F(x;\, \theta, \eta) = F_0\left(\frac{x - \eta}{\theta}\right) \tag{3.4.5}$$

In other words, the pdf has the form

$$f(x;\, \theta, \eta) = \frac{1}{\theta}\, f_0\left(\frac{x - \eta}{\theta}\right) \tag{3.4.6}$$

The normal distribution is the most commonly encountered location-scale distribution, but there are other important examples.

Example 3.4.3 Consider a pdf of the form

$$f_0(z) = \frac{1}{\pi}\, \frac{1}{1 + z^2} \qquad -\infty < z < \infty \tag{3.4.7}$$

If X has pdf of the form $(1/\theta)f_0[(x - \eta)/\theta]$, with $f_0(z)$ given by equation (3.4.7), then X is said to have the **Cauchy distribution** with location-scale parameters η and θ, denoted

$$X \sim \text{CAU}(\theta, \eta) \tag{3.4.8}$$

It is easy to show that the mean and variance of X do not exist, so η and θ cannot be related to a mean and standard deviation. We still can interpret η as either the median or mode.

Another location-scale distribution, which is frequently encountered in life-testing applications, has pdf

$$f(x; \theta, \eta) = \frac{1}{\theta} \exp\left(-\frac{x - \eta}{\theta}\right) \qquad x > \eta \tag{3.4.9}$$

and zero otherwise. This is called the **two-parameter exponential distribution**, denoted by

$$X \sim \text{EXP}(\theta, \eta) \tag{3.4.10}$$

A location-scale distribution based on the pdf $f_0(z)$ of Example 3.4.2 is called the **Laplace** or **double-exponential distribution**, denoted by

$$X \sim \text{DE}(\theta, \eta) \tag{3.4.11}$$

It also is possible to define three-parameter models if we replace $f_0(z)$ with a pdf, $f_Z(z)$, that depends on another parameter, say β. For example, if $Z \sim \text{WEI}(1, \beta)$, then X has the **three-parameter Weibull distribution**, with location-scale parameters η and θ and shape parameter β if its pdf is of the form $f(x; \theta, \eta, \beta) = (1/\theta)f_Z[(x - \eta)/\theta]$. Similarly, if $Z \sim \text{GAM}(1, \kappa)$, then X has the **three-parameter gamma distribution**. These are denoted, respectively, by $X \sim \text{WEI}(\theta, \eta, \beta)$ and $X \sim \text{GAM}(\theta, \eta, \kappa)$.

SUMMARY

The purpose of this chapter was to develop special probability distributions. Special discrete distributions—such as the binomial, hypergeometric, negative binomial, and Poisson distributions—provide useful models for experiments that involve counting or other integer-value responses. Special continuous distributions—such as the uniform, exponential, gamma, Weibull, and normal distributions—provide useful models when experiments involve measurements on a continuous scale such as time, length, or weight.

EXERCISES

1. An office has 10 dot matrix printers. Each requires a new ribbon approximately every seven weeks. If the stock clerk finds at the beginning of a certain week that there are only five ribbons in stock, what is the probability that the supply will be exhausted during that week?

2. In a 10-question true–false test:
   (a) What is the probability of getting all answers correct by guessing?
   (b) What is the probability of getting eight correct by guessing?

3. A basketball player shoots 10 shots and the probability of hitting is 0.5 on each shot.
   (a) What is the probability of hitting eight shots?
   (b) What is the probability of hitting eight shots if the probability on each shot is 0.6?
   (c) What are the expected value and variance of the number of shots hit if $p = 0.5$?

4. A four-engine plane can fly if at least two engines work.
   (a) If the engines operate independently and each malfunctions with probability q, what is the probability that the plane will fly safely?
   (b) A two-engined plane can fly if at least one engine works. If an engine malfunctions with probability q, what is the probability that the plane will fly safely?
   (c) Which plane is the safest?

5. (a) The Chevalier de Mere used to bet that he would get at least one 6 in four rolls of a die. Was this a good bet?
   (b) He also bet that he would get at least one pair of 6's in 24 rolls of two dice. What was his probability of winning this bet?
   (c) Compare the probability of at least one 6 when six dice are rolled with the probability of at least two 6's when 12 dice are rolled.

6. If the probability of picking a winning horse in a race is 0.2, and if X is the number of winning picks out of 20 races, what is:
   (a) $P[X = 4]$.
   (b) $P[X \leqslant 4]$.
   (c) $E(X)$ and $\text{Var}(X)$.

7. If $X \sim \text{BIN}(n, p)$, derive $E(X)$ using Definition 2.2.3.

8. A jar contains 30 green jelly beans and 20 purple jelly beans. Suppose 10 jelly beans are selected at random from the jar.
   (a) Find the probability of obtaining exactly five purple jelly beans if they are selected with replacement.
   (b) Find the probability of obtaining exactly five purple beans if they are selected without replacement.

9. An office has 10 employees, three men and seven women. The manager chooses four at random to attend a short course on quality improvement.

(a) What is the probability that an equal number of men and women are chosen?
(b) What is the probability that more women are chosen?

10. Five cards are drawn without replacement from a regular deck of 52 cards. Give the probability of each of the following events:
(a) exactly two aces.
(b) exactly two kings.
(c) less than two aces.
(d) at least two aces.

11. A shipment of 50 mechanical devices consists of 42 good ones and eight defective. An inspector selects five devices at random without replacement.
(a) What is the probability that exactly three are good?
(b) What is the probability that at most three are good?

12. Repeat Exercise 10 if cards are drawn with replacement.

13. A man pays \$1 a throw to try to win a \$3 Kewpie doll. His probability of winning on each throw is 0.1.
(a) What is the probability that two throws will be required to win the doll?
(b) What is the probability that x throws will be required to win the doll?
(c) What is the probability that more than three throws will be required to win the doll?
(d) What is the expected number of throws needed to win a doll?

14. Three men toss coins to see who pays for coffee. If all three match, they toss again. Otherwise, the "odd man" pays for coffee.
(a) What is the probability that they will need to do this more than once?
(b) What is the probability of tossing at most twice?

15. The man in Exercise 13 has three children, and he must win a Kewpie doll for each one.
(a) What is the probability that 10 throws will be required to win the three dolls?
(b) What is the probability that at least four throws will be required?
(c) What is the expected number of throws needed to win three dolls?

16. Consider a seven-game world series between team A and team B, where for each game $P(A \text{ wins}) = 0.6$.
(a) Find $P(A \text{ wins series in } x \text{ games})$.
(b) You hold a ticket for the seventh game. What is the probability that you will get to use it?
(c) If $P(A \text{ wins a game}) = p$, what value of p maximizes your chance in (b)?
(d) What is the most likely number of games to be played in the series for $p = 0.6$?

17. The probability of a successful missile launch is 0.9. Test launches are conducted until three successful launches are achieved. What is the probability of each of the following?
(a) Exactly six launches will be required.
(b) Fewer than six launches will be required.
(c) At least four launches will be required.

18. Let $X \sim \text{GEO}(p)$.

(a) Derive the MGF of X.

(b) Find the FMGF of X.

(c) Find $E(X)$.

(d) Find $E[X(X-1)]$.

(e) Find $\text{Var}(X)$.

19. Let $X \sim \text{NB}(r, p)$.

(a) Derive the MGF of X.

(b) Find $E(X)$.

(c) Find $\text{Var}(X)$.

20. Suppose an ordinary six-sided die is rolled repeatedly, and the outcome—1, 2, 3, 4, 5 or 6—is noted on each roll.

(a) What is the probability that the third 6 occurs on the seventh roll?

(b) What is the probability that the number of rolls until the first 6 occurs is at most 10?

21. The number of calls that arrive at a switchboard during one hour is Poisson distributed with mean $\mu = 10$. Find the probability of occurrence during an hour of each of the following events:

(a) Exactly seven calls arrive.

(b) At most seven calls arrive.

(c) Between three and seven calls (inclusive) arrive.

22. If X has a Poisson distribution and if $P[X = 0] = 0.2$, find $P[X > 4]$.

23. A certain assembly line produces electronic components, and defective components occur independently with probability .01. The assembly line produces 500 components per hour.

(a) For a given hour, what is the probability that the number of defective components is at most two?

(b) Give the Poisson approximation for (a).

24. The probability that a certain type of electronic component will fail during the first hour of operation is 0.005. If 400 components are tested independently, find the Poisson approximation of the probability that at most two will fail during the first hour.

25. Suppose that 3% of the items produced by an assembly line are defective. An inspector selects 100 items at random from the assembly line. Approximate the probability that exactly five defectives are selected.

26. The number of vehicles passing a certain intersection in the time interval $[0, t]$ is a Poisson process $X(t)$ with mean $E[X(t)] = 3t$, where the unit of time is minutes.

(a) Find the probability that at least two vehicles will pass during a given minute.

(b) Define the events A = at least four vehicles pass during the first minute and B = at most two vehicles pass during the second minute. Find the probability that both A and B occur.

27. Let $X \sim POI(\mu)$.

(a) Find the factorial moment generating function (FMGF) of X, $G_X(t)$.

(b) Use $G_X(t)$ to find $E(X)$.

(c) Use $G_X(t)$ to find $E[X(X-1)]$.

28. Suppose the $X \sim POI(10)$.

(a) Find $P[5 < X < 15]$.

(b) Use the Chebychev Inequality to find a lower bound for $P[5 < X < 15]$

(c) Find a lower bound for $P[1 - k < X/\mu < 1 + k]$ for arbitrary $k > 0$.

29. A 20-sided (icosahedral) die has each face marked with a different integer from 1 through 20. Assuming that each face is equally likely to occur on a single roll, the outcome is a random variable $X \sim DU(20)$.

(a) If the die is rolled twice, find the pdf of the smallest value obtained, say Y.

(b) If the die is rolled three times, find the probability that the largest value is 3.

(c) Find $E(X)$ and Var (X).

30. Let $X \sim DU(N)$. Derive the MGF of X. *Hint:* Make use of the identity $s + s^2 + \cdots + s^N = \dfrac{s(1 - s^N)}{1 - s}$ for $s \neq 1$.

31. Let $X \sim UNIF(a, b)$. Derive the MGF of X.

32. The hardness of a certain alloy (measured on the Rockwell scale) is a random variable X. Assume that $X \sim UNIF(50, 75)$.

(a) Give the CDF of X.

(b) Find $P[60 < X < 70]$.

(c) Find $E(X)$.

(d) Find Var(X).

33. If $Q \sim UNIF(0, 3)$, find the probability that the roots of the equation $g(t) = 0$ are real, where $g(t) = 4t^2 + 4Qt + Q + 2$.

34. Suppose a value x is chosen "at random" in the interval [0, 10]. In other words, x is an observed value of a random variable $X \sim UNIF(0, 10)$. The value x divides the interval [0, 10] into two subintervals.

(a) Find the CDF of the length of the shorter subinterval.

(b) What is the probability that the ratio of lengths of the shorter to the longer subinterval is less than 1/4?

35. Prove that $\Gamma(1/2) = \sqrt{\pi}$. *Hint:* Use the following steps.

(1) Make the substitution $x = \sqrt{t}$ in the integral

$$\Gamma(1/2) = \int_0^\infty t^{-1/2} e^{-t}\, dt$$

(2) Change to polar coordinates in the double integral

$$\left[\Gamma\left(\frac{1}{2}\right)\right]^2 = \int_0^\infty \int_0^\infty 4 \exp\,[-(x^2 + y^2)]\, dx\, dy$$

36. Use the properties of Theorem 3.3.1 to find each of the following:

(a) $\Gamma(5)$.

(b) $\Gamma(5/2)$.

(c) Give an expression for the binomial coefficient, $\binom{n}{k}$, in terms of the gamma function.

37. The survival time (in days) of a white rat that was subjected to a certain level of X-ray radiation is a random variable $X \sim \text{GAM}(5, 4)$. Use Theorem 3.3.2 to find:

(a) $P[X \leqslant 15]$.

(b) $P[15 < X < 20]$.

(c) Find the expected survival time, $E(X)$.

38. The time (in minutes) until the third customer of the day enters a store is a random variable $X \sim \text{GAM}(1, 3)$. If the store opens at 8 A.M., find the probability that:

(a) the third customer arrives between 8:05 and 8:10;

(b) the third customer arrives after 8:10;

(c) Sketch the graph of the pdf of X.

39. Suppose that for the variable Q of Exercise 33, instead of a uniform distribution we assume $Q \sim \text{EXP}(1.5)$. Find the probability that the roots of $g(t) = 0$ are real.

40. Assume that the time (in hours) until failure of a transistor is a random variable $X \sim \text{EXP}(100)$.

(a) Find the probability that $X > 15$.

(b) Find the probability that $X > 110$.

(c) It is observed after 95 hours that the transistor still is working. Find the conditional probability that $X > 110$. How does this compare to (a)? Explain this result.

(d) What is $\text{Var}(X)$?

41. If $X \sim \text{GAM}(1, 2)$, find the mode of X.

42. For a switchboard, suppose the time X (in minutes) until the third call of the day arrives is gamma distributed with scale parameter $\theta = 2$ and shape parameter $\kappa = 3$. If the switchboard is activated at 8 A.M. find the probability that the third call arrives before 8:06 A.M.

43. If $X \sim \text{WEI}(\theta, \beta)$, derive $E(X^k)$ assuming that $k > -\beta$.

44. Suppose $X \sim \text{PAR}(\theta, \kappa)$.

(a) Derive $E(X)$; $\kappa > 1$.

(b) Derive $E(X^2)$; $\kappa > 2$.

45. If $X \sim \text{PAR}(100, 3)$, find $E(X)$ and $\text{Var}(X)$.

46. The shear strength (in pounds) of a spot weld is a Weibull distributed random variable, $X \sim \text{WEI}(400, 2/3)$.

(a) Find $P[X > 410]$.

(b) Find the conditional probability $P[X > 410 \mid X > 390]$.
(c) Find $E(X)$.
(d) Find Var(X).

47. The distance (in meters) that a bomb hits from the center of a target area is a random variable $X \sim \text{WEI}(10, 2)$.
(a) Find the probability that the bomb hits at least 20 meters from the center of the target.
(b) Sketch the graph of the pdf of X.
(c) Find $E(X)$ and $\text{Var}(X)$.

48. Suppose that $X \sim \text{PAR}(\theta, \kappa)$.
(a) Derive the $100 \times p$th percentile of X.
(b) Find the median of X if $\theta = 10$ and $\kappa = 2$.

49. Rework Exercise 37 assuming that, rather than being gamma distributed, the survival time is a random variable $X \sim \text{PAR}(4, 1.2)$.

50. Rework Exercise 40 assuming that, rather than being exponential, the failure time has a Pareto distribution $X \sim \text{PAR}(100, 2)$.

51. Suppose that $Z \sim N(0, 1)$. Find the following probabilities:
(a) $P(Z \leqslant 1.53)$.
(b) $P(Z > -0.49)$.
(c) $P(0.35 < Z < 2.01)$.
(d) $P(|Z| > 1.28)$.

Find the values a and b such that:
(e) $P(Z \leqslant a) = 0.648$.
(f) $P(|Z| < b) = 0.95$.

52. Suppose that $X \sim N(3, 0.16)$. Find the following probabilities:
(a) $P(X > 3)$.
(b) $P(X > 3.3)$.
(c) $P(2.8 \leqslant X \leqslant 3.1)$.
(d) Find the 98th percentile of X.
(e) Find the value c such that $P(3 - c < X < 3 + c) = 0.90$.

53. The Rockwell hardness of a metal specimen is determined by impressing the surface of the specimen with a hardened point, and then measuring the depth of penetration. The hardness of a certain alloy is normally distributed with mean of 70 units and standard deviation of 3 units.
(a) If a specimen is acceptable only if its hardness is between 66 and 74 units, what is the probability that a randomly chosen specimen is acceptable?
(b) If the acceptable range is $70 \pm c$, for what value of c would 95% of all specimens be acceptable?

54. Suppose that $X \sim N(10, 16)$. Find:

(a) $P[X \leqslant 14]$.

(b) $P[4 \leqslant X \leqslant 18]$.

(c) $P[2X - 10 \leqslant 18]$.

(d) $x_{0.95}$, the 95th percentile of X.

55. Assume the amount of light X (in lumens) produced by a certain type of light bulb is normally distributed with mean $\mu = 350$ and variance $\sigma^2 = 400$.

(a) Find $P[325 < X < 363]$.

(b) Find the value c such that the amount of light produced by 90% of the light bulbs will exceed c lumens.

56. Suppose that $X \sim N(1, 2)$.

(a) Find $E(X - 1)^4$.

(b) Find $E(X^4)$.

57. Suppose the computer store in Exercise 26 of Chapter 2 expands its marketing operation and orders 10 copies of the software package. As before, the annual demand is a random variable, X, and unsold copies are discarded; but assume now that $X \sim \text{BIN}(10, p)$.

(a) Find the expected net profit to the store as a function of p.

(b) How large must p be to produce a positive expected net profit?

(c) If instead $X \sim \text{POI}(2)$, would the store make a greater expected net profit by ordering more copies of the software?

58. Consider the following continuous analog of Exercise 57. Let X represent the annual demand for some commodity that is measured on a continuous scale, such as a liquid pesticide which can be measured in gallons (or fractions thereof). At the beginning of the year, a farm-supply store orders c gallons at d_1 dollars per gallon and sells it to customers at d_2 dollars per gallon. The pesticide loses effectiveness if it is stored during the off-season, so any amount unsold at the end of the year is a loss.

(a) If S is the amount sold, show that $E(S) = \int_0^c x f(x)\, dx + c[1 - F(c)]$.

(b) Show that the amount c that maximizes the expected net profit is the $100 \times p$th percentile of X with $p = (d_2 - d_1)/d_2$.

(c) If $d_1 = 6$, $d_2 = 14$, and $X \sim \text{UNIF}(980, 1020)$, find the optimum choice for c.

(d) Rework (c) if, instead, $X \sim N(1000, 100)$.

59. The solution of Exercise 58 can be extended to the discrete case. Suppose now that X is discrete as in Exercise 57, and the store pays d_1 dollars per copy, and charges each customer d_2 dollars per copy. Furthermore, let the demand X be an arbitrary nonnegative integer-valued random variable, with pdf $f(x)$ and CDF $F(x)$. Again, let c be the number of copies ordered by the store.

(a) Show that $E(S) = \sum_{x=0}^{c} x f(x) + c[1 - F(c)]$.

(b) Express the net profit Y as a linear function of S, and find $E(Y)$.

(c) Verify that the solution that maximizes $E(Y)$ is the smallest integer c such that $F(c) \geqslant (d_2 - d_1)/d_2$. *Hint:* Note that the expected net profit is a function of c, say $g(c) = E(Y)$, and the optimum solution will be the smallest c such that $g(c + 1) - g(c) \leqslant 0$.

(d) If $X \sim \text{BIN}(10, 1/2)$, $d_1 = 10$, and $d_2 = 35$, find the optimum solution.

(e) Rework (*d*) if, instead, $X \sim \text{POI}(5)$.

CHAPTER 4

JOINT DISTRIBUTIONS

4.1 INTRODUCTION

In many applications there will be more than one random variable of interest, say $X_1, X_2, \ldots, X_k$. It is convenient mathematically to regard these variables as components of a k-dimensional vector, $\mathbf{X} = (X_1, X_2, \ldots, X_k)$, which is capable of assuming values $\mathbf{x} = (x_1, x_2, \ldots, x_k)$ in a k-dimensional Euclidean space. Note, for example, that an observed value $\mathbf{x}$ may be the result of measuring k characteristics once each, or the result of measuring one characteristic k times. That is, in the latter case $\mathbf{x}$ could represent the outcomes on k repeated trials of an experiment concerning a single variable.

As before, we will develop the discrete and continuous cases separately.

4.2

JOINT DISCRETE DISTRIBUTIONS

Definition 4.2.1

The **joint probability density function** (joint pdf) of the k-dimensional discrete random variable $X = (X_1, X_2, \ldots, X_k)$ is defined to be

$$f(x_1, x_2, \ldots, x_k) = P[X_1 = x_1, X_2 = x_2, \ldots, X_k = x_k] \quad \textbf{(4.2.1)}$$

for all possible values $x = (x_1, x_2, \ldots, x_k)$ of X.

In this context, the notation $[X_1 = x_1, X_2 = x_2, \ldots, X_k = x_k]$ represents the intersection of k events $[X_1 = x_1] \cap [X_2 = x_2] \cap \cdots \cap [X_k = x_k]$. Another notation for the joint pdf involves subscripts, namely $f_{X_1, X_2, \ldots, X_k}(x_1, x_2, \ldots, x_k)$. This notation is a bit more cumbersome, and we will use it only when necessary.

Example 4.2.1 Recall in Example 3.2.6 that a bin contained 1000 flower seeds and 400 were red flowering seeds. Of the remaining seeds, 400 are white flowering and 200 are pink flowering. If 10 seeds are selected at random without replacement, then the number of red flowering seeds, X_1, and the number of white flowering seeds, X_2, in the sample are jointly distributed discrete random variables.

The joint pdf of the pair (X_1, X_2) is obtained easily by the methods of Section 1.6. Specifically,

$$f(x_1, x_2) = \frac{\binom{400}{x_1}\binom{400}{x_2}\binom{200}{10 - x_1 - x_2}}{\binom{1000}{10}} \quad \textbf{(4.2.2)}$$

for all $0 \leqslant x_1$, $0 \leqslant x_2$, and $x_1 + x_2 \leqslant 10$. The probability of obtaining exactly two red, five white, and three pink flowering seeds is $f(2, 5) = 0.0331$. Notice that once the values of x_1 and x_2 are specified, the number of pink is also determined, namely $10 - x_1 - x_2$, so it suffices to consider only two variables.

This is a special case of a more general type of hypergeometric distribution.

EXTENDED HYPERGEOMETRIC DISTRIBUTION

The hypergeometric distribution of equation (3.2.10) can be generalized to apply in cases where there are more than two types of outcomes of interest.

Suppose that a collection consists of a finite number of items, N, and that there are $k + 1$ different types; M_1 of type 1, M_2 of type 2, and so on. Select n items at random *without* replacement, and let X_i be the number of items of type i that are selected. The vector $X = (X_1, X_2, \ldots, X_k)$ has an **extended hypergeometric distribution** and a joint pdf of the form

$$f(x_1, x_2, \ldots, x_k) = \frac{\binom{M_1}{x_1}\binom{M_2}{x_2}\cdots\binom{M_k}{x_k}\binom{M_{k+1}}{x_{k+1}}}{\binom{N}{n}} \tag{4.2.3}$$

for all $0 \leqslant x_i \leqslant M_i$, where $M_{k+1} = N - \sum_{i=1}^{k} M_i$ and $x_{k+1} = n - \sum_{i=1}^{k} x_i$. A special notation for this is

$$X \sim \text{HYP}(n, M_1, M_2, \ldots, M_k, N) \tag{4.2.4}$$

Note that only k random variables are here, and x_{k+1} is used only as a notational convenience. The corresponding problem when the items are selected *with replacement* can be solved with a more general form of the binomial distribution known as the multinomial distribution.

MULTINOMIAL DISTRIBUTION

Suppose that there are $k + 1$ mutually exclusive and exhaustive events, say $E_1, E_2, \ldots, E_k, E_{k+1}$, which can occur on any trial of an experiment, and let $p_i = P(E_i)$ for $i = 1, 2, \ldots, k + 1$. On n independent trials of the experiment, we let X_i be the number of occurrences of the event E_i. The vector $X = (X_1, X_2, \ldots, X_k)$ is said to have the **multinomial distribution**, which has a joint pdf of the form

$$f(x_1, x_2, \ldots, x_k) = \frac{n!}{x_1!\,x_2!\cdots x_{k+1}!}\, p_1^{x_1} p_2^{x_2} \cdots p_{k+1}^{x_{k+1}} \tag{4.2.5}$$

for all $0 \leqslant x_i \leqslant n$, where $x_{k+1} = n - \sum_{i=1}^{k} x_i$ and $p_{k+1} = 1 - \sum_{i=1}^{k} p_i$.

A special notation for this is

$$X \sim \text{MULT}(n, p_1, p_2, \ldots, p_k) \tag{4.2.6}$$

The rationale for equation (4.2.5) is similar to that of the binomial distribution. To have exactly x_i occurrences of E_i, it is necessary to have some permutation of x_1 E_1's, x_2 E_2's, and so on. The total number of such permutations is

$$n!/(x_1!)(x_2!)\cdots(x_{k+1}!),$$

and each permutation occurs with probability $p_1^{x_1} p_2^{x_2} \cdots p_{k+1}^{x_{k+1}}$.

Just as the binomial provides an approximation to the hypergeometric distribution, under certain conditions equation (4.2.5) approximates equation (4.2.3). In Example 4.2.1, let us approximate the value of $f(2, 5)$ with (4.2.5), where $p_1 = p_2 = 0.4$ and $p_3 = 0.2$. This yields an approximate value of 0.0330, which agrees with the exact answer to three decimal places. Actually, this corresponds to the situation of sampling *with* replacement. In other words, if n is small, relative to N and to the values of M_i, then the effect of replacement or nonreplacement is negligible.

Example 4.2.2 The four-sided die of Example 2.1.1 is rolled 20 times, and the number of occurrences of each side is recorded. The probability of obtaining four 1's, six 2's, five 3's, and five 4's can be computed from (3.2.5) with $p_i = 0.25$, namely

$$[20!/(4!)(6!)(5!)(5!)](0.25)^{20} = 0.0089$$

If we were concerned only with recording 1's, 3's, and even numbers then equation (4.2.5) would apply with $p_1 = p_3 = 0.25$ and $1 - p_1 - p_3 = 0.5$. The probability of four 1's, five 3's, and 11 even numbers would be

$$[20!/(4!)(5!)(11!)](0.25)^9(0.5)^{11} = 0.0394$$

The functions defined by equations (4.2.3) and (4.2.5) both sum to 1 when summed over all possible values of $\boldsymbol{x} = (x_1, x_2, \ldots, x_k)$, and both are nonnegative. This is necessary to define a discrete pdf.

Theorem 4.2.1 A function $f(x_1, x_2, \ldots, x_k)$ is the joint pdf for some vector-valued random variable $\boldsymbol{X} = (X_1, X_2, \ldots, X_k)$ if and only if the following properties are satisfied:

$$f(x_1, x_2, \ldots, x_k) \geq 0 \quad \text{for all possible values } (x_1, x_2, \ldots, x_k) \tag{4.2.7}$$

and

$$\sum_{x_1} \cdots \sum_{x_k} f(x_1, x_2, \ldots, x_k) = 1 \tag{4.2.8}$$

■

In some two-dimensional problems it is convenient to present the joint pdf in a tabular form, particularly if a simple functional form for the joint pdf $f(x_1, x_2)$ is not known. For the purpose of illustration, let X_1 and X_2 be discrete random variables with joint probabilities $f(x_1, x_2)$ as given in Table 4.1. These values represent probabilities from a multinomial distribution $(X_1, X_2) \sim \text{MULT}(3; 0.4, 0.4)$. For example, this model would apply to Example 4.2.1 if the sampling had been with replacement, or it would be an approximation to the extended hypergeometric model for the without-replacement case. First notice that

$$\sum_{x_1=0}^{3} \sum_{x_2=0}^{3} f(x_1, x_2) = 1$$

as shown in the table. It is convenient to include impossible outcomes such as (3, 3) in the table and assign them probability zero. Care must be taken with the limits of the summations so that the points with zero probability are not included inadvertently when the nonzero portion of the pdf is summed.

TABLE 4.1 **Values of the discrete pdf of MULT(3; 0.4, 0.4)**

		x_2				
		0	1	2	3	
	0	0.008	0.048	0.096	0.064	0.216
x_1	1	0.048	0.192	0.192	0.000	0.432
	2	0.096	0.192	0.000	0.000	0.288
	3	0.064	0.000	0.000	0.000	0.064
		0.216	0.432	0.288	0.064	1.000

Now we are interested in the "marginal" probability, say $P[X_1 = 0]$, without regard to what value X_2 may assume. Relative to the joint sample space, X_2 has the effect of partitioning the event, say A, that $X_1 = 0$; computing the marginal probability $P(A) = P[X_1 = 0]$ is equivalent to computing a total probability as discussed in Section 1.5. That is, if B_j denotes the event that $X_2 = j$, then

$$A = (A \cap B_0) \cup (A \cap B_1) \cup (A \cap B_2) \cup (A \cap B_3)$$

and

$$\begin{aligned} P(A) &= \sum_{j=0}^{3} P(A \cap B_j) \\ &= \sum_{j=0}^{3} P[X_1 = 0, X_2 = j] \\ &= \sum_{j=0}^{3} f(0, j) \\ &= \sum_{x_2} f(0, x_2) \\ &= 0.216 \end{aligned}$$

as shown in the right margin of the top row of the table. Similarly, we could compute $P[X_1 = 1]$, $P[X_1 = 2]$, and so on.

The numerical values of $f_1(x_1) = P[X_1 = x_1]$ are given in the right margin of the table for each possible value of x_1. Clearly,

$$\sum_{x_1} f_1(x_1) = \sum_{x_1} \sum_{x_2} f(x_1, x_2) = 1 \tag{4.2.9}$$

so $f_1(x_1)$ is a legitimate pdf and is referred to as the marginal pdf of X_1 relative to the original joint sample space. Similarly, numerical values of the function

$$f_2(x_2) = P[X_2 = x_2] = \sum_{i=0}^{3} f(i, x_2)$$

are given in the bottom margin of the table, and this provides a means of finding the pdf's of X_1 and X_2 from $f(x_1, x_2)$.

Definition 4.2.2

If the pair (X_1, X_2) of discrete random variables has the joint pdf $f(x_1, x_2)$, then the **marginal pdf's** of X_1 and X_2 are

$$f_1(x_1) = \sum_{x_2} f(x_1, x_2) \tag{4.2.10}$$

and

$$f_2(x_2) = \sum_{x_1} f(x_1, x_2) \tag{4.2.11}$$

Because (4.2.10) and (4.2.11) are the pdf's of X_1 and X_2, another notation would be $f_{X_1}(x_1)$ and $f_{X_2}(x_2)$.

Although the marginal pdf's were motivated by means of a tabled distribution, it often is possible to derive formulas for $f_1(x_1)$ and $f_2(x_2)$ if an analytic expression for $f(x_1, x_2)$ is available.

For example, if $(X_1, X_2) \sim \text{MULT}(n, p_1, p_2)$, then the marginal pdf of X_1 is

$$\begin{aligned}
f_1(x_1) &= \sum_{x_2} f(x_1, x_2) \\
&= \sum_{x_2=0}^{n-x_1} f(x_1, x_2) \\
&= \frac{n!}{x_1!(n-x_1)!} p_1^{x_1} \sum_{x_2=0}^{n-x_1} \frac{(n-x_1)!\, p_2^{x_2}[(1-p_1)-p_2]^{(n-x_1)-x_2}}{x_2![(n-x_1)-x_2]!} \\
&= \frac{n!}{x_1!(n-x_1)!} p_1^{x_1} \sum_{x_2=0}^{n-x_1} \binom{n-x_1}{x_2} p_2^{x_2}[(1-p_1)-p_2]^{(n-x_1)-x_2} \\
&= \binom{n}{x_1} p_1^{x_1}[p_2 + (1-p_1) - p_2]^{n-x_1} \\
&= \binom{n}{x_1} p_1^{x_1}(1-p_1)^{n-x_1}
\end{aligned}$$

That is, $X_1 \sim \text{BIN}(n, p_1)$. This is what we would expect, because X_1 is counting the number of occurrences of some event E_1 on n independent trials.

For the flower seed example, $n = 3$, $p_1 = 0.4$, and $p_2 = 0.4$. If we are interested only in X_1, the number of red flowering seeds obtained, then we can lump the white and pink flowering seeds together and reduce the problem to a binomial-type problem. Specifically, $X_1 \sim \text{BIN}(3, 0.4)$. Similarly, the marginal distribution of X_2 is $\text{BIN}(n, p_2) = \text{BIN}(3, 0.4)$. Of course, probabilities other than marginal probabilities can be computed from joint pdf's.

If X is a k-dimensional discrete random variable and A is an event, then

$$P[X \in A] = \sum_{x \in A} \cdots \sum f(x_1, x_2, \ldots, x_k) \tag{4.2.12}$$

For example, in the flower seed problem, we may want to know the probability that the number of red flowering seeds is less than the number of white flowering seeds. In other words, the event of interest is $[X_1 < X_2]$. If the sampling is done *with* replacement, Table 4.1 applies. The possible pairs corresponding to the event $[X_1 < X_2]$ are $(x_1, x_2) = (0, 1), (0, 2), (0, 3)$, and $(1, 2)$.

It also is possible to express this event as $[X \in A]$, where A is a region of the plane, which in this example is $A = \{(x_1, x_2) | x_1 < x_2\}$. In this setting, the procedure would be to sum $f(x_1, x_2)$ over all pairs that are enclosed within the boundary of the region A. In this example, we would sum over the pairs above the graph of the line $x_2 = x_1$, which is the boundary of A, as shown in Figure 4.1, and thus

$$\begin{aligned} P[X_1 < X_2] &= f(0, 1) + f(0, 2) + f(0, 3) + f(1, 2) \\ &= 0.048 + 0.096 + 0.064 + 0.192 = 0.400 \end{aligned}$$

Marginal probabilities can be evaluated as discussed in Section 2.2 by summing over the marginal pdf's.

A joint probability of special importance involves sets of the form $A = (-\infty, x_1] \times \cdots \times (-\infty, x_k]$; in other words, Cartesian products of intervals of the type $(-\infty, x_i]$, $i = 1, 2, \ldots, k$.

FIGURE 4.1 Region corresponding to the event $[X_1 < X_2]$

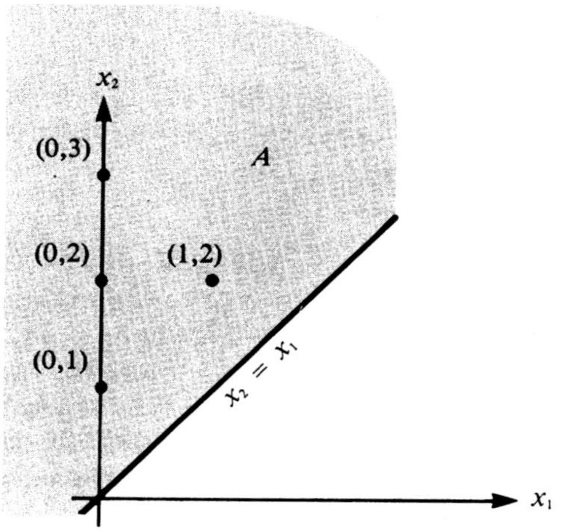

Definition 4.2.3

Joint CDF The **joint cumulative distribution function** of the k random variables $X_1, X_2, \ldots, X_k$ is the function defined by

$$F(x_1, \ldots, x_k) = P[X_1 \leqslant x_1, \ldots, X_k \leqslant x_k] \tag{4.2.13}$$

That is, $F(x_1, \ldots, x_k)$ denotes the probability that the random vector $\boldsymbol{X}$ will assume a value in the indicated k-dimensional rectangle, A. As in the one-dimensional case, other events can be expressed in terms of events of type A so that the CDF completely specifies the probability model.

As in the one-dimensional case, the only requirement that a function must satisfy to qualify as a joint CDF is that it must assign probabilities to events in such a way that Definition 1.3.1 will be satisfied. In particular, a joint CDF must satisfy properties analogous to the properties given in Theorem 2.2.3 for the one-dimensional case. Properties of a bivariate CDF are listed below, and properties of a k-dimensional CDF would be similar.

Theorem 4.2.2 A function $F(x_1, x_2)$ is a bivariate CDF if and only if

$$\lim_{x_1 \to -\infty} F(x_1, x_2) = F(-\infty, x_2) = 0 \quad \text{for all } x_2 \tag{4.2.14}$$

$$\lim_{x_2 \to -\infty} F(x_1, x_2) = F(x_1, -\infty) = 0 \quad \text{for all } x_1 \tag{4.2.15}$$

$$\lim_{\substack{x_1 \to \infty \\ x_2 \to \infty}} F(x_1, x_2) = F(\infty, \infty) = 1 \tag{4.2.16}$$

$$F(b, d) - F(b, c) - F(a, d) + F(a, c) \geqslant 0 \quad \text{for all } a < b \text{ and } c < d \tag{4.2.17}$$

$$\lim_{h \to 0^+} F(x_1 + h, x_2) = \lim_{h \to 0^+} F(x_1, x_2 + h) = F(x_1, x_2) \quad \text{for all } x_1 \text{ and } x_2 \tag{4.2.18}$$

■

Note that property (4.2.17) is a monotonicity condition that is the two-dimensional version of equation (2.2.11). This is needed to prevent the assignment of negative values as probabilities to events of the form $A = (a, b] \times (c, d]$. In particular,

$$P[a < X_1 \leqslant b, c < X_2 \leqslant d] = F(b, d) - F(b, c) - F(a, d) + F(a, c)$$

which is the value on the left of inequality (4.2.17).

Property (4.2.18) asserts that $F(x_1, x_2)$ is continuous from the right in each variable separately. Also note that (4.2.17) is something more than simply requiring that $F(x_1, x_2)$ be nondecreasing in each variable separately.

Example 4.2.3 Consider the function defined as follows:

$$F(x_1, x_2) = \begin{cases} 0 & \text{if } x_1 + x_2 < -1 \\ 1 & \text{if } x_1 + x_2 \geqslant -1 \end{cases} \tag{4.2.19}$$

If we let $a = c = -1$ and $b = d = 1$ in (4.2.17), then

$$F(1, 1) - F(1, -1) - F(-1, 1) + F(-1, -1) = 1 - 1 - 1 + 0 = -1$$

which means that (4.2.17) is not satisfied. However, it is not hard to verify that (4.2.19) is nondecreasing in each variable separately, and all of the other properties are satisfied. Thus, a set function based on (4.2.19) would violate property (1.3.1) of the definition of probability.

4.3

JOINT CONTINUOUS DISTRIBUTIONS

The joint CDF provides a means for defining a joint continuous distribution.

Definition 4.3.1

A k-dimensional vector-valued random variable $X = (X_1, X_2, \ldots, X_k)$ is said to be **continuous** if there is a function $f(x_1, x_2, \ldots, x_k)$, called the **joint probability density function** (joint pdf), of X, such that the joint CDF can be written as

$$F(x_1, \ldots, x_k) = \int_{-\infty}^{x_k} \cdots \int_{-\infty}^{x_1} f(t_1, \ldots, t_k)\, dt_1 \cdots dt_k \tag{4.3.1}$$

for all $x = (x_1, \ldots, x_k)$.

As in the one-dimensional case, the joint pdf can be obtained from the joint CDF by differentiation. In particular,

$$f(x_1, \ldots, x_k) = \frac{\partial^k}{\partial x_1 \cdots \partial x_k} F(x_1, \ldots, x_k) \tag{4.3.2}$$

wherever the partial derivatives exist. To serve the purpose of a joint pdf, two properties must be satisfied.

Theorem 4.3.1 Any function $f(x_1, x_2, \ldots, x_k)$ is a joint pdf of a k-dimensional random variable if and only if

$$f(x_1, \ldots, x_k) \geqslant 0 \quad \text{for all } x_1, \ldots, x_k \tag{4.3.3}$$

and

$$\int_{-\infty}^{\infty} \cdots \int_{-\infty}^{\infty} f(x_1, \ldots, x_k)\, dx_1 \cdots dx_k = 1 \tag{4.3.4}$$

■

Numerous applications can be modeled by joint continuous variables.

Example 4.3.1 Let X_1 denote the concentration of a certain substance in one trial of an experiment, and X_2 the concentration of the substance in a second trial of the experiment. Assume that the joint pdf is given by $f(x_1, x_2) = 4x_1 x_2$; $0 < x_1 < 1$, $0 < x_2 < 1$, and zero otherwise. The joint CDF is given by

$$\begin{aligned} F(x_1, x_2) &= \int_{-\infty}^{x_2} \int_{-\infty}^{x_1} f(t_1, t_2)\, dt_1\, dt_2 \\ &= \int_0^{x_2} \int_0^{x_1} 4t_1 t_2\, dt_1\, dt_2 \\ &= x_1^2 x_2^2 \qquad 0 < x_1 < 1, 0 < x_2 < 1 \end{aligned}$$

This defines $F(x_1, x_2)$ over the region $(0, 1) \times (0, 1)$, but there are four other regions of the plane where it must be defined. In particular, see Figure 4.2 for the definition of $F(x_1, x_2)$ on the five regions.

FIGURE 4.2 Values of a joint CDF

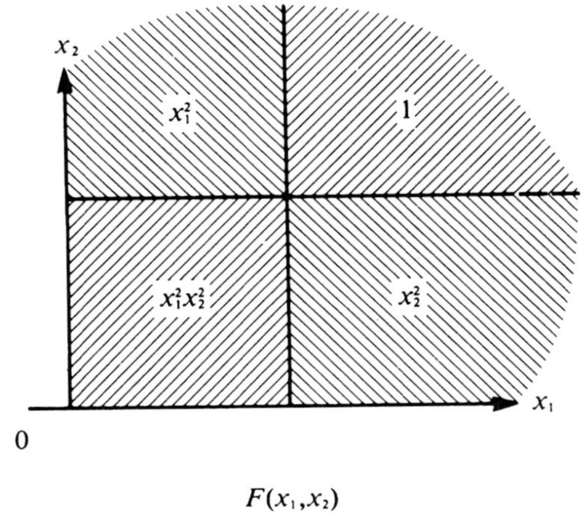

It also is possible to evaluate joint probabilities by integrating the joint pdf over the appropriate region. For example, we will find the probability that for both trials of the experiment, the "average concentration is less than 0.5." This event can be represented by $[(X_1 + X_2)/2 < 0.5]$, or more generally by $[(X_1, X_2) \in A]$, where $A = \{(x_1, x_2) | (x_1 + x_2)/2 < 0.5]$. Thus,

$$\begin{aligned} P[(X_1 + X_2)/2 < 0.5] &= P[(X_1, X_2) \in A] \\ &= \iint_A f(x_1, x_2)\, dx_1\, dx_2 \\ &= \int_0^1 \int_0^{1-x_2} 4x_1x_2\, dx_1\, dx_2 \\ &= \int_0^1 2x_2(1 - x_2)^2\, dx_2 \\ &= \frac{1}{6} \end{aligned}$$

The region A is illustrated in Figure 4.3.

For a general k-dimensional continuous random variable $X = (X_1, \ldots, X_k)$ and a k-dimensional event A we have

$$P[X \in A] = \int \cdots \int_A f(x_1, \ldots, x_k)\, dx_1 \ldots dx_k \tag{4.3.5}$$

Earlier in the section the notion of marginal distributions was discussed for joint discrete random variables. A similar concept can be developed for joint

FIGURE 4.3 Region corresponding to the event $[(X_1 + X_2)/2 < 0.5]$

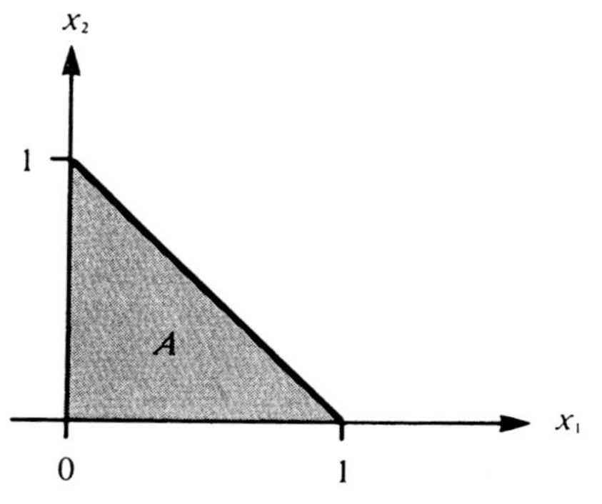

continuous random variables, but the approach is slightly different. In particular, consider the joint CDF, $F(x_1, x_2)$, of a pair of random variables $X = (X_1, X_2)$. The CDF of X_1 is

$$\begin{aligned}
F_1(x_1) &= P[X_1 \leqslant x_1] \\
&= P[X_1 \leqslant x_1, X_2 < \infty] \\
&= F(x_1, \infty) \\
&= \int_{-\infty}^{x_1} \left(\int_{-\infty}^{\infty} f_{X_1, X_2}(t_1, t_2)\, dt_2 \right) dt_1 \\
&= \int_{-\infty}^{x_1} f_1(t_1)\, dt_1
\end{aligned}$$

Thus, for the continuous case, the distribution function that we can interpret as the marginal CDF of X_1 is given by $F(x_1, \infty)$ and the pdf associated with $F_1(x_1)$ is the quantity enclosed in parentheses. That is,

$$\begin{aligned}
f_1(x_1) &= \frac{d}{dx_1} F_1(x_1) \\
&= \frac{d}{dx_1} \int_{-\infty}^{\infty} \int_{-\infty}^{x_1} f(t_1, x_2)\, dt_1\, dx_2 \\
&= \int_{-\infty}^{\infty} f(x_1, x_2)\, dx_2
\end{aligned}$$

Similar results can be obtained for X_2, which suggests the following definition.

Definition 4.3.2

If the pair (X_1, X_2) of continuous random variables has the joint pdf $f(x_1, x_2)$, then the **marginal pdf's** of X_1 and X_2 are

$$f_1(x_1) = \int_{-\infty}^{\infty} f(x_1, x_2)\, dx_2 \tag{4.3.6}$$

$$f_2(x_2) = \int_{-\infty}^{\infty} f(x_1, x_2)\, dx_1 \tag{4.3.7}$$

It follows from the preceding argument that $f_1(x_1)$ and $f_2(x_2)$ are the pdf's of X_1 and X_2, and consequently, another possible notation is $f_{X_1}(x_1)$ and $f_{X_2}(x_2)$. Consider the joint pdf in Example 4.3.1. The marginal pdf of X_1 is

$$\begin{aligned} f_1(x_1) &= \int_0^1 4x_1 x_2 \, dx_2 \\ &= 4x_1 \int_0^1 x_2 \, dx_2 \\ &= 2x_1 \end{aligned}$$

for any $0 < x_1 < 1$, and zero otherwise. Similarly, $f_2(x_2) = 2x_2$ for any $0 < x_2 < 1$.

Actually, the argument preceding Definition 4.3.2 provides a general approach to defining marginal distributions.

Definition 4.3.3

If $X = (X_1, X_2, \ldots, X_k)$ is a k-dimensional random variable with joint CDF $F(x_1, x_2, \ldots, x_k)$, then the **marginal CDF** of X_j is

$$F_j(x_j) = \lim_{\substack{x_i \to \infty \\ \text{all } i \neq j}} F(x_1, \ldots, x_j, \ldots, x_k) \tag{4.3.8}$$

Furthermore, if X is discrete, the marginal pdf is

$$f_j(x_j) = \sum_{\text{all } i \neq j} \cdots \sum f(x_1, \ldots, x_j, \ldots, x_k) \tag{4.3.9}$$

and if X is continuous, the marginal pdf is

$$f_j(x_j) = \underbrace{\int \cdots \int}_{\text{all } i \neq j} f(x_1, \ldots, x_j, \ldots, x_k) \, dx_1 \ldots dx_k \tag{4.3.10}$$

Example 4.3.2 Let X_1, X_2, and X_3 be continuous with a joint pdf of the form $f(x_1, x_2, x_3) = c$; $0 < x_1 < x_2 < x_3 < 1$, and zero otherwise, where c is a constant. First, note that $c = 6$, which follows from (4.3.4). Suppose it is desired to find the marginal of X_3. From (4.3.10) we obtain

$$\begin{aligned} f_3(x_3) &= \int_0^{x_3} \int_0^{x_2} 6 \, dx_1 \, dx_2 \\ &= 6 \int_0^{x_3} x_2 \, dx_2 \\ &= 3x_3^2 \end{aligned}$$

if $0 < x_3 < 1$, and zero otherwise.

Notice that a similar procedure can be used to obtain the joint pdf of any subset of the original set of random variables. For example, the joint pdf of the pair (X_1, X_2) is obtained by integrating $f(x_1, x_2, x_3)$ with respect to x_3 as follows:

$$f(x_1, x_2) = \int_{-\infty}^{\infty} f(x_1, x_2, x_3)\, dx_3$$

$$= \int_{x_2}^{1} 6\, dx_3$$

$$= 6(1 - x_2)$$

if $0 < x_1 < x_2 < 1$, and zero otherwise.

A general formula for the joint pdf of a subset of an arbitrary collection of random variables would involve a rather complicated expression, and we will not attempt to give such a formula. However, the procedure described above, which involves integrating with respect to the "unwanted" variables, provides a general approach to this problem.

4.4 INDEPENDENT RANDOM VARIABLES

Suppose that X_1 and X_2 are discrete random variables with joint probabilities as given in Table 4.2. Note that $f(1, 1) = 0.2 = f_1(1)f_2(1)$. That is, $P[X_1 = 1$ and $X_2 = 1] = P[X_1 = 1]P[X_2 = 1]$, and we would say that the events $[X_1 = 1]$ and $[X_2 = 1]$ are independent events in the usual sense of Chapter 1. However, for example, $f(1, 2) = 0.1 \neq f_1(1)f_2(2)$, so these events are not independent. Thus there is in general a dependence between the random variables X_1 and X_2, although certain events are independent. If $f(x_1, x_2) = f_1(x_1)f_2(x_2)$ for all possible (x_1, x_2), then it would be reasonable to say in general that the random variables X_1 and X_2 are independent.

TABLE 4.2 **Values of the joint pdf of two dependent random variables**

		x_2: 0	1	2	$f_1(x_1)$
x_1	0	0.1	0.2	0.1	0.4
	1	0.1	0.2	0.1	0.4
	2	0.1	0.1	0.0	0.2
$f_2(x_2)$		0.3	0.5	0.2	

Similarly, for continuous random variables, suppose that $f(x_1, x_2) = f_1(x_1)f_2(x_2)$ for all x_1 and x_2. It follows that if $a < b$ and $c < d$, then

$$\begin{aligned} P[a \leqslant X_1 \leqslant b, c \leqslant X_2 \leqslant d] &= \int_c^d \int_a^b f(x_1, x_2)\, dx_1\, dx_2 \\ &= \int_c^d \int_a^b f_1(x_1)f_2(x_2)\, dx_1\, dx_2 \\ &= \int_a^b f_1(x_1)\, dx_1 \int_c^d f_2(x_2)\, dx_2 \\ &= P[a \leqslant X_1 \leqslant b]P[c \leqslant X_2 \leqslant d] \end{aligned}$$

Thus, the events $A = [a \leqslant X_1 \leqslant b]$ and $B = [c \leqslant X_2 \leqslant d]$ are independent in the usual sense, where in set notation this was expressed as $P(A \cap B) = P(A)P(B)$. It is natural to extend the concept of independence of events to the independence of random variables, where random variables X_1 and X_2 would be said to be independent if all events of type A and B are independent. Note that these concepts apply to both discrete and continuous random variables.

Definition 4.4.1

Independent Random Variables Random variables $X_1, \ldots, X_k$ are said to be **independent** if for every $a_i < b_i$,

$$P[a_1 \leqslant X_1 \leqslant b_1, \ldots, a_k \leqslant X_k \leqslant b_k] = \prod_{i=1}^{k} P[a_i \leqslant X_i \leqslant b_i] \quad \textbf{(4.4.1)}$$

The expression on the right of equation (4.4.1) is the product of the marginal probabilities $P[a_1 \leqslant X_1 \leqslant b_1], \ldots, P[a_k \leqslant X_k \leqslant b_k]$. The terminology **stochastically independent** also is often used in this context. If (4.4.1) does not hold for all $a_i < b_i$, the random variables are called **dependent**. Some properties that are equivalent to independence are stated in the following theorem.

Theorem 4.4.1 Random variables $X_1, \ldots, X_k$ are independent if and only if the following properties holds:

$$F(x_1, \ldots, x_k) = F_1(x_1) \cdots F_k(x_k) \quad \textbf{(4.4.2)}$$

$$f(x_1, \ldots, x_k) = f_1(x_1) \cdots f_k(x_k) \quad \textbf{(4.4.3)}$$

where $F_i(x_i)$ and $f_i(x_i)$ are the marginal CDF and pdf of X_i, respectively. ■

Clearly, the random variables in Example 4.3.1 are independent. Indeed, anytime the limits of the variables are not functionally related and the joint pdf can be factored into a function of x_1 times a function of x_2, say $f(x_1, x_2) = g(x_1)h(x_2)$, then it can be factored into a product of the marginal pdf's, say $f(x_1, x_2) = f_1(x_1)f_2(x_2)$, by adjusting the constants properly. This is formalized as follows.

Theorem 4.4.2 Two random variables X_1 and X_2 with joint pdf $f(x_1, x_2)$ are independent if and only if:

1. the "support set," $\{(x_1, x_2) \mid f(x_1, x_2) > 0\}$, is a Cartesian product, $A \times B$, and
2. the joint pdf can be factored into the product of functions of x_1 and x_2, $f(x_1, x_2) = g(x_1)h(x_2)$. ■

Example 4.4.1 The joint pdf of a pair X_1 and X_2 is

$$f(x_1, x_2) = 8x_1x_2 \qquad 0 < x_1 < x_2 < 1$$

and zero otherwise. This function can clearly be factored according to part (2) of the theorem, but the support set, $\{(x_1, x_2) \mid 0 < x_1 < x_2 < 1\}$, is a triangular region that cannot be represented as a Cartesian product. Thus, X_1 and X_2 are dependent.

Example 4.4.2 Consider now a pair X_1 and X_2 with joint pdf

$$f(x_1, x_2) = x_1 + x_2 \qquad 0 < x_1 < 1, 0 < x_2 < 1$$

and zero otherwise. In this case the support set is $\{(x_1, x_2) \mid 0 < x_1 < 1$ and $0 < x_2 < 1\}$, which can be represented as $A \times B$, where A and B are both the open interval (0, 1). However, part (2) of the theorem is not satisfied because $x_1 + x_2$ cannot be factored as $g(x_1)h(x_2)$. Thus, X_1 and X_2 are dependent.

Many interesting problems can be modeled in terms of independent random variables.

Example 4.4.3 Two components in a rocket operate independently, and the probability that each component fails on a launch is p. Let X denote the number of launches required to have a failure of component 1, and let Y denote the number of launches required to have a failure of component 2. Assuming the launch trials are independent, each variable would follow a geometric distribution, $X, Y \sim \text{GEO}(p)$, and if the components are assumed to operate independently, a reasonable model for the pair of variables (X, Y) would be

$$\begin{aligned} f_{X,Y}(x, y) &= f_X(x)f_Y(y) \\ &= pq^{x-1}pq^{y-1} = p^2q^{x+y-2}; \quad x = 1, 2, \ldots; \quad y = 1, 2, \ldots \end{aligned}$$

where $q = 1 - p$. The joint CDF of X and Y is

$$\begin{aligned} F_{X,Y}(x, y) &= \sum_{j=1}^{y} \sum_{i=1}^{x} f_{X,Y}(i, j) \\ &= \sum_{j=1}^{y} pq^{j-1} \sum_{i=1}^{x} pq^{i-1} \\ &= (1 - q^x)(1 - q^y) \\ &= F_X(x)F_Y(y) \end{aligned}$$

which also could be obtained directly from (4.4.1) of Theorem 4.4.1.

We also are interested in the random variables X and T, where $T = X + Y$ is the number of trials needed to get a failure in both components. Now, to have $X = x$ and $T = t$, it is necessary to have $X = x$ and $Y = t - x$, so the joint pdf of X and T can be given by

$$\begin{aligned} f_{X,T}(x, t) &= P[X = x, T = t] \\ &= P[X = x, Y = t - x] \\ &= f_{X,Y}(x, t - x) \\ &= pq^{x-1}pq^{t-x-1} \\ &= p^2q^{t-2} \qquad t = x + 1, x + 2, \ldots, ; x = 1, \ldots, t - 1 \end{aligned}$$

It is clear from Theorem 4.4.2 that X and T are dependent because the support set cannot be a Cartesian product. This is also reasonable intuitively, because a large value of X implies an even larger value of T.

As noted earlier, the probability model is specified completely by either the pdf or the CDF. For example, the joint CDF of X and T is

$$\begin{aligned} F_{X,T}(x, t) &= \sum_{k=1}^{x} \sum_{i=k+1}^{t} p^2q^{i-2} \\ &= 1 - q^x - pxq^{t-1} \end{aligned}$$

for $t = x + 1, x + 2, \ldots;\ x = 1, 2, \ldots, t - 1$.

The marginal CDFs of X and T can be obtained by taking limits as in equation (4.3.8). Specifically,

$$F_X(x) = \lim_{t \to \infty} F_{X,T}(x, t) = 1 - q^x \qquad x = 1, 2, \ldots$$

$$\begin{aligned} F_T(t) &= \lim_{x \to \infty} F_{X,T}(x, t) = F_{X,T}(t - 1, t) \\ &= 1 - q^{t-1} - p(t - 1)q^{t-1} \qquad t = 2, 3, \ldots \end{aligned}$$

The marginal pdf of T could be obtained by taking differences, $f_T(t) = F_T(t) - F_T(t-1)$, or directly from Definition 4.2.2:

$$f_T(t) = \sum_{x=1}^{t-1} f_{X,T}(x, t)$$
$$= (t-1)p^2q^{t-2} \qquad t = 2, 3, \ldots$$

This is the pdf of a negative binomial variable with parameters $r = 2$ and p, which should not be surprising because T is the number of trials required to obtain two failures. Thus, $T \sim \text{NB}(2, p)$. It also should come as no surprise that $X \sim \text{GEO}(p)$; this was assumed at the outset.

As noted earlier, the variables X and T are dependent and thus $f_{X,T}(x, t) \neq f_X(x)f_T(t)$ for some values of x and t. It might be worth noting that verification of independence by seeing that the joint pdf factors into the product of the marginal pdf's requires verification for every pair of values. To see this, suppose we take $p = 1/2$ on the above example. It turns out that the events $[X = 1]$ and $[T = 3]$ are independent, because $f_{X,T}(1, 3) = 1/8 = f_X(1)f_T(3)$. However, X and T are not independent unless all such events are independent. This is not true in this case because, for example, $f_{X,T}(2, 3) = 1/8$ but $f_X(2)f_T(3) = 1/16$. To show dependence, it suffices to find only one such pair of values.

4.5 CONDITIONAL DISTRIBUTIONS

Recall that independence also is related to the concept of conditional probability, and this suggests that the definition of conditional probability of events could be extended to the concept of conditional random variables. In the previous example, one may be interested in a general formula for expressing conditional probabilities of the form

$$P[T = t \mid X = x] = \frac{P[X = x, T = t]}{P[X = x]} = \frac{f_{X,T}(x, t)}{f_X(x)}$$

which suggests the following definition.

Definition 4.5.1

Conditional pdf If X_1 and X_2 are discrete or continuous random variables with joint pdf $f(x_1, x_2)$, then the **conditional probability density function** (conditional pdf) of X_2 given $X_1 = x_1$ is defined to be

$$f(x_2 \mid x_1) = \frac{f(x_1, x_2)}{f_1(x_1)} \qquad \textbf{(4.5.1)}$$

for values x_1 such that $f_1(x_1) > 0$, and zero otherwise.

Similarly, the conditional pdf of X_1 given $X_2 = x_2$ is

$$f(x_1 \mid x_2) = \frac{f(x_1, x_2)}{f_2(x_2)} \tag{4.5.2}$$

for x_2 such that $f_2(x_2) > 0$, and zero otherwise.

As noted in the previous example, for discrete variables a conditional pdf is actually a conditional probability. For example, if X_1 and X_2 are discrete, then $f(x_2 \mid x_1)$ is the conditional probability of the event $[X_2 = x_2]$ given the event $[X_1 = x_1]$. In the case of continuous random variables, the interpretation of the conditional pdf is not as obvious because $P[X_1 = x_1] = 0$. Although $f(x_2 \mid x_1)$ cannot be interpreted as a conditional probability in this case, it can be thought of as assigning conditional "probability density" to arbitrarily small intervals $[x_2, x_2 + \Delta x_2]$, in much the same way that the marginal pdf, $f_2(x_2)$, assigns marginal probability density. Thus, in the continuous case, the conditional probability of an event of the form $[a \leqslant X_2 \leqslant b]$ given $X_1 = x_1$ is

$$\begin{aligned} P[a \leqslant X_2 \leqslant b \mid X_1 = x_1] &= \int_a^b f(x_2 \mid x_1)\, dx_2 \\ &= \frac{\int_a^b f(x_1, x_2)\, dx_2}{\int_{-\infty}^{\infty} f(x_1, x_2)\, dx_2} \end{aligned} \tag{4.5.3}$$

That is, the denominator is the total area under the joint pdf at $X_1 = x_1$, and the numerator is the amount of that area for which $a \leqslant X_2 \leqslant b$ (see Figure 4.4). This could be regarded as a way of assigning probability to an event $[a \leqslant X_2 \leqslant b]$ over a "slice," $X_1 = x_1$, of the joint sample space of the pair (X_1, X_2).

For this to be a valid way of assigning probability, $f(x_2 \mid x_1)$ must satisfy the usual properties of a pdf in the variable x_2 with x_1 fixed. The fact that $f(x_2 \mid x_1) \geqslant 0$ follows from (4.5.2); also note that

$$\begin{aligned} \int_{-\infty}^{\infty} f(x_2 \mid x_1)\, dx_2 &= \frac{1}{f_1(x_1)} \int_{-\infty}^{\infty} f(x_1, x_2)\, dx_2 \\ &= \frac{1}{f_1(x_1)} f_1(x_1) = 1 \end{aligned}$$

for the continuous case. The discrete case is similar.

The concept of conditional distribution can be extended to vectors of random variables. Suppose, for example, that $\boldsymbol{X} = (X_1, \ldots, X_r, \ldots, X_k)$ has joint pdf $f(\boldsymbol{x})$ and $\boldsymbol{X}_1 = (X_1, \ldots, X_r)$ has joint pdf $f_1(\boldsymbol{x}_1)$. If $\boldsymbol{X}_2 = (X_{r+1}, \ldots, X_k)$, then the conditional pdf of $\boldsymbol{X}_2$ given $\boldsymbol{X}_1 = \boldsymbol{x}_1$ is $f(\boldsymbol{x}_2 \mid \boldsymbol{x}_1) = f(\boldsymbol{x})/f_1(\boldsymbol{x}_1)$ for values $\boldsymbol{x}_1$ such that $f_1(\boldsymbol{x}_1) > 0$. As an illustration, consider the random variables X_1, X_2, and X_3 of

FIGURE 4.4 Conditional distribution of probability

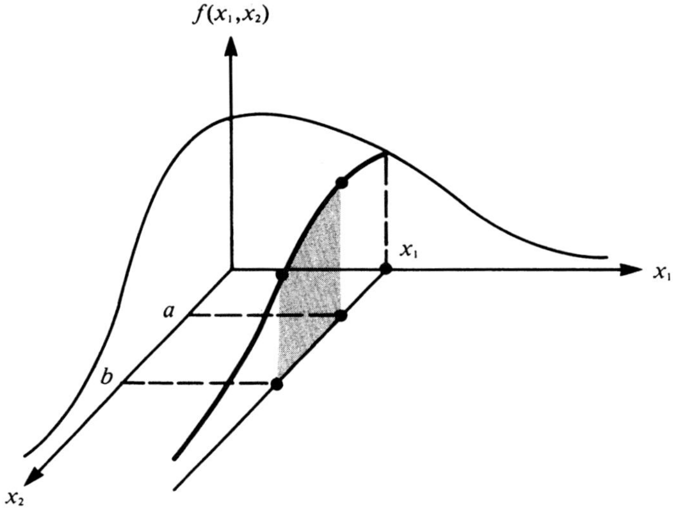

Example 4.3.2. The conditional pdf of X_3 given $(X_1, X_2) = (x_1, x_2)$ is

$$\begin{aligned} f(x_3 \mid x_1, x_2) &= \frac{f(x_1, x_2, x_3)}{f(x_1, x_2)} \\ &= \frac{6}{6(1 - x_2)} \\ &= \frac{1}{1 - x_2} \qquad 0 < x_1 < x_2 < x_3 < 1 \end{aligned}$$

and zero otherwise.

Some properties of conditional pdf's, which correspond to similar properties of conditional probability, are stated in the following theorem.

Theorem 4.5.1 If X_1 and X_2 are random variables with joint pdf $f(x_1, x_2)$ and marginal pdf's $f_1(x_1)$ and $f_2(x_2)$, then

$$f(x_1, x_2) = f_1(x_1)f(x_2 \mid x_1) = f_2(x_2)f(x_1 \mid x_2) \tag{4.5.4}$$

and if X_1 and X_2 are independent, then

$$f(x_2 \mid x_1) = f_2(x_2) \tag{4.5.5}$$

and

$$f(x_1 \mid x_2) = f_1(x_1) \tag{4.5.6}$$

■

Other notations often are used for conditional pdf's. For example, if X and Y are jointly distributed random variables, the conditional pdf of Y given $X = x$ also could be written as $f_{Y|X}(y|x)$ or possibly $f_{Y|x}(y)$. In most applications, there will be no confusion if we suppress the subscripts and use the simpler notation $f(y|x)$. It is also common practice to speak of "conditional random variables" denoted by $Y|X$ or $Y|X = x$.

Example 4.5.1 Consider the variables X and T of Example 4.4.3. The conditional pdf of T given $X = x$ is

$$f(t|x) = \frac{p^2q^{t-2}}{pq^{x-1}} = pq^{t-x-1} \qquad t = x+1, x+2, \ldots$$

Notice that this means that for any $x = 1, 2, \ldots,$ the conditional pdf of $U = T - X$ given $X = x$ is

$$\begin{aligned} f_{U|X}(u|x) &= P[T - x = u \mid X = x] \\ &= P[T = u + x \mid X = x] \\ &= f(u + x|x) \\ &= pq^{u-1} \qquad u = 1, 2, 3, \ldots \end{aligned}$$

Thus, conditional on $X = x$, $U \sim \text{GEO}(p)$. The conditional pdf of X given $T = t$ is

$$\begin{aligned} f(x|t) &= \frac{p^2q^{t-2}}{(t-1)p^2q^{t-2}} \\ &= \frac{1}{t-1} \qquad x = 1, 2, \ldots, t-1 \end{aligned}$$

Thus, conditional on $T = t$, $X \sim \text{DU}(t - 1)$, the discrete uniform distribution on the integers $1, 2, \ldots, t - 1$.

Example 4.5.2 A piece of flat land is in the form of a right triangle with a southern boundary two miles long and eastern boundary one mile long (see Figure 4.5).

The point at which an airborne seed lands is of interest. The following assumption will be made: Given that the seed lands within the boundaries, its coordinates X and Y are "uniformly" distributed over the surface of the triangle. In other words, the pair (X, Y) has a constant joint pdf $f(x, y) = c > 0$ for points within the boundaries, and zero otherwise. By property (4.3.4), we have $c = 1$. The marginal pdf's of X and Y are

$$f_1(x) = \int_0^{x/2} dy = \frac{x}{2} \qquad 0 < x < 2$$

$$f_2(y) = \int_{2y}^{2} dx = 2(1 - y) \qquad 0 < y < 1$$

FIGURE 4.5 A triangular region

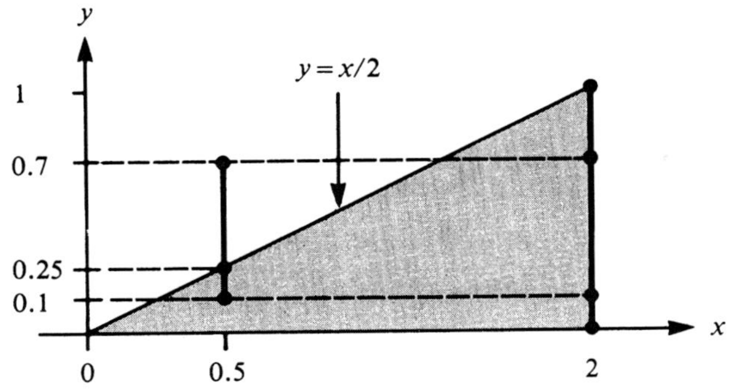

The conditional pdf of Y given $X = x$ is

$$f(y \mid x) = \frac{1}{x/2} = \frac{2}{x} \qquad 0 < y < \frac{x}{2}$$

and zero otherwise. This means that conditional on $X = x$, $Y \sim \text{UNIF}(0, x/2)$.

One possible interpretation of this result is the following. Suppose we are able to observe the x-coordinate but not the y-coordinate. If we observe that $X = x$, then it is sensible to use this information in assigning probability to events relative to Y. Thus, if we observe $X = 0.5$, then the conditional probability that Y is in some interval, say $[0.1, 0.7]$, is

$$\begin{aligned} P[0.1 \leqslant Y \leqslant 0.7 \mid X = 0.5] &= \int_{0.1}^{0.7} f(y \mid 0.5)\, dy \\ &= \int_{0.1}^{0.25} 4\, dy \\ &= 0.6 \end{aligned}$$

where the change in the upper limit is because $f(y \mid x) = 0$ if $x = 0.5$ and $y > 0.25$ (see Figure 4.5). A conditional probability can take this sort of information into account, whereas a marginal probability cannot. For comparison, the marginal probability of this event is

$$\begin{aligned} P[0.1 \leqslant Y \leqslant 0.7] &= \int_{0.1}^{0.7} \int_{2y}^{2} f(x, y)\, dx\, dy \\ &= \int_{0.1}^{0.7} f_2(y)\, dy \\ &= \int_{0.1}^{0.7} 2(1 - y)\, dy \\ &= 0.72 \end{aligned}$$

Mathematically, there is nothing wrong with the marginal probability, but it cannot take information about related variables into account.

In this example, the conditional pdf's turned out to be uniform, but this is not always the case.

Example 4.5.3 Consider the joint density discussed in Example 4.4.2,

$$f(x, y) = x + y \qquad 0 < x < 1,\ 0 < y < 1$$

In this case, for any x between 0 and 1,

$$f(y \mid x) = \frac{f(x, y)}{f_1(x)} = \frac{x + y}{x + 0.5} \qquad 0 < y < 1$$

For example,

$$P[0 < Y < 0.5 \mid X = 0.25] = \int_0^{0.5} \frac{0.25 + y}{0.25 + 0.5}\,dy = \frac{1}{3}$$

4.6 RANDOM SAMPLES

If X represents a random variable of interest, say the lifetime of a certain type of light bulb, then $f(x)$ represents the population probability density function of this variable. If one light bulb is selected "at random" from the population and placed in operation, then $f(x)$ provides the failure density for that light bulb. This process might be described as one trial of an experiment. If one assumes that conceptually there is an infinite population of light bulbs of this type, or if the actual population of light bulbs is sufficiently large so that the population may be assumed to remain essentially the same after a finite number, n, of light bulbs is drawn from it, then it would be reasonable to assume that the population probability density function also is applicable to the lifetime of the second light bulb drawn. Indeed, we could conduct n trials of the experiment, and to distinguish the n trials we may let X_i denote the lifetime of the light bulb obtained on the ith trial, where each $X_i \sim f(x_i)$. That is, each X_i is distributed according to the common parent population density. In addition, if the items are sampled (or the trials are conducted) in such a way that the outcome on one trial does not affect the probability distribution of the variable on a different trial, then the variables may be assumed to be independent. Ordinarily, we will assume that the trials of the experiment or the sampling are conducted in such a way that these two conditions are satisfied, and we will refer to this as "random sampling."

Definition 4.6.1

Random Sample The set of random variables $X_1, \ldots, X_n$ is said to be a **random sample** of size n from a population with density function $f(x)$ if the joint pdf has the form

$$f(x_1, x_2, \ldots, x_n) = f(x_1)f(x_2) \cdots f(x_n) \tag{4.6.1}$$

That is, random sampling assumes that the sample is taken in such a way that the random variables for each trial are independent and follow the common population density function. In this case, the joint density function is the product of the common marginal densities. It also is common practice to refer to the set of observed values, or data, $x_1, x_2, \ldots, x_n$, obtained from the experiment as a random sample.

In many cases it is necessary to obtain actual observed data from a population to help validate an assumed model or to help select an appropriate model. If the data can be assumed to represent a random sample, then equation (4.6.1) provides the connecting link between the observed data and the mathematical model.

Example 4.6.1 The lifetime of a certain type of light bulb is assumed to follow an "exponential" population density function given by

$$f(x) = e^{-x} \qquad 0 < x < \infty \tag{4.6.2}$$

where the lifetime is measured in years. If a random sample of size two is obtained from this population, then we would have

$$f(x_1, x_2) = e^{-(x_1 + x_2)} \qquad 0 < x_i < \infty \tag{4.6.3}$$

Now suppose that the total lifetime of the two light bulbs turned out to be $x_1 + x_2 = 0.5$ years. One may wonder whether this sample result is reasonable when the population density is given by equation (4.6.2). If not, then it may be that a different population model is more appropriate. Questions of this type can be answered by using equation (4.6.3). In particular, consider

$$\begin{aligned} P[X_1 + X_2 \leqslant c] &= \int_0^c \int_0^{c - x_2} e^{-(x_1 + x_2)}\, dx_1\, dx_2 \\ &= 1 - ce^{-c} - e^{-c} \end{aligned}$$

For $c = 0.5$, $P[X_1 + X_2 \leqslant 0.5] = 0.09$; thus it would be unlikely to find the total lifetime of the two bulbs to be 0.5 years or less, if the true population model is given by equation (4.6.2).

Specific techniques for making decisions or drawing inferences based on sample data will be emphasized in later chapters, and we now are interested primarily in developing the mathematical properties that will be needed in carrying out those statistical procedures.

EMPIRICAL DISTRIBUTIONS

It is possible to use some of the work on discrete distributions to study the adequacy of specific continuous models. For example, let $X_1, X_2, \ldots, X_n$ be a random sample of size n, each distributed as $X \sim f(x)$, where, respectively, $f(x)$ and $F(x)$ are the population pdf and CDF. For each real x, let W be the number of variables, X_i, in the random sample that are less than or equal to x. We can regard the occurrence of the event $[X_i \leqslant x]$ for some i as "success," and because the variables in a random sample are independent, W simply counts the number of successes on n independent Bernoulli trials with probability of success $p = P[X \leqslant x]$. Thus, $W \sim \text{BIN}(n, p)$ with $p = F(x)$. The relative frequency of a success on n trials of the experiment would be W/n, which we will denote by $F_n(x)$, referred to as the empirical CDF. The property of statistical regularity, as discussed in Section 1.2, suggests that $F_n(x)$ should be close to $F(x)$ for large n. Thus, for any proposed model, the corresponding CDF, $F(x)$, should be consistent with the empirical CDF, $F_n(x)$, based on data from $F(x)$.

We now take a set of data $x_1, x_2, \ldots, x_n$ from a random sample of size n from $f(x)$, and let $y_1 < y_2 < \cdots < y_n$ be the **ordered** values of the data. Then the **empirical CDF** based on this data can be represented as

$$F_n(x) = \begin{cases} 0 & x < y_1 \\ \dfrac{i}{n} & y_i \leqslant x < y_{i+1} \\ 1 & y_n \leqslant x \end{cases} \tag{4.6.4}$$

Example 4.6.2 Consider the following data from a simulated sample of size $n = 10$ from the distribution of Example 2.3.2, where the CDF has the form $F(x) = 1 - (1 + x)^{-2}$; $x > 0$:

x_i: 0.85, 1.08, 0.35, 3.28, 1.24, 2.58, 0.02, 0.13, 0.22, 0.52

y_i: 0.02, 0.13, 0.22, 0.35, 0.52, 0.85, 1.08, 1.24, 2.58, 3.28

The graphs of $F(x)$ and $F_{10}(x)$ are shown in Figure 4.6. Although the graph of an empirical CDF, $F_n(x)$, is a step function, it generally should be possible to get at least a rough idea of the shape of the corresponding CDF $F(x)$. In this example, the sample size $n = 10$ is probably too small to conclude much, but the graphs in Figure 4.6 show a fairly good agreement.

FIGURE 4.6 Comparison of a CDF with an empirical CDF

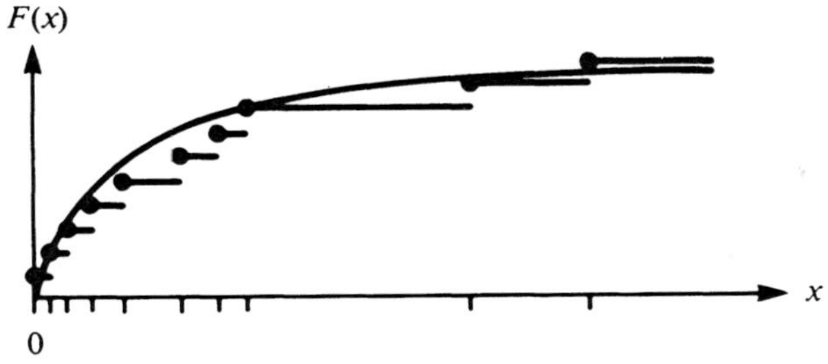

To select a suitable distribution for the population, it is helpful to have numerical estimates for unknown population parameters. For example, suppose it is desired to estimate the population mean μ. The mean of the empirical distribution, computed from the sample data, is a natural choice as an estimate of μ. Corresponding to the empirical CDF is a discrete distribution with pdf, say $f_n(x)$, which assigns the value $1/n$ at each of the data values $x = x_1, x_2, \ldots, x_n$ and zero otherwise. The mean of this distribution is $\sum_{i=1}^{n} x_i\, f_n(x_i) = x_1(1/n) + \cdots + x_n(1/n)$, which provides the desired estimate of μ. This estimate, which is simply the arithmetic average of the data, is called the **sample mean**, denoted by $\bar{x} = \sum_{i=1}^{n} x_i/n$. This rationale also can be used to obtain estimates of other unknown parameter values. For example, an estimate of the population variance σ^2 would be the variance of the empirical distribution, $\sum_{i=1}^{n} (x_i - \bar{x})^2 f_n(x_i) = (x_1 - \bar{x})^2/n + \cdots + (x_n - \bar{x})^2/n$. However, it turns out that such a procedure tends to underestimate the population variance. This point will be discussed in Chapter 8, where it will be shown that the following modified version does not suffer from this problem. The **sample variance** is defined as $s^2 = \sum_{i=1}^{n} (x_i - \bar{x})^2/(n - 1)$. Another illustration involves the estimation of a population proportion. For example, suppose a study is concerned with whether individuals in a population have been exposed to a particular contagious disease. The proportion, say p, who have been exposed is another parameter of the population. If n individuals are selected at random from the population and y is the number who have been exposed, then the **sample proportion** is defined as $\hat{p} = y/n$. This also can be treated as a special type of sample mean. For the ith individual in the sample, define $x_i = 1$ if he has been exposed and zero otherwise. The data $x_1, \ldots, x_n$ correspond to observed values of a random sample from a Bernoulli distribution with parameter p, and $\bar{x} = (x_1 + \cdots + x_n)/n = y/n = \hat{p}$.

HISTOGRAMS

It usually is easier to study the distribution of probability in terms of the pdf, $f(x)$, rather than the CDF. This leads us to consider a different type of empirical distribution, known as a **histogram**. Although this concept generally is considered as a purely descriptive method in lower-level developments of probability and statistics, it is possible to provide a rationale in terms of the multinomial distribution, which was presented in Section 4.2.

Suppose the data can be sorted into k disjoint intervals, say $I_j = (a_j, a_{j+1}]$; $j = 1, 2, \ldots, k$. Then the relative frequency, f_j, with which an observation falls into I_j gives at least a rough indication of what range of values the pdf, $f(x)$, might have over that interval. This can be made more precise by considering $k + 1$ events $E_1, E_2, \ldots, E_{k+1}$, where E_j occurs if and only if some variable, X_i, from the random sample is in the interval I_j if $j = 1, 2, \ldots, k$ and E_{k+1} occurs if an X_i is not in any I_j. If Y_j is the number of variables from the random sample that fall into I_j, and $Y = (Y_1, Y_2, \ldots, Y_k)$, then $Y \sim \text{MULT}(n, p_1, \ldots, p_k)$, where

$$p_j = F(a_{j+1}) - F(a_j) = \int_{a_j}^{a_{j+1}} f(x)\,dx \qquad \textbf{(4.6.5)}$$

Again, because of statistical regularity, we would expect the observed relative frequency, $f_j = y_j/n$, to be close to p_j for large n. It usually is possible to choose the intervals so that the probability of E_{k+1} is negligible. Actually, in practice, this is accomplished by choosing the intervals after the data have been obtained. This is a convenient practice, although theoretically incorrect.

Example 4.6.3 The following observations represent the observed lifetimes in months of a random sample of 40 electrical parts, which have been ordered from smallest to largest:

0.15	2.37	2.90	7.39	7.99	12.05	15.17	17.56
22.40	34.84	35.39	36.38	39.52	41.07	46.50	50.52
52.54	58.91	58.93	66.71	71.48	71.84	77.66	79.31
80.90	90.87	91.22	96.35	108.92	112.26	122.71	126.87
127.05	137.96	167.59	183.53	282.49	335.33	341.19	409.97

We will use k = nine intervals of length 50, $I_1 = (0, 50]$, $I_2 = (50, 100]$, and so on. The distribution of the data is summarized in Table 4.3.

It is seen, for example, that the proportion $15/40 = 0.375$ of the sample values fall below 50 months, so one would expect approximately 37.5% of the total population to fall before 50 months, and so on. Of course, the accuracy of these estimates or approximations would depend primarily on the sample size n. To plot this information so that it directly approximates the population pdf, place a

TABLE 4.3 **Frequency distribution of lifetimes of 40 electrical parts**

Limits of I_j	f_j	f_j/n	Height
0–50	15	0.375	0.0075
50–100	13	0.325	0.0065
100–150	6	0.150	0.0030
150–200	2	0.050	0.0010
200–250	0	0.000	0.0000
250–300	1	0.025	0.0005
300–350	2	0.050	0.0010
350–400	0	0.000	0.0000
400–450	1	0.025	0.0005

rectangle over the interval (0, 50] with area 0.375, a rectangle over the interval (50, 100] with area 0.325, and so on. To achieve this, the height of the rectangles should be taken as the fraction desired divided by the length of the interval. Thus the height of the rectangle over (0, 50] should be $0.375/50 = 0.0075$, the height over (50, 100] should be $0.325/50 = 0.0065$, and so on. This results in Figure 4.7, which sometimes is referred to as a **modified relative frequency histogram.**

FIGURE 4.7 Comparison of an exponential pdf with a modified relative frequency histogram

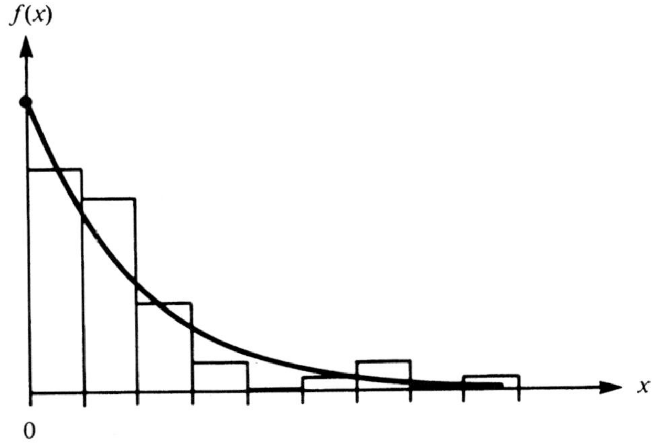

A smooth curve through the tops of the rectangles then would provide a direct approximation to the population pdf. The number and length of the intervals can be adjusted as desired, taking into account such factors as sample size or range of the data. Such a decision is purely subjective, however, and there is no universally accepted rule for doing this.

The shape of the histogram of Figure 4.7 appears to be consistent with an exponential pdf, say

$$f(x) = (1/100)e^{-x/100} \qquad 0 < x < \infty$$

and the graph of this pdf also is shown in the figure. In this case, the CDF is

$$F(x) = 1 - e^{-x/100} \qquad 0 < x < \infty$$

According to this model, $P[X \leqslant 50] = F(50) = 0.393$, $P[50 < X < 100] = F(100) - F(50) = 0.239$, and so on. These probabilities are given in Table 4.4 and are compared with the observed frequencies for the sample of size 40.

TABLE 4.4 **Observed and fitted probabilities**

	Interval probabilities		Cumulative probabilities	
Limits of I_j	**Observed**	**Exponential**	**Observed**	**Exponential**
0–50	0.375	0.393	0.375	0.393
50–100	0.325	0.239	0.700	0.632
100–150	0.150	0.145	0.850	0.777
150–200	0.050	0.088	0.900	0.865
200–250	0.000	0.053	0.900	0.918
250–300	0.025	0.032	0.925	0.950
300–350	0.050	0.020	0.975	0.970
350–400	0.000	0.012	0.975	0.982
400–450	0.025	0.007	1.000	0.989

The data in this example were simulated from an EXP(100) model, and the discrepancy between the empirical distribution and the true model results from the natural "sampling error" involved. As the sample size increases, the histogram or empirical distribution should approach the true model. Of course, in practice the true model would be unknown.

Perhaps it should be noted that definition 4.6.1 of a random sample does not apply to the case of "random sampling without replacement" from a finite population. The "random sampling" terminology is used in this case to reflect the fact that on each trial the elements remaining in the population are equally likely to be selected, but the trials are not independent and the usual counting techniques discussed earlier would apply here. The random sample terminology also may refer here to the idea that each subset of size n elements of the population is equally likely to be selected as the sample. Definition 4.6.1 is applicable to sampling from finite populations if the sampling is with replacement (or it would be approximately suitable if the population is quite large).

SUMMARY

The purpose of this chapter was to further develop the concept of a random variable to include experiments that involve two or more numerical responses. The joint pdf and CDF provide ways to express the joint probability distribution. When the random variables are considered individually, the marginal pdf's and CDFs express their probability distributions.

When the joint pdf can be factored into the product of the marginal pdf's, then the random variables are independent. If the jointly distributed random variables are dependent, then the conditional pdf's provide a valuable way to express this dependence.

Information about the nature of the true probability model may be obtained by conducting n independent trials of an experiment, and obtaining n observed values of the random variable of interest. These observations constitute a random sample from a real or conceptual population. Useful information about the population distribution can be obtained by descriptive methods such as the empirical CDF or a histogram.

EXERCISES

1. For the discrete random variables defined in Exercise 1 of Chapter 3, tabulate:

(a) the joint pdf of Y and Z.

(b) the joint pdf of Z and W.

2. In Exercise 2 of Chapter 3, a game consisted of rolling a die and tossing a coin. If X denotes the number of spots showing on the die plus the number of heads showing on the coin, and if Y denotes just the number of spots showing on the die, tabulate the joint pdf of X and Y.

3. Five cards are drawn *without* replacement from a regular deck of 52 cards. Let X represent the number of aces, Y the number of kings, and Z the number of queens obtained. Give the probability of each of the following events:

(a) $A = [X = 2]$.

(b) $B = [Y = 2]$.

(c) $A \cap B$.

(d) $A \cup B$.

(e) A given B.

(f) $[X = x]$.

(g) $[X < 2]$.

(h) $[X \geqslant 2]$.

(i) $[X = 2, Y = 2, Z = 1)$.

(j) Write an expression for the joint pdf of X, Y, and Z.

4. In reference to Example 4.2.1, find the probability for each of the following events.
(a) Exactly five red and two white.
(b) Exactly five red and two pink.

5. Rework Exercise 3, assuming that the cards were drawn *with* replacement.

6. An ordinary six-sided die is rolled 12 times. If X_1 is the number of 1's, X_2 is the number of 2's, and so on, give the probability for each of the following events:
(a) $[X_1 = 2, X_2 = 3, X_3 = 1, X_4 = 0, X_5 = 4, X_6 = 2]$.
(b) $[X_1 = X_2 = X_3 = X_4 = X_5 = X_6]$.
(c) $[X_1 = 1, X_2 = 2, X_3 = 3, X_4 = 4]$.
(d) Write an expression for the joint pdf of X_1, X_3, and X_5.

7. Suppose that X_1 and X_2 are discrete random variables with joint pdf of the form

$$f(x_1, x_2) = c(x_1 + x_2) \qquad x_1 = 0, 1, 2;\ x_2 = 0, 1, 2$$

and zero otherwise. Find the constant c.

8. If X and Y are discrete random variables with joint pdf

$$f(x, y) = c\,\frac{2^{x+y}}{x!\,y!} \qquad x = 0, 1, 2, \ldots;\ y = 0, 1, 2, \ldots,$$

and zero otherwise.
(a) Find the constant c.
(b) Find the marginal pdf's of X and Y.
(c) Are X and Y independent? Why or why not?

9. Let X_1 and X_2 be discrete random variables with joint pdf $f(x_1, x_2)$ given by the following table:

		x_2		
		1	2	3
x_1	1	1/12	1/6	0
	2	0	1/9	1/5
	3	1/18	1/4	2/15

(a) Find the marginal pdf's of X_1 and X_2.
(b) Are X_1 and X_2 independent? Why or why not?
(c) Find $P[X_1 \leqslant 2]$.
(d) Find $P[X_1 \leqslant X_2]$.
(e) Tabulate the conditional pdf's, $f(x_2 \mid x_1)$ and $f(x_1 \mid x_2)$.

10. Two cards are drawn at random *without* replacement from an ordinary deck. Let X be the number of hearts and Y the number of black cards obtained.
(a) Write an expression for the joint pdf, $f(x, y)$.

(b) Tabulate the joint CDF, $F(x, y)$.
(c) Find the marginal pdf's, $f_1(x)$ and $f_2(y)$.
(d) Are X and Y independent?
(e) Find $P[Y = 1 \mid X = 1]$.
(f) Find $P[Y = y \mid X = 1]$.
(g) Find $P[Y = y \mid X = x]$.
(h) Tabulate $P[X + Y \leqslant z]$; $z = 0, 1, 2$.

11. Rework Exercise 10, assuming that the cards are drawn *with* replacement.

12. Consider the function $F(x_1, x_2)$ defined as follows:

$$F(x_1, x_2) = \begin{cases} 0.25(x_1 + x_2)^2 & \text{if } 0 \leqslant x_1 < 1 \text{ and } 0 \leqslant x_2 < 1 \\ 0 & \text{if } x_1 < 0 \text{ or } x_2 < 0 \\ 1 & \text{otherwise} \end{cases}$$

Is $F(x_1, x_2)$ a bivariate CDF? *Hint:* Check the properties of Theorem 4.2.2.

13. Prove Theorem 4.4.2.

14. Suppose the joint pdf of lifetimes of a certain part and a spare is given by

$$f(x, y) = e^{-(x+y)} \qquad 0 < x < \infty,\ 0 < y < \infty$$

and zero otherwise. Find each of the following:
(a) The marginal pdf's, $f_1(x)$ and $f_2(y)$.
(b) The joint CDF, $F(x, y)$.
(c) $P[X > 2]$.
(d) $P[X < Y]$.
(e) $P[X + Y > 2]$.
(f) Are X and Y independent?

15. Suppose X_1 and X_2 are the survival times (in days) of two white rats that were subjected to different levels of radiation. Assume that X_1 and X_2 are independent,

$$X_1 \sim \text{PAR}(1, 1) \text{ and } X_2 \sim \text{PAR}(1, 2)$$

(a) Give the joint pdf of X_1 and X_2.
(b) Find the probability that the second rat outlives the first rat, $P[X_1 < X_2]$.

16. Assume that X and Y are independent with $X \sim \text{UNIF}(-1, 1)$ and $Y \sim \text{UNIF}(0, 1)$. Find the probability that the roots of the equation $h(t) = 0$ are real, where

$$h(t) = t^2 + 2Xt + Y$$

17. For the random variables X_1, X_2, and X_3 of Example 4.3.2:
(a) Find the marginal pdf $f_1(x_1)$.
(b) Find the marginal pdf $f_2(x_2)$.
(c) Find the joint pdf of the pair (X_1, X_2).

18. Consider a pair of continuous random variables X and Y with a joint CDF of the form

$$F(x, y) = \begin{cases} 0.5xy(x + y) & \text{if } 0 < x < 1,\ 0 < y < 1 \\ 0.5x(x + 1) & \text{if } 0 < x < 1,\ 1 \leqslant y \\ 0.5y(y + 1) & \text{if } 1 \leqslant x,\ 0 < y < 1 \\ 1 & \text{if } 1 \leqslant x,\ 1 \leqslant y \end{cases}$$

and zero otherwise. Find each of the following:

(a) The joint pdf, $f(x, y)$.
(b) $P[X \leqslant 0.5, Y \leqslant 0.5]$.
(c) $P[X < Y]$.
(d) $P[X + Y \leqslant 0.5]$.
(e) $P[X + Y \leqslant 1.5]$.
(f) $P[X + Y \leqslant z]$; $0 < z$.

19. Let X and Y be continuous random variables with a joint pdf of the form

$$f(x, y) = k(x + y) \qquad 0 \leqslant x \leqslant y \leqslant 1$$

and zero otherwise.

(a) Find k so that $f(x, y)$ is a joint pdf.
(b) Find the marginals, $f_1(x)$ and $f_2(y)$.
(c) Find the joint CDF, $F(x, y)$.
(d) Find the conditional pdf $f(y \mid x)$.
(e) Find the conditional pdf $f(x \mid y)$.

20. Suppose that X and Y have the joint pdf

$$f(x, y) = 8xy \qquad 0 \leqslant x \leqslant y \leqslant 1$$

and zero otherwise. Find each of the following:

(a) The joint CDF $F(x, y)$.
(b) $f(y \mid x)$.
(c) $f(x \mid y)$.
(d) $P[X \leqslant 0.5 \mid Y = 0.75]$.
(e) $P[X \leqslant 0.5 \mid Y \leqslant 0.75]$.

21. Suppose that X and Y have the joint pdf

$$f(x, y) = (2/3)(x + 1) \qquad 0 < x < 1,\ 0 < y < 1$$

and zero otherwise. Find each of the following:

(a) $f_1(x)$.
(b) $f_2(y)$.
(c) $f(y \mid x)$.
(d) $P[X + Y \leqslant 1]$.
(e) $P[X < 2Y < 3X]$.
(f) Are X and Y independent?

22. Let $X_1, X_2, \ldots, X_n$ denote a random sample from a population with pdf $f(x) = 3x^2$; $0 < x < 1$, and zero otherwise.

(a) Write down the joint pdf of $X_1, X_2, \ldots, X_n$.

(b) Find the probability that the first observation is less than 0.5, $P[X_1 < 0.5]$.

(c) Find the probability that all of the observations are less than 0.5.

23. Rework Exercise 22 if the random sample is from a Weibull population, $X_i \sim \text{WEI}(1, 2)$.

24. The following set of data consists of weight measurements (in ounces) for 60 major league baseballs:

5.09	5.08	5.21	5.17	5.07	5.24	5.12	5.16	5.18	5.19
5.26	5.10	5.28	5.29	5.27	5.09	5.24	5.26	5.17	5.13
5.27	5.26	5.17	5.19	5.28	5.28	5.18	5.27	5.25	5.26
5.26	5.18	5.13	5.08	5.25	5.17	5.09	5.16	5.24	5.23
5.28	5.24	5.23	5.23	5.27	5.22	5.26	5.27	5.24	5.27
5.25	5.28	5.24	5.26	5.24	5.24	5.27	5.26	5.22	5.09

(a) Construct a frequency distribution by sorting the data into five intervals of length 0.05, starting at 5.05.

(b) Based on the result of (a), graph a modified relative frequency histogram.

(c) Construct a table that compares observed and fitted probabilities based on the intervals of (a), and for pdf $f(x)$ that is uniform on the interval [5.05, 5.30].

25. For the first 10 observations in Exercise 24, graph the empirical CDF, $F_{10}(x)$, and also graph the CDF, $F(x)$, of a uniform distribution on the interval [5.05, 5.30].

26. Consider the jointly distributed random variables X_1, X_2, and X_3 of Example 4.3.2.

(a) Find the joint pdf of X_1 and X_3.

(b) Find the joint pdf of X_2 and X_3.

(c) Find the conditional pdf of X_2 given $(X_1, X_3) = (x_1, x_3)$.

(d) Find the conditional pdf of X_1 given $(X_2, X_3) = (x_2, x_3)$.

(e) Find the conditional pdf of (X_1, X_2) given $X_3 = x_3$.

27. Suppose X_1, X_2 is a random sample of size $n = 2$ from a discrete distribution with pdf given by $f(1) = f(3) = .2$ and $f(2) = .6$.

(a) Tabulate the values of the joint pdf of X_1 and X_2.

(b) Tabulate the values of the joint CDF of X_1 and X_2, $F(x_1, x_2)$.

(c) Find $P[X_1 + X_2 \leqslant 4]$.

28. Suppose X and Y are continuous random variables with joint pdf $f(x, y) = 4(x - xy)$ if $0 < x < 1$ and $0 < y < 1$, and zero otherwise.

(a) Are X and Y independent? Why or why not?

(b) Find $P[X < Y]$.

29. Suppose X and Y are continuous random variables with joint pdf given by $f(x, y) = 24xy$ if $0 < x$, $0 < y$, $x + y < 1$, and zero otherwise.

(a) Are X and Y independent? Why or why not?

(b) Find $P[Y > 2X]$.

(c) Find the marginal pdf of X.

30. Suppose X and Y are continuous random variables with joint pdf $f(x, y) = 60x^2y$ if $0 < x,\ 0 < y,\ x + y < 1$, and zero otherwise.

(a) Find the marginal pdf of X.

(b) Find the conditional pdf of Y given $X = x$.

(c) Find $P[Y > .1 \mid X = .5]$.

31. Suppose X_1 and X_2 are continuous random variables with joint pdf given by $f(x_1, x_2) = 2(x_1 + x_2)$ if $0 < x_1 < x_2 < 1$, and zero otherwise.

(a) Find $P[X_2 > 2X_1]$.

(b) Find the marginal pdf of X_2.

(c) Find the conditional pdf of X_1 given $X_2 = x_2$.

CHAPTER 5

PROPERTIES OF RANDOM VARIABLES

5.1 INTRODUCTION

The use of a random variable and its probability distribution has been discussed as a way of expressing a mathematical model for a nondeterministic physical phenomenon. The random variable may be associated with some numerical characteristic of a real or conceptual population of items, and the pdf represents the distribution of the population over the possible values of the characteristic. Quite often the true population density may be unknown. One possibility in this case is to consider a family of density functions indexed by an unknown parameter as a possible model, and then concentrate on selecting a value for the parameter.

A major emphasis in statistics is to develop estimates of unknown parameters based on sample data. In some cases a parameter may represent a physically meaningful quantity, such as an average or mean value of the population. Thus, it is worthwhile to define and study various properties of random variables that may be useful in representing and interpreting the original population, as well as useful in estimating or selecting an appropriate model. In some cases, special properties of a model (such as the no-memory property of the exponential distribution) may be quite helpful in indicating the type of physical assumptions that would be consistent with that model, although the implications of a model usually are less clear. In such a case, more reliance must be placed on basic descriptive measures such as the mean and variance of a distribution. In this chapter, additional descriptive measures and further properties of random variables will be developed.

5.2 PROPERTIES OF EXPECTED VALUES

As noted in Chapter 2, it often is necessary to consider the expected value of some function of one or more random variables. For example, a study might involve a vector of k random variables, $\mathbf{X} = (X_1, X_2, \ldots, X_k)$, and we would wish to know the expected value of some function of $\mathbf{X}$, say $Y = u(\mathbf{X})$. We could use the standard notation $E(Y)$, or another possibility would be $E[u(\mathbf{X})]$, or $E_{\mathbf{X}}[u(\mathbf{X})]$, where the subscript emphasizes that the sum or integral used to evaluate this expected value is taken relative to the joint pdf of $\mathbf{X}$. The following theorem asserts that both approaches yield the same result.

Theorem 5.2.1 If $\mathbf{X} = (X_1, \ldots, X_k)$ has a joint pdf $f(x_1, \ldots, x_k)$, and if $Y = u(X_1, \ldots, X_k)$ is a function of $\mathbf{X}$, then $E(Y) = E_{\mathbf{X}}[u(X_1, \ldots, X_k)]$, where

$$E_{\mathbf{X}}[u(X_1, \ldots, X_k)] = \sum_{x_1} \cdots \sum_{x_k} u(x_1, \ldots, x_k) f(x_1, \ldots, x_k) \tag{5.2.1}$$

if $\mathbf{X}$ is discrete, and

$$E_{\mathbf{X}}[u(X_1, \ldots, X_k)] = \int_{-\infty}^{\infty} \cdots \int_{-\infty}^{\infty} u(x_1, \ldots, x_k) f(x_1, \ldots, x_k)\, dx_1 \ldots dx_k \tag{5.2.2}$$

if $\mathbf{X}$ is continuous. ■

The proof will be omitted here, but the method of proof will be discussed in Chapter 6. We use the results of the theorem to derive some additional properties of expected values.

Theorem 5.2.2 If X_1 and X_2 are random variables with joint pdf $f(x_1, x_2)$, then

$$E(X_1 + X_2) = E(X_1) + E(X_2) \tag{5.2.3}$$

Proof

Note that the expected value on the left side of equation (5.2.3) is relative to the joint pdf of $X = (X_1, X_2)$, while the terms on the right side could be relative to either the joint or the marginal pdf's. Thus, a more precise statement of equation (5.2.3) would be

$$\begin{aligned} E_X(X_1 + X_2) &= E_X(X_1) + E_X(X_2) \\ &= E_{X_1}(X_1) + E_{X_2}(X_2) \end{aligned}$$

We will show this for the continuous case:

$$\begin{aligned} E(X_1 + X_2) &= E_X(X_1 + X_2) \\ &= \int_{-\infty}^{\infty} \int_{-\infty}^{\infty} (x_1 + x_2) f(x_1, x_2)\, dx_1\, dx_2 \\ &= \int_{-\infty}^{\infty} \int_{-\infty}^{\infty} x_1 f(x_1, x_2)\, dx_1\, dx_2 \\ &\quad + \int_{-\infty}^{\infty} \int_{-\infty}^{\infty} x_2 f(x_1, x_2)\, dx_1\, dx_2 \\ &= \int_{-\infty}^{\infty} x_1 \int_{-\infty}^{\infty} f(x_1, x_2)\, dx_2\, dx_1 \\ &\quad + \int_{-\infty}^{\infty} x_2 \int_{-\infty}^{\infty} f(x_1, x_2)\, dx_1\, dx_2 \\ &= \int_{-\infty}^{\infty} x_1 f_{X_1}(x_1)\, dx_1 + \int_{-\infty}^{\infty} x_2 f_{X_2}(x_2)\, dx_2 \\ &= E_{X_1}(X_1) + E_{X_2}(X_2) \\ &= E(X_1) + E(X_2) \end{aligned}$$

The discrete case is similar. ■

It is possible to combine the preceding theorems to show that if $a_1, a_2, \ldots, a_k$ are constants and $X_1, X_2, \ldots, X_k$ are jointly distributed random variables, then

$$E\left(\sum_{i=1}^{k} a_i X_i\right) = \sum_{i=1}^{k} a_i E(X_i) \tag{5.2.4}$$

Another commonly encountered function of random variables is the product.

Theorem 5.2.3 If X and Y are independent random variables and $g(x)$ and $h(y)$ are functions, then

$$E[g(X)h(Y)] = E[g(X)]E[h(Y)] \tag{5.2.5}$$

Proof

In the continuous case,

$$\begin{aligned} E[g(X)h(Y)] &= \int_{-\infty}^{\infty}\int_{-\infty}^{\infty} g(x)h(y)f(x, y)\,dx\,dy \\ &= \int_{-\infty}^{\infty}\int_{-\infty}^{\infty} g(x)h(y)f_1(x)f_2(y)\,dx\,dy \\ &= \left[\int_{-\infty}^{\infty} g(x)f_1(x)\,dx\right]\left[\int_{-\infty}^{\infty} h(y)f_2(y)\,dy\right] \\ &= E[g(X)]E[h(Y)] \end{aligned}$$

■

It is possible to generalize this theorem to more than two variables. Specifically, if $X_1, \ldots, X_k$ are independent random variables, and $u_1(x_1), \ldots, u_k(x_k)$ are functions, then

$$E[u_1(X_1) \cdots u_k(X_k)] = E[u_1(X_1)] \cdots E[u_k(X_k)] \quad \textbf{(5.2.6)}$$

Certain expected values provide information about the relationship between two variables.

Definition 5.2.1

The **covariance** of a pair of random variables X and Y is defined by

$$\text{Cov}(X, Y) = E[(X - \mu_X)(Y - \mu_Y)] \quad \textbf{(5.2.7)}$$

Another common notation for covariance is σ_{XY}.

Some properties that are useful in dealing with covariances are given in the following theorems.

Theorem 5.2.4 If X and Y are random variables and a and b are constants, then

$$\text{Cov}(aX, bY) = ab\,\text{Cov}(X, Y) \quad \textbf{(5.2.8)}$$

$$\text{Cov}(X + a, Y + b) = \text{Cov}(X, Y) \quad \textbf{(5.2.9)}$$

$$\text{Cov}(X, aX + b) = a\,\text{Var}(X) \quad \textbf{(5.2.10)}$$

Proof

See Exercise 26.

■

Theorem 5.2.5 If X and Y are random variables, then

$$\text{Cov}(X, Y) = E(XY) - E(X)E(Y) \tag{5.2.11}$$

and $\text{Cov}(X, Y) = 0$ whenever X and Y are independent.

Proof

See Exercise 27. ■

Theorem 5.2.2 dealt with the expected value of a sum of two random variables. The following theorem deals with the variance of a sum.

Theorem 5.2.6 If X_1 and X_2 are random variables with joint pdf $f(x_1, x_2)$, then

$$\text{Var}(X_1 + X_2) = \text{Var}(X_1) + \text{Var}(X_2) + 2\ \text{Cov}(X_1, X_2) \tag{5.2.12}$$

and

$$\text{Var}(X_1 + X_2) = \text{Var}(X_1) + \text{Var}(X_2) \tag{5.2.13}$$

whenever X_1 and X_2 are independent.

Proof

For convenience, denote the expected values of X_1 and X_2 by $\mu_i = E(X_i)$; $i = 1, 2$.

$$\begin{aligned}
\text{Var}(X_1 + X_2) &= E[(X_1 + X_2) - (\mu_1 + \mu_2)]^2 \\
&= E[(X_1 - \mu_1) + (X_2 - \mu_2)]^2 \\
&= E[(X_1 - \mu_1)^2] + E[(X_2 - \mu_2)^2] \\
&\quad + 2E[(X_1 - \mu_1)(X_2 - \mu_2)] \\
&= \text{Var}(X_1) + \text{Var}(X_2) + 2\ \text{Cov}(X_1, X_2)
\end{aligned}$$

which establishes equation (5.2.12). Equation (5.2.13) follows from Theorem 5.2.5. ■

It also can be verified that if $X_1, \ldots, X_k$ are random variables and $a_1, \ldots, a_k$ are constants, then

$$\text{Var}\left(\sum_{i=1}^{k} a_i X_i\right) = \sum_{i=1}^{k} a_i^2\ \text{Var}(X_i) + 2 \sum_{i<j}\sum a_i a_j\ \text{Cov}(X_i, X_j) \tag{5.2.14}$$

and if $X_1, \ldots, X_k$ are independent, then

$$\text{Var}\left(\sum_{i=1}^{k} a_i X_i\right) = \sum_{i=1}^{k} a_i^2\ \text{Var}(X_i) \tag{5.2.15}$$

Example 5.2.1 Suppose that $Y \sim \text{BIN}(n, p)$. Because binomial random variables represent the number of successes in n independent trials of a Bernoulli experiment, Y is distributed as a sum, $Y = \sum_{i=1}^{n} X_i$, of independent Bernoulli variables, $X_i \sim \text{BIN}(1, p)$. It follows from equations (5.2.4) and (5.2.15) that the mean and variance of Y are $E(Y) = np$ and $\text{Var}(Y) = npq$, because the mean and variance of a Bernoulli variable are $E(X_i) = p$ and $\text{Var}(X_i) = pq$. This is somewhat easier than the approach used in Chapter 3.

Example 5.2.2 The approach of the previous examples also can be used if $Y \sim \text{HYP}(n, M, N)$, but the derivation is more difficult because draws are dependent if sampling is done *without* replacement. For example, suppose that a set of N components contains M defective ones, and n components are drawn at random without replacement. If X_i represents the number of defective components (either 0 or 1) obtained on the ith draw, then $X_1, X_2, \ldots, X_n$ are *dependent* Bernoulli variables. Consider the pair (X_1, X_2). It is not difficult to see that the marginal distribution of the first variable is $X_1 \sim \text{BIN}(1, p_1)$ where $p_1 = M/N$, and conditional on $X_1 = x_1$, $X_2 \sim \text{BIN}(1, p_2)$, where $p_2 = (M - x_1)/(N - 1)$. Thus, the joint pdf of (X_1, X_2) is

$$f(x_1, x_2) = p_1^{x_1} q_1^{1-x_1} p_2^{x_2} q_2^{1-x_2} \qquad x_i = 0, 1 \tag{5.2.16}$$

from which we can obtain the covariance,

$$\text{Cov}(X_1, X_2) = -\frac{M}{N}\left(1 - \frac{M}{N}\right)\frac{1}{N-1} \tag{5.2.17}$$

Actually, it can be shown that for any pair (X_i, X_j) with $i \neq j$, $\text{Cov}(X_i, X_j)$ is given by equation (5.2.17), and for any i,

$$E(X_i) = \frac{M}{N} \tag{5.2.18}$$

and

$$\text{Var}(X_i) = \frac{M}{N}\left(1 - \frac{M}{N}\right) \tag{5.2.19}$$

It follows from equations (5.2.4) and (5.2.14) that

$$E(Y) = n\,\frac{M}{N} \tag{5.2.20}$$

$$\text{Var}(Y) = n\,\frac{M}{N}\left(1 - \frac{M}{N}\right)\left(\frac{N-n}{N-1}\right) \tag{5.2.21}$$

In the case of equation (5.2.21), there are n terms of the form (5.2.19) and $n(n-1)$ of the form (5.2.17), and the result follows after simplification.

Note that the mean and variance of the hypergeometric distribution have forms similar to the mean and variance of the binomial, with p replaced by M/N, except for the last factor in the variance, $(N-n)/(N-1)$, which is referred to as the **finite multiplier** term. Because the hypergeometric random variable, Y, represents the number of defects obtained when sampling from a finite population without replacement, it is clear that $\text{Var}(Y)$ must approach zero as n approaches N. That is, $Y = M$ when $n = N$, with variance zero, whereas this effect does not occur for the binomial case, which corresponds to sampling with replacement.

APPROXIMATE MEAN AND VARIANCE

Chapter 2 discussed a method for approximating the mean and variance of a function of a random variable X. Similar results can be developed for a function of more than one variable. For example, consider a pair of random variables (X, Y) with means μ_1 and μ_2, variances σ_1^2 and σ_2^2, and covariance σ_{12}; further suppose that the function $H(x, y)$ has partial derivatives in an open rectangle containing (μ_1, μ_2). Using Taylor approximations, we obtain the following approximate formulas for the mean and variance of $H(X, Y)$:

$$E[H(X, Y)] \doteq H(\mu_1, \mu_2) + \frac{\partial^2 H}{\partial x^2}\sigma_1^2 + \frac{\partial^2 H}{\partial y^2}\sigma_2^2$$

$$\text{Var}[H(X, Y)] \doteq \left(\frac{\partial H}{\partial x}\right)^2 \sigma_1^2 + \left(\frac{\partial H}{\partial y}\right)^2 \sigma_2^2 + 2\left(\frac{\partial H}{\partial x}\right)\left(\frac{\partial H}{\partial y}\right)\sigma_{12}$$

where the partial derivatives are evaluated at the means (μ_1, μ_2).

5.3 CORRELATION

The importance of the mean and variance in characterizing the distribution of a random variable was discussed earlier, and the covariance was described as a useful measure of dependence between two random variables.

It was shown in Theorem 5.2.5 that $\text{Cov}(X, Y) = 0$ whenever X and Y are independent. The converse, in general, is not true.

Example 5.3.1 Consider a pair of discrete random variables X and Y with joint pdf $f(x, y) = 1/4$ if $(x, y) = (0, 1), (1, 0), (0, -1)$, or $(-1, 0)$. The marginal pdf of X is $f_X(\pm 1) = 1/4$, $f_X(0) = 1/2$, and $f_X(x) = 0$ otherwise. The pdf of Y is similar. Thus, $E(X) = -1(1/4) + 0(1/2) + 1(1/4) = 0$. Because $xy = 0$ whenever $f(x, y) > 0$, it follows that

$E(XY) = 0$. Consequently, $\text{Cov}(X, Y) = E(XY) - E(X)E(Y) = 0$. However, $f(0, 0) = 0 \neq f_X(0)f_Y(0)$, so X and Y are dependent. In general, we can conclude that X and Y are dependent if $\text{Cov}(X, Y) \neq 0$, but $\text{Cov}(X, Y) = 0$ does not necessarily imply that X and Y are independent.

Definition 5.3.1

If X and Y are random variables with variances σ_X^2 and σ_Y^2 and covariance $\sigma_{XY} = \text{Cov}(X, Y)$, then the **correlation coefficient** of X and Y is

$$\rho = \frac{\sigma_{XY}}{\sigma_X \sigma_Y} \tag{5.3.1}$$

The random variables X and Y are said to be **uncorrelated** if $\rho = 0$; otherwise they are said to be **correlated**.

A subscripted notation, ρ_{XY}, also is sometimes used. The following theorem gives some important properties of the correlation coefficient.

Theorem 5.3.1 If ρ is the correlation coefficient of X and Y, then

$$-1 \leqslant \rho \leqslant 1 \tag{5.3.2}$$

and

$$\rho = \pm 1 \text{ if and only if } Y = aX + b \text{ with probability 1 for some } a \neq 0 \text{ and } b \tag{5.3.3}$$

Proof

For convenience we will use the simplified notation $\mu_1 = \mu_X$, $\mu_2 = \mu_Y$, $\sigma_1 = \sigma_X$, $\sigma_2 = \sigma_Y$, and $\sigma_{12} = \sigma_{XY}$.

To show equation (5.3.2), let

$$W = \frac{Y}{\sigma_2} - \rho \frac{X}{\sigma_1}$$

so that

$$\begin{aligned}\text{Var}(W) &= \left(\frac{1}{\sigma_2}\right)^2 \sigma_2^2 + \left(\frac{\rho}{\sigma_1}\right)^2 \sigma_1^2 - 2\rho \frac{\sigma_{12}}{\sigma_1 \sigma_2} \\ &= 1 + \rho^2 - 2\rho^2 \\ &= 1 - \rho^2 \geqslant 0\end{aligned}$$

because $\text{Var}(W) \geqslant 0$.

To show (5.3.3), notice that $\rho = \pm 1$ implies $\text{Var}(W) = 0$, which by Theorem 2.4.7 implies that $P[W = \mu_W] = 1$, so that with probability 1, $Y/\sigma_2 - X\rho/\sigma_1 = \mu_2/\sigma_2 - \mu_1 \rho/\sigma_1$, or $Y = aX + b$ where $a = \rho\sigma_2/\sigma_1$ and $b = \mu_2 - \mu_1 \rho\sigma_2/\sigma_1$. On the other hand, if $Y = aX + b$, then by Theorems 2.4.3 and 5.2.4, $\sigma_2^2 = a^2\sigma_1^2$ and $\sigma_{12} = a\sigma_1^2$, in which case $\rho = a/|a|$, so that $\rho = 1$ if $a > 0$ and $\rho = -1$ if $a < 0$. ■

Example 5.3.2 Consider random variables X and Y with joint pdf of the form $f(x, y) = 1/20$ if $(x, y) \in C$, and zero otherwise, where

$$C = \{(x, y) \mid 0 < x < 10,\ x - 1 < y < x + 1]$$

which is shown in Figure 5.1.

FIGURE 5.1 Region corresponding to $0 < x < 10$ and $x - 1 < y < x + 1$

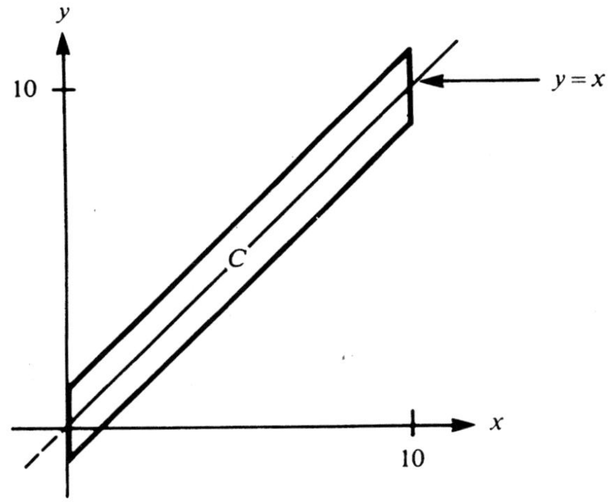

Although Y is not a function of X, the joint distribution of X and Y is "clustered" around the line $y = x$, and we should expect ρ to be near 1.

The variance of X is $\sigma_1^2 = (10)^2/12 = 25/3$, the variance of Y is $\sigma_2^2 = (10)^2/12 + (2)^2/12 = 26/3$, and the covariance is $\sigma_{12} = E(XY) - E(X)E(Y) = 25/3$. Thus, the correlation coefficient is

$$\rho = \frac{25/3}{\sqrt{25/3}\sqrt{26/3}} = \sqrt{25/26} \doteq 0.981$$

which, as expected, is close to 1.

5.4

CONDITIONAL EXPECTATION

It is possible to extend the notions of expectation and variance to the conditional framework.

Definition 5.4.1

If X and Y are jointly distributed random variables, then the **conditional expectation** of Y given $X = x$ is given by

$$E(Y \mid x) = \sum_y y f(y \mid x) \qquad \text{if } X \text{ and } Y \text{ are discrete} \tag{5.4.1}$$

$$E(Y \mid x) = \int_{-\infty}^{\infty} y f(y \mid x)\, dy \qquad \text{if } X \text{ and } Y \text{ are continuous} \tag{5.4.2}$$

Other common notations for conditional expectation are $E_{Y|x}(Y)$ and $E(Y \mid X = x)$.

Example 5.4.1 Consider Example 4.5.2, where a certain airborne particle lands at a random point (X, Y) on a triangular region. For this example, the conditional pdf of Y given $X = x$ is

$$f(y \mid x) = \frac{2}{x} \qquad 0 < y < \frac{x}{2}$$

The conditional expectation is

$$\begin{aligned} E(Y \mid x) &= \int_0^{x/2} y\left(\frac{2}{x}\right) dy \\ &= \frac{(2/x)(x/2)^2}{2} \\ &= \frac{x}{4} \qquad 0 < x < 2 \end{aligned}$$

If we are trying to "predict" the value of the vertical coordinate of (X, Y), then $E(Y \mid x)$ should be more useful than the marginal expectation, $E(Y)$, because it uses information about the horizontal coordinate. Of course, this assumes that such information is available.

Notice that the conditional expectation of Y given $X = x$ is a function of x, say $u(x) = E(Y \mid x)$. The following theorem says that, in general, the random variable $u(X) = E(Y \mid X)$ has the marginal expectation of Y, $E(Y)$.

Theorem 5.4.1 If X and Y are jointly distributed random variables, then

$$E[E(Y \mid X)] = E(Y) \tag{5.4.3}$$

Proof

Consider the continuous case:

$$\begin{aligned}
E[E(Y \mid X)] &= \int_{-\infty}^{\infty} E(Y \mid x) f_1(x)\, dx \\
&= \int_{-\infty}^{\infty} \int_{-\infty}^{\infty} y f(y \mid x) f_1(x)\, dy\, dx \\
&= \int_{-\infty}^{\infty} y \int_{-\infty}^{\infty} f(x, y)\, dx\, dy \\
&= \int_{-\infty}^{\infty} y f_2(y)\, dy \\
&= E(Y)
\end{aligned}$$

■

In the previous example, suppose that we wish to know the marginal expectation, $E(Y)$. One possibility would be to use the previous theorem, with $E(Y \mid x) = x/4$ and $f_1(x) = x/2$ for $0 < x < 2$. Specifically,

$$E(Y) = E[E(Y \mid X)] = \int_0^2 \left(\frac{x}{4}\right)\left(\frac{x}{2}\right) dx = \frac{1}{3}$$

Example 5.4.2 Suppose the number of misspelled words in a student's term paper is a Poisson-distributed random variable X with mean $E(X) = 20$. The student's roommate is asked to proofread the paper in hopes of finding the spelling errors. On average, the roommate finds 85% of such errors, and if x errors are present in the paper, then it is reasonable to consider the number of spelling errors found to be a binomial variable with parameters $n = x$ and $p = .85$. In other words, conditional on $X = x$, $Y \sim \text{BIN}(x, .85)$, and because, in general, the mean of a binomial distribution is np, the conditional expectation is $E(Y \mid x) = .85x$. Thus, the expected number of spelling errors that are found by the roommate would be $E(Y) = E[E(Y \mid X)] = E(.85X) = .85E(X) = 17$.

An interesting situation occurs when X and Y are independent.

Theorem 5.4.2 If X and Y are independent random variables, then $E(Y \mid x) = E(Y)$ and $E(X \mid y) = E(X)$.

Proof

If X and Y are independent, then $f(x, y) = f_1(x)f_2(y)$, so that $f(y \mid x) = f_2(y)$ and $f(x \mid y) = f_1(x)$. In the continuous case

$$\begin{aligned} E(Y \mid x) &= \int_{-\infty}^{\infty} y f(y \mid x)\, dy \\ &= \int_{-\infty}^{\infty} y f_2(y)\, dy \\ &= E(Y) \end{aligned}$$

The discrete case is similar. ■

It also is useful to study the variance of a conditional distribution, which usually is referred to as the conditional variance.

> ***Definition 5.4.2***
>
> The **conditional variance** of Y given $X = x$ is given by
>
> $$\operatorname{Var}(Y \mid x) = E\{[Y - E(Y \mid x)]^2 \mid x\} \tag{5.4.4}$$

An equivalent form, which is analogous to equation (2.4.8), is

$$\operatorname{Var}(Y \mid x) = E(Y^2 \mid x) - [E(Y \mid x)]^2 \tag{5.4.5}$$

Theorem 5.4.3 If X and Y are jointly distributed random variables, then

$$\operatorname{Var}(Y) = E_X[\operatorname{Var}(Y \mid X)] + \operatorname{Var}_X[E(Y \mid X)] \tag{5.4.6}$$

Proof

$$\begin{aligned} E_X[\operatorname{Var}(Y \mid X)] &= E_X\{E(Y^2 \mid X) - [E(Y \mid X)]^2\} \\ &= \{E(Y^2) - E_X[E(Y \mid X)]^2\} \\ &= E(Y^2) - [E(Y)]^2 - \{E_X[E(Y \mid X)]^2 - [E(Y)]^2\} \\ &= \operatorname{Var}(Y) - \operatorname{Var}_X[E(Y \mid X)] \end{aligned}$$

■

This theorem indicates that, on the average (over X), the conditional variance will be smaller than the unconditional variance. Of course, they would be equal if X and Y are independent, because $E(Y \mid X)$ then would not be a function of X, and $\operatorname{Var}[E(Y \mid X)]$ would be zero. If one is interested in estimating the mean height of an individual, $E(Y)$, then the theorem suggests that it might be easier to estimate the individual's (conditional) height if you know the person's weight, because the unconditional population of heights would have greater variance than the reduced population of individuals all with fixed weight, x. This fact leads to the important area of regression analysis, where information about one variable is used to aid in understanding a related variable.

An argument similar to that of Theorem 5.4.1 yields the following, more general, theorem.

Theorem 5.4.4 If X and Y are jointly distributed random variables and $h(x, y)$ is a function, then

$$E[h(X, Y)] = E_X\{E[h(X, Y) \mid X]\} \tag{5.4.7}$$

■

This theorem says that a joint expectation, such as the left side of equation (5.4.7), can be solved by first finding the conditional expectation $E[h(x, Y) \mid x]$, and then finding its expectation relative to the marginal distribution of X. This theorem is very useful in certain applications, when used in conjunction with the following theorem, which is stated without proof.

Theorem 5.4.5 If X and Y are jointly distributed random variables, and $g(x)$ is a function, then

$$E[g(X)Y \mid x] = g(x)E(Y \mid x) \tag{5.4.8}$$

■

Example 5.4.3 If $(X, Y) \sim \text{MULT}(n, p_1, p_2)$, then by straightforward derivation it follows that $X \sim \text{BIN}(n, p_1)$, $Y \sim \text{BIN}(n, p_2)$, and conditional on $X = x$, $Y \mid x \sim \text{BIN}(n - x, p)$, where $p = p_2/(1 - p_1)$. The means and variances of X and Y follow from the results of Section 5.2. Note also that $E(Y \mid x) = (n - x)p_2/(1 - p_1)$.

The two previous theorems can be used to find the covariance of X and Y. Specifically,

$$\begin{aligned}
E(XY) &= E[E(XY \mid X)] \\
&= E[XE(Y \mid X)] \\
&= E\left[\frac{X(n - X)p_2}{1 - p_1}\right] \\
&= \left[\frac{p_2}{1 - p_1}\right][nE(X) - E(X^2)]
\end{aligned}$$

If we substitute $E(X) = np_1$ and $E(X^2) = \text{Var}(X) + (np_1)^2 = np_1[1 + (n - 1)p_1]$ and simplify, then the result is

$$E(XY) = n(n - 1)p_1 p_2$$

Thus,

$$\operatorname{Cov}(X, Y) = n(n-1)p_1 p_2 - (np_1)(np_2)$$
$$= -np_1 p_2$$

As in Theorem 5.3.1, we adopt the convenient notation $\mu_1 = E(X)$, $\mu_2 = E(Y)$, $\sigma_1^2 = \operatorname{Var}(X)$, and $\sigma_2^2 = \operatorname{Var}(Y)$.

Theorem 5.4.6 If $E(Y \mid x)$ is a linear function of x, then

$$E(Y \mid x) = \mu_2 + \rho \frac{\sigma_2}{\sigma_1}(x - \mu_1) \tag{5.4.9}$$

and

$$E_X[\operatorname{Var}(Y \mid X)] = \sigma_2^2(1 - \rho^2) \tag{5.4.10}$$

Proof

Consider (5.4.9). If $E(Y \mid x) = ax + b$, then

$$\mu_2 = E(Y) = E_X[E(Y \mid X)] = E_X(aX + b) = a\mu_1 + b$$

and

$$\begin{aligned}
\sigma_{XY} &= E[(X - \mu_1)(Y - \mu_2)] \\
&= E[(X - \mu_1)Y] - 0 \\
&= E_X\{E[(X - \mu_1)Y \mid X]\} \\
&= E_X[(X - \mu_1)E(Y \mid X)] \\
&= E_X[(X - \mu_1)(aX + b)] \\
&= a\sigma_1^2
\end{aligned}$$

Thus,

$$a = \frac{\sigma_{XY}}{\sigma_1^2} = \rho \frac{\sigma_2}{\sigma_1} \quad \text{and} \quad b = \mu_2 - \rho \frac{\sigma_2}{\sigma_1} \mu_1$$

Equation (5.4.10) follows from Theorem 5.4.3,

$$\begin{aligned}
E_X[\operatorname{Var}(Y \mid X)] &= \operatorname{Var}(Y) - \operatorname{Var}_X\left[\mu_2 + \rho \frac{\sigma_2}{\sigma_1}(X - \mu_1)\right] \\
&= \operatorname{Var}(Y) - \rho^2 \frac{\sigma_2^2 \sigma_1^2}{\sigma_1^2} \\
&= \sigma_2^2(1 - \rho^2)
\end{aligned}$$

Note that if the conditional variance does not depend on x, then

$$\text{Var}(Y \mid x) = E_X[\text{Var}(Y \mid X)] = \sigma_2^2(1 - \rho^2)$$ ■

Thus, the amount of decrease in the variance of the conditional population compared to the unconditional population depends on the correlation, ρ, between the variables.

An important case will now be discussed in which $E(Y \mid x)$ is a linear function of x and $\text{Var}(Y \mid x)$ does not depend on x.

BIVARIATE NORMAL DISTRIBUTION

A pair of continuous random variables X and Y is said to have a **bivariate normal distribution** if it has a joint pdf of the form

$$f(x, y) = \frac{1}{2\pi\sigma_1\sigma_2\sqrt{1-\rho^2}} \times \exp\left\{-\frac{1}{2(1-\rho^2)}\left[\left(\frac{x-\mu_1}{\sigma_1}\right)^2 - 2\rho\left(\frac{x-\mu_1}{\sigma_1}\right)\left(\frac{y-\mu_2}{\sigma_2}\right) + \left(\frac{y-\mu_2}{\sigma_2}\right)^2\right]\right\} \quad -\infty < x < \infty, \; -\infty < y < \infty \tag{5.4.11}$$

A special notation for this is

$$(X, Y) \sim \text{BVN}(\mu_1, \mu_2, \sigma_1^2, \sigma_2^2, \rho) \tag{5.4.12}$$

which depends on five parameters, $-\infty < \mu_1 < \infty$, $-\infty < \mu_2 < \infty$, $\sigma_1 > 0$, $\sigma_2 > 0$, and $-1 < \rho < 1$.

The following theorem says that the marginal pdf's are normal and the notation used for the parameters is appropriate.

Theorem 5.4.7 If $(X, Y) \sim \text{BVN}(\mu_1, \mu_2, \sigma_1^2, \sigma_2^2, \rho)$, then

$$X \sim N(\mu_1, \sigma_1^2) \quad \text{and} \quad Y \sim N(\mu_2, \sigma_2^2)$$

and ρ is the correlation coefficient of X and Y.

Proof

See Exercise 23. ■

Strictly speaking, first we should have established that

$$\int_{-\infty}^{\infty}\int_{-\infty}^{\infty} f(x, y)\, dx\, dy = 1$$

but this would follow from Theorem 5.4.7.

It was noted in Section 5.2 that independent random variables are uncorrelated. In other words, if X and Y are independent, then $\rho = 0$. Notice that the above joint pdf, (5.4.11), factors into the product of the marginal pdf's if $\rho = 0$. Thus, for bivariate normal variables, the terms "uncorrelated" and "independent" can be used interchangeably, although this is not true for general bivariate distributions.

Theorem 5.4.8 If $(X, Y) \sim \text{BVN}(\mu_1, \mu_2, \sigma_1^2, \sigma_2^2, \rho)$, then

1. conditional on $X = x$,

$$Y \mid x \sim N\left[\mu_2 + \rho \frac{\sigma_2}{\sigma_1}(x - \mu_1), \sigma_2^2(1 - \rho^2)\right]$$

2. conditional on $Y = y$,

$$X \mid y \sim N\left[\mu_1 + \rho \frac{\sigma_1}{\sigma_2}(y - \mu_2), \sigma_1^2(1 - \rho^2)\right]$$

Proof

See Exercise 24. ■

This theorem shows that both conditional expectations are linear functions of the conditional variable, and both conditional variances are constant. If ρ is close to zero, then the conditional variance is close to the marginal variance; and if ρ is close to ± 1, then the conditional variance is close to zero.

As noted earlier in the chapter, a conditional expectation, say $E(Y \mid x)$, is a function of x that, when applied to X, has the same expected value as Y. This function sometimes is referred to as a **regression function**, and the graph of $E(Y \mid x)$ is called the **regression curve** of Y on $X = x$. The previous theorem asserts that for bivariate normal variables we have a linear regression function. It follows from Theorem 5.4.3 that the variance of $E(Y \mid X)$ is less than or equal to that of Y, and thus the regression function, in general, should be a better estimator of $\mu_2 = E(Y)$ than is Y itself. Of course, this could be explained by the fact that the marginal distribution of Y makes no use of information about X, whereas the conditional distribution of Y given $X = x$ does.

5.5

JOINT MOMENT GENERATING FUNCTIONS

The moment generating function concept can be generalized to k-dimensional random variables.

Definition 5.5.1

The joint MGF of $X = (X_1, \ldots, X_k)$, if it exists, is defined to be

$$M_X(t) = E\left[\exp\left(\sum_{i=1}^{k} t_i X_i\right)\right] \tag{5.5.1}$$

where $t = (t_1, \ldots, t_k)$ and $-h < t_i < h$ for some $h > 0$.

The joint MGF has properties analogous to those of the univariate MGF. Mixed moments such as $E[X_i^r X_j^s]$ may be obtained by differentiating the joint MGF r times with respect to t_i and s times with respect to t_j and then setting all $t_i = 0$. The joint MGF also uniquely determines the joint distribution of the variables $X_1, \ldots, X_k$.

Note that it also is possible to obtain the MGF of the marginal distributions from the joint MGF. For example,

$$M_X(t_1) = M_{X,Y}(t_1, 0) \tag{5.5.2}$$

$$M_Y(t_2) = M_{X,Y}(0, t_2) \tag{5.5.3}$$

Theorem 5.5.1 If $M_{X,Y}(t_1, t_2)$ exists, then the random variables X and Y are independent if and only if

$$M_{X,Y}(t_1, t_2) = M_X(t_1)M_Y(t_2)$$ ■

Example 5.5.1 Suppose that

$$X = (X_1, \ldots, X_k) \sim \text{MULT}(n, p_1, \ldots, p_k)$$

We have discused earlier that the marginal distributions are binomial, $X_i \sim \text{BIN}(n, p_i)$.

The joint MGF of the multinomial distribution may be evaluated along the lines followed for the binomial distribution to obtain

$$M_X(t) = E\left[\exp\left(\sum_{i=1}^{k} t_i X_i\right)\right]$$

$$= \sum \cdots \sum \frac{n}{x_1! \cdots x_{k+1}!} (p_1 e^{t_1})^{x_1} \cdots (p_k e^{t_k})^{x_k} p_{k+1}^{x_{k+1}}$$

$$= (p_1 e^{t_1} + \cdots + p_k e^{t_k} + p_{k+1})^n \tag{5.5.4}$$

where $p_{k+1} = 1 - p_1 - \cdots - p_k$.

Clearly, the joint marginal distributions also are multinomial. For example, if $(X_1, X_2, X_3) \sim \text{MULT}(n, p_1, p_2, p_3)$, then

$$\begin{aligned} M_{X_1, X_2}(t_1, t_2) &= M_{X_1, X_2, X_3}(t_1, t_2, 0) \\ &= [p_1 e^{t_1} + p_2 e^{t_2} + p_3 + (1 - p_1 - p_2 - p_3)]^n \\ &= [p_1 e^{t_1} + p_2 e^{t_2} + 1 - p_1 - p_2]^n \end{aligned}$$

so

$$(X_1, X_2) \sim \text{MULT}(n, p_1, p_2)$$

Example 5.5.2 Consider a pair of bivariate normal random variables with means μ_1 and μ_2, variances σ_1^2 and σ_2^2, and correlation coefficient ρ. In other words, $(X, Y) \sim \text{BVN}(\mu_1, \mu_2, \sigma_1^2, \sigma_2^2, \rho)$. The joint MGF of X and Y can be evaluated directly by integration: $\int_{-\infty}^{\infty}\int_{-\infty}^{\infty} \exp(t_1 x + t_2 y) f(x, y)\, dx\, dy$ with $f(x, y)$ given by equation (5.4.11). The direct approach is somewhat tedious, so we will make use of some of the results on conditional expectations. Specifically, from Theorems 5.4.4 and 5.4.5, it follows that

$$\begin{aligned} M_{X,Y}(t_1, t_2) &= E[\exp(t_1 X + t_2 Y)] \\ &= E_X\{E[\exp(t_1 X + t_2 Y) \mid X]\} \\ &= E_X\{\exp(t_1 X) E[\exp(t_2 Y) \mid X]\} \end{aligned} \tag{5.5.5}$$

Furthermore, by Theorem 5.4.8

$$Y \mid x \sim N\left[\mu_2 + \rho \frac{\sigma_2}{\sigma_1}(x - \mu_1), \sigma_2^2(1 - \rho^2)\right]$$

so that

$$E[\exp(t_2 Y) \mid X = x] = \exp\left\{\left[\mu_2 + \rho \frac{\sigma_2}{\sigma_1}(x - \mu_1)\right] t_2 + \sigma_2^2(1 - \rho^2) t_2^2/2\right\}$$

After substitution into equation (5.5.5) and some simplification, we obtain

$$M_{X,Y}(t_1, t_2) = \exp[\mu_1 t_1 + \mu_2 t_2 + \tfrac{1}{2}(\sigma_1^2 t_1^2 + \sigma_2^2 t_2^2 + 2\rho\sigma_1\sigma_2 t_1 t_2)] \tag{5.5.6}$$

SUMMARY

The main purpose of this chapter was to develop general properties involving both expected values of and functions of random variables. Sums and products are important functions of random variables that are given special attention. For example, it is shown that the expected value of a sum is the sum of the (marginal) expected values. If the random variables are independent, then the expected value of a product is the product of the (marginal) expected values, and the variance of a sum is the sum of the (marginal) variances.

The correlation coefficient provides a measure of dependence between two random variables. When the correlation coefficient is zero, the random variables are said to be uncorrelated. For two random variables to be independent, it is necessary, but not sufficient, that they be uncorrelated. When the random variables are dependent, the conditional expectation is useful in attempting to predict the value of one variable given an observed value of the other variable.

EXERCISES

1. Let X_1, X_2, X_3, and X_4 be independent random variables, each having the same distribution with mean 5 and standard deviation 3, and let $Y = X_1 + 2X_2 + X_3 - X_4$.

(a) Find $E(Y)$.

(b) Find $\text{Var}(Y)$.

2. Suppose the weight (in ounces) of a major league baseball is a random variable X with mean $\mu = 5$ and standard deviation $\sigma = 2/5$. A carton contains 144 baseballs. Assume that the weights of individual baseballs are independent, and let T represent the total weight of all the baseballs in the carton.

(a) Find the expected total weight, $E(T)$.

(b) Find the variance, $\text{Var}(T)$.

3. Suppose X and Y are continuous random variables with joint pdf $f(x, y) = 24xy$ if $0 < x$, $0 < y$, and $x + y < 1$, and zero otherwise.

(a) Find $E(XY)$.

(b) Find the covariance of X and Y.

(c) Find the correlation coefficient of X and Y.

(d) Find $\text{Cov}(3X, 5Y)$.

(e) Find $\text{Cov}(X + 1, Y - 2)$.

(f) Find $\text{Cov}(X + 1, 5Y - 2)$.

(g) Find $\text{Cov}(3X + 5, X)$.

4. Let X and Y be discrete random variables with joint pdf $f(x, y) = 4/(5xy)$ if $x = 1, 2$ and $y = 2, 3$, and zero otherwise. Find:

(a) $E(X)$.

(b) $E(Y)$.

(c) $E(XY)$.

(d) $\text{Cov}(X, Y)$.

5. Let X and Y be continuous random variables with joint pdf $f(x, y) = x + y$ if $0 < x < 1$ and $0 < y < 1$, and zero otherwise. Find:

(a) $E(X)$.

(b) $E(X + Y)$.

(c) $E(XY)$.

(d) $\text{Cov}(2X, 3Y)$.

(e) $E(Y \mid x)$.

6. If X, Y, Z, and W are random variables, then show that:

(a) $\text{Cov}(X \pm Y, Z) = \text{Cov}(X, Z) \pm \text{Cov}(Y, Z)$.

(b) $\text{Cov}(X + Y, Z + W) = \text{Cov}(X, Z) + \text{Cov}(X, W) + \text{Cov}(Y, Z) + \text{Cov}(Y, W)$.

(c) $\text{Cov}(X + Y, X - Y) = \text{Var}(X) - \text{Var}(Y)$.

7. Suppose X and Y are independent random variables with $E(X) = 2$, $E(Y) = 3$, $\text{Var}(X) = 4$, and $\text{Var}(Y) = 16$.

(a) Find $E(5X - Y)$.

(b) Find $\text{Var}(5X - Y)$.

(c) Find $\text{Cov}(3X + Y, Y)$.

(d) Find $\text{Cov}(X, 5X - Y)$.

8. If $X_1, X_2, \ldots, X_k$ and $Y_1, Y_2, \ldots, Y_m$ are jointly distributed random variables, and if $a_1, a_2, \ldots, a_k$ and $b_1, b_2, \ldots, b_m$ are constants, show that

$$\text{Cov}\left(\sum_{i=1}^{k} a_i X_i, \sum_{j=1}^{m} b_j Y_j\right) = \sum_{i=1}^{k} \sum_{j=1}^{m} a_i b_j \text{Cov}(X_i, Y_j)$$

9. Use the result of Exercise 8 to verify equations (5.2.14) and (5.2.15).

10. For the random variables in Exercise 5, find the approximate mean and variance of $W = XY$.

11. Let $f(x, y) = 6x$, $0 < x < y < 1$, and zero otherwise. Find:

(a) $f_1(x)$.

(b) $f_2(y)$.

(c) $\text{Cov}(X, Y)$.

(d) ρ.

(e) $f(y \mid x)$.

(f) $E(Y \mid x)$.

12. Suppose X and Y are continuous random variables with joint pdf $f(x, y) = 4(x - xy)$ if $0 < x < 1$ and $0 < y < 1$, and zero otherwise.

(a) Find $E(X^2 Y)$.

(b) Find $E(X - Y)$.

(c) Find $\text{Var}(X - Y)$.

(d) What is the correlation coefficient of X and Y?

(e) What is $E(Y \mid x)$?

13. Let $f(x, y) = 1$ if $0 < y < 2x$, $0 < x < 1$, and zero otherwise. Find:

(a) $f(y \mid x)$.

(b) $E(Y \mid x)$.
(c) ρ.

14. For the joint pdf of Exercise 30 in Chapter 4 (page 170):
(a) Find the correlation coefficient of X and Y.
(b) Find the conditional expectation, $E(Y \mid x)$.
(c) Find the conditional variance, $\mathrm{Var}(Y \mid x)$.

15. (a) Determine $E(Y \mid x)$ in Exercise 4.
(b) Determine $\mathrm{Var}(Y \mid x)$ in Exercise 4.

16. Let X and Y have joint pdf $f(x, y) = e^{-y}$ if $0 < x < y < \infty$ and zero otherwise. Find $E(X \mid y)$.

17. Suppose that the conditional distribution of Y given $X = x$ is Poisson with mean $E(Y \mid x) = x$, $Y \mid x \sim \mathrm{POI}(x)$, and that $X \sim \mathrm{EXP}(1)$.
(a) Find $E(Y)$.
(b) Find $\mathrm{Var}(Y)$.

18. One box contains five red and six black marbles. A second box contains 10 red and five black marbles. One marble is drawn from box 1 and placed in box 2. Two marbles then are drawn from box 2 without replacement. What is the expected number of red marbles obtained on the second draw?

19. The number of times a batter gets to bat in a game follows a binomial distribution $N \sim \mathrm{BIN}(6, 0.8)$. Given the number of times at bat, n, that the batter has, the number of hits he gets conditionally follows a binomial distribution, $X \mid n \sim \mathrm{BIN}(n, 0.3)$.
(a) Find $E(X)$.
(b) Find $\mathrm{Var}(X)$.
(c) Find $E(X^2)$.

20. Let X be the number of customers arriving in a given minute at the drive-up window of a local bank, and let Y be the number who make withdrawals. Assume that X is Poisson distributed with expected value $E(X) = 3$, and that the conditional expectation and variance of Y given $X = x$ are $E(Y \mid x) = x/2$ and $\mathrm{Var}(Y \mid x) = (x + 1)/3$.
(a) Find $E(Y)$.
(b) $\mathrm{Var}(Y)$.
(c) Find $E(XY)$.

21. Suppose that Y_1 and Y_2 are continuous with joint pdf $f(y_1, y_2) = 2e^{-y_1 - y_2}$ if $0 < y_1 < y_2 < \infty$ and zero otherwise. Derive the joint MGF of Y_1 and Y_2.

22. Find the joint MGF of the continuous random variables X and Y with joint pdf $f(x, y) = e^{-y}$ if $0 < x < y < \infty$ and zero otherwise.

23. Prove Theorem 5.4.7. *Hint:* Use the joint MGF of Example 5.5.2.

24. Prove Theorem 5.4.8.

25. Let X_1 and X_2 be independent normal random variables, $X_i \sim N(\mu_i, \sigma_i^2)$, and let $Y_1 = X_1$ and $Y_2 = X_1 + X_2$.

(a) Show that Y_1 and Y_2 are bivariate normal.

(b) What are the means, variances, and correlation coefficient of Y_1 and Y_2?

(c) Find the conditional distribution of Y_2 given $Y_1 = y_1$. *Hint:* Use Theorem 5.4.8.

26. Prove Theorem 5.2.4.

27. Prove Theorem 5.2.5.

CHAPTER 6

FUNCTIONS OF RANDOM VARIABLES

6.1

INTRODUCTION

In Chapter 1, probability was defined first in a set theoretic framework. The concept of a random variable then was introduced so that events could be associated with sets of real numbers in the range space of the random variable. This makes it possible to mathematically express the probability model for the population or characteristic of interest in the form of a pdf or a CDF for the associated random variable, say X. In this case, X represents the initial characteristic of interest, and the pdf, $f_X(x)$, may be referred to as the population pdf. It often may be the case that some function of this variable also is of interest. Thus, if X represents the age in weeks of some component, another experimenter may be expressing the age, Y, in days, so that $Y = 7X$. Similarly, $W = \ln X$ or some other function of X may be of interest. Any function of a random variable X is itself a random variable, and the probability distribution of a function of X is determined by the probability distribution of X. For example, for Y above,

$P[14 \leqslant Y \leqslant 21] = P[2 \leqslant X \leqslant 3]$, and so on. Clearly, probabilities concerning functions of a random variable may be of interest, and it is useful to be able to express the pdf or CDF of a function of a random variable in terms of the pdf or CDF of the original variable. Such pdf's sometimes are referred to as "derived" distributions. Of course, a certain pdf may represent a population pdf in one application, but correspond to a derived distribution in a different application.

In Example 4.6.1, the total lifetime, $X_1 + X_2$, of two light bulbs was of interest, and a method of deriving the distribution of this variable was suggested. General techniques for deriving the pdf of a function of random variables will be discussed in this chapter.

6.2 THE CDF TECHNIQUE

We will assume that a random variable X has CDF $F_X(x)$, and that some function of X is of interest, say $Y = u(X)$. The idea behind the CDF technique is to express the CDF of Y in terms of the distribution of X. Specifically, for each real y, we can define a set $A_y = \{x \mid u(x) \leqslant y\}$. It follows that $[Y \leqslant y]$ and $[X \in A_y]$ are equivalent events, in the sense discussed in Section 2.1, and consequently

$$F_Y(y) = P[u(X) \leqslant y] \tag{6.2.1}$$

which also can be expressed as $P[X \in A_y]$. This probability can be expressed as the integral of the pdf, $f_X(x)$, over the set A_y if X is continuous, or the summation of $f_X(x)$ over x in A_y if X is discrete.

For example, it often is possible to express $[u(X) \leqslant y]$ in terms of an equivalent event $[x_1 \leqslant X \leqslant x_2]$, where one or both of the limits x_1 and x_2 depend on y.

In the continuous case,

$$\begin{aligned} F_Y(y) &= \int_{x_1}^{x_2} f_X(x)\,dx \\ &= F_X(x_2) - F_X(x_1) \end{aligned} \tag{6.2.2}$$

and, of course, the pdf is $f_Y(y) = (d/dy)F_Y(y)$.

Example 6.2.1 Suppose that $F_X(x) = 1 - e^{-2x}$, $0 < x < \infty$, and consider $Y = e^X$. We have

$$\begin{aligned} F_Y(y) &= P[Y \leqslant y] \\ &= P[e^X \leqslant y] \\ &= P[X \leqslant \ln y] \\ &= F_X(\ln y) \\ &= 1 - y^{-2} \qquad 1 < y < \infty \end{aligned}$$

In this case, $x_1 = -\infty$ and $x_2 = \ln y$, and the pdf of Y is

$$f_Y(y) = \frac{d}{dy} F_Y(y) = 2y^{-3} \qquad 1 < y < \infty$$

Example 6.2.2 Consider a continuous random variable X, and let $Y = X^2$. It follows that

$$\begin{aligned} F_Y(y) &= P[X^2 \leqslant y] \\ &= P[-\sqrt{y} \leqslant X \leqslant \sqrt{y}] \\ &= F_X(\sqrt{y}) - F_X(-\sqrt{y}) \end{aligned} \tag{6.2.3}$$

The pdf of Y can be expressed easily in terms of the pdf of X in this case, because

$$\begin{aligned} f_Y(y) &= \frac{d}{dy} [F_X(\sqrt{y}) - F_X(-\sqrt{y})] \\ &= f_X(\sqrt{y}) \frac{d}{dy} \sqrt{y} - f_X(-\sqrt{y}) \frac{d}{dy} (-\sqrt{y}) \\ &= \frac{1}{2\sqrt{y}} [f_X(\sqrt{y}) + f_X(-\sqrt{y})] \end{aligned} \tag{6.2.4}$$

for $y > 0$.

Evaluation of equation (6.2.1) can be more complicated than the form given by equation (6.2.2), because a union of intervals may occur.

Example 6.2.3 A signal is sent to a two-sided rotating antenna, and the angle of the antenna at the time the signal is received can be assumed to be uniformly distributed from 0 to 2π, $\Theta \sim \text{UNIF}(0, 2\pi)$. The signal can be received if $Y = \tan \Theta > y_0$. For example, $y \geqslant 1$ corresponds to the angle $45° < \Theta < 90°$ and $225° < \Theta < 270°$. The CDF of Y when $y < 0$ is

$$\begin{aligned} F_Y(y) &= P[\tan(\Theta) \leqslant y] \\ &= P[\pi/2 < \Theta \leqslant \pi + \tan^{-1}(y)] + P[3\pi/2 < \Theta \leqslant 2\pi + \tan^{-1}(y)] \\ &= (1/2\pi)[\pi + \tan^{-1}(y) - \pi/2 + 2\pi + \tan^{-1}(y) - 3\pi/2] \\ &= 1/2 + (1/\pi)\tan^{-1}(y) \end{aligned}$$

By symmetry, $P[Y > y] = P[Y < -y]$. Thus, for $y > 0$,

$$\begin{aligned} F_Y(y) &= 1 - P[Y > y] = 1 - P[Y < -y] \\ &= 1 - [1/2 + (1/\pi)\tan^{-1}(-y)] \\ &= 1/2 + (1/\pi)\tan^{-1}(y) \end{aligned}$$

It is interesting that

$$f_Y(y) = \frac{dF_Y(y)}{dy} = \frac{1}{\pi}\frac{1}{1+y^2} \qquad -\infty < y < \infty$$

This is the pdf of a Cauchy distribution (defined in Chapter 3), $Y \sim \text{CAU}(1, 0)$.

The CDF technique also can be extended to apply to a function of several variables, although the analysis generally is more complicated.

Theorem 6.2.1 Let $\mathbf{X} = (X_1, X_2, \ldots, X_k)$ be a k-dimensional vector of continuous random variables, with joint pdf $f(x_1, x_2, \ldots, x_k)$. If $Y = u(\mathbf{X})$ is a function of $\mathbf{X}$, then

$$F_Y(y) = P[u(\mathbf{X}) \leqslant y] = \int \cdots \int_{A_y} f(x_1, \ldots, x_k)\, dx_1 \ldots dx_k \tag{6.2.5}$$

where $A_y = \{\mathbf{x} \mid u(\mathbf{x}) \leqslant y\}$. ■

Of course, the limits of the integral (6.2.5) are functions of y, and the convenience of this method will depend on the complexity of the resulting limits.

Example 6.2.4 In Example 4.6.1 we considered the sum of two independent random variables, say $Y = X_1 + X_2$, where $X_i \sim \text{EXP}(1)$. The set required in (6.2.5) is, as shown in Figure 6.1,

$$A_y = \{(x_1, x_2) \mid 0 \leqslant x_1 \leqslant y - x_2;\quad 0 \leqslant x_2 \leqslant y\}$$

FIGURE 6.1 Region A_y such that $x_1 + x_2 \leqslant y$

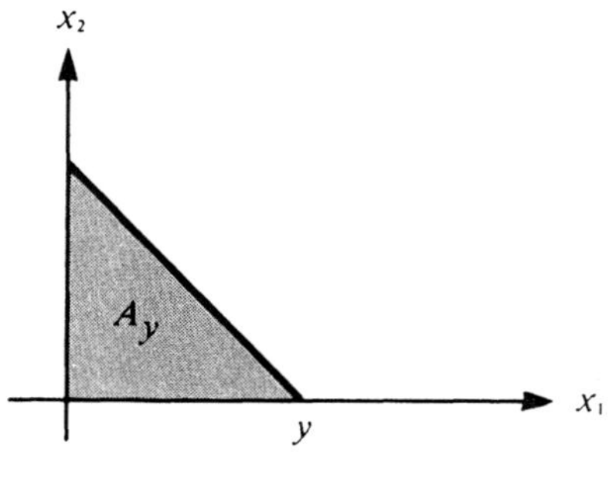

and consequently

$$F_Y(y) = \int_0^y \int_0^{y-x_2} e^{-(x_1+x_2)}\, dx_1\, dx_2$$

$$= 1 - e^{-y} - ye^{-y}$$

and

$$f_Y(y) = \frac{d}{dy} F_Y(y)$$

$$= ye^{-y} \qquad y > 0$$

It is possible in many cases to derive the pdf directly and without first deriving the CDF.

6.3 TRANSFORMATION METHODS

First we will consider transformations of variables in one dimension. Let $u(x)$ be a real-valued function of a real variable x. If the equation $y = u(x)$ can be solved uniquely, say $x = w(y)$, then we say the transformation is **one-to-one**.

It will be necessary to consider discrete and continuous cases separately, and also whether the function is one-to-one.

ONE-TO-ONE TRANSFORMATIONS

Theorem 6.3.1 **Discrete Case** Suppose that X is a discrete random variable with pdf $f_X(x)$ and that $Y = u(X)$ defines a one-to-one transformation. In other words, the equation $y = u(x)$ can be solved uniquely, say $x = w(y)$. Then the pdf of Y is

$$f_Y(y) = f_X(w(y)) \qquad y \in B \tag{6.3.1}$$

where $B = \{y \mid f_Y(y) > 0\}$.

Proof

This follows because $f_Y(y) = P[Y = y] = P[u(X) = y] = P[X = w(y)] = f_X(w(y))$. ■

Example 6.3.1 Let $X \sim \text{GEO}(p)$, so that

$$f_X(x) = pq^{x-1} \qquad x = 1, 2, 3, \ldots$$

Another frequently encountered random variable that also is called geometric is of the form $Y = X - 1$, so that $u(x) = x - 1$, $w(y) = y + 1$, and

$$f_Y(y) = f_X(y + 1)$$
$$= pq^y \qquad y = 0, 1, 2, \ldots$$

which is nothing more than the pdf of the number of failures before the first success.

Theorem 6.3.2 **Continuous Case** Suppose that X is a continuous random variable with pdf $f_X(x)$, and assume that $Y = u(X)$ defines a one-to-one transformation from $A = \{x \mid f_X(x) > 0\}$ on to $B = \{y \mid f_Y(y) > 0\}$ with inverse transformation $x = w(y)$. If the derivative $(d/dy)w(y)$ is continuous and nonzero on B, then the pdf of Y is

$$f_Y(y) = f_X(w(y)) \left| \frac{d}{dy} w(y) \right| \qquad y \in B \tag{6.3.2}$$

Proof

If $y = u(x)$ is one-to-one, then it is either monotonic increasing or monotonic decreasing. If we first assume that it is increasing, then $u(x) \leqslant y$ if and only if $x \leqslant w(y)$. Thus,

$$F_Y(y) = P[u(X) \leqslant y] = P[X \leqslant w(y)] = F_X(w(y))$$

and, consequently,

$$f_Y(y) = \frac{d}{dy} F_X(w(y)) = \frac{d}{dw(y)} F_X(w(y)) \frac{d}{dy} w(y)$$
$$= f_X(w(y)) \left| \frac{d}{dy} w(y) \right|$$

because $(d/dy)w(y) > 0$ in this case.

In the decreasing case, $u(x) \leqslant y$ if and only if $w(y) \leqslant x$, and thus

$$F_Y(y) = P[u(X) \leqslant y] = P[X \geqslant w(y)] = 1 - F_X(w(y))$$

and

$$f_Y(y) = -f_X(w(y)) \frac{d}{dy} w(y)$$
$$= f_X(w(y)) \left| \frac{d}{dy} w(y) \right|$$

because $(d/dy)w(y) < 0$ in this case. ∎

In this context, the derivative of $w(y)$ is usually referred to as the **Jacobian** of the transformation, and denoted by $J = (d/dy)w(y)$. Note also that transforming a continuous random variable is equivalent to the problem of making a change of variables in an integral. This should not be surprising, because a continuous pdf is simply the function that is integrated over events to obtain probabilities.

Example 6.3.2 We wish to use Theorem 6.3.2 to determine the pdf of $Y = e^X$ in Example 6.2.1. We obtain the inverse transformation $x = w(y) = \ln y$, and the Jacobian $J = w'(y) = 1/y$, so that

$$
\begin{aligned}
f_Y(y) &= f_X(\ln y)\left|\frac{1}{y}\right| \\
&= 2e^{-2\ln y}\left(\frac{1}{y}\right) \qquad 1 < y < \infty \\
&= 2y^{-3} \qquad y \in B = (1, \infty)
\end{aligned}
$$

In a transformation problem, it is always important to identify the set B where $f_Y(y) > 0$, which in this example is $B = (1, \infty)$, because $e^x > 1$ when $x > 0$.

The transformation $Y = e^X$, when applied to a normally distributed random variable, yields a positive-valued random variable with an important special distribution.

Example 6.3.3 A distribution that is related to the normal distribution, but for which the random variable assumes only positive values, is the **lognormal distribution**, which is defined by the pdf

$$f_Y(y) = \frac{1}{y\sigma\sqrt{2\pi}} e^{-(\ln y - \mu)^2/2\sigma^2} \qquad 0 < y < \infty \tag{6.3.3}$$

with parameters $-\infty < \mu < \infty$; $0 < \sigma < \infty$. This will be denoted by $Y \sim \text{LOGN}(\mu, \sigma^2)$, and it is related to the normal distribution by the relationship $Y \sim \text{LOGN}(\mu, \sigma^2)$ if and only if $X = \ln Y \sim N(\mu, \sigma^2)$.

In some cases, the lognormal distribution is reparameterized by letting $\mu = \ln \theta$, which gives

$$f_Y(y) = \frac{1}{y\sigma\sqrt{2\pi}} e^{-[\ln(y/\theta)]^2/2\sigma^2} \tag{6.3.4}$$

and in this notation θ becomes a scale parameter.

It is clear that cumulative lognormal probabilities can be expressed in terms of normal probabilities, because if $Y \sim \text{LOGN}(\mu, \sigma^2)$, then

$$
\begin{aligned}
F_Y(y) = P[Y \leqslant y] &= P[\ln Y \leqslant \ln y] \\
&= P[X \leqslant \ln y] \\
&= \Phi\left(\frac{\ln y - \mu}{\sigma}\right)
\end{aligned}
\tag{6.3.5}
$$

Another important special distribution is obtained by a log transformation applied to a Pareto variable.

Example 6.3.4 If $X \sim \text{PAR}(1, 1)$, then the pdf of $Z = \ln X$ is

$$f_Z(z) = \frac{e^z}{(1 + e^z)^2}$$

and the CDF is

$$F_Z(z) = \frac{e^z}{1 + e^z}$$

for all real z. If we introduce location and scale parameters ξ and θ, respectively, by the transformation $y = u(z) = \xi + \theta z$, then the pdf of $Y = u(Z)$ is

$$f_Y(y) = \frac{1}{\theta} \frac{\exp[(y - \xi)/\theta]}{\{1 + \exp[(y - \xi)/\theta]\}^2} \tag{6.3.6}$$

for all real y. The distribution of Y is known as the **logistic distribution**, denoted by $Y \sim \text{LOG}(\theta, \xi)$. This is another example of a symmetric distribution, which follows by noting that

$$f_Z(-z) = \frac{(e^{2z}/e^{2z})e^{-z}}{(1 + e^{-z})^2} = \frac{e^z}{(e^z + 1)^2} = f_Z(z)$$

The transformation $y = \xi + \theta z$ provides a general approach to introducing location and scale parameters into a model

Recall that Theorem 2.4.1 was stated without proof. A proof for the special case of a continuous random variable X, under the conditions of Theorem 6.3.2, now will be provided.

Consider the case in which $u(x)$ is an increasing function, so that the inverse transformation, $x = w(y)$, also is increasing:

$$\begin{aligned} E[u(X)] &= \int_{-\infty}^{\infty} u(x) f_X(x)\, dx \\ &= \int_{-\infty}^{\infty} u[w(y)] f_X[w(y)] \frac{d}{dy} w(y)\, dy \\ &= \int_{-\infty}^{\infty} y f_Y(y)\, dy \\ &= E(Y) \end{aligned}$$

The case in which $u(x)$ is decreasing is similar.

A very useful special transformation is given by the following theorem.

Theorem 6.3.3 **Probability Integral Transformation** If X is continuous with CDF $F(x)$, then $U = F(X) \sim \text{UNIF}(0, 1)$.

Proof

We will prove the theorem in the case where $F(x)$ is one-to-one, so that the inverse, $F^{-1}(u)$, exists:

$$\begin{aligned} F_U(u) &= P[F(X) \leqslant u] \\ &= P[X \leqslant F^{-1}(u)] \\ &= F(F^{-1}(u)) \\ &= u \end{aligned}$$

Because $0 \leqslant F(x) \leqslant 1$, we have $F_U(u) = 0$ if $u \leqslant 0$ and $F_U(u) = 1$ if $u \geqslant 1$. ■

A more general proof is obtained if $F^{-1}(u)$ is replaced by the function $G(u)$ that assigns to each value u the minimum value of x such that $u \leqslant F(x)$,

$$G(u) = \min \{x \mid u \leqslant F(x)\} \qquad 0 \leqslant u \leqslant 1 \tag{6.3.7}$$

The function $G(u)$ exists for any CDF, $F(x)$, and it agrees with $F^{-1}(u)$ if $F(x)$ is a one-to-one function.

The following example involves a continuous distribution with a CDF that is not one-to-one.

Example 6.3.5 Let X be a continuous random variable with pdf

$$f(x) = \begin{cases} 1/2 & \text{if } 1 < |x - 2| < 2 \\ 0 & \text{otherwise} \end{cases}$$

The CDF of X, whose graph is shown in Figure 6.2, is not one-to-one, because it assumes the value 1/2 for all $1 \leqslant x \leqslant 3$.

FIGURE 6.2 A CDF that is continuous but not one-to-one

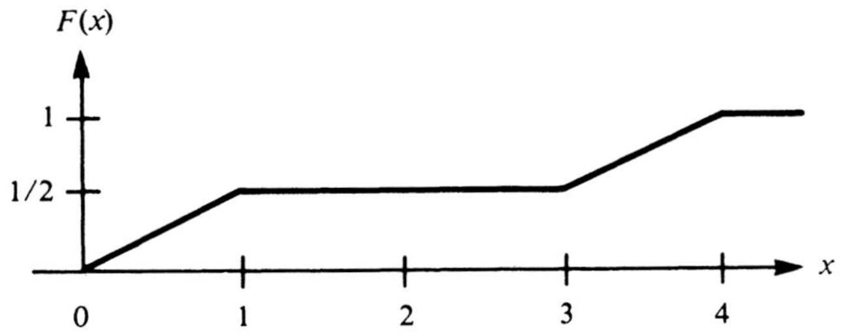

The function $G(u)$, for this example, is

$$G(u) = \begin{cases} 2u & \text{if } 0 \leqslant u \leqslant 1/2 \\ 2(u+1) & \text{if } 1/2 < u \leqslant 1 \end{cases}$$

The function $G(u)$ has another important application, which follows from the next theorem.

Theorem 6.3.4 Let $F(x)$ be a CDF and let $G(u)$ be the function defined by (6.3.7). If $U \sim \text{UNIF}(0, 1)$, then $X = G(U) \sim F(x)$.

Proof

See Exercise 5. ∎

An important application of the preceding theorem is the generation of "pseudo" random variables from some specified distribution using a computer. In other words, the computer generates data that are distributed as the observations of a random sample from some specified distribution with CDF $F(x)$. Specifically, if n "random numbers," say $u_1, u_2, \ldots, u_n$, are generated by a random number generator, these represent a "simulated" random sample of size n from UNIF(0, 1). It follows, then, that $x_1, x_2, \ldots, x_n$ where

$$x_i = G(u_i) \qquad i = 1, 2, \ldots, n \tag{6.3.8}$$

corresponds to a simulated random sample from a distribution with CDF $F(x)$. Of course, in many examples the CDF is one-to-one, and we could use $x_i = F^{-1}(u_i)$. Equation (6.3.8) also can be used with discrete distributions.

Example 6.3.6 If $X \sim \text{BIN}(1, 1/2)$, then

$$F(x) = \begin{cases} 0 & \text{if } x < 0 \\ 1/2 & \text{if } 0 \leqslant x < 1 \\ 1 & \text{if } 1 \leqslant x \end{cases}$$

and

$$G(u) = \begin{cases} 0 & \text{if } 0 \leqslant u \leqslant 1/2 \\ 1 & \text{if } 1/2 < u \leqslant 1 \end{cases}$$

TRANSFORMATIONS THAT ARE NOT ONE-TO-ONE

Suppose that the function $u(x)$ is not one-to-one over $A = \{x \mid f_X(x) > 0\}$. Although this means that no unique solution to the equation $y = u(x)$ exists, it usually is possible to partition A into disjoint subsets $A_1, A_2, \ldots$ such that $u(x)$ is one-to-one over each A_j. Then, for each y in the range of $u(x)$, the equation $y = u(x)$ has a unique solution $x_j = w_j(y)$ over the set A_j. In the discrete case, it follows that Theorem 6.3.1 can be extended to functions that are not one-to-one by replacing equation (6.3.1) with

$$f_Y(y) = \sum_j f_X(w_j(y)) \tag{6.3.9}$$

That is, $f_Y(y) = \sum_j f_X(x_j)$ where the sum is over all x_j such that $u(x_j) = y$.

Example 6.3.7 Let $f_X(x) = \frac{4}{31}(\frac{1}{2})^x$; $x = -2, -1, 0, 1, 2$, and consider $Y = |X|$. Clearly, $B = \{0, 1, 2\}$ and

$$f_Y(0) = f_X(0) = \frac{4}{31}$$

$$f_Y(1) = f_X(-1) + f_X(1) = \frac{8}{31} + \frac{2}{31} = \frac{10}{31}$$

$$f_Y(2) = f_X(-2) + f_X(2) = \frac{17}{31}$$

Another way to express this is

$$\begin{aligned} f_y(y) &= \frac{4}{31} & y &= 0 \\ &= \frac{4}{31}\left[\left(\frac{1}{2}\right)^{-y} + \left(\frac{1}{2}\right)^{y}\right] & y &= 1, 2 \end{aligned}$$

An expression analogous to equation (6.3.9) for the discrete case is obtained for continuous functions that are not one-to-one by extending equation (6.3.2) to

$$f_Y(y) = \sum_j f_X(w_j(y)) \left| \frac{d}{dy} w_j(y) \right| \tag{6.3.10}$$

That is, for the transformation $Y = u(X)$, the summation is again over all the values of j for which $u(x_j) = y$, although the Jacobian enters into the equation for the continuous case. We found by the cumulative method in Example 6.2.2 that if $Y = X^2$, then

$$F_Y(y) = F_X(\sqrt{y}) - F_X(-\sqrt{y})$$

and by taking derivatives

$$f_Y(y) = \frac{1}{2\sqrt{y}} [f_X(\sqrt{y}) + f_X(-\sqrt{y})] \tag{6.3.11}$$

Equation (6.3.11) also follows directly by the transformation method now by applying equation (6.3.10).

Example 6.3.8 Suppose that $X \sim \text{UNIF}(-1, 1)$ and $Y = X^2$. If we partition $A = (-1, 1)$ into $A_1 = (-1, 0)$ and $A_2 = (0, 1)$, then $y = x^2$ has unique solutions $x_1 = w_1(y) = -\sqrt{y}$ and $x_2 = w_2(y) = \sqrt{y}$ over these intervals. We can neglect the point $x = 0$ in this partition, because X is continuous. The pdf of Y is thus

$$f_Y(y) = f_X(-\sqrt{y})\left|\frac{-1}{2\sqrt{y}}\right| + f_X(\sqrt{y})\left|\frac{1}{2\sqrt{y}}\right| = \frac{1}{2\sqrt{y}} \qquad y \in B = (0, 1)$$

If the limits of the function $u(x)$ are not the same over each set A_j of the partition, then greater care must be exercised in applying the equations. This is illustrated in the following example.

Example 6.3.9 With $f_X(x) = x^2/3$, $-1 < x < 2$, and zero otherwise, consider the transformation $Y = X^2$. In general, we still have the inverse transformations $x_1 = w_1(y) = -\sqrt{y}$ for $x < 0$ and $x_2 = w_2(y) = \sqrt{y}$ for $x > 0$; however, for $0 < y < 1$ there are two points with nonzero pdf, namely $x_1 = -\sqrt{y}$ and $x_2 = \sqrt{y}$, that map into y, whereas for $1 < y < 4$ there is only one point with nonzero pdf, $x_2 = \sqrt{y}$, that maps into y. Thus the pdf of Y is

$$f_Y(y) = \frac{1}{2\sqrt{y}}\left[f_X(\sqrt{y}) + f_X(-\sqrt{y})\right]$$

$$= \begin{cases} \dfrac{1}{2\sqrt{y}}\left[\dfrac{(-\sqrt{y})^2}{3} + \dfrac{(\sqrt{y})^2}{3}\right] & 0 < y < 1 \\[2ex] \dfrac{1}{2\sqrt{y}}\left[0 + \dfrac{(\sqrt{y})^2}{3}\right] & 1 < y < 4 \end{cases}$$

In the previous example, notice that it is possible to solve the problem without explicitly using the functional notation $u(x)$ or $w_i(y)$. Thus, as suggested earlier, a simpler way of expressing equations (6.3.9) and (6.3.10), respectively, is

$$f_Y(y) = \sum_j f_X(x_j) \tag{6.3.12}$$

and

$$f_Y(y) = \sum_j f_X(x_j)\left|\frac{dx_j}{dy}\right| \tag{6.3.13}$$

where it must be kept in mind that $x_j = w_j(y)$ is a function of y.

This simpler notation will be convenient in expressing the results of joint transformations of several random variables.

JOINT TRANSFORMATIONS

The preceding theorems can be extended to apply to functions of several random variables.

Example 6.3.10 Consider the geometric variables $X, Y \sim \text{GEO}(p)$ of Example 4.4.3. As we found in this example, the joint pdf of X and $T = X + Y$ can be expressed in terms of the joint pdf of X and Y, namely

$$f_{X,T}(x, t) = f_{X,Y}(x, t - x)$$

where x and $t - x$ are the solutions of the joint transformation $x = u_1(x, y)$ and $t = u_2(x, y) = x + y$.

This can be generalized as follows. Consider a k-dimensional vector $\boldsymbol{X} = (X_1, X_2, \ldots, X_k)$ of random variables, and suppose that $u_1(\boldsymbol{x}), u_2(\boldsymbol{x}), \ldots, u_k(\boldsymbol{x})$ are k functions of $\boldsymbol{x}$, so that $Y_i = u_i(\boldsymbol{X})$ for $i = 1, \ldots, k$ defines another vector of random variables, $\boldsymbol{Y} = (Y_1, Y_2, \ldots, Y_k)$. A more concise way to express this is $\boldsymbol{Y} = \boldsymbol{u}(\boldsymbol{X})$.

In the discrete case, we can state a k-dimensional version of Theorem 6.3.1.

Theorem 6.3.5 If $\boldsymbol{X}$ is a vector of discrete random variables with joint pdf $f_{\boldsymbol{X}}(\boldsymbol{x})$ and $\boldsymbol{Y} = \boldsymbol{u}(\boldsymbol{X})$ defines a one-to-one transformation, then the joint pdf of $\boldsymbol{Y}$ is

$$f_{\boldsymbol{Y}}(y_1, y_2, \ldots, y_k) = f_{\boldsymbol{X}}(x_1, x_2, \ldots, x_k) \tag{6.3.14}$$

where $x_1, x_2, \ldots, x_k$ are the solutions of $\boldsymbol{y} = \boldsymbol{u}(\boldsymbol{x})$, and consequently depend on $y_1, y_2, \ldots, y_k$.

If the transformation is not one-to-one, and if a partition exists, say $A_1, A_2, \ldots$, such that the equation $\boldsymbol{y} = \boldsymbol{u}(\boldsymbol{x})$ has a unique solution $\boldsymbol{x} = \boldsymbol{x}_j$ or

$$\boldsymbol{x}_j = (x_{1j}, x_{2j}, \ldots, x_{kj}) \tag{6.3.15}$$

over A_j, then the pdf of $\boldsymbol{Y}$ is

$$f_{\boldsymbol{Y}}(y_1, \ldots, y_k) = \sum_j f_{\boldsymbol{X}}(x_{1j}, \ldots, x_{kj}) \tag{6.3.16}$$

■

Joint transformations of continuous random variables can be accomplished, although the notion of the Jacobian must be generalized. Suppose, for example, that $u_1(x_1, x_2)$ and $u_2(x_1, x_2)$ are functions, and x_1 and x_2 are unique solutions to the transformation $y_1 = u_1(x_1, x_2)$ and $y_2 = u_2(x_1, x_2)$. Then the **Jacobian** of the transformation is the determinant

$$J = \begin{vmatrix} \dfrac{\partial x_1}{\partial y_1} & \dfrac{\partial x_1}{\partial y_2} \\ \dfrac{\partial x_2}{\partial y_1} & \dfrac{\partial x_2}{\partial y_2} \end{vmatrix} \tag{6.3.17}$$

Example 6.3.11 It is desired to transform x_1 and x_2 into x_1 and the product $x_1 x_2$. Specifically, let $y_1 = x_1$ and $y_2 = x_1 x_2$. The solution is $x_1 = y_1$ and $x_2 = y_2/y_1$, and the Jacobian is

$$J = \begin{vmatrix} 1 & 0 \\ -y_2/y_1^2 & 1/y_1 \end{vmatrix} = 1/y_1$$

For a transformation of k variables $\boldsymbol{y} = \boldsymbol{u}(\boldsymbol{x})$, with a unique solution $\boldsymbol{x} = (x_1, x_2, \ldots, x_k)$, the Jacobian is the determinant of the $k \times k$ matrix of partial derivatives:

$$J = \begin{vmatrix} \dfrac{\partial x_1}{\partial y_1} & \dfrac{\partial x_1}{\partial y_2} & \cdots & \dfrac{\partial x_1}{\partial y_k} \\ \dfrac{\partial x_2}{\partial y_1} & & & \vdots \\ \vdots & & & \\ \dfrac{\partial x_k}{\partial y_1} & \cdots & & \dfrac{\partial x_k}{\partial y_k} \end{vmatrix} \tag{6.3.18}$$

Theorem 6.3.2 can be generalized as follows.

Theorem 6.3.6 Suppose that $\boldsymbol{X} = (X_1, X_2, \ldots, X_k)$ is a vector of continuous random variables with joint pdf $f_{\boldsymbol{X}}(x_1, x_2, \ldots, x_k) > 0$ on A, and $\boldsymbol{Y} = (Y_1, Y_2, \ldots, Y_k)$ is defined by the one-to-one transformation

$$Y_i = u_i(X_1, X_2, \ldots, X_k) \qquad i = 1, 2, \ldots, k$$

If the Jacobian is continuous and nonzero over the range of the transformation, then the joint pdf of $\boldsymbol{Y}$ is

$$f_{\boldsymbol{Y}}(y_1, \ldots, y_k) = f_{\boldsymbol{X}}(x_1, \ldots, x_k)\,|J| \tag{6.3.19}$$

where $\boldsymbol{x} = (x_1, \ldots, x_k)$ is the solution of $\boldsymbol{y} = \boldsymbol{u}(\boldsymbol{x})$.

Proof

As noted earlier, the problem of finding the pdf of a function of a random variable is related to a change of variables in an integral. This approach extends readily to transformations of k variables. Denote by B the range of a transformation $\boldsymbol{y} = \boldsymbol{u}(\boldsymbol{x})$ with inverse $\boldsymbol{x} = \boldsymbol{w}(\boldsymbol{y})$. Assume $D \subset B$, and let C be the set of all points $\boldsymbol{x} = (x_1, \ldots, x_k)$ that map into D under the transformation. We have

$$\begin{aligned} P[\boldsymbol{Y} \in D] &= \int_D \cdots \int f_{\boldsymbol{Y}}(y_1, \ldots, y_k)\, dy_1 \cdots dy_n \\ &= \int_C \cdots \int f_{\boldsymbol{X}}(x_1, \ldots, x_k)\, dx_1 \cdots dx_n. \end{aligned}$$

But this also can be written as

$$\int_D \cdots \int f_{\mathbf{X}}[w_1(y_1, \ldots, y_k), \ldots, w_k(y_1, \ldots, y_k)]\,|J|\,dy_1 \cdots dy_k,$$

as the result of a standard theorem on change of variables in an integral. Because this is true for arbitrary $D \subset B$, equation (6.3.19) follows. ■

Example 6.3.12 Let X_1 and X_2 be independent and exponential, $X_i \sim \text{EXP}(1)$. Thus, the joint pdf is

$$f_{X_1, X_2}(x_1, x_2) = e^{-(x_1 + x_2)} \qquad (x_1, x_2) \in A$$

where $A = \{(x_1, x_2) \mid 0 < x_1,\ 0 < x_2\}$. Consider the random variables $Y_1 = X_1$ and $Y_2 = X_1 + X_2$. This corresponds to the transformation $y_1 = x_1$ and $y_2 = x_1 + x_2$, which has a unique solution, $x_1 = y_1$ and $x_2 = y_2 - y_1$. The Jacobian is

$$J = \begin{vmatrix} 1 & 0 \\ -1 & 1 \end{vmatrix} = 1$$

and thus

$$\begin{aligned} f_{Y_1, Y_2}(y_1, y_2) &= f_{X_1, X_2}(y_1, y_2 - y_1) \\ &= e^{-y_2} \qquad (y_1, y_2) \in B \end{aligned}$$

and zero otherwise. The set B is obtained by transforming the set A, and this corresponds to $y_1 = x_1 > 0$ and $y_2 - y_1 = x_2 > 0$. Thus, $B = \{(y_1, y_2) \mid 0 < y_1 < y_2 < \infty\}$, which is a triangular region in the plane with boundaries $y_1 = 0$ and $y_2 = y_1$. The regions A and B are shown in Figure 6.3.

FIGURE 6.3 Regions corresponding to the transformation $y_1 = x_1$ and $y_2 = x_1 + x_2$

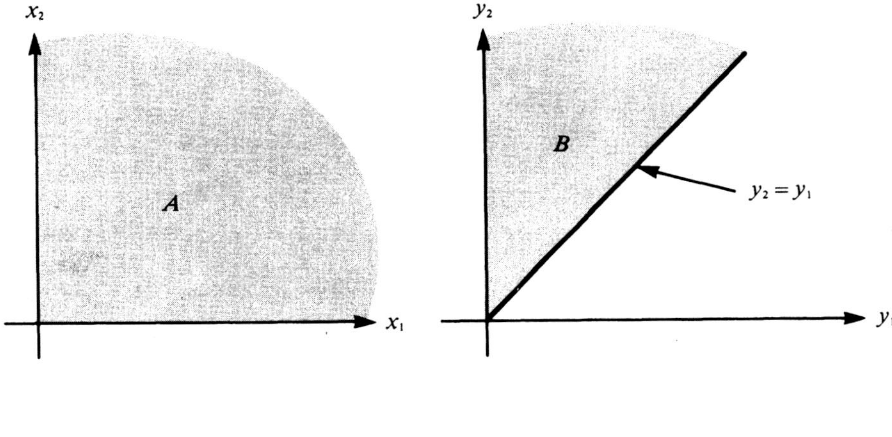

The marginal pdf's of Y_1 and Y_2 are given as follows:

$$f_{Y_1}(y_1) = \int_{y_1}^{\infty} e^{-y_2}\, dy_2$$
$$= e^{-y_1} \qquad y_1 > 0$$
$$f_{Y_2}(y_2) = \int_{0}^{y_2} e^{-y_2}\, dy_1$$
$$= y_2 e^{-y_2} \qquad y_2 > 0$$

Note that $Y_2 \sim \text{GAM}(1, 2)$.

Example 6.3.13 Suppose that, instead of the transformation of the previous example, we consider a different transformation, $y_1 = x_1 - x_2$ and $y_2 = x_1 + x_2$.

The solution is $x_1 = (y_1 + y_2)/2$ and $x_2 = (y_2 - y_1)/2$, so the Jacobian is

$$J = \begin{vmatrix} 1/2 & 1/2 \\ -1/2 & 1/2 \end{vmatrix} = 1/2$$

The joint pdf is given by

$$f_{Y_1, Y_2}(y_1, y_2) = \frac{1}{2} e^{-y_2} \qquad (y_1, y_2) \in B$$

where, in this example, $B = \{(y_1, y_2) \mid -y_2 < y_1 < y_2,\ y_2 > 0\}$ with boundaries $y_2 = -y_1$ and $0 < y_2 = y_1$. The region A is the same as in Figure 6.3, but B is different, as shown in Figure 6.4.

FIGURE 6.4 Region corresponding to the transformation $y_1 = x_1 - x_2$ and $y_2 = x_1 + x_2$

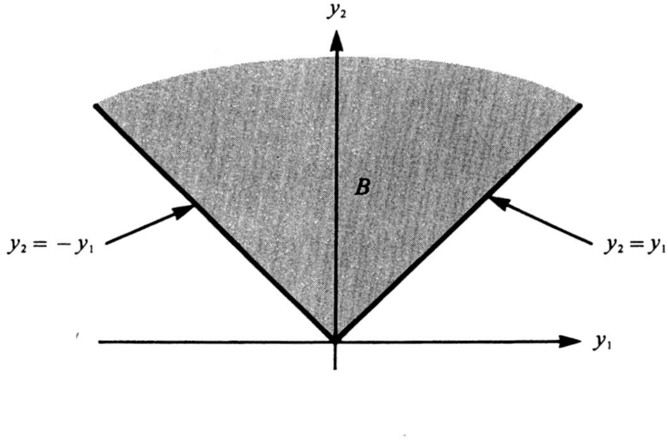

The marginal pdf's of Y_1 and Y_2 are

$$f_{Y_1}(y_1) = \int_{-y_1}^{\infty} \frac{1}{2} e^{-y_2}\, dy_2 = \frac{1}{2} e^{y_1} \qquad y_1 < 0$$

$$= \int_{y_1}^{\infty} \frac{1}{2} e^{-y_2}\, dy_2 = \frac{1}{2} e^{-y_1} \qquad y_1 > 0$$

$$= \frac{1}{2} e^{-|y_1|} \qquad -\infty < y_1 < \infty$$

$$f_{Y_2}(y_2) = \int_{-y_2}^{y_2} \frac{1}{2} e^{-y_2}\, dy_1 = y_2\, e^{-y_2} \qquad y_2 > 0$$

As in the previous example, $Y_2 \sim \text{GAM}(1, 2)$, and Y_1 has the double exponential distribution DE(1, 0).

It is possible to extend Theorem 6.3.6 to transformations that are not one-to-one in a manner similar to equation (6.3.13). Specifically, if the equation $\mathbf{y} = \mathbf{u}(\mathbf{x})$ can be solved uniquely over each set in a partition $A_1, A_2, \ldots$, to yield solutions such as in equation (6.3.15), and if these solutions have nonzero continuous Jacobians, then

$$f_{\mathbf{Y}}(y_1, \ldots, y_k) = \sum_i f_{\mathbf{X}}(x_{1i}, \ldots, x_{ki})\,|J_i| \tag{6.3.20}$$

where J_i is the Jacobian of the solution over A_i.

An important application of equation (6.3.20) will be considered in Section 6.5, but first we will consider methods for dealing with sums.

6.4

SUMS OF RANDOM VARIABLES

Special methods are provided here for dealing with the important special case of sums of random variables.

CONVOLUTION FORMULA

If one is interested only in the pdf of a sum $S = X_1 + X_2$, where X_1 and X_2 are continuous with joint pdf $f(x_1, x_2)$, then a general formula can be derived using the approach of Example 6.3.12, namely

$$f_S(s) = \int_{-\infty}^{\infty} f(t, s - t)\, dt \tag{6.4.1}$$

If X_1 and X_2 are independent, then this usually is referred to as the **convolution formula,**

$$f_S(s) = \int_{-\infty}^{\infty} f_1(t) f_2(s - t)\, dt \tag{6.4.2}$$

Example 6.4.1 Let X_1 and X_2 be independent and uniform, $X_i \sim \text{UNIF}(0, 1)$, and let $S = X_1 + X_2$. The region B corresponding to the transformation $t = x_1$ and $s = x_1 + x_2$ is $B = \{(s, t) \mid 0 < t < s < t + 1 < 2\}$, and this is shown in Figure 6.5.

FIGURE 6.5 Regions corresponding to the transformation $t = x_1$ and $s = x_1 + x_2$

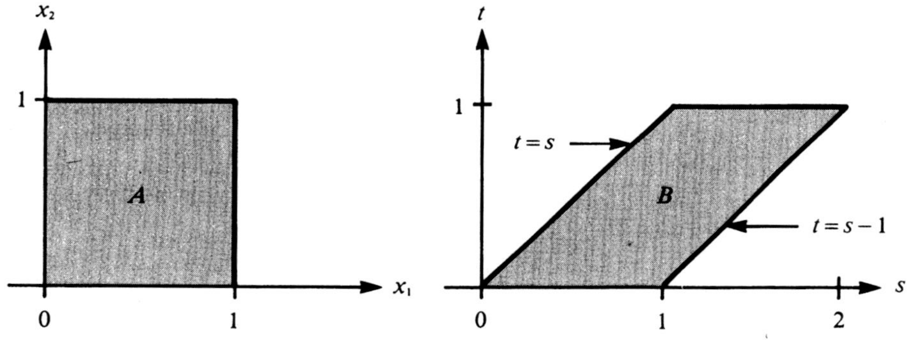

Thus, from equation (6.4.2) we have

$$\begin{aligned} f_S(s) &= \int_0^s dt = s & 0 < s < 1 \\ &= \int_{s-1}^1 dt = 2 - s & 1 \leqslant s < 2 \\ &= 1 - |s - 1| & 0 < s < 2 \end{aligned}$$

and zero otherwise.

In some cases it may be necessary to consider points other than just the boundaries in determining the new range space B. Care must be exercised in determining the appropriate limits of integration, depending on B.

Example 6.4.2 Suppose that X_1 and X_2 are independent gamma variables,

$$f(x_1, x_2) = \frac{1}{\Gamma(\alpha)\Gamma(\beta)} x_1^{\alpha-1} x_2^{\beta-1} e^{-x_1-x_2} \qquad 0 < x_i < \infty$$

Let $Y_1 = X_1 + X_2$ and $Y_2 = X_1/(X_1 + X_2)$, with inverse transformations $x_1 = y_1 y_2$ and $x_2 = y_1(1 - y_2)$. We have

$$J = \begin{vmatrix} y_2 & y_1 \\ 1 - y_2 & -y_1 \end{vmatrix} = -y_1$$

and

$$f_{Y_1, Y_2}(y_1, y_2) = \frac{(y_1 y_2)^{\alpha - 1}}{\Gamma(\alpha)\Gamma(\beta)} [y_1(1 - y_2)]^{\beta - 1} e^{-y_1} |-y_1|$$

$$= \frac{y_2^{\alpha - 1}(1 - y_2)^{\beta - 1} y_1^{\alpha + \beta - 1} e^{-y_1}}{\Gamma(\alpha)\Gamma(\beta)} \tag{6.4.3}$$

if $(y_1, y_2) \in B = \{(y_1, y_2) | 0 < y_1 < \infty, 0 < y_2 < 1\}$ and zero otherwise. The boundary composed of the positive segment of the line $x_1 = 0$ maps into the positive segment of the line $y_2 = 0$, and the positive segment of the line $x_2 = 0$ maps into the positive segment of the line $y_2 = 1$, because $y_2 = x_1(x_1 + x_2) = x_1/x_1 = 1$ and $y_2 = x_1 > 0$ when $x_2 = 0$. This does not completely bound B, but consider the positive segment $x_2 = kx_1$. As k takes on values between 0 and ∞, all points in A are included, and such a line maps into the line $y_2 = x_1/(x_1 + kx_1) = 1/(1 + k)$, where $y_1 = x_1 + kx_1 = (1 + k)x_1$ goes between 0 and ∞ as $0 < x_1 < \infty$. Thus, as k goes from 0 to ∞, B is composed of all the parallel lines between $y_2 = 0$ and $y_2 = 1$.

It is interesting to observe from equation (6.4.3) that Y_1 and Y_2 are independent random variables and $Y_1 \sim \text{GAM}(1, \alpha + \beta)$.

Example 6.4.3 Assume that X_1, X_2, and X_3 are independent gamma variables, $X_i \sim \text{GAM}(1, \alpha_i)$; $i = 1, 2, 3$. The joint pdf is

$$f_{X_1, X_2, X_3}(x_1, x_2, x_3) = \prod_{i=1}^{3} \frac{1}{\Gamma(\alpha_i)} x_i^{\alpha_i - 1} e^{-x_i} \qquad 0 < x_i < \infty$$

Let $Y_i = X_i \Big/ \sum_{j=1}^{3} X_j$, $i = 1, 2$; and $Y_3 = \sum_{j=1}^{3} X_j$ with inverse transformation

$$x_1 = y_1 y_3, \; x_2 = y_2 y_3, \text{ and } x_3 = y_3(1 - y_1 - y_2)$$

We have

$$J = \begin{vmatrix} y_3 & 0 & y_1 \\ 0 & y_3 & y_2 \\ -y_3 & -y_3 & 1 - y_1 - y_2 \end{vmatrix} = y_3^2$$

and

$$f_{Y_1, Y_2, Y_3}(y_1, y_2, y_3) = \frac{y_3^{\alpha_1+\alpha_2+\alpha_3-1}e^{-y_3}y_1^{\alpha_1-1}y_2^{\alpha_2-1}(1-y_1-y_2)^{\alpha_3-1}}{\Gamma(\alpha_1)\Gamma(\alpha_2)\Gamma(\alpha_3)}$$

$$(y_1, y_2, y_3) \in B$$

where

$$B = \{(y_1, y_2, y_3) \mid 0 < y_1 < \infty,\ 0 < y_2 < \infty,\ y_1 + y_2 < 1,\ 0 < y_3 < \infty\}$$

We see in this case that Y_3 is again a gamma variable and is independent of Y_1 and Y_2, although Y_1 and Y_2 are not independent of each other. The joint density of Y_1 and Y_2 is known as a **Dirichlet distribution.** A similar pattern will hold if this transformation is extended to k variables. In particular, if $X_i \sim \text{GAM}(1, \alpha_i)$, $i = 1, \ldots, k$, and are independent, then

$$Y_k = \sum_{i=1}^{k} X_i \sim \text{GAM}\left(1, \sum_{i=1}^{k} \alpha_i\right)$$

Sums of independent random variables often arise in practice. A technique based on moment generating functions usually is much more convenient than using transformations for determining the distribution of sums of independent random variables; this approach will be discussed next.

MOMENT GENERATING FUNCTION METHOD

Theorem 6.4.1 If $X_1, \ldots, X_n$ are independent random variables with MGFs $M_{X_i}(t)$, then the MGF of $Y = \sum_{i=1}^{n} X_i$ is

$$M_Y(t) = M_{X_1}(t) \cdots M_{X_n}(t) \tag{6.4.4}$$

Proof

Notice that $e^{tY} = e^{t(X_1 + \cdots + X_n)} = e^{tX_1} \cdots e^{tX_n}$ so by property (5.2.6),

$$\begin{aligned} M_Y(t) &= E(e^{tY}) \\ &= E(e^{tX_1} \cdots e^{tX_n}) \\ &= E(e^{tX_1}) \cdots E(e^{tX_n}) \\ &= M_{X_1}(t) \cdots M_{X_n}(t) \end{aligned}$$

■

This has a special form when $X_1, \ldots, X_n$ represents a random sample from a population with common pdf $f(x)$ and MGF $M(t)$, namely

$$M_Y(t) = [M(t)]^n \tag{6.4.5}$$

As noted in Chapter 2, the MGF of a random variable uniquely determines its distribution.

It is clear that the MGF can be used as a technique for determining the distribution of a function of a random variable, and it is undoubtedly more important for this purpose than for computing moments. The MGF approach is particularly useful for determining the distribution of a sum of independent random variables, and it often will be much more convenient than trying to carry out a joint transformation. If the MGF of a variable is ascertained, then it is necessary to recognize what distribution has that MGF. The MGFs of many of the most common distributions have been included in Appendix B.

Example 6.4.4 Let $X_1, \ldots, X_k$ be independent binomial random variables with respective parameters n_i and p, $X_i \sim \text{BIN}(n_i, p)$, and let $Y = \sum_{i=1}^{k} X_i$. It follows that

$$\begin{aligned} M_Y(t) &= M_{X_1}(t) \cdots M_{X_k}(t) \\ &= (pe^t + q)^{n_1} \cdots (pe^t + q)^{n_k} \\ &= (pe^t + q)^{n_1 + \cdots + n_k} \end{aligned}$$

We recognize that this is the binomial MGF with parameters $n_1 + \cdots + n_k$ and p, and thus, $Y \sim \text{BIN}(n_1 + \cdots + n_k, p)$.

Example 6.4.5 Let $X_1, \ldots, X_n$ be independent Poisson-distributed random variables, $X_i \sim \text{POI}(\mu_i)$, and let $Y = X_1 + \cdots + X_n$. The MGF of X_i is $M_{X_i}(t) = \exp[\mu_i(e^t - 1)]$, and consequently the MGF of Y is

$$\begin{aligned} M_Y(t) &= \exp[\mu_1(e^t - 1)] \cdots \exp[\mu_n(e^t - 1)] \\ &= \exp[(\mu_1 + \cdots + \mu_n)(e^t - 1)] \end{aligned}$$

which shows that $Y \sim \text{POI}(\mu_1 + \cdots + \mu_n)$.

Example 6.4.6 Suppose that $X_1, \ldots, X_n$ are independent gamma-distributed random variables with respective shape parameters $\kappa_1, \kappa_2, \ldots, \kappa_n$ and common scale parameter θ, $X_i \sim \text{GAM}(\theta, \kappa_i)$ for $i = 1, \ldots, n$. The MGF of X_i is

$$M_{X_i}(t) = (1 - \theta t)^{-\kappa_i} \qquad t < 1/\theta$$

If $Y = \sum_{i=1}^{n} X_i$, then the MGF of Y is

$$\begin{aligned} M_Y(t) &= (1 - \theta t)^{-\kappa_1} \cdots (1 - \theta t)^{-\kappa_n} \\ &= (1 - \theta t)^{-(\kappa_1 + \cdots + \kappa_n)} \end{aligned}$$

and consequently, $Y \sim GAM(\theta, \kappa_1 + \cdots + \kappa_n)$. Of course, this is consistent with some earlier examples.

Example 6.4.7 Let $X_1, \ldots, X_n$ be independent normally distributed random variables, $X_i \sim N(\mu_i, \sigma_i^2)$, and let $Y = \sum_{i=1}^{n} X_i$. The MGF of X_i is

$$M_{X_i}(t) = \exp(\mu_i t + \sigma_i^2 t^2/2)$$

and thus the MGF of Y is

$$\begin{aligned} M_Y(t) &= \exp(\mu_1 t + \sigma_1^2 t^2/2) \cdots \exp(\mu_n t + \sigma_n^2 t^2/2) \\ &= \exp(\mu_1 t + \sigma_1^2 t^2/2 + \cdots + \mu_n t + \sigma_n^2 t^2/2) \\ &= \exp[(\mu_1 + \cdots + \mu_n)t + (\sigma_1^2 + \cdots + \sigma_n^2)t^2/2] \end{aligned}$$

which shows that $Y \sim N(\mu_1 + \cdots + \mu_n, \sigma_1^2 + \cdots + \sigma_n^2)$.

This includes the special case of a random sample $X_1, \ldots, X_n$ from a normally distributed population, say $X_i \sim N(\mu, \sigma^2)$. In this case, $\mu = \mu_i$ and $\sigma^2 = \sigma_i^2$ for all $i = 1, \ldots, n$, and consequently $\sum_{i=1}^{n} X_i \sim N(n\mu, n\sigma^2)$. It also follows readily in this case that the sample mean, $\bar{X} = \sum_{i=1}^{n} X_i/n$ is normally distributed, $\bar{X} \sim N(\mu, \sigma^2/n)$.

An important application of the transformation method that involves ordered random variables is discussed next.

6.5

ORDER STATISTICS

The concept of a random sample of size n was discussed earlier, and the joint density function of the associated n independent random variables, say $X_1, \ldots, X_n$, is given by

$$f(x_1, \ldots, x_n) = f(x_1) \cdots f(x_n) \tag{6.5.1}$$

For example, if a random sample of five light bulbs is tested, the observed failure times might be (in months) $(x_1, \ldots, x_5) = (5, 11, 4, 100, 17)$. Now, the actual observations would have taken place in the order $x_3 = 4$, $x_1 = 5$, $x_2 = 11$, $x_5 = 17$, and $x_4 = 100$. It often is useful to consider the "ordered" random sample of size n, denoted by $(x_{1:n}, x_{2:n}, \ldots, x_{n:n})$. That is, in this example $x_{1:5} = x_3 = 4$, $x_{2:5} = x_1 = 5$, $x_{3:5} = x_2 = 11$, $x_{4:5} = x_5 = 17$, and $x_{5:5} = x_4 = 100$. Because we do not really care which bulbs happened to be labeled number 1, number 2, and so on, one could equivalently record the ordered data as it was taken without keeping track on the initial labeling. In some cases one may desire to stop after the r smallest ordered observations out of n have been observed, because this could result in a great saving of time. In the example, 100 months were required before all five light bulbs failed, but the first four failed in 17 months.

The joint distribution of the ordered variables is not the same as the joint density of the unordered variables. For example, the 5! different permutations of a sample of five observations would correspond to just one ordered result. This suggests the result of the following theorem. We will consider a transformation that orders the values $x_1, x_2, \ldots, x_n$. For example,

$$y_1 = u_1(x_1, x_2, \ldots, x_n) = \min(x_1, x_2, \ldots, x_n)$$

$$y_n = u_n(x_1, x_2, \ldots, x_n) = \max(x_1, x_2, \ldots, x_n)$$

and in general $y_i = u_i(x_1, x_2, \ldots, x_n)$ represents the ith smallest of $x_1, x_2, \ldots, x_n$. For an example of this transformation, see the above light-bulb data. Sometimes we will use the notation $x_{i:n}$ for $u_i(x_1, x_2, \ldots, x_n)$, but ordinarily we will use the simpler notation y_i. Similarly, when this transformation is applied to a random sample $X_1, X_2, \ldots, X_n$ we will obtain a set of ordered random variables, called the **order statistics** and denoted by either $X_{1:n}, X_{2:n}, \ldots, X_{n:n}$ or $Y_1, Y_2, \ldots, Y_n$.

Theorem 6.5.1 If $X_1, X_2, \ldots, X_n$ is a random sample from a population with continuous pdf $f(x)$, then the joint pdf of the order statistics $Y_1, Y_2, \ldots, Y_n$ is

$$g(y_1, y_2, \ldots, y_n) = n!f(y_1)f(y_2)\cdots f(y_n) \tag{6.5.2}$$

if $y_1 < y_2 < \cdots < y_n$, and zero otherwise. ■

This is an example of a transformation of continuous random variables that is not one-to-one, and it may be carried out by partitioning the domain into subsets $A_1, A_2, \ldots$ such that the transformation is one-to-one on each subset, and then summing as suggested by equation (6.3.20).

Rather than attempting a general proof of the theorem, we will illustrate it for the case $n = 3$. In this case, the sample space can be partitioned into the following $3! = 6$ disjoint sets:

$$A_1 = \{(x_1, x_2, x_3) \mid x_1 < x_2 < x_3\}$$

$$A_2 = \{(x_1, x_2, x_3) \mid x_2 < x_1 < x_3\}$$

$$A_3 = \{(x_1, x_2, x_3) \mid x_1 < x_3 < x_2\}$$

$$A_4 = \{(x_1, x_2, x_3) \mid x_2 < x_3 < x_1\}$$

$$A_5 = \{(x_1, x_2, x_3) \mid x_3 < x_1 < x_2\}$$

$$A_6 = \{(x_1, x_2, x_3) \mid x_3 < x_2 < x_1\}$$

and the range of the transformation is $B = \{(y_1, y_2, y_3) \mid y_1 < y_2 < y_3\}$.

In transforming to the ordered random sample, we have the one-to-one transformation

$$Y_1 = X_1,\ Y_2 = X_2,\ Y_3 = X_3 \quad \text{with } J_1 = 1 \text{ on } A_1$$
$$Y_1 = X_2,\ Y_2 = X_1,\ Y_3 = X_3 \quad \text{with } J_2 = -1 \text{ on } A_2$$
$$Y_1 = X_1,\ Y_2 = X_3,\ Y_3 = X_2 \quad \text{with } J_3 = -1 \text{ on } A_3$$

and so forth. Notice that in each case $|J_i| = 1$. Furthermore, for each region, the joint pdf is the product of factors $f(y_i)$ multiplied in some order, but it can be written as $f(y_1)f(y_2)f(y_3)$ regardless of the order. If we sum over all $3! = 6$ subsets, then the joint pdf of Y_1, Y_2, and Y_3 is

$$\begin{aligned} g(y_1, y_2, y_3) &= \sum_{i=1}^{6} f(y_1)f(y_2)f(y_3) \\ &= 3!f(y_1)f(y_2)f(y_3) \qquad y_1 < y_2 < y_3 \end{aligned}$$

and zero otherwise. The argument for a sample of size n, as given by equation (6.5.2), is similar.

Example 6.5.1 Suppose that X_1, X_2, and X_3 represent a random sample of size 3 from a population with pdf

$$f(x) = 2x \qquad 0 < x < 1$$

It follows that the joint pdf of the order statistics Y_1, Y_2, and Y_3 is

$$\begin{aligned} g(y_1, y_2, y_3) &= 3!(2y_1)(2y_2)(2y_3) \\ &= 48y_1 y_2 y_3 \qquad 0 < y_1 < y_2 < y_3 < 1 \end{aligned}$$

and zero otherwise.

Quite often one may be interested in the marginal density of a single order statistic, say Y_k, and this density can be obtained in the usual fashion by integrating over the other variables. In this example, let us find the marginal pdf of the smallest order statistic, Y_1:

$$\begin{aligned} g_1(y_1) &= \int_{y_1}^{1}\int_{y_2}^{1} 48y_1 y_2 y_3 \, dy_3 \, dy_2 \\ &= 6y_1(1 - y_1^2)^2 \qquad 0 < y_1 < 1 \end{aligned}$$

If we want to know the probability that the smallest observation is below some value, say 0.1, it follows that

$$P[Y_1 < 0.1] = \int_0^{0.1} g_1(y_1)\, dy_1 = 0.030$$

It is possible to derive an explicit general formula for the distribution of the kth order statistic in terms of the pdf, $f(x)$, and CDF, $F(x)$, of the population random variable X. If X is a continuous random variable with $f(x) > 0$ on $a < x < b$ (a may be $-\infty$ and b may be ∞), then, for example, for $n = 3$,

$$g_1(y_1) = \int_{y_1}^{b} \int_{y_2}^{b} 3!f(y_1)f(y_2)f(y_3)\,dy_3\,dy_2$$

$$= 3!f(y_1) \int_{y_1}^{b} f(y_2)[F(b) - F(y_2)]\,dy_2$$

$$= -3!f(y_1) \left. \frac{[1 - F(y_2)]^2}{2} \right|_{y_1}^{b}$$

$$= 3f(y_1)[1 - F(y_1)]^2 \qquad a < y_1 < b$$

Similarly,

$$g_2(y_2) = \int_{y_2}^{b} \int_{a}^{y_2} 3!f(y_1)f(y_2)f(y_3)\,dy_1\,dy_3$$

$$= 3!f(y_2)[F(y_2) - F(a)] \int_{y_2}^{b} f(y_3)\,dy_3$$

$$= 3!f(y_2)[1 - F(y_2)]F(y_2) \qquad a < y_2 < b$$

where $F(a) = 0$ and $F(b) = 1$.

These results may be generalized to the n-dimensional case to obtain the following theorem.

Theorem 6.5.2 Suppose that $X_1, \ldots, X_n$ denotes a random sample of size n from a continuous pdf, $f(x)$, where $f(x) > 0$ for $a < x < b$. Then the pdf of the kth order statistic Y_k is given by

$$g_k(y_k) = \frac{n!}{(k-1)!(n-k)!} [F(y_k)]^{k-1}[1 - F(y_k)]^{n-k} f(y_k) \qquad \textbf{(6.5.3)}$$

if $a < y_k < b$, and zero otherwise. ■

An interesting heuristic argument can be given, based on the notion that the "likelihood" of an observation is assigned by the pdf. To have $Y_k = y_k$, one must have $k - 1$ observations less than y_k, one at y_k, and $n - k$ observations greater than y_k, where $P[X \leqslant y_k] = F(y_k)$, $P[X \geqslant y_k] = 1 - F(y_k)$, and the likelihood of an observation at y_k is $f(y_k)$. There are $n!/(k-1)!\,1!(n-k)!$ possible orderings of the n independent observations, and $g_k(y_k)$ is given by the multinomial expression (6.5.3). This is illustrated in Figure 6.6.

A similar argument can be used to easily give the joint pdf of any set of order statistics. For example, consider a pair of order statistics Y_i and Y_j where $i < j$. To

FIGURE 6.6 The kth ordered observation

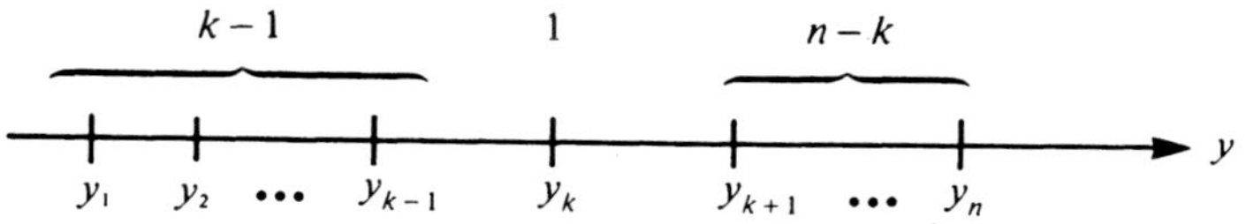

have $Y_i = y_i$ and $Y_j = y_j$, one must have $i - 1$ observations less than y_i, one at y_i, $j - i - 1$ between y_i and y_j, one at y_j, and $n - j$ greater than y_j. Applying the multinomial form gives the joint pdf for Y_i and Y_j as

$$g_{ij}(y_i, y_j) = \frac{n!}{(i-1)!(j-i-1)!(n-j)!}[F(y_i)]^{i-1}f(y_i) \times [F(y_j) - F(y_i)]^{j-i-1}[1 - F(y_j)]^{n-j}f(y_j) \tag{6.5.4}$$

if $a < y_i < y_j < b$, and zero otherwise. This is illustrated by Figure 6.7.

The smallest and largest order statistics are of special importance, as are certain functions of order statistics known as the sample median and range. If n is odd, then the **sample median** is the middle observation, Y_k where $k = (n + 1)/2$; if n is even, then it is considered to be any value between the two middle observations Y_k and Y_{k+1} where $k = n/2$, although it is often taken to be their average. The **sample range** is the difference of the smallest from the largest, $R = Y_n - Y_1$. For continuous random variables, the pdf's of the minimum and maximum, Y_1 and Y_n, which are special cases of equation (6.5.3), are

$$g_1(y_1) = n[1 - F(y_1)]^{n-1}f(y_1) \qquad a < y_1 < b \tag{6.5.5}$$

and

$$g_n(y_n) = n[F(y_n)]^{n-1}f(y_n) \qquad a < y_n < b \tag{6.5.6}$$

For *discrete* and *continuous* random variables, the CDF of the minimum or maximum of the sample can be derived directly by following the CDF technique. For the minimum.

$$\begin{aligned} G_1(y_1) &= P[Y_1 \leqslant y_1] \\ &= 1 - P[Y_1 > y_1] \\ &= 1 - P[\text{all } X_i > y_1] \\ &= 1 - [1 - F(y_1)]^n \end{aligned} \tag{6.5.7}$$

FIGURE 6.7 The ith and jth ordered observations

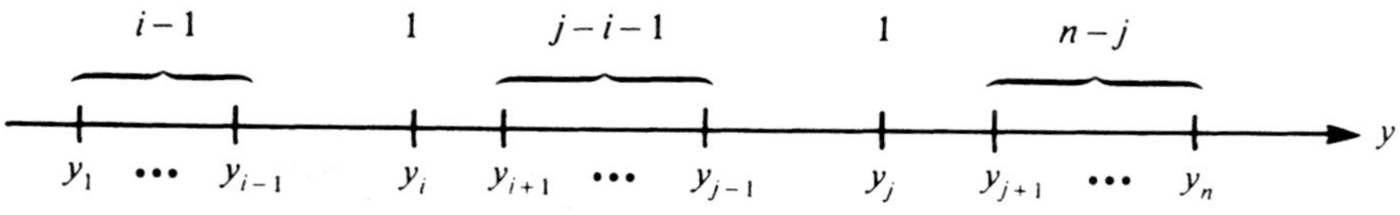

For the maximum

$$G_n(y_n) = P[Y_n \leqslant y_n] = P[\text{all } X_i \leqslant y_n] = [F(y_n)]^n \quad (6.5.8)$$

Following similar arguments, it is possible to express the CDF of the kth order statistic. In this case we have $Y_k \leqslant y_k$ if k or more X_i are at most y_k, where the number of X_i that are at most y_k follows a binomial distribution with parameters n and $p = F(y_k)$.

That is, let A_j denote the event that exactly j X_i's are less than or equal to y_k and let B denote the event that $Y_k \leqslant y_k$; then

$$B = \bigcup_{j=k}^{n} A_j$$

where the A_j are disjoint and $P(A_j) = \binom{n}{j} p^j (1-p)^{n-j}$. It follows that $P(B) = \sum_{j=k}^{n} P(A_j)$, which gives the result stated in the following theorem.

Theorem 6.5.3 For a random sample of size n from a discrete or continuous CDF, $F(x)$, the marginal CDF of the kth order statistic is given by

$$G_k(y_k) = \sum_{j=k}^{n} \binom{n}{j} [F(y_k)]^j [1 - F(y_k)]^{n-j} \quad (6.5.9)$$

■

Example 6.5.2 Consider the result of two rolls of the four-sided die in Example 2.1.1. The graph of the CDF of the maximum is shown in Figure 2.3. Although this function was obtained numerically from a table of the pdf, we can obtain an analytic expression using equation (6.5.8). Specifically, let X_1 and X_2 represent a random sample of size 2 from the discrete uniform distribution, $X_i \sim \text{DU}(4)$. The CDF of X_i is $F(x) = [x]/4$ for $1 \leqslant x \leqslant 4$, where $[x]$ is the greatest integer not exceeding x. If $Y_2 = \max(X_1, X_2)$, then $G_2(y_2) = ([y_2]/4)^2$ for $1 \leqslant y_2 \leqslant 4$, according to equation (6.5.8). The CDF of the minimum, $Y_1 = \min(X_1, X_2)$, would be given by $G_1(y_1) = 1 - (1 - [y_1]/4)^2$ for $1 \leqslant y_1 \leqslant 4$, according to equation (6.5.7).

Example 6.5.3 Consider a random sample of size n from a distribution with pdf and CDF given by $f(x) = 2x$ and $F(x) = x^2$; $0 < x < 1$. From equations (6.5.5) and (6.5.6), we have that

$$g_1(y_1) = 2ny_1(1 - y_1^2)^{n-1} \qquad 0 < y_1 < 1$$

and

$$g_n(y_n) = 2ny_n(y_n^2)^{n-1}$$
$$= 2ny_n^{2n-1} \qquad 0 < y_n < 1$$

The corresponding CDFs may be obtained by integration or directly from equations (6.5.7) and (6.5.8).

Example 6.5.4 Suppose that in Example 6.5.3 we are interested in the density of the range of the sample, $R = Y_n - Y_1$. From expression (6.5.4), we have

$$g_{1,n}(y_1, y_n) = \frac{n!}{(n-2)!}(2y_1)[y_n^2 - y_1^2]^{n-2}(2y_n) \qquad 0 < y_1 < y_n < 1$$

Making the transformation $R = Y_n - Y_1$, $S = Y_1$, yields the inverse transformation $y_1 = s$, $y_n = r + s$, and $|J| = 1$. Thus, the joint pdf of R and S is

$$h(r, s) = \frac{4n!}{(n-2)!} s(r + s)[r^2 + 2rs]^{n-2} \qquad 0 < s < 1 - r,\ 0 < r < 1$$

The regions A and B of the transformation are shown in Figure 6.8.

FIGURE 6.8 Regions corresponding to the transformation $r = y_n - y_1$ and $s = y_1$

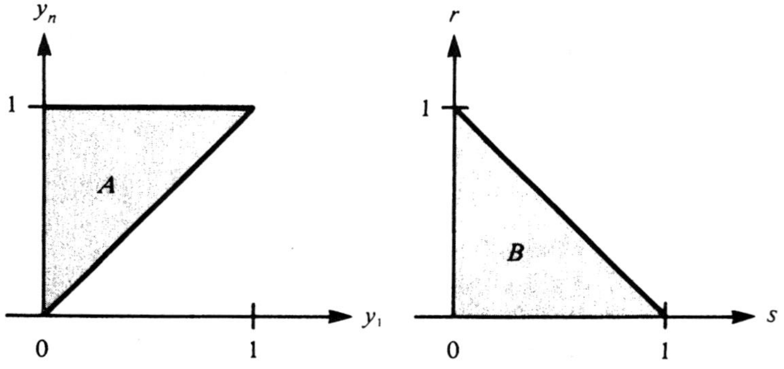

The marginal density of the range then is given by

$$h_1(r) = \int_0^{1-r} h(r, s)\, ds$$

For example, for the case $n = 2$, we have

$$h_1(r) = \int_0^{1-r} 8s(r + s)\, ds$$
$$= (4/3)(r + 2)(1 - r)^2 \qquad \textbf{(6.5.10)}$$

for $0 < r < 1$.

An interesting general expression can be obtained for the marginal CDF of R, because

$$H_1(r) = \int_{-\infty}^{r} \int_{-\infty}^{\infty} g_{1,n}(s, x + s)\, ds\, dx$$

$$= \int_{-\infty}^{\infty} \int_{-\infty}^{r} \frac{n!}{(n-2)!} f(s)[F(x+s) - F(s)]^{n-2} f(x+s)\, dx\, ds$$

$$= \int_{-\infty}^{\infty} n f(s)[F(r+s) - F(s)]^{n-1}\, ds \qquad \textbf{(6.5.11)}$$

Note, however, that great care must be taken in applying this formula to an example where the region with $f(x) > 0$ has finite limits.

Example 6.5.5 Again consider Example 6.5.4. In that case $F(s + r) = 1$ if $s > 1 - r$, so equation (6.5.11) becomes

$$H_1(r) = \int_0^{1-r} n(2s)[(r+s)^2 - s^2]^{n-1}\, ds + \int_{1-r}^{1} n(2s)[1 - s^2]^{n-1}\, ds$$

For the case $n = 2$,

$$H_1(r) = \int_0^{1-r} 4s(r^2 + 2rs)\, ds + \int_{1-r}^{1} 4s(1 - s^2)\, ds$$

$$= \frac{8r}{3} - 2r^2 + \frac{r^4}{3}$$

which is consistent with the pdf given by equation (6.5.10).

CENSORED SAMPLING

As mentioned earlier, in certain types of problems such as life-testing experiments, the ordered observations may occur naturally. In such cases a great savings in time and cost may be realized by terminating the experiment after only the first r ordered observations have occurred, rather than waiting for all n failures to occur. This usually is referred to as **Type II censored sampling**. In this case, the joint marginal density function of the first r order statistics may be obtained by integrating over the remaining variables. Censored sampling is applicable to many different types of problems, but for convenience the variable will be referred to as "time" in the following discussion.

Theorem 6.5.4 **Type II Censored Sampling** The joint marginal density function of the first r order statistics from a random sample of size n from a continuous pdf, $f(x)$, is given by

$$g(y_1, \ldots, y_r) = \frac{n!}{(n-r)!} [1 - F(y_r)]^{n-r} \prod_{i=1}^{r} f(y_i) \qquad \textbf{(6.5.12)}$$

if $-\infty < y_1 < \cdots < y_r < \infty$ and zero otherwise. ■

In Type II censored sampling the number of observations, r, is fixed but the length of the experiment, Y_r, is a random variable. If one terminates the experiment after a fixed time, t_0, this procedure is referred to as **Type I censored sampling**. In this case the number of observations, R, is a random variable. The probability that a failure occurs before time t_0 for any given trial is $p = F(t_0)$, so for a random sample of size n the random variable R follows a binomial distribution:

$$R \sim \text{BIN}(n, F(t_0)) \tag{6.5.13}$$

Type I censored sampling is related to the concept of truncated sampling and truncated distributions. Consider a random variable X with pdf $f(x)$ and CDF $F(x)$. If it is given that a random variable from this distribution has a value less than t_0, then the CDF of X given $X \leqslant t_0$ is referred to as the *truncated distribution* of X, truncated on the right at t_0, and is given by

$$F(x \mid x \leqslant t_0) = \frac{P[X \leqslant x, X \leqslant t_0]}{P[X \leqslant t_0]}$$

$$= \frac{F(x)}{F(t_0)} \qquad 0 < x < t_0 \tag{6.5.14}$$

and

$$f(x \mid x \leqslant t_0) = \frac{f(x)}{F(t_0)} \qquad 0 < x < t_0$$

Distributions truncated on the left are defined similarly.

Now, consider a random sample of size n from $f(x)$, and suppose it is given that r observations occur before the truncation point t_0; then, given $R = r$, the joint conditional density function of these values, say $x_1, \ldots, x_r$, is given by

$$h(x_1, \ldots, x_r \mid r) = \prod_{i=1}^{r} f(x_i \mid x_i \leqslant t_0)$$

$$= \frac{1}{[F(t_0)]^r} \prod_{i=1}^{r} f(x_i) \tag{6.5.15}$$

if all $x \leqslant t_0$ and zero otherwise.

Equation (6.5.15) also would be the density function of a random sample of size r, when the parent population density function is assumed to be the truncated density function $f(x)/F(t_0)$. Thus, (6.5.15) may arise either when the pdf of the population sampled originally is in the form of a truncated density or when the original population density is not truncated but the observed sample values are restricted or truncated. This restriction could result from any of several reasons, including limitations of measuring devices.

Thus, one could have a sample of size r from a truncated density or a truncated sample from a regular density. In the first case, equation (6.5.15) provides the usual density function for a random sample of size r, and in the second case it provides the conditional density function for the r observations that were given to have values less than t_0. It is interesting to note that equation (6.5.15) does not involve the original sample size n. Indeed, truncated sampling may occur in two slightly different ways. Suppose that the failure time of a unit follows the density $f(x)$, and that the unit is guaranteed for t_0 years. If a unit fails under warranty, then it is returned to a certain repair center, and the failure times of these units are recorded until r failure times are observed. The conditional density function of these r failure times then would follow equation (6.5.15), which does not depend on n, and the original number of units, n, placed in service may be known or unknown. Also note that the data would again naturally occur as ordered data and the original labeling of the original random units placed in service would be unimportant or unknown. Thus, it again would be reasonable to consider directly the joint density of the ordered observations given by

$$g(y_1, \dots, y_r \mid r) = \frac{r!}{[F(t_0)]^r} \prod_{i=1}^{r} f(y_i) \tag{6.5.16}$$

if $y_1 < \cdots < y_r < t_0$ and zero otherwise.

In a slightly different setup, one may place a known number of units, say n, in service and record failure times until time t_0. Again, as mentioned, the conditional density function still is given by equation (6.5.15) [or by equation (6.5.16) for the ordered data], but in this case additional information is available, namely that $n - r$ items survived longer than time t_0. This information is not ignored if the unconditional joint density of $Y_1, \dots, Y_R$ is considered, and this usually is a preferred approach when the sample size n is known. This situation usually is referred to as Type I censored sampling (on the right), rather than truncated sampling.

Theorem 6.5.5 **Type I Censored Sampling** If $Y_1, \dots, Y_r$ denote the observed values of a random sample of size n from $f(x)$ that is Type I censored on the right at t_0, then the joint pdf of $Y_1, \dots, Y_R$ is given by

$$f_{Y_1, \dots, Y_R}(y_1, \dots, y_r) = \frac{n!}{(n-r)!} [1 - F(t_0)]^{n-r} \prod_{i=1}^{r} f(y_i) \tag{6.5.17}$$

if $y_1 < \cdots < y_r < t_0$ and $r = 1, 2, \dots, n$, and

$$P[R = 0] = [1 - F(t_0)]^n$$

Proof

This follows by factoring the joint pdf into the product of the marginal pdf of R with the conditional pdf of $Y_1, \dots, Y_R$ given $R = r$. Specifically,

$$f_{Y_1, \dots, Y_R}(y_1, \dots, y_r) = g(y_1, \dots, y_r \mid r) b(r; n, F(t_0))$$

$$= \frac{r! \prod_{i=1}^{r} f(y_i)}{[F(t_0)]^r} \frac{n!}{r!(n-r)!} [F(t_0)]^r [1 - F(t_0)]^{n-r}$$

which simplifies to equation (6.5.17). ■

Note that the forms of equations (6.5.12) and (6.5.17) are quite similar, with t_0 replacing y_r.

As suggested earlier, we will wish to use sample data to make statistical inferences about the probability model for a given experiment. The joint density function or "likelihood function" of the sample data is the connecting link between the observed data and the mathematical model, and indeed many statistical procedures are expressed directly in terms of the likelihood function of the data. In the case of censored data, equations (6.5.12), (6.5.16), or (6.5.17) give the likelihood function or joint density function of the available ordered data, and statistical or probabilistic results must be based on these equations. Thus it is clear that the type of data available and the methods of sampling can affect the likelihood function of the observed data.

Example 6.5.6 We will assume that failure times of airplane air conditioners follow an exponential model EXP(θ). We will study properties of random variables in the next chapter that will help us characterize a distribution and interpret the physical meaning of parameters such as θ. However, for illustration purposes, suppose the manufacturer claims that an exponential distribution with $\theta = 200$ provides a good model for the failure times of such air conditioners, but the mechanics feel $\theta = 150$ provides a better model. Thirteen airplanes were placed in service, and the first 10 air conditioner failure times were as follows (Proschan, 1963):

23, 50, 50, 55, 74, 90, 97, 102, 130, 194

For Type II censored sampling, the likelihood function for the exponential distribution is given by equation (6.5.12) as

$$g(y_1, \ldots, y_r; \theta) = \frac{n!}{(n-r)!} \exp\left[-\frac{(n-r)y_r}{\theta}\right] \frac{1}{\theta^r} \exp\left[-\sum_{i=1}^{r} \frac{y_i}{\theta}\right]$$

$$= \frac{n!}{(n-r)!\,\theta^r} \exp\left[-\left(\sum_{i=1}^{r} y_i + (n-r)y_r\right)\Big/\theta\right]$$

For the above data, $r = 10$, $n = 13$, and

$$T = \sum_{i=1}^{10} y_i + (13-10)y_{10} = 1447$$

It would be interesting to compare the likelihoods of the observed data assuming $\theta = 200$ and $\theta = 150$. The ratio of the likelihoods is

$$\frac{g(y_1, \ldots, y_{10}; 200)}{g(y_1, \ldots, y_{10}; 150)} = \left(\frac{150}{200}\right)^{10} \exp\left[-1447\left(\frac{1}{200} - \frac{1}{150}\right)\right]$$

$$= 0.628$$

Thus we see that the observed data values are more likely under the assumption $\theta = 150$ than when $\theta = 200$. Based on these data, it would be reasonable to infer that the exponential model with $\theta = 150$ provides the better model. Indeed, it is possible to show that the value of θ that yields the maximum value of the likelihood is the value

$$\theta = \frac{T}{r} = \frac{1447}{10} = 144.7$$

Thus, if one wished to choose a value of θ based on these data, the value $\theta = 144.7$ seems reasonable.

For illustration purposes, suppose that Type I censoring had been used and that the experiment had been conducted for 200 flying hours for each plane to obtain the preceding data. The likelihood function now is given by equation (6.5.17):

$$f(y_1, \ldots, y_n; \theta, t_0) = \frac{n!}{(n-r)!\,\theta^r} \exp\left[-\left(\sum_{i=1}^{r} y_i + (n-r)t_0\right)\Big/\theta\right]$$

For our example, $r = 10$, $n = 13$, and $t_0 = 200$.

It is interesting that the likelihood function is maximized in this case by the value of θ given by

$$\theta = \left(\sum_{i=1}^{r} y_i + (n-r)t_0\right)\Big/r = 146.5$$

As a final illustration, suppose that a large fleet of planes is placed in service and a repair depot decides to record the failure times that occur before 200 hours. However, some units in service may be taken to a different depot for repair, so it is unknown how many units have not failed after 200 hours. That is, the sample size n is unknown. Given that r ordered observations have been recorded, the conditional likelihood is given by equation (6.5.16):

$$h(y_1, \ldots, y_r; \theta, t_0 \mid r) = \frac{r!\exp\left(-\sum_{i=1}^{r} y_i/\theta\right)}{\theta^r[1 - \exp(-t_0/\theta)]^r}$$

where $r = 10$ and $t_0 = 200$.

The value of θ that maximizes this joint pdf cannot be expressed in closed form; however, the approximate value for this case based on the given data is $\theta \doteq 245$. This value is not too close to the other values obtained, but of course the data were not actually obtained under this mode of sampling. If two different assumptions are made about the same data, then one cannot expect to always get similar results (although the Type I and Type II censoring formulas are quite similar).

SUMMARY

The main purpose of this chapter was to develop methods for deriving the distribution of a function of one or more random variables. The CDF technique is a general method that involves expressing the CDF of the "new" random variable in terms of the distribution of the "old" random variable (or variables). When one k-dimensional vector of random variables (new variables) is defined as a function of another k-dimensional vector of random variables (old variables) by means of a set of equations, transformation methods make it possible to express the joint pdf of the new random variables in terms of the joint pdf of the old random variables. The continuous case also involves multiplying by a function called the Jacobian of the transformation. A special transformation, called the probability integral transformation, and its inverse are useful in applications such as computer simulation of data.

The transformation that orders the values in a random sample from smallest to largest can be used to define the order statistics. A set of order statistics in which a specified subset is not observed is termed a censored sample. This concept is useful in applications such as life-testing of manufactured components, where it is not feasible to wait for all components to fail before analyzing the data.

EXERCISES

1. Let X be a random variable with pdf $f(x) = 4x^3$ if $0 < x < 1$ and zero otherwise. Use the cumulative (CDF) technique to determine the pdf of each of the following random variables:
   (a) $Y = X^4$.
   (b) $W = e^X$.
   (c) $Z = \ln X$.
   (d) $U = (X - 0.5)^2$.

2. Let X be a random variable that is uniformly distributed, $X \sim \text{UNIF}(0, 1)$. Use the CDF technique to determine the pdf of each of the following:
   (a) $Y = X^{1/4}$.
   (b) $W = e^{-X}$.
   (c) $Z = 1 - e^{-X}$.
   (d) $U = X(1 - X)$.

3. The measured radius of a circle, R, has pdf $f(r) = 6r(1 - r)$, $0 < r < 1$.
   (a) Find the distribution of the circumference.
   (b) Find the distribution of the area of the circle.

4. If X is Weibull distributed, $X \sim \text{WEI}(\theta, \beta)$, find both the CDF and pdf of each of the following:

(a) $Y = (X/\theta)^\beta$.

(b) $W = \ln X$.

(c) $Z = (\ln X)^2$.

5. Prove Theorem 6.3.4, assuming that the CDF $F(x)$ is a one-to-one function.

6. Let X have the pdf given in Exercise 1. Find the transformation $y = u(x)$ such that $Y = u(X) \sim \text{UNIF}(0, 1)$.

7. Let $X \sim \text{UNIF}(0, 1)$. Find transformations $y = G_1(u)$ and $w = G_2(u)$ such that

(a) $Y = G_1(U) \sim \text{EXP}(1)$.

(b) $W = G_2(U) \sim \text{BIN}(3, 1/2)$.

8. Rework Exercise 1 using transformation methods.

9. Rework Exercise 2 using transformation methods.

10. Suppose X has pdf $f(x) = (1/2) \exp(-|x|)$ for all $-\infty < x < \infty$.

(a) Find the pdf of $Y = |X|$.

(b) Let $W = 0$ if $X \leq 0$ and $W = 1$ if $X > 0$. Find the CDF of W.

11. If $X \sim \text{BIN}(n, p)$, then find the pdf of $Y = n - X$.

12. If $X \sim \text{NB}(r, p)$, then find the pdf of $Y = X - r$.

13. Let X have pdf $f(x) = x^2/24$; $-2 < x < 4$ and zero otherwise. Find the pdf of $Y = X^2$.

14. Let X and Y have joint pdf $f(x, y) = 4e^{-2(x+y)}$; $0 < x < \infty$, $0 < y < \infty$, and zero otherwise.

(a) Find the CDF of $W = X + Y$.

(b) Find the joint pdf of $U = X/Y$ and $V = X$.

(c) Find the marginal pdf of U.

15. If X_1 and X_2 denote a random sample of size 2 from a Poisson distribution, $X_i \sim \text{POI}(\lambda)$, find the pdf of $Y = X_1 + X_2$.

16. Let X_1 and X_2 denote a random sample of size 2 from a distribution with pdf $f(x) = 1/x^2$; $1 \leq x < \infty$ and zero otherwise.

(a) Find the joint pdf of $U = X_1 X_2$ and $V = X_1$.

(b) Find the marginal pdf of U.

17. Suppose that X_1 and X_2 denote a random sample of size 2 from a gamma distribution, $X_i \sim \text{GAM}(2, 1/2)$.

(a) Find the pdf of $Y = \sqrt{X_1 + X_2}$.

(b) Find the pdf of $W = X_1/X_2$.

18. Let X and Y have joint pdf $f(x, y) = e^{-y}$; $0 < x < y < \infty$ and zero otherwise.

(a) Find the joint pdf of $S = X + Y$ and $T = X$.

(b) Find the marginal pdf of T.

(c) Find the marginal pdf of S.

19. Suppose that $X_1, X_2, \ldots, X_k$ are independent random variables and let $Y_i = u_i(X_i)$ for $i = 1, 2, \ldots, k$. Show that $Y_1, Y_2, \ldots, Y_k$ are independent. Consider only the case where X_i is continuous and $y_i = u_i(x_i)$ is one-to-one. *Hint:* If $x_i = w_i(y_i)$ is the inverse transformation, then the Jacobian has the form

$$J = \prod_{i=1}^{k} \frac{d}{dy_i} w_i(y_i)$$

20. Prove Theorem 5.4.5 in the case of discrete random variables X and Y. *Hint:* Use the transformation $s = x$ and $t = g(x)y$.

21. Suppose X and Y are continuous random variables with joint pdf $f(x, y) = 2(x + y)$ if $0 < x < y < 1$ and zero otherwise.

(a) Find the joint pdf of the variables $S = X$ and $T = XY$.

(b) Find the marginal pdf of T.

22. As in Exercise 2 of Chapter 5 (page 189), assume the weight (in ounces) of a major league baseball is a random variable, and recall that a carton contains 144 baseballs. Assume now that the weights of individual baseballs are independent and normally distributed with mean $\mu = 5$ and standard deviation $\sigma = 2/5$, and let T represent the total weight of all baseballs in the carton. Find the probability that the total weight of baseballs in a carton is at most 725 ounces.

23. Suppose that $X_1, X_2, \ldots, X_k$ are independent random variables, and let $Y = X_1 + X_2 + \cdots + X_k$. If $X_i \sim \text{GEO}(p)$, then find the MGF of Y. What is the distribution of Y?

24. Let $X_1, X_2, \ldots, X_{10}$ be a random sample of size $n = 10$ from an exponential distribution with mean 2, $X_i \sim \text{EXP}(2)$.

(a) Find the MGF of the sum $Y = \sum_{i=1}^{10} X_i$.

(b) What is the pdf of Y?

25. Let X_1, X_2, X_3, and X_4 be independent random variables. Assume that X_2, X_3, and X_4 each are Poisson distributed with mean 5, and suppose that $Y = X_1 + X_2 + X_3 + X_4 \sim \text{POI}(25)$.

(a) What is the distribution of X_1?

(b) What is the distribution of $W = X_1 + X_2$?

26. Let X_1 and X_2 be independent negative binomial random variables, $X_1 \sim \text{NB}(r_1, p)$ and $X_2 \sim \text{NB}(r_2, p)$.

(a) Find the MGF of $Y = X_1 + X_2$.

(b) What is the distribution of Y?

27. Recall that $Y \sim \text{LOGN}(\mu, \sigma^2)$ if $\ln Y \sim N(\mu, \sigma^2)$. Assume that $Y_i \sim \text{LOGN}(\mu_i, \sigma_i^2)$, $i = 1, \ldots, n$ are independent. Find the distribution of:

(a) $\prod_{i=1}^{n} Y_i$.

(b) $\prod_{i=1}^{n} Y_i^a$.

(c) Y_1/Y_2.

(d) Find $E\left[\prod_{i=1}^{n} Y_i\right]$.

28. Let X_1 and X_2 be a random sample of size $n = 2$ from a continuous distribution with pdf of the form $f(x) = 2x$ if $0 < x < 1$ and zero otherwise.

(a) Find the marginal pdfs of the smallest and largest order statistics, Y_1 and Y_2.

(b) Find the joint pdf of Y_1 and Y_2.

(c) Find the pdf of the sample range $R = Y_2 - Y_1$.

29. Consider a random sample of size n from a distribution with pdf $f(x) = 1/x^2$ if $1 \leqslant x < \infty$; zero otherwise.

(a) Give the joint pdf of the order statistics.

(b) Give the pdf of the smallest order statistic, Y_1.

(c) Give the pdf of the largest order statistic, Y_n.

(d) Derive the pdf of the sample range, $R = Y_n - Y_1$, for $n = 2$.

(e) Give the pdf of the sample median, Y_r, assuming that n is odd so that $r = (n + 1)/2$.

30. Consider a random sample of size $n = 5$ from a Pareto distribution, $X_i \sim \text{PAR}(1, 2)$.

(a) Give the joint pdf of the second and fourth order statistics, Y_2 and Y_4.

(b) Give the joint pdf of the first three order statistics, Y_1, Y_2, and Y_3.

(c) Give the CDF of the sample median, Y_3.

31. Consider a random sample of size n from an exponential distribution, $X_i \sim \text{EXP}(1)$. Give the pdf of each of the following:

(a) The smallest order statistic, Y_1.

(b) The largest order statistic, Y_n.

(c) The sample range, $R = Y_n - Y_1$.

(d) The first r order statistics, $Y_1, \ldots, Y_r$.

32. A system is composed of five independent components connected in series.

(a) If the pdf of the time to failure of each component is exponential, $X_i \sim \text{EXP}(1)$, then give the pdf of the time to failure of the system.

(b) Repeat (a), but assume that the components are connected in parallel.

(c) Suppose that the five-component system fails when at least three components fail. Give the pdf of the time to failure of the system.

(d) Suppose that n independent components are not distributed identically, but rather $X_i \sim \text{EXP}(\theta_i)$. Give the pdf of the time to failure of a series system in this case.

33. Consider a random sample of size n from a geometric distribution, $X_i \sim \text{GEO}(p)$. Give the CDF of each of the following:

(a) The minimum, Y_1.

(b) The kth smallest, Y_k.

(c) The maximum, Y_n.

(d) Find $P[Y_1 \leqslant 1]$.

34. Suppose X_1 and X_2 are continuous random variables with joint pdf $f(x_1, x_2)$. Prove Theorem 5.2.1 assuming the transformation $y_1 = u(x_1, x_2)$, $y_2 = x_2$ is one-to-one. *Hint:* First derive the marginal pdf of $Y_1 = u(X_1, X_2)$ and show that $E(Y_1) = \int y_1 f_{Y_1}(y_1)\, dy_1 = \iint u(x_1, x_2) f(x_1, x_2)\, dx_1\, dx_2$. Use a similar proof in the case of discrete random variables. Notice that proofs for the cases of k variables and transformations that are not one-to-one are similar but more complicated.

35. Suppose X_1, X_2 are independent exponentially distributed random variables, $X_i \sim \text{EXP}(\theta)$, and let $Y = X_1 - X_2$.

(a) Find the MGF of Y.

(b) What is the distribution of Y?

36. Show that if $X_1, \ldots, X_n$ are independent random variables with FMGFs $G_1(t), \ldots, G_n(t)$, and $Y = X_1 + \cdots + X_n$, respectively, then the FMGF of Y is $G_Y(t) = G_1(t) \cdots G_n(t)$.

CHAPTER 7

LIMITING DISTRIBUTIONS

7.1 INTRODUCTION

In Chapter 6, general methods were discussed for deriving the distribution of a function of n random variables, say $Y_n = u(X_1, \ldots, X_n)$. In some cases, the pdf of Y_n is obtained easily, but there are many important cases where the derivation is not tractable. In many of these, it is possible to obtain useful approximate results that apply when n is large. These results are based on the notions of convergence in distribution and limiting distribution.

7.2

SEQUENCES OF RANDOM VARIABLES

Consider a sequence of random variables $Y_1, Y_2, \ldots$ with a corresponding sequence of CDFs $G_1(y), G_2(y), \ldots$ so that for each $n = 1, 2, \ldots$

$$G_n(y) = P[Y_n \leqslant y] \tag{7.2.1}$$

Definition 7.2.1

If $Y_n \sim G_n(y)$ for each $n = 1, 2, \ldots$, and if for some CDF $G(y)$,

$$\lim_{n\to\infty} G_n(y) = G(y) \tag{7.2.2}$$

for all values y at which $G(y)$ is continuous, then the sequence $Y_1, Y_2, \ldots$ is said to **converge in distribution** to $Y \sim G(y)$, denoted by $Y_n \xrightarrow{d} Y$. The distribution corresponding to the CDF $G(y)$ is called the **limiting distribution** of Y_n.

Example 7.2.1 Let $X_1, \ldots, X_n$ be a random sample from a uniform distribution, $X_i \sim \text{UNIF}(0, 1)$, and let $Y_n = X_{n:n}$, the largest order statistic. From the results of Chapter 6, it follows that the CDF of Y_n is

$$G_n(y) = y^n \qquad 0 < y < 1 \tag{7.2.3}$$

zero if $y \leqslant 0$ and one if $y \geqslant 1$. Of course, when $0 < y < 1$, y^n approaches 0 as n approaches ∞, and when $y \leqslant 0$ or $y \geqslant 1$, $G_n(y)$ is a sequence of constants, with

FIGURE 7.1 Comparison of CDFs $G_n(y)$ with limiting degenerate CDF $G(y)$

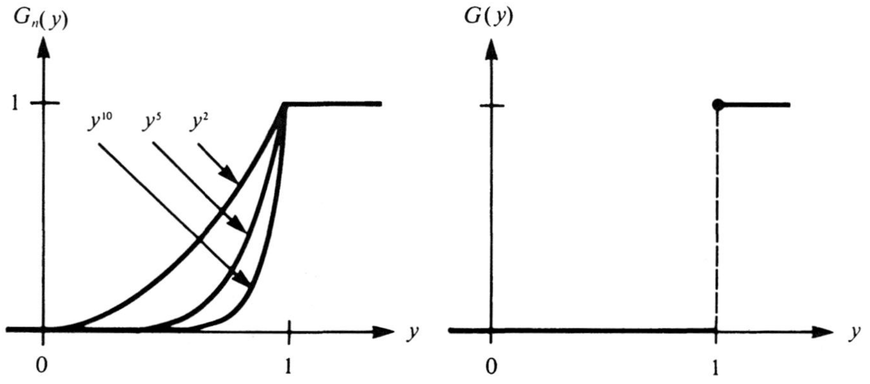

respective limits 0 or 1. Thus, $\lim_{n \to \infty} G_n(y) = G(y)$ where

$$G(y) = \begin{cases} 0 & y < 1 \\ 1 & y \geqslant 1 \end{cases} \tag{7.2.4}$$

This situation is illustrated in Figure 7.1, which shows $G(y)$ and $G_n(y)$ for $n = 2$, 5, and 10.

The function defined by equation (7.2.4) is the CDF of a random variable that is concentrated at one value, $y = 1$. Such distributions occur often as limiting distributions.

Definition 7.2.2

The function $G(y)$ is the CDF of a **degenerate distribution** at the value $y = c$ if

$$G(y) = \begin{cases} 0 & y < c \\ 1 & y \geqslant c \end{cases} \tag{7.2.5}$$

In other words, $G(y)$ is the CDF of a discrete distribution that assigns probability one at the value $y = c$ and zero otherwise.

Example 7.2.2 Let $X_1, X_2, \ldots, X_n$ be a random sample from an exponential distribution, $X_i \sim \text{EXP}(\theta)$, and let $Y_n = X_{1:n}$ be the smallest order statistic. It follows that the CDF of Y_n is

$$G_n(y) = 1 - e^{-ny/\theta} \qquad y > 0 \tag{7.2.6}$$

and zero otherwise. We have $\lim_{n \to \infty} G_n(y) = 1$ if $y > 0$ because $e^{-y/\theta} < 1$ in this case. Thus, the limit is zero if $y < 0$ and one if $y > 0$, which corresponds to a degenerate distribution at the value $y = 0$. Notice that the limit at $y = 0$ is zero, which means that the limiting function is not only discontinuous at $y = 0$ but also not even continuous from the right at $y = 0$, which is a requirement of a CDF. This is not a problem, because Definition 7.2.1 requires only that the limiting function agrees with a CDF at its points of continuity.

Definition 7.2.3

A sequence of random variables, $Y_1, Y_2, \ldots$, is said to **converge stochastically** to a constant c if it has a limiting distribution that is degenerate at $y = c$.

An alternative formulation of stochastic convergence will be considered in Section 7.6, and a more general concept called convergence in probability will be discussed in Section 7.7.

Not all limiting distributions are degenerate, as seen in the next example. The following limits are useful in many problems:

$$\lim_{n\to\infty}\left(1+\frac{c}{n}\right)^{nb} = e^{cb} \tag{7.2.7}$$

$$\lim_{n\to\infty}\left[1+\frac{c}{n}+\frac{d(n)}{n}\right]^{nb} = e^{cb} \quad \text{if } \lim_{n\to\infty} d(n) = 0 \tag{7.2.8}$$

These are obtained easily from expansions involving the natural logarithm. For example, limit (7.2.7) follows from the expansion $nb \ln(1 + c/n) = nb(c/n + \cdots) = cb + \cdots$, where the rest of the terms approach zero as $n \to \infty$.

Example 7.2.3 Suppose that $X_1, \ldots, X_n$ is a random sample from a Pareto distribution, $X_i \sim \text{PAR}(1, 1)$, and let $Y_n = nX_{1:n}$. The CDF of X_i is $F(x) = 1 - (1 + x)^{-1}$; $x > 0$, so the CDF of Y_n is

$$G_n(y) = 1 - \left(1 + \frac{y}{n}\right)^{-n} \qquad y > 0 \tag{7.2.9}$$

Using limit (7.2.7), we obtain the limit $G(y) = 1 - e^{-y}$; $y > 0$ and zero otherwise, which is the CDF of an exponential distribution, EXP(1). This is illustrated in Figure 7.2, which shows the graphs of $G(y)$ and $G_n(y)$ for $n = 1$, 2, and 5.

FIGURE 7.2 Comparison of CDFs $G_n(y)$ with limiting CDF $G(y)$

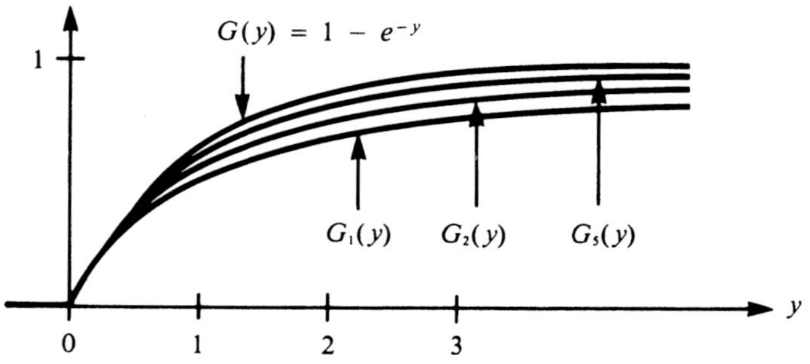

The following example shows that a sequence of random variables need not have a limiting distribution.

Example 7.2.4 For the random sample of the previous example, let us consider the largest order statistic, $Y_n = X_{n:n}$. The CDF of Y_n is

$$G_n(y) = \left(\frac{y}{1 + y}\right)^n \qquad y > 0 \tag{7.2.10}$$

and zero otherwise. Because $y/(1 + y) < 1$, we have $\lim_{n \to \infty} G_n(y) = G(y) = 0$ for all y, which is not a CDF because it does not approach one as $y \to \infty$.

Example 7.2.5 In the previous example, suppose that we consider a rescaled variable, $Y_n = (1/n)X_{n:n}$, which has CDF

$$G_n(y) = \left(1 + \frac{1}{ny}\right)^{-n} \qquad y > 0 \tag{7.2.11}$$

and zero otherwise. Using limit (7.2.7), we obtain the CDF $G(y) = e^{-1/y}$; $y > 0$.

Example 7.2.6 For the random sample of Example 7.2.2, consider the modified sequence $Y_n = (1/\theta)X_{n:n} - \ln n$. The CDF is

$$G_n(y) = \left[1 - \left(\frac{1}{n}\right)e^{-y}\right]^n \qquad y > -\ln n \tag{7.2.12}$$

and zero otherwise. Following from limit (7.2.7), the limiting CDF is $G(y) = \exp(-e^{-y})$; $-\infty < y < \infty$.

We now illustrate the accuracy when this limiting CDF is used as an approximation to $G_n(y)$ for large n. Suppose that the lifetime in months of a certain type of component is a random variable $X \sim \text{EXP}(1)$, and suppose that 10 independent components are connected in a parallel system. The time to failure of the system is $T = X_{10:10}$, and the CDF is $F_T(t) = (1 - e^{-t})^{10}$; $t > 0$. This CDF is evaluated at $t = 1$, 2, 5, and 7 months in the table at the top of page 236. To approximate these probabilities with the limiting distribution, then

$$\begin{aligned} F_T(t) &= P[T \leqslant t] \\ &= P[Y_{10} + \ln 10 \leqslant t] \\ &\doteq G(t - \ln 10) \\ &= \exp(-e^{-(t - \ln 10)}) \\ &= \exp(-10e^{-t}) \end{aligned}$$

The approximate probabilities are given in the table for comparison

t:	1	2	5	7
$F_T(t)$:	0.010	0.234	0.935	0.9909
$G(t - \ln 10)$:	0.025	0.258	0.935	0.9909

The approximation should improve as n increases.

Example 7.2.7 Consider the sample mean of a random sample from a normal distribution, $N(\mu, \sigma^2)$, $Y_n = \bar{X}_n$. From the results of the previous chapter, $Y_n \sim N(\mu, \sigma^2/n)$, and

$$G_n(y) = \Phi\left[\frac{\sqrt{n}(y - \mu)}{\sigma}\right] \tag{7.2.13}$$

The limiting CDF is degenerate at $y = \mu$, because $\lim_{n \to \infty} G_n(y) = 0$ if $y < \mu$, 1/2 if $y = \mu$, and 1 if $y > \mu$, so that the sample mean converges stochastically to μ. We will show this in a more general setting in a later section.

Certain limiting distributions are easier to derive by using moment generating functions.

7.3

THE CENTRAL LIMIT THEOREM

In the previous examples, the exact CDF was known for each finite n, and the limiting distribution was obtained directly from this sequence. One advantage of limiting distributions is that it often may be possible to determine the limiting distribution without knowing the exact form of the CDF for finite n. The limiting distribution then may provide a useful approximation when the exact probabilities are not available. One method of accomplishing this result is to make use of MGFs. The following theorem is stated without proof.

Theorem 7.3.1 Let $Y_1, Y_2, \ldots$ be a sequence of random variables with respective CDFs $G_1(y)$, $G_2(y), \ldots$ and MGFs $M_1(t), M_2(t), \ldots$. If $M(t)$ is the MGF of a CDF $G(y)$, and if $\lim_{n \to \infty} M_n(t) = M(t)$ for all t in an open interval containing zero, $-h < t < h$, then $\lim_{n \to \infty} G_n(y) = G(y)$ for all continuity points of $G(y)$. ■

Example 7.3.1 Let $X_1, \ldots, X_n$ be a random sample from a Bernoulli distribution, $X_i \sim \text{BIN}(1, p)$, and consider $Y_n = \sum_{i=1}^{n} X_i$. If we let $p \to 0$ as $n \to \infty$ in such a way that $np = \mu$, for fixed $\mu > 0$, then

$$M_n(t) = (pe^t + q)^n = \left(\frac{\mu e^t}{n} + 1 - \frac{\mu}{n}\right)^n = \left[1 + \frac{\mu(e^t - 1)}{n}\right]^n \tag{7.3.1}$$

and from limit (7.2.7) we have

$$\lim_{n \to \infty} M_n(t) = e^{\mu(e^t - 1)} \tag{7.3.2}$$

which is the MGF of the Poisson distribution with mean μ. This is consistent with the result of Theorem 3.2.3 and is somewhat easier to verify. We conclude that $Y_n \xrightarrow{d} Y \sim \text{POI}(\mu)$.

Example 7.3.2 **Bernoulli Law of Large Numbers** Suppose now that we keep p fixed and consider the sequence of sample proportions, $W_n = \hat{p}_n = Y_n/n$. By using the series expansion $e^u = 1 + u + u^2/2 + \cdots$ with $u = t/n$, we obtain

$$M_{W_n}(t) = (pe^{t/n} + q)^n = \left[p\left(1 + \frac{t}{n} + \cdots\right) + q\right]^n = \left[1 + \frac{pt}{n} + \frac{d(n)}{n}\right]^n \tag{7.3.3}$$

where $d(n)/n$ involves the disregarded terms of the series expansion, and $d(n) \to 0$ as $n \to \infty$. From limit (7.2.8) we have

$$\lim_{n \to \infty} M_{W_n}(t) = e^{pt} \tag{7.3.4}$$

which is the MGF of a degenerate distribution at $y = p$, and thus $\hat{p}_n$ converges stochastically to p as n approaches infinity.

Note that this example provides an approach to answering the question that was raised in Chapter 1 about statistical regularity. If, in a sequence of M independent trials of an experiment, Y_M represents the number of occurrences of an event A, then $f_A = Y_M/M$ is the relative frequency of occurrence of A. Because the

Bernoulli parameter has the value $p = P(A)$ in this case, it follows that f_A converges stochastically to $P(A)$ as $M \to \infty$. For example, if a coin is tossed repeatedly, and $A = \{H\}$, then the successive relative frequencies of A correspond to a sequence of random variables that will converge stochastically to $p = 1/2$ for an unbiased coin. Even though different sequences of tosses generally produce different observed numerical sequences of f_A, in the long run they all tend to stabilize near 1/2.

Example 7.3.3 Now we consider the sequence of "standardized" variables:

$$Z_n = \frac{Y_n - np}{\sqrt{npq}} \tag{7.3.5}$$

With the simplified notation $\sigma_n = \sqrt{npq}$, we have $Z_n = Y_n/\sigma_n - np/\sigma_n$. Using the series expansion of the previous example,

$$\begin{aligned} M_{Z_n}(t) &= e^{-npt/\sigma_n}(pe^{t/\sigma_n} + q)^n \\ &= [e^{-pt/\sigma_n}(pe^{t/\sigma_n} + q)]^n \\ &= \left[\left(1 - \frac{pt}{\sigma_n} + \frac{p^2t^2}{2\sigma_n^2} + \cdots\right)\left(1 + \frac{pt}{\sigma_n} + \frac{pt^2}{2\sigma_n^2} + \cdots\right)\right]^n \\ &= \left[1 + \frac{t^2}{2n} + \frac{d(n)}{n}\right]^n \end{aligned} \tag{7.3.6}$$

where $d(n) \to 0$ as $n \to \infty$. Thus,

$$\lim_{n \to \infty} M_{Z_n}(t) = e^{t^2/2} \tag{7.3.7}$$

which is the MGF of the standard normal distribution, and so $Z_n \xrightarrow{d} Z \sim \text{N}(0, 1)$. This is an example of a special limiting result known as the Central Limit Theorem.

Theorem 7.3.2 **Central Limit Theorem (CLT)** If $X_1, \ldots, X_n$ is a random sample from a distribution with mean μ and variance $\sigma^2 < \infty$, then the limiting distribution of

$$Z_n = \frac{\sum_{i=1}^{n} X_i - n\mu}{\sqrt{n}\,\sigma} \tag{7.3.8}$$

is the standard normal, $Z_n \xrightarrow{d} Z \sim \text{N}(0, 1)$ as $n \to \infty$.

Proof

This limiting result holds for random samples from any distribution with finite mean and variance, but the proof will be outlined under the stronger assumption that the MGF of the distribution exists. The proof can be modified for the more general case by using a more general concept called a characteristic function, which we will not consider here.

Let $m(t)$ denote the MGF of $X - \mu$, $m(t) = M_{X-\mu}(t)$, and note that $m(0) = 1$, $m'(0) = E(X - \mu) = 0$, and $m''(0) = E(X - \mu)^2 = \sigma^2$. Expanding $m(t)$ by the Taylor series formula about 0 gives, for ξ between 0 and t,

$$\begin{aligned} m(t) &= m(0) + m'(0)t + \frac{m''(\xi)t^2}{2} \\ &= 1 + \frac{m''(\xi)t^2}{2} \\ &= 1 + \frac{\sigma^2 t^2}{2} + \frac{(m''(\xi) - \sigma^2)t^2}{2} \end{aligned} \tag{7.3.9}$$

by adding and subtracting $\sigma^2 t^2/2$.

Now we may write

$$Z_n = \frac{\sum (X_i - \mu)}{\sqrt{n}\,\sigma}$$

and

$$\begin{aligned} M_{Z_n}(t) &= M_{\Sigma(X_i - \mu)}\left(\frac{t}{\sqrt{n}\,\sigma}\right) \\ &= \prod_{i=1}^{n} M_{X_i - \mu}\left(\frac{t}{\sqrt{n}\,\sigma}\right) \\ &= \left[M_{X-\mu}\left(\frac{t}{\sqrt{n}\,\sigma}\right)\right]^n \\ &= \left[m\left(\frac{t}{\sqrt{n}\,\sigma}\right)\right]^n \\ &= \left[1 + \frac{\sigma^2 t^2}{2n\sigma^2} + \frac{(m''(\xi) - \sigma^2)t^2}{2n\sigma^2}\right]^n \qquad 0 < |\xi| < \frac{|t|}{\sqrt{n}\,\sigma} \end{aligned}$$

As $n \to \infty$ $t/\sqrt{n}\,\sigma \to 0$, $\xi \to 0$, and $m''(\xi) - \sigma^2 \to 0$, so

$$M_{Z_n}(t) = \left[1 + \frac{t^2}{2n} + \frac{d(n)}{n}\right]^n \tag{7.3.10}$$

where $d(n) \to 0$ as $n \to \infty$. It follows that

$$\lim_{n \to \infty} M_{Z_n}(t) = e^{t^2/2} \tag{7.3.11}$$

or

$$\lim_{n \to \infty} F_{Z_n}(z) = \Phi(z) \tag{7.3.12}$$

which means that $Z_n \xrightarrow{d} Z \sim N(0, 1)$. ∎

Note that the variable in limit (7.3.8) also can be related to the sample mean,

$$Z_n = \frac{\bar{X}_n - \mu}{\sigma/\sqrt{n}} \tag{7.3.13}$$

The major application of the CLT is to provide an approximate distribution in cases where the exact distribution is unknown or intractable.

Example 7.3.4 Let $X_1, \ldots, X_n$ be a random sample from a uniform distribution, $X_i \sim \text{UNIF}(0, 1)$, and let $Y_n = \sum_{i=1}^{n} X_i$. Because $E(X_i) = 1/2$ and $\text{Var}(X_i) = 1/12$, we have the approximation

$$Y_n \sim N\left(\frac{n}{2}, \frac{n}{12}\right)$$

For example, if $n = 12$, then approximately

$$Y_{12} - 6 \sim N(0, 1)$$

This approximation is so close that it often is used to simulate standard normal random numbers in computer applications. Of course, this requires 12 uniform random numbers to be generated to obtain one normal random number.

7.4 APPROXIMATIONS FOR THE BINOMIAL DISTRIBUTION

Examples 7.3.1 through 7.3.3 demonstrated that various limiting distributions apply, depending on how the sequence of binomial variables is standardized and also on assumptions about the behavior of p as $n \to \infty$.

Example 7.3.1 suggests that for a binomial variable $Y_n \sim \text{BIN}(n, p)$, if n is large and p is small, then approximately $Y_n \sim \text{POI}(np)$. This was discussed in a different context and an illustration was given in Example 3.2.9 of Chapter 3.

Example 7.3.3 considered a fixed value of p, and a suitably standardized sequence was found to have a standard normal distribution, suggesting a normal approximation. In particular, it suggests that for large n and fixed p, approximately $Y_n \sim N(np, npq)$. This approximation works best when p is close to 0.5, because the binomial distribution is symmetric when $p = 0.5$. The accuracy required in any approximation depends on the application. One guideline is to use the normal approximation when $np \geqslant 5$ and $nq \geqslant 5$, but again this would depend on the accuracy required.

Example 7.4.1 The probability that a basketball player hits a shot is $p = 0.5$. If he takes 20 shots, what is the probability that he hits at least nine? The exact probability is

$$P[Y_{20} \geqslant 9] = 1 - P[Y_{20} \leqslant 8]$$
$$= 1 - \sum_{y=0}^{8} \binom{20}{y} 0.5^y 0.5^{20-y} = 0.7483$$

A normal approximation is

$$P[Y_{20} \geqslant 9] = 1 - P[Y_{20} \leqslant 8] \doteq 1 - \Phi\left(\frac{8 - 10}{\sqrt{5}}\right)$$
$$= 1 - \Phi(-0.89) = 0.8133$$

Because the binomial distribution is discrete and the normal distribution is continuous, the approximation can be improved by making a **continuity correction**. In particular, each binomial probability, $b(y; n, p)$, has the same value as the area of a rectangle of height $b(y; n, p)$ and with the interval $[y - 0.5, y + 0.5]$ as its base, because the length of the base is one unit. The area of this rectangle can be approximated by the area under the pdf of $Y \sim N(np, npq)$, which corresponds to fitting a normal distribution with the same mean and variance as $Y_n \sim BIN(n, p)$. This is illustrated for the case of $n = 20$, $p = 0.5$, and $y = 7$ in Figure 7.3, where the exact probability is $b(7;\ 20,\ 0.5) = \binom{20}{7}(0.5)^7(0.5)^{13}$ $= 0.0739$. The approximation, which is the shaded area in the figure, is

FIGURE 7.3 Continuity correction for normal approximation of a binomial probability

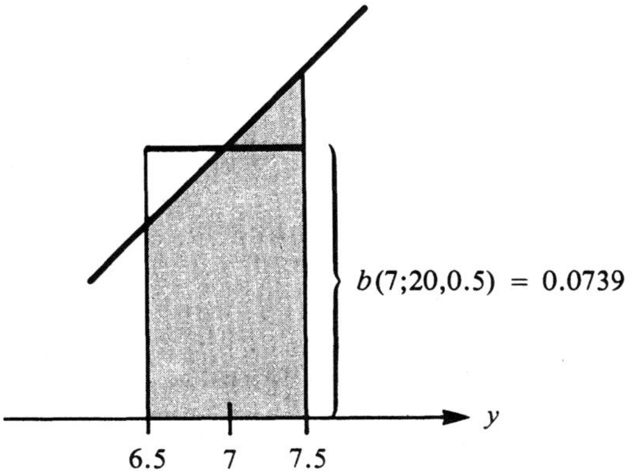

$$\Phi\left(\frac{7.5 - 10}{\sqrt{5}}\right) - \Phi\left(\frac{6.5 - 10}{\sqrt{5}}\right) = \Phi(-1.12) - \Phi(-1.57) = 0.0732$$

The same idea can be used with other binomial probabilities, such as

$$\begin{aligned} P[Y_{20} \geqslant 9] &= 1 - P[Y_{20} \leqslant 8] \\ &\doteq 1 - \Phi\left(\frac{8.5 - 10}{\sqrt{5}}\right) \\ &= 1 - \Phi(-0.67) \\ &= 0.7486 \end{aligned}$$

which is much closer to the exact value than without the continuity correction. The situation is shown in Figure 7.4.

In general, if $Y_n \sim \text{BIN}(n, p)$ and $a \leqslant b$ are integers, then

$$P[a \leqslant Y_n \leqslant b] \doteq \Phi\left(\frac{b + 0.5 - np}{\sqrt{npq}}\right) - \Phi\left(\frac{a - 0.5 - np}{\sqrt{npq}}\right) \tag{7.4.1}$$

Continuity corrections also are useful with other discrete distributions that can be approximated by the normal distribution.

FIGURE 7.4 The normal approximation for a binomial distribution

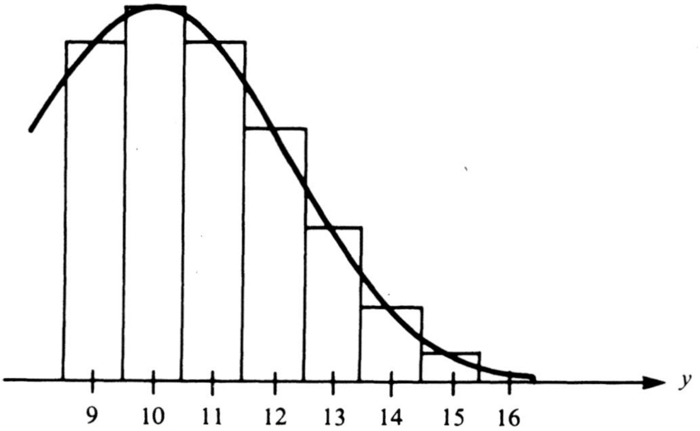

Example 7.4.2 Suppose that $Y_n \sim \text{POI}(n)$, where n is a positive integer. From the results of Chapter 6, we know that Y_n has the same distribution as a sum $\sum_{i=1}^{n} X_i$, where $X_1, \ldots, X_n$ are independent, $X_i \sim \text{POI}(1)$. According to the CLT, $Z_n = (Y_n - n)/\sqrt{n} \xrightarrow{d} Z \sim \text{N}(0, 1)$, which suggests the approximation $Y_n \sim \text{N}(n, n)$ for large n. For example, $n = 20$, we desire to find $P[10 \leqslant Y_{20} \leqslant 30]$. The exact value is $\sum_{y=10}^{30} e^{-20}(20)^y/y! = 0.982$, and the approximate value is

$$\Phi\left(\frac{30.5 - 20}{\sqrt{20}}\right) - \Phi\left(\frac{9.5 - 20}{\sqrt{20}}\right) = \Phi(2.35) - \Phi(-2.35) = 0.981$$

which is quite close to the exact value.

7.5 ASYMPTOTIC NORMAL DISTRIBUTIONS

From the CLT it follows that when the sample mean is standardized according to equation (7.3.13), the corresponding sequence $Z_n \xrightarrow{d} Z \sim \text{N}(0, 1)$.

It would not be unreasonable to consider the distribution of the sample mean $\bar{X}_n$ as approximately N $(\mu, \sigma^2/n)$ for large n. This is an example of a more general notion.

Definition 7.5.1

If $Y_1, Y_2, \ldots$ is a sequence of random variables and m and c are constants such that

$$Z_n = \frac{Y_n - m}{c/\sqrt{n}} \xrightarrow{d} Z \sim \text{N}(0, 1) \tag{7.5.1}$$

as $n \to \infty$, then Y_n is said to have an **asymptotic normal distribution** with **asymptotic mean** m and **asymptotic variance** c^2/n.

Example 7.5.1 Consider the random sample of Example 4.6.3, which involved $n = 40$ lifetimes of electrical parts, $X_i \sim \text{EXP}(100)$. By the CLT, $\bar{X}_n$ has an asymptotic normal distribution with mean $m = 100$ and variance $c^2/n = (100)^2/40 = 250$.

ASYMPTOTIC DISTRIBUTION OF CENTRAL ORDER STATISTICS

In Section 7.1 we showed several examples that involved extreme order statistics, such as the largest and smallest, with limiting distributions that were not normal. Under certain conditions, it is possible to show that "central" order statistics are asymptotically normal.

Theorem 7.5.1 Let $X_1, \ldots, X_n$ be a random sample from a continuous distribution with a pdf $f(x)$ that is continuous and nonzero at the pth percentile, x_p, for $0 < p < 1$. If $k/n \to p$ (with $k - np$ bounded), then the sequence of kth order statistics, $X_{k:n}$, is asymptotically normal with mean x_p and variance c^2/n, where

$$c^2 = \frac{p(1-p)}{[f(x_p)]^2} \tag{7.5.2}$$

■

Example 7.5.2 Let $X_1, \ldots, X_n$ be a random sample from an exponential distribution, $X_i \sim \text{EXP}(1)$, so that $f(x) = e^{-x}$ and $F(x) = 1 - e^{-x}$; $x > 0$. For odd n, let $k = (n+1)/2$, so that $Y_k = X_{k:n}$ is the sample median. If $p = 0.5$, then the median is $x_{0.5} = -\ln(0.5) = \ln 2$ and

$$c^2 = \frac{0.5(1-0.5)}{[f(\ln 2)]^2} = \frac{0.25}{(0.5)^2} = 1$$

Thus, $X_{k:n}$ is asymptotically normal with asymptotic mean $x_{0.5} = \ln 2$ and asymptotic variance $c^2/n = 1/n$.

Example 7.5.3 Suppose that $X_1, \ldots, X_n$ is a random sample from a uniform distribution, $X_i \sim \text{UNIF}(0, 1)$, so that $f(x) = 1$ and $F(x) = x$; $0 < x < 1$. Also assume that n is odd and $k = (n+1)/2$, so that $Y_k = X_{k:n}$ is the middle order statistic or sample median. Formula (6.5.3) gives the pdf of Y_k, which has a special form because $k - 1 = n - k = (n-1)/2$ in this example. The pdf is

$$g_n(y) = \frac{n!}{\{[(n-1)/2]!\}^2} [y(1-y)]^{(n-1)/2} \qquad 0 < y < 1 \tag{7.5.3}$$

According to the theorem, with $p = 0.5$, the pth percentile is $x_{0.5} = 0.5$ and $c^2 = 0.5(1-0.5)/[1]^2 = 0.25$, so that $Z_n = \sqrt{n}(X_{k:n} - 0.5)/0.5 \xrightarrow{d} Z \sim N(0, 1)$. Actually, this is strongly suggested by the pdf (7.5.3) after the transformation $z = \sqrt{n}(y - 0.5)/0.5$, which has inverse transformation $y = 0.5 + 0.5z/\sqrt{n}$ and Jacobian $J = 1/\sqrt{n}$. The resulting pdf is

$$f_n(z) = \frac{n!(0.5)^{n-1}}{\sqrt{n}\{[(n-1)/2]!\}^2}\left(1 - \frac{z^2}{n}\right)^{(n-1)/2} \qquad |z| < \sqrt{n} \tag{7.5.4}$$

It follows from limit (7.2.7), and the fact that $(1 - z^2/n)^{-1/2} \to 1$, that

$$\lim_{n\to\infty}\left(1 - \frac{z^2}{n}\right)^{(n-1)/2} = e^{-z^2/2}$$

and it is also possible to show that the constant in (7.5.4) approaches $1/\sqrt{2\pi}$ as $n \to \infty$.

Thus, in the example, the sequence of pdf's corresponding to Z_n converges to a standard normal pdf. It is not obvious that this will imply that the CDFs also converge, but this can be proved. However, we will not pursue this point.

7.6

PROPERTIES OF STOCHASTIC CONVERGENCE

We encountered several examples in which a sequence of random variables converged stochastically to a constant. For instance, in Example 7.3.2 we discovered that the sample proportion converges stochastically to the population proportion. Clearly, this is a useful general concept for evaluating estimators of unknown population parameters, and it would be reasonable to require that a good estimator should have the property that it converges stochastically to the parameter value as the sample size approaches infinity.

The following theorem, stated without proof, provides an alternate criterion for showing stochastic convergence.

Theorem 7.6.1 The sequence $Y_1, Y_2, \ldots$ converges stochastically to c if and only if for every $\varepsilon > 0$,

$$\lim_{n\to\infty} P[|Y_n - c| < \varepsilon] = 1 \quad \textbf{(7.6.1)}$$

■

A sequence of random variables that satisfies Theorem (7.6.1) is also said to **converge in probability** to the constant c, denoted by $Y_n \xrightarrow{P} c$. The notion of convergence in probability will be discussed in a more general context in the next section.

Example 7.6.1 Example 7.3.2 verified the so-called Bernoulli Law of Large Numbers with the MGF approach. It also can be verified with the previous theorem and the Chebychev inequality. Specifically, the mean and variance of $\hat{p}_n$ are $E(\hat{p}_n) = p$ and

$\text{Var}(\hat{p}_n) = pq/n$, so that

$$P[|\hat{p}_n - p| < \varepsilon] \geqslant 1 - \frac{pq}{\varepsilon^2 n} \tag{7.6.2}$$

for any $\varepsilon > 0$, so $\lim_{n \to \infty} P[|\hat{p}_n - p| < \varepsilon] = 1$.

This same approach can be used to prove a more general result, usually referred to as the **Law of Large Numbers** (LLN).

Theorem 7.6.2 If $X_1, \ldots, X_n$ is a random sample from a distribution with finite mean μ and variance σ^2, then the sequence of sample means converges in probability to μ,

$\bar{X}_n \xrightarrow{P} \mu.$

Proof

This follows from the fact that $E(\bar{X}_n) = \mu$, $\text{Var}(\bar{X}_n) = \sigma^2/n$, and thus

$$P[|\bar{X}_n - \mu| < \varepsilon] \geqslant 1 - \frac{\sigma^2}{\varepsilon^2 n} \tag{7.6.3}$$

so that $\lim_{n \to \infty} P[|\bar{X}_n - \mu| < \varepsilon] = 1$. ■

These results further illustrate that the sample mean provides a good estimate of the population mean, in the sense that the probability approaches 1 that $\bar{X}_n$ is arbitrarily close to μ as $n \to \infty$.

Actually, the right side of inequality (7.6.3) provides additional information. Namely, for any $\varepsilon > 0$ and $0 < \delta < 1$, if $n > \sigma^2/(\varepsilon^2 \delta)$, then

$$P[\mu - \varepsilon < \bar{X}_n < \mu + \varepsilon] \geqslant 1 - \delta$$

The following theorem, which is stated without proof, asserts that a sequence of asymptotically normal variables converges in probability to the asymptotic mean.

Theorem 7.6.3 If $Z_n = \sqrt{n}(Y_n - m)/c \xrightarrow{d} Z \sim \text{N}(0, 1)$, then $Y_n \xrightarrow{P} m$. ■

Example 7.6.2 We found in Examples 7.5.2 and 7.5.3 that the sample median $X_{k:n}$ is asymptotically normal with asymptotic mean $x_{0.5}$, the distribution median. It follows from the theorem that $X_{k:n} \xrightarrow{P} x_{0.5}$ as $n \to \infty$, with $k/n \to 0.5$.

Similarly, under the conditions of Theorem 7.5.1, it follows that if $k/n \to p$, then the kth smallest order statistic converges stochastically to the pth percentile, $X_{k:n} \xrightarrow{P} x_p$.

7.7

ADDITIONAL LIMIT THEOREMS

Definition 7.7.1

Convergence in Probability The sequence of random variables Y_n is said to **converge in probability** to Y, written $Y_n \xrightarrow{P} Y$, if

$$\lim_{n \to \infty} P[|Y_n - Y| < \varepsilon] = 1 \tag{7.7.1}$$

It follows from equation (7.5.2) that stochastic convergence is equivalent to convergence in probability to the constant c, and for the most part we will restrict attention to this special case. Note that convergence in probability is a stronger property than convergence in distribution. This should not be surprising, because convergence in distribution does not impose any requirement on the joint distribution of Y_n and Y, whereas convergence in probability does. The following theorem is stated without proof.

Theorem 7.7.1 For a sequence of random variables, if

$$Y_n \xrightarrow{P} Y$$

then

$$Y_n \xrightarrow{d} Y$$

For the special case $Y = c$, the limiting distribution is the degenerate distribution $P[Y = c] = 1$. This was the condition we initially used to define stochastic convergence. ∎

Theorem 7.7.2 If $Y_n \xrightarrow{P} c$, then for any function $g(y)$ that is continuous at c,

$$g(Y_n) \xrightarrow{P} g(c)$$

Proof

Because $g(y)$ is continuous at c, it follows that for every $\varepsilon > 0$ a $\delta > 0$ exists such that $|y - c| < \delta$ implies $|g(y) - g(c)| < \varepsilon$. This, in turn, implies that

$$P[|g(Y_n) - g(c)| < \varepsilon] \geqslant P[|Y_n - c| < \delta]$$

because $P(B) \geqslant P(A)$ whenever $A \subset B$. But because $Y_n \xrightarrow{P} c$, it follows for every $\delta > 0$ that

$$\lim_{n\to\infty} P[|g(Y_n) - g(c)| < \varepsilon] \geqslant \lim_{n\to\infty} P[|Y_n - c| < \delta] = 1$$

The left-hand limit cannot exceed 1, so it must equal 1, and $g(Y_n) \xrightarrow{P} g(c)$. ■

Theorem 7.7.2 is also valid if Y_n and c are k-dimensional vectors. Thus this theorem is very useful, and examples of the types of results that follow are listed in the next theorem.

Theorem 7.7.3 If X_n and Y_n are two sequences of random variables such that $X_n \xrightarrow{P} c$ and $Y_n \xrightarrow{P} d$, then:

1. $aX_n + bY_n \xrightarrow{P} ac + bd$.
2. $X_n Y_n \xrightarrow{P} cd$.
3. $X_n/c \xrightarrow{P} 1$, for $c \neq 0$.
4. $1/X_n \xrightarrow{P} 1/c$ if $P[X_n \neq 0] = 1$ for all n, $c \neq 0$.
5. $\sqrt{X_n} \xrightarrow{P} \sqrt{c}$ if $P[X_n \geqslant 0] = 1$ for all n. ■

Example 7.7.1 Suppose that $Y \sim \text{BIN}(n, p)$. We know that $\hat{p} = Y/n \xrightarrow{P} p$. Thus it follows that $\hat{p}(1 - \hat{p}) \xrightarrow{P} p(1 - p)$.

The following theorem is helpful in determining limits in distribution.

Theorem 7.7.4 **Slutsky's Theorem** If X_n and Y_n are two sequences of random variables such that $X_n \xrightarrow{P} c$ and $Y_n \xrightarrow{d} Y$, then:

1. $X_n + Y_n \xrightarrow{d} c + Y$.
2. $X_n Y_n \xrightarrow{d} cY$.
3. $Y_n/X_n \xrightarrow{d} Y/c$; $c \neq 0$.

Note that as a special case X_n could be an ordinary numerical sequence such as $X_n = n/(n-1)$. ■

Example 7.7.2 Consider a random sample of size n from a Bernoulli distribution, $X_i \sim \text{BIN}(1, p)$. We know that

$$\frac{\hat{p} - p}{\sqrt{p(1-p)/n}} \xrightarrow{d} Z \sim N(0, 1)$$

We also know that $\hat{p}(1-\hat{p}) \xrightarrow{P} p(1-p)$, so dividing by $[\hat{p}(1-\hat{p})/p(1-p)]^{1/2}$ gives

$$\frac{\hat{p} - p}{\sqrt{\hat{p}(1-\hat{p})/n}} \xrightarrow{d} Z \sim N(0, 1) \tag{7.7.2}$$

Theorem 7.7.2 also may be generalized.

Theorem 7.7.5 If $Y_n \xrightarrow{d} Y$, then for any continuous function $g(y)$, $g(Y_n) \xrightarrow{d} g(Y)$.
Note that $g(y)$ is assumed not to depend on n. ■

Theorem 7.7.6 If $\sqrt{n}(Y_n - m)/c \xrightarrow{d} Z \sim N(0, 1)$, and if $g(y)$ has a nonzero derivative at $y = m$, $g'(m) \neq 0$, then

$$\frac{\sqrt{n}[g(Y_n) - g(m)]}{|cg'(m)|} \xrightarrow{d} Z \sim N(0, 1)$$

Proof

Define $u(y) = [g(y) - g(m)]/(y - m) - g'(m)$ if $y \neq m$, and let $u(m) = 0$. It follows that $u(y)$ is continuous at m with $u(m) = 0$, and thus $g'(m) + u(Y_n) \xrightarrow{P} g'(m)$. Furthermore,

$$\frac{\sqrt{n}[g(Y_n) - g(m)]}{[cg'(m)]} = \left[\frac{\sqrt{n}(Y_n - m)}{c}\right]\frac{[g'(m) + u(Y_n)]}{g'(m)}$$

From Theorem 7.7.3, we have $[g'(m) + u(Y_n)]/g'(m) \xrightarrow{P} 1$, and the result follows from Theorem 7.7.4. ■

According to our earlier interpretation of an asymptotic normal distribution, we conclude that for large n, if $Y_n \sim N(m, c^2/n)$, then approximately

$$g(Y_n) \sim N\left\{g(m), \frac{c^2[g'(m)]^2}{n}\right\} \tag{7.7.3}$$

Note the similarities between this result and the approximate mean and variance formulas given in Section 2.4.

Example 7.7.3 The Central Limit Theorem says that the sample mean is asymptotically normally distributed,

$$\frac{\sqrt{n}(\bar{X}_n - \mu)}{\sigma} \xrightarrow{d} Z \sim N(0, 1)$$

or, approximately for large n,

$$\bar{X}_n \sim N\left(\mu, \frac{\sigma^2}{n}\right)$$

We now know from Theorem 7.7.6 that differentiable functions of $\bar{X}_n$ also will be asymptotically normally distributed. For example, if $g(\bar{X}_n) = \bar{X}_n^2$, then $g'(\mu) = 2\mu$, and approximately,

$$\bar{X}_n^2 \sim N\left[\mu^2, \frac{(2\mu)^2\sigma^2}{n}\right]$$

7.8*

ASYMPTOTIC DISTRIBUTIONS OF EXTREME ORDER STATISTICS

As noted in Section 7.5, the central order statistics, $X_{k:n}$, are asymptotically normally distributed as $n \to \infty$ and $k/n \to p$. If extreme order statistics such as $X_{1:n}$, $X_{2:n}$, and $X_{n:n}$ are standardized so that they have a nondegenerate limiting distribution, this limiting distribution will not be normal. Examples of such limiting distributions were given earlier. It can be shown that the nondegenerate limiting distribution of an extreme variable must belong to one of three possible types of distributions. Thus, these three types of distributions are useful when studying extremes, analogous to the way the normal distribution is useful when studying means through the Central Limit Theorem.

For example, in studying floods, the variable of interest may be the maximum flood stage during the year. This variable may behave approximately like the

* Advanced (or optional) topic

maximum of a large number of independent flood levels attained through the year. Thus, one of the three limiting types may provide a good model for this variable. Similarly, the strength of a chain is equal to that of its weakest link, or the strength of a ceramic may be the strength at its weakest flaw, where the number of flaws, n, may be quite large. Also, the lifetime of a system of independent and identically distributed components connected in series is equal to the minimum lifetime of the components. Again, one of the limiting distributions may provide a good approximation for the lifetime of the system, even though the distribution of the lifetimes of the individual components may not be known. Similarly, the lifetime of a system of components connected in parallel is equal to the maximum lifetime of the components.

The following theorems, which are stated without proof, are useful in studying the asymptotic behavior of extreme order statistics.

Theorem 7.8.1 If the limit of a sequence of CDFs is a continuous CDF, $F(y) = \lim_{n\to\infty} F_n(y)$, then for any $a_n > 0$ and b_n,

$$\lim_{n\to\infty} F_n(a_n y + b_n) = F(ay + b) \tag{7.8.1}$$

if and only if $\lim_{n\to\infty} a_n = a > 0$ and $\lim_{n\to\infty} b_n = b$. ■

Theorem 7.8.2 If the limit of a sequence of CDFs is a continuous CDF, and if $\lim_{n\to\infty} F_n(a_n y + b_n) = G(y)$ for all $a_n > 0$ and all real y, then $\lim F_n(\alpha_n y + \beta_n) = G(y)$ for $\alpha_n > 0$, if and only if $\alpha_n/a_n \to 1$ and $(\beta_n - b_n)/a_n \to 0$ as $n \to \infty$. ■

LIMITING DISTRIBUTIONS OF MAXIMUMS

Let $X_{1:n}, \ldots, X_{n:n}$ denote an ordered random sample of size n from a distribution with CDF $F(x)$. In the context of extreme-value theory, the maximum $X_{n:n}$ is said to have a (nondegenerate) limiting distribution $G(y)$ if there exist sequences of standardizing constants $\{a_n\}$ and $\{b_n\}$ with $a_n > 0$ such that the standardized variable, $Y_n = (X_{n:n} - b_n)/a_n$, converges in distribution to $G(y)$,

$$Y_n = \frac{X_{n:n} - b_n}{a_n} \xrightarrow{d} Y \sim G(y) \tag{7.8.2}$$

That is, if we say that $X_{n:n}$ has a limiting distribution of type G, we will mean that the limiting distribution of the standardized variable Y_n is a nondegenerate distribution $G(y)$. As suggested by Theorems 7.8.1 and 7.8.2, if $G(y)$ is continuous, the sequence of standardizing constants will not be unique; however, it is not possible to obtain a limiting distribution of a different type by changing the standardizing constants.

Recall that the exact distribution of $X_{n:n}$ is given by

$$F_{n:n}(x) = [F(x)]^n \tag{7.8.3}$$

If we consider $Y_n = (X_{n:n} - b_n)/a_n$, then the exact distribution of Y_n is

$$\begin{aligned} G_n(y) = P[Y_n \leqslant y] &= F_{n:n}(a_n y + b_n) \\ &= [F(a_n y + b_n)]^n \end{aligned} \tag{7.8.4}$$

Thus, the limiting distribution of $X_{n:n}$ (or more correctly Y_n) is given by

$$G(y) = \lim_{n\to\infty} G_n(y) = \lim_{n\to\infty} [F(a_n y + b_n)]^n \tag{7.8.5}$$

Thus, equation (7.8.5) provides a direct approach for determining a limiting extreme-value distribution, if sequences $\{a_n\}$ and $\{b_n\}$ can be found that result in a nondegenerate limit.

Recall from Example 7.2.6 that if $X \sim \text{EXP}(1)$, then we may let $a_n = 1$ and $b_n = \ln n$. Thus,

$$G_n(y) = [F(y + \ln n)]^n = \left[1 - \left(\frac{1}{n}\right)e^{-y}\right]^n \tag{7.8.6}$$

and thus,

$$G(y) = \lim_{n\to\infty} \left[1 - \left(\frac{1}{n}\right)e^{-y}\right]^n = \exp(-e^{-y}) \tag{7.8.7}$$

The three possible types of limiting distributions are provided in the following theorem, which is stated without proof.

Theorem 7.8.3 If $Y_n = (X_{n:n} - b_n)/a_n$ has a limiting distribution $G(y)$, then $G(y)$ must be one of the following three types of extreme-value distributions:

1. Type I (for maximums) (**Exponential type**)

$$G^{(1)}(y) = \exp(-e^{-y}) \qquad -\infty < y < \infty \tag{7.8.8}$$

2. Type II (for maximums) (**Cauchy type**)

$$G^{(2)}(y) = \exp(-y^{-\gamma}) \qquad y > 0,\ \gamma > 0 \tag{7.8.9}$$

3. Type III (for maximums) (**Limited type**)

$$G^{(3)}(y) = \begin{cases} \exp[-(-y)^{\gamma}] & y < 0,\ \gamma > 0 \\ 1 & y \geqslant 0 \end{cases} \tag{7.8.10}$$

■

The limiting distribution of the maximum from densities such as the normal, lognormal, logistic, and gamma distributions is a Type I extreme-value distribution. Generally speaking, such densities have tails no thicker than the exponential distribution. This class includes a large number of the most common distributions, and the Type I extreme-value distribution (for maximum) should provide a useful model for many types of variables related to maximums. Of course, a location parameter and a scale parameter would need to be introduced into the model when applied directly to the nonstandardized variable.

The Type II limiting distribution results for maximums from densities with thicker tails, such as the Cauchy distribution. The Type III case may arise from densities with finite upper limits on the range of the variables.

The following theorem provides an alternative form to equation (7.8.5), which is sometimes more convenient for carrying out the limit.

Theorem 7.8.4 **Gnedenko** In determining the limiting distribution of $Y_n = (X_{n:n} - b_n)/a_n$,

$$\lim_{n\to\infty} G_n(y) = \lim [F(a_n y + b_n)]^n = G(y) \tag{7.8.11}$$

if and only if

$$\lim_{n\to\infty} n[1 - F(a_n y + b_n)] = -\ln G(y) \tag{7.8.12}$$

■

In many cases the greatest difficulty involves determining suitable standardizing sequences so that a nondegenerate limiting distribution will result. For a given CDF, $F(x)$, it is possible to use Theorem 7.8.4 to solve for a_n and b_n in terms of $F(x)$ for each of the three possible types of limiting distributions. Thus, if the limiting type for $F(x)$ is known, then a_n and b_n can be computed. If the type is not known, then a_n and b_n can be computed for each type and then applied to see which type works out. One property of a CDF that is useful in expressing the standardizing constants is its "characteristic largest value."

Definition 7.8.1

The **characteristic largest value,** u_n, of a CDF $F(x)$ is defined by the equation

$$n[1 - F(u_n)] = 1 \tag{7.8.13}$$

For a random sample of size n from $F(x)$, the expected number of observations that will exceed u_n is 1. The probability that one observation will exceed u_n is

$$p = P[X > u_n] = 1 - F(u_n)$$

and the expected number for n independent observations is

$$np = n[1 - F(u_n)]$$

Theorem 7.8.5 Let $X \sim F(x)$, and assume that $Y_n = (X_{n:n} - b_n)/a_n$ has a limiting distribution.

1. If $F(x)$ is continuous and strictly increasing, then the limiting distribution of Y_n is of exponential type if and only if

$$\lim_{n \to \infty} n[1 - F(a_n y + b_n)] = e^{-y} \qquad -\infty < y < \infty \tag{7.8.14}$$

where $b_n = u_n$ and a_n is the solution of

$$F(a_n + u_n) = 1 - (ne)^{-1}$$

2. $G(y)$ is of Cauchy type if and only if

$$\lim_{y \to \infty} \frac{1 - F(y)}{1 - F(ky)} = k^{\gamma} \qquad k > 0, \gamma > 0 \tag{7.8.15}$$

and in this case, $a_n = u_n$ and $b_n = 0$.

3. $G(y)$ is of limited type if and only if

$$\lim_{y \to 0-} \frac{1 - F(ky + x_0)}{1 - F(y + x_0)} = k^{\gamma} \qquad k > 0 \tag{7.8.16}$$

where $x_0 = \max \{x \mid F(x) < 1\}$, the upper limit of x. Also $b_n = x_0$ and $a_n = x_0 - u_n$. ■

Example 7.8.1 Suppose again that $X \sim \text{EXP}(\theta)$, and we are interested in the maximum of a random sample of size n. The characteristic largest value u_n is obtained from

$$n[1 - F(u_n)] = n[1 - (1 - e^{-u_n/\theta})] = 1$$

which gives

$$u_n = \theta \ln n$$

We happen to know that the exponential density falls in the Type I case, so we will try that case first. We have $b_n = u_n = \theta \ln n$, and a_n is determined from

$$F(a_n + u_n) = 1 - e^{-(a_n + \theta \ln n)/\theta} = 1 - (1/n)e^{-a_n/\theta}$$
$$= 1 - 1/(ne)$$

which gives $a_n = \theta$.

Thus, if the exponential density is in the Type I case, we know that

$$Y_n = \frac{X_{n:n} - \theta \ln n}{\theta} \xrightarrow{d} Y \sim G^{(1)}(y) \tag{7.8.17}$$

This is verified easily by using condition 1 of Theorem 7.8.5, because

$$\begin{aligned}\lim_{n\to\infty} n[1 - F(a_n y + b_n)] &= \lim_{n\to\infty} n[e^{-(y+\ln n)}] \\ &= \lim_{n\to\infty} e^{-y} \\ &= e^{-y} \qquad -\infty < y < \infty\end{aligned}$$

Example 7.8.2 The density for the CDF $F(x) = 1 - x^{-\theta}$, $x \geqslant 1$, has a thick upper tail, so one would expect the limiting distribution to be of Cauchy type. If we check the Cauchy-type condition given in Theorem 7.8.5, we find

$$\lim_{y\to\infty} \frac{1 - F(y)}{1 - F(ky)} = \lim_{y\to\infty} \frac{y^{-\theta}}{(ky)^{-\theta}} = k^{\theta} \tag{7.8.18}$$

so the limiting distribution is of Cauchy type with $\gamma = \theta$. Also, we have $n[1 - F(u_n)] = nu_n^{-\theta} = 1$, which gives $u_n = n^{1/\theta} = a_n$, and we let $b_n = 0$ in this case. Thus we know that

$$Y_n = \frac{X_{n:n}}{n^{1/\theta}} \xrightarrow{d} Y \sim G^{(2)}(y) \tag{7.8.19}$$

Now that we know how to standardize the variable, we also can verify this result directly by Theorem 7.8.4. We have

$$\lim_{n\to\infty} n[1 - F(a_n y + b_n)] = \lim_{n\to\infty} y^{-\theta} = -\ln G(y) \tag{7.8.20}$$

so $G(y) = \exp(-y^{-\theta})$, which is the Cauchy type with $\gamma = \theta$.

Example 7.8.3 For $X \sim \text{UNIF}(0, 1)$, where $F(x) = x$, $0 < x < 1$, we should expect a Type III limiting distribution. We have

$$n[1 - F(u_n)] = n(1 - u_n) = 1$$

which gives $u_n = 1 - 1/n$. Thus, $b_n = x_0 = 1$ and $a_n = x_0 - u_n = 1/n$. Checking condition 3 of Theorem 7.8.5,

$$\lim_{y\to 0^-} \frac{1 - F(ky + x_0)}{1 - F(y + x_0)} = \lim_{y\to 0^-} \frac{1 - (ky + x_0)}{1 - (y + x_0)} = \lim_{y\to 0^-} \frac{ky}{y} = k$$

so the limiting distribution of $Y_n = n(X_{n:n} - 1)$ is Type III with $\gamma = 1$. Again, if we look directly at Theorem 7.8.4 to further illustrate, we have

$$\begin{aligned}\lim_{n\to\infty} n[1 - F(a_n y + b_n)] &= \lim_{n\to\infty} n\left[1 - \left(\frac{y}{n} + 1\right)\right] \\ &= -y \\ &= -\ln G(y)\end{aligned} \tag{7.8.21}$$

and $Y_n = n(X_{n:n} - 1) \xrightarrow{d} Y \sim G(y)$, where

$$G(y) = G^{(3)}(y) = \begin{cases} e^y & y < 0 \\ 1 & 0 \leqslant y \end{cases}$$

LIMITING DISTRIBUTIONS OF MINIMUMS

If a nondegenerate limiting distribution exists for the minimum of a random sample, then it also will be one of three possible types. Indeed, the distribution of a minimum can be related to the distribution of a maximum, because

$$\min(x_1, \ldots, x_n) = -\max(-x_1, \ldots, -x_n) \tag{7.8.22}$$

Thus, all the results on maximums can be modified to apply to minimums if the details can be sorted out.

Let X be continuous, $X \sim F_X(x)$, and let $Z = -X \sim F_Z(z) = 1 - F_X(-z)$. Note also that $X_{1:n} = -Z_{n:n}$.

Now consider $W_n = (X_{1:n} + b_n)/a_n$. We have

$$\begin{aligned} G_{W_n}(w) &= P\left[\frac{X_{1:n} + b_n}{a_n} \leqslant w\right] \\ &= P\left[\frac{-Z_{n:n} + b_n}{a_n} \leqslant w\right] \\ &= P\left[\frac{Z_{n:n} - b_n}{a_n} \geqslant -w\right] \\ &= P[Y_n \geqslant -w] \\ &= 1 - G_{Y_n}(-w) \end{aligned}$$

The limiting distribution of W_n, say $H(w)$, then is given by

$$\begin{aligned} H(w) = \lim_{n \to \infty} G_{W_n}(w) &= \lim_{n \to \infty} [1 - G_{Y_n}(-w)] \\ &= 1 - G(-w) \end{aligned}$$

where $G(y)$ now denotes the limiting distribution of $Y_n = (Z_{n:n} - b_n)/a_n$. Thus to find $H(w)$, the limiting distribution for a minimum, the first step is to determine

$$F_Z(z) = 1 - F_X(-z)$$

then determine a_n, b_n, and the limiting distribution $G(y)$ by the methods described for maximums as applied to $F_Z(z)$. Then the limiting distribution for W_n is

$$H(w) = 1 - G(-w) \tag{7.8.23}$$

Note that if $F_X(x)$ belongs to one limiting type, it is possible that $F_Z(z)$ will belong to a different type. For example, maximums from EXP(θ) have a Type I limiting distribution, whereas $F_Z(z)$ in this case has a Type III limiting distribution, so the limiting distribution of the minimum will be a transformed Type III distribution.

In summary, a straightforward procedure for determining a_n, b_n, and $H(w)$ is first to find $F_Z(z)$ and apply the methods for maximums to determine $G(y)$ for $Y_n = (Z_{n:n} - b_n)/a_n$, and then to use equation (7.8.23) to obtain $H(w)$. It also is possible to express the results directly in terms of the original distribution $F_X(x)$.

> ***Definition 7.8.2***
>
> The **smallest characteristic value** is the value s_n defined by
>
> $$nF(s_n) = 1 \qquad \textbf{(7.8.24)}$$

It follows from equation (7.8.22) that $s_n(x) = -u_n(z)$. Similarly, the condition $F_Z(a_n + u_n(z)) = 1 - 1/(ne)$ becomes $F_X(-a_n + s_n) = 1/(ne)$, and so on.

Theorem 7.8.6 If $W_n = (X_{1:n} + b_n)/a_n$ has a limiting distribution $H(w)$, then $H(w)$ must be one of the following three types of extreme-value distributions:

1. Type I (for minimums) (**Exponential type**)
   In this case, $b_n = -s_n$, a_n is defined by

   $$F(s_n - a_n) = \frac{1}{ne} \qquad W_n = \frac{X_{1:n} - s_n}{a_n}$$

   and

   $$H_W^{(1)}(w) = 1 - G^{(1)}(-w) = 1 - \exp(-e^w) \qquad -\infty < w < \infty$$

   if and only if $\lim_{n \to \infty} nF(-a_n y + s_n) = e^{-y}$.

2. Type II (for minimums) (**Cauchy type**)
   In this case, $a_n = -s_n$, $b_n = 0$, $W_n = -X_{1:n}/s_n$, and

   $$H_W^{(2)}(w) = 1 - G^{(2)}(-w) = 1 - \exp[-(-w)^{-\gamma}] \quad w < 0, \gamma > 0$$

   if and only if

   $$\lim_{y \to \infty} \frac{F(-y)}{F(-ky)} = k^{\gamma} \qquad k > 0, \gamma > 0$$

   or

   $$\lim_{n \to \infty} nF(s_n y) = y^{-\gamma} \qquad y > 0$$

3. Type III (for minimums) (**Limited type**)
   If $x_1 = \min \{x \mid F(x) > 0\}$ denotes the lower limit for x (that is, $x_1 = -x_0$), then

$$b_n = -x_1 \qquad a_n = -x_1 + s_n \qquad w_n = \frac{X_{1:n} - x_1}{s_n - x_1}$$

and

$$H_W^{(3)}(w) = 1 - G^{(3)}(-w) = 1 - \exp(-w^\gamma) \qquad w > 0, \gamma > 0$$

if and only if

$$\lim_{y \to 0} \frac{F(ky + x_1)}{F(y + x_1)} = k^\gamma \qquad k > 0$$

or

$$\lim_{n \to \infty} nF[(x_1 - s_n)y + x_1] = (-y)^\gamma$$

■

Note that the Type I distribution for minimums is known as the **Type I extreme-value distribution**. Also, the Type III distribution for minimums is a Weibull distribution. Recall that the limiting distribution for maximums is Type I for many of the common densities. In determining the type of limiting distribution of the minimum, it is necessary to consider the thickness of the right-hand tail of $F_Z(z)$, where $Z = -X$. Thus the limiting distribution of the minimum for some of these common densities, such as the exponential and gamma, belongs to Type III. This may be one reason that the Weibull distribution often is encountered in applications.

Example 7.8.4 We now consider the minimum of a random sample of size n from EXP(θ). We already know in this case that $X_{1:n} \sim \text{EXP}(\theta/n)$, and so $nX_{1:n}/\theta \sim \text{EXP}(1)$. Thus the limiting distribution of $nX_{1:n}/\theta$ is also EXP(1), which is the Type III case with $\gamma = 1$. If we did not know the answer, then we would guess that the limiting distribution was Type III, because the range of the variable $Z = -X$ is limited on the right. Checking condition 3 in Theorem 7.8.6, we have $x_1 = 0$ and

$$\lim_{y \to 0+} \frac{F(ky + x_1)}{F(y + x_1)} = \lim_{y \to 0+} \frac{1 - \exp(-ky)}{1 - \exp(-y)} = \lim_{y \to 0+} \frac{k \exp(-ky)}{\exp(-y)} = k$$

Thus, we know that $H_W(w) = 1 - e^{-w}$, where

$$W_n = \frac{X_{1:n} - x_1}{s_n - x_1} = \frac{X_{1:n}}{s_n}$$

In this case, s_n is given by

$$F(s_n) = 1 - e^{-s_n/\theta} = \frac{1}{n} \quad \text{or} \quad s_n = -\theta \ln\left(1 - \frac{1}{n}\right)$$

This does not yield identically the same standardizing constant as suggested earlier; however, the results are consistent because

$$\frac{-\ln(1 - 1/n)}{1/n} \to 1$$

SUMMARY

The purpose of this chapter was to introduce and develop the notions of convergence in distribution, limiting distributions, and convergence in probability. These concepts are important in studying the asymptotic behavior of sequences of random variables and their distributions.

The Law of Large Numbers (LLN) and the Central Limit Theorem (CLT) deal with the limiting behavior of certain functions of the sample mean as the sample size approaches infinity. Specifically, the LLN asserts that a sequence of sample means converges stochastically to the population mean under certain mild conditions. This type of convergence is also equivalent to convergence in probability in this case, because the limit is constant. Under certain conditions, the CLT asserts that a suitably transformed sequence of sample means has a normal limiting distribution. These theorems have important theoretical implications in probability and statistics, and they also provide useful approximations in many applied situations. For example, the CLT yields a very good approximation for the binomial distribution.

EXERCISES

1. Consider a random sample of size n from a distribution with CDF $F(x) = 1 - 1/x$ if $1 \leq x < \infty$, and zero otherwise.

(a) Derive the CDF of the smallest order statistic, $X_{1:n}$.

(b) Find the limiting distribution of $X_{1:n}$.

(c) Find the limiting distribution of $X_{1:n}^n$.

2. Consider a random sample of size n from a distribution with CDF $F(x) = (1 + e^{-x})^{-1}$ for all real x.

(a) Does the largest order statistic, $X_{n:n}$, have a limiting distribution?

(b) Does $X_{n:n} - \ln n$ have a limiting distribution? If so, what is it?

3. Consider a random sample of size n from a distribution with CDF $F(x) = 1 - x^{-2}$ if $x > 1$, and zero otherwise. Determine whether each of the following sequences has a limiting distribution; if so, then give the limiting distribution.

(a) $X_{1:n}$.

(b) $X_{n:n}$.

(c) $n^{-1/2}X_{n:n}$.

4. Let $X_1, X_2, \ldots$ be independent Bernoulli random variables, $X_i \sim \text{BIN}(1, p_i)$, and let $Y_n = \sum_{i=1}^{n} (X_i - p_i)/n$. Show that the sequence $Y_1, Y_2, \ldots$ converges stochastically to $c = 0$ as $n \to \infty$. *Hint:* Use the Chebychev inequality.

5. Suppose that $Z_i \sim N(0, 1)$ and that $Z_1, Z_2, \ldots$ are independent. Use moment generating functions to find the limiting distribution of $\sum_{i=1}^{n} (Z_i + 1/n)/\sqrt{n}$.

6. Show that the limit in equation (7.3.2) is still correct if the assumption $np = \mu$ is replaced by the weaker assumption $np \to \mu$ as $n \to \infty$.

7. Consider a random sample from a Weibull distribution, $X_i \sim \text{WEI}(1, 2)$. Find approximate values a and b such that for $n = 35$:

(a) $P[a < \bar{X} < b] \doteq 0.95$.

(b) $P[a < \tilde{X} < b] \doteq 0.95$, where $\tilde{X} = X_{18:35}$ is the sample median.

8. In Exercise 2 of Chapter 5, a carton contains 144 baseballs, each of which has a mean weight of 5 ounces and a standard deviation of 2/5 ounces. Use the Central Limit Theorem to approximate the probability that the total weight of the baseballs in the carton is a maximum of 725 ounces.

9. Let $X_1, X_2, \ldots, X_{100}$ be a random sample from an exponential distribution, $X_i \sim \text{EXP}(1)$, and let $Y = X_1 + X_2 + \cdots + X_{100}$.

(a) Give an approximation for $P[Y > 110]$.

(b) If $\bar{X}$ is the sample mean, then approximate $P[1.1 < \bar{X} < 1.2]$.

10. Assume $X_n \sim \text{GAM}(1, n)$ and let $Z_n = (X_n - n)/\sqrt{n}$. Show that $Z_n \xrightarrow{d} Z \sim N(0, 1)$. *Hint:* Show that $M_{Z_n}(t) = \exp(-\sqrt{n}\,t - \ln(1 - t/\sqrt{n}))$ and then use the expansion $\ln(1 - s) = -s - (1 + \xi)s^2/2$ where $\xi \to 0$ as $s \to 0$. Does the above limiting distribution also follow as a result of the CLT? Explain your answer.

11. Let $X_i \sim \text{UNIF}(0, 1)$, where $X_1, X_2, \ldots, X_{20}$ are independent. Find normal approximations for each of the following:

(a) $P\left[\sum_{i=1}^{20} X_i \leqslant 12\right]$.

(b) The 90th percentile of $\sum_{i=1}^{20} X_i$.

12. A certain type of weapon has probability p of working successfully. We test n weapons, and the stockpile is replaced if the number of failures, X, is at least one. How large must n

be to have $P[X \geqslant 1] \doteq 0.99$ when $p = 0.95$?

(a) Use exact binomial.
(b) Use normal approximation.
(c) Use Poisson approximation.
(d) Rework (a) through (c) with $p = 0.90$.

13. Suppose that $Y_n \sim \text{NB}(n, p)$. Give a normal approximation for $P[Y_n \leqslant y]$ for large n. *Hint:* Y_n is distributed as the sum of n independent geometric random variables.

14. For the sequence Y_n of Exercise 13:

(a) Show that Y_n/n converges stochastically to $1/p$, using Theorem 7.6.2.
(b) Rework (a) using Theorem 7.6.3.

15. Let W_i be the weight of the ith airline passenger's luggage. Assume that the weights are independent, each with pdf

$$f(w) = \theta B^{-\theta} w^{\theta - 1} \qquad \text{if } 0 < w < B$$

and zero otherwise.

(a) For $n = 100$, $\theta = 3$, and $B = 80$, approximate $P\left[\sum_{i=1}^{100} W_i > 6025\right]$.
(b) If $W_{1:n}$ is the smallest value out of n, then show that $W_{1:n} \xrightarrow{P} 0$ as $n \to \infty$.
(c) If $W_{n:n}$ is the largest value out of n, then show that $W_{n:n} \xrightarrow{P} B$ as $n \to \infty$.
(d) Find the limiting distribution of $(W_{n:n}/B)^n$.
(e) Find the asymptotic normal distribution of the median, $W_{k:n}$, where $k/n \to 0.5$ with $k - 0.5n$ bounded.
(f) To what does $W_{k:n}$ of (e) converge stochastically?
(g) What is the limiting distribution of $n^{1/\theta} W_{1:n}/B$?

16. Consider a random sample from a Poisson distribution, $X_i \sim \text{POI}(\mu)$.

(a) Show that $Y_n = e^{-\bar{X}_n}$ converges stochastically to $P[X = 0] = e^{-\mu}$.
(b) Find the asymptotic normal distribution of Y_n.
(c) Show that $\bar{X}_n \exp(-\bar{X}_n)$ converges stochastically to $P[X = 1] = \mu e^{-\mu}$.

17. Let $X_1, X_2, \ldots, X_n$ be a random sample of size n from a normal distribution, $X_i \sim \text{N}(\mu, \sigma^2)$, and let $\tilde{X}_n$ be the sample median. Find constants m and c such that $\tilde{X}_n$ is asymptotically normal $\text{N}(m, c^2/n)$.

18. In Exercise 1, find the limiting distribution of $n \ln X_{1:n}$.

19. In Exercise 2, find the limiting distribution of $(1/n) \exp(X_{n:n})$.

20. Under the assumptions of Theorem 7.5.1:

(a) Show that $X_{k:n}$ converges stochastically to x_p.
(b) Show that $F(X_{k:n}) \xrightarrow{P} p$ if $F(x)$ is continuous.

21. As noted in the chapter, convergence of a sequence of real numbers to a limit can be regarded as a special case of stochastic convergence. That is, if $P[Y_n = c_n] = 1$ for each n, and $c_n \to c$, then $Y_n \xrightarrow{P} c$. Use this along with Theorem 7.7.3 to show that if $a_n \to a$, $b_n \to b$ and $Y_n \xrightarrow{P} c$, then $a_n + b_n Y_n \xrightarrow{P} a + bc$.

22. Consider the sequence of independent Bernoulli variables in Exercise 4. Show that if $\sum_{i=1}^{n} p_i/n \to l$ as $n \to \infty$, then $\sum_{i=1}^{n} X_i/n \xrightarrow{P} l$.

23. For the sequence of random variables Z_n of Exercise 10, if Y_n is another sequence such that $Y_n \xrightarrow{P} c$ and if $W_n = Y_n Z_n$ does W_n have a limiting distribution? If so, what is it? What is the limiting distribution of Z_n/Y_n?

24. Use the normal approximation to work Exercise 6 of Chapter 3.

25. Use the theorems of Section 7.8 to determine the standardizing constants and the appropriate extreme-value distributions for each of the following:

(a) $X_{1:n}$ and $X_{n:n}$ where $F(x) = (1 + e^{-x})^{-1}$.
(b) $X_{1:n}$ and $X_{n:n}$ where $X_i \sim \text{WEI}(\theta, \beta)$.
(c) $X_{1:n}$ and $X_{n:n}$ where $X_i \sim \text{EV}(\theta, \eta)$.
(d) $X_{1:n}$ and $X_{n:n}$ where $X_i \sim \text{PAR}(\theta, \kappa)$.

26. Consider the CDF of Exercise 1. Find the limiting extreme-value distribution of $X_{1:n}$ and compare this result to the results of Exercise 1.

27. Consider a random sample from a gamma distribution, $X_i \sim \text{GAM}(\theta, \kappa)$. Determine the limiting extreme-value distribution of $X_{1:n}$.

28. Consider a random sample from a Cauchy distribution, $X_i \sim \text{CAU}(1, 0)$. Determine the type of limiting extreme-value distribution of $X_{n:n}$.

CHAPTER 8

STATISTICS AND SAMPLING DISTRIBUTIONS

8.1 INTRODUCTION

In Chapter 4, the notion of random sampling was presented. The empirical distribution function was used to provide a rationale for the sample mean and sample variance as intuitive estimates of the mean and variance of the population distribution. The purpose of this chapter is to introduce the concept of a statistic, which includes the sample mean and sample variance as special cases, and to derive properties of certain statistics that play an important role in later chapters.

8.2 STATISTICS

Consider a set of observable random variables $X_1, \ldots, X_n$. For example, suppose the variables are a random sample of size n from a population.

Definition 8.2.1

A function of observable random variables, $T = \ell(X_1, \ldots, X_n)$, which does not depend on any unknown parameters, is called a **statistic**.

In our notation, script ℓ is the function that we apply to $X_1, \ldots, X_n$ to define the statistic, which is denoted by capital T.

It is required that the variables be observable because of the intended use of a statistic. The intent is to make inferences about the distribution of the set of random variables, and if the variables are not observable or if the function $\ell(x_1, \ldots, x_n)$ depends on unknown parameters, then T would not be useful in making such inferences. For example, consider the data of Example 4.6.3, which were obtained by observing the lifetimes of 40 randomly selected electrical parts. It is reasonable to assume that they are the observed values of a random sample of size 40 from the population of all such parts. Typically, such a population will have one or more unknown parameters, such as an unknown population mean, say μ. To make an inference about the population, suppose it is necessary to numerically evaluate some function of the data that also depends on the unknown parameter, such as $\ell(x_1, \ldots, x_{40}) = (x_1 + \cdots + x_{40})/\mu$ or $x_{1:40} - \mu$. Of course, such computations would be impossible, because μ is unknown, and these functions would not be suitable for defining statistics.

Also note that, in general, the set of observable random variables need not be a random sample. For example, the set of ordered random variables $Y_1, \ldots, Y_{10}$ of Example 6.5.6 is not a random sample. However, a function of these variables that does not depend on unknown parameters, such as $\ell(y_1, \ldots, y_{10}) = (y_1 + \cdots + y_{10}) + 3y_{10}$, would be a statistic.

Most of the discussion in the chapters that follow will involve random samples.

Example 8.2.1 Let $X_1, \ldots, X_n$ represent a random sample from a population with pdf $f(x)$. The **sample mean**, as defined in Chapter 4, provides an example of a statistic with the function $\ell(x_1, \ldots, x_n) = (x_1 + \cdots + x_n)/n$.

This statistic usually is denoted by

$$\bar{X} = \sum_{i=1}^{n} \frac{X_i}{n} \tag{8.2.1}$$

When a random sample is observed, the value of $\bar{X}$, computed from the data, usually is denoted by lower case $\bar{x}$. As noted in Chapter 4, $\bar{x}$ is useful as an estimate of the population mean, $\mu = E(X)$.

The following theorem provides important properties of the sample mean.

Theorem 8.2.1 If $X_1, \ldots, X_n$ denotes a random sample from $f(x)$ with $E(X) = \mu$ and $\text{Var}(X) = \sigma^2$, then

$$E(\bar{X}) = \mu \tag{8.2.2}$$

and

$$\text{Var}(\bar{X}) = \frac{\sigma^2}{n} \tag{8.2.3}$$

■

Property (8.2.2) indicates that if the sample mean is used to estimate the population mean, then the values of sample estimates will, on the average, be the population mean μ. Of course, for any one sample the value $\bar{x}$ may differ substantially from μ. A statistic with this property is said to be **unbiased** for the parameter it is intended to estimate. This property will receive more attention in the next chapter.

An important special case of the theorem occurs when the population distribution is Bernoulli.

Example 8.2.2 Consider the random variables $X_1, \ldots, X_n$ of Example 5.2.1, which we can regard as a random sample of size n from a Bernoulli distribution, $X_i \sim \text{BIN}(1, p)$. The Bernoulli distribution provides a model for a dichotomous or two-valued population. The mean and variance of such a population are $\mu = p$ and $\sigma^2 = pq$, respectively, where, as usual, $q = 1 - p$. The sample mean in this case is $\bar{X} = Y/n$, where Y is the binomial variable of Example 5.2.1, and it usually is called the **sample proportion**, denoted $\hat{p} = Y/n$.

It is rather straightforward to show that $\hat{p}$ is an unbiased estimate of p,

$$E(\hat{p}) = p \tag{8.2.4}$$

and that

$$\text{Var}(\hat{p}) = \frac{pq}{n} \tag{8.2.5}$$

As noted earlier, the binomial distribution provides a model for the situation of sampling *with* replacement. In Example 5.2.2, the comparable result for sampling *without* replacement was considered, with $Y \sim \text{HYP}(n, M, N)$.

In that example, suppose that we want to estimate M/N, the proportion of defective components in the population, based on the sample proportion, Y/n. We know the mean and variance of Y from equations (5.2.20) and (5.2.21). Specifically,

$$E\left(\frac{Y}{n}\right) = \frac{M}{N} = p \tag{8.2.6}$$

which means that Y/n is unbiased for p, and $\text{Var}(Y/n)$ is shown easily to approach zero as n increases. Actually, in this example, it is possible for the variance to attain the value 0 if $n = N$, which means that the entire population has been inspected.

Example 8.2.3 The function $\ell(x_1, \ldots, x_n) = [(x_1 - \bar{x})^2 + \cdots + (x_n - \bar{x})^2]/(n - 1)$ when applied to data corresponds to the sample variance which was discussed in Chapter 4. Specifically, the **sample variance** is given by

$$S^2 = \frac{\sum_{i=1}^{n} (X_i - \bar{X})^2}{n - 1} \tag{8.2.7}$$

The following alternate forms may be obtained by expanding the square:

$$S^2 = \frac{\sum_{i=1}^{n} X_i^2 - \left(\sum_{i=1}^{n} X_i\right)^2 \Big/ n}{n - 1} \tag{8.2.8}$$

$$= \frac{\sum_{i=1}^{n} X_i^2 - n\bar{X}^2}{n - 1} \tag{8.2.9}$$

The following theorem provides important properties of the sample variance.

Theorem 8.2.2 If $X_1, \ldots, X_n$ denotes a random sample of size n from $f(x)$ with $E(X) = \mu$, $\text{Var}(X) = \sigma^2$, then

$$E(S^2) = \sigma^2 \tag{8.2.10}$$

$$\text{Var}(S^2) = \left(\mu_4 - \frac{n - 3}{n - 1}\sigma^4\right)\Big/ n \qquad n > 1 \tag{8.2.11}$$

Proof

Consider property (8.2.10). Based on equation (8.2.9), we have

$$\begin{aligned}
E(S^2) &= E\left[\sum_{i=1}^{n} X_i^2 - n\bar{X}^2\right]\Big/(n - 1) \\
&= \frac{1}{n - 1}\left[\sum_{i=1}^{n} E(X_i^2) - nE(\bar{X}^2)\right] \\
&= \frac{1}{n - 1}\left[n(\mu^2 + \sigma^2) - n\left(\mu^2 + \frac{\sigma^2}{n}\right)\right] \\
&= \frac{1}{n - 1}[(n - 1)\sigma^2] \\
&= \sigma^2
\end{aligned}$$

The proof of equation (8.2.11) is omitted. ■

According to property (8.2.10), the sample variance provides another example of an unbiased statistic, and this is the principal reason for using the divisor $n - 1$ rather than n.

8.3

SAMPLING DISTRIBUTIONS

A statistic is also a random variable, the distribution of which depends on the distribution of a random sample and on the form of the function $\ell(x_1, x_2, \ldots, x_n)$. The distribution of a statistic sometimes is referred to as a **derived distribution** or **sampling distribution**, in contrast to the population distribution.

Many important statistics can be expressed as linear combinations of independent normal random variables.

LINEAR COMBINATIONS OF NORMAL VARIABLES

Theorem 8.3.1 If $X_i \sim N(\mu_i, \sigma_i^2)$; $i = 1, \ldots, n$ denote independent normal variables, then

$$Y = \sum_{i=1}^{n} a_i X_i \sim N\left(\sum_{i=1}^{n} a_i \mu_i, \sum_{i=1}^{n} a_i^2 \sigma_i^2\right) \tag{8.3.1}$$

Proof

$$\begin{aligned} M_Y(t) &= \prod_{i=1}^{n} M_{X_i}(a_i t) \\ &= \prod_{i=1}^{n} e^{a_i \mu_i t + a_i^2 t^2 \sigma_i^2/2} \\ &= \exp\left[t \sum_{i=1}^{n} a_i \mu_i + t^2 \sum_{i=1}^{n} a_i^2 \sigma_i^2 \Big/ 2\right] \end{aligned}$$

which is the MGF of a normal variable with mean $\sum a_i \mu_i$ and variance $\sum a_i^2 \sigma_i^2$. ■

Corollary 8.3.1 If $X_1, \ldots, X_n$ denotes a random sample from $N(\mu, \sigma^2)$, then $\bar{X} \sim N(\mu, \sigma^2/n)$. ■

Example 8.3.1 In the situation of Example 3.3.4, we wish to investigate the claim that $X \sim N(60, 36)$, so 25 batteries are life-tested and the average of the survival times of the 25 batteries is computed. If the claim is true, the average life of 25 batteries should exceed what value 95% of the time? We have $E(\bar{X}) = 60$, $\text{Var}(\bar{X}) = 36/25$, and

$$P[\bar{X} > c] = 1 - \Phi\left(\frac{c - 60}{6/5}\right)$$
$$= 0.95$$

so $\dfrac{c - 60}{6/5} = z_{0.05} = -1.645$ and $c = 58.026$ months.

In general, for a prescribed probability level $1 - \alpha$, one would have

$$c = \mu + \frac{z_\alpha \sigma}{\sqrt{n}} \tag{8.3.2}$$

or in terms of percentiles directly available in Table 3 (Appendix C) for small α, one could write

$$c = \mu - \frac{z_{1-\alpha} \sigma}{\sqrt{n}} \tag{8.3.3}$$

Thus a reasonable procedure would be to accept the claim if the observed $\bar{x} \geqslant 58.026$, but to disbelieve or reject the claim if $\bar{x} < 58.026$, because that should happen with very small probability (less than 0.05) if the claim is true. If one wished to be more certain before rejecting, then a smaller α, say $\alpha = 0.01$, could be used to determine the critical value c. This test procedure favors the consumer, because it does not reject when a large mean life is indicated. An appropriate test for the other direction (or both directions) also could be constructed.

Example 8.3.2 Consider two independent random samples $X_1, X_2, \ldots, X_{n_1}$, and $Y_1, Y_2, \ldots, Y_{n_2}$, with respective sample sizes n_1 and n_2, from normally distributed populations, $X_i \sim N(\mu_1, \sigma_1^2)$ and $Y_j \sim N(\mu_2, \sigma_2^2)$, and denote by $\bar{X}$ and $\bar{Y}$ the sample means. It follows from Theorem 8.3.1 that the difference also is normally distributed, $\bar{X} - \bar{Y} \sim N(\mu_1 - \mu_2, \sigma_1^2/n_1 + \sigma_2^2/n_2)$. It is clear that the first n_1 terms of the difference have coefficient $a_i = 1/n_1$, and the last n_2 terms have coefficient $a_i = -1/n_2$. Consequently, the mean of the difference is $n_1(1/n_1)\mu_1 + n_2(-1/n_2)\mu_2 = \mu_1 - \mu_2$, and the variance is $n_1(1/n_1)^2\sigma_1^2 + n_2(-1/n_2)^2\sigma_2^2 = \sigma_1^2/n_1 + \sigma_2^2/n_2$.

Certain additional properties involve a special case of the gamma distribution.

CHI-SQUARE DISTRIBUTION

Consider a special gamma distribution with $\theta = 2$ and $\kappa = \nu/2$. The variable Y is said to follow a **chi-square distribution** with ν **degrees of freedom** if

$Y \sim \text{GAM}(2, \nu/2)$. A special notation for this is

$$Y \sim \chi^2(\nu) \tag{8.3.4}$$

Theorem 8.3.2 If $Y \sim \chi^2(\nu)$, then

$$M_Y(t) = (1 - 2t)^{-\nu/2} \tag{8.3.5}$$

$$E(Y^r) = 2^r \frac{\Gamma(\nu/2 + r)}{\Gamma(\nu/2)} \tag{8.3.6}$$

$$E(Y) = \nu \tag{8.3.7}$$

$$\text{Var}(Y) = 2\nu \tag{8.3.8}$$

Proof

These results follow from the corresponding properties of the gamma distribution. ■

The cumulative chi-square distribution has been extensively tabulated in the literature. In most cases percentiles, $\chi^2_\gamma(\nu)$, are provided for particular γ levels of interest and for different values of ν. Specifically, if $Y \sim \chi^2(\nu)$, then $\chi^2_\gamma(\nu)$ is the value such that

$$P[Y \leqslant \chi^2_\gamma(\nu)] = \gamma \tag{8.3.9}$$

Values of $\chi^2_\gamma(\nu)$ are provided in Table 4 (Appendix C) for various values of γ and ν. These values also can be used to obtain percentiles for the gamma distribution.

Theorem 8.3.3 If $X \sim \text{GAM}(\theta, \kappa)$, then $Y = 2X/\theta \sim \chi^2(2\kappa)$.

Proof

$$\begin{aligned} M_Y(t) &= M_{2X/\theta}(t) \\ &= M_X(2t/\theta) \\ &= (1 - 2t)^{-2\kappa/2} \end{aligned}$$

which is the MGF of a chi-square distribution with 2κ degrees of freedom. ■

The gamma CDF also can be expressed in terms of the chi-square notation. If $X \sim \text{GAM}(\theta, \kappa)$, and if $H(y;\ \nu)$ denotes a chi-square CDF with ν degrees of freedom, then

$$F_X(x) = H(2x/\theta;\ 2\kappa) \tag{8.3.10}$$

Cumulative chi-square probabilities $H(c; \nu)$ are provided in Table 5 (Appendix C) for various values of c and ν.

Example 8.3.3 The time to failure (in years) of a certain type of component follows a gamma distribution with $\theta = 3$ and $\kappa = 2$. It is desired to determine a guarantee period for which 90% of the components will survive. That is, the 10th percentile, $x_{0.10}$, is desired such that $P[X \leq x_{0.10}] = 0.10$. We find

$$P[X \leq x_{0.10}] = H(2x_{0.10}/\theta; 2\kappa) = 0.10$$

Thus setting

$$\frac{2x_{0.10}}{\theta} = \chi^2_{0.10}(2\kappa)$$

gives

$$x_{0.10} = \frac{\theta\chi^2_{0.10}(2\kappa)}{2}$$

For $\theta = 3$ and $\kappa = 2$,

$$x_{0.10} = \frac{3\chi^2_{0.10}(4)}{2} = \frac{(3)(1.06)}{2} = 1.59 \text{ years}$$

It is clear in general that the pth percentile of the gamma distribution may be expressed as

$$x_p = \frac{\theta\chi^2_p(2\kappa)}{2} \tag{8.3.11}$$

The following theorem states the useful property that the sum of independent chi-square variables also is chi-square distributed.

Theorem 8.3.4 If $Y_i \sim \chi^2(\nu_i)$; $i = 1, \ldots, n$ are independent chi-square variables, then

$$V = \sum_{i=1}^{n} Y_i \sim \chi^2\left(\sum_{i=1}^{n} \nu_i\right) \tag{8.3.12}$$

Proof

$$M_V(t) = (1 - 2t)^{-\nu_1/2} \cdots (1 - 2t)^{-\nu_n/2}$$
$$= (1 - 2t)^{-\sum \nu_i/2}$$

which is the MGF of $\chi^2(\sum \nu_i)$. ■

The following theorem establishes a connection between standard normal and chi-square variables.

Theorem 8.3.5 If $Z \sim N(0, 1)$, then $Z^2 \sim \chi^2(1)$.

Proof

$$\begin{aligned}
M_{Z^2}(t) &= E[e^{tZ^2}] \\
&= \int_{-\infty}^{\infty} \frac{1}{\sqrt{2\pi}} e^{tz^2 - z^2/2}\, dz \\
&= \frac{1}{\sqrt{1-2t}} \int_{-\infty}^{\infty} \frac{\sqrt{1-2t}}{\sqrt{2\pi}} e^{-z^2(1-2t)/2}\, dz \\
&= (1-2t)^{-1/2}
\end{aligned}$$

which is the MGF of a chi-square distribution with one degree of freedom. ■

Corollary 8.3.2 If $X_1, \ldots, X_n$ denotes a random sample from $N(\mu, \sigma^2)$, then

$$\sum_{i=1}^{n} \frac{(X_i - \mu)^2}{\sigma^2} \sim \chi^2(n) \tag{8.3.13}$$

$$\frac{n(\bar{X} - \mu)^2}{\sigma^2} \sim \chi^2(1) \tag{8.3.14}$$

■

The sample variance was discussed previously, and for a sample from a normal population its distribution can be related to a chi-square distribution. The sampling distribution of S^2 does not follow directly from the previous corollary because the terms, $X_i - \bar{X}$, are not independent. Indeed, they are functionally dependent because $\sum_{i=1}^{n} (X_i - \bar{X}) = 0$.

Theorem 8.3.6 If $X_1, \ldots, X_n$ denotes a random sample from $N(\mu, \sigma^2)$, then

1. $\bar{X}$ and the terms $X_i - \bar{X}$; $i = 1, \ldots, n$ are independent.
2. $\bar{X}$ and S^2 are independent.
3. $(n - 1)S^2/\sigma^2 \sim \chi^2(n - 1)$. **(8.3.15)**

Proof

To obtain part 1, first note that by adding and subtracting $\bar{x}$ and then expanding, we obtain the relationship

$$\sum_{i=1}^{n} \frac{(x_i - \mu)^2}{\sigma^2} = \sum_{i=1}^{n} \frac{(x_i - \bar{x})^2}{\sigma^2} + \frac{n(\bar{x} - \mu)^2}{\sigma^2} \tag{8.3.16}$$

Thus the joint density of $X_1, \ldots, X_n$ may be expressed as

$$f(x_1, \ldots, x_n) = \frac{1}{(2\pi)^{n/2}\sigma^n} \exp\left[-\frac{1}{2}\sum_{i=1}^{n}\left(\frac{x_i - \mu}{\sigma}\right)^2\right]$$

$$= \frac{1}{(2\pi)^{n/2}\sigma^n} \exp\left[-\frac{1}{2\sigma^2}\left(\sum_{i=1}^{n}(x_i - \bar{x})^2 + n(\bar{x} - \mu)^2\right)\right]$$

Now consider the joint transformation

$$y_1 = \bar{x}, \ y_i = x_i - \bar{x} \qquad i = 2, \ldots, n$$

We know that

$$x_1 - \bar{x} = -\sum_{i=2}^{n}(x_i - \bar{x}) = -\sum_{i=2}^{n} y_i$$

so

$$\sum_{i=1}^{n}(x_i - \bar{x})^2 = \left(-\sum_{i=2}^{n} y_i\right)^2 + \sum_{i=2}^{n} y_i^2$$

and

$$g(y_1, \ldots, y_n) = \frac{|J|}{(2\pi)^{n/2}\sigma^n}$$
$$\times \exp\left\{-\frac{1}{2\sigma^2}\left[\left(-\sum_{i=2}^{n} y_i\right)^2 + \sum_{i=2}^{n} y_i^2 + n(y_1 - \mu)^2\right]\right\}$$

It is easy to see that the Jacobian, J, is a constant, and in particular it can be shown that $|J| = n$. Therefore, the joint density function factors into the marginal density function of y_1 times a function of $y_2, \ldots, y_n$ only, which shows that $Y_1 = \bar{X}$ and the terms $Y_i = X_i - \bar{X}$ for $i = 2, \ldots, n$ are independent. Because $X_1 - \bar{X} = -\sum_{i=2}^{n}(X_i - \bar{X})$, it follows that $\bar{X}$ and $X_1 - \bar{X}$ also are independent.

Part 2 follows from part 1, because S^2 is a function only of the $X_i - \bar{X}$.

To obtain part 3, consider again equation (8.3.16) applied to the random sample,

$$V_1 = \sum_{i=1}^{n} \frac{(X_i - \mu)^2}{\sigma^2} = \frac{(n-1)S^2}{\sigma^2} + \frac{n(\bar{X} - \mu)^2}{\sigma^2}$$

$$= V_2 + V_3$$

From Corollary 8.3.2, $V_1 \sim \chi^2(n)$ and $V_3 \sim \chi^2(1)$. Also, V_2 and V_3 are independent, so

$$M_{V_1}(t) = M_{V_2}(t)M_{V_3}(t)$$

and

$$M_{V_2}(t) = \frac{M_{V_1}(t)}{M_{V_3}(t)} = \frac{(1-2t)^{-n/2}}{(1-2t)^{-1/2}}$$
$$= (1-2t)^{-(n-1)/2}$$

Thus, $V_2 = [(n-1)S^2]/\sigma^2 \sim \chi^2(n-1)$.

Thus, if c_γ is the γth percentile of the distribution of S^2, then

$$c_\gamma = \frac{\sigma^2\chi^2_\gamma(n-1)}{n-1} \tag{8.3.17}$$

■

Consider again Example 8.3.1, where it was assumed that $X \sim N(60, 36)$. Suppose that it was decided to sample 25 batteries, and to reject the claim that $\sigma^2 = 36$ if $s^2 \geqslant 54.63$, and not reject the claim if $s^2 < 54.63$. Under this procedure, what would be the probability of rejecting the claim when in fact $\sigma^2 = 36$? We see that

$$P[S^2 \geqslant 54.63] = P[24S^2/36 \geqslant 36.42]$$
$$= 1 - H(36.42; 24)$$
$$= 0.05$$

If instead one wished to be wrong only 1% of the time when rejecting the claim, then the procedure would be to reject if $s^2 \geqslant c_{0.99}$, where $c_{0.99} = \sigma^2\chi^2_{0.99}(n-1)/(n-1) = 36(42.98)/24 = 64.47$.

8.4

THE *t*, *F*, AND BETA DISTRIBUTIONS

Certain functions of normal samples are very important in statistical analysis of populations.

STUDENT'S t *DISTRIBUTION*

We noticed that S^2 can be used to make inferences about the parameter σ^2 in a normal distribution. Similarly, $\bar{X}$ is useful concerning the parameter μ; however, the distribution of $\bar{X}$ also depends on the parameter σ^2. This makes it impossible to use $\bar{X}$ for certain types of statistical procedures concerning the mean when σ^2 is unknown. It turns out that if σ is replaced by S in the quantity $\sqrt{n}(\bar{X} - \mu)/\sigma$, then the resulting distribution is no longer standard normal, but it does not depend on σ, and it can be derived using transformation methods.

Theorem 8.4.1 If $Z \sim N(0, 1)$ and $V \sim \chi^2(\nu)$, and if Z and V are independent, then the distribution of

$$T = \frac{Z}{\sqrt{V/\nu}} \tag{8.4.1}$$

is referred to as **Student's *t* distribution** with ν degrees of freedom, denoted by $T \sim t(\nu)$. The pdf is given by

$$f(t; \nu) = \frac{\Gamma\left(\dfrac{\nu+1}{2}\right)}{\Gamma\left(\dfrac{\nu}{2}\right)} \frac{1}{\sqrt{\nu\pi}} \left(1 + \frac{t^2}{\nu}\right)^{-(\nu+1)/2} \tag{8.4.2}$$

Proof

The joint density of Z and V is given by

$$f_{Z,V}(z, v) = \frac{v^{\nu/2-1}e^{-v/2}e^{-z^2/2}}{\sqrt{2\pi}\,\Gamma(\nu/2)2^{\nu/2}} \qquad 0 < v < \infty,\ -\infty < z < \infty$$

Consider the transformation $T = Z/\sqrt{V/\nu}$, $W = V$, with inverse transformation $v = w$, $z = t\sqrt{w/\nu}$. The Jacobian is $J = \sqrt{w/\nu}$ and

$$f_{T,W}(t, w) = \frac{(w/\nu)^{1/2}w^{\nu/2-1}e^{-w/2}e^{-t^2w/2\nu}}{\sqrt{2\pi}\,\Gamma(\nu/2)2^{\nu/2}} \qquad -\infty < t < \infty,\ 0 < w < \infty$$

After some simplification, the marginal pdf $f(t; \nu) = \int_0^\infty f_{T,W}(t, w)\, dw$ yields equation (8.4.2). ■

The t distribution is symmetric about zero, and its general shape is similar to that of the standard normal distribution. Indeed, the t distribution approaches the standard normal distribution as $\nu \to \infty$. For smaller ν the t distribution is flatter with thicker tails and, in fact, $T \sim CAU(1, 0)$ when $\nu = 1$.

Percentiles, $t_\gamma(\nu)$, are provided in Table 6 (Appendix C) for selected values of γ and for $\nu = 1, \ldots, 30, 40, 60, 120, \infty$.

Theorem 8.4.2 If $T \sim t(\nu)$, then for $\nu > 2r$,

$$E(T^{2r}) = \frac{\Gamma((2r+1)/2)\Gamma((\nu-2r)/2)}{\Gamma(1/2)\Gamma(\nu/2)}\nu^r \tag{8.4.3}$$

$$E(T^{2r-1}) = 0 \qquad r = 1, 2, \ldots \tag{8.4.4}$$

$$\text{Var}(T) = \frac{\nu}{\nu-2} \qquad 2 < \nu \tag{8.4.5}$$

Proof

The 2*r*th moment is

$$E(T^{2r}) = E(Z^{2r})E[(V/\nu)^{-r}]$$

where $Z \sim \text{N}(0, 1)$ and $V \sim \chi^2(\nu)$. Substitution of the normal and chi-square moments gives the required result. ■

As suggested earlier, one application of the *t* distribution arises when sampling from a normal distribution, as illustrated by the following theorem.

Theorem 8.4.3 If $X_1, \ldots, X_n$ denotes a random sample from $\text{N}(\mu, \sigma^2)$, then

$$\frac{\bar{X} - \mu}{S/\sqrt{n}} \sim t(n-1) \tag{8.4.6}$$

Proof

This follows from Theorem 8.4.1, because $Z = \sqrt{n}(\bar{X} - \mu)/\sigma \sim \text{N}(0, 1)$ and, by Theorem 8.3.6, $V = (n-1)S^2/\sigma^2 \sim \chi^2(n-1)$, and $\bar{X}$ and S^2 are independent. ■

SNEDECOR'S F DISTRIBUTION

Another derived distribution of great importance in statistics is called Snedecor's *F* distribution.

Theorem 8.4.4 If $V_1 \sim \chi^2(\nu_1)$ and $V_2 \sim \chi^2(\nu_2)$ are independent, then the random variable

$$X = \frac{V_1/\nu_1}{V_2/\nu_2} \tag{8.4.7}$$

has the following pdf for $x > 0$:

$$g(x; \nu_1, \nu_2) = \frac{\Gamma\left(\frac{\nu_1 + \nu_2}{2}\right)}{\Gamma\left(\frac{\nu_1}{2}\right)\Gamma\left(\frac{\nu_2}{2}\right)} \left(\frac{\nu_1}{\nu_2}\right)^{\nu_1/2} x^{(\nu_1/2)-1}\left(1 + \frac{\nu_1}{\nu_2} x\right)^{-(\nu_1+\nu_2)/2} \tag{8.4.8}$$

■

This is known as **Snedecor's *F* distribution** with ν_1 and ν_2 degrees of freedom, and is denoted by $X \sim F(\nu_1, \nu_2)$. Some authors use the notation F rather than X for the ratio (8.4.7).

The pdf (8.4.8) can be derived in a manner similar to that of the t distribution as in Theorem 8.4.1.

Theorem 8.4.5 If $X \sim F(\nu_1, \nu_2)$, then

$$E(X^r) = \frac{\left(\frac{\nu_2}{\nu_1}\right)^r \Gamma\left(\frac{\nu_1}{2} + r\right)\Gamma\left(\frac{\nu_2}{2} - r\right)}{\Gamma\left(\frac{\nu_1}{2}\right)\Gamma\left(\frac{\nu_2}{2}\right)} \qquad \nu_2 > 2r \tag{8.4.9}$$

$$E(X) = \frac{\nu_2}{\nu_2 - 2} \qquad 2 < \nu_2 \tag{8.4.10}$$

$$\text{Var}(X) = \frac{2\nu_2^2(\nu_1 + \nu_2 - 2)}{\nu_1(\nu_2 - 2)^2(\nu_2 - 4)} \qquad 4 < \nu_2 \tag{8.4.11}$$

Proof

These results follow from the fact that V_1 and V_2 are independent, and from chi-square moments (8.3.6). Specifically, they can be obtained from

$$E(X^r) = \left(\frac{\nu_2}{\nu_1}\right)^r E(V_1^r)E(V_2^{-r}) \tag{8.4.12}$$

■

Percentiles $f_\gamma(\nu_1, \nu_2)$ of $X \sim F(\nu_1, \nu_2)$ such that

$$P[X \leqslant f_\gamma(\nu_1, \nu_2)] = \gamma \tag{8.4.13}$$

are provided in Table 7 (Appendix C) for selected values of γ, ν_1, and ν_2. Percentiles for small values of γ can be obtained by using the fact that if $X \sim F(\nu_1, \nu_2)$, then $Y = 1/X \sim F(\nu_2, \nu_1)$. Thus,

$$\begin{aligned} 1 - \gamma &= P[X < f_{1-\gamma}(\nu_1, \nu_2)] \\ &= P\left[Y > \frac{1}{f_{1-\gamma}(\nu_1, \nu_2)}\right] \\ &= 1 - P\left[Y \leqslant \frac{1}{f_{1-\gamma}(\nu_1, \nu_2)}\right] \end{aligned}$$

so that

$$\frac{1}{f_{1-\gamma}(\nu_1, \nu_2)} = f_\gamma(\nu_2, \nu_1)$$

$$f_{1-\gamma}(\nu_1, \nu_2) = \frac{1}{f_\gamma(\nu_2, \nu_1)} \tag{8.4.14}$$

■

Example 8.4.1 Let $X_1, \ldots, X_{n_1}$ and $Y_1, \ldots, Y_{n_2}$ be independent random samples from populations with respective distributions $X_i \sim N(\mu_1, \sigma_1^2)$ and $Y_j \sim N(\mu_2, \sigma_2^2)$. If $\nu_1 = n_1 - 1$ and $\nu_2 = n_2 - 1$, then $\nu_1 S_1^2/\sigma_1^2 \sim \chi^2(\nu_1)$ and $\nu_2 S_2^2/\sigma_2^2 \sim \chi^2(\nu_2)$, so that

$$\frac{S_1^2 \sigma_2^2}{S_2^2 \sigma_1^2} \sim F(\nu_1, \nu_2)$$

and thus

$$P\left[\frac{S_1^2 \sigma_2^2}{S_2^2 \sigma_1^2} \leqslant f_{0.95}(\nu_1, \nu_2)\right] = 0.95$$

and

$$P\left[\frac{S_1^2}{S_2^2 f_{0.95}(\nu_1, \nu_2)} \leqslant \frac{\sigma_1^2}{\sigma_2^2}\right] = 0.95$$

If $n_1 = 16$ and $n_2 = 21$, then $f_{0.95}(15, 20) = 2.20$, and for two such samples it usually is said that we are 95% "confident" that the ratio $\sigma_1^2/\sigma_2^2 > s_1^2/[s_2^2 f_{0.95}(15, 20)]$. This notion will be further developed in a later chapter.

BETA DISTRIBUTION

An F variable can be transformed to have the beta distribution. If $X \sim F(\nu_1, \nu_2)$ then the random variable

$$Y = \frac{(\nu_1/\nu_2)X}{1 + (\nu_1/\nu_2)X} \tag{8.4.15}$$

has the pdf

$$f(y; a, b) = \frac{\Gamma(a+b)}{\Gamma(a)\Gamma(b)} y^{a-1}(1-y)^{b-1} \qquad 0 < y < 1 \tag{8.4.16}$$

where $a = \nu_1/2$ and $b = \nu_2/2$. This pdf defines the **beta distribution** with parameters $a > 0$ and $b > 0$, denoted $Y \sim \text{BETA}(a, b)$.

The mean and variance of Y easily are shown to be

$$E(Y) = \frac{a}{a+b} \tag{8.4.17}$$

$$\text{Var}(Y) = \frac{ab}{(a+b+1)(a+b)^2} \tag{8.4.18}$$

The γth percentile of a beta distribution can be expressed in terms of a percentile of the F distribution as a result of equation (8.4.15), namely

$$y_\gamma(a, b) = \frac{a f_\gamma(2a, 2b)}{b + a f_\gamma(2a, 2b)} \tag{8.4.19}$$

If a and b are positive integers, then successive integration by parts leads to a relationship between the CDFs of beta and binomial distributions. If $X \sim \text{BIN}(n, p)$ and $Y \sim \text{BETA}(n - i + 1, i)$, then $F_X(i - 1) = F_Y(1 - p)$.

The beta distribution arises in connection with distributions of order statistics. For a continuous random variable $X \sim f(x)$, the pdf of the kth order statistic from a random sample of size n is given by

$$g_k(x_{k:n}) = \frac{n!}{(k-1)!(n-k)!} [F(x_{k:n})]^{k-1}[1 - F(x_{k:n})]^{n-k} f(x_{k:n})$$

Making the change of variable $U_{k:n} = F(X_{k:n})$ gives

$$U_{k:n} \sim \text{BETA}(k, n - k + 1)$$

Because $U = F(X) \sim \text{UNIF}(0, 1)$, it also follows that $U_{k:n}$ represents the kth largest ordered uniform random variable. The CDF of $X_{k:n}$ can be expressed in terms of a beta CDF, because

$$\begin{aligned} G_k(x_{k:n}) &= P[X_{k:n} \leqslant x_{k:n}] \\ &= P[F(X_{k:n}) \leqslant F(x_{k:n})] \\ &= H(F(x_{k:n}); k, n - k + 1) \end{aligned}$$

where $H(y; a, b)$ denotes the CDF of $Y \sim \text{BETA}(a, b)$.

Example 8.4.2 Suppose that $X \sim \text{EXP}(\theta)$, and one wishes to compute probabilities concerning $X_{k:n}$. We have

$$F(x) = 1 - e^{-x/\theta} \qquad U_{k:n} = F(X_{k:n}) \sim \text{BETA}(k, n - k + 1)$$

and

$$P[X_{k:n} \leqslant c] = P[F(X_{k:n}) \leqslant F(c)]$$
$$= P[U_{k:n} \leqslant F(c)]$$
$$= P\left[\frac{(n-k+1)U_{k:n}}{k(1-U_{k:n})} \leqslant \frac{(n-k+1)F(c)}{k(1-F(c))}\right]$$

where this last probability involves a variable distributed as $F(2k, 2(n-k+1))$. Thus for specified values of θ, c, k, and n, this probability can be obtained from a cumulative beta table, or from a cumulative F table if the proper α level is available. For the purpose of illustration, we wish to know c_γ such that

$$P[X_{k:n} < c_\gamma] = \gamma$$

then

$$f_\gamma(2k, 2(n-k+1)) = \frac{(n-k+1)F(c_\gamma)}{k[1-F(c_\gamma)]}$$
$$= \frac{(n-k+1)(1-\exp(-c_\gamma/\theta))}{k \exp(-c_\gamma/\theta)}$$

and

$$c_\gamma = \theta \ln\left[1 + \frac{k f_\gamma(2k, 2(n-k+1))}{n-k+1}\right] \qquad \textbf{(8.4.20)}$$

If $n = 11$, $k = 6$, and $\gamma = 0.95$, then

$$c_\gamma = \theta \ln\left[1 + \frac{6(2.69)}{6}\right] = 1.31\theta$$

and

$$P[X_{6:11} \leqslant 1.31\theta] = 0.95$$

or

$$P\left[\frac{X_{6:11}}{1.31} \leqslant \theta\right] = 0.95$$

where $X_{6:11}$ is the median of the sample, and θ is the mean of the population.

We have defined the beta distribution and we have seen its relationship to the F distribution and the binomial CDF, as well as its application to the distribution of ordered uniform random variables. The beta distribution represents a generalization of the uniform distribution, and provides a rather flexible two-parameter model for various types of variables that must lie between 0 and 1.

8.5

LARGE-SAMPLE APPROXIMATIONS

The sampling distributions discussed in the earlier sections have approximations that apply for large sample sizes.

Theorem 8.5.1 If $Y_\nu \sim \chi^2(\nu)$, then

$$Z_\nu = \frac{Y_\nu - \nu}{\sqrt{2\nu}} \xrightarrow{d} Z \sim N(0, 1)$$

as $\nu \to \infty$.

Proof

This follows from the CLT, because Y_ν is distributed as a sum, $\sum_{i=1}^{\nu} X_i$, where $X_1, \ldots, X_\nu$ are independent, and $X_i \sim \chi^2(1)$, so that $E(X_i) = 1$ and $\text{Var}(X_i) = 2$. ■

We also would expect the pdf's of Z_ν to closely approximate the pdf of Z for large ν. This is illustrated in Figure 8.1, which shows the pdf's for $\nu = 20$, 80, and 200.

FIGURE 8.1 Comparison of pdf's of standardized chi-square and standard normal distributions

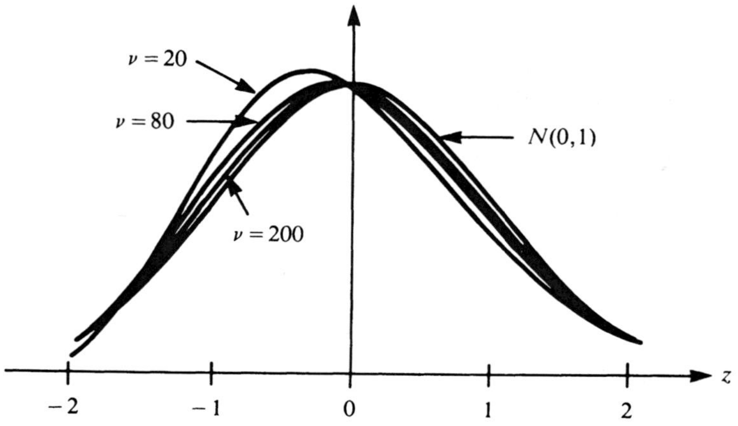

It follows that chi-square percentiles can be approximated in terms of standard normal percentiles for large ν. Specifically,

$$\gamma = P[Y_\nu \leqslant \chi^2_\gamma(\nu)]$$

$$\doteq \Phi\left(\frac{\chi^2_\gamma(\nu) - \nu}{\sqrt{2\nu}}\right)$$

so that

$$\frac{\chi^2_\gamma(\nu) - \nu}{\sqrt{2\nu}} \doteq z_\gamma$$

$$\chi^2_\gamma(\nu) \doteq \nu + z_\gamma\sqrt{2\nu} \tag{8.5.1}$$

For example, for $\nu = 30$ and $\gamma = 0.95$,

$$\chi^2_{0.95}(30) \doteq 30 + 1.645\sqrt{60} = 42.74$$

compared to the exact value $\chi^2_{0.95}(30) = 43.77$. A more accurate approximation, known as the **Wilson-Hilferty** approximation, is given by

$$\chi^2_\gamma(\nu) \doteq \nu\left[1 - \frac{2}{9\nu} + z_\gamma\sqrt{\frac{2}{9\nu}}\right]^3 \tag{8.5.2}$$

This gives approximate values of $\chi^2_\gamma(\nu)/\nu$ within 0.01 of exact values for $\nu \geqslant 3$ and $0.01 \leqslant \gamma \leqslant 0.99$. For example, if $\nu = 30$ and $\gamma = 0.95$, approximation (8.5.2) gives $\chi^2_{0.95}(30) \doteq 43.77$, which agrees to two decimal places with the exact value.

It also is possible to derive asymptotic normal distributions directly for S_n^2 and S_n.

Example 8.5.1 Let S_n^2 denote the sample variance from a random sample of size n from $N(\mu, \sigma^2)$. We know that

$$V_n = \frac{(n-1)S_n^2}{\sigma^2} \sim \chi^2(n-1)$$

and from Theorem 8.5.1,

$$\frac{V_n - (n-1)}{\sqrt{2(n-1)}} \xrightarrow{d} Z \sim N(0, 1)$$

That is,

$$\frac{\sqrt{n-1}\,[S_n^2 - \sigma^2]}{\sigma^2\sqrt{2}} \xrightarrow{d} Z \tag{8.5.3}$$

or approximately,

$$S_n^2 \sim N\left(\sigma^2, \frac{2\sigma^4}{n-1}\right) \tag{8.5.4}$$

If $Y_n = S_n^2$, and $g(y) = \sqrt{y}$, then $g'(y) = 1/2\sqrt{y}$, $g'(\sigma^2) = 1/2\sigma$, and approximately,

$$S_n \sim N\left[\sigma, \frac{\sigma^2}{2(n-1)}\right] \tag{8.5.5}$$

It also is possible to show that a t-distributed variable has a limiting standard normal distribution as the degrees of freedom ν increases. To see this, consider a variable $T_\nu \sim t(\nu)$, where

$$T_\nu = \frac{Z}{\sqrt{\chi_\nu^2/\nu}}$$

We know that $E(\chi_\nu^2/\nu) = 1$, $\text{Var}(\chi_\nu^2/\nu) = 2/\nu$, and by Chebychev's inequality, $P[|\chi_\nu^2/\nu - 1| < \varepsilon] \geqslant 1 - 2/\nu\varepsilon^2$, so that $\chi_\nu^2/\nu \xrightarrow{P} 1$ as $\nu \to \infty$.

Thus Student's t distribution has a limiting standard normal distribution by Theorem 7.7.4, part 3:

$$T_\nu = \frac{Z}{\sqrt{\chi_\nu^2/\nu}} \xrightarrow{d} Z \sim N(0, 1) \tag{8.5.6}$$

FIGURE 8.2 Comparison of pdf's of t and standard normal distributions

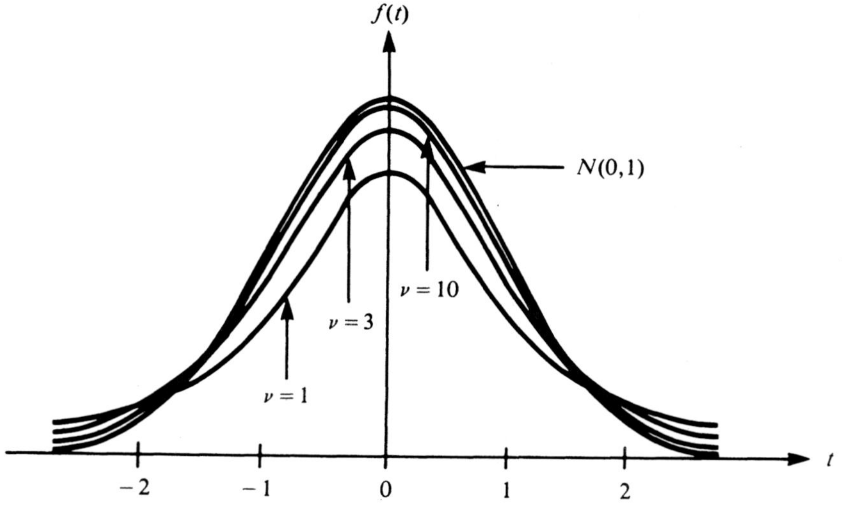

This is illustrated by Figure 8.2, which shows the pdf's of N(0, 1) and $t(\nu)$ for $\nu = 1, 3$, and 10.

This suggests that the t percentile, $t_\gamma(\nu)$, is approximately equal to z_γ for large ν, and it leads to the last row of values in Table 6 (Appendix C) that correspond to the standard normal percentiles.

A similar rationale yields approximate percentiles for an F distribution. Suppose $X_{\nu_1, \nu_2} = (V_1/\nu_1)/(V_2/\nu_2)$. As noted in the above discussion, $V_2/\nu_2 \to 1$ as $\nu_2 \to \infty$. Thus, if ν_1 is kept fixed, $X_{\nu_1, \nu_2} \xrightarrow{d} V_1/\nu_1$ as $\nu_2 \to \infty$. The resulting approximation for an F percentile is $f_\gamma(\nu_1, \nu_2) \doteq \chi^2_\gamma(\nu_1)/\nu_1$ for large ν_2. A similar argument leads to an approximation for large ν_1, namely, $f_\gamma(\nu_1, \nu_2) \doteq \nu_2/\chi^2_{1-\gamma}(\nu_2)$. These also provide the limiting entries in Table 7 (Appendix C).

SUMMARY

Our purpose in this chapter was to study properties of the normal distribution and to derive other related distributions that arise in the statistical analysis of data from normally distributed populations.

An important property of the normal distribution is that linear combinations of independent normal random variables are also normally distributed, which means, among other things, that the sample mean is normally distributed. A certain function of the sample variance is shown to be chi-square distributed, and the sample mean and sample variance are shown to be independent random variables. This is important in the development of statistical methods for the analysis of the population mean when the population variance is unknown. This corresponds to Student's t distribution, which is obtained as the distribution of a standard normal variable divided by the square root of an independent chi-square variable over its degrees of freedom. Another example involves Snedecor's F distribution, which is obtained as the distribution of the ratio of two independent chi-square variables over their respective degrees of freedom. Variables of the latter type are important in statistical analyses that compare the variances of two normally distributed populations.

EXERCISES

1. Let X denote the weight in pounds of a bag of feed, where $X \sim N(101, 4)$. What is the probability that 20 bags will weigh at least a ton?

2. S denotes the diameter of a shaft and B the diameter of a bearing, where S and B are independent with $S \sim N(1, 0.0004)$ and $B \sim N(1.01, 0.0009)$.

(a) If a shaft and a bearing are selected at random, what is the probability that the shaft diameter will exceed the bearing diameter?

(b) Assume equal variances $\sigma_1^2 = \sigma_2^2 = \sigma^2$, and find the value of σ that will yield a probability of noninterference of 0.95.

3. Let $X_1, X_2, \ldots, X_n$ be a random sample of size n from a normal distribution, $X_i \sim N(\mu, \sigma^2)$, and define $U = \sum_{i=1}^{n} X_i$ and $W = \sum_{i=1}^{n} X_i^2$.

(a) Find a statistic that is a function of U and W and unbiased for the parameter $\theta = 2\mu - 5\sigma^2$.

(b) Find a statistic that is unbiased for $\sigma^2 + \mu^2$.

(c) Let c be a constant, and define $Y_i = 1$ if $X_i \leqslant c$ and zero otherwise. Find a statistic that is a function of $Y_1, Y_2, \ldots, Y_n$ and also unbiased for $F_X(c) = \Phi\left(\frac{c-\mu}{\sigma}\right)$.

4. Assume that X_1 and X_2 are independent normal random variables, $X_i \sim N(\mu, \sigma^2)$, and let $Y_1 = X_1 + X_2$ and $Y_2 = X_1 - X_2$. Show that Y_1 and Y_2 are independent and normally distributed.

5. A new component is placed in service and nine spares are available. The times to failure in days are independent exponential variables, $T_i \sim EXP(100)$.

(a) What is the distribution of $\sum_{i=1}^{10} T_i$?

(b) What is the probability that successful operation can be maintained for at least 1.5 years? *Hint:* Use Theorem 8.3.3 to transform to a chi-square variable.

(c) How many spares would be needed to be 95% sure of successful operation for at least two years?

6. Repeat Exercise 5 assuming $T_i \sim GAM(100, 1.2)$.

7. Five independent tasks are to be performed, where the time in hours to complete the ith task is given by $T_i \sim GAM(100, \kappa_i)$, where $\kappa_i = 3 + i/3$. What is the probability that it will take less than 2600 hours to complete all five tasks?

8. Suppose that $X \sim \chi^2(m)$, $Y \sim \chi^2(n)$, and X and Y are independent. Is $Y - X \sim \chi^2$ if $n > m$?

9. Suppose that $X \sim \chi^2(m)$, $S = X + Y \sim \chi^2(m + n)$, and X and Y are independent. Use MGFs to show that $S - X \sim \chi^2(n)$.

10. A random sample of size $n = 15$ is drawn from EXP(θ). Find c so that $P[c\bar{X} < \theta] = 0.95$, where $\bar{X}$ is the sample mean.

11. Let $Z \sim N(0, 1)$.

(a) Find $P[Z^2 < 3.84]$ using tabled values of the standard normal distribution.

(b) Find $P[Z^2 < 3.84]$ using tabled values of the chi-square distribution.

12. The distance in feet by which a parachutist misses a target is $D = \sqrt{X_1^2 + X_2^2}$, where X_1 and X_2 are independent with $X_i \sim N(0, 25)$. Find $P[D \leqslant 12.25 \text{ feet}]$.

13. Consider independent random variables $Z_i \sim N(0, 1)$, $i = 1, \ldots, 16$, and let $\bar{Z}$ be the sample mean. Find:

(a) $P[\bar{Z} < \frac{1}{2}]$.

(b) $P[Z_1 - Z_2 < 2]$.

(c) $P[Z_1 + Z_2 < 2]$.

(d) $P\left[\sum_{i=1}^{16} Z_i^2 < 32\right]$.

(e) $P\left[\sum_{i=1}^{16} (Z_i - \bar{Z})^2 < 25\right]$.

14. If $T \sim t(\nu)$, give the distribution of T^2.

15. Suppose that $X_i \sim N(\mu, \sigma^2)$, $i = 1, \ldots, n$ and $Z_i \sim N(0, 1)$, $i = 1, \ldots, k$, and all variables are independent. State the distribution of each of the following variables if it is a "named" distribution or otherwise state "unknown."

(a) $X_1 - X_2$

(b) $X_2 + 2X_3$

(c) $\dfrac{X_1 - X_2}{\sigma S_Z \sqrt{2}}$

(d) Z_1^2

(e) $\dfrac{\sqrt{n}(\bar{X} - \mu)}{\sigma S_z}$

(f) $Z_1^2 + Z_2^2$

(g) $Z_1^2 - Z_2^2$

(h) $\dfrac{Z_1}{\sqrt{Z_2^2}}$

(i) $\dfrac{Z_1^2}{Z_2^2}$

(j) $\dfrac{Z_1}{Z_2}$

(k) $\dfrac{\bar{X}}{\bar{Z}}$

(l) $\dfrac{\sqrt{nk}(\bar{X} - \mu)}{\sigma \sqrt{\sum_{i=1}^{k} Z_i^2}}$

(m) $\dfrac{\sum_{i=1}^{n} (X_i - \mu)^2}{\sigma^2} + \sum_{i=1}^{k} (Z_i - \bar{Z})^2$

(n) $\dfrac{\bar{X}}{\sigma^2} + \dfrac{\sum_{i=1}^{k} Z_i}{k}$

(o) $k\bar{Z}^2$

(p) $\dfrac{(k-1)\sum_{i=1}^{n} (X_i - \bar{X})^2}{(n-1)\sigma^2 \sum_{i=1}^{k} (Z_i - \bar{Z})^2}$

16. Let $X_1, X_2, \ldots, X_9$ be a random sample from a normal distribution, $X_i \sim N(6, 25)$, and denote by $\bar{X}$ and S^2 the sample mean and sample variance. Use tables from Appendix C to find each of the following:

(a) $P[3 < \bar{X} < 7]$.

(b) $P[1.860 < 3(\bar{X} - 6)/S]$.

(c) $P[S^2 \leqslant 31.9375]$

17. Use tabled values from Appendix C to find the following:

(a) $P[7.26 < Y < 22.31]$ if $Y \sim \chi^2(15)$.

(b) The value b such that $P[Y \leqslant b] = 0.75$ if $Y \sim \chi^2(23)$.

(c) $P\left[\dfrac{Y}{1+Y} > \dfrac{11}{16}\right]$ if $Y \sim \chi^2(6)$.

(d) $P[0.87 < T < 2.65]$ if $T \sim t(13)$.

(e) The value b such that $P[T < b] = 0.60$ if $T \sim t(26)$.

(f) The value c such that $P[|T| \geqslant c] = 0.02$ if $T \sim t(23)$.

(g) $P[2.91 < X < 5.52]$ if $X \sim F(7, 12)$.

(h) $P\left[\dfrac{1}{X} > 0.25\right]$ if $X \sim F(20, 8)$.

18. Assume that Z, V_1, and V_2 are independent random variables with $Z \sim N(0, 1)$, $V_1 \sim \chi^2(5)$, and $V_2 \sim \chi^2(9)$. Find the following:

(a) $P[V_1 + V_2 < 8.6]$.

(b) $P[Z/\sqrt{V_1/5} < 2.015]$.

(c) $P[Z > 0.611\sqrt{V_2}]$.

(d) $P[V_1/V_2 < 1.450]$.

(e) The value b such that $P\left[\dfrac{V_1}{V_1 + V_2} < b\right] = 0.90$.

19. If $T \sim t(1)$, then show the following:

(a) The CDF of T is $F(t) = \dfrac{1}{2} + \dfrac{1}{\pi}\arctan(t)$.

(b) The $100 \times \gamma$th percentile is $t_\gamma(1) = \tan[\pi(\gamma - 1/2)]$.

20. Show that if $X \sim F(2, 2b)$, then

(a) $P[X > x] = \left(1 + \dfrac{x}{b}\right)^{-b}$ for all $x > 0$.

(b) The $100 \times \gamma$th percentile is $f_\gamma(2, 2b) = b[(1 - \gamma)^{-1/b} - 1]$.

21. Show that if $F(x; \mu)$ is the CDF of $X \sim POI(\mu)$, and if $H(y; v)$ is the CDF of a chi-square distribution with v degrees of freedom, then $F(x; \mu) = 1 - H[2\mu; 2(x + 1)]$. *Hint:* Use Theorem 3.3.2 and the fact that $Y \sim \chi^2(v)$ corresponds to $Y \sim GAM(2, v/2)$.

22. If $X \sim BETA(p, q)$, derive $E(X^n)$.

23. Consider a random sample from a beta distribution, $X_i \sim BETA(1, 2)$. Use the CLT (Theorem 7.3.2) to approximate $P[\bar{X} \leqslant 0.5]$ for $n = 12$.

24. Let $Y_n \sim \chi^2(n)$. Find the limiting distribution of $(Y_n - n)/\sqrt{2n}$ as $n \to \infty$, using moment generating functions.

25. Rework Exercise 5(b) and (c) using normal approximation, and compare to the exact results.

26. Let $X_1, X_2, \ldots, X_n$ be a random sample from a distribution whose first four moments exist, and let

$$S_n^2 = \sum_{i=1}^{n} (X_i - \bar{X})^2/(n-1)$$

Show that $S_n^2 \xrightarrow{P} \sigma^2$ as $n \to \infty$. *Hint:* Use Theorem 8.2.2 and the Chebychev inequality.

27. Compare the Wilson-Hilferty approximation (Equation 8.5.2) to the exact tabled values of $\chi^2_{0.95}(10)$ and $\chi^2_{0.05}(10)$.

CHAPTER 9

POINT ESTIMATION

9.1 INTRODUCTION

The previous chapters were concerned with developing the concepts of probability and random variable to build mathematical models of nondeterministic physical phenomena. A certain numerical characteristic of the physical phenomenon may be of interest, but its value cannot be computed directly. Instead, it is possible to observe one or more random variables, the distribution of which depends on the characteristic of interest. Our main objective in the next few chapters will be to develop methods to analyze the observed values of random variables in order to gain information about the unknown characteristic.

As mentioned earlier, the process of obtaining an observed result of a physical phenomenon is called an experiment. Suppose that the result of the experiment is a random variable X, and $f(x; \theta)$ represents its pdf. It is common practice to consider X as a measurement value obtained from an individual chosen at random from a population. In this context, $f(x; \theta)$ will be referred to as the

population pdf, and it reflects the distribution of individual measurements in the population. Complete specification of $f(x; \theta)$ achieves the goal of identifying the distribution of the response of interest.

In some cases it is possible to arrive at a specified model based on axiomatic assumptions or other knowledge about the population, as was the case in certain counting problems discussed earlier. More often, the experimenter is not able to specify the pdf completely, but it may be possible to assume that $f(x; \theta)$ is a member of some known family of distributions (such as normal, gamma, Weibull, or Poisson), and that θ is an unknown parameter such as the mean or variance of the distribution. The objective of point estimation is to assign an appropriate value for θ based on observed data from the population. The observed results of repeated trials of an experiment can be modeled mathematically as a random sample from the population pdf. In other words, it is assumed that a set of n independent random variables, $X_1, X_2, \ldots, X_n$, each with pdf $f(x; \theta)$, will be observed, resulting in a set of data $x_1, x_2, \ldots, x_n$. Of course, it is possible to represent the joint pdf of the random sample as a product:

$$f(x_1, x_2, \ldots, x_n; \theta) = f(x_1; \theta)f(x_2; \theta) \cdots f(x_n; \theta) \tag{9.1.1}$$

This joint pdf provides the connection between the observed data and the mathematical model for the population. In this chapter we will be concerned with ways to make use of such data in estimating the unknown value of the parameter θ.

In subsequent chapters, other kinds of analyses will be developed. For example, the data not only can provide information about the parameter value, but also can provide information about more basic questions, such as what family of pdf's should be considered to begin with. This notion, which is generally referred to as goodness-of-fit, will be considered in a later chapter. It also is possible to answer certain questions about the population without assuming a functional form for $f(x; \theta)$. Such methods, known as nonparametric methods as well as other types of analyses, such as confidence intervals and tests of hypotheses about the value of θ, also will be considered later.

In this chapter we will assume that the distribution of a population of interest can be represented by a member of some specified family of pdf's, $f(x; \theta)$, indexed by a **parameter** θ. In some cases, the parameter will be vector-valued, and we will use boldfaced $\boldsymbol{\theta}$ to denote this.

We will let Ω, called the **parameter space**, denote the set of all possible values that the parameter θ could assume. If $\boldsymbol{\theta}$ is a vector, then Ω will be a subset of a Euclidean space of the same dimension, and the dimension of Ω will correspond to the number of unknown real parameters.

In what follows, we will assume that $X_1, X_2, \ldots, X_n$ is a random sample from $f(x; \theta)$ and that $\tau(\theta)$ is some function of θ.

Definition 9.1.1

A statistic, $T = \ell(X_1, X_2, \ldots, X_n)$, that is used to estimate the value of $\tau(\theta)$ is called an **estimator** of $\tau(\theta)$, and an observed value of the statistic, $t = \ell(x_1, x_2, \ldots, x_n)$, is called an **estimate** of $\tau(\theta)$.

Of course, this includes the case of estimating the parameter value itself, if we let $\tau(\theta) = \theta$.

Notice that we are using three different kinds of letters in our notation. The capital T represents the statistic that we use as an estimator, the lower case t is an observed value or estimate, and the script ℓ represents the function that we apply to the random sample.

Another fairly suggestive notation involves the use of a circumflex (also called a "hat") above the parameter, $\hat{\theta}$, to distinguish between the unknown parameter value and its estimator. Yet another common notation involves the use of a tilde, $\tilde{\theta}$. The practice of using capital and lowercase letters to distinguish between estimators and estimates usually is not followed with notations such as $\hat{\theta}$ and $\tilde{\theta}$.

Two of the most frequently used approaches to the problem of estimation are given in the next section.

9.2

SOME METHODS OF ESTIMATION

In some cases, reasonable estimators can be found on the basis of intuition, but various general methods have been developed for deriving estimators.

METHOD OF MOMENTS

The sample mean, $\bar{X}$, was proposed in Chapter 8 as an estimator of the population mean μ. A more general approach, which produces estimators known as the **method of moments estimators** (MMEs), can be developed.

Consider a population pdf, $f(x; \theta_1, \ldots, \theta_k)$, depending on one or more parameters $\theta_1, \ldots, \theta_k$. In Chapter 2, the moments about the origin, μ'_j, were defined. Generally, these will depend on the parameters, say

$$\mu'_j(\theta_1, \ldots, \theta_k) = E(X^j) \qquad j = 1, 2, \ldots, k \tag{9.2.1}$$

It is possible to define estimators of these distribution moments.

Definition 9.2.1

If $X_1, \ldots, X_n$ is a random sample from $f(x; \theta_1, \ldots, \theta_k)$, the first k **sample moments** are given by

$$M'_j = \frac{\sum_{i=1}^{n} X_i^j}{n} \qquad j = 1, 2, \ldots, k \tag{9.2.2}$$

As noted in Chapter 2, the first moment is the mean, $\mu'_1 = \mu$. Similarly, the first sample moment is the sample mean.

Consider the simple case of one unknown parameter, say $\theta = \theta_1$. That $\bar{X} = M'_1$ is generally a reasonable estimator of $\mu = \mu'_1(\theta)$ suggests using the solution $\hat{\theta}$ of the equation $M'_1 = \mu'_1(\hat{\theta})$ as an estimator of θ. In other words, because M'_1 tends to be close to $\mu'_1(\theta)$, under certain conditions we might expect that $\hat{\theta}$ will tend to be close to θ.

More generally, the method of moments principle is to choose as estimators of the parameters $\theta_1, \ldots, \theta_k$ the values $\hat{\theta}_1, \ldots, \hat{\theta}_k$ that render the population moments equal to the sample moments. In other words, $\hat{\theta}_1, \ldots, \hat{\theta}_k$ are solutions of the equations

$$M'_j = \mu'_j(\hat{\theta}_1, \ldots, \hat{\theta}_k) \qquad j = 1, 2, \ldots, k \tag{9.2.3}$$

Example 9.2.1 Consider a random sample from a distribution with two unknown parameters, the mean μ and the variance σ^2. We know from earlier considerations that $\mu = \mu'_1$ and $\sigma^2 = E(X^2) - \mu^2 = \mu'_2 - (\mu'_1)^2$, so that the MMEs are solutions of the equations $M'_1 = \hat{\mu}$ and $M'_2 = \hat{\sigma}^2 + (\hat{\mu})^2$, which are $\hat{\mu} = \bar{X}$ and

$$\hat{\sigma}^2 = \sum_{i=1}^{n} \frac{X_i^2}{n} - \bar{X}^2 = \sum_{i=1}^{n} \frac{(X_i - \bar{X})^2}{n}$$

Notice that the MME of σ^2 is closely related to the sample variance that was defined in Chapter 8, namely $\hat{\sigma}^2 = [(n-1)/n]S^2$.

Example 9.2.2 Consider a random sample from a two-parameter exponential distribution, $X_i \sim \text{EXP}(1, \eta)$. We know that the mean is $\mu = \mu(\eta) = 1 + \eta$, and if we set $\bar{X} = 1 + \hat{\eta}$, then $\hat{\eta} = \bar{X} - 1$ is the MME of η.

Example 9.2.3 Consider now a random sample from an exponential distribution, $X_i \sim \text{EXP}(\theta)$, and suppose we wish to estimate the probability $p(\theta) = P(X \geqslant 1) = e^{-1/\theta}$. Notice that $\mu'_1(\theta) = \mu = \theta$, so the MME of θ is $\hat{\theta} = \bar{X}$. If we reparameterize the model with $p = p(\theta) = e^{-1/\theta} = e^{-1/\mu}$, then $\mu = \mu(p) = -1/\ln p$, and if we equate $\bar{X} = \mu(\hat{p}) = -1/\ln \hat{p}$, then the MME of p is $\hat{p} = e^{-1/\bar{X}}$. Thus, in this case we see that

$\hat{p} = p(\hat{\theta})$. If a class of estimators has this property, it is said to have an "invariance" property.

Thus, to estimate $\tau(\theta)$, one might first solve $\bar{X} = \mu(\hat{\theta})$ to obtain the MME of θ and then use $\tau(\hat{\theta})$, or else one might express μ directly in terms of τ and solve $\bar{X} = \mu(\hat{\tau})$ for the MME of τ. It is not clear that both approaches will always give the same result, but if $\hat{\theta}$ is a MME of θ, then we also will refer to $\tau(\hat{\theta})$ as an MME of $\tau(\theta)$. In general, if the MMEs of the natural parameters $\theta_1, \ldots, \theta_k$ are obtained, then $\hat{\tau}_j(\theta_1, \ldots, \theta_k) = \tau_j(\hat{\theta}_1, \ldots, \hat{\theta}_k)$ will be used to estimate other functions of the natural parameters, rather than require that the moment equations be expressed directly in terms of the τ_j's.

Example 9.2.4 Consider a random sample from a gamma distribution, $X_i \sim \text{GAM}(\theta, \kappa)$. Because $\mu_1' = \mu = \kappa\theta$ and $\mu_2' = \sigma^2 + \mu^2 = \kappa\theta^2 + \kappa^2\theta^2 = \kappa(1 + \kappa)\theta^2$, so that

$$\kappa\theta = \bar{X} \quad \text{and} \quad \kappa(1 + \kappa)\theta^2 = \sum_{i=1}^{n} \frac{X_i^2}{n}$$

The resulting MMEs are

$$\hat{\theta} = \sum_{i=1}^{n} \frac{(X_i - \bar{X})^2}{n\bar{X}} = \frac{[(n-1)/n]S^2}{\bar{X}} \quad \text{and} \quad \hat{\kappa} = \frac{\bar{X}}{\hat{\theta}}$$

METHOD OF MAXIMUM LIKELIHOOD

We now will consider a method that quite often leads to estimators possessing desirable properties, particularly large sample properties. The idea is to use a value in the parameter space that corresponds to the largest "likelihood" for the observed data as an estimate of an unknown parameter.

Example 9.2.5 Suppose that a coin is biased, and it is known that the average proportion of heads is one of the three values $p = 0.20$, 0.30, or 0.80. An experiment consists of tossing the coin twice and observing the number of heads. This could be modeled mathematically as a random sample X_1, X_2 of size $n = 2$ from a Bernoulli distribution, $X_i \sim \text{BIN}(1, p)$, where the parameter space is $\Omega = \{0.20, 0.30, 0.80\}$. Notice that the MME of p, which is $\bar{X}$, does not produce reasonable estimates in this example, because $\bar{x} = 0, 0.5$, or 1 are the only possibilities, and these are not values in Ω.

Consider now the joint pdf of the random sample,

$$f(x_1, x_2; p) = p^{x_1 + x_2}(1 - p)^{2 - x_1 - x_2}$$

for $x_i = 0$ or 1. The values of $f(x_1, x_2; p)$ are provided in Table 9.1 for the various pairs (x_1, x_2) and values of p.

TABLE 9.1 Joint pdf of the numbers of heads for two tosses of a biased coin

	(x_1, x_2)			
p	(0, 0)	(0, 1)	(1, 0)	(1, 1)
0.20	0.64	0.16	0.16	0.04
0.30	0.49	0.21	0.21	0.09
0.80	0.04	0.16	0.16	0.64

Suppose that the experiment results in the observed pair $(x_1, x_2) = (0, 0)$. From Table 9.1, it would seem more likely that $p = 0.20$ rather than the other two values. Similarly, $(x_1, x_2) = (0, 1)$ or $(1, 0)$ would correspond to $p = 0.30$, and $(x_1, x_2) = (1, 1)$ would correspond to $p = 0.80$. Thus, the estimate that maximizes the "likelihood" for an observed pair (x_1, x_2) is

$$\hat{p} = \begin{cases} 0.20 & \text{if } (x_1, x_2) = (0, 0) \\ 0.30 & \text{if } (x_1, x_2) = (0, 1), (1, 0) \\ 0.80 & \text{if } (x_1, x_2) = (1, 1) \end{cases}$$

More generally, for a set of discrete random variables, the joint density function of a random sample evaluated at a particular set of sample data, say $f(x_1, \ldots, x_n; \theta)$, represents the probability that the observed set of data $x_1, \ldots, x_n$ will occur. For continuous random variables, $f(x_1, \ldots, x_n; \theta)$ is not a probability but it still reflects the relative "likelihood" that the set of data will occur, and this likelihood depends on the true value of the parameter.

Definition 9.2.2

Likelihood Function The joint density function of n random variables $X_1, \ldots, X_n$ evaluated at $x_1, \ldots, x_n$, say $f(x_1, \ldots, x_n; \theta)$, is referred to as a **likelihood function**. For fixed $x_1, \ldots, x_n$ the likelihood function is a function of θ and often is denoted by $L(\theta)$.

If $X_1, \ldots, X_n$ represents a random sample from $f(x; \theta)$, then

$$L(\theta) = f(x_1; \theta) \cdots f(x_n; \theta) \quad \textbf{(9.2.4)}$$

For a given observed set of data, $L(\theta)$ gives the likelihood of that set occurring as a function of θ. The maximum likelihood principle of estimation is to choose as the estimate of θ, for a given set of data, that value for which the observed set of data would have been most likely to occur. That is, if the likelihood of observing a given set of observations is much higher when $\theta = \theta_1$ than when $\theta = \theta_2$, then it is reasonable to choose θ_1 as an estimate of θ rather than θ_2.

Definition 9.2.3

Maximum Likelihood Estimator Let $L(\theta) = f(x_1, \ldots, x_n; \theta)$, $\theta \in \Omega$, be the joint pdf of $X_1, \ldots, X_n$. For a given set of observations, $(x_1, \ldots, x_n)$, a value $\hat{\theta}$ in Ω at which $L(\theta)$ is a maximum is called a **maximum likelihood estimate** (MLE) of θ. That is, $\hat{\theta}$ is a value of θ that satisfies

$$f(x_1, \ldots, x_n; \hat{\theta}) = \max_{\theta \in \Omega} f(x_1, \ldots, x_n; \theta) \tag{9.2.5}$$

Notice that if each set of observations $(x_1, \ldots, x_n)$ corresponds to a unique value $\hat{\theta}$, then this procedure defines a function, $\hat{\theta} = \ell(x_1, \ldots, x_n)$. This same function, when applied to the random sample, $\hat{\theta} = \ell(X_1, \ldots, X_n)$, is called the **maximum likelihood estimator**, also denoted MLE. Usually, the same notation, $\hat{\theta}$, is used for both the ML estimate and the ML estimator.

In most applications, $L(\theta)$ represents the joint pdf of a random sample, although the maximum likelihood principle also applies to other cases such as sets of order statistics.

If Ω is an open interval, and if $L(\theta)$ is differentiable and assumes a maximum on Ω, then the MLE will be a solution of the equation (maximum likelihood equation)

$$\frac{d}{d\theta} L(\theta) = 0 \tag{9.2.6}$$

Strictly speaking, if one or more solutions of equation (9.2.6) exist, then it should be verified which, if any, maximize $L(\theta)$. Note also that any value of θ that maximizes $L(\theta)$ also will maximize the log-likelihood, $\ln L(\theta)$, so for computational convenience the alternate form of the maximum likelihood equation,

$$\frac{d}{d\theta} \ln L(\theta) = 0 \tag{9.2.7}$$

often will be used.

Example 9.2.6 Consider a random sample from a Poisson distribution, $X_i \sim \text{POI}(\theta)$. The likelihood function is

$$L(\theta) = \prod_{i=1}^{n} f(x_i; \theta) = \frac{e^{-n\theta}\theta^{\sum_{i=1}^{n} x_i}}{\prod_{i=1}^{n} x_i!}$$

and the log-likelihood is

$$\ln L(\theta) = -n\theta + \sum_{i=1}^{n} x_i \ln \theta - \ln \left(\prod_{i=1}^{n} x_i!\right)$$

The maximum likelihood equation is

$$\frac{d}{d\theta} \ln L(\theta) = -n + \sum_{i=1}^{n} \frac{x_i}{\theta} = 0$$

which has the solution $\hat{\theta} = \sum_{i=1}^{n} \frac{x_i}{n} = \bar{x}$. It is possible to verify that this is a maximum by use of the second derivative,

$$\frac{d^2}{d\theta^2} \ln L(\theta) = -\sum_{i=1}^{n} \frac{x_i}{\theta^2}$$

which is negative when evaluated at $\bar{x}$, $-n/\bar{x} < 0$.

In subsequent examples, unless otherwise indicated, $\sum$ will denote a summation from 1 to n. Similarly, $\prod$ will denote a product from 1 to n.

Suppose now that we wish to estimate

$$\tau = \tau(\theta) = P[X = 0] = e^{-\theta}$$

We may reparameterize the model in terms of τ by letting $\theta = -\ln \tau$. We obtain

$$f^*(x; \tau) = \frac{\tau^n(-\ln \tau)^{\sum x_i}}{\prod x_i!}$$

If $L^*(\tau)$ represents the likelihood function relative to τ, then

$$\ln L^*(\tau) = n \ln \tau + \sum x_i \ln(-\ln \tau) - \ln \prod x_i!$$

$$\frac{d \ln L^*(\tau)}{d\tau} = \frac{n}{\tau} + \frac{\sum x_i}{-\ln \tau}\left(\frac{-1}{\tau}\right)$$

and setting the derivative equal to zero gives

$$-\ln \hat{\tau} = \bar{x} \quad \hat{\tau} = e^{-\bar{x}}$$

In this example, it follows that $\hat{\tau} = \hat{\tau}(\theta) = \tau(\hat{\theta})$.

We could have maximized $L^*(\tau)$ relative to τ in this case directly by the chain rule without carrying out the reparameterization. Specifically,

$$\frac{d}{d\theta} \ln L(\theta) = \frac{d}{d\tau} \ln L^*(\tau) \frac{d\tau}{d\theta}$$

and if $d\tau/d\theta \neq 0$, then $d/d\tau[\ln L^*(\tau)] = 0$ whenever $d/d\theta[\ln L(\theta)] = 0$, so that the maximum with respect to τ occurs at $\tau(\hat{\theta})$.

It should be noted that we are using the notation τ to represent both a function of θ and a value in the range of the function. This is a slight misuse of standard mathematical notation, but it is a convenient practice that often is used in problems involving reparameterization.

In general, if u is some one-to-one function with inverse u^{-1}, and if $\tau = u(\theta)$, then we can define $L^*(\tau) = L(u^{-1}(\tau))$. It follows that $\hat{\tau}$ will maximize $L^*(\tau)$ if $u^{-1}(\hat{\tau}) = \hat{\theta}$, or equivalently, $\hat{\tau} = u(\hat{\theta})$. When u is not one-to-one, there is no unique solution to $\tau = u(\theta)$ for every value of τ. The usual approach in this case is to extend the definition of the function $L^*(\tau)$. For example, if for every value τ, $L(\theta)$ attains a maximum over the subset of Ω such that $\tau = u(\theta)$, then we define $L^*(\tau)$ to be this maximum value. This generalizes the reparameterized likelihood function to cases where u is not one-to-one, and it follows that $\hat{\tau} = u(\hat{\theta})$ maximizes $L^*(\tau)$ when $\hat{\theta}$ maximizes $L(\theta)$. (See Exercise 43.) These results are summarized in the following theorem.

Theorem 9.2.1 **Invariance Property** If $\hat{\theta}$ is the MLE of θ and if $u(\theta)$ is a function of θ, then $u(\hat{\theta})$ is an MLE of $u(\theta)$. ■

In other words, if we reparameterize by $\tau = \tau(\theta)$, then the MLE of τ is $\hat{\tau} = \tau(\hat{\theta})$.

Example 9.2.7 Consider a random sample from an exponential distribution, $X_i \sim \text{EXP}(\theta)$. The likelihood function for a sample of size n is

$$L(\theta) = \frac{1}{\theta^n} e^{-\sum x_i/\theta} \qquad 0 < x_i < \infty$$

Thus,

$$\ln L(\theta) = -n \ln \theta - \frac{\sum x_i}{\theta} \quad \text{and} \quad \frac{d}{d\theta} \ln L(\theta) = \frac{-n}{\theta} + \frac{\sum x_i}{\theta^2}$$

Equating this derivative to zero gives $\hat{\theta} = \bar{x}$.

If we wish to estimate $p(\theta) = P(X \geqslant 1) = e^{-1/\theta}$, then we know from Theorem 9.2.1 that the MLE is $p(\hat{\theta}) = e^{-1/\bar{x}}$.

There are cases where the MLE exists but cannot be obtained as a solution of the ML equation.

Example 9.2.8 Consider a random sample from a two-parameter exponential distribution, $X_i \sim \text{EXP}(1, \eta)$. The likelihood function is $L(\eta) = \exp\left[-\sum (x_i - \eta)\right]$ if all $x_i \geqslant \eta$ and zero otherwise. If we denote the minimum of $x_1, \ldots, x_n$ by $x_{1:n}$, then we can write $L(\eta) = \exp\left[n(\eta - \bar{x})\right]$ if $x_{1:n} \geqslant \eta$ and zero otherwise. The graph of $L(\eta)$ is shown in Figure 9.1.

FIGURE 9.1 The likelihood function for a random sample from EXP(1, η)

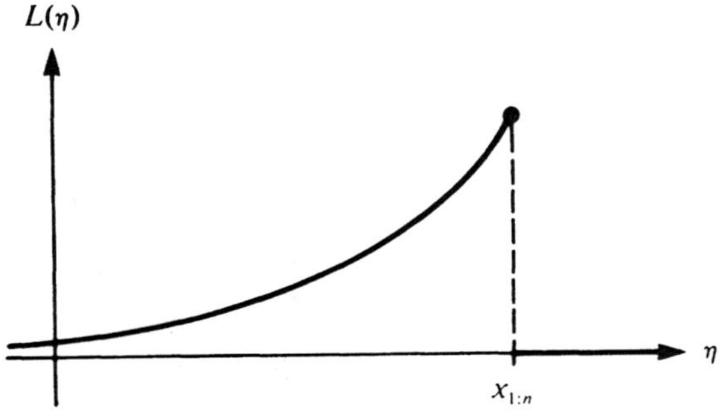

It is clear from the figure that $L(\eta)$ is maximized at $\hat{\eta} = x_{1:n}$, and the ML estimator is the first order statistic. This is an example where the MLE and the MME are different. (See Example 9.2.2).

As noted earlier, the ML principle can be used in situations where the observed variables are not independent or identically distributed.

Example 9.2.9 The lifetime of a certain component follows an exponential distribution, EXP(θ). Suppose that n components are randomly selected and placed on test, and the first r observed failure times are denoted by $x_{1:n}, \ldots, x_{r:n}$. From equation (6.5.12) the joint density of $X_{1:n}, \ldots, X_{r:n}$ is given by

$$L(\theta) = f(x_{1:n}, \ldots, x_{r:n}; \theta)$$

$$= \frac{n!}{(n-r)!} \exp\left(\frac{-\sum_{i=1}^{r} x_{i:n}}{\theta}\right) \exp\left[\frac{-(n-r)x_{r:n}}{\theta}\right]\left(\frac{1}{\theta}\right)^r$$

$$= \frac{n!}{(n-r)!\,\theta^r} \exp \frac{-\left[\sum_{i=1}^{r} x_{i:n} + (n-r)x_{r:n}\right]}{\theta}$$

Note that $T = \sum_{i=1}^{r} X_{i:n} + (n-r)X_{r:n}$ represents the total survival time of the n items on test until the experiment is terminated. To obtain the MLE of θ based on these data, we have

$$\ln L(\theta) = \text{const} - r \ln \theta - \frac{T}{\theta} \qquad \frac{d}{d\theta} \ln L(\theta) = \frac{-r}{\theta} + \frac{T}{\theta^2}$$

Setting the derivative equal to zero gives $\hat{\theta} = T/r$.

If the complete sample is observed, then $r = n$ and $\hat{\theta} = \bar{x}$ as before.

The previous examples have involved the distributions with one unknown parameter. The definitions of likelihood function and maximum likelihood estimator can be applied in the case of more than one unknown parameter if $\boldsymbol{\theta}$ represents a vector of parameters, say $\boldsymbol{\theta} = (\theta_1, \ldots, \theta_k)$. Although Ω could, in general, be almost any sort of k-dimensional set, in most examples it is a Cartesian product of k intervals. When Ω is of this form and if the partial derivatives of $L(\theta_1, \ldots, \theta_k)$ exist, and the MLEs do not occur on the boundary of Ω, then the MLEs will be solutions of the simultaneous equations

$$\frac{\partial}{\partial \theta_j} \ln L(\theta_1, \ldots, \theta_k) = 0 \tag{9.2.8}$$

for $j = 1, \ldots, k$. These are called the **maximum likelihood equations**, and the solutions are denoted by $\hat{\theta}_1, \ldots, \hat{\theta}_k$. As in the one-parameter case, it generally is necessary to verify that the solutions of the ML equations maximize $L(\theta_1, \ldots, \theta_k)$.

Theorem 9.2.2 **Invariance Property** If $\hat{\boldsymbol{\theta}} = (\hat{\theta}_1, \ldots, \hat{\theta}_k)$ denotes the MLE of $\boldsymbol{\theta} = (\theta_1, \ldots, \theta_k)$, then the MLE of $\boldsymbol{\tau} = (\tau_1(\boldsymbol{\theta}), \ldots, \tau_r(\boldsymbol{\theta}))$ is $\hat{\boldsymbol{\tau}} = (\hat{\tau}_1, \ldots, \hat{\tau}_r) = (\tau_1(\hat{\boldsymbol{\theta}}), \ldots, \tau_r(\hat{\boldsymbol{\theta}}))$ for $1 \leqslant r \leqslant k$. ■

The situation here is similar to the case of a single parameter. If $\boldsymbol{\tau}$ represents a one-to-one transformation, then a reparameterized likelihood function can be defined, and the MLE of $\boldsymbol{\tau}$ is obtained as the transformation of the MLE of $\boldsymbol{\theta}$. In the case of a transformation that is not one-to-one, the likelihood function relative to $\boldsymbol{\tau}$ must be extended in a manner similar to the single-parameter case.

Note that the multiparameter estimators often are not the same as the individual estimators when the other parameters are assumed to be known. This is illustrated by the following example.

Example 9.2.10 For a set of random variables $X_i \sim N(\mu, \sigma^2)$, the MLEs of μ and $\theta = \sigma^2$ based on a random sample of size n are desired. We have

$$f(x; \mu, \theta) = \frac{1}{\sqrt{2\pi\theta}} e^{-(x-\mu)^2/2\theta}$$

$$L(\mu, \theta) = (2\pi\theta)^{-n/2} \exp\left[\frac{-\sum (x_i - \mu)^2}{2\theta}\right]$$

$$\ln L(\mu, \theta) = \text{const} - \frac{n}{2} \ln \theta - \frac{\sum (x_i - \mu)^2}{2\theta}$$

$$\frac{\partial \ln L(\mu, \theta)}{\partial \mu} = \frac{2 \sum (x_i - \mu)}{2\theta}$$

and

$$\frac{\partial \ln L(\mu, \theta)}{\partial \theta} = \frac{-n}{2\theta} + \frac{\sum (x_i - \mu)^2}{2\theta^2}$$

Setting these derivatives equal to zero and solving simultaneously for the solution values $\hat{\mu}$ and $\hat{\theta}$ yields the MLEs

$$\hat{\mu} = \bar{x} \qquad \hat{\theta} = \hat{\sigma}^2 = \frac{\sum (x_i - \bar{x})^2}{n}$$

The ML method enjoys the invariance property in the two-parameter normal case. For example, if the likelihood function is maximized with respect to μ and σ, then one obtains $\hat{\mu} = \bar{x}$ and $\hat{\sigma} = \sqrt{\hat{\theta}}$, and similarly for other functions of the parameters. Notice that if $\theta = \theta_0$ is known, then from the first ML equation

$$\frac{2 \sum (x_i - \hat{\mu})}{2\theta_0} = 0$$

yields $\hat{\mu} = \bar{x}$ as before, but if $\mu = \mu_0$ is assumed known, then from the second ML equation

$$\frac{-n}{2\hat{\theta}} + \frac{\sum (x_i - \mu_0)^2}{2\hat{\theta}^2} = 0$$

yields

$$\hat{\theta} = \frac{\sum (x_i - \mu_0)^2}{n}$$

Example 9.2.11 Consider a random sample from a two-parameter distribution with both parameters unknown, $X_i \sim \text{EXP}(\theta, \eta)$. The population pdf is

$$f(x; \theta, \eta) = \left(\frac{1}{\theta}\right) e^{-(x-\eta)/\theta} \qquad \eta \leqslant x$$

The likelihood function is

$$L(\theta, \eta) = \left(\frac{1}{\theta}\right)^n \exp\left[\frac{-\sum (x_i - \eta)}{\theta}\right] \qquad \text{all } x_i \geqslant \eta$$

and the log-likelihood is

$$\ln L(\theta, \eta) = -n \ln \theta - \frac{\sum x_i - n\eta}{\theta} \qquad x_{1:n} \geqslant \eta$$

where $x_{1:n}$ is the minimum of $x_1, \ldots, x_n$. As in Example 9.2.8, the likelihood is maximized with respect to η by taking $\hat{\eta} = x_{1:n}$. To maximize relative to θ, we may differentiate $\ln L(\theta, \hat{\eta})$ with respect to θ, and solve the resulting equation,

$$\frac{d \ln L(\theta, \hat{\eta})}{d\theta} = \frac{-n}{\theta} + \frac{\sum (x_i - \hat{\eta})}{\theta^2} = 0$$

which yields

$$\hat{\theta} = \frac{\sum (x_i - \hat{\eta})}{n} = \bar{x} - \hat{\eta} = \bar{x} - x_{1:n}$$

The α percentile, x_α, such that $F(x_\alpha) = \alpha$ is given by $x_\alpha = -\theta \ln (1 - \alpha) + \eta$, and the MLE of x_α is $\hat{x}_\alpha = -\hat{\theta} \ln (1 - \alpha) + \hat{\eta}$ by Theorem 9.2.2.

Example 9.2.12 Let us consider ML estimation for the parameters of a gamma distribution, GAM(θ, κ), based on a random sample of size n. We have

$$L(\theta, \kappa) = \frac{1}{\theta^{n\kappa}[\Gamma(\kappa)]^n} \left[\prod x_i\right]^{\kappa - 1} \exp \left[-\sum x_i/\theta\right]$$

and

$$\ln L(\theta, \kappa) = -n\kappa \ln \theta - n \ln \Gamma(\kappa) + (\kappa - 1) \ln \prod x_i - \sum x_i/\theta$$

The partial derivatives are

$$\frac{\partial \ln L(\theta, \kappa)}{\partial \theta} = \frac{-n\kappa}{\theta} + \frac{\sum x_i}{\theta^2}$$

$$\frac{\partial \ln L(\theta, \kappa)}{\partial \kappa} = -n \ln \theta - n\Gamma'(\kappa)/\Gamma(\kappa) + \ln \prod x_i$$

If we let $\tilde{x} = (\prod x_i)^{1/n}$ denote the geometric mean of the sample and let $\Psi(\kappa) = \Gamma'(\kappa)/\Gamma(\kappa)$ denote the psi function, then setting the derivatives equal to zero gives the equations

$$\hat{\theta} = \bar{x}/\hat{\kappa}$$

$$\ln (\hat{\kappa}) - \Psi(\hat{\kappa}) - \ln (\bar{x}/\tilde{x}) = 0$$

This provides an example where the ML equations cannot be solved in closed form, although a numerical solution for $\hat{\kappa}$ can be obtained from the last equation; for example, by using tables of the psi function. We see that $\hat{\kappa}$ is a function of $\bar{x}/\tilde{x}$ and is not a function of $\bar{x}$, $\tilde{x}$, and n separately. Thus it is convenient to provide tables of $\hat{\kappa}$ in terms of $\bar{x}/\tilde{x}$. Perhaps the best approach to ML estimation for the gamma distribution is the use of the following rational approximation [Greenwood and Durand (1960)]:

$$\hat{\kappa} = \frac{0.5000876 + 0.1648852M - 0.0544274M^2}{M} \qquad 0 < M \leqslant 0.5772$$

$$\hat{\kappa} = \frac{8.898919 + 9.059950M + 0.9775373M^2}{M(17.79728 + 11.968477M + M^2)} \qquad 0.5772 < M \leqslant 17$$

$$\hat{\kappa} = \frac{1}{M} \qquad M > 17$$

where $M = \ln (\bar{x}/\tilde{x})$.

The MLEs are not the same as the MMEs, but the ML estimate of the mean is $\hat{\mu} = \hat{\theta}\hat{\kappa} = \bar{x}$.

It is also possible to have solutions to the ML equations that can be obtained only by numerical methods.

Example 9.2.13 Consider a random sample of size n from a Weibull distribution with both scale and shape parameters unknown, $X_i \sim \text{WEI}(\theta, \beta)$. The population pdf is

$$f(x; \theta, \beta) = (\beta/\theta)(x/\theta)^{\beta - 1} \exp\left[-(x/\theta)^{\beta}\right]$$

for $x > 0$; $\theta > 0$, and $\beta > 0$, and the log-likelihood function is

$$\ln L(\theta, \beta) = n \ln(\beta/\theta) + (\beta - 1) \sum \ln(x_i/\theta) - \sum (x_i/\theta)^{\beta}$$

which leads to the ML equations

$$\frac{\partial}{\partial\theta} \ln L(\theta, \beta) = -n\beta/\theta + (\beta/\theta) \sum (x_i/\theta)^{\beta} = 0$$

$$\frac{\partial}{\partial\beta} \ln L(\theta, \beta) = n/\beta + \sum \ln(x_i/\theta) - \sum (x_i/\theta)^{\beta} \ln(x_i/\theta) = 0$$

After some algebra, the MLEs are the solutions $\beta = \hat{\beta}$ and $\theta = \hat{\theta}$ of the equations

$$g(\beta) = \frac{\sum x_i^{\beta} \ln x_i}{\sum x_i^{\beta}} - \frac{1}{\beta} - \frac{\sum \ln x_i}{n} = 0$$

$$\theta = \left(\sum x_i^{\beta}/n\right)^{1/\beta}$$

The equation $g(\beta) = 0$ cannot be solved explicitly as a function of the data, but for a given set of data it is possible to solve for $\hat{\beta}$ by an iterative numerical method such as the Newton-Raphson method. Specifically, we can define a sequence $\beta_1, \beta_2, \ldots$ such that

$$\hat{\beta}_m = \hat{\beta}_{m-1} - \frac{g(\hat{\beta}_{m-1})}{g'(\hat{\beta}_{m-1})}$$

where $\hat{\beta}_0 > 0$ is an initial value, $g'(\beta)$ is the derivative of $g(\beta)$, and $\hat{\beta}_m \to \hat{\beta}$ as $m \to \infty$.

Some large-sample properties of MLEs will be discussed in Section 9.4, and additional methods of estimation will be presented in Section 9.5.

9.3

CRITERIA FOR EVALUATING ESTIMATORS

Several properties of estimators would appear to be desirable, including unbiasedness.

> ***Definition 9.3.1***
>
> **Unbiased Estimator** An estimator T is said to be an **unbiased estimator** of $\tau(\theta)$ if
>
> $$E(T) = \tau(\theta) \tag{9.3.1}$$
>
> for all $\theta \in \mathbf{\Omega}$. Otherwise, we say that T is a **biased estimator** of $\tau(\theta)$.

If an unbiased estimator is used to assign a value of $\tau(\theta)$, then the correct value of $\tau(\theta)$ may not be achieved by any given estimate, t, but the "average" value of T will be $\tau(\theta)$.

Example 9.3.1 Consider a random sample from a distribution $f(x; \boldsymbol{\theta})$ with $\boldsymbol{\theta} = (\mu, \sigma^2)$, where μ and σ^2 are the mean and variance of the population. It was shown in Section 8.2 that the sample mean and variance, $\bar{X}$ and S^2, are unbiased estimators of μ and σ^2, respectively. If both μ and σ^2 are unknown, then the appropriate parameter space is a subset of two-dimensional Euclidean space. In particular, $\mathbf{\Omega}$ is the Cartesian product of the intervals $(-\infty, \infty)$ and $(0, \infty)$; $\mathbf{\Omega} = (-\infty, \infty) \times (0, \infty)$. If only one parameter is unknown, then Ω will consist of the corresponding one-dimensional set. For example, suppose the population is normal with unknown mean, μ, but known variance $\sigma^2 = 9$. The appropriate parameter space is $\mathbf{\Omega} = (-\infty, \infty)$, because in general for the mean of a normal distribution, $-\infty < \mu < \infty$.

We may desire to estimate a percentile, say the 95th percentile, of the distribution $N(\mu, 9)$. This is an example of a function of the parameter, because $\tau(\mu) = \mu + \sigma z_{0.95} = \mu + 4.95$. It follows that $T = \bar{X} + 4.95$ is an unbiased estimator of $\tau(\mu)$, because $E(T) = E(\bar{X} + 4.95) = E(\bar{X}) + 4.95 = \mu + 4.95$, regardless of the value of μ.

It is possible to have a reasonable estimator that is biased, and often an estimator can be adjusted to make it unbiased.

Example 9.3.2 Consider a random sample of size n from an exponential distribution, $X_i \sim \text{EXP}(\theta)$, with parameter θ. Because θ is the mean of the distribution, we know that the MLE, $\bar{X}$, is unbiased for θ. If we wish to estimate the reciprocal of the mean, $\tau(\theta) = 1/\theta$, then by the invariance property the MLE is $T_1 = 1/\bar{X}$. However, T_1 is a biased estimator of $1/\theta$, which follows from results in Chapter 8. In particular, from Theorems 8.3.3 and 8.3.4, we know that

$$Y = \frac{2n\bar{X}}{\theta} = \sum_{i=1}^{n} \frac{2X_i}{\theta} \sim \chi^2(2n)$$

Furthermore, it follows from equation (8.3.6) with $r = -1$ that $E(Y^{-1}) = 1/[2(n-1)]$, and consequently $E(T_1) = [n/(n-1)](1/\theta)$. Although this shows that T_1 is a biased estimator of $1/\theta$, it also follows that an adjusted estimator of the form cT_1, where $c = (n-1)/n$ is unbiased for $1/\theta$. This also suggests that while the unadjusted estimator, T_1, is biased, it still might be reasonable because the amount of bias, $1/[(n-1)\theta]$, is small for large n.

It is not always possible to adjust a biased estimator in this manner. For example, suppose that it is desired to estimate $1/\theta$ using only the smallest observation, which would correspond to observing the first order statistic, $X_{1:n}$. It was shown in Example 7.2.2 that $X_{1:n} \sim \text{EXP}(\theta/n)$, and consequently $nX_{1:n}$ is another example of an unbiased estimator of θ. This suggests that $T_2 = 1/(nX_{1:n})$ also could be used to estimate $1/\theta$. The statistic T_2 cannot be adjusted in the above manner to be unbiased for $1/\theta$, because $E(T_2)$ does not even exist.

The statistics T_1 and T_2 illustrate a possible flaw in the concept of unbiasedness as a general principle. In particular, if $\hat{\theta}$ is an unbiased estimator of θ, then $\tau(\hat{\theta})$ will not necessarily be an unbiased estimator $\tau(\theta)$. Yet $\tau(\hat{\theta})$ may be a reasonable estimator of $\tau(\theta)$, such as the case when $\hat{\theta} = \bar{X}$ and $\tau(\hat{\theta}) = 1/\bar{X}$.

It often is possible to derive several different potential estimators of a parameter. For example, in some cases the MLEs and the MMEs have basically different forms. This raises the obvious question of how to decide which estimators are "best" in some sense, and this question will be discussed next.

A very general idea is to select the estimator that tends to be closest or "most concentrated" around the true value of the parameter. It might be reasonable to say that T_1 is **more concentrated** than T_2 about $\tau(\theta)$ if

$$P[\tau(\theta) - \varepsilon < T_1 < \tau(\theta) + \varepsilon] \geq P[\tau(\theta) - \varepsilon < T_2 < \tau(\theta) + \varepsilon] \tag{9.3.2}$$

for all $\varepsilon > 0$, and that an estimator is **most concentrated** if it is more concentrated than any other estimator.

The idea of a more concentrated estimator is illustrated in Figure 9.2, which shows the pdf's of two estimators T_1 and T_2.

It is not clear how to obtain an estimator that is most concentrated, but some other concepts will be discussed that may partially achieve this goal. For

FIGURE 9.2 The concept of "more concentrated"

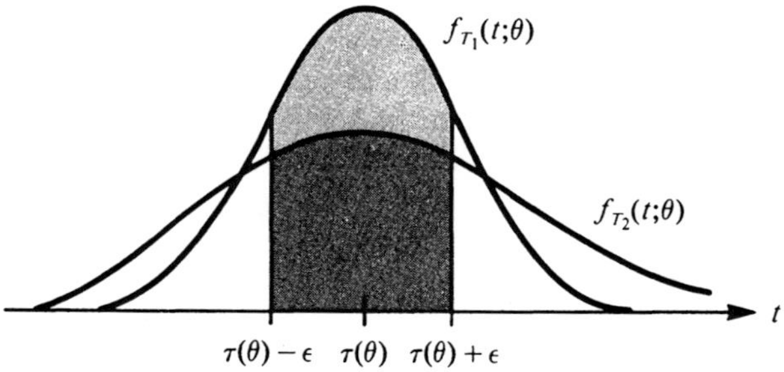

example, if T is an unbiased estimator of $\tau(\theta)$, it follows from the Chebychev inquality that

$$P[\tau(\theta) - \varepsilon < T < \tau(\theta) + \varepsilon] \geqslant 1 - \operatorname{Var}(T)/\varepsilon^2 \tag{9.3.3}$$

for all $\varepsilon > 0$. This suggests that for unbiased estimators, one with a smaller variance will tend to be more concentrated and thus may be preferable.

Example 9.3.3 Let us reconsider Example 9.3.2, where we are interested only in estimating the mean, θ. If $\hat{\theta}_1 = \bar{X}$ and $\hat{\theta}_2 = nX_{1:n}$, then both estimators are unbiased for θ, but $\operatorname{Var}(\hat{\theta}_1) = \theta^2/n$ and $\operatorname{Var}(\hat{\theta}_2) = \theta^2$. Thus, for $n > 1$, $\operatorname{Var}(\hat{\theta}_1) < \operatorname{Var}(\hat{\theta}_2)$ for all $\theta > 0$, and $\hat{\theta}_1$ is the better estimator by this criterion.

In some cases one estimator may have a smaller variance for some values of θ and a larger variance for other values of θ. In such a case neither estimator can be said to be better than the other in general. In certain cases it is possible to show that a particular unbiased estimator has the smallest possible variance among all possible unbiased estimators for all values of θ. In such a case one could restrict attention to that particular estimator.

UNIFORMLY MINIMUM VARIANCE UNBIASED ESTIMATORS

Definition 9.3.2

Let $X_1, X_2, \ldots, X_n$ be a random sample of size n from $f(x; \theta)$. An estimator T^* of $\tau(\theta)$ is called a **uniformly minimum variance unbiased estimator** (UMVUE) of $\tau(\theta)$ if

1. T^* is unbiased for $\tau(\theta)$, and
2. for any other unbiased estimator T of $\tau(\theta)$, $\operatorname{Var}(T^*) \leqslant \operatorname{Var}(T)$ for all $\theta \in \mathbf{\Omega}$.

In some cases, lower bounds can be derived for the variance of unbiased estimators. If an unbiased estimator can be found that attains such a lower bound, then it follows that the estimator is a UMVUE. In the following discussion, if the appropriate derivatives exist and can be passed under the integral sign (or summation), then a lower bound for the variance of an unbiased estimator can be established. Among other things, this will require that the domain of the integrand must not depend on θ.

If T is an unbiased estimator of $\tau(\theta)$, then the **Cramer-Rao lower bound** (CRLB), based on a random sample, is

$$\operatorname{Var}(T) \geqslant \frac{[\tau'(\theta)]^2}{nE\left[\frac{\partial}{\partial\theta} \ln f(X; \theta)\right]^2} \tag{9.3.4}$$

Assuming differentiability conditions as mentioned earlier, the CRLB can be developed as follows. We will assume the case of sampling from a continuous distribution. The discrete case is similar.

Consider the function defined by

$$u(x_1, \ldots, x_n; \theta) = \frac{\partial}{\partial\theta} \ln f(x_1, \ldots, x_n; \theta)$$

which also can be written

$$u(x_1, \ldots, x_n; \theta) = \frac{1}{f(x_1, \ldots, x_n; \theta)} \frac{\partial}{\partial\theta} f(x_1, \ldots, x_n; \theta) \tag{9.3.5}$$

If we define a random variable $U = u(X_1, \ldots, X_n; \theta)$, then

$$\begin{aligned}
E(U) &= \int \cdots \int u(x_1, \ldots, x_n; \theta) f(x_1, \ldots, x_n; \theta)\, dx_1 \cdots dx_n \\
&= \int \cdots \int \frac{\partial}{\partial\theta} f(x_1, \ldots, x_n; \theta)\, dx_1 \cdots dx_n \\
&= \frac{d}{d\theta} \int \cdots \int f(x_1, \ldots, x_n; \theta)\, dx_1 \cdots dx_n \\
&= \frac{d}{d\theta} 1 \\
&= 0
\end{aligned}$$

Note also that if $T = t(X_1, \ldots, X_n)$ is unbiased for $\tau(\theta)$, then

$$\tau(\theta) = E(T) = \int \cdots \int t(x_1, \ldots, x_n) f(x_1, \ldots, x_n; \theta)\, dx_1 \cdots dx_n$$

If we differentiate with respect to θ, then

$$\begin{aligned}\tau'(\theta) &= \int \cdots \int \ell(x_1, \ldots, x_n) \frac{\partial}{\partial \theta} f(x_1, \ldots, x_n; \theta)\, dx_1 \cdots dx_n \\ &= \int \cdots \int \ell(x_1, \ldots, x_n) u(x_1, \ldots, x_n; \theta) f(x_1, \ldots, x_n; \theta)\, dx_1 \cdots dx_n \\ &= E(TU)\end{aligned}$$

It also follows, from equations (2.4.6) and (5.2.11) and the fact that $E(U) = 0$, that $\mathrm{Var}(U) = E(U^2)$ and $\mathrm{Cov}(T, U) = E(TU)$.

Because the correlation coefficient is always between ± 1 [see equation (5.3.2)], it follows that $[\mathrm{Cov}(T, U)]^2 \leqslant \mathrm{Var}(T)\,\mathrm{Var}(U)$, and consequently $\mathrm{Var}(T)E(U^2) \geqslant [\tau'(\theta)]^2$, so that

$$\mathrm{Var}(T) \geqslant \frac{[\tau'(\theta)]^2}{E\left[\frac{\partial}{\partial \theta} \ln f(X_1, \ldots, X_n; \theta)\right]^2} \tag{9.3.6}$$

When $X_1, \ldots, X_n$ represent a random sample,

$$f(x_1, \ldots, x_n; \theta) = f(x_1; \theta) \cdots f(x_n; \theta)$$

so that

$$u(x_1, \ldots, x_n; \theta) = \sum_{i=1}^{n} \frac{\partial}{\partial \theta} \ln f(x_i; \theta)$$

in which case

$$E(U^2) = \mathrm{Var}(U) = n\,\mathrm{Var}\left[\frac{\partial}{\partial \theta} \ln f(X; \theta)\right] = nE\left[\frac{\partial}{\partial \theta} \ln f(X; \theta)\right]^2$$

which yields inequality (9.3.4).

Note that if the proper differentiability conditions hold, as mentioned earlier, it can be shown that

$$E\left[\frac{\partial}{\partial \theta} \ln f(X; \theta)\right]^2 = -E\left[\frac{\partial^2}{\partial \theta^2} \ln f(X; \theta)\right]$$

Example 9.3.4 Consider a random sample from an exponential distribution, $X_i \sim \mathrm{EXP}(\theta)$. Because

$$\ln f(x; \theta) = -x/\theta - \ln \theta$$

$$\begin{aligned}\frac{\partial}{\partial \theta} \ln f(x; \theta) &= x/\theta^2 - 1/\theta \\ &= (x - \theta)/\theta^2\end{aligned}$$

Thus,

$$E\left[\frac{\partial}{\partial\theta}\ln f(X;\theta)\right]^2 = E[(X-\theta)^2/\theta^4]$$
$$= \theta^2/\theta^4$$
$$= 1/\theta^2$$

and the CRLB for $\tau(\theta) = \theta$ is $1/[n(1/\theta^2)] = \theta^2/n$. Because $\text{Var}(\bar{X}) = \theta^2/n$, it follows that $\bar{X}$ is the UMVUE of θ.

It is possible to obtain more information about the type of estimator, the variance of which can attain the CRLB, by further considering the derivation of inequality (9.3.6). The lower bound is attained only when the correlation coefficient of T and U is ± 1. It follows from Theorem 5.3.1 that this occurs if and only if T and U are linearly related, say $T = aU + b$ with probability 1 for some constants $a \neq 0$ and b. Thus, for T to attain the CRLB of $\tau(\theta)$, it must be a linear function of

$$\sum_{i=1}^{n}\frac{\partial}{\partial\theta}\ln f(X_i;\theta)$$

Example 9.3.5 We take a random sample $X_1, \ldots, X_n$ from a geometric distribution with parameter $\theta = p$, $X_i \sim \text{GEO}(\theta)$, and we wish to find a UMVUE for $\tau(\theta) = 1/\theta$. Because

$$\ln f(x;\theta) = \ln\theta + (x-1)\ln(1-\theta)$$
$$\frac{\partial}{\partial\theta}\ln f(x;\theta) = \frac{1}{\theta} - \frac{x-1}{1-\theta}$$
$$= \frac{x - 1/\theta}{\theta - 1}$$

For the variance of an unbiased estimator T to attain the CRLB, it must have the form

$$T = a\sum_{i=1}^{n}(X_i - 1/\theta)/(\theta - 1) + b$$

which also can be expressed as a linear function of the sample mean, say $T = c\bar{X} + d$ for constants c and d. Because $\bar{X}$ is unbiased for $1/\theta$, then necessarily $c = 1$ and $d = 0$, so that $T = \bar{X}$ is the only such estimator. The variance of $\bar{X}$ is $\text{Var}(\bar{X}) = (1-\theta)/(n\theta^2)$, which also can be shown to be the CRLB for this case.

This discussion also suggests that only certain types of functions will admit an unbiased estimator, the variance of which can attain the CRLB.

Theorem 9.3.1 If an unbiased estimator for $\tau(\theta)$ exists, the variance of which achieves the CRLB, then only a linear function of $\tau(\theta)$ will admit an unbiased estimator, the variance of which achieves the corresponding CRLB. ∎

Thus, in the previous example, there is no unbiased estimator, the variance of which attains the CRLB for unbiased estimators of θ because θ is not a linear function of $1/\theta$. It cannot be concluded from this that an UMVUE of θ does not exist, only that it cannot be found by the CRLB approach. In the next chapter we will study a method that often works when the present approach fails.

Comparisons involving the variances of estimators often are used to decide which method makes more efficient use of the data.

Definition 9.3.3

Efficiency The **relative efficiency** of an unbiased estimator T of $\tau(\theta)$ to another unbiased estimator T^* of $\tau(\theta)$ is given by

$$\text{re}(T, T^*) = \frac{\text{Var}(T^*)}{\text{Var}(T)} \tag{9.3.7}$$

An unbiased estimator T^* of $\tau(\theta)$ is said to be **efficient** if $\text{re}(T, T^*) \leqslant 1$ for all unbiased estimators T of $\tau(\theta)$, and all $\theta \in \boldsymbol{\Omega}$. The efficiency of an unbiased estimator T of $\tau(\theta)$ is given by

$$\text{e}\,(T) = \text{re}(T, T^*) \tag{9.3.8}$$

if T^* is an efficient estimator of $\tau(\theta)$.

Notice that in this terminology an efficient estimator is just a UMVUE.

The notion of relative efficiency can be interpreted in terms of the sample sizes required for two types of estimators to estimate a parameter with comparable accuracy. Specifically, suppose that T_1 and T_2 are unbiased estimators of $\tau(\theta)$, and that the variances are of the form $\text{Var}(T_1) = k_1/n$ and $\text{Var}(T_2) = k_2/n$. The relative efficiency in this case is of the form $\text{re}(T_1, T_2) = k_2/k_1$. If it is desired to choose sample sizes, say n_1 and n_2, to achieve the same variance by either method, then $k_1/n_1 = k_2/n_2$, which implies $n_2/n_1 = \text{re}(T_1, T_2)$. In other words, if T_1 is less efficient than T_2, one could choose a larger sample size, by a factor of k_1/k_2, to achieve equal variances.

Some authors define the efficiency of T to be the ratio CRLB/Var(T), which allows the possibility that a UMVUE could exist but not be efficient by this definition. However, it does follow that if CRLB/Var(T) = 1, then T is an efficient estimator by Definition 9.3.3. At this point, use of the CRLB is the only convenient means we have to verify that an estimator is efficient.

Example 9.3.6 Recall in Example 9.3.2 that the estimator $T = (n-1)/(n\bar{X})$ is unbiased for $\tau(\theta) = 1/\theta$. In this case $\tau'(\theta) = -1/\theta^2$ and the CRLB is $[-1/\theta^2]^2/[n(1/\theta^2)] = 1/(n\theta^2)$. It was found in Example 9.3.4 that the variance of $\bar{X}$ attains the CRLB of unbiased estimators of θ. Because $\tau(\theta) = 1/\theta$ is not a linear function of θ, there is no unbiased estimator of $1/\theta$ whose variance equals $1/(n\theta^2)$. In terms of the random variable $Y = 2n\bar{X}/\theta$, we can express T as $T = [2(n-1)/\theta]Y^{-1}$. From this and equation (8.3.6) we can show that $\text{Var}(T) = 1/[(n-2)\theta^2]$. Even though $\text{Var}(T)$ does not attain the CRLB, it is quite close to it for large n. It often is possible to obtain an unbiased estimator, the variance of which is close to the CRLB even though it does not achieve it exactly, so the CRLB can be useful in evaluating a proposed estimator, whether an UMVUE exists or not. Actually, we will be able to show in the next chapter that the estimator T is an UMVUE for $1/\theta$. This means that T is an example of an efficient estimator that cannot be obtained by the CRLB method.

Example 9.3.7 Recall in Example 9.3.3 that we had unbiased estimators $\hat{\theta}_1 = \bar{X}$ and $\hat{\theta}_2 = nX_{1:n}$ of θ. It was later found that $\hat{\theta}_1$ is a UMVUE. Thus, $\hat{\theta}_1$ is an efficient estimator of θ and the efficiency of $\hat{\theta}_2$ is

$$\text{e}(\hat{\theta}_2) = \text{re}(\hat{\theta}_2, \hat{\theta}_1) = \frac{\theta^2/n}{\theta^2} = \frac{1}{n}$$

and thus $\hat{\theta}_2$ is a very poor estimator of θ because its efficiency is small for large n.

A slightly biased estimator that is highly concentrated about the parameter of interest may be preferable to an unbiased estimator that is less concentrated. Thus, it is desirable to have more general criteria that allow for both biased and unbiased estimators to be compared.

Definition 9.3.4

If T is an estimator of $\tau(\theta)$, then the **bias** is given by

$$\text{b}(T) = E(T) - \tau(\theta) \tag{9.3.9}$$

and the **mean squared error** (MSE) of T is given by

$$\text{MSE}(T) = E[T - \tau(\theta)]^2 \tag{9.3.10}$$

Theorem 9.3.2 If T is an estimator of $\tau(\theta)$, then

$$\text{MSE(T)} = \text{Var}(T) + [\text{b}(T)]^2 \tag{9.3.11}$$

Proof

$$\begin{aligned}\text{MSE}(T) &= E[T - \tau(\theta)]^2\\ &= E[T - E(T) + E(T) - \tau(\theta)]^2\\ &= E[T - E(T)]^2 + 2[E(T) - \tau(\theta)]\\ &\quad \times [E(T) - E(T)] + [E(T) - \tau(\theta)]^2\\ &= \text{Var}(T) + [\text{b}(T)]^2\end{aligned}$$

■

The MSE is a reasonable criterion that considers both the variance and the bias of an estimator, and it agrees with the variance criterion if attention is restricted to unbiased estimators. It provides a useful means for comparing two or more estimators, but it is not possible to obtain an estimator that has uniformly minimum MSE for all $\theta \in \mathbf{\Omega}$ and all possible estimators.

Example 9.3.8 Consider a family of pdf's $f(x; \theta)$ where the parameter space $\mathbf{\Omega}$ contains at least two values. If no restrictions are placed on the type of estimators under consideration, then constant estimators, $\hat{\theta}_c = c$ for $c \in \mathbf{\Omega}$, cannot be excluded. Such estimators clearly are not desirable from a practical point of view, because they do not even depend on the sample, yet each such estimator has a small MSE for values of θ near c. In particular, $\text{MSE}(\hat{\theta}_c) = (c - \theta)^2$, which is zero if $\theta = c$. This means that for a uniformly minimum MSE estimator, $\hat{\theta}$, necessarily $\text{MSE}(\hat{\theta}) = 0$ for all $\theta \in \mathbf{\Omega}$. This would mean that $\hat{\theta}$ is constant, say $\hat{\theta} = c^*$ (with probability 1). Now, if $\theta \in \mathbf{\Omega}$ and $\theta \neq c^*$, then $\text{MSE}(\hat{\theta}) = (c^* - \theta)^2 > 0$, in which case $\hat{\theta}$ does not have uniformly minimum MSE.

If the class of estimators under consideration can be restricted to a smaller class, then it may be possible to find a uniformly minimum MSE estimator. For example, restriction to unbiased estimators eliminates estimators of the constant type, because $\hat{\theta}_c = c$ is not an unbiased estimator of θ.

Example 9.3.9 Consider a random sample from a two-parameter exponential distribution with known scale parameter, say $\theta = 1$, and unknown location parameter η. In other words, $X_i \sim \text{EXP}(1, \eta)$.

We wish to compare the MME and the MLE, $\hat{\eta}_1$ and $\hat{\eta}_2$, respectively. Specifically, let $\hat{\eta}_1 = \bar{X} - 1$ and $\hat{\eta}_2 = X_{1:n}$. It is easy to show that $\bar{X} - \eta \sim \text{GAM}(1/n, n)$ and $X_{1:n} - \eta \sim \text{EXP}(1/n)$, and it follows that

$$E(\hat{\eta}_1) = E(\bar{X} - 1) = E(\bar{X}) - 1 = 1 + \eta - 1 = \eta$$

and

$$E(\hat{\eta}_2) = E(X_{1:n}) = E(X_{1:n} - \eta + \eta) = E(X_{1:n} - \eta) + \eta = 1/n + \eta$$

Thus, $\hat{\eta}_1$ is unbiased and $\hat{\eta}_2$ is biased with bias term $b(\hat{\eta}_2) = 1/n$. Their MSEs are

$$\text{MSE}(\hat{\eta}_1) = \text{Var}(\bar{X} - 1) = \text{Var}(\bar{X}) = 1/n$$

and

$$\begin{aligned}\text{MSE}(\hat{\eta}_2) &= \text{Var}(\hat{\eta}_2) + (1/n)^2 = \text{Var}(X_{1:n}) + (1/n)^2 \\ &= \text{Var}(X_{1:n} - \eta) + (1/n)^2 = (1/n)^2 + (1/n)^2 = 2/n^2\end{aligned}$$

Thus, for $n > 2$ the biased estimator has a much smaller MSE than does the unbiased estimator.

It also is possible to adjust $\hat{\eta}_2$ to be unbiased, say $\hat{\eta}_3 = X_{1:n} - 1/n$, so that $E(\hat{\eta}_3) = E(X_{1:n}) - 1/n = \eta + 1/n - 1/n = \eta$ and $\text{MSE}(\hat{\eta}_3) = \text{Var}(X_{1:n}) = \text{Var}(X_{1:n} - \eta) = 1/n^2$. Thus, for $n > 1$, $\hat{\eta}_3$ has the smallest MSE of the three.

It is interesting to note that in Example 9.3.3, when the sample was assumed to be from an exponential distribution with unknown scale parameter θ, the MLE of θ, which is $\bar{X}$, was much superior to the one based on $X_{1:n}$. In the present example, where the distribution is exponential with a known scale but unknown location parameter, the result is just the reverse.

9.4 LARGE-SAMPLE PROPERTIES

We have discussed properties of estimators such as unbiasedness and uniformly minimum variance. These are defined for any fixed sample size n, and are examples of "small-sample" properties. It also is useful to consider asymptotic or "large-sample" properties of a particular type of estimator. An estimator may have undesirable properties for small n, but still be a reasonable estimator in certain applications if it has good asymptotic properties as the sample size increases. It also is possible quite often to evaluate the asymptotic properties of an estimator when small-sample properties are difficult to determine.

Definition 9.4.1

Simple Consistency Let $\{T_n\}$ be a sequence of estimators of $\tau(\theta)$. These estimators are said to be **consistent** estimators of $\tau(\theta)$ if for every $\varepsilon > 0$,

$$\lim_{n\to\infty} P[|T_n - \tau(\theta)| < \varepsilon] = 1 \qquad \textbf{(9.4.1)}$$

for every $\theta \in \boldsymbol{\Omega}$.

In the terminology of Chapter 7, T_n converges stochastically to $\tau(\theta)$, $T_n \xrightarrow{P} \tau(\theta)$ as $n \to \infty$. Sometimes this also is referred to as **simple** consistency. One interpretation of consistency is that for larger sample sizes the estimator tends to be more concentrated about $\tau(\theta)$, and by making n sufficiently large T_n can be made as concentrated as desired.

Another slightly stronger type of consistency is based on the MSE.

Definition 9.4.2

MSE Consistency If $\{T_n\}$ is a sequence of estimators of $\tau(\theta)$, then they are called **mean squared error consistent** if

$$\lim_{n \to \infty} E[T_n - \tau(\theta)]^2 = 0 \tag{9.4.2}$$

for every $\theta \in \boldsymbol{\Omega}$.

Another desirable property is asymptotic unbiasedness.

Definition 9.4.3

Asymptotic Unbiased A sequence $\{T_n\}$ is said to be **asymptotically unbiased** for $\tau(\theta)$ if

$$\lim_{n \to \infty} E(T_n) = \tau(\theta) \tag{9.4.3}$$

for all $\theta \in \boldsymbol{\Omega}$.

It can be shown that a MSE consistent sequence also is asymptotically unbiased and simply consistent.

Theorem 9.4.1 A sequence $\{T_n\}$ of estimators of $\tau(\theta)$ is mean squared error consistent if and only if it is asymptotically unbiased and $\lim\limits_{n \to \infty} \mathrm{Var}(T_n) = 0$.

Proof

This follows immediately from Theorem 9.3.2, because

$$\mathrm{MSE}(T_n) = \mathrm{Var}(T_n) + [E(T_n) - \tau(\theta)]^2$$

Because both terms on the right are nonnegative, $\mathrm{MSE}(T_n) \to 0$ implies both $\mathrm{Var}(T_n) \to 0$ and $E(T_n) \to \tau(\theta)$. The converse is obvious. ■

Example 9.4.1 In Example 9.3.2 we considered the reciprocal of the sample mean, which we now denote by $T_n = 1/\bar{X}$, as an estimator of $\tau(\theta) = 1/\theta$. As noted earlier, $Y = 2n\bar{X}/\theta \sim \chi^2(2n)$. It follows from equation (8.3.6) that $E(T_n) = [n/(n-1)](1/\theta)$ and $\text{Var}(T_n) = [n/(n-1)]^2/[(n-2)\theta^2]$, so that $E(T_n) \to 1/\theta$ and $\text{Var}(T_n) \to 0$ as $n \to \infty$. Thus, even though T_n is not unbiased, it is asymptotically unbiased and MSE consistent for $\tau(\theta) = 1/\theta$.

As mentioned earlier, MSE consistency is a stronger property than simple consistency.

Theorem 9.4.2 If a sequence $\{T_n\}$ is mean squared error consistent, it also is simply consistent.

Proof

This follows from the Markov inequality (2.4.11), with $X = T_n - \tau(\theta)$, $r = 2$, and $c = \varepsilon$, so that

$$P[|T_n - \tau(\theta)| < \varepsilon] \geqslant 1 - \text{MSE}(T_n)/\varepsilon^2$$

which approaches 1 as $n \to \infty$. ∎

Example 9.4.2 Let $X_1, \ldots, X_n$ be a random sample from a distribution with finite mean μ and variance σ^2. It was shown in Chapter 7 that the sample mean, $\bar{X}_n$, converges stochastically to μ, and if the fourth moment, μ_4', is finite, then the sample variance, S_n^2, converges stochastically to σ^2. Actually, because $\bar{X}_n$ and S_n^2 are unbiased and their respective variances approach zero as $n \to \infty$, it follows that they are both simply and MSE consistent.

If the distribution is exponential, $X_i \sim \text{EXP}(\theta)$, then it follows that $\bar{X}_n$ is MSE consistent for θ, but the estimator $\hat{\theta}_n = nX_{1:n}$ is not even simply consistent, because $nX_{1:n} \sim \text{EXP}(\theta)$.

If the distribution is the two-parameter exponential distribution, EXP(1, η), as in Example 9.3.10, then the unbiased estimator $\hat{\eta}_3 = X_{1:n} - 1/n$ is MSE consistent, because $\text{MSE}(\hat{\eta}_3) = 1/n^2 \to 0$. However, $\hat{\eta}_4 = c(X_{1:n} - 1/n)$ is not MSE consistent for fixed $c \neq 1$ and $\eta \neq 0$, because $\text{MSE}(\hat{\eta}_4) = c^2/n^2 + (c-1)^2\eta^2 \to (c-1)^2\eta^2 \neq 0$. The choice of c that minimized MSE when $\eta = 1$ was $c = n^2/(1 + n^2)$, which has limit 1. In general, if $c \to 1$ as $n \to \infty$, then $\text{MSE}(\hat{\eta}_4) \to 0$, and $\hat{\eta}_4$ is MSE consistent, and also asymptotically unbiased for η.

Theorem 9.4.3 If $\{T_n\}$ is simply consistent for $\tau(\theta)$ and if $g(t)$ is continuous at each value of $\tau(\theta)$, then $g(T_n)$ is simply consistent for $g(\tau(\theta))$.

Proof

This follows immediately from Theorem 7.7.2 with $Y_n = T_n$ and $c = \tau(\theta)$. ■

A special application of this theorem is that if $\tau(\theta)$ is a continuous function of θ and $\hat{\theta}_n$ is simply consistent for θ, then $\tau(\hat{\theta}_n)$ is simply consistent for $\tau(\theta)$.

It also is possible to formulate an asymptotic version of efficiency.

Definition 9.4.4

Asymptotic Efficiency Let $\{T_n\}$ and $\{T_n^*\}$ be two asymptotically unbiased sequences of estimators for $\tau(\theta)$. The **asymptotic relative efficiency** of T_n relative to T_n^* is given by

$$\text{are}(T_n, T_n^*) = \lim_{n \to \infty} \frac{\text{Var}(T_n^*)}{\text{Var}(T_n)} \tag{9.4.4}$$

The sequence $\{T_n^*\}$ is said to be **asymptotically efficient** if $\text{are}(T_n, T_n^*) \leqslant 1$ for all other asymptotically unbiased sequences $\{T_n\}$, and all $\theta \in \mathbf{\Omega}$, The **asymptotic efficiency** of an asymptotically unbiased sequence $\{T_n\}$ is given by

$$\text{ae}(T_n) = \text{are}(T_n, T_n^*) \tag{9.4.5}$$

if $\{T_n^*\}$ is asymptotically efficient.

The CRLB is not always attainable for fixed n, but it often is attainable asymptotically, in which case it can be quite useful in determining asymptotic efficiency.

Example 9.4.3 Recall that in Example 9.4.1, which involved sampling from EXP(θ), the sequence $T_n = 1/\bar{X}$ was shown to be asymptotically unbiased for $1/\theta$. The variance is $\text{Var}(T_n) = [n/(n-1)]^2/[(n-2)\theta^2]$ and the CRLB is $[-1/\theta^2]^2/[n(1/\theta^2)] = 1/[n\theta^2]$. Because

$$\lim_{n \to \infty} \frac{\text{CRLB}}{\text{Var}(T_n)} = \lim_{n \to \infty} \frac{1/[n\theta^2]}{[n/(n-1)]^2/[(n-2)\theta^2]} = 1$$

it follows that T_n is asymptotically efficient for estimating $1/\theta$.

Example 9.4.4 Consider again Example 9.3.9, where the population pdf is a two-parameter exponential, $X_i \sim \text{EXP}(1, \eta)$. Because the range of X_i depends on η, the CRLB cannot be used here. The estimators $\hat{\eta}_2 = X_{1:n}$ and $\hat{\eta}_3 = X_{1:n} - 1/n$ are both asymptotically unbiased, and both have the same variance, $\text{Var}(\hat{\eta}_2) = \text{Var}(\hat{\eta}_3) = 1/n^2$, and thus

$$\text{are}(\hat{\eta}_3, \hat{\eta}_2) = \lim_{n \to \infty} \frac{1/n^2}{1/n^2} = 1$$

We will see later that $\hat{\eta}_3$ is a UMVUE for η, and thus $\hat{\eta}_2$ also is asymptotically efficient for estimating η.

Another idea might be to compare MSEs rather than variance. The present example suggests that this generally will not provide the same criterion as comparing variances, because

$$\frac{\text{MSE}(\hat{\eta}_3)}{\text{MSE}(\hat{\eta}_4)} = \frac{1/n^2}{1/n^2 + 1/n^2} = \frac{1}{2}$$

This results from the fact that the bias squared and the variance are both the same power of $1/n$, namely $1/n^2$. Thus, some limitation on the bias must be considered or the asymptotic relative efficiency may be misleading.

In Example 9.3.9, another unbiased estimator, $\hat{\eta}_1 = \bar{X} - 1$, also was considered, and $\text{Var}(\hat{\eta}_1) = 1/n$. Thus

$$\text{are}(\hat{\eta}_1, \hat{\eta}_3) = \lim_{n \to \infty} \frac{1/n^2}{1/n} = 0$$

so $\hat{\eta}_1$ is not as desirable as $\hat{\eta}_3$.

It should be noted that this is an unusual example; in most cases the variance of an estimator is of the form c/n, whereas in the case of $\hat{\eta}_3$ it is $1/n^2$, which is a higher power of $1/n$.

An estimator with variance of order $1/n^2$ usually is referred to as a **superefficient** estimator.

It often is difficult to obtain an exact expression for the variance of a proposed estimator. Another approach, which sometimes is used, restricts attention only to estimators that are asymptotically normal and replaces the exact variances with asymptotic variances in the definition of asymptotic relative efficiency. Specifically, if $\{T_n\}$ and $\{T_n^*\}$ are asymptotically normal with asymptotic mean $\tau(\theta)$ and respective asymptotic variances $k(\theta)/n$ and $k^*(\theta)/n$, then the alternative definition is

$$\text{are}(T_n, T_n^*) = \frac{k^*(\theta)}{k(\theta)} \tag{9.4.6}$$

Of course, this approach is appropriate only for comparing asymptotically normal estimators, but it is somewhat simpler to use in many cases. An estimator T_n^* is at least as good as T_n if $k^*(\theta) \leqslant k(\theta)$ for all $\theta \in \boldsymbol{\Omega}$, and T_n^* is asymptotically efficient, in this sense, if this inequality holds for all asymptotically normal estimators T_n. Such an estimator is often referred to as **best asymptotically normal** (BAN).

Example 9.4.5 A random sample of size n is drawn from an exponential distribution, $X_i \sim \text{EXP}(\theta)$, and it is desired to estimate the distribution median, $\tau(\theta) = (\ln 2)\theta$. One possible estimator is the sample median, $T_n = X_{r:n}$ with $r/n \to 1/2$ as $n \to \infty$. Under the conditions of Theorem 7.5.1 with $p = 1/2$, T_n is asymptotically normal with asymptotic mean $\tau(\theta)$ and asymptotic variance θ^2/n. Another possibility is based on the sample mean, $T_n^* = (\ln 2)\bar{X}_n$. It follows from the central limit theorem that $Z_n = \sqrt{n}(\bar{X}_n - \theta)/\theta \xrightarrow{d} \text{N}(0, 1)$ as $n \to \infty$, and consequently T_n^* is asymptotically normal with asymptotic mean and variance, respectively $\tau(\theta)$ and $(\ln 2)^2\theta^2/n$. Thus, $k(\theta) = \theta^2$ and $k^*(\theta) = (\ln 2)^2\theta^2$, and by definition (9.4.6) the asymptotic relative efficiency is $(\ln 2)^2\theta^2/\theta^2 = (\ln 2)^2 = 0.48$, and T_n^* is the better estimator, Actually, it is possible to show, by comparison with the CRLB, that T_n^* is efficient, but it still might be useful in some applications to know that for large samples, a method based on the sample median is 48% as efficient as one based on the sample mean.

ASYMPTOTIC PROPERTIES OF MLEs

Under certain circumstances, it can be shown that the MLEs have very desirable properties. Specifically, if certain regularity conditions are satisfied, then the solutions, $\hat{\theta}_n$, of the maximum likelihood equations have the following properties:

1. $\hat{\theta}_n$ exists and is unique,
2. $\hat{\theta}_n$ is a consistent estimator of θ,
3. $\hat{\theta}_n$ is asymptotically normal with asymptotic mean θ and variance

$$1/nE\left[\frac{\partial}{\partial\theta}\ln f(X;\theta)\right]^2, \quad \text{and}$$

4. $\hat{\theta}_n$ is asymptotically efficient.

Of course, for an MLE to result from solving the ML equation (9.2.6), it is necessary that the partial derivative of $\ln f(x; \theta)$ with respect to θ exists, and also that the set $A = \{x\colon f(x; \theta) > 0\}$ does not depend on θ. Additional conditions involving the derivatives of $\ln f(x; \theta)$ and $f(x; \theta)$ also are required, but we will not discuss them here. Different sets of regularity conditions are discussed by Wasan (1970, p. 158) and Bickel and Doksum (1977, p. 150).

Notice that the asymptotic efficiency of $\hat{\theta}_n$ follows from the fact that the asymptotic variance is the same as the CRLB for unbiased estimators of θ. Thus, for large n, approximately

$$\hat{\theta}_n \sim \text{N}(\theta, \text{CRLB})$$

It also follows from Theorem 7.7.6 that if $\tau(\theta)$ is a function with nonzero derivative, then $\hat{\tau}_n = \tau(\hat{\theta}_n)$ also is asymptotically normal with asymptotic mean $\tau(\theta)$ and variance $[\tau'(\theta)]^2\text{CRLB}$. Notice also that the asymptotic variance of $\hat{\tau}_n$ is the CRLB for variances of unbiased estimators $\tau = \tau(\theta)$, so that $\hat{\tau}_n$ also is asymptotically efficient.

Example 9.4.6 Recall in Example 9.2.7 that the MLE of the mean θ of an exponential distribution is the sample mean, $\hat{\theta}_n = \bar{X}_n$. It is possible to infer the same asymptotic properties either from the preceding discussion or from the Central Limit Theorem. In particular, $\hat{\theta}_n$ is asymptotically normal with asymptotic mean θ and variance θ^2/n. It also was shown in Example 9.3.4 that CRLB $= \theta^2/n$. We also know the exact distribution of $\hat{\theta}_n$ in this case, because

$$\frac{2n\hat{\theta}_n}{\theta} \sim \chi^2(2n)$$

which is consistent with the asymptotic normal result

$$\sqrt{n}\,\frac{\hat{\theta}_n - \theta}{\theta} \sim \mathrm{N}(0, 1)$$

because a properly standardized chi-square variable has a standard normal limiting distribution.

Suppose that now we are interested in estimating

$$R = R(t;\, \theta) = P(X > t) = \exp\,(-t/\theta)$$

An approximation for the variance of $\hat{R} = \exp\,(-t/\hat{\theta})$ is given by the asymptotic variance

$$\begin{aligned}\mathrm{Var}(\hat{R}) &\doteq \left[\frac{\partial}{\partial\theta} R(t;\, \theta)\right]^2 (\theta^2/n)\\ &= [\exp\,(-t/\theta)(t/\theta^2)]^2(\theta^2/n)\\ &= [\exp\,(-t/\theta)(t/\theta)]^2/n\\ &= [R(\ln R)]^2/n\end{aligned}$$

and thus for large n, approximately

$$\hat{R} \sim \mathrm{N}(R,\, R(\ln R)^2/n)$$

Example 9.4.7 Consider a random sample from a Pareto distribution, $X_i \sim \mathrm{PAR}(1, \kappa)$ where κ is unknown. Because

$$f(x;\, \kappa) = \kappa(1 + x)^{-\kappa - 1} \qquad x > 0$$

it follows that

$$\ln L(\kappa) = n \ln \kappa - (\kappa + 1) \sum \ln\,(1 + x_i)$$

and the ML equation is

$$\frac{d}{d\kappa} \ln L(\kappa) = n/\kappa - \sum \ln\,(1 + x_i) = 0$$

which yields the MLE

$$\hat{\kappa} = n/\sum \ln(1 + x_i)$$

To find the CRLB, note that

$$\ln f(x; \kappa) = \ln \kappa - (\kappa + 1) \ln(1 + x)$$

$$\frac{\partial}{\partial \kappa} \ln f(x; \kappa) = 1/\kappa - \ln(1 + x)$$

and thus,

$$\text{CRLB} = 1/nE[1/\kappa - \ln(1 + X)]^2$$

To evaluate this last expression, it is convenient to consider the transformation

$$Y = \ln(1 + X) \sim \text{EXP}(1/\kappa)$$

so that

$$E[\ln(1 + X)] = 1/\kappa$$

$$E[1/\kappa - \ln(1 + X)]^2 = \text{Var}[\ln(1 + X)] = 1/\kappa^2$$

$$\text{Var}(\hat{\kappa}) \doteq \text{CRLB} = \kappa^2/n$$

and approximately

$$\hat{\kappa} \sim \text{N}(\kappa, \kappa^2/n)$$

Example 9.4.8 Consider a random sample from the two-parameter exponential distribution, $X_i \sim \text{EXP}(1, \eta)$. We recall from Example 9.2.8 that the MLE, $\hat{\eta} = X_{1:n}$, cannot be obtained as a solution to the ML equation (9.2.7), because $\ln L(\eta)$ is not differentiable over the whole parameter space. Of course, the difficulty results from the fact that the set $A = \{x: f(x; \eta) > 0\} = [\eta, \infty)$ depends on the parameter η. Thus, we have an example where the asymptotic normality of the MLE is not expected to hold.

As a matter of fact, we know from the results of Chapter 7 that the first order statistic, $X_{1:n}$, is not asymptotically normal; rather, for a suitable choice of norming constants, the corresponding limiting distribution is an extreme-value type for minimums.

Asymptotic properties such as those discussed earlier in the section exist for MLEs in the multiparameter case, but they cannot be expressed conveniently without matrix notation. Consequently, we will not consider them here.

9.5

BAYES AND MINIMAX ESTIMATORS

When an estimate differs from the true value of the parameter being estimated, one may consider the loss involved to be a function of this difference. If it is assumed that the loss increases as the square of the difference, then the MSE criterion simply considers the average squared error loss associated with the estimator. Clearly the MSE criterion can be generalized to other types of loss functions besides squared error.

Definition 9.5.1

Loss Function If T is an estimator of $\tau(\theta)$, then a **loss function** is any real-valued function, $L(t; \theta)$, such that

$$L(t; \theta) \geqslant 0 \quad \text{for every } t \tag{9.5.1}$$

and

$$L(t; \theta) = 0 \quad \text{when } t = \tau(\theta) \tag{9.5.2}$$

Definition 9.5.2

Risk Function The **risk function** is defined to be the expected loss,

$$R_T(\theta) = E[L(T; \theta)] \tag{9.5.3}$$

Thus, if a parameter or a function of a parameter is being estimated, one may choose an appropriate loss function depending on the problem, and then try to find an estimator, the average loss or risk function of which is small for all possible values of the parameter. If the loss function is taken to be squared error loss, then the risk becomes the MSE as considered previously. Another reasonable loss function is absolute error, which gives the risk function $R_T(\theta) = E|T - \tau(\theta)|$.

As for the MSE, it usually will not be possible to determine for other risk functions an estimator that has smaller risk than all other estimators uniformly for all θ. When comparing two specific estimators, it is possible that one may have a smaller risk than the other for all θ.

Definition 9.5.3

Admissible Estimator An estimator T_1 is a **better estimator** than T_2 if and only if

$$R_{T_1}(\theta) \leqslant R_{T_2}(\theta) \quad \text{for all } \theta \in \mathbf{\Omega}$$

and

$$R_{T_1}(\theta) < R_{T_2}(\theta) \quad \text{for at least one } \theta$$

An estimator T is **admissible** if and only if there is no better estimator.

Thus, if one estimator has uniformly smaller risk than another, we will retain the first estimator for consideration and eliminate the latter as not admissible. Typically, some estimators will have smallest risk for some values of θ but not for others. As mentioned earlier, one possible approach to select a best estimator is to restrict the class of estimators. This was discussed in connection with the class of unbiased estimators with MSE risk. There is no guarantee that this approach will work in every case.

Example 9.5.1 In Example 9.3.9 we found an unbiased estimator, $\hat{\eta}_3 = X_{1:n} - 1/n$, which appeared to be reasonable for estimating the location parameter η. We now consider a class of estimators of the form $\hat{\eta}_4 = c\hat{\eta}_3$ for some constant $c > 0$. Such estimators will be biased except in the case $c = 1$, and the MSE risk would be $\text{MSE}(\hat{\eta}_4) = \text{Var}(c\hat{\eta}_3) + [b(c\hat{\eta}_3)]^2 = c^2/n^2 + (c-1)^2\eta^2$.

Let us attempt to find a member of this class of estimators with minimum MSE. This corresponds to choosing $c = n^2\eta^2/(1 + n^2\eta^2)$. Unfortunately, this depends upon the unknown parameter that we are trying to estimate. However, this suggests the possibility of choosing c to obtain an estimator that will have smaller risk, at least over a portion of the parameter space. For example, if it is suspected that η is somewhere close to 1, then the appropriate constant would be $c = n^2/(1 + n^2)$. For this choice of c, $\text{MSE}(\hat{\eta}_4) < \text{MSE}(\hat{\eta}_3)$ if and only if η satisfies this inequality $c^2/n^2 + (c-1)^2\eta^2 < 1/n^2$, which corresponds to $\eta^2 < 2 + 1/n^2$. For a sample of size $n = 3$, $c = 0.9$, $\text{MSE}(\hat{\eta}_4) = 0.09 + 0.01\eta^2$, and $\text{MSE}(\hat{\eta}_4) < \text{MSE}(\hat{\eta}_3)$ if $-1.45 < \eta < 1.45$ and $\text{MSE}(\hat{\eta}_4) \geqslant \text{MSE}(\hat{\eta}_3)$ otherwise. Thus, the MSE is not uniformly smaller for either estimator, and one could not choose between them solely by comparing MSEs. The comparison between the MSEs for these estimators is provided in Figure 9.3.

If it could be determined a priori that $-1.45 < \eta < 1.45$, then $\hat{\eta}_4$ would be preferable, but otherwise it might be best to use the unbiased estimator $\hat{\eta}_3$.

FIGURE 9.3 Comparison of MSEs for two different estimators of the threshold parameter

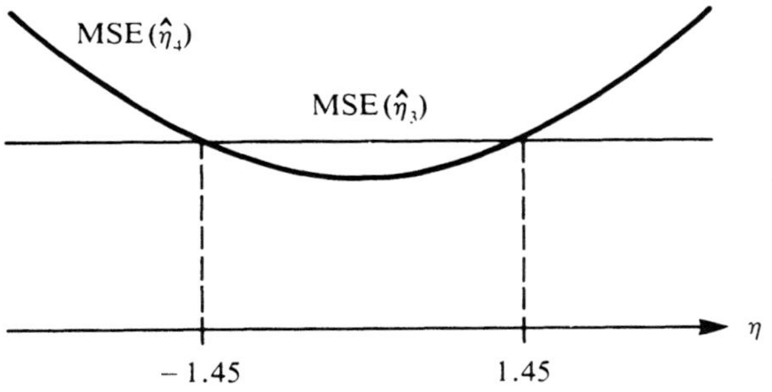

Another criterion that sometimes is used to select an estimator from the class of admissible estimators is the minimax criterion.

> ***Definition 9.5.4***
>
> **Minimax Estimator** An estimator T_1 is a **minimax estimator** if
>
> $$\max_{\theta} R_{T_1}(\theta) \leqslant \max_{\theta} R_T(\theta) \tag{9.5.4}$$
>
> for every estimator T.

In other words, T_1 is an estimator that minimizes the maximum risk, or

$$\max_{\theta} R_{T_1}(\theta) = \min_{T} \max_{\theta} R_T(\theta) \tag{9.5.5}$$

Of course, this assumes that the risk function attains a maximum value for some θ and that such maximum values attain a minimum for some T. In a more general treatment of the topic, the maximum and minimum could be replaced with the more general concepts of least upper bound and greatest lower bound, respectively.

The minimax principle is a conservative approach, because it attempts to protect against the worst risk that can occur.

Example 9.5.2 Consider the class of estimators of the form $\hat{\eta}_4 = c(X_{1:n} - 1/n)$ discussed in Example 9.3.10, and MSE risk. Recall that $\text{MSE}(\hat{\eta}_4) = c^2/n^2 + (c - 1)^2\eta^2$, which depends on η except when $c = 1$. This last case corresponds to the unbiased estimate $\hat{\eta}_3$. If $0 < c < 1$, then neither $\hat{\eta}_3$ nor $\hat{\eta}_4$ has uniformly smaller MSE for all η, so we might consider using the minimax principle. Because

$$\max_{\eta} \text{MSE}(\hat{\eta}_3) = 1/n^2$$

and

$$\max_{\eta} \text{MSE}(\hat{\eta}_4) = \max_{\eta} [c^2/n^2 + (c-1)^2\eta^2] = \infty$$

the unbiased estimator, $\hat{\eta}_3$, is the minimax estimator within this class of estimators. It is not clear, at this point, whether $\hat{\eta}_3$ is minimax for any larger class of estimators.

One possible flaw of the minimax principle is illustrated by the graph of two possible risk functions in Figure 9.4. The minimax principle would choose T_1 over T_2, yet T_2 is much better than T_1 for most values of θ.

In Example 9.3.10 it was suggested that an experimenter might have some prior knowledge about where, at least approximately, the parameter may be located. More generally, one might want to use an estimator that has small risk for values of θ that are "most likely" to occur in a given experiment. This can be modeled mathematically by treating θ as a random variable, say $\theta \sim p(\theta)$, where $p(\theta)$ is a function that has the usual properties (2.3.4) and (2.3.5) of a pdf in the variable θ. A reasonable approach then would be to compute the "average" or expected risk of an estimator, averaged over values of θ with respect to the pdf $p(\theta)$, and choose an estimator with smallest average risk.

Definition 9.5.5

Bayes Risk For a random sample from $f(x; \theta)$, the **Bayes risk** of an estimator T relative to a risk function $R_T(\theta)$ and pdf $p(\theta)$ is the average risk with respect to $p(\theta)$,

$$A_T = E_\theta[R_T(\theta)] = \int_\Omega R_T(\theta)p(\theta)\, d\theta \qquad \textbf{(9.5.6)}$$

If an estimator has the smallest Bayes risk, then it is referred to as a Bayes estimator.

FIGURE 9.4 Comparison of risk functions for two estimators

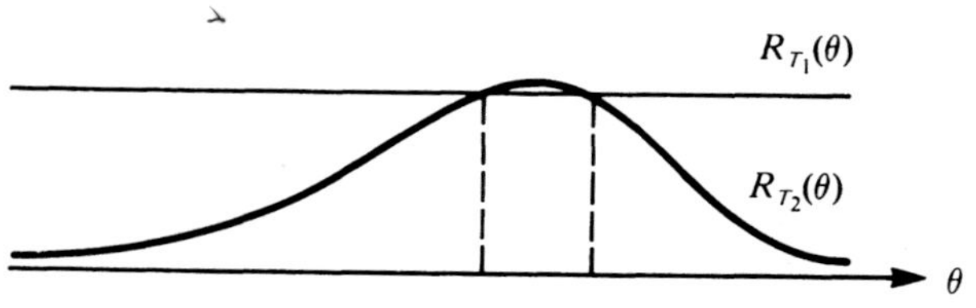

Definition 9.5.6

Bayes Estimator For a random sample from $f(x; \theta)$, the **Bayes estimator** T^* relative to the risk function $R_T(\theta)$ and pdf $p(\theta)$ is the estimator with minimum expected risk,

$$E_\theta[R_{T^*}(\theta)] \leqslant E_\theta[R_T(\theta)] \tag{9.5.7}$$

for every estimator T.

In some kinds of problems it is reasonable to assume that the parameter varies for different cases, and it may be proper to treat θ as a random variable. In other cases, $p(\theta)$ may reflect prior information or belief as to what the true value of the parameter may be. In either case, introduction of the pdf $p(\theta)$, which usually is called a **prior density** for the parameter θ, constitutes an additional assumption that may be helpful or harmful depending on its correctness. In any event, averaging the risk relative to a pdf $p(\theta)$ is a procedure that provides a possible way to discriminate between two estimators when neither of their risk functions is uniformly smaller than the other for all θ.

A whole class of estimators can be produced by considering different pdf's $p(\theta)$. It is useful to have a class of estimators in a problem, although if there is some physical reason to justify choosing a particular $p(\theta)$, then the estimator associated with that $p(\theta)$ would presumably be the best one to use in that problem. There are different philosophies involved with choosing prior densities $p(\theta)$, but we will not be too concerned with how the $p(\theta)$ is chosen in this work. In some cases θ may indeed act like a random variable, and $p(\theta)$ would reflect this fact. Alternatively, $p(\theta)$ may represent a degree of belief concerning the value of θ arrived at from previous sampling information, or by other means. In any event, potentially useful estimators can be developed through this structure. The subject of choosing a prior pdf is discussed in books by DeGroot (1970) and Zellner (1971).

Example 9.5.3 Consider again the estimators $\hat{\eta}_3 = X_{1:n} - 1/n$ and $\hat{\eta}_4 = 0.9(X_{1:n} - 1/n)$ of Example 9.3.10. With squared error loss we found that $\hat{\eta}_3$ is better by the minimax principle, but $\hat{\eta}_4$ is better if it is known that $\eta^2 < 2 + 1/n^2$, because it has smaller MSE for η in this subset of $\boldsymbol{\Omega}$. We now assume a standard normal prior density, $\eta \sim \text{N}(0, 1)$, and compare the Bayes risk. It follows that $E_\eta[R_{\hat{\eta}_3}(\eta)] = E_\eta(1/n^2) = 1/n^2$ and $E_\eta[R_{\hat{\eta}_4}(\eta)] = E_\eta[0.81/n^2 - 0.01\eta^2] = 0.81/n^2 + 0.01$. According to this criterion, $\hat{\eta}_3$ is better if $n \geqslant 5$ and $\hat{\eta}_4$ is better if $n \leqslant 4$.

A few results now are considered that are useful in determining a Bayes estimator. Note that in this framework the density function $f(x; \theta)$ is interpreted as a conditional density function $f(x \mid \theta)$.

Definition 9.5.7

Posterior Distribution The conditional density of θ given the sample observations $\boldsymbol{x} = (x_1, \ldots, x_n)$ is called the **posterior density** or **posterior pdf**, and is given by

$$f_{\theta|\boldsymbol{x}}(\theta) = \frac{f(x_1, \ldots, x_n \mid \theta)p(\theta)}{\int f(x_1, \ldots, x_n \mid \theta)p(\theta)\, d\theta} \tag{9.5.8}$$

The Bayes estimator is the estimator that minimizes the average risk over θ, $E_\theta[R_T(\theta)]$. However,

$$\begin{aligned} E_\theta[R_T(\theta)] &= E_\theta\{E_{\mathbf{X}|\theta}[L(T;\, \theta)]\} \\ &= E_{\mathbf{X}}\{E_{\theta|\mathbf{X}}[L(T;\, \theta)]\} \end{aligned} \tag{9.5.9}$$

and an estimator T that minimizes $E_{\theta|\boldsymbol{x}}[L(T;\, \theta)]$ for each $\boldsymbol{x}$ also minimizes the average over $\boldsymbol{X}$. Thus the Bayes estimator may be obtained by minimizing the expected loss relative to the posterior distribution.

Theorem 9.5.1 If $X_1, \ldots, X_n$ denotes a random sample from $f(x \mid \theta)$, then the Bayes estimator is the estimator that minimizes the expected loss relative to the posterior distribution of $\theta \mid \boldsymbol{x}$,

$$E_{\theta|\boldsymbol{x}}[L(T;\, \theta)]$$

■

For certain types of loss functions, expressions for the Bayes estimator can be determined more explicitly in terms of the posterior distribution.

Theorem 9.5.2 The Bayes estimator, T, of $\tau(\theta)$ under the squared error loss function,

$$L(T;\, \theta) = [T - \tau(\theta)]^2 \tag{9.5.10}$$

is the conditional mean of $\tau(\theta)$ relative to the posterior distribution,

$$T = E_{\theta|\mathbf{x}}[\tau(\theta)] = \int \tau(\theta) f_{\theta|\mathbf{x}}(\theta)\, d\theta \tag{9.5.11}$$

Proof

See Exercise 41.

■

Example 9.5.4 Consider a random sample of size n from a Bernoulli distribution,

$$f(x \mid \theta) = \theta^x(1 - \theta)^{1-x} \qquad x = 0, 1$$

and let $\Theta \sim \text{UNIF}(0, 1)$. Equation (9.5.8) yields the posterior density

$$f_{\Theta|x}(\theta) = \frac{\theta^{\Sigma x_i}(1-\theta)^{n-\Sigma x_i}}{\int_0^1 \theta^{\Sigma x_i}(1-\theta)^{n-\Sigma x_i}\, d\theta} \qquad 0 < \theta < 1$$

It is convenient to express this using the notation of the beta distribution. As noted in Chapter 8, a random variable Y has a beta distribution with parameters a and b, denoted $Y \sim \text{BETA}(a, b)$, if it has pdf of the form $f(y; a, b) = y^{a-1}(1-y)^{b-1}/B(a, b)$ when $0 < y < 1$, and zero otherwise, with constant $B(a, b) = \Gamma(a)\Gamma(b)/\Gamma(a+b)$. In this notation recall that the mean of the distribution is $E(Y) = a/(a+b)$. It also is possible to express the posterior distribution in terms of this notation. Specifically,

$$f_{\Theta|x}(\theta) = \frac{\theta^{\Sigma x_i}(1-\theta)^{n-\Sigma x_i}}{B(\sum x_i + 1, n - \sum x_i + 1)} \qquad 0 < \theta < 1 \tag{9.5.12}$$

In other words, $\Theta | x \sim \text{BETA}(\sum x_i + 1, n - \sum x_i + 1)$. Consequently, $E_{\theta|x}(\Theta) = (\sum x_i + 1)/[(\sum x_i + 1) + (n - \sum x_i + 1)] = (\sum x_i + 1)/(n + 2)$. With squared error loss, we have by Theorem 9.5.2 that the Bayes estimator of θ is $T = (\sum X_i + 1)/(n + 2)$.

Example 9.5.5 Suppose that $X_i \sim \text{POI}(\theta)$, and we are interested in a Bayes estimator of θ assuming squared error loss. We choose to consider the class of gamma prior densities, $\theta \sim \text{GAM}(\beta, \kappa)$,

$$p(\theta) = \frac{1}{\beta^\kappa \Gamma(\kappa)} \theta^{\kappa-1} e^{-\theta/\beta}$$

where β and κ are known arbitrary constants. The posterior distribution is given by

$$f_{\theta|x}(\theta) = \left[\frac{e^{-n\theta}\theta^{\Sigma x_i}\theta^{\kappa-1}e^{-\theta/\beta}}{\prod (x_i!)\beta^\kappa\Gamma(\kappa)}\right] \Bigg/ \left[\int \frac{e^{-n\theta}\theta^{\Sigma x}\theta^{\kappa-1}e^{-\theta/\beta}\, d\theta}{\prod (x_i!)\beta^\kappa\Gamma(\kappa)}\right]$$

That is,

$$\theta | x \sim \text{GAM}[(n + 1/\beta)^{-1}, \sum x_i + \kappa] \tag{9.5.13}$$

The Bayes estimator of θ is therefore

$$T = E(\theta | X) = \frac{\sum X_i + \kappa}{n + 1/\beta}$$

A prior density with large β and small κ makes this estimator close to the MLE, $\hat{\theta} = \bar{x}$.

The risk in this case is

$$R_T(\theta) = E[T - \theta]^2 = \text{Var}(T) + [E(T) - \theta]^2$$
$$= \frac{n\ \text{Var}(X)}{(n + 1/\beta)^2} + \left[\frac{n\theta + \kappa}{n + 1/\beta} - \theta\right]^2$$
$$= \frac{n\theta + [\kappa - \theta/\beta]^2}{(n + 1/\beta)^2}$$

Some authors define the Bayes estimator to be the mean of the posterior distribution, which (according to Theorem 9.5.2) results from squared error loss. However, it is desirable in some applications to use other loss functions.

Theorem 9.5.3 The Bayes estimator, $\hat{\theta}$, of θ under absolute error loss,

$$L(\hat{\theta}; \theta) = |\hat{\theta} - \theta| \tag{9.5.14}$$

is the median of the posterior distribution $f_{\theta|\mathbf{x}}(\theta)$.

Proof

See Exercise 42. ∎

The Bayes estimator structure sometimes is helpful in finding a minimax estimator.

Theorem 9.5.4 If T^* is a Bayes estimator with constant risk, $R_{T^*}(\theta) = c$, then T^* is a minimax estimator.

Proof

We have $\max_\theta R_{T^*}(\theta) = \max_\theta c = c = R_{T^*}(\theta)$, but because $R_{T^*}(\theta)$ is constant over θ,

$$R_{T^*}(\theta) = E_\theta[R_{T^*}(\theta)] \leqslant E_\theta[R_T(\theta)]$$

for every T because T^* is the Bayes estimator. Now the average of a variable is not larger than the maximum value of a variable, so

$$E_\theta[R_T(\theta)] \leqslant \max_\theta R_T(\theta)$$

and

$$\max_\theta R_{T^*}(\theta) \leqslant \max_\theta R_T(\theta)$$

which shows that T^* is a minimax estimator. ∎

It follows that if an appropriate prior pdf, $p(\theta)$, can be found that will yield a Bayes estimator with constant risk, then the Bayes estimator also will be the minimax estimator.

Example 9.5.6 Recall the prior and posterior distributions of Example 9.5.4, but now consider a "weighted" squared error loss,

$$L(t; \theta) = \frac{(t - \theta)^2}{\theta(1 - \theta)} \tag{9.5.15}$$

which gives more weight to the values of θ that are closer to zero or one. Note that

$$E_{\Theta|x}[L(t; \Theta)] = \int_0^1 \frac{(t - \theta)^2}{\theta(1 - \theta)} \frac{\theta^{\sum x_i}(1 - \theta)^{n - \sum x_i}}{B(\sum x_i + 1, n - \sum x_i + 1)} \, d\theta$$

$$= C(x) \int_0^1 (t - \theta)^2 \frac{\theta^{\sum x_i - 1}(1 - \theta)^{n - \sum x_i - 1}}{B(\sum x_i, n - \sum x_i)} \, d\theta$$

with $C(x) = B(\sum x_i, n - \sum x_i)/B(\sum x_i + 1, n - \sum x_i + 1)$, which means that the expression $E_{\Theta|x}[L(t; \Theta)]$ is minimized when the latter integral is minimized. Notice that this integral corresponds to the conditional expectation of ordinary squared error loss $(t - \theta)^2$ relative to the posterior distribution BETA $(\sum x_i, n - \sum x_i)$. By Theorem 9.5.2, this integral is minimized when t is the mean of BETA $(\sum x_i, n - \sum x_i)$. This mean is $t = \sum x_i/(\sum x_i + n - \sum x_i) = \bar{x}$. It follows that $t^*(x) = \bar{x}$, and the Bayes estimator is $T^* = \bar{X}$. Furthermore, the risk is

$$R_{T^*}(\theta) = \frac{E[(\bar{X} - \theta)^2]}{\theta(1 - \theta)} = \frac{\theta(1 - \theta)/n}{\theta(1 - \theta)} = \frac{1}{n}$$

which is constant with respect to θ. By Theorem 9.5.4, $\bar{X}$ is a minimax estimator in this example.

SUMMARY

Our purpose in this chapter was to provide general methods for estimating unknown parameters and to present criteria for evaluating the properties of estimators. The two methods receiving the most attention were the method of moments and the method of maximum likelihood. The MLEs were found to have desirable asymptotic properties under certain conditions. For example, the MLEs in certain cases are asymptotically efficient and asymptotically normally distributed.

In general, it is desirable to have the distribution of the estimator highly concentrated about the true value of the parameter being estimated. This concentration may be reflected by an appropriate loss function, but most attention is centered on squared error loss and MSE risk. If the estimator is unbiased, then the MSE simply becomes the variance of the estimator. Within the class of unbiased estimators there may exist an estimator with uniformly minimum variance for all possible values of the parameter. This estimator is referred to as the UMVUE. At this point a direct method for finding a UMVUE has not been provided; however, if an estimator satisfies the CRLB, then we know it is a UMVUE. The concepts of sufficiency and completeness discussed in the next chapter will provide a more systematic approach for attempting to find a UMVUE.

In lieu of finding an estimator that has uniformly minimum variance over the parameter, θ, we considered the principle of minimizing the maximum variance (risk) over θ (minimax estimator), and minimizing the average variance (or more generally risk) over θ, giving the Bayes estimators. Bayes estimation requires specifying an additional prior density $p(\theta)$. Information was provided on how to compute a Bayes estimator, but very little on how to find a minimax estimator.

EXERCISES

1. Find method of moments estimators (MMEs) of θ based on a random sample $X_1, \ldots, X_n$ from each of the following pdf's:

(a) $f(x; \theta) = \theta x^{\theta - 1}$; $0 < x < 1$, zero otherwise; $0 < \theta$.

(b) $f(x; \theta) = (\theta + 1)x^{-\theta - 2}$; $1 < x$, zero otherwise; $0 < \theta$.

(c) $f(x; \theta) = \theta^2 x e^{-\theta x}$; $0 < x$, zero otherwise; $0 < \theta$.

2. Find the MMEs based on a random sample of size n from each of the following distributions (see Appendix B):

(a) $X_i \sim \text{NB}(3, p)$.

(b) $X_i \sim \text{GAM}(2, \kappa)$.

(c) $X_i \sim \text{WEI}(\theta, 1/2)$.

(d) $X_i \sim \text{DE}(\theta, \eta)$ with both θ and η unknown.

(e) $X_i \sim \text{EV}(\theta, \eta)$ with both θ and η unknown.

(f) $X_i \sim \text{PAR}(\theta, \kappa)$ with both θ and κ unknown.

3. Find maximum likelihood estimators (MLEs) for θ based on a random sample of size n for each of the pdf's in Exercise 1.

4. Find the MLEs based on a random sample $X_1, \ldots, X_n$ from each of the following distributions:

(a) $X_i \sim \text{BIN}(1, p)$.

(b) $X_i \sim \text{GEO}(p)$.

(c) $X_i \sim \text{NB}(3, p)$.

(d) $X_i \sim \text{N}(0, \theta)$.

(e) $X_i \sim \text{GAM}(\theta, 2)$.
(f) $X_i \sim \text{DE}(\theta, 0)$.
(g) $X_i \sim \text{WEI}(\theta, 1/2)$.
(h) $X_i \sim \text{PAR}(1, \kappa)$.

5. Find the MLE for θ based on a random sample of size n from a distribution with pdf

$$f(x;\theta) = \begin{cases} 2\theta^2 x^{-3} & \theta \leqslant x \\ 0 & x < \theta;\ 0 < \theta \end{cases}$$

6. Find the MLEs based on a random sample $X_1, \ldots, X_n$ from each of the following pdf's:
(a) $f(x; \theta_1, \theta_2) = 1/(\theta_2 - \theta_1)$; $\theta_1 \leqslant x \leqslant \theta_2$, zero otherwise;
(b) $f(x; \theta, \eta) = \theta\eta^\theta x^{-\theta-1}$; $\eta \leqslant x$ zero otherwise; $0 < \theta, 0 < \eta < \infty$.

7. Let $X_1, \ldots, X_n$ be a random sample from a geometric distribution, $X \sim \text{GEO}(p)$. Find the MLEs of the following quantities:
(a) $E(X) = 1/p$.
(b) $\text{Var}(X) = (1 - p)/p^2$.
(c) $P[X > k] = (1 - p)^k$ for arbitrary $k = 1, 2, \ldots$
Hint: Use the invariance property of MLEs.

8. Based on a random sample of size n from a normal distribution, $X \sim N(\mu, \sigma^2)$, find the MLEs of the following:
(a) $P[X > c]$ for arbitrary c.
(b) The 95th percentile of X.

9. Suppose that $x_{1:n}$ and $x_{n:n}$ are the smallest and largest observed values of a random sample of size n from a distribution with pdf $f(x; \theta)$; $0 < \theta$.
(a) If $f(x; \theta) = 1$ for $\theta - 0.5 \leqslant x \leqslant \theta + 0.5$, zero otherwise, then show that any value $\hat{\theta}$ such that $x_{n:n} - 0.5 \leqslant \hat{\theta} \leqslant x_{1:n} + 0.5$ is an ML estimate of θ.
(b) If $f(x; \theta) = 1/\theta$ for $\theta \leqslant x \leqslant 2\theta$, zero otherwise, then show that $\hat{\theta} = 0.5x_{n:n}$ is an ML estimate of θ.

10. Consider a random sample of size n from a double exponential distribution, $X_i \sim \text{DE}(\theta, \eta)$.
(a) Find the MLE of η when $\theta = 1$. *Hint:* Show first that if $x_1, \ldots, x_n$ are observed values, then the sum $\sum_{i=1}^{n} |x_i - a|$ is minimized when a is the sample median.
(b) Find the MLEs when both θ and η are unknown.

11. Consider a random sample of size n from a Pareto distribution, $X_i \sim \text{PAR}(\theta, 2)$.
(a) Find the ML equation (9.2.7).
(b) From the data of Example 4.6.2, compute the ML estimate, $\hat{\theta}$, to three decimal places. *Note:* The ML equation cannot be solved explicitly for $\hat{\theta}$, but it can be solved numerically, by an iterative method, or by trial and error.

12. Let $Y_1, \ldots, Y_n$ be a random sample from a log normal distribution, $Y \sim \text{LOGN}(\mu, \sigma^2)$.

(a) Find the MLEs of μ and σ^2.

(b) Find the MLE of $E(Y)$.

Hint: Recall that $Y \sim \text{LOGN}(\mu, \sigma^2)$ if $\ln(Y) \sim \text{N}(\mu, \sigma^2)$, or, equivalently, $Y = \exp(X)$ where $X \sim \text{N}(\mu, \sigma^2)$.

13. Consider independent random samples $X_1, \ldots, X_n$ and $Y_1, \ldots, Y_m$ from normal distributions with a common mean μ, but possibly different variances, σ_1^2 and σ_2^2, so that $X_i \sim \text{N}(\mu, \sigma_1^2)$ and $Y_j \sim \text{N}(\mu, \sigma_2^2)$. Find the MLEs of μ, σ_1^2, and σ_2^2.

14. Let X be the number of independent trials of some component until it fails, where $1 - p$ is the probability of failure on each trial. We record the exact number of trials, $Y = X$, if $X \leqslant r$; otherwise we record $Y = r + 1$, where r is a fixed positive integer.

(a) Show that the discrete pdf of Y is

$$f(y; p) = \begin{cases} (1-p)p^{y-1} & y = 1, \ldots, r \\ p^r & y = r + 1 \end{cases}$$

(b) Let $Y_1, \ldots, Y_n$ be a random sample from $f(y; p)$. Find the MLE of p.

Hint: $f(y; p) = c(y; p)p^{y-1}$, where $c(y; p) = 1 - p$ if $y = 1, \ldots, r$ and $c(r + 1; p) = 1$. It follows that $\prod_{i=1}^{n} c(y_i; p) = (1 - p)^m$ where m is the number of observed y_i that are less than $r + 1$.

15. Let $X \sim \text{BIN}(n, p)$ and $\hat{p} = X/n$.

(a) Find a constant c so that $E[c\hat{p}(1 - \hat{p})] = p(1 - p)$.

(b) Find an unbiased estimator of $\text{Var}(X)$.

(c) Consider a random sample of size N from $\text{BIN}(n, p)$. Find unbiased estimators of p and $\text{Var}(X)$ based on the random sample.

16. (Truncated Poisson.) Let $X \sim \text{POI}(\mu)$, and suppose we cannot observe $X = 0$, so the observed random variable, Y, has discrete pdf

$$f(y; \mu) = \begin{cases} \dfrac{e^{-\mu}\mu^y}{y!(1 - e^{-\mu})} & y = 1, 2, \ldots \\ 0 & \text{otherwise} \end{cases}$$

We desire to estimate $P[X > 0] = 1 - e^{-\mu}$. Show that an unbiased (but unreasonable) estimator of $1 - e^{-\mu}$ is given by $u(Y)$ where $u(y) = 0$ if y is odd, and $u(y) = 2$ if y is even.
Hint: Consider the power series expansion of $(e^{\mu} + e^{-\mu})/2$.

17. Let $X_1, \ldots, X_n$ be a random sample from a uniform distribution, $X_i \sim \text{UNIF}(\theta - 1, \theta + 1)$.

(a) Show that the sample mean, $\bar{X}$, is an unbiased estimator of θ.

(b) Show that the "midrange," $(X_{1:n} + X_{n:n})/2$, is an unbiased estimator of θ.

18. Suppose that X is continuous and its pdf, $f(x; \mu)$, is symmetric about μ. That is, $f(\mu + c; \mu) = f(\mu - c; \mu)$ for all $c > 0$.

(a) Show that for a random sample of size n, where n is odd ($n = 2k - 1$), the sample median, $X_{k:n}$, is an unbiased estimator of μ.

(b) Show that $Z = X_{1:n} - \mu$ and $W = \mu - X_{n:n}$ have the same distribution, and thus that the midrange, $(X_{1:n} + X_{n:n})/2$, is unbiased for μ.

19. Consider a random sample of size n from a uniform distribution, $X_i \sim \text{UNIF}(-\theta, \theta)$; $\theta > 0$. Find a constant c so that $c(X_{n:n} - X_{1:n})$ is an unbiased estimator of θ.

20. Let S be the sample standard deviation, based on a random sample of size n from a distribution with pdf $f(x; \mu, \sigma^2)$ with mean μ and variance σ^2.

(a) Show that $E(S) \leqslant \sigma$, where equality holds if and only if $f(x; \mu, \sigma^2)$ is degenerate at μ, $P[X = \mu] = 1$. *Hint:* Consider Var(S).

(b) If $X_i \sim \text{N}(\mu, \sigma^2)$, find a constant c such that cS is an unbiased estimator of σ. *Hint:* Use the fact that $(n-1)S^2/\sigma^2 \sim \chi^2(n-1)$ and $S = (S^2)^{1/2}$.

(c) Relative to (b), find a function of $\bar{X}$ and S that is unbiased for the 95th percentile of $X \sim \text{N}(\mu, \sigma^2)$.

21. Consider a random sample of size n from a Bernoulli distribution, $X_i \sim \text{BIN}(1, p)$.

(a) Find the CRLB for the variances of unbiased estimators of p.

(b) Find the CRLB for the variances of unbiased estimators of $p(1-p)$.

(c) Find a UMVUE of p.

22. Consider a random sample of size n from a normal distribution, $X_i \sim \text{N}(\mu, 9)$.

(a) Find the CRLB for variances of unbiased estimators of μ.

(b) Is the MLE, $\hat{\mu} = \bar{X}$, a UMVUE of μ?

(c) Is the MLE of the 95th percentile a UMVUE?

23. Let $X_1, \ldots, X_n$ be a random sample from a normal distribution, $\text{N}(0, \theta)$.

(a) Is the MLE, $\hat{\theta}$, an unbiased estimator of θ?

(b) Is $\hat{\theta}$ a UMVUE of θ?

24. Let $X \sim \text{POI}(\mu)$, and let $\theta = P[X = 0] = e^{-\mu}$.

(a) Is $\hat{\theta} = e^{-X}$ an unbiased estimator of θ?

(b) Show that $\tilde{\theta} = u(X)$ is an unbiased estimator of θ, where $u(0) = 1$ and $u(x) = 0$ if $x = 1, 2, \ldots$

(c) Compare the MSEs of $\hat{\theta}$ and $\tilde{\theta}$ for estimating $\theta = e^{-\mu}$ when $\mu = 1$ and $\mu = 2$.

25. Consider the estimator $T_1 = 1/\bar{X}$ of Example 9.3.2. Compare the MSEs of T_1 and cT_1 for estimating $1/\theta$, where $c = (n-1)/n$.

26. Consider a random sample of size n from a distribution with pdf $f(x; \theta) = 1/\theta$ if $0 < x \leqslant \theta$, and zero otherwise; $0 < \theta$.

(a). Find the MLE $\hat{\theta}$.

(b) Find the MME $\tilde{\theta}$.

(c) Is $\hat{\theta}$ unbiased?

(d) Is $\tilde{\theta}$ unbiased?

(e) Compare the MSEs of $\hat{\theta}$ and $\tilde{\theta}$.

27. Consider a random sample of size $n = 2$ from a normal distribution, $X_i \sim N(\theta, 1)$, where $\Omega = \{\theta \mid 0 \leqslant \theta \leqslant 1\}$. Define estimators as follows: $\hat{\theta}_1 = (1/2)X_1 + (1/2)X_2$, $\hat{\theta}_2 = (1/4)X_1 + (3/4)X_2$, $\hat{\theta}_3 = (2/3)X_1$, and $\hat{\theta}_4 = (2/3)\hat{\theta}_1$. Consider squared error loss $L(t; \theta) = (t - \theta)^2$.

(a) Compare the risk functions for these estimators.

(b) Compare the estimators by the minimax principle.

(c) Find the Bayes risk of the estimators, using $\theta \sim$ UNIF(0, 1).

(d) Find the Bayes risk of the estimators, using $\theta \sim$ BETA(2, 1).

28. Let $X_1, \ldots, X_n$ be a random sample from EXP(θ), and define $\hat{\theta}_1 = \bar{X}$ and $\hat{\theta}_2 = n\bar{X}/(n + 1)$.

(a) Find the variances of $\hat{\theta}_1$ and $\hat{\theta}_2$.

(b) Find the MSEs of $\hat{\theta}_1$ and $\hat{\theta}_2$.

(c) Compare the variances of $\hat{\theta}_1$ and $\hat{\theta}_2$ for $n = 2$.

(d) Compare the MSEs of $\hat{\theta}_1$ and $\hat{\theta}_2$ for $n = 2$.

(e) Find the Bayes risk of $\hat{\theta}_1$ using $\theta \sim$ EXP(2).

29. Consider a random sample of size n from a Bernoulli distribution, $X_i \sim$ BIN(1, p). For a uniform prior density, $p \sim$ UNIF(0, 1), and squared error loss, find the following:

(a) Bayes estimator of p.

(b) Bayes estimator of $p(1 - p)$.

(c) Bayes risk for the estimator in (a).

30. Let $X \sim$ POI(μ), and consider the loss function $L(\hat{\mu}, \mu) = (\hat{\mu} - \mu)^2/\mu$. Assume a gamma prior density, $\mu \sim$ GAM(θ, κ), where θ and κ are known.

(a) Find the Bayes estimator of μ.

(b) Show that $\hat{\mu} = X$ is the minimax estimator.

31. Let $\hat{\theta}$ and $\tilde{\theta}$ be the MLE and MME, respectively, for θ in Exercise 26.

(a) Show that $\hat{\theta}$ is MSE consistent.

(b) Show that $\tilde{\theta}$ is MSE consistent.

32. Show that the MLE of θ in Exercise 5 is simply consistent.

33. Consider a random sample of size n from a Poisson distribution, $X_i \sim$ POI(μ).

(a) Find the CRLB for the variances of unbiased estimators of μ.

(b) Find the CRLB for the variances of unbiased estimators of $\theta = e^{-\mu}$.

(c) Find a UMVUE of μ.

(d) Find the MLE $\hat{\theta}$ of θ.

(e) Is $\hat{\theta}$ an unbiased estimator of θ?

(f) Is $\hat{\theta}$ asymptotically unbiased?

(g) Show that $\tilde{\theta} = [(n - 1)/n]^{\Sigma X_i}$ is an unbiased estimator of θ.

(h) Find Var($\tilde{\theta}$) and compare to the CRLB of (b).

Hint: Note that $Y = \sum X_i \sim$ POI($n\mu$), and that $E(\tilde{\theta})$ and Var($\tilde{\theta}$) are related to the MGF of Y.

34. Consider a random sample of size n from a distribution with discrete pdf $f(x; p) = p(1-p)^x$; $x = 0, 1, \ldots$, zero otherwise.

(a) Find the MLE of p.

(b) Find the MLE of $\theta = (1-p)/p$.

(c) Find the CRLB for variances of unbiased estimators of θ.

(d) Is the MLE of θ a UMVUE?

(e) Is the MLE of θ MSE consistent?

(f) Find the asymptotic distribution of the MLE of θ.

(g) Let $\tilde{\theta} = n\bar{X}/(n+1)$. Find risk functions of both $\tilde{\theta}$ and $\bar{X}$ using the loss function $L(t; \theta) = (t-\theta)^2/(\theta^2 + \theta)$.

35. Find the asymptotic distribution of the MLE of p in Exercise 4(a).

36. Find the asymptotic distribution of the MLE of θ in Exercise 4(d).

37. Let $X_1, \ldots, X_n$ be a random sample with an odd sample size ($n = 2k - 1$).

(a) If $X_i \sim \text{DE}(1, \eta)$, find the asymptotic relative efficiency, as defined by equation (9.4.6), of $\hat{\eta}_n = \bar{X}_n$ relative to $\tilde{\eta}_n = X_{k:n}$.

(b) If $X_i \sim \text{N}(\mu, 1)$, find the asymptotic relative efficiency, as defined by equation (9.4.6), of $\tilde{\mu}_n = X_{k:n}$ relative to $\hat{\mu}_n = \bar{X}_n$.

38. An estimator $\hat{\theta}$ is said to be **median unbiased** if $P[\hat{\theta} < \theta] = P[\hat{\theta} > \theta]$. Consider a random sample of size n from an exponential distribution, $X_i \sim \text{EXP}(\theta)$.

(a) Find a median unbiased estimator of θ that has the form $\hat{\theta} = c\bar{X}$.

(b) Find the relative efficiency of $\hat{\theta}$ compared to $\bar{X}$.

(c) Compare the MSEs of $\hat{\theta}$ and $\bar{X}$ when $n = 5$.

39. Suppose that $\hat{\theta}_i$, $i = 1, \ldots, n$ are independent unbiased estimators of θ with $\text{Var}(\hat{\theta}_i) = \sigma_i^2$. Consider a combined estimator $\hat{\theta}_c = \sum_{i=1}^{n} a_i \hat{\theta}_i$ where $\sum_{i=1}^{n} a_i = 1$.

(a) Show that $\hat{\theta}_c$ is unbiased.

(b) It can be shown that $\text{Var}(\hat{\theta}_c)$ is minimized by letting $a_i = (1/\sigma_i^2) \Big/ \sum_{i=1}^{n} (1/\sigma_i^2)$. Verify this for the case $n = 2$.

40. Let X be a random variable with CDF $F(x)$.

(a) Show that $E[(X-c)^2]$ is minimized by the value $c = E(X)$.

(b) Assuming that X is continuous, show that $E[|X - c|]$ is minimized if c is the median, that is, the value such that $F(c) = 1/2$.

41. Prove Theorem 9.5.2. *Hint:* Use Exercise 40(a) applied to the posterior distribution for fixed x.

42. Prove Theorem 9.5.3. *Hint:* Use Exercise 40(b).

43. Consider the functions $L(\theta)$, $L^*(\tau)$, and $u(\theta)$ in the discussion preceding Theorem 9.2.1. Show that $\hat{\tau} = u(\hat{\theta})$ maximizes $L^*(\tau)$ if $\hat{\theta}$ maximizes $L(\theta)$. *Hint:* Note that $L^*(\tau) \leqslant L(\hat{\theta})$ for all τ in the range of the function $u(\theta)$, and $L(\hat{\theta}) = L^*(\hat{\tau})$ if $\hat{\tau} = u(\hat{\theta})$.

44. Let $X_1, X_2, \ldots, X_n$ be a random sample from an exponential distribution with mean $1/\theta$, $f(x \mid \theta) = \theta \exp(-\theta x)$ for $x > 0$, and assume that the prior density of θ also is exponential with mean $1/\beta$ where β is known.

(a) Show that the posterior distribution is $\theta \mid \mathbf{x} \sim \text{GAM}[(\beta + \sum x_i)^{-1}, n + 1]$.

(b) Using squared error loss, find the Bayes estimator of θ.

(c) Using squared error loss, find the Bayes estimator of $\mu = 1/\theta$.

(d) Using absolute error loss, find the Bayes estimator of θ. Use chi-square notation to express the solution.

(e) Using absolute error loss, find the Bayes estimator of $\mu = 1/\theta$.

CHAPTER 10

SUFFICIENCY AND COMPLETENESS

10.1

INTRODUCTION

Chapter 9 presented methods for deriving point estimators based on a random sample to estimate unknown parameters of the population distribution. In some cases, it is possible to show, in a certain sense, that a particular statistic or set of statistics contains all of the "information" in the sample about the parameters. It then would be reasonable to restrict attention to such statistics when estimating or otherwise making inferences about the parameters.

More generally, the idea of sufficiency involves the reduction of a data set to a more concise set of statistics with no loss of information about the unknown parameter. Roughly, a statistic S will be considered a "sufficient" statistic for a parameter θ if the conditional distribution of any other statistic T given the value of S does not involve θ. In other words, once the value of a sufficient statistic is known, the observed value of any other statistic does not contain any further information about the parameter.

Example 10.1.1 A coin is tossed n times, and the outcome is recorded for each toss. As usual, this process could be modeled in terms of a random sample $X_1, \ldots, X_n$ from a Bernoulli distribution. Suppose that it is not known whether the coin is fair, and we wish to estimate $\theta = P(\text{head})$. It would seem that the total number of heads, $S = \sum X_i$, should provide as much information about the value θ as the actual outcomes. To check this out, consider

$$f(x_1, \ldots, x_n; \theta) = \theta^{\sum x_i}(1-\theta)^{n-\sum x_i} \qquad x_i = 0, 1$$

We also know that $S \sim \text{BIN}(n, \theta)$, so that

$$f_S(s; \theta) = \binom{n}{s}\theta^s(1-\theta)^{n-s} \qquad s = 0, 1, \ldots, n$$

If $\sum x_i = s$, then the events $[X_1 = x_1, \ldots, X_n = x_n, S = s]$ and $[X_1 = x_1, \ldots, X_n = x_n]$ are equivalent, and

$$\begin{aligned} f_{X_1, \ldots, X_n|s}(x_1, \ldots, x_n) &= \frac{P[X_1 = x_1, \ldots, X_n = x_n, S = s]}{P[S = s]} \\ &= \frac{f(x_1, \ldots, x_n; \theta)}{f_S(s; \theta)} \\ &= \frac{\theta^{\sum x_i}(1-\theta)^{n-\sum x_i}}{\binom{n}{s}\theta^s(1-\theta)^{n-s}} \\ &= \frac{1}{\binom{n}{s}} \end{aligned}$$

If $\sum x_i \neq s$, then the conditional pdf is zero. In either case, it does not involve θ. Furthermore, let $T = \ell(X_1, \ldots, X_n)$ be any other statistic, and define the set $C_t = \{(x_1, \ldots, x_n): \ell(x_1, \ldots, x_n) = t\}$. The conditional pdf of T given $S = s$ is

$$\begin{aligned} f_{T|s}(t) &= P[T = t \mid S = s] \\ &= \sum_{C_t} f_{X_1, \ldots, X_n|s}(x_1, \ldots, x_n) \end{aligned}$$

which also does not involve θ.

It will be desirable to have a more precise definition of sufficiency for more than one parameter, and a set of jointly sufficient statistics.

10.2

SUFFICIENT STATISTICS

As in the previous chapter, a set of data $x_1, \ldots, x_n$ will be modeled mathematically as observed values of a set of random variables $X_1, \ldots, X_n$. For convenience, we will use vector notation, $\mathbf{X} = (X_1, \ldots, X_n)$ and $\mathbf{x} = (x_1, \ldots, x_n)$, to refer to the observed random variables and their possible values. We also will allow the possibility of a vector-valued parameter $\boldsymbol{\theta}$ and vector-valued statistics $\mathbf{S}$ and $\mathbf{T}$.

Definition 10.2.1

Jointly Sufficient Statistics Let $\mathbf{X} = (X_1, \ldots, X_n)$ have joint pdf $f(\mathbf{x}, \boldsymbol{\theta})$, and let $\mathbf{S} = (S_1, \ldots, S_k)$ be a k-dimensional statistic. Then $S_1, \ldots, S_k$ is a set of **jointly sufficient statistics** for $\boldsymbol{\theta}$ if for any other vector of statistics, $\mathbf{T}$, the conditional pdf of $\mathbf{T}$ given $\mathbf{S} = \mathbf{s}$, denoted by $f_{\mathbf{T}|\mathbf{s}}(\mathbf{t})$, does not depend on $\boldsymbol{\theta}$. In the one-dimensional case, we simply say that S is a **sufficient statistic** for θ.

Again, the idea is that if $\mathbf{S}$ is observed, then additional information about $\boldsymbol{\theta}$ cannot be obtained from $\mathbf{T}$ if the conditional distribution of $\mathbf{T}$ given $\mathbf{S} = \mathbf{s}$ is free of θ. We usually will assume that $X_1, \ldots, X_n$ is a random sample from a population pdf $f(x; \boldsymbol{\theta})$, and for convenience we often will refer to the vector $\mathbf{X} = (X_1, \ldots, X_n)$ as the random sample. However, in general, $\mathbf{X}$ could represent some other vector of observed random variables, such as a censored sample or some other set of order statistics.

The primary purpose is to reduce the sample to the smallest set of sufficient statistics, referred to as a "minimal set" of sufficient statistics. If k unknown parameters are present in the model, then quite often there will exist a set of k sufficient statistics. In some cases, the number of sufficient statistics will exceed the number of parameters, and indeed in some cases no reduction in the number of statistics is possible. The whole sample is itself a set of sufficient statistics, but when we refer to sufficient statistics we ordinarily will be thinking of some smaller set of sufficient statistics.

Definition 10.2.2

A set of statistics is called a **minimal sufficient** set if the members of the set are jointly sufficient for the parameters and if they are a function of every other set of jointly sufficient statistics.

For example, the order statistics will be shown to be jointly sufficient. In a sense, this does represent a reduction of the sample, although the number of statistics in this case is not reduced. In some cases, the order statistics may be a minimal sufficient set, but of course we hope to reduce the sample to a small number of jointly sufficient statistics.

Clearly one cannot actually consider all possible statistics, T, in attempting to use Definition 10.2.1 to verify that $\mathbf{S}$ is a sufficient statistic. However, because T may be written as a function of the sample, $\mathbf{X} = (X_1, \ldots, X_n)$, one possible approach would be to show that $f_{\mathbf{X}|\mathbf{s}}(\mathbf{x})$ is free of $\boldsymbol{\theta}$. Actually, this approach was used in Example 10.1.1, where $\mathbf{X}$ was a random sample from a Bernoulli distribution. Essentially the same derivation could be used in the more general situation where $\mathbf{X}$ is discrete and $\mathbf{S}$ and $\boldsymbol{\theta}$ are vector-valued. Suppose that $\mathbf{S} = (S_1, \ldots, S_k)$ where $S_j = \mathscr{s}_j(X_1, \ldots, X_n)$ for $j = 1, \ldots, k$, and denote by $\mathscr{s}(x_1, \ldots, x_n)$ the vector-valued function whose jth coordinate is $\mathscr{s}_j(x_1, \ldots, x_n)$. In a manner analogous to Example 10.1.1, the conditional pdf of $\mathbf{X} = (X_1, \ldots, X_n)$ given $\mathbf{S} = \mathbf{s}$ can be written as

$$f_{\mathbf{X}|\mathbf{s}}(x_1, \ldots, x_n) = \begin{cases} \dfrac{f(x_1, \ldots, x_n; \boldsymbol{\theta})}{f_{\mathbf{s}}(\mathbf{s}; \boldsymbol{\theta})} & \text{if } \mathscr{s}(x_1, \ldots, x_n) = \mathbf{s} \\ 0 & \text{otherwise} \end{cases} \tag{10.2.1}$$

This would not be a standard situation for continuous random variables, because we have an n-dimensional vector of random variables with the distribution of probability restricted to an $n - k$ dimensional subspace. Consequently, care must be taken with regard to the meaning of an expression such as (10.2.1) in the continuous case.

In general, we can say that $S_1, \ldots, S_k$ are jointly sufficient for $\boldsymbol{\theta}$ if equation (10.2.1) is free of $\boldsymbol{\theta}$.

Some authors avoid any concern with technical difficulties by directly defining $S_1, \ldots, S_k$ to be jointly sufficient for $\boldsymbol{\theta}$ if $f(x_1, \ldots, x_n; \boldsymbol{\theta})/f_{\mathbf{S}}(\mathbf{s}; \boldsymbol{\theta})$ is free of $\boldsymbol{\theta}$. In any event, equation (10.2.1) will be used here without resorting to a more careful mathematical development.

Example 10.2.1 Consider a random sample from an exponential distribution, $X_i \sim \text{EXP}(\theta)$. It follows that

$$f(x_1, \ldots, x_n; \theta) = \frac{1}{\theta^n} \exp\left(-\frac{\sum x_i}{\theta}\right) \qquad x_i > 0$$

which suggests checking the statistic $S = \sum X_i$. We also know that $S \sim \text{GAM}(\theta, n)$, so that

$$f_S(s; \theta) = \frac{1}{\theta^n \Gamma(n)} s^{n-1} e^{-s/\theta} \qquad s > 0$$

If $s = \sum x_i$, then

$$\frac{f(x_1, \ldots, x_n; \theta)}{f_S(s; \theta)} = \frac{\Gamma(n)}{s^{n-1}}$$

which is free of θ, and thus by equation (10.2.1) S is sufficient for θ.

A slightly simpler criterion also can be derived. In particular, if $S_1, \ldots, S_k$ are jointly sufficient for $\boldsymbol{\theta}$, then

$$\begin{aligned} f(x_1, \ldots, x_n; \boldsymbol{\theta}) &= f_{\mathbf{S}}(\mathbf{s}; \boldsymbol{\theta}) f_{\mathbf{X}|\mathbf{s}}(x_1, \ldots, x_n) \\ &= g(\mathbf{s}; \boldsymbol{\theta}) h(x_1, \ldots, x_n) \end{aligned} \tag{10.2.2}$$

That is, the joint pdf of the sample can be factored into a function of $\mathbf{s}$ and $\boldsymbol{\theta}$ times a function of $\mathbf{x} = (x_1, \ldots, x_n)$ that does not involve $\boldsymbol{\theta}$. Conversely, suppose that $f(x_1, \ldots, x_n; \boldsymbol{\theta}) = g(\mathbf{s}; \boldsymbol{\theta})h(x_1, \ldots, x_n)$, where it is assumed that for fixed $\mathbf{s}$, $h(x_1, \ldots, x_n)$ does not depend on $\boldsymbol{\theta}$. Note that this means that if the joint pdf of $X_1, \ldots, X_n$ is zero over some region of the x_i's, then it must be possible to identify this region in terms of $\mathbf{s}$ and $\boldsymbol{\theta}$, and in terms of $\mathbf{x}$ and $\mathbf{s}$ without otherwise involving the $\mathbf{x}$ with the $\boldsymbol{\theta}$. If this is not possible, then the joint pdf really is not completely specified in the form stated. Basically, then, if equation (10.2.2) holds for some functions g and h, then the marginal pdf of $\mathbf{S}$ must be in the form

$$f_{\mathbf{s}}(\mathbf{s}; \boldsymbol{\theta}) = g(\mathbf{s}; \boldsymbol{\theta})c(\mathbf{s})$$

because for fixed $\mathbf{s}$ integrating or summing over the remaining variables cannot bring $\boldsymbol{\theta}$ into the function. Thus

$$f(x_1, \ldots, x_n; \boldsymbol{\theta}) = f_{\mathbf{S}}(\mathbf{s}; \boldsymbol{\theta})h(x_1, \ldots, x_n)/c(\mathbf{s})$$

and

$$\frac{f(x_1, \ldots, x_n; \boldsymbol{\theta})}{f_{\mathbf{S}}(\mathbf{s}; \boldsymbol{\theta})} = h(x_1, \ldots, x_n)/c(\mathbf{s})$$

which is independent of $\boldsymbol{\theta}$. This provides the outline of the proof of the following theorem.

Theorem 10.2.1 **Factorization Criterion** If $X_1, \ldots, X_n$ have joint pdf $f(x_1, \ldots, x_n; \boldsymbol{\theta})$, and if $\mathbf{S} = (S_1, \ldots, S_k)$, then $S_1, \ldots, S_k$ are jointly sufficient for $\boldsymbol{\theta}$ if and only if

$$f(x_1, \ldots, x_n; \boldsymbol{\theta}) = g(\mathbf{s}; \boldsymbol{\theta})h(x_1, \ldots, x_n) \tag{10.2.3}$$

where $g(\mathbf{s}; \boldsymbol{\theta})$ does not depend on $x_1, \ldots, x_n$, except through $\mathbf{s}$, and $h(x_1, \ldots, x_n)$ does not involve $\boldsymbol{\theta}$. ■

Example 10.2.2 Consider the random sample of Example 10.1.1, where $X_i \sim \text{BIN}(1, \theta)$. In that example, we conjectured that $S = \sum X_i$ would be sufficient, and then verified it directly by deriving the conditional pdf of $\mathbf{X}$ given $S = s$. The procedure is somewhat simpler if we use the factorization criterion. In particular, we have

$$\begin{aligned} f(x_1, \ldots, x_n; \theta) &= \theta^{\Sigma x_i}(1 - \theta)^{n - \Sigma x_i} \\ &= \theta^s(1 - \theta)^{n-s} \\ &= g(s; \theta)h(x_1, \ldots, x_n) \end{aligned}$$

where $s = \sum x_i$ and, in this case, we define $h(x_1, \ldots, x_n) = 1$ if all $x_i = 0$ or 1, and zero otherwise. It should be noted that the sample proportion, $\hat{\theta} = S/n$, also is sufficient for θ. In general, if a statistic S is sufficient for θ, then any one-to-one function of S also is sufficient for θ.

It is important to specify completely the functions involved in the factorization criterion, including the identification of regions of zero probability. The following example shows that care must be exercised in this matter.

Example 10.2.3 Consider a random sample from a uniform distribution, $X_i \sim \text{UNIF}(0, \theta)$, where θ is unknown. The joint pdf of $X_1, \ldots, X_n$ is

$$f(x_1, \ldots, x_n; \theta) = \frac{1}{\theta^n} \qquad 0 < x_i < \theta \qquad i = 1, \ldots, n$$

and zero otherwise. It is easier to specify this pdf in terms of the minimum, $x_{1:n}$, and maximum, $x_{n:n}$, of $x_1, \ldots, x_n$. In particular,

$$f(x_1, \ldots, x_n; \theta) = \frac{1}{\theta^n} \qquad 0 < x_{1:n} \qquad x_{n:n} < \theta$$

which means that

$$f(x_1, \ldots, x_n; \theta) = g(x_{n:n}; \theta)h(x_1, \ldots, x_n)$$

where $g(s; \theta) = 1/\theta^n$ if $s < \theta$ and zero otherwise, and $h(x_1, \ldots, x_n) = 1$ if $0 < x_{1:n}$ and zero otherwise. It follows from the factorization criterion that the largest order statistic, $S = X_{n:n}$, is sufficient for θ.

This type of problem is made more clear by using "indicator function" notation, which allows the conditions on the limits of the variables to be incorporated directly into the functional form of the pdf.

Definition 10.2.3

If A is a set, then the **indicator function** of A, denoted by I_A, is defined as

$$I_A(x) = \begin{cases} 1 & \text{if } x \in A \\ 0 & \text{if } x \notin A \end{cases}$$

In the previous example, we let $A = (0, \theta)$, then

$$f(x; \theta) = (1/\theta)I_{(0,\,\theta)}(x)$$

so that

$$f(x_1, \ldots, x_n; \theta) = (1/\theta^n)\prod_{i=1}^{n} I_{(0,\,\theta)}(x_i)$$

Because $\prod_{i=1}^{n} I_{(0,\,\theta)}(x_i) = 1$ if and only if $0 < x_{1:n}$ and $0 < x_{n:n} < \theta$, equation (10.2.3) is satisfied with $s = x_{n:n}$, $g(s; \theta) = (1/\theta^n)I_{(0,\,\theta)}(s)$ and $h(x_1, \ldots, x_n) = I_{(0,\,\infty)}(x_{1:n})$.

Example 10.2.4 Consider a random sample from a normal distribution, $X_i \sim N(\mu, \sigma^2)$, where both μ and σ^2 are unknown. It follows that

$$f(x_1, \ldots, x_n; \mu, \sigma^2) = \frac{1}{(2\pi\sigma^2)^{n/2}} \exp\left[-\frac{1}{2\sigma^2}\sum (x_i - \mu)^2\right]$$

Because $\sum (x_i - \mu)^2 = \sum x_i^2 - 2\mu \sum x_i + n\mu^2$, it follows that equation (10.2.3) holds with $s_1 = \sum x_i$, $s_2 = \sum x_i^2$,

$$g(s_1, s_2; \mu, \sigma^2) = \frac{1}{(2\pi\sigma^2)^{n/2}} \exp\left[-\frac{1}{2\sigma^2}(s_2 - 2\mu s_1 + n\mu^2)\right]$$

and $h(x_1, \ldots, x_n) = 1$. Thus, by the factorization criterion, $S_1 = \sum X_i$ and $S_2 = \sum X_i^2$ are jointly sufficient for $\boldsymbol{\theta} = (\mu, \sigma^2)$. Notice also that the MLEs, $\hat{\mu} = \bar{X} = S_1/n$ and $\hat{\sigma}^2 = \sum (X_i - \bar{X})^2/n = S_2/n - (S_1/n)^2$, correspond to a one-to-one transformation of S_1 and S_2, so that $\hat{\mu}$ and $\hat{\sigma}^2$ also are jointly sufficient for μ and σ^2.

In the next section, the general connection between MLEs and sufficient statistics will be established.

When a minimal set of sufficient statistics exists, we might expect the number of sufficient statistics to be equal to the number of unknown parameters. In the following example, two statistics are required to obtain sufficiency for a single parameter.

Example 10.2.5 Consider a random sample from a uniform distribution, $X_i \sim \text{UNIF}(\theta, \theta + 1)$. Notice that the length of the interval is one unit, but the endpoints are assumed to be unknown. The pdf of X is the indicator function of the interval, $f(x; \theta) = I_{(\theta, \theta+1)}(x)$, so the joint pdf of $X_1, \ldots, X_n$ is

$$f(x_1, \ldots, x_n; \theta) = \prod_{i=1}^{n} I_{(\theta, \theta+1)}(x_i)$$

This function assumes the value 1 if and only if $\theta < x_i$ and $x_i < \theta + 1$ for all x_i, so that

$$f(x_1, \ldots, x_n; \theta) = \prod_{i=1}^{n} I_{(\theta, \infty)}(x_i) \prod_{i=1}^{n} I_{(-\infty, \theta+1)}(x_i)$$

$$= I_{(\theta, \infty)}(x_{1:n}) I_{(-\infty, \theta+1)}(x_{n:n})$$

which shows, by the factorization criterion, that the smallest and largest order statistics $S_1 = X_{1:n}$ and $S_2 = X_{n:n}$ are jointly sufficient for θ. Actually, it can be shown that S_1 and S_2 are minimal sufficient.

Methods for verifying whether a set of statistics is minimal sufficient are discussed by Wasan (1970), but we will not elaborate on them here.

10.3

FURTHER PROPERTIES OF SUFFICIENT STATISTICS

It is possible to relate sufficiency to several of the concepts that were discussed in earlier chapters.

Theorem 10.3.1 If $S_1, \ldots, S_k$ are jointly sufficient for $\boldsymbol{\theta}$ and if $\hat{\boldsymbol{\theta}}$ is a unique maximum likelihood estimator of $\boldsymbol{\theta}$, then $\hat{\boldsymbol{\theta}}$ is a function of $\boldsymbol{S} = (S_1, \ldots, S_k)$.

Proof

By the factorization criterion,

$$L(\boldsymbol{\theta}) = f(x_1, \ldots, x_n; \boldsymbol{\theta}) = g(\boldsymbol{s}; \boldsymbol{\theta}) h(x_1, \ldots, x_n)$$

which means that a value that maximizes the likelihood function must depend on $\boldsymbol{s}$, say $\hat{\boldsymbol{\theta}} = \ell(\boldsymbol{s})$. If the MLE is unique, this defines a function of $\boldsymbol{s}$. ■

Actually, the result can be stated more generally: If there exist jointly sufficient statistics, and if there exists an MLE, then there exists an MLE that is a function of the sufficient statistics.

It also follows that if the MLEs, $\hat{\theta}_1, \ldots, \hat{\theta}_k$ are unique and jointly sufficient, then they are a minimal sufficient set, because the factorization criterion applies for every set of jointly sufficient statistics.

The following simple example shows that it is possible to have a sufficient statistic S and an MLE that is not a function of S.

Example 10.3.1 Suppose that X is discrete with pdf $f(x; \theta)$ and $\mathbf{\Omega} = \{0, 1\}$, where, with the use of indicator functions,

$$f(x; \theta) = \frac{1}{4} I_{\{1, 4\}}(x) + \left(\frac{1-\theta}{4} + \frac{\theta}{6}\right) I_{\{2\}}(x) + \left(\frac{1-\theta}{4} + \frac{\theta}{3}\right) I_{\{3\}}(x)$$

If $\mathscr{s}(x) = 3I_{\{1, 4\}}(x) + 2I_{\{2\}}(x) + 4I_{\{3\}}(x)$, then $S = \mathscr{s}(X)$ is sufficient for θ, which can be seen from the factorization criterion with

$$h(x) = I_{\{1, 2, 3, 4\}}(x) \quad \text{and} \quad g(s; \theta) = \frac{1-\theta}{4} + \frac{\theta}{12} s$$

Furthermore, more than one MLE exists. For example, the functions $\ell_1(x) = I_{\{1, 3\}}(x)$ and $\ell_2(x) = I_{\{1, 3, 4\}}(x)$ both produce MLEs, $\hat{\theta}_1 = \ell_1(x)$ and $\hat{\theta}_2 = \ell_2(x)$, because the corresponding estimates maximize $f(x; \theta)$ for each fixed x. Clearly, $\hat{\theta}_1$ is not a function of S because $\mathscr{s}(1) = \mathscr{s}(4) = 3$, but $\ell_1(1) = 1$ while $\ell_1(4) = 0$. However, $\hat{\theta}_2 = \ell(s)$ where $\ell(s) = I_{\{3, 4\}}(s)$.

This shows that some care must be taken in stating the relationship between sufficient statistics and MLEs. If the MLE is unique, however, then the situation is rather straightforward.

Example 10.3.2 Consider a random sample of size n from a Bernoulli distribution $X_i \sim \text{BIN}(1, p)$. We know that $S = \sum X_i$ is sufficient for p, and that $S \sim \text{BIN}(n, p)$. Thus, we may determine the MLE of p directly from the pdf of S, giving $\hat{p} = S/n$ as before.

Theorem 10.3.2 If S is sufficient for θ, then any Bayes estimator will be a function of S.

Proof

Because the function $h(x_1, \ldots, x_n)$ in the factorization criterion does not depend on θ, it can be eliminated in equation (9.5.8), and the posterior density $f_{\theta|\mathbf{x}}(\theta)$ can be replaced by

$$f_{\theta|\mathbf{x}}(\theta) = \frac{g(s;\, \theta)p(\theta)}{\displaystyle\int g(s;\, \theta)p(\theta)\, d\theta}$$ ■

As mentioned earlier, the order statistics are jointly sufficient.

Theorem 10.3.3 If $X_1, \ldots, X_n$ is a random sample from a continuous distribution with pdf $f(x;\, \boldsymbol{\theta})$, then the order statistics form a jointly sufficient set for $\boldsymbol{\theta}$.

Proof

For fixed $x_{1:n}, \ldots, x_{n:n}$, and associated $x_1, \ldots, x_n$

$$\frac{f(x_1;\, \boldsymbol{\theta}) \cdots f(x_n;\, \boldsymbol{\theta})}{n!\, f(x_{1:n};\, \boldsymbol{\theta}) \cdots f(x_{n:n};\, \boldsymbol{\theta})} = \frac{1}{n!} \qquad x_{1:n} = \min(x_i), \ldots, x_{n:n} = \max(x_i)$$

and zero otherwise. ■

Generally, sufficient statistics are involved in the construction of UMVUEs.

Theorem 10.3.4 **Rao-Blackwell** Let $X_1, \ldots, X_n$ have joint pdf $f(x_1, \ldots, x_n;\, \boldsymbol{\theta})$, and let $\mathbf{S} = (S_1, \ldots, S_k)$ be a vector of jointly sufficient statistics for $\boldsymbol{\theta}$. If T is any unbiased estimator of $\tau(\boldsymbol{\theta})$, and if $T^* = E(T \mid \mathbf{S})$, then

1. T^* is an unbiased estimator of $\tau(\boldsymbol{\theta})$,
2. T^* is a function of $\mathbf{S}$, and
3. $\text{Var}(T^*) \leqslant \text{Var}(T)$ for every $\boldsymbol{\theta}$, and $\text{Var}(T^*) < \text{Var}(T)$ for some $\boldsymbol{\theta}$ unless $T^* = T$ with probability 1.

Proof

By sufficiency, $f_{T|\mathbf{s}}(t)$ does not involve $\boldsymbol{\theta}$, and thus the function $\ell^*(\mathbf{s}) = E(T \mid \mathbf{s})$ does not depend on $\boldsymbol{\theta}$. Thus, $T^* = \ell^*(\mathbf{S}) = E(T \mid \mathbf{S})$ is an estimator that is a function of $\mathbf{S}$, and furthermore,

$$\begin{aligned} E(T^*) &= E_{\mathbf{S}}(T^*) \\ &= E_{\mathbf{S}}[E(T \mid \mathbf{S})] \\ &= E(T) \\ &= \tau(\boldsymbol{\theta}) \end{aligned}$$

by Theorem 5.4.1. From Theorem 5.4.3,

$$\begin{aligned}\text{Var}(T) &= \text{Var}[E(T \mid S)] + E[\text{Var}(T \mid S)] \\ &\geqslant \text{Var}[E(T \mid S)] \\ &= \text{Var}(T^*)\end{aligned}$$

with equality if and only if $E[\text{Var}(T \mid S)] = 0$, which occurs if and only if $\text{Var}(T \mid S) = 0$ with probability 1, or equivalently $T = E(T \mid S) = T^*$. ■

It is clear from the Rao-Blackwell theorem that if we are searching for an unbiased estimator with small variance, we may as well restrict attention to functions of sufficient statistics. If any unbiased estimator exists, then there will be one that is a function of sufficient statistics, namely $E(T \mid S)$, which also is unbiased and has variance at least as small or smaller. In particular, we still are interested in knowing how to find a UMVUE for a parameter, and the above theorem narrows our problem down somewhat. For example, consider a one-parameter model $f(x; \theta)$, and assume that a single sufficient statistic, S, exists. We know we must consider only unbiased functions of S in searching for a UMVUE. In some cases it may be possible to show that only one function of S is unbiased, and in that case we would know that it is a UMVUE. The concept of "completeness" is helpful in determining unique unbiased estimators, and this concept is defined in the next section.

10.4

COMPLETENESS AND EXPONENTIAL CLASS

Definition 10.4.1

Completeness A family of density functions $\{f_T(t; \boldsymbol{\theta}); \boldsymbol{\theta} \in \boldsymbol{\Omega}\}$, is called **complete** if $E[u(T)] = 0$ for all $\boldsymbol{\theta} \in \boldsymbol{\Omega}$ implies $u(T) = 0$ with probability 1 for all $\boldsymbol{\theta} \in \boldsymbol{\Omega}$.

This sometimes is expressed by saying that there are no nontrivial unbiased estimators of zero. In particular, it means that two different functions of $\boldsymbol{T}$ cannot have the same expected value. For example, if $E[u_1(\boldsymbol{T})] = \tau(\boldsymbol{\theta})$ and $E[u_2(\boldsymbol{T})] = \tau(\boldsymbol{\theta})$, then $E[u_1(\boldsymbol{T}) - u_2(\boldsymbol{T})] = 0$, which implies $u_1(\boldsymbol{T}) - u_2(\boldsymbol{T}) = 0$, or $u_1(\boldsymbol{T}) = u_2(\boldsymbol{T})$ with probability 1, if the family of density functions is complete. That is, any unbiased estimator is unique in this case. We primarily are interested in knowing that the family of density functions of a sufficient statistic is complete, because in that case an unbiased function of the sufficient statistic will be unique, and it must be a UMVUE by the Rao-Blackwell theorem.

A sufficient statistic the density of which is a member of a complete family of density functions will be referred to as a **complete sufficient statistic.**

Theorem 10.4.1 **Lehmann-Scheffe** Let $X_1, \ldots, X_n$ have joint pdf $f(x_1, \ldots, x_n; \boldsymbol{\theta})$, and let $\boldsymbol{S}$ be a vector of jointly complete sufficient statistics for $\boldsymbol{\theta}$. If $T^* = t^*(\boldsymbol{S})$ is a statistic that is unbiased for $\tau(\boldsymbol{\theta})$ and a function of $\boldsymbol{S}$, then T^* is a UMVUE of $\tau(\boldsymbol{\theta})$.

Proof

It follows by completeness that any statistic that is a function of $\boldsymbol{S}$ and an unbiased estimator of $\tau(\boldsymbol{\theta})$ must be equal to T^* with probability 1. If T is any other statistic that is an unbiased estimator of $\tau(\boldsymbol{\theta})$, then by the Rao-Blackwell theorem $E(T \mid \boldsymbol{S})$ also is unbiased for $\tau(\boldsymbol{\theta})$ and a function of $\boldsymbol{S}$, so by uniqueness, $T^* = E(T \mid \boldsymbol{S})$ with probability 1. Furthermore, $\text{Var}(T^*) \leqslant \text{Var}(T)$ for all $\boldsymbol{\theta}$. Thus, T^* is a UMVUE of $\tau(\boldsymbol{\theta})$. ∎

Example 10.4.1 Let $X_1, \ldots, X_n$ denote a random sample from a Poisson distribution, $X_i \sim \text{POI}(\mu)$, so that

$$f(x_1, \ldots, x_n; \mu) = \frac{e^{-n\mu}\mu^{\Sigma x}}{\prod (x_i!)}$$

By the factorization criterion, $S = \sum X_i$ is a sufficient statistic. We know that $S \sim \text{POI}(n\mu)$, and we can show that a Poisson family is complete. For convenience let $\theta = n\mu$, and consider any function $u(s)$. We have

$$E[u(S)] = \sum_{s=0}^{\infty} \frac{u(s)e^{-\theta}\theta^s}{s!}$$

Because $e^{-\theta} \neq 0$, setting $E[u(S)] = 0$ requires all the coefficients, $u(s)/s!$, of θ^s to be zero. But $u(S)/s! = 0$ implies $u(s) = 0$. By completeness, $\bar{X} = S/n$ is the unique function of S that is unbiased for $E(\bar{X}) = \mu$, and by Theorem 10.4.1 it must be a UMVUE of μ.

This particular result also can be verified by comparing $\text{Var}(\bar{X})$ to the CRLB; however, the CRLB approach will not work for a nonlinear function of S. The present approach, on the other hand, can be used to find the UMVUE of $\tau(\theta) = E[u(S)]$, for any function $u(s)$ for which the expected value exists. For example, in the Poisson case, $E(\bar{X}^2) = \mu^2 + \mu/n$, so that $\bar{X}^2 = (S/n)^2$ is the UMVUE of $\mu^2 + \mu/n$. It also follows that $\bar{X}^2 - \bar{X}/n = (S/n)^2 - S/n^2$ is the UMVUE of μ^2. If a UMVUE is desired for any specified $\tau(\mu)$, it is only necessary to find some function of S that is unbiased for $\tau(\mu)$; then that will be the UMVUE. If there is difficulty in finding a $u(s)$ such that $E[u(S)] = \tau(\mu)$, one possibility is to find any function, $h(X_1, \ldots, X_n) = T$, that is unbiased, and then $E(T \mid S)$ will be an unbiased estimator that is a function of S. Thus, use of complete sufficient statistics and the Rao-Blackwell theorem provides one possible systematic approach for attempting to find UMVUEs in certain cases.

Note that completeness is a property of a family of densities, and the family must be large enough or "complete" to enjoy this property. That is, there may be a nonzero $u(S)$ whose mean is zero for some densities, but this situation may not hold if more densities are added to the family. If one considers a single Poisson distribution, say $\mu = 1$, then $E(S - n) = 0$, and a family consisting of this single Poisson density function is not complete because $u(s) = s - n \neq 0$ if $s \neq n$.

If the range of the random variable does not depend on parameters, then one may essentially restrict attention to families of densities that fall in the form of the "exponential class" when considering complete sufficient statistics, so we need not consider these families individually in detail.

Definition 10.4.2

Exponential Class A density function is said to be a member of the **regular exponential class** if it can be expressed in the form

$$f(x; \boldsymbol{\theta}) = c(\boldsymbol{\theta})h(x) \exp\left[\sum_{j=1}^{k} q_j(\boldsymbol{\theta})t_j(x)\right] \qquad x \in A \qquad \textbf{(10.4.1)}$$

and zero otherwise, where $\boldsymbol{\theta} = (\theta_1, \ldots, \theta_k)$ is a vector of k unknown parameters, if the parameter space has the form

$$\boldsymbol{\Omega} = \{\boldsymbol{\theta} \mid a_i \leqslant \theta_i \leqslant b_i,\ i = 1, \ldots, k\}$$

(note that $a_i = -\infty$ and $b_i = \infty$ are permissible values), and if it satisfies regularity conditions 1, 2, and 3a or 3b given by:

1. The set $A = \{x : f(x; \boldsymbol{\theta}) > 0\}$ does not depend on $\boldsymbol{\theta}$.
2. The functions $q_j(\boldsymbol{\theta})$ are nontrivial, functionally independent, continuous functions of the θ_i.

3a. For a continuous random variable, the derivatives $t'_j(x)$ are linearly independent continuous functions of x over A.

3b. For a discrete random variable, the $t_j(x)$ are nontrivial functions of x on A, and none is a linear function of the others.

For convenience, we will write that $f(x; \boldsymbol{\theta})$ is a member of REC$(q_1, \ldots, q_k)$ or simply REC.

Example 10.4.2 Consider a Bernoulli distribution, $X \sim \text{BIN}(1, p)$. It follows that

$$f(x; p) = p^x(1-p)^{1-x}$$
$$= (1-p) \exp\{x \ln[p/(1-p)]\} \qquad x \in A = \{0, 1\}$$

which is REC(q_1) with $q_1(p) = \ln[p/(1-p)]$ and $t_1(x) = x$.

Note that the notion of REC, with slightly modified regularity conditions, can be extended to the case where X is a vector.

It can be shown that the REC is a complete family for the special case when $t_j(x) = x$. Many of the common density functions such as binomial, Poisson, exponential, gamma, and normal pdf's are in the form of the REC, but we are particularly interested in knowing that the pdf's of the sufficient statistics from these models are complete. If a random sample is considered from a member of the REC, then a set of joint sufficient statistics is identified readily by the factorization criterion; moreover, the pdf of these sufficient statistics also turns out to be in the special form of a (possibly multivariate) REC, and therefore they are complete sufficient statistics.

Theorem 10.4.2 If $X_1, \ldots, X_n$ is a random sample from a member of the regular exponential class REC $(q_1, \ldots, q_k)$, then the statistics

$$S_1 = \sum_{i=1}^{n} t_1(X_i), \ldots, S_k = \sum_{i=1}^{n} t_k(X_i)$$

are a minimal set of complete sufficient statistics for $\theta_1, \ldots, \theta_k$. ■

Example 10.4.3 Consider the previous example, $X \sim \text{BIN}(1, p)$. For a random sample of size n, $t(x_i) = x_i$ and $S = \sum_{i=1}^{n} X_i$ is a complete sufficient statistic for p.

If we desire a UMVUE of $\text{Var}(X) = p(1-p)$, we might try $\bar{X}(1-\bar{X})$. Now

$$E[\bar{X}(1-\bar{X})] = E(\bar{X}) - E(\bar{X}^2)$$
$$= p - [p^2 + \text{Var}(\bar{X})]$$
$$= p - p^2 - p(1-p)/n$$
$$= p(1-p)(1-1/n)$$

and thus $E[n\bar{X}(1-\bar{X})/(n-1)] = p(1-p)$, and this gives the UMVUE of $p(1-p)$ as $c\bar{X}(1-\bar{X})$ where $c = n/(n-1)$.

Example 10.4.4 If $X \sim N(\mu, \sigma^2)$, then

$$f(x; \mu, \sigma) = \frac{1}{\sigma\sqrt{2\pi}} \exp\left[-(x-\mu)^2/2\sigma^2\right]$$

$$= \frac{1}{\sigma\sqrt{2\pi}} \exp\left[-\frac{x^2}{2\sigma^2} + \frac{2\mu x}{2\sigma^2} - \frac{\mu^2}{2\sigma^2}\right]$$

For a random sample of size n, it is clear that $S_1 = \sum X_i^2$ and $S_2 = \sum X_i$ are jointly complete and sufficient statistics of μ and σ^2. Because MLEs are one-to-one functions of S_1 and S_2, they also could be used as the jointly complete sufficient statistics here.

It can be shown that under mild regularity conditions, families of density functions that admit k-dimensional sufficient statistics for all sample sizes must be k-parameter RECs.

Thus, for the regular case, for most practical purposes Theorem 10.4.2 covers all of the models that admit complete sufficient statistics, and there is no point in attempting to find complete sufficient statistics in the regular case for models that are not in the REC form.

We have seen that a close connection exists among the REC, complete sufficient statistics, and UMVUEs. Also, MLEs are functions of minimal sufficient statistics, and the MLEs are asymptotically efficient with asymptotic variance being the CRLB. If we call an estimator whose variance achieves the CRLB a CRLB estimator, then the following theorems can be stated.

Theorem 10.4.3 If a CRLB estimator T exists for $\tau(\theta)$, then a single sufficient statistic exists, and T is a function of the sufficient statistic. Conversely, if a single sufficient statistic exists and the CRLB exists, then a CRLB estimator exists for some $\tau(\theta)$. ■

Theorem 10.4.4 If the CRLB exists, then a CRLB estimator will exist for some function $\tau(\theta)$ if and only if the density function is a member of the REC. Furthermore, the CRLB estimator of $\tau(\theta)$ will be $\tau(\hat{\theta})$, where $\hat{\theta}$ is the MLE of θ. ■

Most pdf's of practical interest that are not included in the REC belong to another general class, which allows the range of X, denoted by $A = \{x : f(x; \boldsymbol{\theta}) > 0\}$, to depend on $\boldsymbol{\theta}$.

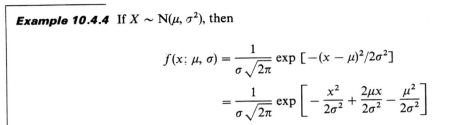

Definition 10.4.3

A density function is said to be a member of the **range-dependent exponential class**, denoted by RDEC $(q_1, \ldots, q_k)$, if it satisfies regularity conditions 2 and 3a or 3b of Definition 10.4.2, for $j = 3, \ldots, k$, and if it has the form

$$f(x; \boldsymbol{\theta}) = c(\boldsymbol{\theta})h(x) \exp\left[\sum_{j=3}^{k} q_j(\theta_3, \ldots, \theta_k)\ell_j(x)\right] \tag{10.4.2}$$

where $A = \{x \mid q_1(\theta_1, \theta_2) < x < q_2(\theta_1, \theta_2)\}$ and $\boldsymbol{\theta} \in \boldsymbol{\Omega}$.

We will include as special cases the following:

1. The one-parameter case, where

$$f(x; \theta) = c(\theta)h(x) \tag{10.4.3}$$

with

$$A = \{x \mid q_1(\theta) < x < q_2(\theta)\}$$

2. The two-parameter case, where

$$f(x; \theta_1, \theta_2) = c(\theta_1, \theta_2)h(x) \tag{10.4.4}$$

with

$$A = \{x \mid q_1(\theta_1, \theta_2) < x < q_2(\theta_1, \theta_2)\}$$

The following theorem is useful in identifying sufficient statistics in the range-dependent case.

Theorem 10.4.5 Let $X_1, \ldots, X_n$ be a random sample from a member of the RDEC $(q_1, \ldots, q_k)$.

1. If $k > 2$, then $S_1 = X_{1:n}$, $S_2 = X_{n:n}$ and $S_3, \ldots, S_k$ where $S_j = \sum_{i=1}^{n} \ell_j(X_i)$ are jointly sufficient for $\boldsymbol{\theta} = (\theta_1, \ldots, \theta_k)$.
2. In the two-parameter case, $S_1 = X_{1:n}$ and $S_2 = X_{n:n}$ are jointly sufficient for $\boldsymbol{\theta} = (\theta_1, \theta_2)$.
3. In the one-parameter case, $S_1 = X_{1:n}$ and $S_2 = X_{n:n}$ are jointly sufficient for θ. If $q_1(\theta)$ is increasing and $q_2(\theta)$ is decreasing, then $T_1 = \min[q_1^{-1}(X_{1:n}), q_2^{-1}(X_{n:n})]$ is a single sufficient statistic for θ. If $q_1(\theta)$ is decreasing and $q_2(\theta)$ is increasing, then $T_2 = \max[q_1^{-1}(X_{1:n}), q_2^{-1}(X_{n:n})]$ is a single sufficient statistic for θ. ■

If one of the limits is constant and the other depends on a single parameter, say θ, then the following theorem can be stated.

Theorem 10.4.6 Suppose that $X_1, \ldots, X_n$ is a random sample from a member of the RDEC.

1. If $k > 2$ and the lower limit is constant, say $q_1(\theta) = a$, then $X_{n:n}$ and the statistics $\sum_{i=1}^{n} t_j(X_i)$ are jointly sufficient for θ and θ_j; $j = 3, \ldots, k$. If the upper limit is constant, say $q_2(\theta) = b$, then $X_{1:n}$ and the statistics $\sum_{i=1}^{n} t_j(X_i)$ are jointly sufficient for θ and θ_j; $j = 3, \ldots, k$.
2. In the one-parameter case, if $q_1(\theta)$ does not depend on θ, then $S_2 = X_{n:n}$ is sufficient for θ, and if $q_2(\theta)$ does not depend on θ, then $S_1 = X_{1:n}$ is sufficient for θ. ■

Example 10.4.5 Consider the pdf

$$f(x; \theta) = \frac{1}{2\theta} \qquad -\theta < x < \theta$$

and zero otherwise. We have $q_1(\theta) = -\theta$, a decreasing function, and $q_2(\theta) = \theta$, an increasing function of θ. Thus, by Theorem 10.4.5, $T_2 = \max\,[-X_{1:n}, X_{n:n}]$ is a single sufficient statistic for θ.

Example 10.4.6 Consider a two-parameter exponential distribution, $X \sim \text{EXP}(\theta, \eta)$.

$$\begin{aligned} f(x; \theta, \eta) &= (1/\theta) \exp\,[-(x - \eta)/\theta] && \eta < x < \infty \\ &= (1/\theta) \exp\,(\eta/\theta) \exp\,(-x/\theta) && \eta < x < \infty \end{aligned}$$

If $X_1, \ldots, X_n$ is a random sample, then it follows from Theorem 10.4.6 that $X_{1:n}$ and $\sum X_i$ are joint sufficient statistics for (θ, η). Because $q_2(\eta) = \infty$ is not a function of parameters, $X_{n:n}$ is not involved.

Suppose that θ is known, say $\theta = 1$. Then

$$\begin{aligned} f(x; \eta) &= e^{-(x+\eta)} && \eta < x < \infty \\ &= e^{-x} e^{\eta} && \eta < x < \infty \end{aligned}$$

We see that $X_{1:n}$ is sufficient for η. This is consistent with earlier results, where we found that estimators of η based on $X_{1:n}$ were better than estimators based on other statistics, such as $\bar{X}$, for this model.

Example 10.4.7 Consider a random sample of size n from a uniform distribution, $X_i \sim \text{UNIF}(\theta_1, \theta_2)$. Because

$$f(x; \theta_1, \theta_2) = \frac{1}{\theta_2 - \theta_1} \qquad \theta_1 < x < \theta_2$$

it follows from Theorem 10.4.5 that $X_{1:n}$ and $X_{n:n}$ are jointly sufficient for (θ_1, θ_2).

The previous results deal only with finding sufficient statistics for members of the RDEC. These statistics also may be complete, but this must be verified by a separate argument.

Example 10.4.8 Consider a random sample of size n from a uniform distribution $X_i \sim \text{UNIF}(0, \theta)$. It follows from the previous theorem that $X_{n:n}$ is sufficient for θ. The pdf of $S = X_{n:n}$ is

$$f(s; \theta) = ns^{n-1}/\theta^n \qquad 0 < s < \theta$$

and zero otherwise.

To verify completeness, assume that $E[u(S)] = 0$ for all $\theta > 0$, which means that

$$\int_0^\theta [u(s)ns^{n-1}/\theta^n] \, ds = 0$$

If we multiply by θ^n and differentiate with respect to θ, then $u(\theta)n\theta^{n-1} = 0$ for all $\theta > 0$, which implies $u(s) = 0$ for all $s > 0$, and thus S is a complete sufficient statistic for θ.

The following interesting theorem sometimes is useful for establishing certain distributional results.

Theorem 10.4.7 **Basu** Let $X_1, \ldots, X_n$ have joint pdf $f(x_1, \ldots, x_n; \boldsymbol{\theta})$; $\boldsymbol{\theta} \in \boldsymbol{\Omega}$. Suppose that $\boldsymbol{S} = (S_1, \ldots, S_k)$ where $S_1, \ldots, S_k$ are jointly complete sufficient statistics for $\boldsymbol{\theta}$, and suppose that $\boldsymbol{T}$ is any other statistic. If the distribution of $\boldsymbol{T}$ does not involve $\boldsymbol{\theta}$, then $\boldsymbol{S}$ and $\boldsymbol{T}$ are stochastically independent.

Proof

We will consider the discrete case. Denote by $f(\mathbf{t}), f(\mathbf{s}; \boldsymbol{\theta})$, and $f(\mathbf{t} \mid \mathbf{s})$ the pdf's of $\boldsymbol{T}$, $\boldsymbol{S}$, and the conditional pdf of $\boldsymbol{T}$ given $\boldsymbol{S} = \mathbf{s}$, respectively. Consider the following expected value relative to the distribution of $\boldsymbol{S}$:

$$\begin{aligned} E_{\boldsymbol{S}}[f(\mathbf{t}) - f(\mathbf{t} \mid \boldsymbol{S})] &= f(\mathbf{t}) - \sum_{\mathbf{s}} f(\mathbf{t} \mid \mathbf{s}) f(\mathbf{s}; \boldsymbol{\theta}) \\ &= f(\mathbf{t}) - \sum_{\mathbf{s}} f(\mathbf{s}, \mathbf{t}; \boldsymbol{\theta}) \\ &= f(\mathbf{t}) - f(\mathbf{t}) = 0 \end{aligned}$$

Because $\boldsymbol{S}$ is a complete sufficient statistic, $f(\mathbf{t} \mid \mathbf{s}) = f(\mathbf{t})$, which means that $\boldsymbol{S}$ and $\boldsymbol{T}$ are stochastically independent.

The continuous case is similar. ∎

Example 10.4.9 Consider a random sample of size n from a normal distribution, $X_i \sim N(\mu, \sigma^2)$, and consider the MLEs, $\hat{\mu} = \bar{X}$ and $\hat{\sigma}^2 = \sum (X_i - \bar{X})^2/n$. It is easy to verify that $\bar{X}$ is a complete sufficient statistic for μ, for fixed values of σ^2. Also

$$\frac{n\hat{\sigma}^2}{\sigma^2} \sim \chi^2(n-1)$$

which does not depend on μ. It follows that $\bar{X}$ and $\hat{\sigma}^2$ are independent random variables. Also, $\bar{X}$ and $\hat{\sigma}^2$ are jointly complete sufficient statistics for μ and σ, and quantities of the form $(X_i - \bar{X})/\hat{\sigma}$ are distributed independently of μ and σ, so these quantities are stochastically independent of $\bar{X}$ and $\hat{\sigma}^2$.

SUMMARY

Our purpose in this chapter was to introduce the concepts of sufficiency and completeness. Generally speaking, a statistic provides a reduction of a set of data from some distribution to a more concise form. If a statistic is sufficient, then it contains, in a certain sense, all of the "information" in the data concerning an unknown parameter of the distribution. Although sufficiency can be verified directly from the definition, at least theoretically, this usually can be accomplished more easily by using the factorization criterion.

If a statistic is sufficient and a unique MLE exists, then the MLE is a function of the sufficient statistic. Sufficient statistics also are important in the construction of UMVUEs. If a statistic is complete as well as sufficient for a parameter, and if an unbiased estimator of the parameter (or a function of the parameter) exists, then a UMVUE exists and it is a function of the complete sufficient statistic. It often is difficult to verify completeness directly from the definition, but a special class of pdf's, known as the exponential class, provides a convenient way to identify complete sufficient statistics.

EXERCISES

1. Let $X_1, \ldots, X_n$ be a random sample from a Poisson distribution, $X_i \sim POI(\mu)$. Verify that $S = \sum_{i=1}^{n} X_i$ is sufficient for μ by using equation (10.2.1).

2. Consider a random sample of size n from a geometric distribution, $X_i \sim GEO(p)$. Use equation (10.2.1) to show that $S = \sum X_i$ is sufficient for p.

3. Suppose that $X_1, \ldots, X_n$ is a random sample from a normal distribution, $X_i \sim N(0, \theta)$. Show that equation (10.2.1) does not depend on θ if $S = \sum X_i^2$.

4. Consider a random sample of size n from a two-parameter exponential distribution, $X_i \sim \text{EXP}(1, \eta)$. Show that $S = X_{1:n}$ is sufficient for η by using equation (10.2.1).

5. Let $X_1, \ldots, X_n$ be a random sample from a gamma distribution, $X_i \sim \text{GAM}(\theta, 2)$. Show that $S = \sum X_i$ is sufficient for θ

(a) by using equation (10.2.1),

(b) by the factorization criterion of equation (10.2.3).

6. Suppose that $X_1, X_2, \ldots, X_n$ are independent, with $X_i \sim \text{BIN}(m_i, p)$; $i = 1, 2, \ldots, n$. Show that $S = \sum_{i=1}^{n} X_i$ is sufficient for p by the factorization criterion.

7. Let $X_1, X_2, \ldots, X_n$ be independent with $X_i \sim \text{NB}(r_i, p)$. Find a sufficient statistic for p.

8. Rework Exercise 4 using the factorization criterion.

9. Consider a random sample of size n from a Weibull distribution, $X_i \sim \text{WEI}(\theta, \beta)$.

(a) Find a sufficient statistic for θ with β known, say $\beta = 2$.

(b) If β is unknown, can you find a single sufficient statistic for β?

10. Let $X_1, \ldots, X_n$ be a random sample from a normal distribution, $X_i \sim N(\mu, \sigma^2)$.

(a) Find a single sufficient statistic for μ with σ^2 known.

(b) Find a single sufficient statistic for σ^2 with μ known.

11. Consider a random sample of size n from a uniform distribution, $X_i \sim \text{UNIF}(\theta_1, \theta_2)$.

(a) Show that $X_{1:n}$ is sufficient for θ_1, if θ_2 is known.

(b) Show that $X_{1:n}$ and $X_{n:n}$ are jointly sufficient for θ_1 and θ_2.

12. Let $X_1, \ldots, X_n$ be a random sample from a two-parameter exponential distribution, $X_i \sim \text{EXP}(\theta, \eta)$. Show that $X_{1:n}$ and $\bar{X}$ are jointly sufficient for θ and η.

13. Suppose that $X_1, \ldots, X_n$ is a random sample from a beta distribution, $X_i \sim \text{BETA}(\theta_1, \theta_2)$. Find joint sufficient statistics for θ_1 and θ_2.

14. Consider a random sample of size n from a uniform distribution, $X_i \sim \text{UNIF}(\theta, 2\theta)$; $\theta > 0$. Can you find a single sufficient statistic for θ? Can you find a pair of jointly sufficient statistics for θ?

15. For the random sample of Exercise 2, find the estimator of p obtained by maximizing the pdf of $S = \sum X_i$, and compare this with the usual MLE of p.

16. For the random variables $X_1, \ldots, X_n$ in Exercise 7, find the MLE of p by maximizing the pdf of the sufficient statistic. Is this the same as the usual MLE? Explain why this result is expected.

17. Consider the sufficient statistic, $S = X_{1:n}$, of Exercise 4.

(a) Show that S also is complete.

(b) Verify that $X_{1:n} - 1/n$ is the UMVUE of η.

(c) Find the UMVUE of the pth percentile.

18. Let $X \sim \mathrm{N}(0, \theta)$; $\theta > 0$.
   (a) Show that X^2 is complete and sufficient for θ.
   (b) Show that $\mathrm{N}(0, \theta)$ is not a complete family.

19. Show that $\mathrm{N}(\mu, \mu^2)$ does not belong to the regular exponential class.

20. Show that the following families of distributions belong to the regular exponential class, and for each case use this information to find complete sufficient statistics based on a random sample $X_1, \ldots, X_n$.
   (a) $\mathrm{BIN}(1, p)$; $0 < p < 1$.
   (b) $\mathrm{POI}(\mu)$; $\mu > 0$.
   (c) $\mathrm{NB}(r, p)$; r known, $0 < p < 1$.
   (d) $\mathrm{N}(\mu, \sigma^2)$; $-\infty < \mu < \infty$, $\sigma^2 > 0$.
   (e) $\mathrm{EXP}(\theta)$; $\theta > 0$.
   (f) $\mathrm{GAM}(\theta, \kappa)$; $\theta > 0$, $\kappa > 0$.
   (g) $\mathrm{BETA}(\theta_1, \theta_2)$; $\theta_1 > 0$, $\theta_2 > 0$.
   (h) $\mathrm{WEI}(\theta, \beta)$; β known, $\theta > 0$.

21. Let $X_1, \ldots, X_n$ be a random sample from a Bernoulli distribution, $X_i \sim \mathrm{BIN}(1, p)$; $0 < p < 1$.
   (a) Find the UMVUE of $\mathrm{Var}(X) = p(1 - p)$.
   (b) Find the UMVUE of p^2.

22. Consider a random sample of size n from a Poisson distribution, $X_i \sim \mathrm{POI}(\mu)$; $\mu > 0$. Find the UMVUE of $P[X = 0] = e^{-\mu}$. *Hint:* Recall Exercise 33(g) of Chapter 9.

23. Suppose that $X_1, \ldots, X_n$ is a random sample from a normal distribution, $X_i \sim \mathrm{N}(\mu, 9)$.
   (a) Find the UMVUE of the 95th percentile.
   (b) Find the UMVUE of $P[X \leqslant c]$ where c is a known constant.
   Hint: Find the conditional distribution of X_1 given $\bar{X} = \bar{x}$ and apply the Rao-Blackwell theorem with $T = u(X_1)$, where $u(x_1) = 1$ if $x_1 \leqslant c$, and zero otherwise.

24. If $X \sim \mathrm{POI}(\mu)$, show that $S = (-1)^X$ is the UMVUE of $e^{-2\mu}$. Is this a reasonable estimator?

25. Consider a random sample of size n from a distribution with pdf $f(x; \theta) = \theta x^{\theta - 1}$ if $0 < x < 1$ and zero otherwise; $\theta > 0$.
   (a) Find the UMVUE of $1/\theta$. *Hint:* $E[-\ln X] = 1/\theta$.
   (b) Find the UMVUE of θ.

26. For the random sample of Exercise 11, show that the jointly sufficient statistics $X_{1:n}$ and $X_{n:n}$ also are complete. Suppose that it is desired to estimate the mean $\mu = (\theta_1 + \theta_2)/2$. Find the UMVUE of μ. *Hint:* First find the expected values $E(X_{1:n})$ and $E(X_{n:n})$ and show that $(X_{1:n} + X_{n:n})/2$ is unbiased for the mean.

27. Let $X_1, \ldots, X_n$ be a random sample from a normal distribution, $X_i \sim \mathrm{N}(\mu, \sigma^2)$.

(a) Find the UMVUE of σ^2.

(b) Find the UMVUE of the 95th percentile.

Hint: Recall Exercise 20 of Chapter 9.

28. Use Theorems 10.4.5 and 10.4.6 to find sufficient statistics for the parameters of the distributions in Exercises 5 and 6(b) of Chapter 9.

29. Consider a random sample of size n from a gamma distribution, $X_i \sim \mathrm{GAM}(\theta, \kappa)$, and let $\bar{X} = (1/n) \sum X_i$ and $\tilde{X} = (\prod X_i)^{1/n}$ be the sample mean and geometric mean, respectively.

(a) Show that $\bar{X}$ and $\tilde{X}$ are jointly complete and sufficient for θ and κ.

(b) Find the UMVUE of $\mu = \theta\kappa$.

(c) Find the UMVUE of μ^n.

(d) Show that the distribution of $T = \bar{X}/\tilde{X}$ does not depend on θ.

(e) Show that $\bar{X}$ and T are stochastically independent random variables.

(f) Show that the conditional pdf of $\bar{X}$ given $\tilde{X} = \tilde{x}$ does not depend on κ.

30. Consider a random sample of size n from a two-parameter exponential distribution, $X_i \sim \mathrm{EXP}(\theta, \eta)$. Recall from Exercise 12 that $X_{1:n}$ and $\bar{X}$ are jointly sufficient for θ and η. Because $X_{1:n}$ is complete and sufficient for η for each fixed value of θ, argue from Theorem 10.4.7 that $X_{1:n}$ and $T = X_{1:n} - \bar{X}$ are stochastically independent.

(a) Find the MLE $\hat{\theta}$ of θ.

(b) Find the UMVUE of η.

(c) Show that the conditional pdf of $X_{1:n}$ given $\bar{X}$ does not depend on θ.

(d) Show that the distribution of $Q = (X_{1:n} - \eta)/\hat{\theta}$ is free of η and θ.

31. Let $X_1, \ldots, X_n$ be a random sample of size n from a distribution with pdf

$$f(x; \theta) = \begin{cases} \theta(1 + x)^{-(1+\theta)} & 0 < x \\ 0 & x \leqslant 0 \end{cases}$$

(a) Find the MLE $\hat{\theta}$ of θ.

(b) Find a complete sufficient statistic for θ.

(c) Find the CRLB for $1/\theta$.

(d) Find the UMVUE of $1/\theta$.

(e) Find the asymptotic normal distribution for $\hat{\theta}$ and also for $\tau(\hat{\theta}) = 1/\hat{\theta}$.

(f) Find the UMVUE of θ.

32. Consider a random sample of size n from a distribution with pdf

$$f(x; \theta) = \begin{cases} \dfrac{(\ln \theta)^x}{\theta x!} & x = 0, 1, \ldots; \theta > 1 \\ 0 & \text{otherwise} \end{cases}$$

(a) Find a complete sufficient statistic for θ.

(b) Find the MLE of θ.

(c) Find the CRLB for θ.

(d) Find the UMVUE of $\ln \theta$.

(e) Find the UMVUE of $(\ln \theta)^2$.

(f) Find the CRLB for $(\ln \theta)^2$.

33. Suppose that only the first r order statistics are observed, based on a random sample of size n from an exponential distribution, $X_i \sim \text{EXP}(\theta)$. In other words, we have a Type II censored sample.

(a) Find the MLE of θ based only on $X_{1:n}, \ldots, X_{r:n}$.

(b) Relative to these order statistics, find a complete sufficient statistic for θ.

CHAPTER 11

INTERVAL ESTIMATION

11.1 INTRODUCTION

The problem of point estimation was discussed in Chapter 9. Along with a point estimate of the value of a parameter, we want to have some understanding of how close we can expect our estimate to be to the true value. Some information on this question is provided by knowing the variance or the MSE of the estimator. Another approach would be to consider interval estimates; one then could consider the probability that such an interval will contain the true parameter value. Indeed, one could adjust the interval to achieve some prescribed probability level, and thus a measure of its accuracy would be incorporated automatically into the interval estimate.

Example 11.1.1 In Example 4.6.3, the observed lifetimes (in months) of 40 electrical parts were given, and we argued that an exponential distribution of lifetimes might be reasonable. Consequently, we will assume that the data are the observed values of a random sample of size $n = 40$ from an exponential distribution, $X_i \sim \text{EXP}(\theta)$, where θ is the mean lifetime. Recall that in Example 9.3.4 we found that the sample mean, $\bar{X}$, is the UMVUE of θ. For the given set of data, the estimate of θ is $\bar{x} = 93.1$ months. Although we know that this estimate is based on an estimator with optimal properties, a point estimate in itself does not provide information about accuracy. Our solution to this problem will be to derive an interval whose endpoints are random variables that include the true value of θ between them with probability near 1, for example, 0.95.

It was noted in Example 9.3.2 that $2n\bar{X}/\theta \sim \chi^2(2n)$, and we know that percentiles of the chi-square distribution are given in Table 4 (Appendix C). For example, with $n = 40$ and $\nu = 80$, we find that $\chi^2_{0.025}(80) = 57.15$ and $\chi^2_{0.975}(80) = 106.63$. It follows that $P[57.15 < 80\bar{X}/\theta < 106.63] = 0.975 - 0.025$, and consequently

$$P[80\bar{X}/106.63 < \theta < 80\bar{X}/57.15] = 0.95$$

In general, an interval with random endpoints will be called a **random interval**. In particular, the interval $(80\bar{X}/106.63,\ 80\bar{X}/57.15)$ is a random interval that contains the true value of θ with probability 0.95. If we now replace $\bar{X}$ with the estimate $\bar{x} = 93.1$, then the resulting interval is (69.9, 130.3). We will refer to this interval as a 95% **confidence interval** for θ. Because the estimated interval has known endpoints, it is not appropriate to say that it contains the true value of θ with probability 0.95. That is, the parameter θ, although unknown, is a constant, and this particular interval either does or does not contain θ. However, the fact that the associated random interval had probability 0.95 prior to estimation might lead us to assert that we are "95% confident" that $69.9 < \theta < 130.3$.

The rest of the chapter will include a formal definition of confidence intervals and a discussion of general methods for deriving confidence intervals.

11.2 CONFIDENCE INTERVALS

Let $X_1, \ldots, X_n$ have joint pdf $f(x_1, \ldots, x_n;\ \theta)$; $\theta \in \mathbf{\Omega}$, where $\mathbf{\Omega}$ is an interval. Suppose that L and U are statistics, say $L = \ell(X_1, \ldots, X_n)$ and $U = u(X_1, \ldots, X_n)$. If an experiment yields data $x_1, \ldots, x_n$, then we have observed values $\ell(x_1, \ldots, x_n)$ and $u(x_1, \ldots, x_n)$.

Definition 11.2.1

Confidence Interval An interval $(\ell(x_1, \ldots, x_n), u(x_1, \ldots, x_n))$ is called a **100γ% confidence interval** for θ if

$$P[\ell(X_1, \ldots, X_n) < \theta < u(X_1, \ldots, X_n)] = \gamma \quad (11.2.1)$$

where $0 < \gamma < 1$. The observed values $\ell(x_1, \ldots, x_n)$ and $u(x_1, \ldots, x_n)$ are called **lower** and **upper confidence limits**, respectively.

Other notations that often are encountered in the statistical literature are θ_L and θ_U for lower and upper confidence limits, respectively. We also sometimes will use the abbreviated notations $\ell(\mathbf{x}) = \ell(x_1, \ldots, x_n)$ and $u(\mathbf{x}) = u(x_1, \ldots, x_n)$ to denote the observed limits.

Strictly speaking, a distinction should be made between the random interval (L, U) and the observed interval $(\ell(\mathbf{x}), u(\mathbf{x}))$ as mentioned previously. This situation is analogous to the distinction in point estimation between an estimator and an estimate. Other terminology, which is useful in maintaining this distinction, is to call (L, U) an **interval estimator** and $(\ell(\mathbf{x}), u(\mathbf{x}))$ an **interval estimate.** The probability level, γ, also is called the **confidence coefficient** or **confidence level.**

Perhaps the most common interpretation of a confidence interval is based on the relative frequency property of probability. Specifically, if such interval estimates are computed from many different samples, then in the long run we would expect approximately $100\gamma\%$ of the intervals to include the true value of θ. That is, our confidence is in the method, and because of Definition (11.2.1), the confidence level reflects the long-term frequency interpretation of probability.

It often is desirable to have either a lower or an upper confidence limit, but not both.

Definition 11.2.2

One-Sided Confidence Limits

1. If

$$P[\ell(X_1, \ldots, X_n) < \theta] = \gamma \quad (11.2.2)$$

then $\ell(\mathbf{x}) = \ell(x_1, \ldots, x_n)$ is called a **one-sided lower 100γ% confidence limit** for θ.

2. If

$$P[\theta < u(X_1, \ldots, X_n)] = \gamma \quad (11.2.3)$$

then $u(\mathbf{x}) = u(x_1, \ldots, x_n)$ is called a **one-sided upper 100γ% confidence limit** for θ.

It may not always be clear how to obtain confidence limits that satisfy Definitions 11.2.1 or 11.2.2. The concept of sufficiency often offers some aid in this problem. If a single sufficient statistic S exists, then one might consider finding confidence limits that are functions of S. Otherwise, another reasonable statistic, such as an MLE, might be considered.

Example 11.2.1 We take a random sample of size n from an exponential distribution, $X_i \sim \text{EXP}(\theta)$, and we wish to derive a one-sided lower $100\gamma\%$ confidence limit for θ. We know that $\bar{X}$ is sufficient for θ and also that $2n\bar{X}/\theta \sim \chi^2(2n)$. As mentioned in Chapter 8, γth percentiles, $\chi^2_\gamma(v)$, are provided in Table 4 (Appendix C). Thus,

$$\begin{aligned}\gamma &= P[2n\bar{X}/\theta < \chi^2_\gamma(2n)] \\ &= P[2n\bar{X}/\chi^2_\gamma(2n) < \theta]\end{aligned}$$

If $\bar{x}$ is observed, then a one-sided lower $100\gamma\%$ confidence limit is given by

$$\ell(\mathbf{x}) = 2n\bar{x}/\chi^2_\gamma(2n) \tag{11.2.4}$$

Similarly, a one-sided upper $100\gamma\%$ confidence limit is given by

$$u(\mathbf{x}) = 2n\bar{x}/\chi^2_{1-\gamma}(2n) \tag{11.2.5}$$

Notice that in the case of an upper limit we must use the value $1 - \gamma$ rather than γ when we read Table 4. For example, if a one-sided upper 90% confidence limit is desired $1 - 0.90 = 0.10$. For a sample of size $n = 40$, the required percentile is $\chi^2_{0.10}(80) = 64.28$, and the desired upper confidence limit has the form $u(\mathbf{x}) = 80\bar{x}/64.28$.

Suppose that we want a $100\gamma\%$ confidence interval for θ. If we choose values $\alpha_1 > 0$ and $\alpha_2 > 0$ such that $\alpha_1 + \alpha_2 = \alpha = 1 - \gamma$, then it follows that

$$P[\chi^2_{\alpha_1}(2n) < 2n\bar{X}/\theta < \chi^2_{1-\alpha_2}(2n)] = 1 - \alpha_2 - \alpha_1$$

and thus

$$P[2n\bar{X}/\chi^2_{1-\alpha_2}(2n) < \theta < 2n\bar{X}/\chi^2_{\alpha_1}(2n)] = \gamma$$

It is common in practice to let $\alpha_1 = \alpha_2$, which is known as the **equal tailed** choice, and this would imply $\alpha_1 = \alpha_2 = \alpha/2$. The corresponding confidence interval has the form

$$(2n\bar{x}/\chi^2_{1-\alpha/2}(2n),\ 2n\bar{x}/\chi^2_{\alpha/2}(2n)) \tag{11.2.6}$$

Generally speaking, for a prescribed confidence level, we want to use a method that produces an interval with some optimal property such as minimal length. Actually, the length, $U - L$, of the corresponding random interval generally will be a random variable, so a criterion such as **minimum expected length** might be more appropriate. For some problems, the equal tailed choice of α_1 and α_2 will provide the minimum expected length, but for others it will not. For example, interval (11.2.6) of the previous example does not have this property (see Exercise 26).

Example 11.2.2 Consider a random sample from a normal distribution, $X_i \sim N(\mu, \sigma^2)$, where σ^2 is assumed to be known. In this case $\bar{X}$ is sufficient for μ, and it is known that $Z = \sqrt{n}(\bar{X} - \mu)/\sigma \sim N(0, 1)$. By symmetry, we also know that $z_{\alpha/2} = -z_{1-\alpha/2}$, and thus

$$\begin{aligned}1 - \alpha &= P[-z_{1-\alpha/2} < \sqrt{n}(\bar{X} - \mu)/\sigma < z_{1-\alpha/2}] \\ &= P[\bar{X} - z_{1-\alpha/2}\,\sigma/\sqrt{n} < \mu < \bar{X} + z_{1-\alpha/2}\,\sigma/\sqrt{n}]\end{aligned}$$

It follows that a $100(1 - \alpha)\%$ confidence interval for μ is given by

$$(\bar{x} - z_{1-\alpha/2}\,\sigma/\sqrt{n},\ \bar{x} + z_{1-\alpha/2}\,\sigma/\sqrt{n}) \tag{11.2.7}$$

For example, for a 95% confidence interval, $1 - \alpha/2 = 0.975$ and the upper and lower confidence limits are $\bar{x} \pm 1.96\sigma/\sqrt{n}$.

Notice that this solution is not acceptable if σ^2 is unknown, because the confidence limits then would depend on an unknown parameter and could not be computed. With a slightly modified derivation if will be possible to obtain a confidence interval for μ, even if σ^2 is an unknown "nuisance parameter." Indeed, a major difficulty in determining confidence intervals arises in multiparameter cases where unknown nuisance parameters are present. A general method that often provides a way of dealing with this problem is presented in the next section.

In multiparameter cases it also may be desirable to have a "joint confidence region" that applies to all parameters simultaneously. Also, a confidence region for a single parameter, in the one-dimensional case, could be some set other than an interval. In general, if $\boldsymbol{\theta} \in \boldsymbol{\Omega}$, then any region $A_\theta(x_1, \ldots, x_n)$ in $\boldsymbol{\Omega}$ is a **$100\gamma\%$ confidence region** if the probability is γ that $A_\theta(X_1, \ldots, X_n)$ contains the true value of $\boldsymbol{\theta}$.

11.3 PIVOTAL QUANTITY METHOD

Suppose that $X_1, \ldots, X_n$ has joint pdf $f(x_1, \ldots, x_n; \theta)$, and we wish to obtain confidence limits for θ where other unknown nuisance parameters also may be present.

Definition 11.3.1

Pivotal Quantity If $Q = \varphi(X_1, \ldots, X_n; \theta)$ is a random variable that is a function only of $X_1, \ldots, X_n$ and θ, then Q is called a **pivotal quantity** if its distribution does not depend on θ or any other unknown parameters.

Example 11.3.1 In Example 11.2.1, we encountered a chi-square distributed random variable, which will be denoted here as $Q = 2n\bar{X}/\theta$, and which clearly satisfies the definition of a pivotal quantity. In that example we were able to proceed from a probability statement about Q to obtain confidence limits for θ.

More generally, if Q is a pivotal quantity for a parameter θ and if percentiles of Q, say q_1 and q_2, are available such that

$$P[q_1 < \varphi(X_1, \ldots, X_n; \theta) < q_2] = \gamma \tag{11.3.1}$$

then for an observed sample, $x_1, \ldots, x_n$, a $100\gamma\%$ confidence region for θ is the set of $\theta \in \mathbf{\Omega}$ that satisfy

$$q_1 < \varphi(x_1, \ldots, x_n; \theta) < q_1 \tag{11.3.2}$$

Such a confidence region will not necessarily be an interval, and in general it might be quite complicated. However, in some rather important situations confidence intervals can be obtained. One general situation that will always yield an interval occurs when, for each fixed set of values $x_1, \ldots, x_n$, the function $\varphi(x_1, \ldots, x_n; \theta)$ is a monotonic increasing (or decreasing) function of θ. It also is possible to identify certain types of distributions that will admit pivotal quantities. Specifically, Chapter 3 included a discussion of location and scale parameter models, which include most of the special distributions we have considered. Recall that a parameter θ is a **location parameter** if the pdf has the form $f(x; \theta) = f_0(x - \theta)$, and it is a **scale parameter** if it has the form $f(x; \theta) = (1/\theta)f_0(x/\theta)$, where $f_0(z)$ is a pdf that is free of unknown parameters (including θ). In the case of **location-scale parameters**, say θ_1 and θ_2, the pdf has the form $f(x; \theta_1, \theta_2) = (1/\theta_2)f_0[x - \theta_1)/\theta_2]$. If MLEs exist in any of these cases, then they can be used to form pivotal quantities.

Theorem 11.3.1 Let $X_1, \ldots, X_n$ be a random sample from a distribution with pdf $f(x; \theta)$ for $\theta \in \mathbf{\Omega}$, and assume that an MLE $\hat{\theta}$ exists.

1. If θ is a location parameter, then $Q = \hat{\theta} - \theta$ is a pivotal quantity.
2. If θ is a scale parameter, then $Q = \hat{\theta}/\theta$ is a pivotal quantity. ■

We already have seen examples of pivotal quantities that are slight variations of the ones suggested in this theorem. Specifically, recall Example 11.2.2, where

$X_i \sim N(\mu, \sigma^2)$. With σ^2 known, μ is a location parameter, and the MLE is $\bar{X}$; thus $\bar{X} - \mu$ is a pivotal quantity. In Example 11.2.1, $X_i \sim EXP(\theta)$, so that θ is a scale parameter and the MLE is $\bar{X}$; thus $\bar{X}/\theta$ is a pivotal quantity. Notice that it sometimes is convenient to make a slight modification, such as multiplying by a known scale factor, so that the pivotal quantity has a known distribution. For example, we know that $2n\bar{X}/\theta \sim \chi^2(2n)$, which has tabulated percentiles, so it might be better to let this be our pivotal quantity rather than $\bar{X}/\theta$.

Theorem 11.3.2 Let $X_1, \ldots, X_n$ be a random sample from a distribution with location-scale parameters

$$f(x; \theta_1, \theta_2) = \frac{1}{\theta_2} f_0\left(\frac{x - \theta_1}{\theta_2}\right)$$

If MLEs $\hat{\theta}_1$ and $\hat{\theta}_2$ exist, then $(\hat{\theta}_1 - \theta_1)/\hat{\theta}_2$ and $\hat{\theta}_2/\theta_2$ are pivotal quantities for θ_1 and θ_2, respectively. ■

We will not prove this theorem here, but details are provided by Antle and Bain (1969).

Notice also that $(\hat{\theta}_1 - \theta_1)/\theta_2$ has a distribution that is free of unknown parameters, but it is not a pivotal quantity unless θ_2 is known.

If sufficient statistics exist, then MLEs can be found that are functions of them, and the method should provide good results.

Example 11.3.2 Consider a random sample from a normal distribution, $X_i \sim N(\mu, \sigma^2)$, where both μ and σ^2 are unknown. If $\hat{\mu}$ and $\hat{\sigma}$ are the MLEs of μ and σ, then $(\hat{\mu} - \mu)/\hat{\sigma}$ and $\hat{\sigma}/\sigma$ are pivotal quantities, which could be used to derive confidence intervals for each parameter with the other considered as an unknown nuisance parameter. It will be convenient to express the results in terms of the unbiased estimator $S^2 = n\hat{\sigma}^2/(n-1)$ to take advantage of some known distributional properties, namely

$$\frac{\bar{X} - \mu}{S/\sqrt{n}} \sim t(n-1) \tag{11.3.3}$$

and

$$\frac{(n-1)S^2}{\sigma^2} \sim \chi^2(n-1) \tag{11.3.4}$$

If $t_{1-\alpha/2} = t_{1-\alpha/2}(n-1)$ is the $(1-\alpha/2)$th percentile of the t distribution with $n-1$ degrees of freedom, then

$$1 - \alpha = P\left[-t_{1-\alpha/2} < \frac{\bar{X} - \mu}{S/\sqrt{n}} < t_{1-\alpha/2}\right]$$
$$= P[\bar{X} - t_{1-\alpha/2} S/\sqrt{n} < \mu < \bar{X} + t_{1-\alpha/2} S/\sqrt{n}]$$

which means that a $100(1-\alpha)\%$ confidence interval for μ is given by

$$(\bar{x} - t_{1-\alpha/2}\, s/\sqrt{n},\ \bar{x} + t_{1-\alpha/2}\, s/\sqrt{n}) \tag{11.3.5}$$

with observed values $\bar{x}$ and s.

Similarly, if $\chi^2_{\alpha/2} = \chi^2_{\alpha/2}(n-1)$ and $\chi^2_{1-\alpha/2} = \chi^2_{1-\alpha/2}(n-1)$ are the $(\alpha/2)$th and $(1-\alpha/2)$th percentiles of the chi-square distribution with $n-1$ degrees of freedom, then

$$1 - \alpha = P[\chi^2_{\alpha/2} < (n-1)S^2/\sigma^2 < \chi^2_{1-\alpha/2}]$$
$$= P\left[\frac{(n-1)S^2}{\chi^2_{1-\alpha/2}} < \sigma^2 < \frac{(n-1)S^2}{\chi^2_{\alpha/2}}\right]$$

and a $100(1-\alpha)\%$ confidence interval for σ^2 is given by

$$((n-1)s^2/\chi^2_{1-\alpha/2},\ (n-1)s^2/\chi^2_{\alpha/2}) \tag{11.3.6}$$

Also, confidence limits for σ are obtained by computing the positive square roots of these limits.

In general, if (θ_L, θ_U) is a $100\gamma\%$ confidence interval for a parameter θ, and if $\tau(\theta)$ is a monotonic increasing function of $\theta \in \mathbf{\Omega}$, then $(\tau(\theta_L), \tau(\theta_U))$ is a $100\gamma\%$ confidence interval for $\tau(\theta)$.

Example 11.3.3 In Example 9.2.13, the computation of MLEs for the parameters of a Weibull distribution $X_i \sim \text{WEI}(\theta, \beta)$, was discussed. Although the Weibull distribution is not a location-scale model, it is not difficult to show that the distribution of $Y_i = \ln X_i$ has an extreme-value distribution that is a location-scale model. Specifically,

$$f(y; \theta_1, \theta_2) = (1/\theta_2) f_0[(y - \theta_1)/\theta_2] \tag{11.3.7}$$

where $f_0(z) = \exp(z - e^z)$. The relationship between parameters is $\theta_1 = \ln \theta$ and $\theta_2 = 1/\beta$, and thus

$$Q_1 = \hat{\beta} \ln(\hat{\theta}/\theta) = \frac{\hat{\theta}_1 - \theta_1}{\hat{\theta}_2} \tag{11.3.8}$$

and

$$Q_2 = \hat{\beta}/\beta = (\hat{\theta}_2/\theta_2)^{-1} \tag{11.3.9}$$

are pivotal quantities for θ and β. Because the MLEs must be computed by iterative methods for this model, there is no known way to derive the exact distributions of Q_1 and Q_2, but percentiles can be obtained by computer simulation. Tables of these percentiles and derivations of the confidence intervals are given by Bain and Engelhardt (1991). Approximate distributions are given in Chapter 16.

It may not always be possible to find a pivotal quantity based on MLEs, but for a sample from a continuous distribution with a single unknown parameter, at least one pivotal quantity can always be derived by use of the probability integral transform.

If $X_i \sim f(x; \theta)$ and if $F(x; \theta)$ is the CDF of X_i, then it follows from Theorem 6.3.3 that $F(X_i; \theta) \sim \text{UNIF}(0, 1)$, and consequently $Y_i = -\ln F(X_i; \theta) \sim \text{EXP}(1)$. For a random sample $X_1, \ldots, X_n$, it follows that

$$-2\sum_{i=1}^{n} \ln F(X_i; \theta) \sim \chi^2(2n) \tag{11.3.10}$$

so that

$$P\left[\chi^2_{\alpha/2}(2n) < -2\sum_{i=1}^{n} \ln F(X_i; \theta) < \chi^2_{1-\alpha/2}(2n)\right] = 1 - \alpha \tag{11.3.11}$$

and inverting this statement will provide a confidence region for θ. If the CDF is not in closed form or if it is too complicated, then the inversion may have to be done numerically. If $F(x; \theta)$ is a monotonic increasing (or decreasing) function of θ, then the resulting confidence region will be an interval. Notice also that $1 - F(X_i; \theta) \sim \text{UNIF}(0, 1)$, and

$$-2\sum_{i=1}^{n} \ln [1 - F(X_i; \theta)] \sim \chi^2(2n) \tag{11.3.12}$$

In general, expressions (11.3.10) and (11.3.12) will give different intervals, and perhaps computational convenience would be a reasonable criterion for choosing between them.

Example 11.3.4 Consider a random sample from a Pareto distribution, $X_i \sim \text{PAR}(1, \kappa)$. The CDF is

$$F(x; \kappa) = 1 - (1 + x)^{-\kappa} \qquad x > 0$$

If we use equation (11.3.12), then $-\ln [1 - F(x; \kappa)] = \kappa \ln (1 + x)$, so

$$2\kappa \sum_{i=1}^{n} \ln (1 + X_i) \sim \chi^2(2n)$$

and a $100(1 - \alpha)\%$ confidence interval has the form

$$\left(\frac{\chi^2_{\alpha/2}(2n)}{2 \sum \ln (1 + x_i)}, \frac{\chi^2_{1-\alpha/2}(2n)}{2 \sum \ln (1 + x_i)}\right)$$

The solution based on equation (11.3.10) would be much harder because the resulting inequality would have to be solved numerically.

For discrete distributions, and for some multiparameter problems, a pivotal quantity may not exist. However, an approximate pivotal quantity often can be obtained based on asymptotic results. The normal approximation to the binomial distribution, as discussed in Chapter 7, is an example.

APPROXIMATE CONFIDENCE INTERVALS

Let $X_1, \ldots, X_n$ be a random sample from a distribution with pdf $f(x; \theta)$. As noted in Chapter 9, MLEs are asymptotically normal under certain conditions.

Example 11.3.5 Consider a random sample from a Bernoulli distribution, $X_i \sim \text{BIN}(1, p)$. The MLE of p is $\hat{p} = \sum X_i/n$. We also know that $\hat{p}$ is sufficient and that $\sum X_i \sim \text{BIN}(n, p)$, but there is no pivotal quantity for p. However, by the CLT,

$$\frac{\hat{p} - p}{\sqrt{p(1 - p)/n}} \xrightarrow{d} Z \sim \text{N}(0, 1) \tag{11.3.13}$$

and consequently for large n,

$$P\left[-z_{1-\alpha/2} < \frac{\hat{p} - p}{\sqrt{p(1 - p)/n}} < z_{1-\alpha/2}\right] \doteq 1 - \alpha \tag{11.3.14}$$

This approximation is enhanced by using the continuity correction, as discussed in Chapter 7, but we will not pursue this point. Limits for an approximate $100(1 - \alpha)\%$ confidence interval (p_0, p_1) for p are obtained by solving for the smaller solution of

$$\frac{\hat{p} - p_0}{\sqrt{p_0(1 - p_0)/n}} = -z_{1-\alpha/2} \tag{11.3.15}$$

and the larger solution of

$$\frac{\hat{p} - p_1}{\sqrt{p_1(1 - p_1)/n}} = z_{1-\alpha/2} \tag{11.3.16}$$

The common practice in this problem is to simplify the limits by using the limiting result that

$$\frac{\hat{p} - p}{\sqrt{\hat{p}(1 - \hat{p})/n}} \xrightarrow{d} Z \sim N(0, 1) \tag{11.3.17}$$

as $n \to \infty$, which was shown in Example 7.7.2. Thus, for large n, we also have the approximate result

$$P\left[-z_{1-\alpha/2} < \frac{\hat{p} - p}{\sqrt{\hat{p}(1 - \hat{p})/n}} < z_{1-\alpha/2}\right] \doteq 1 - \alpha \tag{11.3.18}$$

This statement is much easier to invert, and approximate confidence limits for p are given by

$$\hat{p} \pm z_{1-\alpha/2}\sqrt{\hat{p}(1 - \hat{p})/n} \tag{11.3.19}$$

An important point here is that the random variables defined by expressions (11.2.13) and (11.3.17) are not pivotal quantities for any finite n, because their exact distributions depend on p. However, the limiting distribution is standard normal, which does not involve p, and hence the degree to which the exact distribution depends on p should be small for large n, and the variables could be regarded as "approximate" pivotal quantities.

Other important distributions also admit approximate pivotal quantities.

Example 11.3.6 Consider a random sample of size n from a Poisson distribution, $X_i \sim POI(\mu)$. By the CLT, we know that

$$\frac{\bar{X} - \mu}{\sqrt{\mu/n}} \xrightarrow{d} Z \sim N(0, 1) \tag{11.3.20}$$

and thus by Theorem 7.7.4 that

$$\frac{\bar{X} - \mu}{\sqrt{\bar{X}/n}} \xrightarrow{d} Z \sim N(0, 1) \tag{11.3.21}$$

as $n \to \infty$. Either of these random variables could be used to derive approximate confidence intervals, although expression (11.3.21) would be more convenient.

Actually, it is possible to generalize this approach when MLEs are asymptotically normal (see Exercise 29).

11.4

GENERAL METHOD

If a pivotal quantity is not available, then it is still possible to determine a confidence region for a parameter θ if a statistic exists with a distribution that depends on θ but not on any other unknown nuisance parameters. Specifically, let $X_1, \ldots, X_n$ have joint pdf $f(x_1, \ldots, x_n; \theta)$, and $S = \mathscr{s}(X_1, \ldots, X_n) \sim g(s; \theta)$. Preferably S will be sufficient for θ, or possibly some reasonable estimator such as an MLE, but this is not required.

Now, for each possible value of θ, assume that we can find values $h_1(\theta)$ and $h_2(\theta)$ such that

$$P[h_1(\theta) < S < h_2(\theta)] = 1 - \alpha \tag{11.4.1}$$

If we observe $S = s$, then the set of values $\theta \in \boldsymbol{\Omega}$ that satisfy $h_1(\theta) < s < h_2(\theta)$ form a $100(1 - \alpha)\%$ confidence region. In other words, if θ_0 is the true value of θ, then θ_0 will be in the confidence region if and only if $h_1(\theta_0) < s < h_2(\theta_0)$, which has $100(1 - \alpha)\%$ confidence level because equation (11.4.1) holds with $\theta = \theta_0$ in this case. Quite often $h_1(\theta)$ and $h_2(\theta)$ will be monotonic increasing (or decreasing) functions of θ, and the resulting confidence region will be an interval.

Example 11.4.1 Consider a random sample of size n from the continuous distribution with pdf

$$f(x; \theta) = \begin{cases} (1/\theta^2) \exp\left[-(x - \theta)/\theta^2\right] & x \geqslant \theta \\ 0 & x < \theta \end{cases}$$

with $\theta > 0$. There is no single sufficient statistic, but $X_{1:n}$ and $\sum X_i$ are jointly sufficient for θ. It is desired to derive a 90% confidence interval for θ based on the statistic $S = X_{1:n}$. The CDF of S is

$$G(s; \theta) = \begin{cases} 1 - \exp[-n(s - \theta)/\theta^2] & s \geqslant \theta \\ 0 & s < \theta \end{cases}$$

One possible choice of functions $h_1(\theta)$ and $h_2(\theta)$ that satisfy equation (11.4.1) is obtained by solving

$$G(h_1(\theta); \theta) = 0.05$$

and

$$G(h_2(\theta); \theta) = 0.95$$

FIGURE 11.1 Functions $h_1(\theta)$ and $h_2(\theta)$ for the general method of constructing confidence intervals

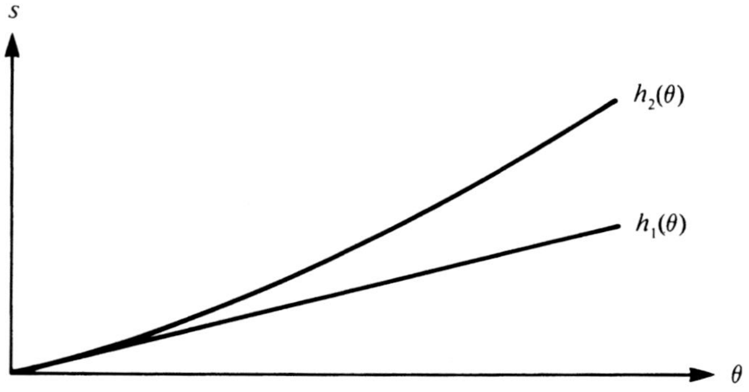

This yields the functions

$$h_1(\theta) = \theta - \ln (0.95)\theta^2/n \doteq \theta + 0.0513\theta^2/n$$

and

$$h_2(\theta) = \theta - \ln (0.05)\theta^2/n \doteq \theta + 2.996\ \theta^2/n$$

The graphs of $h_1(\theta)$ and $h_2(\theta)$ with $n = 10$ are shown in Figure 11.1.

Suppose now that a sample of size $n = 10$ yields a minimum observation $s = x_{1:10} = 2.50$. The solutions of $2.50 = h_1(\theta)$ and $2.50 = h_2(\theta)$ are $\theta_1 = 2.469$ and $\theta_2 = 1.667$. Because $h_1(\theta)$ and $h_2(\theta)$ are increasing, the set of all $\theta > 0$ such that $h_1(\theta) < 2.50 < h_2(\theta)$ is the interval (1.667, 2.469), which is a 90% confidence interval for θ.

Because confidence limits in this approach are values of θ that satisfy $h_1(\theta) = s$ and $h_2(\theta) = s$, a more suggestive notation might be θ_L and θ_U rather than $\ell(\boldsymbol{x})$ and $u(\boldsymbol{x})$.

In general, if $h_1(\theta)$ and $h_2(\theta)$ are both increasing, then the endpoints of the confidence interval can be determined for any observed s by solving for the lower limit θ_L such that $h_2(\theta_L) = s$, and for the upper limit θ_U such that $h_1(\theta_U) = s$. The argument that (θ_L, θ_U) is a $100(1 - \alpha)\%$ confidence interval is illustrated graphically by Figure 11.2.

If θ_0 is the true value of θ, then $P[h_1(\theta_0) < S < h_2(\theta_0)] = 1 - \alpha$, and whenever $h_1(\theta_0) < s < h_2(\theta_0)$, then (θ_L, θ_U) contains θ_0. Also, when s falls outside the interval $(h_1(\theta_0), h_2(\theta_0))$, the resulting limits will not contain θ_0, and the associated probability is α. If $h_1(\theta)$ and $h_2(\theta)$ are both decreasing, then the argument is similar, but in this case $h_1(\theta_L) = s$ and $h_2(\theta_U) = s$. These results can be conveniently formulated in terms of the CDF of S.

FIGURE 11.2 A confidence interval based on the general method

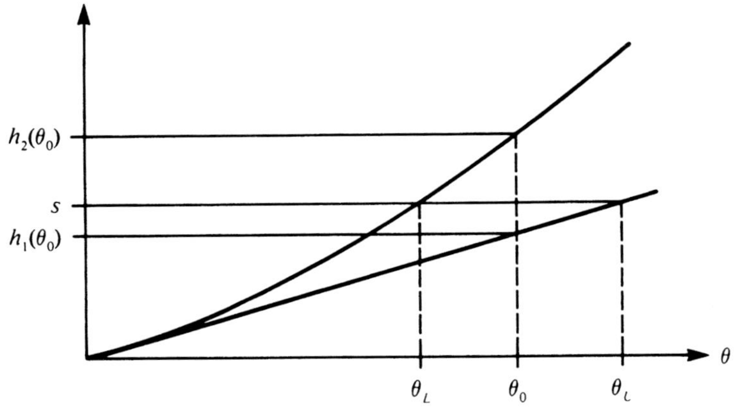

Theorem 11.4.1 Let the statistic S be continuous with CDF $G(s; \theta)$, and suppose that $h_1(\theta)$ and $h_2(\theta)$ are functions that satisfy

$$G(h_1(\theta); \theta) = \alpha_1 \tag{11.4.2}$$

and

$$G(h_2(\theta); \theta) = 1 - \alpha_2 \tag{11.4.3}$$

for each $\theta \in \mathbf{\Omega}$, where $0 < \alpha_1 < 1$ and $0 < \alpha_2 < 1$. Let s be an observed value of S. If $h_1(\theta)$ and $h_2(\theta)$ are increasing functions of θ, then the following statements hold:

1. A one-sided lower $100(1 - \alpha_2)\%$ confidence limit, θ_L, is the solution of
$$h_2(\theta_L) = s \tag{11.4.4}$$
2. A one-sided upper $100(1 - \alpha_1)\%$ confidence limit, θ_U, is the solution of
$$h_1(\theta_U) = s \tag{11.4.5}$$
3. If $\alpha = \alpha_1 + \alpha_2$ and $0 < \alpha < 1$, then (θ_L, θ_U) is a $100(1 - \alpha)\%$ confidence interval for θ. ■

The theorem is modified easily for the case where $h_1(\theta)$ and $h_2(\theta)$ are decreasing. In particular, if $h_1(\theta_L) = s$, then θ_L is a one-sided lower $100(1 - \alpha_1)\%$ confidence limit, and if $h_2(\theta_U) = s$, then θ_U is a one-sided upper $100(1 - \alpha_2)\%$ confidence limit.

Example 11.4.2 Consider again a random sample from an exponential distribution, $X_i \sim \text{EXP}(\theta)$, and recall that $S = \sum X_i$ is sufficient for θ. Because $2S/\theta \sim \chi^2(2n)$, we have

$$\begin{aligned}\alpha &= P[S \leqslant h_1(\theta)] \\ &= P[2S/\theta \leqslant 2h_1(\theta)/\theta]\end{aligned}$$

which implies

$$2h_1(\theta)/\theta = \chi^2_\alpha(2n)$$

and

$$h_1(\theta) = \theta\chi^2_\alpha(2n)/2$$

This is an increasing function of θ, and the solution of $h_1(\theta) = s$ provides a one-sided upper $100(1 - \alpha)\%$ confidence limit $\theta_U = 2s/\chi^2_\alpha(2n)$, which we also obtained by the pivotal quantity approach. The function $h_2(\theta)$ and θ_L are obtained in a similar manner.

There exist examples where $h_1(\theta)$ and $h_2(\theta)$ are not monotonic, and where the resulting confidence region is not an interval (see Exercise 25).

Also note that in practice it is not necessary to know $h_1(\theta)$ and $h_2(\theta)$ for all θ, but it will be necessary to know the values of θ for which $h_1(\theta) = s$ and $h_2(\theta) = s$, and whether these functions are increasing or decreasing, in order to know which function gives an upper limit and which gives a lower limit. It can be shown that if $G(s; \theta)$ is a decreasing function of θ for each fixed s, then both $h_1(\theta)$ and $h_2(\theta)$ are increasing functions of θ. This suggests the following theorem.

Theorem 11.4.2 Suppose that the statistic S is continuous with CDF $G(s; \theta)$, and let s be an observed value of S. If $G(s; \theta)$ is a decreasing function of θ, then the following statements hold:

1. A one-sided lower $100(1 - \alpha_2)\%$ confidence limit, θ_L, is provided by a solution of

$$G(s; \theta_L) = 1 - \alpha_2 \tag{11.4.6}$$

2. A one-sided upper $100(1 - \alpha_1)\%$ confidence limit, θ_U, is provided by a solution

$$G(s; \theta_U) = \alpha_1 \tag{11.4.7}$$

3. If $\alpha = \alpha_1 + \alpha_2$ and $0 < \alpha < 1$, then (θ_L, θ_U) is a $100(1 - \alpha)\%$ confidence interval for θ. ■

A similar theorem can be stated for the case where $G(s;\ \theta)$ is increasing in θ. In particular, if $G(s;\ \theta_U) = 1 - \alpha_2$, then θ_U is a one-sided upper $100(1 - \alpha_2)\%$ confidence limit; and if $G(s;\ \theta_L) = \alpha_1$, then θ_L is a one-sided lower $100(1 - \alpha_1)\%$ confidence limit.

Example 11.4.3 Consider the statistic $S = X_{1:n}$ of Example 11.4.1. For any fixed s, with the substitution $t = s/\theta$, $G(s;\ \theta)$ can be written as $1 - \exp[-(n/s)(t - 1)t]$ for $t \geq 1$. The derivative with respect to t is $(n/x)(2t - 1) \exp[-(n/s)(t - 1)t]$, which is positive because $2t - 1 > 0$ when $t \geq 1$. Consequently, $G(s;\ \theta)$ is an increasing function of t and thus a decreasing function of $\theta = s/t$. It follows from the theorem that one-sided lower and upper 95% confidence limits for observed $s = 2.50$ and $n = 10$ are obtained by solving $G(2.50;\ \theta_L) = 0.95$ and $G(2.50;\ \theta_U) = 0.05$. These solutions are $\theta_L = 1.667$ and $\theta_U = 2.469$, as before.

It also is possible to state a more general theorem that includes discrete cases, but it is not always possible to achieve a prescribed confidence level when the observed statistic is discrete. However, "conservative" confidence intervals, in general, can be obtained.

Definition 11.4.1

An observed confidence interval (θ_L, θ_U) is called a **conservative** $100(1 - \alpha)\%$ confidence interval for θ if the corresponding random interval contains the true value of θ with probability **at least** $1 - \alpha$.

Conservative one-sided confidence limits can be defined similarly.

Theorem 11.4.3 Let S be a statistic with CDF $G(s;\ \theta)$, and let $h_1(\theta)$ and $h_2(\theta)$ be functions that satisfy $G(h_1(\theta);\ \theta) = \alpha_1$ and

$$P[S < h_2(\theta);\ \theta] = 1 - \alpha_2 \tag{11.4.8}$$

where $0 < \alpha_1 < 1$ and $0 < \alpha_2 < 1$.

1. If $h_1(\theta)$ and $h_2(\theta)$ are increasing functions, then a conservative one-sided lower $100(1 - \alpha_2)\%$ confidence limit for θ, based on an observed value s of S, is a solution of $h_2(\theta_L) = s$, or $\theta = \theta_L$ such that

$$P[S < s;\ \theta_L] = 1 - \alpha_2 \tag{11.4.9}$$

A conservative one-sided upper $100(1 - \alpha_1)\%$ confidence limit is a solution of $h_1(\theta_U) = s$, or $G(s;\ \theta_U) = \alpha_1$.

2. If $h_1(\theta)$ and $h_2(\theta)$ are decreasing functions, then a conservative one-sided lower $100(1-\alpha_1)\%$ confidence limit is a solution of $h_1(\theta_L) = s$, or $G(s;\ \theta_L) = \alpha_1$. A conservative one-sided upper $100(1-\alpha_2)\%$ confidence limit is a solution of $h_2(\theta_U) = s$, or $\theta = \theta_U$ such that

$$P[S < s;\ \theta_U] = 1 - \alpha_2 \tag{11.4.10}$$

3. In either case, if $\alpha = \alpha_1 + \alpha_2$ and $0 < \alpha < 1$, then (θ_L, θ_U) is a conservative $100(1-\alpha)\%$ confidence interval for θ. ■

An exact prescribed confidence level may not be achievable if S is discrete, but the confidence levels will be at least the stated levels. This requires keeping the strict inequality in conditions (11.4.8), (11.4.9), and (11.4.10). Of course, if S is continuous, then $P[S < s;\ \theta] = G(s;\ \theta)$, and the previous theorems apply, yielding exact confidence levels.

Consider the case of a discrete distribution, $G(s;\ \theta)$, where $h_i(\theta)$ is an increasing function and $G(s;\ \theta)$ is a decreasing function of θ. Let S assume the discrete values $s_1, s_2, \ldots$, and suppose that there are parameter values $\theta_1, \theta_2, \ldots$ such that $G(s_i;\ \theta_i) = \alpha$. If $S = s_i$ is observed, then let $\theta_U = \theta_i$ be the upper $100(1-\alpha)\%$ confidence limit. The confidence level will be greater than $1 - \alpha$ for the intermediate values of θ. If $\theta_{i-1} < \theta < \theta_i$, then the confidence interval will contain θ if the observed value of S is greater than or equal to s_i, which will occur with probability

$$\begin{aligned} P[S \geqslant s_i \mid \theta_{i-1} < \theta < \theta_i] &= 1 - G(s_{i-1};\ \theta) \\ &\geqslant 1 - G(s_{i-1};\ \theta_{i-1}) \\ &= 1 - \alpha \end{aligned}$$

Similarly, suppose that $G(s_i;\ \theta_i) = 1 - \alpha$, and suppose $\theta_L = \theta_{i-1}$. That is, if $S = s_i$, then θ_L is the solution of $G(s_{i-1};\ \theta_L) = 1 - \alpha$. Now consider a value $\theta_{i-1} < \theta < \theta_i$. This value will be in the confidence interval if $S \leqslant s_i$, which occurs with probability

$$\begin{aligned} P[S \leqslant s_i \mid \theta_{i-1} < \theta < \theta_i] &= G(s_i;\ \theta) \\ &\geqslant G(s_i;\ \theta_i) = 1 - \alpha \end{aligned}$$

Example 11.4.4 In Example 11.3.5, two approaches were presented for obtaining approximate confidence intervals for the binomial parameter p, based on large sample approximations.

We now desire to derive a conservative one-sided $(1-\alpha)100\%$ confidence limit for p. We know that $S = \sum X_i$ is a sufficient statistic, and $S \sim \text{BIN}(n, p)$. We will not find explicit expressions for $h_1(p)$ and $h_2(p)$ in this example, but note that

$G(s; p) = B(s; n, p)$ is a decreasing function of p. Thus, for an observed value s, a solution of

$$\alpha_1 = B(s; n, p_U) = \sum_{y=0}^{s} \binom{n}{y} p_U^y (1 - p_U)^{n-y}$$

is a conservative one-sided upper limit, and a solution of

$$1 - \alpha_2 = B(s - 1; n, p_L) = \sum_{y=0}^{s-1} \binom{n}{y} p_L^y (1 - p_L)^{n-y}$$

is a conservative one-sided lower limit. If $\alpha = \alpha_1 + \alpha_2$, then a conservative $100(1 - \alpha)\%$ confidence interval is given by (p_L, p_U).

For a specified value of s, p_L and p_U can be determined by interpolation from a cumulative binomial table such as Table 1 (Appendix C), or it can be obtained by numerical methods applied to the CDF. For example, suppose that $n = 10$ and $s = 2$. If $\alpha_1 = 0.05$, then $B(2; 10, 0.5) = 0.0547$ and $B(2; 10, 0.55) = 0.0274$. Linear interpolation yields $p_U \doteq 0.509$. By trying a few more values, we obtain a closer value, $B(2; 10, 0.507) = 0.05$, so that $p_U = 0.507$. Similarly, if $\alpha_2 = 0.05$, we can find $B(1; 10, 0.037) = 0.95$ and thus $p_L = 0.037$. It follows also that $(0.037, 0.507)$ is a conservative 90% confidence interval for p.

Example 11.4.5 Recall that methods for deriving approximate confidence intervals for the mean, μ, of a Poisson distribution were discussed in Example 11.3.6. For a random sample of size n, $X_i \sim \text{POI}(\mu)$, a sufficient statistic is $S = \sum X_i$, and $S \sim \text{POI}(n\mu)$. Because the CDF of the Poisson distribution is related to the CDF of a chi-square distribution (see Exercise 21, Chapter 8), the confidence limits can be expressed conveniently in terms of chi-square percentiles. If we denote by $H(y; v)$ the CDF of a chi-square variable with v degrees-of-freedom, then a conservative upper $100(1 - \alpha_1)\%$ confidence limit for μ, for an observed value s, is a solution of

$$\alpha_1 = G(s; \mu_U) = 1 - H(2n\mu_U; 2s + 2)$$

which means that

$$2n\mu_U = \chi^2_{1-\alpha_1}(2s + 2)$$

and thus

$$\mu_U = \chi^2_{1-\alpha_1}(2s + 2)/2n$$

Similarly, a conservative lower $100(1 - \alpha_2)\%$ confidence limit for μ is a solution of

$$1 - \alpha_2 = G(s - 1; \mu_L) = 1 - H(2n\mu_L; 2s)$$

so that

$$2n\mu_L = \chi^2_{\alpha_2}(2s)$$

and

$$\mu_L = \chi^2_{\alpha_2}(2s)/2n$$

If $\alpha_1 = \alpha_2 = \alpha/2$, then a conservative $100(1-\alpha)\%$ confidence interval for μ is given by

$$(\chi^2_{\alpha/2}(2 \sum x_i)/2n, \chi^2_{1-\alpha/2}(2 \sum x_i + 2)/2n) \tag{11.4.11}$$

The general method can be applied to any problem that contains a single unknown parameter. The following theorem may be helpful in identifying a statistic that can be used in the presence of location-scale nuisance parameters.

Theorem 11.4.4 Let $X_1, \ldots, X_n$ be a random sample of size n from a distribution with pdf of the form

$$f(x; \theta_1, \theta_2, \kappa) = \frac{1}{\theta_2} f_0\left(\frac{x-\theta_1}{\theta_2}; \kappa\right) \tag{11.4.12}$$

where $-\infty < \theta_1 < \infty$ and $\theta_2 > 0$, and $f_0(z; \kappa)$ is a pdf that depends on κ but not on θ_1 or θ_2. If there exist MLEs $\hat{\theta}_1$, $\hat{\theta}_2$, and $\hat{\kappa}$, then the distributions of $(\hat{\theta}_1 - \theta_1)/\hat{\theta}_2$, $\hat{\theta}_2/\theta_2$, and $\hat{\kappa}$ do not depend on θ_1 and θ_2. ■

It follows that the general method can be used with the statistic $\hat{\kappa}$ to determine confidence limits on κ, with θ_1 and θ_2 being unknown nuisance parameters. Of course, if κ is known, then the pivotal quantities $(\hat{\theta}_1 - \theta_1)/\hat{\theta}_2$ and $\hat{\theta}_2/\theta_2$ can be used to find confidence intervals for θ_1 and θ_2. Theorem 11.3.2 also would apply in this situation, because θ_1 and θ_2 are location-scale parameters when κ is known. It may not be clear how to derive confidence limits for θ_1 and θ_2 if κ is unknown.

Theorem 11.4.5 Let $X_1, \ldots, X_n$ be a random sample of size n from a distribution with CDF $F(x; \theta_1, \theta_2)$ where θ_1 and θ_2 are location-scale parameters, and suppose that MLEs $\hat{\theta}_1$ and $\hat{\theta}_2$ exist. If t is a fixed value, then $F(t; \hat{\theta}_1, \hat{\theta}_2)$ is a statistic whose distribution depends on t, θ_1, and θ_2 only through $F(t; \theta_1, \theta_2)$. ■

Consider the case where $F(x; \theta_1, \theta_2) = F_0[(x - \theta_1)/\theta_2]$ with $F_0(z)$ a one-to-one function. Let

$$c = (t - \theta_1)/\theta_2 = F_0^{-1}[F(t; \theta_1, \theta_2)]$$

which depends on t, θ_1, and θ_2 only through $F(t; \theta_1, \theta_2)$. It follows that

$$\begin{aligned} F(t; \hat{\theta}_1, \hat{\theta}_2) &= F_0[(t - \hat{\theta}_1)/\hat{\theta}_2] \\ &= F_0[c(\theta_2/\hat{\theta}_2) - (\hat{\theta}_1 - \theta_1)/\hat{\theta}_2] \end{aligned}$$

which is a function only of c and the pivotal quantities $(\hat{\theta}_1 - \theta_1)/\hat{\theta}_2$ and $\hat{\theta}_2/\theta_2$. Consequently, its distribution depends on $F(t;\ \theta_1,\ \theta_2)$, but not on any other unknown nuisance parameters.

Example 11.4.6 Consider the quantity

$$R(t) = P[X > t] = 1 - F(t;\ \theta_1,\ \theta_2)$$

This is an important quantity in applications where X represents a failure time or lifetime of some experimental unit. In some applications it is called the "reliability" function, and in others it is called the "survivor" function. It follows from the previous theorem that the distribution of the MLE of reliability, $\hat{R}(t) = 1 - F(t;\ \hat{\theta}_1,\ \hat{\theta}_2)$, depends only on $R(t)$. Thus, the general method can be used to find confidence limits on $R(t)$.

For a more specific example, consider a random sample of size n from a two-parameter exponential distribution, $X_i \sim \text{EXP}(\theta,\ \eta)$. The MLEs are $\hat{\eta} = X_{1:n}$, $\hat{\theta} = \bar{X} - X_{1:n}$, and

$$\hat{R}(t) = 1 - F(t;\ \hat{\theta},\ \hat{\eta}) = \exp\ [-(t - \hat{\eta})/\hat{\theta}] = \exp\ [(\theta/\hat{\theta}) \ln R(t) - (\hat{\eta} - \eta)/\hat{\theta}]$$

If $Y = \hat{R}(t)$, then it can be shown that the CDF, $G(y;\ R(t))$, is decreasing in $R(t)$ for each fixed y. Thus, by Theorem 11.4.2, a one-sided lower $(1 - \alpha)100\%$ confidence limit, $R_L(t)$, is obtained by solving $G(\hat{R}(t);\ R_L(t)) = 1 - \alpha$. The CDF of $\hat{R}(t)$ is rather complicated in this case, and we will not attempt to derive it.

11.5 TWO-SAMPLE PROBLEMS

Quite often random samples are taken for the purpose of comparing two or more populations. One may be interested in comparing the mean yields of two processes or the relative variation in yields of two processes. Confidence intervals are quite informative in making such comparisons.

TWO-SAMPLE NORMAL PROCEDURES

Consider independent random samples of sizes n_1 and n_2 from two normally distributed populations $X_i \sim \text{N}(\mu_1,\ \sigma_1^2)$ and $Y_j \sim \text{N}(\mu_2,\ \sigma_2^2)$, respectively. Denote by $\bar{X}$, $\bar{Y}$, S_1^2, and S_2^2 the sample means and sample variances.

Suppose we wish to know whether one population has a smaller variance than the other. For example, two methods of producing baseballs might be compared to see which method produces baseballs with a smaller variation in their elasticity. The unbiased point estimators S_1^2 and S_2^2 can be computed, but if there is only a small difference in the estimates, then it may not be clear whether this difference results from a true difference in the variances or whether the difference results from random sampling error. In other words, if the distribution variances are the same, then the difference in sample variances, based on two more samples, might be just as likely to be in the other direction. Note also that if the sample sizes are large, then even a small difference between estimates may indicate a difference between parameter values; but if the sample sizes are small, then fairly large differences might result from chance. The confidence interval approach incorporates this kind of information into the interval estimate.

PROCEDURE FOR VARIANCES

A confidence interval for the ratio σ_2^2/σ_1^2 can be derived by using Snedecor's F distribution as suggested in Example 8.4.1. In particular, we know that

$$\frac{S_1^2 \sigma_2^2}{S_2^2 \sigma_1^2} \sim F(n_1 - 1, n_2 - 1) \tag{11.5.1}$$

which provides a pivotal quantity for σ_2^2/σ_1^2. Percentiles for the F distribution with $\nu_1 = n_1 - 1$ and $\nu_2 = n_2 - 1$ can be obtained from Table 7 (Appendix C), so

$$P\left[f_{\alpha/2}(\nu_1, \nu_2) < \frac{S_1^2}{S_2^2} \frac{\sigma_2^2}{\sigma_1^2} < f_{1-\alpha/2}(\nu_1, \nu_2) \right] = 1 - \alpha \tag{11.5.2}$$

and thus, if s_1^2 and s_2^2 are estimates, then $(1 - \alpha)100\%$ confidence interval for σ_2^2/σ_1^2 is given by

$$\left(\frac{s_2^2}{s_1^2} f_{\alpha/2}(n_1 - 1, n_2 - 1), \frac{s_2^2}{s_1^2} f_{1-\alpha/2}(n_1 - 1, n_2 - 1) \right) \tag{11.5.3}$$

Example 11.5.1 Random samples of size $n_1 = 16$ and $n_2 = 21$ yield estimates $s_1^2 = 0.60$ and $s_2^2 = 0.20$, and a 90% confidence interval is desired. From Table 7 (Appendix C), $f_{0.95}(15, 20) = 2.20$ and $f_{0.05}(15, 20) = 1/f_{0.95}(20, 15) = 1/2.33 = 0.429$. It follows that (0.143, 0.733) is a 90% confidence interval for σ_2^2/σ_1^2. Because the interval does not contain the value 1, we might conclude that $\sigma_2^2 \neq \sigma_1^2$ (or $\sigma_2^2/\sigma_1^2 \neq 1$), and that the two populations have different variances. Because the confidence level is 90%, only 10% of such conclusions, on the average, will be incorrect.

This type of reasoning will be developed more formally in the next chapter.

PROCEDURE FOR MEANS

If the variances, σ_1^2 and σ_2^2, are known, then a pivotal quantity for the difference, $\mu_2 - \mu_1$, is easily obtained. Specifically, because

$$\bar{Y} - \bar{X} \sim \mathrm{N}(\mu_2 - \mu_1, \sigma_1^2/n_1 + \sigma_2^2/n_2) \tag{11.5.4}$$

it follows that

$$Z = \frac{\bar{Y} - \bar{X} - (\mu_2 - \mu_1)}{\sqrt{\sigma_1^2/n_1 + \sigma_2^2/n_2}} \sim \text{N}(0, 1) \tag{11.5.5}$$

With this choice of Z, the statement $P[-z_{1-\alpha/2} < Z < z_{1-\alpha/2}] = 1 - \alpha$ can be solved easily to obtain $100(1 - \alpha)\%$ confidence limits, namely

$$\bar{y} - \bar{x} \pm z_{1-\alpha/2}\sqrt{\sigma_1^2/n_1 + \sigma_2^2/n_2} \tag{11.5.6}$$

In most cases the variances will not be known, but in some cases it will be reasonable to assume that the variances are unknown but equal. For example, one may wish to study the effect on mean yields when an additive or other modification is introduced in an existing method. In some cases it might be reasonable to assume that the additive could affect the mean yield but would not affect the variation in the process. If $\sigma_1^2 = \sigma_2^2 = \sigma^2$, then the common variance can be eliminated in much the same way as in the one-sample case using a Student's t variable. A pooled estimator of the common variance is the weighted average

$$S_p^2 = \frac{(n_1 - 1)S_1^2 + (n_2 - 1)S_2^2}{n_1 + n_2 - 2} \tag{11.5.7}$$

and if

$$V = \frac{(n_1 + n_2 - 2)S_p^2}{\sigma^2} \tag{11.5.8}$$

then

$$V = \frac{(n_1 - 1)S_1^2}{\sigma^2} + \frac{(n_2 - 1)S_2^2}{\sigma^2} \sim \chi^2(n_1 + n_2 - 2) \tag{11.5.9}$$

It is also true that $\bar{X}$ and $\bar{Y}$ are independent of S_1^2 and S_2^2, so with Z as given by equation (11.5.5), with $\sigma_1^2 = \sigma_2^2 = \sigma^2$, and with V given by equation (11.5.8), it follows from Theorem 8.4.1 that

$$T = \frac{\bar{Y} - \bar{X} - (\mu_2 - \mu_1)}{S_p\sqrt{\dfrac{1}{n_1} + \dfrac{1}{n_2}}} = \frac{Z}{\sqrt{V/(n_1 + n_2 - 2)}} \sim t(n_1 + n_2 - 2) \tag{11.5.10}$$

Limits for a $(1 - \alpha)100\%$ confidence interval for $\mu_2 - \mu_1$ are given by

$$\bar{y} - \bar{x} \pm t_{1-\alpha/2}(n_1 + n_2 - 2)s_p\sqrt{\frac{1}{n_1} + \frac{1}{n_2}} \tag{11.5.11}$$

Example 11.5.2 Random samples of size $n_1 = 16$ and $n_2 = 21$ yield estimates $\bar{x} = 4.31$, $\bar{y} = 5.22$, $s_1^2 = 0.12$, and $s_2^2 = 0.10$. We might first consider a confidence interval for the ratio of variances to check the assumption of equal variances. A 90% confidence interval for σ_1^2/σ_2^2 is (0.358, 1.83), which contains the value 1. Thus, there is not strong evidence that the variances are unequal, and we will assume $\sigma_1^2 = \sigma_2^2$. The pooled estimate of variance is $s_p^2 = [(15)(0.12) + (20)(0.10)]/35 = 0.109$, and $s_p = 0.330$. Suppose that a 95% confidence interval for $\mu_2 - \mu_1$ is desired. By linear interpolation between $t_{0.975}(30) = 2.042$ and $t_{0.975}(40) = 2.021$ in Table 6 (Appendix C), we obtain $t_{0.975}(35) \doteq 2.032$. The desired confidence interval, based on the limits in equation (11.5.11), is (0.688, 1.133).

APPROXIMATE METHODS

It is not easy to eliminate unknown variances to obtain a pivotal quantity for $\mu_2 - \mu_1$ when the variances are unequal. One possible approach would be a large-sample method. Specifically, as n_1 and $n_2 \to \infty$,

$$\frac{\bar{Y} - \bar{X} - (\mu_2 - \mu_1)}{\sqrt{S_1^2/n_1 + S_2^2/n_2}} \xrightarrow{d} Z \sim \mathrm{N}(0, 1) \tag{11.5.12}$$

Thus, for large sample sizes, approximate confidence limits for $\mu_2 - \mu_1$ may be easily obtained from expression (11.5.12).

Note that the above limiting results also hold if the samples are not from normal distributions, so this provides a general large-sample result from differences of means. The size of the samples required to make the limiting approximation close would depend somewhat on the form of the densities.

For small samples from normal distributions, the distribution of the random variable in expression (11.5.12) depends on σ_1^2 and σ_2^2, but good small-sample approximations can be based on Student's t distribution. One such approximation, which comes from Welch (1949), is

$$T = \frac{\bar{Y} - \bar{X} - (\mu_2 - \mu_1)}{\sqrt{S_1^2/n_1 + S_2^2/n_2}} \sim t(v) \tag{11.5.13}$$

where the degrees of freedom are estimated as follows:

$$v = \frac{(s_1^2/n_1 + s_2^2/n_2)^2}{[(s_1^2/n_1)^2/(n_1 - 1)] + [(s_2^2/n_2)^2/(n_2 - 1)]} \tag{11.5.14}$$

Notice that this generally will produce noninteger degrees of freedom, but linear interpolation in Table 6 (Appendix C) can be used to obtain the required percentiles for constructing confidence intervals. The general problem of making inferences about $\mu_2 - \mu_1$ with unequal variances is known as the Behrens-Fisher problem. Welch's solution is just one of many that have been proposed in the statistical literature; it was studied by Wang (1971), who found it to be quite good.

PAIRED-SAMPLE PROCEDURE

All of the above results assume that the random samples are independent. In some cases, such as test–retest experiments, dependent samples are appropriate. For example, to measure the effectiveness of a diet plan, we would select n people at random and weigh them both before and after the diet. The observations would be independent between pairs, but the observations within a pair would not be independent because they were taken on the same individual.

We have a random sample of n pairs, (X_i, Y_i), and we assume that the differences $D_i = Y_i - X_i$ for $i = 1, \ldots, n$ are normally distributed with mean $\mu_D = \mu_2 - \mu_1$ and variance $\sigma_D^2 = \sigma_1^2 + \sigma_2^2 - 2\sigma_{12}$, or

$$D_i \sim \mathrm{N}(\mu_2 - \mu_1, \sigma_D^2)$$

Let

$$\bar{D} = \sum_{i=1}^{n} D_i/n = \bar{Y} - \bar{X} \tag{11.5.15}$$

and

$$S_D^2 = \frac{n\sum_{i=1}^{n} D_i^2 - \left(\sum_{i=1}^{n} D_i\right)^2}{n(n-1)} \tag{11.5.16}$$

It follows from the results of Chapter 6 that

$$T = \frac{\bar{D} - (\mu_2 - \mu_1)}{S_D/\sqrt{n}} \sim t(n-1) \tag{11.5.17}$$

and thus a $(1 - \alpha)100\%$ confidence interval for $\mu_2 - \mu_1$ has limits of the form

$$\bar{d} \pm t_{1-\alpha/2}(n-1)s_D/\sqrt{n} \tag{11.5.18}$$

where $\bar{d}$ and s_D are observed.

Note that this method remains valid if the samples are independent, because in this case $D_i \sim \mathrm{N}(\mu_2 - \mu_1, \sigma_1^2 + \sigma_2^2)$. However, the degrees of freedom in the paired sample procedure is $n - 1$, whereas in the independent sample case with $\sigma_1^2 = \sigma_2^2$ we obtained a t statistic with $2n - 2$ degrees of freedom; so the effective sample size is twice as large in the independent sample case, and consequently the paired-sample method would not be as good. However, if there is a reason for pairing and the pairs are highly correlated, then $\sigma_D^2 = \sigma_1^2 + \sigma_2^2 - 2\sigma_{12}$ may be much smaller than $\sigma_1^2 + \sigma_2^2$, and this could offset the loss in effective sample size. Thus pairing is a useful technique, but it should not be used indiscriminately.

It is interesting to note that if two independent samples have equal sample size, and if the variances are not equal, then the paired sample procedure still can be used to provide an exact t statistic, but the resulting confidence interval would tend to be wider than one based on an approximate t variable such as that of expression (11.5.13).

TWO-SAMPLE BINOMIAL PROCEDURE

Suppose that $X_1 \sim \text{BIN}(n_1, p_1)$ and $X_2 \sim \text{BIN}(n_2, p_2)$. Letting $\hat{p}_1 = X_1/n_1$ and $\hat{p}_2 = X_2/n_2$, from the results of Chapter 7 we have

$$\frac{\hat{p}_2 - \hat{p}_1 - (p_2 - p_1)}{\sqrt{\hat{p}_1(1 - \hat{p}_1)/n_1 + \hat{p}_2(1 - \hat{p}_2)/n_2}} \xrightarrow{d} Z \sim \text{N}(0, 1) \tag{11.5.19}$$

It is clear that approximate large-sample confidence limits for $p_2 - p_1$ can be obtained in a manner similar to the one-sample case, namely

$$\hat{p}_2 - \hat{p}_1 \pm z_{1-\alpha/2} \sqrt{\frac{\hat{p}_1(1 - \hat{p}_1)}{n_1} + \frac{\hat{p}_2(1 - \hat{p}_2)}{n_2}} \tag{11.5.20}$$

11.6

BAYESIAN INTERVAL ESTIMATION

Bayes estimators were discussed briefly in Chapter 9 for the case of point estimation. There the parameter was treated mathematically as a random variable. In certain cases, this may be a physically meaningful assumption. For example, the parameter may behave as a variable over different conditions in the experiment. The prior density, $p(\theta)$, may be considered to reflect prior knowledge or belief about the true value of the parameter, and the Bayesian structure provides a convenient framework for using this prior belief to order the risk functions and select the best (smallest average risk) estimator. In this case, the prior density is not unlike a class of confidence intervals indexed by α. As α varies from 0 to 1, the resulting confidence intervals for θ could be represented as producing a probability distribution for θ. The induced distribution in this case is based on sample data rather than on subjective criteria.

In any event, suppose that a prior density $p(\theta)$ exists or is introduced into the problem and $f(x; \theta)$ is interpreted as a conditional pdf, $f(x \mid \theta)$. Consider again the posterior density of θ given the sample $\boldsymbol{x} = (x_1, \ldots, x_n)$,

$$f_{\theta|\boldsymbol{x}}(\theta) = \frac{f(x_1, \ldots, x_n \mid \theta)p(\theta)}{\int f(x_1, \ldots, x_n \mid \theta)p(\theta)\, d\theta} \tag{11.6.1}$$

The prior density $p(\theta)$ can be interpreted as specifying an initial probability distribution for the possible values of θ, and in this context $f_{\theta|\boldsymbol{x}}(\theta)$ would represent a revised distribution adjusted by the observed random sample. For a particular $1 - \alpha$ level, a Bayesian confidence interval for θ is given by (θ_L, θ_U) where θ_L and

θ_U satisfy

$$\int_{\theta_L}^{\theta_U} f_{\theta|\mathbf{x}}(\theta)\, d\theta = 1 - \alpha \tag{11.6.2}$$

If θ is a true random variable, then the Bayesian interval would have the usual probability interpretation. Of course, in any such problem the results are correct only to the extent that the assumed models are correct. If $p(\theta)$ represents a degree of belief about the values of θ, then presumably the interval (θ_L, θ_U) also would be interpreted in the degree-of-belief sense.

Example 11.6.1 In Example 9.5.5 it was assumed that $X_i \sim \text{POI}(\theta)$ and that $\theta \sim \text{GAM}(\beta, \kappa)$. The posterior distribution was found to be given by

$$\theta_x = \theta \,|\, \mathbf{x} \sim \text{GAM}[(n + 1/\beta)^{-1}, \textstyle\sum x_i + \kappa] \tag{11.6.3}$$

It follows that

$$\frac{2\theta_x}{(n + 1/\beta)^{-1}} = 2(n + 1/\beta)\theta_x \sim \chi^2[2(\textstyle\sum x_i + \kappa)] \tag{11.6.4}$$

and

$$P[\chi^2_{\alpha/2}(\nu) < 2(n + 1/\beta)\theta_x < \chi^2_{1-\alpha/2}(\nu)] = 1 - \alpha \tag{11.6.5}$$

where $\nu = 2(\sum x_i + \kappa)$. Thus, a $100(1 - \alpha)\%$ Bayesian confidence interval for θ is given by (θ_L, θ_U) where $\theta_L = \chi^2_{\alpha/2}(\nu)/2(n + 1/\beta)$, and $\theta_U = \chi^2_{1-\alpha/2}(\nu)/2(n + 1/\beta)$.

SUMMARY

Our purpose in this chapter was to introduce the concept of an interval estimate or confidence interval. A point estimator in itself does not provide direct information about accuracy. An interval estimator gives one possible solution to this problem. The concept involves an interval whose endpoints are statistics that include the true value of the parameter between them with high probability. This probability corresponds to the confidence level of the interval estimator. Ordinarily, the term *confidence interval* (or *interval estimate*) refers to the observed interval that is computed from data.

There are two basic methods for constructing confidence intervals. One method, which is especially useful in certain applications where unknown nuisance parameters are present, involves the notion of a pivotal quantity. This amounts to finding a random variable that is a function of the observed random variables and the parameter of interest, but not of any other unknown parameters. It also is required that the distribution of the pivotal quantity be free of any unknown parameters. In the case of location-scale parameters, pivotal quantities can be expressed in terms of the MLEs if they exist. Approximate large-sample pivotal quantities can be based on asymptotic normal results in some cases.

The other method, which is referred to as the general method, does not require the existence of a pivotal quantity, but has the disadvantage that it cannot be used when a nuisance parameter is present. This method can be applied with any statistic whose distributions can be expressed in terms of the parameter. The percentiles are functions of the parameter, and the limits of the confidence interval are obtained by solving equations that involve certain percentiles and the observed value of the statistic. Interval estimates obtained by either method can be interpreted in terms of the relative frequency with which the true value of the parameter will be included in the interval, which corresponds to the probability that the interval estimator will contain the true value. Another type of interval is based on the Bayesian approach. This approach provides a convenient way to use information or, in some cases, subjective judgment about the unknown parameter, although the relative frequency interpretation may be inappropriate in some instances.

EXERCISES

1. Consider a random sample of size n from a normal distribution, $X_i \sim N(\mu, \sigma^2)$.
   (a) If it is known that $\sigma^2 = 9$, find a 90% confidence interval for μ based on the estimate $\bar{x} = 19.3$ with $n = 16$.
   (b) Based on the information in (a), find a one-sided lower 90% confidence limit for μ. Also, find a one-sided upper 90% confidence limit for μ.
   (c) For a confidence interval of the form given by expression (11.2.7), derive a formula for the sample size required to obtain an interval of specified length λ. If $\sigma^2 = 9$, then what sample size is needed to achieve a 90% confidence interval of length 2?
   (d) Suppose now that σ^2 is unknown. Find a 90% confidence interval for μ if $\bar{x} = 19.3$ and $s^2 = 10.24$ with $n = 16$.
   (e) Based on the data in (d), find a 99% confidence interval for σ^2.

2. Assume that the weight data of Exercise 24, Chapter 4, are observed values of a random sample of size $n = 60$ from a normal distribution.
   (a) Find a 99% confidence interval for the mean weight of major league baseballs.
   (b) Find a 99% confidence interval for the standard deviation.

3. Let $X_1, \ldots, X_n$ be a random sample from an exponential distribution, $X \sim EXP(\theta)$.
   (a) If $\bar{x} = 17.9$ with $n = 50$, then find a one-sided lower 95% confidence limit for θ.
   (b) Find a one-sided lower 95% confidence limit for $P(X > t) = e^{-t/\theta}$ where t is an arbitrary known value.

4. The following data are times (in hours) between failures of air conditioning equipment in a particular airplane: 74, 57, 48, 29, 502, 12, 70, 21, 29, 386, 59, 27, 153, 26, 326. Assume that the data are observed values of a random sample from an exponential distribution, $X_i \sim \text{EXP}(\theta)$.

(a) Find a 90% confidence interval for the mean time between failures, θ.

(b) Find a one-sided lower 95% confidence limit for the 10th percentile of the distribution of time between failures.

5. Consider a random sample of size n from an exponential distribution, $X_i \sim \text{EXP}(1, \eta)$.

(a) Show that $Q = X_{1:n} - \eta$ is a pivotal quantity and find its distribution.

(b) Derive a $100\gamma\%$ equal tailed confidence interval for η.

(c) The following data are mileages for 19 military personnel carriers that failed in service: 162, 200, 271, 320, 393, 508, 539, 629, 706, 777, 884, 1008, 1101, 1182, 1463, 1603, 1984, 2355, 2880. Assuming that these data are observations of a random sample from an exponential distribution, find a 90% confidence interval for η. Assume that $\theta = 850$ is known.

6. Let $X_1, \ldots, X_n$ be a random sample from a two-parameter exponential distribution, $X_i \sim \text{EXP}(\theta, \eta)$.

(a) Assuming it is known that $\eta = 150$, find a pivotal quantity for the parameter θ based on the sufficient statistic.

(b) Using the data of Exercise 5, find a one-sided lower 95% confidence limit for θ.

7. Let $X_1, X_2, \ldots, X_n$ be a random sample from a Weibull distribution, $X \sim \text{WEI}(\theta, 2)$.

(a) Show that $Q = 2\sum_{i=1}^{n} X_i^2/\theta^2 \sim \chi^2(2n)$.

(b) Use Q to derive an equal tailed $100\gamma\%$ confidence interval for θ.

(c) Find a lower $100\gamma\%$ confidence limit for $P(X > t) = \exp[-(t/\theta)^2]$.

(d) Find an upper $100\gamma\%$ confidence limit for the pth percentile of the distribution.

8. Consider a random sample of size n from a uniform distribution, $X_i \sim \text{UNIF}(0, \theta)$, $\theta > 0$, and let $X_{n:n}$ be the largest order statistic.

(a) Find the probability that the random interval $(X_{n:n}, 2X_{n:n})$ contains θ.

(b) Find the constant c such that $(x_{n:n}, cx_{n:n})$ is a $100(1-\alpha)\%$ confidence interval for θ.

9. Use the approach of Example 11.3.4 with the data of Example 4.6.2 to find a 95% confidence interval for κ.

10. Suppose that the exact values of the data $x_1, \ldots, x_{50}$ in Exercise 3(b) are not known, but it is known that 40 of the 50 measurements are larger than t.

(a) Find an approximate one-sided lower 95% confidence limit for $P(X > t)$ based on this information.

(b) Note that under the exponential assumption, $P(X > t) = \exp(-t/\theta)$. If $t = 5$, use the result from (a) to find an approximate one-sided lower 95% confidence limit for θ and compare this to the confidence limit of Exercise 3(a).

11. Let p be the proportion of people in the United States with red hair. In a sample of size 40, five people with red hair were observed. Find an approximate 90% confidence interval for p.

12. Suppose that 45 workers in a textile mill are selected at random in a study of accident rate. The number of accidents per worker is assumed to be Poisson distributed with mean μ. The average number of accidents per worker is $\bar{x} = 1.7$.

(a) Find an approximate one-sided lower 90% confidence limit for μ using equation (11.3.20).

(b) Repeat (a) using equation (11.3.21) instead.

(c) Find a conservative one-sided lower 90% confidence limit for μ using the approach of Example 11.4.5.

13. Consider a random sample of size n from a gamma distribution, $X_i \sim \text{GAM}(\theta, \kappa)$.

(a) Assuming κ is known, derive a $100(1-\alpha)\%$ equal-tailed confidence interval for θ based on the sufficient statistic.

(b) Assuming that $\theta = 1$, and for $n = 1$, find an equal-tailed 90% confidence interval for κ if $x_1 = 10$ is observed. *Hint:* Note that $2X_1 \sim \chi^2(2\kappa)$, and use interpolation in Table 4 (Appendix C).

14. Assume that the number of defects in a piece of wire that is t yards in length is $X \sim \text{POI}(\lambda t)$ for any $t > 0$.

(a) If five defects are found in a 100-yard roll of wire, find a conservative one-sided upper 95% confidence limit for the mean number of defects in such a roll.

(b) If a total of 15 defects are found in five 100-yard rolls of wire, find a conservative one-sided upper 95% confidence limit for λ.

15. Let $X_1, \ldots, X_n$ be a random sample from a Weibull distribution, $X_i \sim \text{WEI}(\theta, \beta)$, where β is known.

(a) Use the general method of Section 11.4 to derive a $100(1-\alpha)\%$ confidence interval for θ based on the statistic $S_1 = X_{1:n}$.

(b) Use the general method to find a $(1-\alpha)\ 100\%$ confidence interval for θ based on the statistic $S_2 = \sum X_i^\beta$.

16. Let $f(x; p) = p f_{X_1}(x) + (1-p) f_{X_2}(x)$, where $X_1 \sim \text{N}(1, 1)$ and $X_2 \sim \text{N}(0, 1)$. Based on a sample of size $n = 1$ from $f(x; p)$, derive a one-sided lower $100\gamma\%$ confidence limit for p.

17. Suppose that $X \sim \text{GEO}(p)$.

(a) Derive a conservative one-sided lower $100\gamma\%$ confidence limit for p based on a single observation x.

(b) If $x = 5$, find a conservative one-sided lower 90% confidence limit for p.

(c) If $X_1, \ldots, X_n$ is a random sample from GEO(p), describe the form of a conservative one-sided lower $100\gamma\%$ confidence limit for p based on the sufficient statistic.

18. Let $X_1, \ldots, X_n$ be a random sample from a normal distribution, $X \sim \text{N}(\mu, \sigma^2)$. If t is a fixed real number, find a statistic that is a function of the sufficient statistics and whose distribution depends on t, μ, and σ^2 only through $F(t; \mu, \sigma^2) = P(X \leqslant t)$.

19. Consider independent random samples from two normal distributions, $X_i \sim N(\mu_1, \sigma_1^2)$ and $Y_j \sim N(\mu_2, \sigma_2^2)$; $i = 1, \ldots, n_1, j = 1, \ldots, n_2$. Assuming that μ_1 and μ_2 are known, derive a $100(1-\alpha)\%$ confidence interval for σ_2^2/σ_1^2 based on sufficient statistics.

20. Consider independent random samples from two exponential distributions, $X_i \sim \text{EXP}(\theta_1)$ and $Y_j \sim \text{EXP}(\theta_2)$; $i = 1, \ldots, n_1, j = 1, \ldots, n_2$.

(a) Show that $(\theta_2/\theta_1)(\bar{X}/\bar{Y}) \sim F(2n_1, 2n_2)$.

(b) Derive a $100\gamma\%$ confidence interval for θ_2/θ_1.

21. Compute a 95% Bayesian confidence interval for μ based on the results of Exercise 12. Use the prior density of Example 11.6.1 with $\beta = 1$ and $\kappa = 2$.

22. Let $X_1, \ldots, X_n$ be a random sample from a Bernoulli distribution, $X_i \sim \text{BIN}(1, \theta)$, and assume a uniform prior $\theta \sim \text{UNIF}(0, 1)$. Derive a $100(1-\alpha)\%$ Bayesian interval estimate of θ. *Hint:* The posterior distribution is given in Example 9.5.4.

23. Using the densities $f(x \mid \theta)$ and $f(\theta)$ of Exercise 44 of Chapter 9:

(a) Derive a $100(1-\alpha)\%$ Bayesian confidence interval for θ.

(b) Derive a $100(1-\alpha)\%$ Bayesian confidence interval for $\mu = 1/\theta$.

(c) Compute a 90% Bayesian confidence interval for θ if $n = 10$, $\bar{x} = 5$, and $\beta = 2$.

24. Suppose that θ_L and θ_U are one-sided lower and upper confidence limits for θ with confidence coefficients $1 - \alpha_1$ and $1 - \alpha_2$, respectively. Show that (θ_L, θ_U) is a conservative $100(1-\alpha)\%$ confidence interval for θ if $\alpha = \alpha_1 + \alpha_2 < 1$. *Hint:* Use Bonferroni's inequality (1.4.7), with

$$[\theta_L < \theta < \theta_U] = [\theta_L < \theta] \cap [\theta < \theta_U]$$

25. Consider a random sample of size n from a distribution with CDF

$$F(x; \theta) = \begin{cases} 1 - \exp[-\theta(x - \theta)] & x \geqslant \theta \\ 0 & x < \theta \end{cases}$$

with $\theta > 0$.

(a) Find the CDF, $G(s; \theta)$, of $S = X_{1:n}$.

(b) Find the function $h(\theta)$ such that $G(h(\theta); \theta) = 1 - \alpha$, and show that it is not monotonic.

(c) Show that $h(\theta) = s$ has two solutions, $\theta_1 = [s - \sqrt{s^2 + 4(\ln \alpha)/n}]/2$ and $\theta_2 = [s + \sqrt{s^2 + 4(\ln \alpha)/n}]/2$, if $s^2 > -4(\ln \alpha)/n$, and that $h(\theta) > s$ if and only if either $\theta < \theta_1$ or $\theta > \theta_2$. Thus, $(0, \theta_1) \cup (\theta_2, \infty)$ is a $100(1-\alpha)\%$ confidence region for θ, but it is not an interval.

26. Consider the equal-tailed confidence interval of equation (11.2.6).

(a) Use the fact that $Q = 2n\bar{X}/\theta \sim \chi^2(2n)$ to derive a formula for the expected length of the corresponding random interval.

(b) More generally, a $100(1-\alpha)\%$ confidence interval for θ has the form $(2n\bar{x}/q_2, 2n\bar{x}/q_1)$ where $F_Q(q_1) = \alpha_1$ and $F_Q(q_2) = 1 - \alpha_2$ with $\alpha = \alpha_1 + \alpha_2 < 1$, and the expected length is proportional to $(1/q_1 - 1/q_2)$. Note that q_2 is an implicit function of q_1, because $F_Q(q_2) - F_Q(q_1) = 1 - \alpha$ (which is fixed), and consequently that $dq_2/dq_1 = f_Q(q_2)$. Use this to show that the values of q_1 and q_2 that minimize

$(1/q - 1/q_2)$ must satisfy $q_1^2 f_Q(q_1) = q_2^2 f_Q(q_2)$, which is not the equal-tailed choice for a chi-square pdf.

27. Consider the equal-tailed confidence interval of equation (11.2.7). More generally, if $Z = \sqrt{n}(\bar{X} - \mu)/\sigma$, then a $100(1 - \alpha)\%$ confidence interval for μ is given by $(\bar{x} - z_2 \sigma/\sqrt{n}, \bar{x} - z_1 \sigma/\sqrt{n})$ where $\Phi(z_1) = \alpha_1$ and $\Phi(z_2) = 1 - \alpha_2$ with $\alpha = \alpha_1 + \alpha_2 < 1$.

(a) Show that the interval of this form with minimum length is given by equation (11.2.7).

(b) In the case when σ^2 is unknown, can a similar claim be made about the expected length of the random interval corresponding to equation (11.3.5)?

(c) In a manner similar to that of Exercise 26, show that the equal-tailed confidence interval given by equation (11.3.6) does not have minimum expected length.

28. Based on the pivotal quantities Q_1 and Q_2 given by equations (11.3.8) and (11.3.9), derive one-sided lower $100\gamma\%$ confidence limits for the parameters θ_1 and θ_2 of the extreme-value distribution. Leave the answer in terms of arbitrary percentiles q_1 and q_2.

29. As noted in Section 9.4, under certain regularity conditions, the MLEs $\hat{\theta}_n$ are asymptotically normal, $N(\theta, c^2(\theta)/n)$, where $c^2(\theta)/n$ is the CRLB. Assuming further that $c(\theta)$ is a continuous function of θ, it follows from Theorem 7.7.2 that $c(\hat{\theta}_n)$ converges stochastically to $c(\theta)$.

(a) Using other results from Chapter 7, show that if $Z_n = \sqrt{n}(\hat{\theta}_n - \theta)/c(\hat{\theta}_n)$, then $Z_n \xrightarrow{d} Z \sim N(0, 1)$ as $n \to \infty$.

(b) From (a), show that limits for an approximate large-sample $100(1 - \alpha)\%$ confidence interval are given by $\hat{\theta}_n \pm z_{1-\alpha/2}\, c(\hat{\theta}_n)/\sqrt{n}$.

(c) Based on the results of Example 9.4.7, derive an approximate $100(1 - \alpha)\%$ confidence interval for the parameter κ where $X_i \sim PAR(1, \kappa)$.

(d) Use the data of Example 4.6.2 to find an approximate 95% confidence interval for κ, and compare this with the confidence interval of Exercise 9. Would you expect a close approximation in this example?

30. Suppose that $\hat{\theta}_n$ is asymptotically normal, $N(\theta, c^2(\theta)/n)$. It sometimes is desirable to consider a function, say $g(\theta)$, such that the asymptotic variance of $g(\hat{\theta}_n)$ does not depend on θ. Such a function is called a **variance-stabilizing transformation.** If we apply Theorem 7.7.6 with $Y_n = \hat{\theta}_n$, $m = \theta$, and $c = c(\theta)$, then $g(\theta)$ would have to satisfy the equation $[c(\theta)g'(\theta)]^2 = k$, a constant.

(a) If $X_1, \ldots, X_n$ is a random sample from a Poisson distribution, $X_i \sim POI(\mu)$, and $\hat{\theta}_n = \bar{X}$, show that $g(\mu) = \sqrt{\mu}$ is a variance-stabilizing transformation.

(b) Derive an approximate, large-sample $100(1 - \alpha)\%$ confidence interval for μ based on $g(\bar{X})$.

(c) Consider a random sample of size n from $EXP(\theta)$, and let $\hat{\theta}_n = \bar{X}$. Find a variance-stabilizing transformation and use it to derive an approximate large-sample confidence interval for θ.

CHAPTER 12

TESTS OF HYPOTHESES

12.1 INTRODUCTION

In scientific activities, much attention is devoted to answering questions about the validity of theories or hypotheses concerning physical phenomena. Is a new drug effective? Does a lot of manufactured items contain an excessive number of defectives? Is the mean lifetime of a component at least some specified amount? Ordinarily, information about such phenomena can be obtained only by performing experiments whose outcomes have some bearing on the hypotheses of interest. The term **hypotheses testing** will refer to the process of trying to decide the truth or falsity of such hypotheses on the basis of experimental evidence.

For instance, we may suspect that a certain hypothesis, perhaps an accepted theory, is false, and an experiment is conducted. An outcome that is inconsistent with the hypothesis will cast doubt on its validity. For example, the hypothesis to be tested may specify that a physical constant has the value μ_0. In general, experimental measurements are subject to random error, and thus any decision about the truth or falsity of the hypothesis, based on experimental evidence, also is subject to error. It will not be possible to avoid an occasional decision error, but it will be possible to construct tests so that such errors occur infrequently and at some prescribed rate.

A simple example illustrates the concept of hypothesis testing.

Example 12.1.1 A theory proposes that the yield of a certain chemical reaction is normally distributed, $X \sim N(\mu, 16)$. Past experience indicates that $\mu = 10$ if a certain mineral is not present, and $\mu = 11$ if the mineral is present. Our experiment would be to take a random sample of size n. On the basis of that sample, we would try to decide which case is true. That is, we wish to test the "null hypothesis" $H_0 : \mu = \mu_0 = 10$ against the "alternative hypothesis" $H_a : \mu = \mu_1 = 11$.

Definition 12.1.1

If $X \sim f(x; \boldsymbol{\theta})$, a **statistical hypothesis** is a statement about the distribution of X. If the hypothesis completely specifies $f(x; \boldsymbol{\theta})$, then it is referred to as a **simple** hypothesis; otherwise it is called **composite**.

Quite often the distribution in question has a known parametric form with a single unknown parameter θ, and the hypothesis consists of a statement about θ. In this framework, a statistical hypothesis corresponds to a subset of the parameter space, and the objective of a test would be to decide whether the true value of the parameter is in the subset. Thus, a null hypothesis would correspond to a subset $\boldsymbol{\Omega}_0$ of $\boldsymbol{\Omega}$, and the alternative hypothesis would correspond to its complement, $\boldsymbol{\Omega} - \boldsymbol{\Omega}_0$. In the case of simple hypotheses, these sets consist of only one element each, $\boldsymbol{\Omega}_0 = \{\theta_0\}$ and $\boldsymbol{\Omega} - \boldsymbol{\Omega}_0 = \{\theta_1\}$, where $\theta_0 \neq \theta_1$.

Most experiments have some goal or research hypothesis that one hopes to support with statistical evidence, and this hypothesis should be taken as the alternative hypothesis. The reason for this will become clear as we proceed. In our example, if we have strong evidence that the mineral is present, then we may wish to spend a large amount of money to begin mining operations, so we associate this case with the alternative hypothesis. We now must consider sample data, and decide on the basis of the data whether we have sufficient statistical evidence to reject H_0 in favor of the alternative H_a, or whether we do not have sufficient evidence. That is, our philosophy will be to divide the sample space into two regions, the "critical region" or "rejection region" C, and the nonrejection region $S - C$. If the observed sample data fall in C, then we will reject H_0, and if they do not fall in C, then we will not reject H_0.

Definition 12.1.2

The **critical region** for a test of hypotheses is the subset of the sample space that corresponds to rejecting the null hypothesis.

In our example, $\bar{X}$ is a sufficient statistic for μ, so we may conveniently express the critical region directly in terms of the univariate variable $\bar{X}$, and we will refer to $\bar{X}$ as the **test statistic**. Because $\mu_1 > \mu_0$, a natural form for the critical region in this problem is to let $C = \{(x_1, \ldots, x_n) \mid \bar{x} \geqslant c\}$, for some appropriate constant c. That is, we will reject H_0 if $\bar{x} \geqslant c$, and we will not reject H_0 if $\bar{x} < c$. There are two possible errors we may make under this procedure. We might reject H_0 when H_0 is true, or we might fail to reject H_0 when H_0 is false. These errors are referred to as follows:

1. Type I error: Reject a true H_0.
2. Type II error: Fail to reject a false H_0.

Occasionally, for convenience, we may refer to the Type II error as "accepting a false H_0" and to $S - C$ as the "acceptance" region, but it should be understood that this is not strictly a correct interpretation. That is, failure to have enough statistical evidence to reject H_0 is not the same as having strong evidence to support H_0.

We hope to choose a test statistic and a critical region so that we would have a small probability of making these two errors. We will adopt the following notations for these error probabilities:

1. $P[\text{Type I error}] = P[\text{TI}] = \alpha$.
2. $P[\text{Type II error}] = P[\text{TII}] = \beta$.

Definition 12.1.3

For a simple null hypothesis, H_0, the probability of rejecting a true H_0, $\alpha = P[\text{TI}]$, is referred to as the **significance level** of the test. For a composite null hypothesis, H_0, the **size** of the test (or size of the critical region) is the maximum probability of rejecting H_0 when H_0 is true (maximized over the values of the parameter under H_0).

Notice that for a simple H_0 the significance level is also the size of the test.

The standard approach is to specify or select some acceptable level of error such as $\alpha = 0.05$ or $\alpha = 0.01$ for the significance level of the test, and then to determine a critical region that will achieve this α. Among all critical regions of size α we would select the one that has the smallest $P[\text{TII}]$. In our example, if $n = 25$, then $\alpha = 0.05$ gives $c = \mu_0 + z_{1-\alpha}\sigma/\sqrt{n} = 10 + 1.645(4)/5 = 11.316$. This is verified easily, because

$$P[\bar{X} \geqslant c \mid \mu = \mu_0 = 10] = P\left[\frac{\bar{X} - \mu_0}{\sigma/\sqrt{n}} \geqslant \frac{c - \mu_0}{\sigma/\sqrt{n}}\right]$$
$$= P\left[Z \geqslant \frac{11.316 - 10}{4/5}\right]$$
$$= P[Z \geqslant 1.645]$$
$$= 0.05$$

Thus, a size 0.05 test of $H_0 : \mu = 10$ against the alternative $H_a : \mu = 11$ is to reject H_0 if the observed value $\bar{x} \geqslant 11.316$. Note that this critical region provides a size 0.05 test for any alternative value $\mu = \mu_1$, but the fact that $\mu_1 > \mu_0$ means that we will get a smaller Type II error by taking the critical region as the right-hand tail of the distribution of $\bar{X}$ rather than as some other region of size 0.05. For an alternative $\mu_1 < \mu_0$ the left-hand tail would be preferable. Thus, the alternative affects our choice for the location of the critical region, but it is otherwise determined under H_0 for specified α.

The probability of Type II error for the critical region C is

$$\beta = P[\text{TII}] = P[\bar{X} < 11.316 \mid \mu = \mu_1 = 11]$$
$$= P\left[\frac{\bar{X} - 11}{4/5} < \frac{11.316 - 11}{4/5} \,\middle|\, \mu = 11\right]$$
$$= P[Z < 0.395] = 0.654$$

These concepts are illustrated in Figure 12.1.

At this point, there is no theoretical reason for choosing a critical region of the form C over any other. For example, the critical region $C_1 = \{(x_1, \ldots, x_n) \mid 10 < \bar{x} < 10.1006\}$ also has size $\alpha = 0.05$ because, under H_0

$$P[10 < \bar{X} < 10.1006] = P\left[0 < \frac{\bar{X} - 10}{4/5} < 0.1257\right]$$
$$= 0.05$$

FIGURE 12.1 Probabilities of Type I and Type II errors

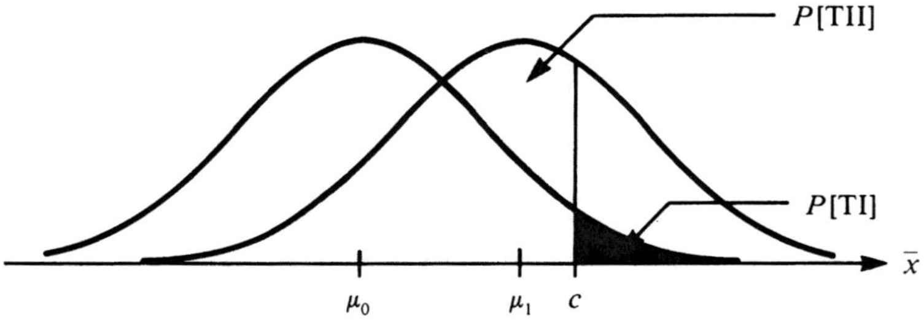

However, $P[\text{TII}]$ for this critical region is

$$\begin{aligned} P[\text{TII}] &= 1 - P[10 < \bar{X} < 10.1006 \mid \mu = 11] \\ &= 1 - P\left[\frac{10 - 11}{4/5} < Z < \frac{10.1006 - 11}{4/5}\right] \\ &= 1 - P[-1.25 < Z < -1.12425] \\ &= 0.9752 \end{aligned}$$

This critical region is much worse than using the right-hand tail of the distribution of $\bar{X}$ under H_0.

In this example there is a good chance of getting a relatively large $\bar{x}$ even when $\mu = 10$, so we must have $\bar{x} \geqslant 11.316$ to be sure (at the $1 - \alpha = 0.95$ level) that we are correct when we reject H_0. Clearly, we may fail to reject quite often when we should reject. This illustrates why we associate the alternative hypothesis with what we hope to establish. If we do reject H_0 and adopt the alternative hypothesis, then we have a controlled error rate α of being wrong. If we do not reject H_0, then H_0 may be true or we may have made a Type II error, which can happen with probability 0.654 in our example. Thus we would not feel secure in advocating H_0 just because we did not have sufficient evidence to reject it. This is also why we prefer to say that we "do not reject H_0" rather than that we "accept H_0" or believe H_0 to be true; we simply would not have sufficient statistical evidence to declare H_0 false at the 0.05 level of significance. Of course, this point is not so important if the Type II error also is known to be small. For example, suppose that a sample size $n = 100$ is available. To maintain a critical region of size $\alpha = 0.05$, we would now use

$$c_2 = \mu_0 + z_{1-\alpha}\sigma/\sqrt{n} = 10 + 1.645(4)/10 = 10.658$$

The value of $P[\text{TII}]$ in this case is

$$\begin{aligned} P[\bar{X} < 10.658 \mid \mu = 11] &= P\left[\frac{\bar{X} - 11}{4/10} < \frac{10.658 - 11}{4/10}\right] \\ &= P[Z < -0.855] \\ &= 0.196 \end{aligned}$$

If the choice of sample size is flexible, then one can specify both α and β in this problem and determine what sample size is necessary.

More generally, we may wish to test $H_0 : \mu = \mu_0$ against $H_a : \mu = \mu_1$ (where $\mu_1 > \mu_0$) at the α significance level. A test based on the test statistic

$$Z_0 = \frac{\bar{X} - \mu_0}{\sigma/\sqrt{n}} \tag{12.1.1}$$

is equivalent to one using $\bar{X}$, so we may conveniently express our test as rejecting H_0 if $z_0 \geqslant z_{1-\alpha}$, where z_0 is the computed value of Z_0. Clearly, under H_0,

$P[Z_0 \geqslant z_{1-\alpha}] = \alpha$, and we have a critical region of size α. The probability of Type II error for the alternative μ_1 is

$$P[\text{TII}] = P\left[\frac{\bar{X} - \mu_0}{\sigma/\sqrt{n}} \leqslant z_{1-\alpha} \,|\, \mu = \mu_1\right]$$
$$= P\left[\frac{\bar{X} - \mu_1}{\sigma/\sqrt{n}} \leqslant z_{1-\alpha} + \frac{\mu_0 - \mu_1}{\sigma/\sqrt{n}} \,\Big|\, \mu = \mu_1\right]$$

so that

$$P[\text{TII}] = \Phi\left(z_{1-\alpha} + \frac{\mu_0 - \mu_1}{\sigma/\sqrt{n}}\right) \tag{12.1.2}$$

The sample size n that will render $P[\text{TII}] = \beta$ is the solution to

$$z_{1-\alpha} + \frac{\mu_0 - \mu_1}{\sigma/\sqrt{n}} = z_\beta = -z_{1-\beta} \tag{12.1.3}$$

This gives

$$n = \frac{(z_{1-\alpha} + z_{1-\beta})^2\sigma^2}{(\mu_0 - \mu_1)^2} \tag{12.1.4}$$

For $\alpha = 0.05$, $\beta = 0.10$, $\mu_0 = 10$, $\mu_1 = 11$, and $\sigma = 4$, we obtain $n = (1.645 + 1.282)^2(16)/1 \doteq 137$.

In considering the error probabilities of a test, it sometimes is convenient to use the "power function" of the test.

Definition 12.1.4

The **power function**, $\pi(\theta)$, of a test of H_0 is the probability of rejecting H_0 when the true value of the parameter is θ.

For simple hypotheses $H_0 : \theta = \theta_0$ versus $H_a : \theta = \theta_1$, we have $\pi(\theta_0) = P[\text{TI}] = \alpha$ and $\pi(\theta_1) = 1 - P[\text{TII}] = 1 - \beta$. For composite hypotheses, say $H_0 : \theta \in \mathbf{\Omega}_0$ versus $H_a : \theta \in \mathbf{\Omega} - \mathbf{\Omega}_0$, the size of the test (or critical region) is

$$\alpha = \max_{\theta \in \mathbf{\Omega}_0} \pi(\theta) \tag{12.1.5}$$

and if the true value θ falls in $\mathbf{\Omega} - \mathbf{\Omega}_0$, then $\pi(\theta) = 1 - P[\text{TII}]$, where we note that $P[\text{TII}]$ depends on θ. In other words, the value of the power function is always the area under the pdf of the test statistic and over the critical region, giving $P[\text{TI}]$ for values of θ in the null hypothesis and $1 - P[\text{TII}]$ for values of θ in the alternative hypothesis. This is illustrated for a test of means in Figure 12.2.

In the next section, tests concerning the mean of a normal distribution will illustrate further the notation of composite hypotheses.

FIGURE 12.2 The relationship of the power function to the probability of a Type II error

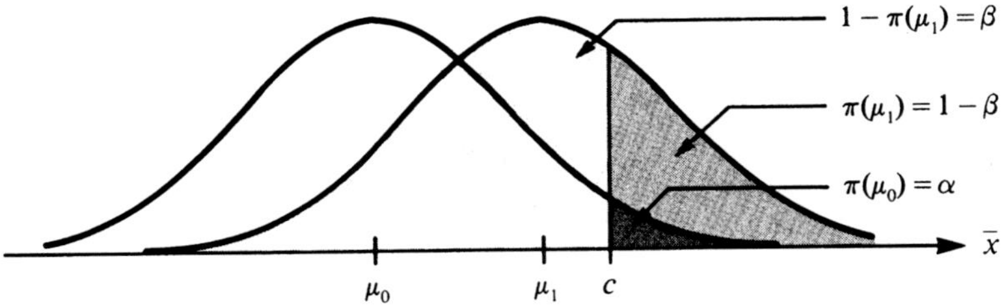

12.2

COMPOSITE HYPOTHESES

Again, we assume that $X \sim N(\mu, \sigma^2)$, where σ^2 is known, and we wish to test $H_0 : \mu = \mu_0$ against the composite alternative $H_a : \mu > \mu_0$. It was suggested in the previous example that the critical region should be located on the right-hand tail for any alternative $\mu_1 > \mu_0$, but the value of the critical value c did not depend on the value of μ_1. Thus, it is clear that the test for the simple alternative also is valid for this composite alternative. A test at significance level α still would reject H_0 if

$$z_0 = \frac{\bar{x} - \mu_0}{\sigma/\sqrt{n}} \geqslant z_{1-\alpha} \tag{12.2.1}$$

The power of this test at any value μ is

$$\pi(\mu) = P\left[\frac{\bar{X} - \mu_0}{\sigma/\sqrt{n}} \geqslant z_{1-\alpha} \,\middle|\, \mu\right]$$

$$= P\left[\frac{\bar{X} - \mu}{\sigma/\sqrt{n}} \geqslant z_{1-\alpha} + \frac{\mu_0 - \mu}{\sigma/\sqrt{n}} \,\middle|\, \mu\right]$$

so that

$$\pi(\mu) = 1 - \Phi\left(z_{1-\alpha} + \frac{\mu_0 - \mu}{\sigma/\sqrt{n}}\right) \tag{12.2.2}$$

For $\mu = \mu_0$ we have $\pi(\mu_0) = \alpha$, and for $\mu > \mu_0$ we have $\pi(\mu) = 1 - P[\text{TII}]$.

We also may consider a composite null hypothesis. Suppose that we wish to test $H_0 : \mu \leqslant \mu_0$ against $H_a : \mu > \mu_0$, and we reject H_0 if inequality (12.2.1) is satisfied. This is still a size α test for the composite null hypothesis. The probability of rejecting H_0 for any $\mu \leqslant \mu_0$ is $\pi(\mu)$, and $\pi(\mu) \leqslant \pi(\mu_0) = \alpha$ for $\mu \leqslant \mu_0$, and thus the size is $\max_{\mu \leqslant \mu_0} \pi(\mu) = \alpha$. That is, if the critical region is chosen to have size α

FIGURE 12.3 Comparison of power functions for tests based on two different sample sizes

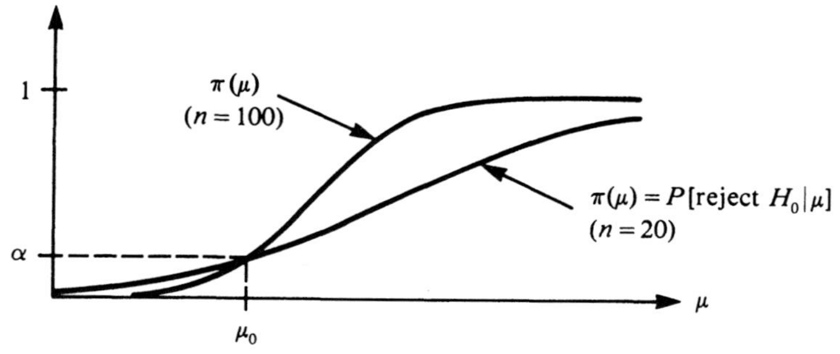

at μ_0, then the Type I error will be less than α for any $\mu < \mu_0$, so the original critical region still is appropriate here. Thus, the α level tests developed for simple null hypotheses often are applicable to the more realistic composite hypotheses, and $P[\text{TI}]$ will be no worse than α. The general shape of the power function is given in Figure 12.3 for $n = 20$ and 100.

From Figure 12.3, it is rather obvious why failure to reject H_0 should not be interpreted as acceptance of H_0. In particular, we always can find an alternative value of μ sufficiently close to μ_0 so that the power of the test, $\pi(\mu)$, is arbitrarily close to α. This would not be a serious problem in particular, if we could determine an indifference region, which is a subset of the alternative on which we are willing to tolerate low power. In other words, it might not be too important to detect alternative values of μ that are close to μ_0. In our example, we may not be too concerned about rejecting H_0 when μ is in some small interval, say (μ_0, μ_1), which we could take as our indifference region. When $\mu \geqslant \mu_1$, a sample size can be determined from equation (12.1.4) that will provide power $\pi(\mu) \geqslant 1 - \beta$. That is, for alternative values outside of the indifference region, a test can be constructed that will achieve or exceed prescribed error rates for both types of error.

P-VALUE

There is not always general agreement about how small α should be for rejection of H_0 to constitute strong evidence in support of H_a. Experimenter 1 may consider $\alpha = 0.05$ sufficiently small, while experimenter 2 insists on using $\alpha = 0.01$. Thus, it would be possible for experimenter 1 to reject when experimenter 2 fails to reject, based on the same data. If the experimenters agree to use the same test statistic, then this problem may be overcome by reporting the results of the experiment in terms of the **observed size** or ***p*-value** of the test, which is defined as the smallest size α at which H_0 can be rejected, based on the observed value of the test statistic.

Example 12.2.1 On the basis of a sample of size $n = 25$ from a normal distribution, $X_i \sim N(\mu, 16)$, we wish to test $H_0 : \mu = 10$ versus $H_a : \mu > 10$. Suppose that we observe $\bar{x} = 11.40$. The p-value is $P[\bar{X} \geqslant 11.40 \mid \mu = 10] = 1 - \Phi(1.75) = 1 - 0.9599 = 0.0401$. Because $0.01 < 0.0401 < 0.05$, the test would reject at the 0.05 level but not at the 0.01 level. If the p-value is reported, then interested readers can apply their own criteria.

To test $H_0 : \mu \geqslant \mu_0$ against $H_a : \mu < \mu_0$, similar results are obtained by rejecting H_0 if

$$z_0 \leqslant -z_{1-\alpha} \tag{12.2.3}$$

That is, the critical region of size α now is taken on the left-hand tail of the distribution of the test statistic. These tests are known as **one-sided tests** of hypotheses. The test with a critical region of form (12.2.1) is called an **upper one-sided test**, and the form (12.2.3) corresponds to a **lower one-sided test.**

Another common type of test involves a **two-sided alternative**. We may wish to test $H_0 : \mu = \mu_0$ against the alternative $H_a : \mu \neq \mu_0$. If we choose the right-hand tail for our critical region, then we will have good power for rejecting H_0 when $\mu > \mu_0$, but we will have poor power when $\mu < \mu_0$. Similarly, if we choose the left-hand tail for the critical region, we will have good power if $\mu < \mu_0$, but poor power if $\mu > \mu_0$. A good compromise is to use a two-sided critical region and reject H_0 if

$$z_0 \leqslant -z_{1-\alpha/2} \quad \text{or} \quad z_0 \geqslant z_{1-\alpha/2} \tag{12.2.4}$$

It is reasonable to use an equal-tailed test (each tail of size $\alpha/2$) in this case because of symmetry considerations, and it is common practice to use equal tails, for the sake of convenience, in most two-sided tests.

The power function for the two-sided test is

$$\pi(\mu) = 1 - P[-z_{1-\alpha/2} < Z_0 < z_{1-\alpha/2} \mid \mu]$$

which gives

$$\pi(\mu) = 1 - \Phi\left(z_{1-\alpha/2} + \frac{\mu_0 - \mu}{\sigma/\sqrt{n}}\right) + \Phi\left(-z_{1-\alpha/2} + \frac{\mu_0 - \mu}{\sigma/\sqrt{n}}\right) \tag{12.2.5}$$

If $\mu > \mu_0$, then, as suggested in Figure 12.4, the last normal probability term in the above power function will be near zero. Similarly, if $\mu < \mu_0$, then one would not expect to reject H_0 by getting a significantly large $\bar{x}$, and the first normal probability term would be near zero. If the appropriate small term is ignored, then the sample size formula for the two-sided test is approximately given by equation (12.1.4) with $z_{1-\alpha}$ replaced by $z_{1-\alpha/2}$ and μ_1 replaced by μ.

At this point, it is convenient to observe a connection between confidence intervals and hypothesis testing. In the two-sided test above, let us determine, for

FIGURE 12.4 Critical region for a two-sided test

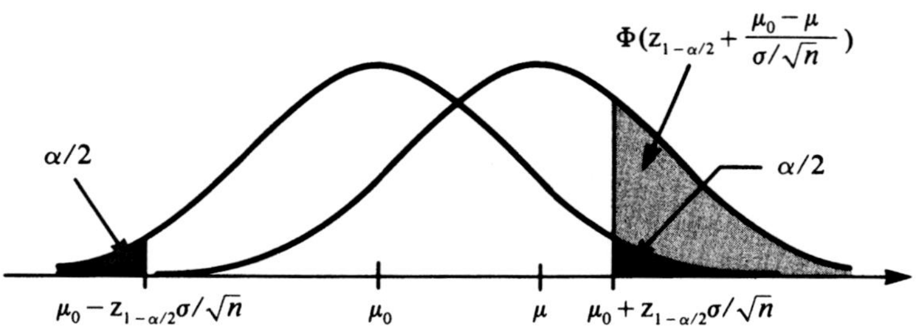

an observed value $\bar{x}$, which hypothesized values of μ_0 would not have been rejected. We see from expression (12.2.4) that these are the values of μ_0 that satisfy

$$\bar{x} - z_{1-\alpha/2}\,\sigma/\sqrt{n} < \mu_0 < \bar{x} + z_{1-\alpha/2}\,\sigma/\sqrt{n} \tag{12.2.6}$$

That is, the set of values of μ_0 that are in the "acceptance region" (nonrejection region) of the test is the same as our earlier $100(1-\alpha)\%$ confidence interval for μ. The values of μ in the "acceptance regions" of the one-sided tests correspond to the one-sided confidence intervals we discussed earlier. Indeed, one could carry out a test of $H_0 : \mu = \mu_0$ by computing the corresponding confidence interval and rejecting H_0 if the interval does not contain μ_0.

12.3

TESTS FOR THE NORMAL DISTRIBUTION

In this section, we will state theorems that summarize the common test procedures for the parameters of a normal distribution. In a later section, we will show that some of these tests have optimal properties.

TESTS FOR THE MEAN (σ^2 KNOWN)

The results discussed in the previous section are summarized in the following theorem.

Theorem 12.3.1 Suppose that $x_1 \ldots, x_n$ is an observed random sample from $N(\mu, \sigma^2)$, where σ^2 is known, and let

$$z_0 = \frac{\bar{x} - \mu_0}{\sigma/\sqrt{n}} \tag{12.3.1}$$

1. A size α test of $H_0 : \mu \leqslant \mu_0$ versus $H_a : \mu > \mu_0$ is to reject H_0 if $z_0 \geqslant z_{1-\alpha}$. The power function for this test is

$$\pi(\mu) = 1 - \Phi\left(z_{1-\alpha} + \frac{\mu_0 - \mu}{\sigma/\sqrt{n}}\right) \tag{12.3.2}$$

2. A size α test of $H_0 : \mu \geqslant \mu_0$ versus $H_a : \mu < \mu_0$ is to reject H_0 if $z_0 \leqslant -z_{1-\alpha}$. The power function for this test is

$$\pi(\mu) = \Phi\left(-z_{1-\alpha} + \frac{\mu_0 - \mu}{\sigma/\sqrt{n}}\right) \tag{12.3.3}$$

3. A size α test of $H_0 : \mu = \mu_0$ versus $H_a : \mu \neq \mu_0$ is to reject H_0 if $z_0 \leqslant -z_{1-\alpha/2}$ or $z_0 \geqslant z_{1-\alpha/2}$.
4. The sample size required to achieve a size α test with power $1 - \beta$ for an alternative value μ is given by

$$n = \frac{(z_{1-\alpha} + z_{1-\beta})^2\sigma^2}{(\mu_0 - \mu)^2} \tag{12.3.4}$$

for a one-sided test, and

$$n \doteq \frac{(z_{1-\alpha/2} + z_{1-\beta})^2\sigma^2}{(\mu_0 - \mu)^2} \tag{12.3.5}$$

for a two-sided test. ■

TESTS FOR THE MEAN (σ^2 UNKNOWN)

In most practical applications it is not possible to assume that the variance is known. It is clear that the pivotal quantities and other statistics considered in developing confidence intervals can be applied to the associated hypothesis-testing problems. Later we will discuss general techniques for deriving statistical tests.

Tests for μ with σ unknown can be based on Student's t statistic, which will be similar to the tests based on the standard normal test statistic for the case in which the variance is known, with σ^2 replaced by the observed sample variance s^2.

Theorem 12.3.2 Let $x_1, \ldots, x_n$ be an observed random sample from $N(\mu, \sigma^2)$, where σ^2 is unknown, and let

$$t_0 = \frac{\bar{x} - \mu_0}{s/\sqrt{n}} \tag{12.3.6}$$

1. A size α test of $H_0 : \mu \leqslant \mu_0$ versus $H_a : \mu > \mu_0$ is to reject H_0 if $t_0 \geqslant t_{1-\alpha}(n-1)$.
2. A size α test of $H_0 : \mu \geqslant \mu_0$ versus $H_a : \mu < \mu_0$ is to reject H_0 if $t_0 \leqslant -t_{1-\alpha}(n-1)$.
3. A size α test of $H_0 : \mu = \mu_0$ versus $H_a : \mu \neq \mu_0$ is to reject H_0 if $t_0 \leqslant -t_{1-\alpha/2}(n-1)$ or $t_0 \geqslant t_{1-\alpha/2}(n-1)$. ■

The power function and the related sample size problem are more complicated than in the case where σ^2 is known.

For part 1 of the theorem, for an alternative $\mu > \mu_0$, the power function is

$$\pi(\mu) = P\left[\frac{\bar{X} - \mu_0}{S/\sqrt{n}} \geqslant t_{1-\alpha}(\nu) \mid \mu\right]$$

$$= P\left[\frac{\bar{X} - \mu + (\mu - \mu_0)}{S/\sqrt{n}} \geqslant t_{1-\alpha}(\nu) \mid \mu\right]$$

so that

$$\pi(\mu) = P\left[\frac{Z + \delta}{\sqrt{V/v}} \geqslant t_{1-\alpha}(\nu)\right], \qquad \textbf{(12.3.7)}$$

where $\nu = n - 1$, $\delta = \sqrt{n}(\mu - \mu_0)/\sigma$, and Z and V are independent, $Z \sim \mathrm{N}(0,\ 1)$ and $V = (n-1)S^2/\sigma^2 \sim \chi^2(\nu)$. The random variable in equation (12.3.7) is said to have a **noncentral *t* distribution** with ν degrees of freedom and **noncentrality parameter** δ. It has the usual Student's t distribution only when $\delta = 0$. Otherwise, the distribution is rather hard to evaluate. Tables of noncentral t distribution are available, and $\pi(\mu)$ can be determined from these for given values of δ. Similarly, the sample size required to give a desired power can be determined for specified values of δ. This can be quite useful if approximate values or previous estimates of σ are available. Table 8 (Appendix C) gives the sample size required to achieve $\pi(\mu) = 1 - \beta$ in terms of $d = |\delta|/\sqrt{n} = |\mu - \mu_0|/\sigma$ for the tests described in Theorem 12.3.2.

Example 12.3.1 It is desired to test, at the $\alpha = 0.05$ level, $H_0 : \mu = 10$ versus $H_a : \mu > 10$ for a normal distribution, $\mathrm{N}(\mu, \sigma^2)$ with σ^2 unknown, and we wish to have power 0.99 if the true value μ is two standard deviations greater than $\mu_0 = 10$. In other words, $0.99 = \pi(\mu)$ when $d = |\mu - \mu_0|/\sigma = 2$. From Table 8 (Appendix C), we obtain $n = 6$.

Tests also can be constructed for other parameters such as the variance.

TEST FOR VARIANCE

It is possible to construct tests of hypotheses such as $H_0 : \sigma^2 = \sigma_0^2$ versus $H_a : \sigma^2 > \sigma_0^2$ based on the test statistic

$$V_0 = (n-1)S^2/\sigma_0^2 \quad \textbf{(12.3.8)}$$

because $V_0 \sim \chi^2(n-1)$ when H_0 is true. An observed value of the sample variance, s^2, that is large relative to σ_0^2 would support H_a. This suggests choosing the right-hand tail of the distribution of V_0 as the critical region for such a test. This test would be very useful for deciding whether the amount of variability in a population is excessive relative to some standard value, σ_0^2. Similar remarks apply for a composite null hypothesis of the form $H_0 : \sigma^2 \leqslant \sigma_0^2$.

In the following theorem, $H(c; \nu)$ is the CDF of $\chi^2(\nu)$.

Theorem 12.3.3 Let $x_1, \ldots, x_n$ be an observed random sample from $N(\mu, \sigma^2)$, and let

$$v_0 = (n-1)s^2/\sigma_0^2 \quad \textbf{(12.3.9)}$$

1. A size α test of $H_0 : \sigma^2 \leqslant \sigma_0^2$ versus $H_a : \sigma^2 > \sigma_0^2$ is to reject H_0 if $v_0 \geqslant \chi^2_{1-\alpha}(n-1)$. The power function for this test is

$$\pi(\sigma^2) = 1 - H[(\sigma_0^2/\sigma^2)\chi^2_{1-\alpha}(n-1); n-1] \quad \textbf{(12.3.10)}$$

2. A size α test of $H_0 : \sigma^2 \geqslant \sigma_0^2$ versus $H_a : \sigma^2 < \sigma_0^2$ is to reject H_0 if $v_0 \leqslant \chi^2_{\alpha}(n-1)$. The power function for this test is

$$\pi(\sigma^2) = H[(\sigma_0^2/\sigma^2)\chi^2_{\alpha}(n-1); n-1] \quad \textbf{(12.3.11)}$$

3. A size α test of $H_0 : \sigma^2 = \sigma_0^2$ versus $H_a : \sigma^2 \neq \sigma_0^2$ is to reject H_0 if $v_0 \leqslant \chi^2_{\alpha/2}(n-1)$ or $v_0 \geqslant \chi^2_{1-\alpha/2}(n-1)$.

Proof

We will derive the power function for part 1 and leave the other details as an exercise.

$$\begin{aligned}\pi(\sigma^2) &= P[V_0 \geqslant \chi^2_{1-\alpha}(n-1) \mid \sigma^2] \\ &= P[(n-1)S^2/\sigma^2 \geqslant (\sigma_0^2/\sigma^2)\chi^2_{1-\alpha}(n-1) \mid \sigma^2] \\ &= 1 - H[(\sigma_0^2/\sigma^2)\chi^2_{1-\alpha}(n-1); n-1]\end{aligned}$$

Notice th[illegible] in particular, $\pi(\sigma_0^2) = 1 - H[\chi^2_{1-\alpha}(n-1); n-1] = 1 - (1-\alpha) = \alpha$, and bec[illegible] $\pi(\sigma^2)$ is increasing, the size of the critical region is α. ■

In practice, it is convenient to use an equal-tailed two-sided test as described in part 3, but unequal tail values α_1 and α_2 with $\alpha = \alpha_1 + \alpha_2$ may be desirable in some situations.

It is possible to solve for a sample size to achieve prescribed levels of size and power for these tests. For example, for the test of part 1, we may wish to have $\pi(\sigma_1^2) = 1 - \beta$ for some $\sigma_1^2 > \sigma_0^2$. This requires finding the value n such that

$$(\sigma_0^2/\sigma_1^2)\chi^2_{1-\alpha}(n-1) = \chi^2_\beta(n-1) \tag{12.3.12}$$

This cannot be solved explicitly for n, but an iterative procedure based on the percentiles in Table 4 (Appendix C) can be used for specified values of α, β, and σ_0^2/σ_1^2. It also would be desirable to have an approximate formula that gives n explicitly as a function of these values. Such an approximation can be based on the normal approximation $\chi^2_\gamma(\nu) \doteq \nu + z_\gamma\sqrt{2\nu}$, which was given in Chapter 8. If this approximation is used in both sides of equation (12.3.12), then it is possible to derive the approximation

$$n \doteq 1 + 2\left[\frac{z_\beta - (\sigma_0/\sigma_1)^2 z_{1-\alpha}}{1 - (\sigma_0/\sigma_1)^2}\right]^2 \tag{12.3.13}$$

Example 12.3.2 We desire to test $H_0 : \sigma^2 \leqslant 16$ versus $H_a : \sigma^2 > 16$ with size $\alpha = 0.10$ and power $1 - \beta = 0.75$ when $\sigma_1^2 = 32$ is true. Based on approximation (12.3.13) with $z_{1-\alpha} = 1.282$, $z_\beta = -0.674$, and $(\sigma_0/\sigma_1)^2 = 0.5$, we obtain $n \doteq 15$. If we compute both sides of equation (12.3.12) for values of ν close to $\nu = 15 - 1 = 14$, we obtain the best agreement when $\nu = 15$, which corresponds to $n = 16$.

TWO-SAMPLE TESTS

It is possible to construct tests of hypotheses concerning the variances of two normal distributions, such as $H_0 : \sigma_2^2/\sigma_1^2 = d_0$, based on an F statistic. In particular, consider the test statistic

$$F_0 = \frac{S_1^2}{S_2^2}\, d_0 \tag{12.3.14}$$

where $F_0 \sim F(n_1 - 1, n_2 - 1)$ if H_0 is true.

Theorem 12.3.4 Suppose that $x_1, \ldots, x_{n_1}$ and $y_1, \ldots, y_{n_2}$ are observed values of independent random samples from $\mathrm{N}(\mu_1, \sigma_1^2)$ and $\mathrm{N}(\mu_2, \sigma_2^2)$, respectively, and let

$$f_0 = \frac{s_1^2}{s_2^2}\, d_0 \tag{12.3.15}$$

1. A size α test of $H_0 : \sigma_2^2/\sigma_1^2 \leqslant d_0$ versus $H_a : \sigma_2^2/\sigma_1^2 > d_0$ is to reject H_0 if $f_0 \leqslant 1/f_{1-\alpha}(n_2 - 1, n_1 - 1)$.
2. A size α test of $H_0 : \sigma_2^2/\sigma_1^2 \geqslant d_0$ versus $H_a : \sigma_2^2/\sigma_1^2 < d_0$ is to reject H_0 if $f_0 \geqslant f_{1-\alpha}(n_1 - 1, n_2 - 1)$.

3. A size α test of $H_0 : \sigma_2^2/\sigma_1^2 = d_0$ versus $H_a : \sigma_2^2/\sigma_1^2 \neq d_0$ is to reject H_0 if $f_0 \leqslant 1/f_{1-\alpha/2}(n_2 - 1, n_1 - 1)$ or $f_0 \geqslant f_{1-\alpha/2}(n_1 - 1, n_2 - 1)$.

If the variances are unknown but equal, then tests of hypotheses concerning the means such as $H_0 : \mu_2 - \mu_1 = d_0$ can be constructed based on the t distribution. In particular, let

$$t_0 = \frac{\bar{y} - \bar{x} - d_0}{s_p\sqrt{\dfrac{1}{n_1} + \dfrac{1}{n_2}}} \tag{12.3.16}$$

where s_p^2 is the pooled estimate defined by equation (11.5.7).

Theorem 12.3.5 Suppose that $x_1, \ldots, x_{n_1}$ and $y_1, \ldots, y_{n_2}$ are observed values of independent random samples from $N(\mu_1, \sigma_1^2)$ and $N(\mu_2, \sigma_2^2)$, respectively, where $\sigma_1^2 = \sigma_2^2 = \sigma^2$.

1. A size α test of $H_0 : \mu_2 - \mu_1 \leqslant d_0$ versus $H_a : \mu_2 - \mu_1 > d_0$ is to reject H_0 if $t_0 \geqslant t_{1-\alpha}(n_1 + n_2 - 2)$.
2. A size α test of $H_0 : \mu_2 - \mu_1 \geqslant d_0$ versus $H_a : \mu_2 - \mu_1 < d_0$ is to reject H_0 if $t_0 \leqslant -t_{1-\alpha}(n_1 + n_2 - 2)$.
3. A size α test of $H_0 : \mu_2 - \mu_1 = d_0$ versus $H_a : \mu_2 - \mu_1 \neq d_0$ is to reject H_0 if $t_0 \leqslant -t_{1-\alpha/2}(n_1 + n_2 - 2)$ or $t_0 \geqslant t_{1-\alpha/2}(n_1 + n_2 - 2)$. ■

The power functions for these tests involve the noncentral t distribution. It is possible to determine equal sample sizes $n_1 = n_2 = n$ for a one-sided size α test with power $1 - \beta$ by using Table 8 (Appendix C) with $d' = |\mu_2 - \mu_1|/\sigma$. For a two-sided test, the size is 2α. Again, it is necessary to have a value to use for σ or else to express the difference in standard deviations. Of course, the test will have the proper size whatever n is used.

For unequal variances, an approximate test can be constructed based on Welch's approximate t statistic as given by equation (11.5.13). Similarly, tests for other cases such as the paired sample case can be set up easily.

PAIRED-SAMPLE t TEST

All of the above tests assume that the random samples are independent. As noted in Chapter 11, there are situations in which an experiment involves only one set of individuals or experimental objects, and two observations are made on each individual or object. For example, one possible way to test the effectiveness of a diet plan would be to weigh each one of a set of n individuals both before and after the diet period. The result would be paired data $(x_1, y_1), \ldots, (x_n, y_n)$ with x_i and y_i the weight measurements, respectively before and after the diet, for the ith individual in the study. Of course, one might reasonably expect a dependence between observations within a pair because they are measurements on the same individual.

We will assume that such paired data are observations on a set of independent pairs of random variables from a bivariate population, $(X, Y) \sim f(x, y)$, and that the differences $D = Y - X$ are normally distributed, $D \sim N(\mu_D, \sigma_D^2)$, with $\mu_D = E(Y) - E(X) = \mu_2 - \mu_1$. One possibility that leads to this situation is if the pairs are a random sample from a bivariate normal population. That is, if there is independence between pairs, and each has the same bivariate normal distribution, then the differences are independent normal with mean $\mu_2 - \mu_1$. Thus, a test based on the T-variable of equation (11.5.17) applied to the differences $d_i = y_i - x_i$ yields a test of $\mu_2 - \mu_1$ with σ_D^2 unknown.

Theorem 12.3.6 Denote by $(x_1, y_1), \ldots, (x_n, y_n)$ the observed values of n independent pairs of random variables, $(X_1, Y_1), \ldots, (X_n, Y_n)$, and assume that the differences $D_i = Y_i - X_i$ are normally distributed, each with mean $\mu_D = \mu_2 - \mu_1$ and variance σ_D^2. Let $\bar{d}$ and s_d^2 be the sample mean and sample variance based on the differences $d_i = y_i - x_i$, for $i = 1, \ldots, n$, and let

$$t_0 = \frac{\bar{d} - d_0}{s_d/\sqrt{n}} \tag{12.3.17}$$

1. A size α test of $H_0 : \mu_2 - \mu_1 \leqslant d_0$ versus $H_a : \mu_2 - \mu_1 > d_0$ is to reject H_0 if $t_0 \geqslant t_{1-\alpha}(n-1)$.
2. A size α test of $H_0 : \mu_2 - \mu_1 \geqslant d_0$ versus $H_a : \mu_2 - \mu_1 < d_0$ is to reject H_0 if $t_0 \leqslant -t_{1-\alpha}(n-1)$.
3. A size α test of $H_0 : \mu_2 - \mu_1 = d_0$ versus $H_a : \mu_2 - \mu_1 \neq d_0$ is to reject H_0 if either $t_0 \leqslant -t_{1-\alpha/2}(n-1)$ or $t_0 \geqslant t_{1-\alpha/2}(n-1)$. ■

12.4

BINOMIAL TESTS

The techniques used to set confidence intervals for the binomial distribution also can be modified to obtain tests of hypotheses. Suppose that $X_i \sim \text{BIN}(1, p)$, $i = 1, \ldots, n$. Then tests for p will be based on the sufficient statistic $S = \sum X_i \sim \text{BIN}(n, p)$.

Theorem 12.4.1 Let $S \sim \text{BIN}(n, p)$. For large n, an approximate size α test of $H_0 : p \leqslant p_0$ against $H_a : p > p_0$ is to reject H_0 if

$$z_0 = \frac{s - np_0}{\sqrt{np_0(1 - p_0)}} > z_{1-\alpha}$$

An approximate size α test of $H_0: p \geqslant p_0$ against $H_a: p < p_0$ is to reject H_0 if

$$z_0 < -z_{1-\alpha}$$

An approximate size α test of $H_0: p = p_0$ against $H_a: p \neq p_0$ is to reject H_0 if

$$z_0 < -z_{1-\alpha/2} \quad \text{or} \quad z_0 > z_{1-\alpha/2}$$ ■

These results follow from the fact that when $p = p_0$, then

$$\frac{S - np_0}{\sqrt{np_0(1-p_0)}} \xrightarrow{d} Z \sim \text{N}(0, 1) \tag{12.4.1}$$

As in the previous one-sided examples, the probability of rejecting a true H_0 will be less than α for other values of p in the null hypotheses.

Exact tests also can be based on S, analogous to the way exact confidence intervals were obtained using the general method. These tests also may be **conservative** in the sense that the size of the test actually may be less than α for all parameter values under H_0, because of the discreteness of the random variable.

Theorem 12.4.2 Suppose that $S \sim \text{BIN}(n, p)$, and $B(s; n, p)$ denotes a binomial CDF. Denote by s an observed value of S.

1. A conservative size α test of $H_0: p \leqslant p_0$ against $H_a: p > p_0$ is to reject H_0 if

$$1 - B(s-1; n, p_0) \leqslant \alpha$$

2. A conservative size α test of $H_0: p \geqslant p_0$ against $H_a: p < p_0$ is to reject H_0 if

$$B(s; n, p_0) \leqslant \alpha$$

3. A conservative two-sided test of $H_0: p = p_0$ against $H_a: p \neq p_0$ is to reject H_0 if

$$B(s; n, p_0) \leqslant \alpha/2 \quad \text{or} \quad 1 - B(s-1; n, p_0) \leqslant \alpha/2$$ ■

The concept of hypothesis testing is well illustrated by this model. If one is testing $H_0: p \geqslant p_0$, then H_0 is rejected if the observed s is so small that it would have been very unlikely ($\leqslant \alpha$) to have obtained such a small value of s when $p \geqslant p_0$. Also, it is clear that the indicated critical regions have size $\leqslant \alpha$.

In case 2, for example, if p_0 happens to be a value such that $B(s; n, p_0) = \alpha$, then the test will have exact size α; otherwise it is conservative. Note also that $B(s; n, p)$, for example, is the p value or observed size of the test in case 2, if one desires to use that concept.

It also is possible to construct tests of hypotheses about the equality of two population proportions. In particular, suppose that $X \sim \text{BIN}(n_1, p_1)$ and $Y \sim \text{BIN}(n_2, p_2)$, and that X and Y are independent. The MLEs are $\hat{p}_1 = X/n_1$ and

$\hat{p}_2 = Y/n_2$. Under $H_0 : p_1 = p_2$, it would seem appropriate to have a pooled estimate of their common value, say $\hat{p} = (X + Y)/(n_1 + n_2)$. It can be shown by the methods of Chapter 7 that if $p_1 = p_2$, then

$$Z_0 = \frac{\hat{p}_2 - \hat{p}_1}{\sqrt{\hat{p}(1 - \hat{p})\left(\frac{1}{n_1} + \frac{1}{n_2}\right)}} \xrightarrow{d} Z \sim \mathrm{N}(0, 1) \tag{12.4.2}$$

and consequently an approximate size α test of $H_0 : p_1 = p_2$ versus $H_a : p_1 \neq p_2$ would reject H_0 if $z_0 \leqslant -z_{1-\alpha/2}$ or $z_0 \geqslant z_{1-\alpha/2}$.

12.5

POISSON TESTS

Tests for the mean of a Poisson distribution can be based on the sufficient statistic $S = \sum X_i$. In the following theorem, $F(x; \mu)$ is the CDF of $X \sim \mathrm{POI}(\mu)$.

Theorem 12.5.1 Let $x_1, \ldots, x_n$ be an observed random sample from $\mathrm{POI}(\mu)$, and let $s = \sum x_i$.

1. A conservative size α test of $H_0 : \mu \leqslant \mu_0$ versus $H_a : \mu > \mu_0$ is to reject H_0 if $1 - F(s - 1; n\mu_0) \leqslant \alpha$.
2. A conservative size α test of $H_0 : \mu \geqslant \mu_0$ versus $H_a : \mu < \mu_0$ is to reject H_0 if $F(s; n\mu_0) \leqslant \alpha$. ■

Using the results in Exercise 18 of Chapter 8, it is possible to give tests in terms of chi-square percentiles. In particular, H_0 of part 1 is rejected if $2n\mu_0 \leqslant \chi^2_{\alpha}(2s)$, and H_0 of part 2 is rejected if $2n\mu_0 \geqslant \chi^2_{1-\alpha}(2s + 2)$. A two-sided size α test of $H_0 : \mu = \mu_0$ versus $H_a : \mu \neq \mu_0$ would reject H_0 if $2n\mu_0 \leqslant \chi^2_{\alpha/2}(2s)$ or $2n\mu_0 \geqslant \chi^2_{1-\alpha/2}(2s + 2)$. Again the concept of p-value may be useful in this problem, where for an observed s, $1 - F(s - 1; n\mu_0)$ and $F(s; n\mu_0)$ are the p-values for cases 1 and 2, respectively.

12.6

MOST POWERFUL TESTS

In the previous sections, the terminology of hypothesis testing has been developed, and some intuitively appealing tests have been described based on pivotal quantities or appropriate sufficient statistics. In most cases these tests are closely related to analogous confidence intervals discussed in the previous chapter. The tests presented earlier were based on reasonable test statistics, but no rationale was provided to suggest that they are best in any sense. If confronted with choosing between two or more tests of the same size, it would seem reasonable to select the one with the greatest chance of detecting a true alternative value. In other words, the strategy will be to select a test with maximum power for alternative values of the parameter. We will approach this problem by considering a method for deriving critical regions corresponding to tests that are most powerful tests of a given size for testing simple hypotheses.

Let $X_1, \ldots, X_n$ have joint pdf $f(x_1, \ldots, x_n; \theta)$, and consider a critical region C. The notation for the power function corresponding to C is

$$\pi_C(\theta) = P[(X_1, \ldots, X_n) \in C \mid \theta] \tag{12.6.1}$$

Definition 12.6.1

A test of $H_0 : \theta = \theta_0$ versus $H_a : \theta = \theta_1$ based on a critical region C^* is said to be a **most powerful test** of size α if

1. $\pi_{C^*}(\theta_0) = \alpha$, and
2. $\pi_{C^*}(\theta_1) \geqslant \pi_C(\theta_1)$ for any other critical region C of size α [that is, $\pi_C(\theta_0) = \alpha$].

Such a critical region, C^*, is called a **most powerful critical region** of size α.

The following theorem shows how to derive a most powerful critical region for testing simple hypotheses.

Theorem 12.6.1 **Neyman-Pearson Lemma** Suppose that $X_1, \ldots, X_n$ have joint pdf $f(x_1, \ldots, x_n; \theta)$. Let

$$\lambda(x_1, \ldots, x_n; \theta_0, \theta_1) = \frac{f(x_1, \ldots, x_n; \theta_0)}{f(x_1, \ldots, x_n; \theta_1)} \tag{12.6.2}$$

and let C^* be the set

$$C^* = \{(x_1, \ldots, x_n) \mid \lambda(x_1, \ldots, x_n; \theta_0, \theta_1) \leqslant k\} \tag{12.6.3}$$

where k is a constant such that

$$P[(X_1, \ldots, X_n) \in C^* \mid \theta_0] = \alpha \tag{12.6.4}$$

Then C^* is a most powerful critical region of size α for testing $H_0 : \theta = \theta_0$ versus $H_a : \theta = \theta_1$.

Proof

For convenience, we will adopt vector notation, $\mathbf{X} = (X_1, \ldots, X_n)$ and $\mathbf{x} = (x_1, \ldots, x_n)$. Also, if A is an n-dimensional event, let

$$P[\mathbf{X} \in A \mid \theta] = \int_A f(\mathbf{x}; \theta) = \int \cdots_A \int f(x_1, \ldots, x_n; \theta)\, dx_1 \ldots dx_n \qquad \textbf{(12.6.5)}$$

for the continuous case. The discrete case would be similar, with integrals replaced by summations. We also will denote the complement of a set C by $\bar{C}$. Note that if A is a subset of C^*, then

$$P[\mathbf{X} \in A \mid \theta_0] \leqslant kP[\mathbf{X} \in A \mid \theta_1] \qquad \textbf{(12.6.6)}$$

because $\int_A f(\mathbf{x}; \theta_0) \leqslant \int_A kf(\mathbf{x}; \theta_1)$. Similarly, if A is a subset of $\bar{C}^*$, then

$$P[\mathbf{X} \in A \mid \theta_0] \geqslant kP[\mathbf{X} \in A \mid \theta_1] \qquad \textbf{(12.6.7)}$$

Notice that for any critical region C we have

$$C^* = (C^* \cap C) \cup (C^* \cap \bar{C}) \qquad \text{and} \qquad C = (C \cap C^*) \cup (C \cap \bar{C}^*)$$

Thus,

$$\pi_{C^*}(\theta) = P[\mathbf{X} \in C^* \cap C \mid \theta] + P[\mathbf{X} \in C^* \cap \bar{C} \mid \theta]$$

and

$$\pi_C(\theta) = P[\mathbf{X} \in C^* \cap C \mid \theta] + P[\mathbf{X} \in C \cap \bar{C}^* \mid \theta]$$

and the difference is

$$\pi_{C^*}(\theta) - \pi_C(\theta) = P[\mathbf{X} \in C^* \cap \bar{C} \mid \theta] - P[\mathbf{X} \in C \cap \bar{C}^* \mid \theta] \qquad \textbf{(12.6.8)}$$

Combining equation (12.6.8) with $\theta = \theta_1$ and inequalities (12.6.6) and (12.6.7), we have

$$\pi_{C^*}(\theta_1) - \pi_C(\theta_1) \geqslant (1/k)\{P[\mathbf{X} \in C^* \cap \bar{C} \mid \theta_0] - P[\mathbf{X} \in C \cap \bar{C}^* \mid \theta_0]\}$$

Again, using (12.6.8) with $\theta = \theta_0$ in the right side of this inequality, we obtain

$$\pi_{C^*}(\theta_1) - \pi_C(\theta_1) \geqslant (1/k)[\pi_{C^*}(\theta_0) - \pi_C(\theta_0)]$$

If C is a critical region of size α, then $\pi_{C^*}(\theta_0) - \pi_C(\theta_0) = \alpha - \alpha = 0$, and the right side of the last inequality is 0, and thus $\pi_{C^*}(\theta_1) \geqslant \pi_C(\theta_1)$. ■

The general philosophy of the Neyman-Pearson approach to hypothesis testing is to put sample points into the critical region until it reaches size α. To maximize power, points should be put into the critical region that are more likely under H_a than under H_0. In particular, the Neyman-Pearson lemma says that the criterion for choosing sample points to be included should be based on the magnitude of the ratio of the likelihood functions under H_0 and H_a.

Example 12.6.1 Consider a random sample of size n from an exponential distribution, $X_i \sim \text{EXP}(\theta)$. We wish to test $H_0 : \theta = \theta_0$ against $H_a : \theta = \theta_1$ where $\theta_1 > \theta_0$. The Neyman-Pearson lemma says to reject H_0 if

$$\lambda(x, \theta_0, \theta_1) = \frac{\theta_0^{-n} \exp(-\sum x_i/\theta_0)}{\theta_1^{-n} \exp(-\sum x_i/\theta_1)} \leqslant k$$

where k is such that $P[\lambda(X, \theta_0, \theta_1) \leqslant k] = \alpha$, under $\theta = \theta_0$. Now

$$P[\lambda(X; \theta_0, \theta_1) \leqslant k \mid \theta_0] = P[\sum X_i(1/\theta_1 - 1/\theta_0) \leqslant \ln((\theta_0/\theta_1)^n k) \mid \theta_0]$$

so that

$$P[X \in C^* \mid \theta_0] = P[\sum X_i \geqslant k_1 \mid \theta_0] \tag{12.6.9}$$

where $k_1 = \ln((\theta_0/\theta_1)^n k)/(1/\theta_1 - 1/\theta_0)$. Notice that the direction of the inequality changed because $1/\theta_1 - 1/\theta_0 < 0$ in this case. Thus, a most powerful critical region has the form $C^* = \{(x_1, \ldots, x_n) \mid \sum x_i \geqslant k_1\}$. Notice that under $H_0 : \theta = \theta_0$, we have $2 \sum X_i/\theta_0 \sim \chi^2(2n)$, so that $k_1 = \theta_0 \chi^2_{1-\alpha}(2n)/2$ would give a critical region of size α, and an equivalent test would be to reject H_0 if $2 \sum x_i/\theta_0 \geqslant \chi^2_{1-\alpha}(2n)$. The original constant k could be computed if desired, but it is not necessary in order to perform the test.

Similarly, if we wish to test $H_0 : \theta = \theta_0$ versus $H_a : \theta = \theta_1$ with $\theta_1 < \theta_0$, then the most powerful test of size α is to reject H_0 if $2 \sum x_i/\theta_0 \leqslant \chi^2_\alpha(2n)$. The only difference between the two tests comes about because of the difference in the sign of $1/\theta_1 - 1/\theta_0$ in the two cases. In other words, the right side of equation (12.6.9) becomes $P[\sum X_i \leqslant k_1 \mid \theta_0]$ if $\theta_1 < \theta_0$, which corresponds to $1/\theta_1 - 1/\theta_2 > 0$. Note also that this is the only way in which C^* depends on the alternative value in this example. That is, the most powerful test of $H_0 : \theta = \theta_0$ versus $H_a : \theta = \theta_2$ is exactly the same, provided that θ_1 and θ_2 are both greater (or both less) than θ_0. This makes it possible to extend the concept of most powerful test for a simple alternative to a "uniformly most powerful" test for a composite alternative such as $H_a : \theta > \theta_0$.

This concept is considered more fully in the next section.

Example 12.6.2 Consider a random sample of size n from a normal distribution with mean zero, $X_i \sim N(0, \sigma^2)$. We wish to test $H_0 : \sigma^2 = \sigma_0^2$ versus $H_a : \sigma^2 = \sigma_1^2$ with $\sigma_1^2 > \sigma_0^2$. In this case

$$\lambda = \lambda(x_1, \ldots, x_n; \sigma_0^2, \sigma_1^2) = \frac{\left(\dfrac{1}{\sqrt{2\pi}\sigma_0}\right)^n \exp[-\sum x_i^2/2\sigma_0^2]}{\left(\dfrac{1}{\sqrt{2\pi}\sigma_1}\right)^n \exp[-\sum x_i^2/2\sigma_1^2]}$$

Thus, $\lambda \leqslant k$ is equivalent to

$$(1/\sigma_1^2 - 1/\sigma_0^2) \sum x_i^2 \leqslant k_1$$

for a constant k_1. Because $\sigma_1^2 > \sigma_0^2$, we have $1/\sigma_1^2 - 1/\sigma_0^2 < 0$, and the most powerful critical region has the form $C^* = \{(x_1, \ldots, x_n) | \sum x_i^2 \geqslant k_2\}$. Notice also that $\sum X_i^2/\sigma_0^2 \sim \chi^2(n)$ under H_0, so that a size α test would reject H_0 if $\sum x_i^2/\sigma_0^2 \geqslant \chi^2_{1-\alpha}(n)$. Note that if $\sigma_1^2 < \sigma_0^2$, a most powerful test of size α would reject if $\sum x_i^2/\sigma_0^2 \leqslant \chi^2_\alpha(n)$.

The previous examples involve continuous distributions, so that a test with a prescribed size α is possible. For discrete distributions it may not be possible to achieve an exact size α, but one could choose k to give size at most α and as close to α as possible. In this case the Neyman-Pearson test is the most powerful test of size $\pi_{C^*}(\theta_0) = \alpha_1 \leqslant \alpha$, and it would be a conservative test for a prescribed size α.

Example 12.6.3 We wish to determine the form of the most powerful test of $H_0 : p = p_0$ against $H_a : p = p_1 > p_0$ based on the statistic $S \sim \text{BIN}(n, p)$. We have

$$\lambda = \frac{\binom{n}{s} p_0^s (1 - p_0)^{n-s}}{\binom{n}{s} p_1^s (1 - p_1)^{n-s}} \leqslant k$$

so that

$$\left\{\frac{p_0(1 - p_1)}{p_1(1 - p_0)}\right\}^s \leqslant k_1$$

or

$$s \ln \left[\frac{p_0(1 - p_1)}{p_1(1 - p_0)}\right] \leqslant \ln k_1$$

Because $p_0(1 - p_1)/p_1(1 - p_0) \leqslant 1$, the log term is negative and the test is to reject H_0 if $s \geqslant k_2$, which is the same form as the binomial test suggested earlier. Now

$$P[S \geqslant i \mid p = p_0] = 1 - B(i - 1; n, p_0) = \alpha_i$$

so for integer values $i = 1, \ldots, n$, exact most powerful tests of size α_i are achieved by rejecting H_0 if $s \geqslant i$. For other prescribed levels of α the test would be chosen to be conservative as discussed earlier.

The Neyman-Pearson lemma does not claim that the conservative tests would be most powerful. Somewhat artificially, one can increase the power of the conservative test by adding a fraction of a sample point to the critical region in the discrete case so that $P[\text{TI}] = \alpha$. That is, if $P[S \geqslant 7] < \alpha$ and $P[S \geqslant 6] > \alpha$, then one could reject H_0 if $s \geqslant 7$ and some fraction of the time when $s = 6$, depending on what fraction is needed to make the size of the critical region equal α. This is referred to as a **randomized test**, because if $s = 6$ is observed one could flip a coin (appropriately biased) to decide whether to reject. In most cases it seems more reasonable to select some exact α_i close to the level desired, and then use the exact, most powerful test for this α_i significance level, rather than randomize on an additional discrete point to get a prescribed size α.

Note that the Neyman-Pearson principle applies to testing any completely specified $H_0 : f_0(\mathbf{x}; \theta_0)$ against any completely specified alternative $H_a : f_1(\mathbf{x}; \theta_1)$.

In most applications $\mathbf{x}$ will result from a random sample from a density with possibly different values of a parameter, but $\mathbf{x}$ could result from a set of order statistics or some other multivariate variable. Also, the densities need not be of the same type under H_0 and H_a as long as they are completely specified, so that the statistic can be computed and the critical region can be determined with size α under H_0.

Example 12.6.4 We have a random sample of size n, and we wish to test $H_0 : X \sim \text{UNIF}(0, 1)$ against $H_a : X \sim \text{EXP}(1)$. We have

$$\lambda = \frac{f_0(x_1, \ldots, x_n)}{f_1(x_1, \ldots, x_n)} = \frac{1}{e^{-\Sigma x_i}} \leqslant k$$

so we reject H_0 if $\sum x_i \leqslant k_1 = \ln k$. The distribution of a sum of uniform variables is not easy to express, but the central limit theorem can be used to obtain an approximate critical value. We know that if $X \sim \text{UNIF}(0, 1)$, then $E(X) = 1/2$, $\text{Var}(X) = 1/12$, and

$$z_n = \frac{\bar{X} - 0.5}{\sqrt{1/(12n)}} \xrightarrow{d} Z \sim \text{N}(0, 1)$$

Thus, an approximate size α test is to reject H_0 if

$$z_n = \sqrt{12n}(\bar{x} - 0.5) \leqslant -z_{1-\alpha}$$

The concept of a most powerful test now will be extended to the case of composite hypotheses.

12.7

UNIFORMLY MOST POWERFUL TESTS

In the last section we saw that in some cases the same test is most powerful against several different alternative values. If a test is most powerful against every possible value in a composite alternative, then it will be called a uniformly most powerful test.

Definition 12.7.1

Let $X_1, \ldots, X_n$ have joint pdf $f(x_1, \ldots, x_n; \theta)$ for $\theta \in \mathbf{\Omega}$, and consider hypotheses of the form $H_0 : \theta \in \mathbf{\Omega}_0$ versus $H_a : \theta \in \mathbf{\Omega} - \mathbf{\Omega}_0$, where $\mathbf{\Omega}_0$ is a subset of $\mathbf{\Omega}$. A critical region C^*, and the associated test, are said to be **uniformly most powerful** (UMP) of size α if

$$\max_{\theta \in \mathbf{\Omega}_0} \pi_{C^*}(\theta) = \alpha \tag{12.7.1}$$

and

$$\pi_{C^*}(\theta) \geqslant \pi_C(\theta) \tag{12.7.2}$$

for all $\theta \in \mathbf{\Omega} - \mathbf{\Omega}_0$ and all critical regions C of size α.

That is, C^* defines a UMP test of size α if it has size α, and if for all parameter values in the alternative, it has maximum power relative to all critical regions of size α.

A UMP test often exists in the case of a one-sided composite alternative, and a possible technique for determining a UMP test is first to derive the Neyman-Pearson test for a particular alternative value and then show that the test does not depend on the specific alternative value.

Example 12.7.1 Consider a random sample of size n from an exponential distribution, $X_i \sim \text{EXP}(\theta)$. It was found in Example 12.6.1 that the most powerful test of size α of $H_0 : \theta = \theta_0$ versus $H_a : \theta = \theta_1$, when $\theta_1 > \theta_0$, is to reject H_0 if $2n\bar{x}/\theta_0 = 2 \sum x_i/\theta_0 \geqslant \chi^2_{1-\alpha}(2n)$. Because this does not depend on the particular value of θ_1, but only on the fact that $\theta_1 > \theta_0$, it follows that it is a UMP test of $H_0 : \theta = \theta_0$ versus $H_a : \theta > \theta_0$. Note also that the power function for this test can be expressed in terms of a chi-square CDF, $H(c; \nu)$, with $\nu = 2n$. In particular,

$$\pi(\theta) = 1 - H[(\theta_0/\theta)\chi^2_{1-\alpha}(2n); 2n] \tag{12.7.3}$$

because $(\theta_0/\theta)[2 \sum X_i/\theta_0] = 2 \sum X_i/\theta \sim \chi^2(2n)$ when θ is the true value. Because $\pi(\theta)$ is an increasing function of θ, $\max_{\theta \leqslant \theta_0} \pi(\theta) = \pi(\theta_0) = \alpha$, and the test is also a UMP test of size α for the composite hypotheses $H_0 : \theta \leqslant \theta_0$ versus $H_a : \theta > \theta_0$.

Similarly, a UMP test of either $H_0 : \theta = \theta_0$ or $H_0 : \theta \geqslant \theta_0$ versus $H_a : \theta < \theta_0$ is to reject H_0 if $2n\bar{x}/\theta_0 \leqslant \chi^2_{\alpha}(2n)$, and the associated power function is

$$\pi(\theta) = H[(\theta_0/\theta)\chi^2_{\alpha}(2n); 2n] \tag{12.7.4}$$

In Example 4.6.3, observed lifetimes of 40 electrical parts were given, and it was conjectured that these observations might be exponentially distributed with mean lifetime $\theta = 100$ months. In a particular application, suppose that the parts will be unsuitable if the mean is less than 100. We will carry out a size $\alpha = 0.05$ test of

$H_0 : \theta \geqslant 100$ versus $H_a : \theta < 100$. The sample mean is $\bar{x} = 93.1$, and consequently $2n\bar{x}/\theta_0 = (80)(93.1)/100 = 74.48 > 60.39 = \chi^2_{0.05}(80)$, which means that we cannot reject H_0 at the 0.05 level of significance. Suppose that we wish to know $P[\text{TII}]$ if, in fact, the mean is $\theta = 50$ months. According to function (12.7.4), $\pi(50) = H(120.78; 80)$. Table 5 (Appendix C) is generally useful for determining such quantities, although in this case $v = 80$ exceeds the values given in the table. From Table 4 we can determine that the power is between 0.995 and 0.999, because $116.32 < 120.78 < 124.84$; however, an approximate value can be found from the approximation provided with Table 5. Specifically, $H(120.28; 80) \doteq 0.9978$, so $P[\text{TII}] \doteq 1 - 0.9978 = 0.0022$, and Type II error is quite unlikely for this alternative.

Note that for the two-sided composite alternative $H_a : \theta \neq \theta_0$, it is not possible to find a test that is UMP for every alternative value. For an alternative value $\theta_1 > \theta_0$ a right-tail critical region is optimal, but if the true θ is $\theta_2 < \theta_0$ then the right-tail critical region is very poor, and vice versa. As suggested earlier, we could compromise in this case and take a two-sided critical region, but it is not most powerful for any particular alternative. It is possible to extend the concept of unbiasedness to tests of hypotheses, and in the restricted class of unbiased tests there may be a UMP unbiased test for a two-sided composite alternative. This concept will be discussed briefly later.

It is easy to see that the other Neyman-Pearson tests illustrated in Examples 12.6.2 and 12.6.3 also provide UMP tests for the corresponding one-sided composite alternatives. General results along these lines can be stated for any pdf that satisfies a property known as the "monotone likelihood ratio."

For the sake of brevity, most of the distributional results in the rest of the chapter will be stated in vector notation. For example, if $\mathbf{X} = (X_1, \ldots, X_n)$, then $\mathbf{X} \sim f(\mathbf{x}; \theta)$ will mean that $X_1, \ldots, X_n$ have joint pdf $f(x_1, \ldots, x_n; \theta)$.

Definition 12.7.2

A joint pdf $f(\mathbf{x}; \theta)$ is said to have a **monotone likelihood ratio** (MLR) in the statistic $T = \ell(\mathbf{X})$ if for any two values of the parameter, $\theta_1 < \theta_2$, the ratio $f(\mathbf{x}; \theta_2)/f(\mathbf{x}; \theta_1)$ depends on $\mathbf{x}$ only through the function $\ell(\mathbf{x})$, and this ratio is a nondecreasing function of $\ell(\mathbf{x})$.

Notice that the MLR property also will hold for any increasing function of $\ell(\mathbf{x})$.

Example 12.7.2 Consider a random sample of size n from an exponential distribution, $X_i \sim \text{EXP}(\theta)$. Because $f(\mathbf{x}; \theta) = (1/\theta)^n \exp(-\sum x_i/\theta)$, we have

$$\frac{f(\mathbf{x}; \theta_2)}{f(\mathbf{x}; \theta_1)} = (\theta_1/\theta_2)^n \exp\left[-\left(\frac{1}{\theta_2} - \frac{1}{\theta_1}\right) \sum x_i\right]$$

which is a nondecreasing function of $\ell(\mathbf{x}) = \sum x_i$ if $\theta_2 > \theta_1$. Thus, $f(\mathbf{x}; \theta)$ has the MLR property in the statistic $T = \sum X_i$. Notice that the MLR property also holds for the statistic $\bar{X}$, because it is an increasing function of T.

The MLR property is useful in deriving UMP tests.

Theorem 12.7.1 If a joint pdf $f(\mathbf{x}; \theta)$ has a monotone likelihood ratio in the statistic $T = \ell(\mathbf{X})$, then a UMP test of size α for $H_0 : \theta \leqslant \theta_0$ versus $H_a : \theta > \theta_0$ is to reject H_0 if $\ell(\mathbf{x}) \geqslant k$, where $P[\ell(\mathbf{X}) \geqslant k \mid \theta_0] = \alpha$. ■

The dual problem of testing $H_0 : \theta \geqslant \theta_0$ versus $H_a : \theta < \theta_0$ also can be handled by the MLR approach, but the inequalities in Theorem 12.7.1 should be reversed. Also, if the ratio is a nonincreasing function of $\ell(\mathbf{x})$, then H_0 of the theorem can be rejected with the inequalities in $\ell(\mathbf{x})$ reversed.

In many applications, the terms nondecreasing and nonincreasing can be replaced with the terms increasing and decreasing respectively, but not in all applications, as the next example demonstrates.

Example 12.7.3 Consider a random sample of size n from a two-parameter exponential distribution, $X_i \sim \text{EXP}(1, \eta)$. The joint pdf is

$$f(\mathbf{x}; \eta) = \begin{cases} \exp\left[-\sum (x_i - \eta)\right] & \eta < x_{1:n} \\ 0 & x_{1:n} \leqslant \eta \end{cases}$$

If $\eta_1 < \eta_2$, then

$$\frac{f(\mathbf{x}; \eta_2)}{f(\mathbf{x}; \eta_1)} = \begin{cases} 0 & \eta_1 < x_{1:n} \leqslant \eta_2 \\ \exp\left[n(\eta_2 - \eta_1)\right] & \eta_2 < x_{1:n} \end{cases}$$

That this function is not defined for $x_{1:n} \leqslant \eta_1$ is not a problem, because $P[X_{1:n} \leqslant \eta_1] = 0$ when η_1 is the true value of η. Thus, the ratio is a nondecreasing function of $x_{1:n}$, and the MLR property holds for $T = X_{1:n}$. According to Theorem 12.7.1, a UMP test of size α for $H_0 : \eta \leqslant \eta_0$ versus $H_a : \eta > \eta_0$ is to reject H_0 if $x_{1:n} \geqslant k$, where $\alpha = P[X_{1:n} \geqslant k \mid \eta_0] = \exp\left[-n(k - \eta_0)\right]$, and thus $k = \eta_0 - (\ln \alpha)/n$.

Theorem 12.7.2 Suppose that $X_1, \ldots, X_n$ have joint pdf of the form

$$f(\mathbf{x}; \theta) = c(\theta)h(\mathbf{x}) \exp\left[q(\theta)\ell(\mathbf{x})\right] \tag{12.7.5}$$

where $q(\theta)$ is an increasing function of θ.

1. A UMP test of size α for $H_0 : \theta \leqslant \theta_0$ versus $H_a : \theta > \theta_0$ is to reject H_0 if $\ell(\mathbf{x}) \geqslant k$, where $P[\ell(\mathbf{X}) \geqslant k \mid \theta_0] = \alpha$.

2. A UMP test of size α for $H_0 : \theta \geqslant \theta_0$ versus $H_a : \theta < \theta_0$ is to reject H_0 if $\ell(x) \leqslant k$, where $P[\ell(X) \leqslant k \mid \theta_0] = \alpha$.

Proof

If $\theta_1 < \theta_2$, then $q(\theta_1) < q(\theta_2)$, so that

$$\frac{f(x;\, \theta_2)}{f(x;\, \theta_1)} = \frac{c(\theta_2)}{c(\theta_1)} \exp\{[q(\theta_2) - q(\theta_1)]\ell(x)\}$$

which is an increasing function of $\ell(x)$ because $q(\theta_2) - q(\theta_1) > 0$. The theorem follows by the MLR property. ■

An obvious application of the theorem occurs when $X_1, \ldots, X_n$ is a random sample from a member of the regular exponential class, say $f(x;\, \theta) = c(\theta)h(x) \exp[q(\theta)u(x)]$ with $\ell(x) = \sum u(x_i)$ and $q(\theta)$ an increasing function of θ.

Example 12.7.4 Consider a random sample of size n from a Poisson distribution, $X_i \sim \text{POI}(\mu)$. The joint pdf is

$$f(x;\, \mu) = \frac{e^{-n\mu}\mu^{\Sigma x_i}}{x_1! \cdots x_n!} \qquad \text{all } x_i = 0, 1, \ldots$$

$$= e^{-n\mu}(x_1! \cdots x_n!)^{-1} \exp[(\ln \mu) \textstyle\sum x_i]$$

The theorem applies with $q(\mu) = \ln \mu$ and $\ell(x) = \sum x_i$. A UMP test of size α for $H_0 : \mu \leqslant \mu_0$ versus $H_a : \mu > \mu_0$ would reject H_0 if $T = \sum X_i \geqslant k$ where $P[T \geqslant k \mid \mu_0] = \alpha$. Because $T \sim \text{POI}(n\mu)$, we must have $\sum_{t=k}^{\infty} \exp(-n\mu_0)(n\mu_0)^t / t! = \alpha$. We again have a discreteness problem, but the tests described in Theorem 12.7.2 are UMP for the particular values of α that can be achieved.

As mentioned earlier, there is a close relationship between tests and confidence intervals. If one tests $H_0 : \theta = \theta_0$ against $H_a : \theta \neq \theta_0$ at the α significance level, then for a given sample, the set of θ_0 for which H_0 would not be rejected represents a $100(1 - \alpha)\%$ confidence region for θ. Loosely speaking, if the acceptance set of a size α test is an interval, then it is a $100(1 - \alpha)\%$ confidence interval. Thus, one approach to find a confidence interval is first to derive an associated test by one of the techniques discussed earlier. Goodness properties of confidence intervals usually are expressed in terms of the associated test. For example, a confidence region associated with a UMP test is termed **uniformly most accurate** (UMA).

UNBIASED TESTS

It was mentioned earlier that in some cases where a UMP test may not exist, particularly for a two-sided alternative, there may exist a UMP test among the restricted class of "unbiased" tests.

Definition 12.7.3

A test of $H_0 : \theta \in \mathbf{\Omega}_0$ versus $H_a : \theta \in \mathbf{\Omega} - \mathbf{\Omega}_0$ is **unbiased** if

$$\min_{\theta \in \mathbf{\Omega} - \mathbf{\Omega}_0} \pi(\theta) \geqslant \max_{\theta \in \mathbf{\Omega}_0} \pi(\theta) \tag{12.7.6}$$

In other words, the probability of rejecting H_0 when it is false is at least as large as the probability of rejecting H_0 when it is true.

Example 12.7.5 Consider a random sample of size n from a normal distribution with mean zero, $X_i \sim N(0, \sigma^2)$. It is desired to test $H_0 : \sigma^2 = \sigma_0^2$ versus a two-sided alternative, $H_a : \sigma^2 \neq \sigma_0^2$, based on the test statistic $S_0 = \sum X_i^2/\sigma_0^2$. Under H_0 we know that $S_0 \sim \chi^2(n)$, so an equal-tailed critical region, similar to that of part 3 of Theorem 12.3.3, would reject H_0 if $s_0 \leqslant \chi^2_{\alpha/2}(n)$ or $s_0 \geqslant \chi^2_{1-\alpha/2}(n)$. In particular, consider a sample of size $n = 2$ and a test of size $\alpha = 0.05$ for $H_0 : \sigma^2 = 1$. The graph of the power function for this test is given in Figure 12.5.

FIGURE 12.5 The power function of a biased test.

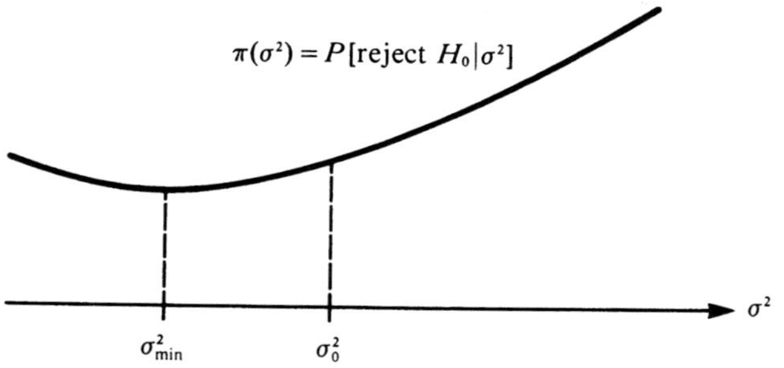

The minimum value of $\pi(\sigma^2)$ occurs at a value $\sigma^2 \neq \sigma_0^2$, and thus the test is less likely to reject H_0 for some values $\sigma^2 \neq \sigma_0^2$ than it is when $\sigma^2 = \sigma_0^2$. Consequently, the test is not unbiased. It is possible to construct an unbiased two-sided test if we abandon the convenient equal-tailed test in favor of one with a particular choice of critical values $\chi^2_{\alpha_1}(n)$ and $\chi^2_{1-\alpha_2}(n)$ with $\alpha_1 + \alpha_2 = \alpha$, but $\alpha_1 \neq \alpha_2$ (see Exercise 26). Such a test is not very convenient to use, and the biased equal-tailed test usually is preferred in practice.

It can be shown that the test described above is a UMP test among the restricted class of unbiased tests of H_0. In fact, it can be shown, under certain conditions, that for joint pdf's of the form given in equation (12.7.5), a uniformly most powerful unbiased (UMPU) test of $H_0: \theta = \theta_0$ versus $H_a: \theta \neq \theta_0$ exists. Methods for deriving UMPU tests are given by Lehmann (1959), but they will not be discussed here.

12.8

GENERALIZED LIKELIHOOD RATIO TESTS

The Neyman-Pearson lemma provides a method for deriving a most powerful test of simple hypotheses, and quite often this test also will be UMP for a one-sided composite alternative. Two theorems that were stated in the previous section also are useful in deriving UMP tests when there is a single unknown parameter.

Methods for deriving tests also are needed when unknown nuisance parameters are present or in other situations where the methods for determining a UMP test do not appear applicable. For example, in a two-sample normal problem, one may test the hypothesis that the means are equal, $H_0: \mu_1 = \mu_2$, without specifying them to be equal to a particular value. We discussed several natural tests in the first five sections, which were mostly suggested by analogous confidence-interval results. Of course, some of these tests also are UMP tests, but it is clear more generally that a test can be based on any statistic for which a critical region of the desired size can be determined. The problem then is to choose a good test statistic and a reasonable form for the critical region to obtain a test with high power. If the distribution depends on a single unknown parameter, then a single sufficient statistic or an MLE may be available, and a test could be based on this statistic. For a multiparameter problem, a test statistic might be some function of joint sufficient statistics or joint MLEs, but it may not always be clear what test statistic would be most suitable. Of course, the distribution of the statistic must be such that the size of the critical region can be computed and not depend on unknown parameters. For example, Student's t statistic may be used to test the mean of a normal distribution when the variance is unknown. Given a suitable test statistic, then, as with the Neyman-Pearson lemma, sample points should be included in the critical region that are less likely to occur under H_0 and more likely to occur under H_a.

The generalized likelihood ratio test is a generalization of the Neyman-Pearson test, and it provides a desirable test in many applications.

Definition 12.8.1

Let $\mathbf{X} = (X_1, \ldots, X_n)$ where $X_1, \ldots, X_n$ have joint pdf $f(\mathbf{x}; \boldsymbol{\theta})$ for $\boldsymbol{\theta} \in \boldsymbol{\Omega}$, and consider the hypothesis $H_0 : \boldsymbol{\theta} \in \boldsymbol{\Omega}_0$ versus $H_a : \boldsymbol{\theta} \in \boldsymbol{\Omega} - \boldsymbol{\Omega}_0$. The **generalized likelihood ratio** (GLR) is defined by

$$\lambda(\mathbf{x}) = \frac{\max\limits_{\boldsymbol{\theta} \in \boldsymbol{\Omega}_0} f(\mathbf{x}; \boldsymbol{\theta})}{\max\limits_{\boldsymbol{\theta} \in \boldsymbol{\Omega}} f(\mathbf{x}; \boldsymbol{\theta})} = \frac{f(\mathbf{x}; \hat{\boldsymbol{\theta}}_0)}{f(\mathbf{x}; \hat{\boldsymbol{\theta}})} \tag{12.8.1}$$

where $\hat{\boldsymbol{\theta}}$ denotes the usual MLE of $\boldsymbol{\theta}$, and $\hat{\boldsymbol{\theta}}_0$ denotes the MLE under the restriction that H_0 is true.

In other words, $\hat{\boldsymbol{\theta}}$ and $\hat{\boldsymbol{\theta}}_0$ are obtained by maximizing $f(\mathbf{x}; \boldsymbol{\theta})$ over $\boldsymbol{\Omega}$ and $\boldsymbol{\Omega}_0$, respectively. The generalized likelihood ratio test is to reject H_0 if $\lambda(\mathbf{x}) \leqslant k$, where k is chosen to provide a size α test.

Another, slightly different approach is to maximize over $\boldsymbol{\theta} \in \boldsymbol{\Omega} - \boldsymbol{\Omega}_0$ in the denominator, but the form in Definition (12.8.1) often is easier to evaluate and yields equivalent results.

Essentially, the GLR principle determines the critical region and associated test, by deciding which points will be included according to the ratio of estimated likelihoods of the observed data, where the numerator is estimated under the restriction that H_0 is true. This is similar to the Neyman-Pearson principle where the likelihoods are completely specified, but it is not, strictly speaking, a complete generalization of the Neyman-Pearson principle, because the unrestricted estimate, $\hat{\boldsymbol{\theta}}$, could possibly be in $\boldsymbol{\Omega}_0$.

We see that $\lambda(\mathbf{X})$ is a valid test statistic that is not a function of unknown parameters; in many cases the distribution of $\lambda(\mathbf{X})$ is free of parameters, and the exact critical value k can be determined. In some cases the distribution of $\lambda(\mathbf{X})$ under H_0 depends on unknown parameters, and an exact size α critical region cannot be determined. If regularity conditions hold, which ensure that the MLEs are asymptotically normally distributed, then it can be shown that the asymptotic distribution of $\lambda(\mathbf{X})$ is free of parameters, and an approximate size α test will be available for large n. In particular, if $\mathbf{X} \sim f(\mathbf{x}; \theta_1, \ldots, \theta_k)$, then under $H_0 : (\theta_1, \ldots, \theta_r) = (\theta_{10}, \ldots, \theta_{r0})$, $r < k$, approximately, for large n,

$$-2 \ln \lambda(\mathbf{X}) \sim \chi^2(r) \tag{12.8.2}$$

Thus, an approximate size α test is to reject H_0 if

$$-2 \ln \lambda(\mathbf{x}) \geqslant \chi^2_{1-\alpha}(r) \tag{12.8.3}$$

Example 12.8.1 Suppose that $X_i \sim N(\mu, \sigma^2)$, where σ^2 is known, and we wish to test $H_0 : \mu = \mu_0$ against $H_a : \mu \neq \mu_0$. The usual (unrestricted) MLE is $\hat{\mu} = \bar{x}$, and the GLR is

$$\lambda(\boldsymbol{x}) = \frac{f(\boldsymbol{x}; \mu_0)}{f(\boldsymbol{x}; \hat{\mu})}$$

$$= \frac{(2\pi\sigma^2)^{-n/2} \exp\left[-\sum (x_i - \mu_0)^2/2\sigma^2\right]}{(2\pi\sigma^2)^{-n/2} \exp\left[-\sum (x_i - \bar{x})^2/2\sigma^2\right]}$$

which, after some simplification, can be expressed as

$$\lambda(\boldsymbol{x}) = \exp\left[-n(\bar{x} - \mu_0)^2/2\sigma^2\right]$$

Rejecting H_0 if $\lambda(\boldsymbol{x}) \leqslant k$ is equivalent to rejecting H_0 if

$$z^2 = \left[\frac{(\bar{x} - \mu_0)}{\sigma/\sqrt{n}}\right]^2 \geqslant k_1$$

where $Z \sim N(0, 1)$ and $Z^2 \sim \chi^2(1)$. Thus, a size α test is to reject H_0 if

$$z^2 \geqslant \chi^2_{1-\alpha}(1)$$

or equivalently, reject H_0 if

$$z \leqslant -z_{1-\alpha/2} \qquad \text{or} \qquad z \geqslant z_{1-\alpha/2}$$

Thus the likelihood ratio test may be reduced to the usual two-sided equal-tailed normal test. It is interesting to note that the asymptotic approximation, $-2 \ln \lambda(\boldsymbol{X}) \sim \chi^2(1)$, is exact in this example.

Example 12.8.2 We now consider the hypotheses $H_0 : \mu = \mu_0$ against $H_a : \mu > \mu_0$. For practical purposes, the GLR test in this case reduces to the one-sided UMP test based on z; however, there is one technical difference. In this case, the MLE relative to $\boldsymbol{\Omega} = [\mu_0, \infty)$ is

$$\hat{\mu} = \begin{cases} \bar{x} & \bar{x} > \mu_0 \\ \mu_0 & \bar{x} \leqslant \mu_0 \end{cases}$$

Because the size of the test, α, usually is quite small, and we will be rejecting H_0 for large $\bar{x}$, ordinarily we will be concerned only with determining a critical region for the GLR statistic for the case when $\bar{x} > \mu_0$.

Specifically, we have

$$\lambda(\boldsymbol{x}) = \begin{cases} \exp\left[-n(\bar{x} - \mu_0)^2/2\sigma_0^2\right] & \bar{x} > \mu_0 \\ 1 & \bar{x} \leqslant \mu_0 \end{cases}$$

but under H_0, $P[\lambda(\boldsymbol{X}) < 1] = P[\bar{X} > \mu_0] = 0.5$. So, for $\alpha < 0.5$, $k < 1$, and the critical region will not contain any $\boldsymbol{x}$ such that $\lambda(\boldsymbol{x}) = 1$. In particular, the

GLR test is to reject H_0 if $\bar{x} > \mu_0$ and $\lambda(\mathbf{x}) \leqslant k$; or $\bar{x} > \mu_0$ and $z^2 = [\sqrt{n}(\bar{x} - \mu_0)/\sigma]^2 \geqslant k_1$ when $\bar{x} > \mu_0$, $z^2 \geqslant k_1$ if and only if $z \geqslant \sqrt{k_1}$. Thus, a size α test (for $\alpha < 0.5$) is to reject H_0 if $z \geqslant z_{1-\alpha}$, which also is the UMP test for these hypotheses.

The GLR test of $H_0 : \mu \leqslant \mu_0$ against $H_a : \mu > \mu_0$ is somewhat similar. In this case, $\hat{\mu} = \bar{x}$, but maximizing the likelihood function over $\mathbf{\Omega}_0$ gives

$$\hat{\mu}_0 = \begin{cases} \bar{x} & \bar{x} \leqslant \mu_0 \\ \mu_0 & \bar{x} > \mu_0 \end{cases}$$

Thus,

$$\lambda(\mathbf{x}) = \frac{f(\mathbf{x}; \hat{\mu}_0)}{f(\mathbf{x}; \hat{\mu})} = \begin{cases} 1 & \bar{x} \leqslant \mu_0 \\ \exp\left[-n(\bar{x} - \mu_0)^2/2\sigma^2\right] & \bar{x} > \mu_0 \end{cases}$$

This is the same result as obtained for testing the simple $H_0 : \mu = \mu_0$ against $H_a : \mu > \mu_0$. The same critical region also gives a size α test for the composite null hypothesis $H_0 : \mu \leqslant \mu_0$, because $z = (\bar{x} - \mu_0)/(\sigma/\sqrt{n})$, $P[Z \geqslant z_{1-\alpha} \mid \mu_0] = \alpha$, and $P[Z \geqslant z_{1-\alpha} \mid \mu] < \alpha$ when $\mu < \mu_0$.

It sometimes is desired to test a hypothesis about one unknown parameter in the presence of another unknown nuisance parameter.

Example 12.8.3 Suppose now that $X_i \sim N(\mu, \sigma^2)$ where σ^2 is assumed unknown, and we wish to test $H_0 : \mu = \mu_0$ against $H_a : \mu \neq \mu_0$. This does not represent a simple null hypothesis, because the distribution is not specified completely under H_0. The parameter space is two-dimensional, $\mathbf{\Omega} = (-\infty, \infty) \times (0, \infty) = \{(\mu, \sigma^2) \mid -\infty < \mu < \infty \text{ and } \sigma^2 > 0\}$, and $H_0 : \mu = \mu_0$ is an abbreviated notation for $H_0 : (\mu, \sigma^2) \in \mathbf{\Omega}_0$ where $\mathbf{\Omega}_0 = \{\mu_0\} \times (0, \infty) = \{(\mu, \sigma^2) \mid \mu = \mu_0 \text{ and } \sigma^2 > 0\}$. These sets are illustrated in Figure 12.6.

FIGURE 12.6 A subset $\mathbf{\Omega}_0$ of hypothesized values μ_0

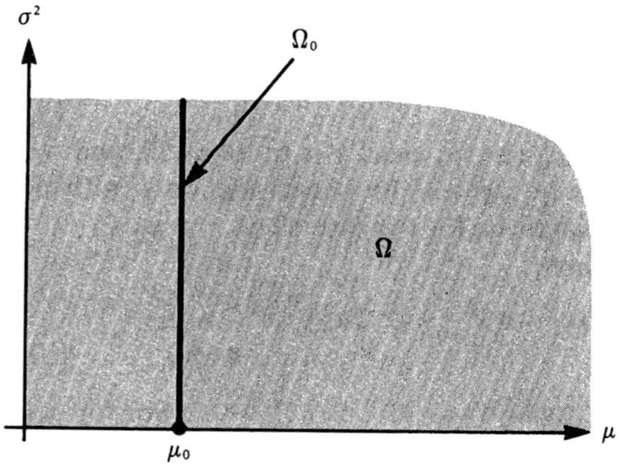

Maximizing $f(\boldsymbol{x};\mu,\sigma^2)$ over $\boldsymbol{\Omega}$ yields the usual MLEs $\hat{\mu}=\bar{x}$ and $\hat{\sigma}^2=\sum(x_i-\bar{x})^2/n$, but over $\boldsymbol{\Omega}_0$ we obtain $\hat{\mu}_0=\mu_0$ and $\hat{\sigma}_0^2=\sum(x_i-\mu_0)^2/n$. Thus,

$$\begin{aligned}
\lambda(\boldsymbol{x}) &= \frac{f(\boldsymbol{x};\mu_0,\hat{\sigma}_0^2)}{f(\boldsymbol{x};\hat{\mu},\hat{\sigma}^2)}\\
&= \frac{(2\pi\hat{\sigma}_0^2)^{-n/2}\exp\left[-\sum(x_i-\mu_0)^2/2\hat{\sigma}_0^2\right]}{(2\pi\hat{\sigma}^2)^{-n/2}\exp\left[-\sum(x_i-\bar{x})^2/2\hat{\sigma}^2\right]}\\
&= [\hat{\sigma}_0^2/\hat{\sigma}^2]^{-n/2}
\end{aligned}$$

and consequently,

$$\begin{aligned}
[\lambda(\boldsymbol{x})]^{-2/n} &= \frac{\sum(x_i-\mu_0)^2}{\sum(x_i-\bar{x})^2}\\
&= \frac{\sum(x_i-\bar{x})^2+n(\bar{x}-\mu_0)^2}{\sum(x_i-\bar{x})^2}\\
&= 1+\frac{n(\bar{x}-\mu_0)^2}{\sum(x_i-\bar{x})^2}\\
&= 1+t^2(\boldsymbol{x})/(n-1)
\end{aligned}$$

where $t(\boldsymbol{x})=\sqrt{n}(\bar{x}-\mu_0)/\sqrt{\sum(x_i-\bar{x})^2/(n-1)}=\sqrt{n}(\bar{x}-\mu_0)/s$. Under $H_0: \mu=\mu_0$, $T=t(\boldsymbol{X})\sim t(n-1)$ and $T^2\sim F(1,n-1)$. Thus, rejecting H_0 when $\lambda(\boldsymbol{x})$ is small is equivalent to rejecting H_0 if T^2 is large, and a size α test is to reject H_0 if $t^2\geqslant f_{1-\alpha}(1,n-1)$, or alternatively if

$$t\leqslant -t_{1-\alpha/2}(n-1) \qquad \text{or} \qquad t\geqslant t_{1-\alpha/2}(n-1)$$

Thus the two-sided test proposed earlier based on Student's t pivotal quantity is now seen to be a GLR test. This two-sided test is not UMP, but it can be shown to be UMPU.

The GLR approach also can be used to derive tests for two-sample problems.

Example 12.8.4 Suppose that $X\sim \text{BIN}(n_1,p_1)$ and $Y\sim \text{BIN}(n_2,p_2)$ with X and Y independent, and we wish to test whether the proportions are equal, $H_0: p_1=p_2=p$ against $H_a: p_1\neq p_2$, where p is unknown. The parameter space is $\boldsymbol{\Omega}=(0,1)\times(0,1)$ $=\{(p_1,p_2)\,|\,0<p_1<1 \text{ and } 0<p_2<1\}$, and the subset corresponding to H_0 is $\boldsymbol{\Omega}_0=\{(p_1,p_2)\,|\,0<p_1=p_2<1\}$. These sets are illustrated in Figure 12.7.

FIGURE 12.7 A subset of hypothesized values $p_1 = p_2$

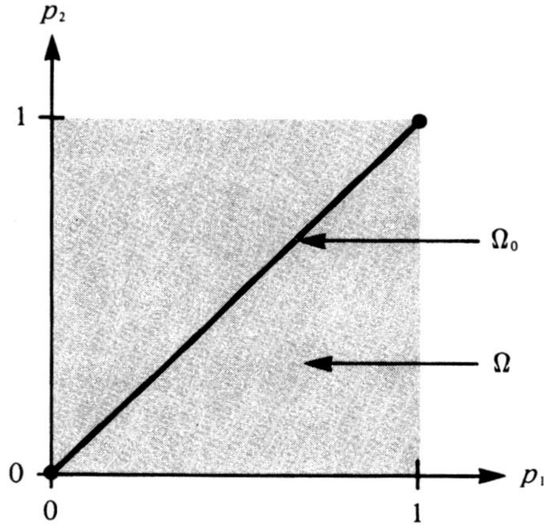

Based on x and y, the MLEs are $\hat{p}_1 = x/n_1$, $\hat{p}_2 = y/n_2$ over $\boldsymbol{\Omega}$, and $\hat{p}_0 = (x + y)/(n_1 + n_2)$ over $\boldsymbol{\Omega}_0$. The GLR statistic is

$$\lambda(x, y) = \frac{f(x; \hat{p}_0)f(y; \hat{p}_0)}{f(x; \hat{p}_1)f(y; \hat{p}_2)}$$

$$= \frac{\binom{n_1}{x}\hat{p}_0^x(1 - \hat{p}_0)^{n_1 - x}\binom{n_2}{y}\hat{p}_0^y(1 - \hat{p}_0)^{n_2 - y}}{\binom{n_1}{x}\hat{p}_1^x(1 - \hat{p}_1)^{n_1 - x}\binom{n_2}{y}\hat{p}_2^y(1 - \hat{p}_2)^{n_2 - y}}$$

Except for the cancellation of the binomial coefficients, this particular GLR statistic does not appear to simplify greatly, but it can be computed easily. The distribution of $\lambda(X, Y)$ will depend on p under H_0, however, so an exact size α critical region cannot be determined. The chi-square approximation should be useful here for large sample sizes. The hypothesis H_0 represents a one-dimensional restriction in the parameter space, because it corresponds to $H_0 : \theta = p_2 - p_1 = 0$; so $r = 1$ in expression (12.8.2), and approximately for a size α test,

$$-2 \ln \lambda(X, Y) \sim \chi^2(1)$$

and H_0 is rejected if $-2 \ln \lambda(x, y) > \chi^2_{1-\alpha}(1)$.

The more common approach in this case is to use the approximate normal test statistic given by expression (12.4.2). It also is interesting to note that an exact conditional test can be constructed in this case, as discussed in the next section.

k-SAMPLE TESTS

The GLR approach also is helpful in deriving important tests involving samples from two or more normal distributions.

Suppose that independent random samples are obtained from k normal distributions. For each $j = 1, \ldots, k$, we denote the mean and variance of the jth distribution by μ_j and σ_j^2, respectively, and we denote by n_j the sample size. Also, we denote by x_{ij} the observed value of the ith random variable from the jth random sample, $X_{ij} \sim N(\mu_j, \sigma_j^2)$. We will denote the individual jth sample means and sample variances, respectively, by $\bar{x}_j = \sum_i x_{ij}/n_j$ and $s_j^2 = \sum_i (x_{ij} - \bar{x}_j)^2/(n_j - 1)$. It also will be convenient to adopt a notation for the mean of the pooled samples, namely $\bar{\bar{x}} = \sum_j \sum_i x_{ij}/N = \sum_j n_j \bar{x}_j/N$ with $N = n_1 + \cdots + n_k$.

A test for the equality of distribution means now will be derived, assuming a common but unknown variance, say $\sigma_j^2 = \sigma^2$. Specifically, we will derive the GLR test of $H_0 : \mu_1 = \cdots = \mu_k$ versus the alternative that $\mu_i \neq \mu_j$ for at least one pair $i \neq j$. The likelihood is

$$L = L(\mu_1, \ldots, \mu_k, \sigma^2) = \prod_j \prod_i \frac{1}{\sqrt{2\pi}\sigma} \exp\left[-\frac{1}{2\sigma^2}(x_{ij} - \mu_j)^2\right]$$

$$= (2\pi\sigma^2)^{-N/2} \exp\left[-\frac{1}{2\sigma^2}\sum_j \sum_i (x_{ij} - \mu_j)^2\right]$$

which yields the following log-likelihood:

$$\ln L = -\frac{N}{2}\ln(2\pi\sigma^2) - \frac{1}{2\sigma^2}\sum_j \sum_i (x_{ij} - \mu_j)^2$$

Relative to the parameter space $\boldsymbol{\Omega} = \{(\mu_1, \ldots, \mu_k, \sigma^2) \mid -\infty < \mu_j < \infty, \sigma^2 > 0\}$, we compute the MLEs by taking partials of $\ln L$ with respect to each of the $k + 1$ parameters and equating them to zero

$$\frac{\partial}{\partial \mu_j}\ln L = -\frac{1}{2\sigma^2}\sum_i 2(x_{ij} - \mu_j)(-1) = 0 \qquad j = 1, \ldots, k$$

$$\frac{\partial}{\partial \sigma^2}\ln L = -\frac{N}{2}\frac{1}{\sigma^2} - \frac{1}{2(\sigma^2)^2}\sum_j \sum_i (x_{ij} - \mu_j)^2(-1) = 0$$

Solving these equations simultaneously we obtain the MLEs

$$\hat{\mu}_j = \sum_i x_{ij}\Big/ n_j = \bar{x}_j \qquad \hat{\sigma}^2 = \sum_j \sum_i (x_{ij} - \bar{x}_j)^2\Big/ N$$

For the subspace $\boldsymbol{\Omega}_0$ under H_0, the μ_j's have a common but unknown value, say μ, and the MLEs are obtained by equating to zero the partials of $\ln L$ with respect to μ and σ^2:

$$\frac{\partial}{\partial \mu} \ln L = -\frac{1}{2\sigma^2} \sum_j \sum_i 2(x_{ij} - \mu)(-1) = 0$$

$$\frac{\partial}{\partial \sigma^2} \ln L = -\frac{N}{2}\frac{1}{\sigma^2} - \frac{1}{2(\sigma^2)^2} \sum_j \sum_i (x_{ij} - \mu)^2(-1) = 0$$

leading to the solutions

$$\hat{\mu}_0 = \sum_j n_j \bar{x}_j \Big/ N = \bar{\bar{x}} \qquad \hat{\sigma}_0^2 = \sum_j \sum_i (x_{ij} - \bar{\bar{x}})^2 \Big/ N$$

Thus, the GLR is

$$\lambda(\boldsymbol{x}) = \frac{L(\hat{\mu}_0, \hat{\sigma}_0^2)}{L(\hat{\mu}_1, \ldots, \hat{\mu}_k, \hat{\sigma}^2)} = \frac{(2\pi\hat{\sigma}_0^2)^{-N/2} \exp\left[-\dfrac{1}{2\hat{\sigma}_0^2} \sum\limits_j \sum\limits_i (x_{ij} - \bar{\bar{x}})^2\right]}{(2\pi\hat{\sigma}^2)^{-N/2} \exp\left[-\dfrac{1}{2\hat{\sigma}^2} \sum\limits_j \sum\limits_i (x_{ij} - \bar{x}_j)^2\right]}$$

$$= \frac{(2\pi\hat{\sigma}_0^2)^{-N/2} \exp(-N/2)}{(2\pi\hat{\sigma}^2)^{-N/2} \exp(-N/2)}$$

$$= (\hat{\sigma}_0^2/\hat{\sigma}^2)^{-N/2}$$

$$= \left[\frac{\sum\limits_j \sum\limits_i (x_{ij} - \bar{\bar{x}})^2}{\sum\limits_j \sum\limits_i (x_{ij} - \bar{x}_j)^2}\right]^{-N/2} \tag{12.8.4}$$

It is possible to express this test in terms of an F-distributed statistic. To show this we first note the following identity (see Exercise 40):

$$\sum_j \sum_i (x_{ij} - \bar{\bar{x}})^2 = \sum_j \sum_i (x_{ij} - \bar{x}_j)^2 + \sum_j n_j(\bar{x}_j - \bar{\bar{x}})^2 \tag{12.8.5}$$

From this and equation (12.8.4), it follows that

$$\lambda(\boldsymbol{x}) = \left[1 + \frac{\sum\limits_j n_j(\bar{x}_j - \bar{\bar{x}})^2}{\sum\limits_j \sum\limits_i (x_{ij} - \bar{x}_j)^2}\right]^{-N/2}$$

which means that the GLR test is equivalent to rejecting H_0 if for some $c > 0$

$$\frac{\sum\limits_j n_j(\bar{x}_j - \bar{\bar{x}})^2}{\sum\limits_j \sum\limits_i (x_{ij} - \bar{x}_j)^2} \geqslant c \tag{12.8.6}$$

Next we define variables $V_1 = \sum_j \sum_i (x_{ij} - \bar{\bar{x}})^2/\sigma^2$, $V_2 = \sum_j \sum_i (x_{ij} - \bar{x}_j)^2/\sigma^2$ and $V_3 = \sum_j n_j(\bar{x}_j - \bar{\bar{x}})^2/\sigma^2$, and note that we can write $V_2 = \sum_j (n_j - 1)s_j^2/\sigma^2$. Thus, V_2

is a function of the sample variances and V_3 is a function of the sample means. Because the $\bar{x}_j$'s and s_j^2's are independent, it follows that V_2 and V_3 also are independent. Furthermore, from identity (12.8.5) we know that $V_1 = V_2 + V_3$, so from the results of Chapter 8 about sums of independent chi-square variables, we have under H_0 that $V_1 \sim \chi^2(N-1)$ and $V_2 \sim \chi^2(N-k)$, from which it follows that $V_3 \sim \chi^2[(N-1)-(N-k)] = \chi^2(k-1)$. Thus, the ratio on the left of equation (12.8.6) is proportional to an F-distributed statistic, namely

$$F = \frac{\sum n_j(\bar{x}_j - \bar{\bar{x}})^2/(k-1)}{\sum_j \sum_i (x_{ij} - \bar{x}_j)^2/(N-k)} \sim F(k-1, N-k) \tag{12.8.7}$$

Finally, we note that the denominator of equation (12.8.7) is a pooled estimate of σ^2, namely $s_p^2 = \sum_j (n_j - 1)s_j^2/(N-k)$.

The above remarks are summarized in the following theorem:

Theorem 12.8.1 For k normal populations with common variance, $N(\mu_j, \sigma^2)$, $j = 1, \ldots, k$, a size α test of $H_0 : \mu_1 = \cdots = \mu_k$ is to reject H_0 if

$$F = \frac{\sum n_j(\bar{x}_j - \bar{\bar{x}})^2/(k-1)}{s_p^2} > f_{1-\alpha}(k-1, N-k)$$

where $s_p^2 = \sum_{j=1}^{k} (n_j - 1)s_j^2/(N-k)$ and $N = \sum n_j$. Furthermore, this test is equivalent to the GLR test. ■

An important application of this theorem involves testing the effects of k different experimental treatments. For example, suppose it is desired to test whether k different brands of plant food are equally effective in promoting growth in garden plants. If μ_j is the mean growth per plant using brand j, then the test of Theorem 12.8.1 would be appropriate for testing whether the different brands of plant food are equivalent in this respect.

This test also is related to a procedure called **analysis of variance**. This terminology is motivated by the identity (12.8.5). The term on the left reflects total variability of the pooled sample data. On the other hand, the first term on the right reflects variation "within" the individual samples, while the second term reflects variation "between" the samples. Strictly speaking, this corresponds to a one-way analysis of variance because it only considers one factor, such as the brand of plant food. It also is possible to consider a second factor in the experiment, such as the amount of water applied to plants. The appropriate procedure in this case is called a two-way analysis of variance, but we will not pursue this point.

The GLR approach also can be used to derive a test of equality of variances, $H_0 : \sigma_1^2 = \cdots = \sigma_k^2$. We will not present this derivation, but an equivalent test and an approximate distribution that is useful in performing such a test are given in the following theorems.

Theorem 12.8.2 Let

$$M = M(v_j, w_j) = N \ln\left[N^{-1} \sum_{j=1}^{k} v_j w_j\right] - \sum_{j=1}^{k} v_j \ln w_j$$

and let

$$c = 1 + \frac{1}{3(k+1)}\left[\sum_{j=1}^{k} \frac{1}{v_j} - \frac{1}{N}\right]$$

where $N = \sum_{j=1}^{k} v_j$ and $v_j w_j \sim \chi^2(v_j)$; then approximately

$$M/c \sim \chi^2(k-1) \qquad \text{if } v_j \geqslant 4$$ ■

For smaller v_j critical values may be determined from Tables 31 and 32 of Pearson and Hartley (1958).

Now for normal samples, $(n_j - 1)s_j^2/\sigma_j^2 \sim \chi^2(n_j - 1)$, and if one lets $w_j = s_j^2/\sigma_j^2$ in M, then the σ_j^2 will cancel out if they are all equal; thus $M(n_j - 1, s_j^2)$ may be used as a test statistic for testing $H_0 : \sigma_1^2 = \cdots = \sigma_k^2$, and its distribution may be expressed in terms of the chi-squared distribution under H_0. This statistic is minimized when all of the observed s_j^2 are equal, and the statistic becomes larger for unequal s_j^2, which favors the alternative.

Theorem 12.8.3 For k normal populations $N(\mu_j, \sigma_j^2)$, $j = 1, \ldots, k$, an approximate size α test of $H_0 : \sigma_1^2 = \sigma_2^2 = \cdots = \sigma_k^2$ against the alternative of at least one inequality is to reject H_0 if

$$M(n_j - 1, s_j^2) > c\chi_{1-\alpha}^2(k-1)$$ ■

12.9

CONDITIONAL TESTS

It sometimes is possible to eliminate unknown nuisance parameters and obtain exact size α tests by considering tests based on conditional variables. For example, if a sufficient statistic S exists for an unknown nuisance parameter θ, then the distribution of $X \mid S$ will not depend on θ. This technique will be illustrated for the two-sample binomial test.

Example 12.9.1 Again let $X \sim \text{BIN}(n_1, p_1)$ and $Y \sim \text{BIN}(n_2, p_2)$ with X and Y independent. We wish a size α test of $H_0: p_1 = p_2 = p$ against $H_a: p_1 < p_2$. Under H_0, the joint density of X and Y is

$$f(x, y) = \binom{n_1}{x}\binom{n_2}{y} p^{x+y}(1-p)^{n_1+n_2-(x+y)}$$

and it is clear that $S = X + Y$ is sufficient for the common unknown p in this density. This suggests considering a test based on the conditional distribution of (X, Y) given $S = s$. Because $Y = S - X$, it suffices to base the test on the conditional distribution of Y given $S = s$. Under H_0, $S \sim \text{BIN}(n_1 + n_2, p)$, and thus

$$\begin{aligned} f_{Y|s}(y) &= \frac{f_{S,Y}(s, y)}{f_S(s)} \\ &= \frac{f_{X,Y}(s-y, y)}{f_S(s)} \\ &= \frac{\binom{n_1}{s-y}\binom{n_2}{y} p^s(1-p)^{n_1+n_2-s}}{\binom{n_1+n_2}{s} p^s(1-p)^{n_1+n_2-s}} \end{aligned}$$

$$= \frac{\binom{n_2}{y}\binom{n_1}{s-y}}{\binom{n_1+n_2}{s}} \qquad y = 0, \ldots, s;\ s = 0, \ldots, n_1 + n_2$$

which is a hypergeometric distribution. This distribution does not involve p, and an exact size α critical region can be determined under H_0 for any given observed value of s. For $H_a: p_1 < p_2$ the best critical region would be for large y. Thus, reject H_0 if $y \geqslant k(s)$, or for a size α test, reject H_0 if

$$\sum_{i=y}^{s} \frac{\binom{n_2}{i}\binom{n_1}{s-i}}{\binom{n_1+n_2}{s}} \leqslant \alpha$$

Tests for other alternatives can be obtained in a similar manner. Except for the discreteness problem, this provides an exact size α test. In other words, it is exact for values of α that the above sum can attain. Otherwise, the test is conservative for the prescribed α.

The following theorem is useful for constructing tests for hypotheses concerning a parameter θ in the presence of nuisance parameters, $\boldsymbol{\kappa} = (\kappa_1, \ldots, \kappa_m)$.

Theorem 12.9.1 Let $X = (X_1, \ldots, X_n)$ where $X_1, \ldots, X_n$ have joint pdf of the form

$$f(x; \theta, \boldsymbol{\kappa}) = c(\theta, \boldsymbol{\kappa})h(x) \exp\left[\theta \ell(x) + \sum_{i=1}^{m} \kappa_i \mathscr{s}_i(x)\right] \tag{12.9.1}$$

If $S_i = \mathscr{s}_i(X)$ for $i = 1, \ldots, m$, and $T = \ell(X)$, then $S_1, \ldots, S_m$ are jointly sufficient for $\kappa_1, \ldots, \kappa_m$ for each fixed θ, and the conditional pdf $f_{T|s}(t; \theta)$ does not depend on $\boldsymbol{\kappa}$. Furthermore,

1. A size α test of $H_0: \theta \leqslant \theta_0$ versus $H_a: \theta > \theta_0$ is to reject H_0 if $\ell(x) \geqslant k(s)$ where $P[T \geqslant k(s) \mid s] = \alpha$ when $\theta = \theta_0$.
2. A size α test of $H_0: \theta \geqslant \theta_0$ versus $H_a: \theta < \theta_0$ is to reject H_0 if $\ell(x) \leqslant k(s)$ where $P[T \leqslant k(s) \mid s] = \alpha$ when $\theta = \theta_0$. ■

Under certain regularity conditions on equation (12.9.1), it also is possible to show that these tests are UMP unbiased tests. For more details, see Lehmann (1959).

Example 12.9.2 Consider a random sample from a gamma distribution, $X_i \sim \text{GAM}(\lambda, \kappa)$. If we reparameterize with $\theta = 1/\lambda$, then the joint pdf is

$$\begin{aligned} f(x; \theta, \kappa) &= \frac{\theta^{n\kappa}}{\Gamma^n(\kappa)} \left(\prod x_i\right)^{\kappa - 1} \exp\left(-\theta \sum x_i\right) \\ &= c(\theta, \kappa)h(x) \exp\left[\theta\left(-\sum x_i\right) + (\kappa - 1) \ln\left(\prod x_i\right)\right] \end{aligned}$$

where $h(x) = 1$ if all $x_i > 0$, and 0 otherwise. According to the theorem, $S = \ln\left(\prod X_i\right)$ is sufficient for any fixed θ, and if $T = -\sum X_i$, then the distribution of T given $S = s$ does not depend on κ. The conditional pdf $f_{T|s}(t)$ is quite complicated in this example, but tables that can be used to perform an equivalent test are given by Engelhardt and Bain (1977).

Note that a conditional size α test also is a size α test unconditionally, because, for example, if $P[T \geqslant k(s) \mid s] = \alpha$, then

$$\begin{aligned} P[T \geqslant k(S)] &= E_S\{P[T \geqslant k(S) \mid S]\} \\ &= E_S(\alpha) \\ &= \alpha \end{aligned}$$

12.10

SEQUENTIAL TESTS

We found earlier that for a fixed sample size n the Neyman-Pearson approach could be used to construct most powerful size α tests for simple hypotheses. Also, in some cases, formulas are available for computing the sample size n that yields a size α test with a specified power, or equivalently, a specified level of Type II error, say β. In this section we consider a sequential test procedure in which the sample size is not fixed in advance.

Example 12.10.1 The force (in pounds) that will cause a certain type of ceramic automobile part to break is normally distributed with mean μ and standard deviation $\sigma = 40$. The original factory parts are known to have a mean breaking strength of 100 pounds. A manufacturer of replacement parts claims that its parts are better than the originals, and that the mean breaking strength of the replacement parts is 120 pounds. To demonstrate its claim, a test is proposed in which the breaking strength is measured for a fixed number n of the replacement parts, yielding data $x_1, \ldots, x_n$. Based on the resulting data, the manufacturer decides to perform a size $\alpha = 0.05$ test of $H_0 : \mu \leqslant 100$ versus $H_a : \mu > 100$. We first consider the question of determining a fixed sample size based on methods discussed earlier in the chapter. It follows from Theorem 12.7.1 that a UMP test of size α for $H_0 : \mu \leqslant \mu_0$ versus $H_a : \mu > \mu_0$ rejects H_0 if $\bar{x} \geqslant c$ for an appropriate choice of c. Furthermore, we have from Theorem 12.3.1 that this is equivalent to rejecting H_0 if $z_0 = \sqrt{n}(\bar{x} - \mu_0)/\sigma \geqslant z_{1-\alpha}$, and that the sample size required to achieve a test with a specified $\beta = P[\text{Type II error}]$ is given by $n = (z_{1-\alpha} + z_{1-\beta})^2\sigma^2/(\mu_0 - \mu_1)^2$. Thus, to have a size $\alpha = 0.05$ test of the replacement parts with $\beta = 0.10$ when $\mu = 120$, a sample of size $n = (1.645 + 1.282)^2(40)^2/(100 - 120)^2 \doteq 34$ is required.

Obviously, the cost of such a project will depend on the number of parts tested, which might lead one to seek a procedure that requires that fewer parts be tested. One possibility is to consider a sequential test. In other words, it might be possible to devise a procedure such that testing of the first few parts would produce sufficient evidence to accept or reject H_0 without the need for further testing.

SEQUENTIAL PROBABILITY RATIO TESTS

Consider the situation of testing a simple null hypothesis $H_0 : \theta = \theta_0$ against a simple alternative hypothesis $H_a : \theta = \theta_1$. If $X_1, \ldots, X_n$ is a random sample of size n from a distribution with pdf $f(x; \theta)$, then we know from the Neyman-Pearson lemma that a most powerful critical region is determined by the inequality

$$\lambda = \frac{f(x_1; \theta_0) \cdots f(x_n; \theta_0)}{f(x_1; \theta_1) \cdots f(x_n; \theta_1)} \leqslant k$$

where k is a positive constant.

A **sequential probability ratio test** (SPRT) is defined in terms of a sequence of such ratios. Specifically, we define

$$\lambda_m = \lambda_m(x_1, \ldots, x_m) = \frac{f(x_1; \theta_0) \cdots f(x_m; \theta_0)}{f(x_1; \theta_1) \cdots f(x_m; \theta_1)}$$

for $m = 1, 2, \ldots$, and adopt the following procedure: Let $k_0 < k_1$ be arbitrary positive numbers, and compute λ_1 based on the first observation x_1. If $\lambda_1 \leqslant k_0$, then reject H_0; if $\lambda_1 \geqslant k_1$, then accept H_0; and if $k_0 < \lambda_1 < k_1$, then take a second observation x_2 and compute λ_2. Similarly, if $\lambda_2 \leqslant k_0$, then reject H_0; if $\lambda_2 \geqslant k_1$, then accept H_0; and if $k_0 < \lambda_2 < k_1$, then take a third observation x_3 and compute λ_3, and so on. The idea is to continue taking x_i's as long as the ratio λ_m remains between k_0 and k_1, and to stop as soon as either $\lambda_m \leqslant k_0$ or $\lambda_m \geqslant k_1$, rejecting H_0 if $\lambda_m \leqslant k_0$ and accepting H_0 if $\lambda_m \geqslant k_1$. The critical region, say C, of the resulting sequential test is the union of the following disjoint sets:

$$C_n = \{(x_1, \ldots, x_n) \mid k_0 < \lambda_j(x_1, \ldots, x_j) < k_1, j = 1, \ldots, n-1, \\ \lambda_n(x_1, \ldots, x_n) \leqslant k_0\}$$

for $n = 1, 2, \ldots$.

In other words, if for some n, a point $(x_1, \ldots, x_n)$ is in C_n, then H_0 is rejected for a sample of size n. On the other hand, H_0 is accepted if such a point is in an acceptance region, say A, which is the union of disjoint sets A_n of the following form:

$$A_n = \{(x_1, \ldots, x_n) \mid k_0 < \lambda_j(x_1, \ldots, x_j) < k_1, j = 1, \ldots, n-1, \\ \lambda_n(x_1, \ldots, x_n) \geqslant k_1\}$$

In the case of the Neyman-Pearson test for fixed sample size n, the constant k was determined so that the size of the test would be some prescribed α. Now it is necessary to find constants k_0 and k_1 so that the SPRT will have prescribed values α and β for the respective probabilities of Type I and Type II error,

$$\alpha = P[\text{reject } H_0 \mid \theta_0] = \sum_{n=1}^{\infty} \int_{C_n} L_n(\theta_0) \tag{12.10.1}$$

and

$$\beta = P[\text{accept } H_0 \mid \theta_1] = \sum_{n=1}^{\infty} \int_{A_n} L_n(\theta_1) \tag{12.10.2}$$

where $L_n(\theta) = f(x_1;\ \theta) \cdots f(x_n;\ \theta)$, and the integral notations are defined as follows:

$$\int_{C_n} L_n(\theta_0) = \int \underset{C_n}{\cdots} \int f(x_1;\ \theta_0) \cdots f(x_n;\ \theta_0)\, dx_1 \cdots dx_n$$

and

$$\int_{A_n} L_n(\theta_1) = \int \underset{A_n}{\cdots} \int f(x_1;\ \theta_1) \cdots f(x_n;\ \theta_1)\, dx_1 \cdots dx_n$$

The constants k_0 and k_1 are solutions of the integral equations (12.10.1) and (12.10.2), and, as might be expected, an exact determination of these constants is not trivial. Fortunately, there is a rather simple approximation available that we will consider shortly.

Before we proceed, there are a number of points to consider about SPRTs. In particular, because the sample size depends on the observed values of the sequence of random variables $X_1, X_2, \ldots$, it is itself a random variable, say N. As one might suspect, the distribution of N is quite complicated, and we will not derive it. Another concern is the possibility that testing could continue indefinitely. Although we will not attempt a proof, it can be shown that a SPRT will terminate in a finite number of steps. Specifically, it can be shown that $N < \infty$ with probability 1. Of course, another point that was raised earlier concerns whether the size of the sample can be reduced by using a sequential test rather than a test with a fixed sample size. We will discuss this latter point after we consider the question of approximations for k_0 and k_1.

APPROXIMATE SEQUENTIAL TESTS

Suppose it is required to perform a sequential test with prescribed probabilities of Type I and Type II errors, α and β, respectively. As noted above, the constants k_0 and k_1 can be obtained by solving the integral equations (12.10.1) and (12.10.2), and exact solutions, in general, will be difficult to achieve. Fortunately, it is possible to obtain approximate solutions that are much easier to compute and rather accurate. If α and β are the exact levels desired, then we define constants $k_0^* = \dfrac{\alpha}{1-\beta}$ and $k_1^* = \dfrac{1-\alpha}{\beta}$.

The following discussion suggests using k_0^* and k_1^* as approximations for k_0 and k_1. Using the above stated property that $N < \infty$ with probability 1 and that $\lambda_n(x_1, \ldots, x_n) \leqslant k_0$ when $(x_1, \ldots, x_n)$ is in C_n, it follows that

$$\begin{aligned}\alpha = P[\text{reject } H_0 \mid \theta_0] &= \sum_{n=1}^{\infty} \int_{C_n} L_n(\theta_0) \leqslant \sum_{n=1}^{\infty} \int_{C_n} k_0 L_n(\theta_1) \\ &= k_0 \sum_{n=1}^{\infty} \int_{C_n} L_n(\theta_1) \\ &= k_0 P[\text{reject } H_0 \mid \theta_1] \\ &= k_0(1-\beta)\end{aligned}$$

and hence $\alpha/(1-\beta) \leqslant k_0$. Similarly, because $\lambda_n(x_1, \ldots, x_n) \geqslant k_1$ when $(x_1, \ldots, x_n)$ is in A_n, it follows that

$$\begin{aligned}1-\alpha = P[\text{accept } H_0 \mid \theta_0] &= \sum_{n=1}^{\infty} \int_{A_n} L_n(\theta_0) \geqslant \sum_{n=1}^{\infty} \int_{A_n} k_1 L_n(\theta_1) \\ &= k_1 P[\text{accept } H_0 \mid \theta_1] \\ &= k_1 \beta\end{aligned}$$

and hence $k_1 \leqslant (1-\alpha)/\beta$. These results imply the inequality $k_0^* \leqslant k_0 < k_1 \leqslant k_1^*$.

A relationship now will be established between the errors for the exact test and those of the approximate test. Denote by α^* and β^* the actual error sizes of the approximate SPRT based on using the constants k_0^* and k_1^*. Also, denote by C_n^* and A_n^* the sets that define, respectively, the critical and acceptance regions for the approximate test based on k_0^* and k_1^*. It follows by an argument similar to that given above that

$$\alpha^* = \sum_{n=1}^{\infty} \int_{C_n^*} L_n(\theta_0) \leqslant \frac{\alpha}{1-\beta} \sum_{n=1}^{\infty} \int_{C_n^*} L_n(\theta_1) = \frac{\alpha}{1-\beta}(1-\beta^*)$$

and

$$1 - \alpha^* = \sum_{n=1}^{\infty} \int_{A_n^*} L_n(\theta_0) \geqslant \frac{1-\alpha}{\beta} \sum_{n=1}^{\infty} \int_{A_n^*} L_n(\theta_1) = \frac{1-\alpha}{\beta} \beta^*$$

It follows that $\alpha^*(1-\beta) \leqslant \alpha(1-\beta^*)$ and $(1-\alpha)\beta^* \leqslant (1-\alpha^*)\beta$, and consequently that $\alpha^*(1-\beta) + (1-\alpha)\beta^* \leqslant \alpha(1-\beta^*) + (1-\alpha^*)\beta$, which, after simplification yields the inequality:

$$\alpha^* + \beta^* \leqslant \alpha + \beta$$

Thus, if the experimenter uses the approximate SPRT based on the constants $k_0^* = \alpha/(1-\beta)$ and $k_1^* = (1-\alpha)/\beta$ rather than the exact SPRT, then based on the constants k_0 and k_1, the sum of the errors of the approximate test are bounded above by the sum of the error of the exact test.

EXPECTED SAMPLE SIZE

We now consider a way of assessing the effectiveness of SPRTs in reducing the amount of sampling relative to tests based on fixed sample sizes. Our criterion involves the expected number of observations required to reach a decision.

As before, we denote by N the number of observations required to reach a decision, either reject H_0 or accept H_0. Theoretically, we might attempt to compute its expectation directly from the definition, but as noted previously the distribution of N is quite complicated and thus we will resort to a different approach. Recall that the test is based on observed values of a sequence of random variables $X_1, X_2, \ldots, X_n$ which are independent and identically distributed with pdf $f(x; \theta)$. Theoretically, we could continue taking observations indefinitely, but according to the sequential procedure defined above, we will terminate as soon as $\lambda_n \leqslant k_0$ or $\lambda_n \geqslant k_1$ for some n, and we define N as the first such value n.

We now define a new random variable, say $Z = \ln f(X; \theta_0) - \ln f(X; \theta_1)$ where $X \sim f(x; \theta)$ for either $\theta = \theta_0$ or θ_1. In a similar manner, we can define a whole sequence of such random variables $Z_1, Z_2, \ldots,$ based on the sequence $X_1, X_2, \ldots$ and we also can define a sequence of sums, $S_m = \sum_{i=1}^{m} Z_i$ for $m \geqslant 1$.

Notice that these sums are related to the likelihood ratios:

$$S_m = \ln [\lambda_m(X_1, \ldots, X_m)] \qquad m = 1, 2, \ldots$$

It follows that N is the subscript of the first sum S_n such that either $S_n \leqslant \ln (k_0)$ or $S_n \geqslant \ln (k_1)$, and we denote the corresponding sum as $S_N = Z_1 + \cdots + Z_N$. It is possible to show that $E(S_N) = E(N)E(Z)$ when $E(N) < \infty$. This relationship, which is known as Wald's equation, is useful in deriving an approximation to the expected sample size. We will not attempt to prove it here.

If the sequential test rejects H_0 at step N, then $S_N \leqslant \ln (k_0)$, and we would expect the sum to be close to $\ln (k_0)$, because it first dropped below this value at the Nth step. Similarly, if the test accepts H_0 at step N, then $S_N \geqslant \ln (k_1)$, and we would expect the sum to be close to $\ln (k_1)$ in this case. These remarks together with Wald's equation suggest the following approximation:

$$E(N) = \frac{E(S_N)}{E(Z)} \cong \frac{\ln (k_0)P[\text{reject } H_0] + \ln (k_1)P[\text{accept } H_0]}{E(Z)}$$

By using the approximations $k_0 \cong k_0^* = \alpha/(1 - \beta)$ and $k_1 \cong k_1^* = (1 - \alpha)/\beta$, we obtain the following approximation to expected sample size when H_0 is true:

$$E(N \mid \theta_0) \cong \frac{\alpha \ln [\alpha/(1 - \beta)] + (1 - \alpha) \ln [(1 - \alpha)/\beta]}{E(Z \mid \theta_0)}$$

Similarly, an approximation when H_a is true is given by

$$E(N \mid \theta_1) \cong \frac{(1 - \beta) \ln [\alpha/(1 - \beta)] + \beta \ln [(1 - \alpha)/\beta]}{E(Z \mid \theta_1)}$$

Example 12.10.2 We consider again the problem of Example 12.10.1, which dealt with the force (in pounds) required to break a certain type of ceramic part whose breaking strength is normally distributed with mean μ and standard deviation $\sigma = 40$. Suppose we wish to test the simple null hypothesis $H_0 : \mu = 100$ versus the simple alternative $H_a : \mu = 120$ with $\alpha = 0.05$ and $\beta = 0.10$. Thus, the approximate critical values for a SPRT are $k_0^* = 0.05/(1 - 0.10) \doteq 0.056$ and $k_1^* = (1 - 0.05)/0.10 = 9.5$, and such a test would reject H_0 as soon as $\lambda_n \leqslant 0.056$, and accept H_0 as soon as $\lambda_n \geqslant 9.5$. In this case, it also is possible to express the test in terms of the sum of the data. Specifically, because $f(x; \mu) = \dfrac{1}{\sqrt{2\pi}\sigma} \exp\left[-\dfrac{1}{2\sigma^2} (x - \mu)^2 \right]$, we can write

$$z_i = \ln f(x_i\,;\mu_0) - \ln f(x_i\,;\mu_1)$$
$$= \frac{1}{2\sigma^2}(x_i - \mu_1)^2 - \frac{1}{2\sigma^2}(x_i - \mu_0)^2$$
$$= \frac{1}{2\sigma^2}[2x_i(\mu_0 - \mu_1) - (\mu_0^2 - \mu_1^2)]$$
$$= \frac{\mu_0 - \mu_1}{2\sigma^2}[2x_i - (\mu_0 + \mu_1)]$$

It follows that

$$s_n = \sum_{i=1}^{n} z_i = \frac{\mu_0 - \mu_1}{2\sigma^2}\left[2\sum_{i=1}^{n} x_i - n(\mu_0 + \mu_1)\right]$$

which is a linear function of $\sum x_i$. Thus, the criterion of stopping the test if $s_n \leqslant \ln(k_0^*)$ or $s_n \geqslant \ln(k_1^*)$ is equivalent to stopping if $\sum x_i \leqslant c_0(n)$ or $\sum x_i \geqslant c_1(n)$ with $c_0(n)$ and $c_1(n)$ determined by n, k_0^*, k_1^*, μ_0, μ_1, and σ^2.

It also would be interesting to approximate the expected sample size and compare this to the sample size required to achieve the corresponding test with a fixed sample size. The expression given above for z_i also can be used to obtain $E(Z)$. Specifically, it follows that

$$E(Z) = \frac{\mu_0 - \mu_1}{2\sigma^2}[2E(X) - (\mu_0 + \mu_1)]$$

and thus,

$$E(Z \mid \mu_i) = \frac{\mu_0 - \mu_1}{2\sigma^2}[2\mu_i - (\mu_0 + \mu_1)] \qquad i = 0, 1$$

As a result, we have that $E(Z \mid \mu_i) = (-1)^i(\mu_0 - \mu_1)^2/(2\sigma^2)$ for $i = 0, 1$, and in our example,

$$E(N \mid \mu_0 = 100)$$
$$\cong \frac{0.05 \ln[0.05/(1-0.10)] + (1-0.05)\ln[(1-0.05)/0.10]}{(100-120)^2/[2(40)^2]}$$
$$\doteq 16$$

and

$$E(N \mid \mu_1 = 120)$$
$$\cong \frac{(1-0.10)\ln[0.05/(1-0.10)] + 0.10 \ln[(1-0.05)/0.10]}{-(100-120)^2/[2(40)^2]}$$
$$\doteq 19$$

For the SPRT, the expected sample sizes 16 and 19 under H_0 and H_1, respectively, compare to the sample size $n = 34$ for the corresponding Neyman-Pearson test considered in Example 12.10.1.

For additional reading about sequential tests, the book by Ghosh (1970) is recommended.

SUMMARY

Our purpose in this chapter was to introduce the concept of hypothesis testing, which corresponds to a process of attempting to determine the truth or falsity of specified statistical hypotheses on the basis of experimental evidence. A statistical hypothesis is a statement about the distribution of the random variable that models the characteristic of interest. If the hypothesis completely specifies the distribution, then it is called a simple hypothesis; otherwise it is called composite. A Type I error occurs when a true null hypothesis is rejected by the test, and a Type II error occurs when a test fails to reject a false null hypothesis. It is not possible to avoid an occasional decision error, but it is possible in many situations to design tests that lead to such errors with a specified relative frequency.

If a test is based on a set of data consisting of n measurements, then the critical region (or rejection region) is a subset of an n-dimensional Euclidean space. The null hypothesis is rejected when the n-dimensional vector of data is contained in the critical region. Often it is possible to express the critical region in terms of a test statistic. For a simple hypothesis, the significance level is the probability of committing a Type I error. In the case of a composite hypothesis, the size of the test (or size of the associated critical region) is the largest probability of a Type I error relative to all distributions specified in the null hypothesis.

The power function gives the probability of rejecting a false null hypothesis for the different alternative values. By means of the Neyman-Pearson lemma, it is possible to derive a most powerful test of a given size, and in some cases this test is a UMP test.

In many cases, a UMP test cannot be obtained, but it is possible to derive reasonable tests by means of the generalized likelihood ratio approach. For example, this approach can be used in many cases to derive tests of hypotheses where nuisance parameters are present, and also in situations involving two-sided alternatives. In a few cases, it is possible to derive UMP unbiased tests that also can be used in nuisance parameter problems and with two-sided alternative hypotheses.

EXERCISES

1. Suppose $X_1, \ldots, X_{16}$ is a random sample of size $n = 16$ from a normal distribution, $X_i \sim \mathrm{N}(\mu, 1)$, and we wish to test $H_0 : \mu = 20$ at significance level $\alpha = 0.05$, based on the sample mean $\bar{X}$.

(a) Determine critical regions of the form $A = \{\bar{x} \mid -\infty < \bar{x} \leqslant a\}$ and $B = \{\bar{x} \mid b \leqslant \bar{x} < \infty\}$.

(b) Find the probability of a Type II error, $\beta = P[\text{TII}]$, for each critical region in (a) for the alternative $H_a : \mu = 21$. Which of these critical regions is unreasonable for this alternative?

(c) Rework (b) for the alternative $H_a : \mu = 19$.

(d) What is the significance level for a test with the critical region $A \cup B$?

(e) What is $\beta = P[\text{TII}]$ for a test with critical region $A \cup B$ if $|\mu - 20| = 1$?

2. Suppose a box contains four marbles, θ white ones and $4 - \theta$ black ones. Test $H_0 : \theta = 2$ against $H_a : \theta \neq 2$ as follows: Draw two marbles with replacement and reject H_0 if both marbles are the same color; otherwise do not reject.

(a) Compute the probability of Type I error.

(b) Compute the probability of Type II error for all possible situations.

(c) Rework (a) and (b) if the two marbles are drawn without replacement.

3. Consider a random sample of size 20 from a normal distribution, $X_i \sim \mathrm{N}(\mu, \sigma^2)$, and suppose that $\bar{x} = 11$ and $s^2 = 16$.

(a) Assuming it is known that $\sigma^2 = 4$, test $H_0 : \mu \geqslant 12$ versus $H_a : \mu < 12$ at the significance level $\alpha = 0.01$.

(b) What is $\beta = P[\text{TII}]$ if in fact $\mu = 10.5$?

(c) What sample size is needed for the power of the test to be 0.90 for the alternative value $\mu = 10.5$?

(d) Test the hypotheses of (a) assuming σ^2 unknown.

(e) Test $H_0 : \sigma^2 \leqslant 9$ against $H_a : \sigma^2 > 9$ with significance level $\alpha = 0.01$.

(f) What sample size is needed for the test of (e) to be 90% certain of rejecting H_0 when in fact $\sigma^2 = 18$? What is $\beta = P[\text{TII}]$ in this case?,

4. Consider the biased coin discussed in Example 9.2.5, where the probability of a head, p, is known to be 0.20, 0.30, or 0.80. The coin is tossed repeatedly, and we let X be the number of tosses required to obtain the first head. To test $H_0 : p = 0.80$, suppose we reject H_0 if $X \geqslant 3$, and do not reject otherwise.

(a) What is the probability of Type I error, $P[\text{TI}]$?

(b) What is the probability of a Type II error, $P[\text{TII}]$, for each of the other two values of p?

(c) For a test of $H_0 : p = 0.30$, suppose we use a critical region of the form $\{1, 14, 15, \ldots\}$. Find $P[\text{TI}]$, and also find $P[\text{TII}]$ for each of the other values of p.

5. It is desired to test the hypothesis that the mean melting point of an alloy is 200 degrees Celsius (°C) so that a difference of 20°C is detected with probability 0.95. Assuming normality, and with an initial guess that $\sigma = 25°C$, how many specimens of the alloy must be tested to perform a t test with $\alpha = 0.01$? Describe the null and alternative hypotheses, and the form of the associated critical region.

6. Let $X_1, \ldots, X_n$ be a random sample of size n from an exponential distribution, $X_i \sim \text{EXP}(1, \eta)$. A test of $H_0 : \eta \leqslant \eta_0$ versus $H_a : \eta > \eta_0$ is desired, based on $X_{1:n}$.
   (a) Find a critical region of size α of the form $\{x_{1:n} \geqslant c\}$.
   (b) Derive the power function for the test of (a).
   (c) Derive a formula to determine the sample size n for a test of size α with $\beta = P[\text{TII}]$ if $\eta = \eta_1$.

7. A coin is tossed 20 times and $x = 6$ heads are observed. Let $p = P(\text{head})$. A test of $H_0 : p \geqslant 0.5$ versus $H_a : p < 0.5$ of size at most 0.10 is desired.
   (a) Perform a test using Theorem 12.4.1.
   (b) Perform a test using Theorem 12.4.2.
   (c) What is the power of a size $\alpha = 0.0577$ test of $H_0 : p \geqslant 0.5$ for the alternative $p = 0.2$?
   (d) What is the p-value for the test in (b)? That is, what is the observed size?

8. Suppose that the number of defects in a piece of wire of length t yards is Poisson distributed, $X \sim \text{POI}(\lambda t)$, and one defect is found in a 100-yard piece of wire.
   (a) Test $H_0 : \lambda \geqslant 0.05$ against $H_a : \lambda < 0.05$ with significance level at most 0.01, by means of Theorem 12.5.1.
   (b) What is the p-value for such a test?
   (c) Suppose a total of two defects are found in two 100-yard pieces of wire. Test $H_0 : \lambda \geqslant 0.05$ versus $H_a : \lambda < 0.05$ at significance level $\alpha = 0.0103$.
   (d) Find the power of the test in (c) if $\lambda = 0.01$.

9. Consider independent random samples from two normal distributions, $X_i \sim N(\mu_1, \sigma_1^2)$ for $i = 1, \ldots, n_1$ and $Y_j \sim N(\mu_2, \sigma_2^2)$ for $j = 1, \ldots, n_2$. Let $n_1 = n_2 = 9$, $\bar{x} = 16$, $\bar{y} = 10$, $s_1^2 = 36$, and $s_2^2 = 45$.
   (a) Assuming equal variances, test $H_0 : \mu_1 = \mu_2$ against $H_a : \mu_1 \neq \mu_2$ at the $\alpha = 0.10$ level of significance.
   (b) Perform an approximate $\alpha = 0.10$ level test of $H_0 : \mu_1 = \mu_2$ against $H_a : \mu_1 \neq \mu_2$ using equation (11.5.13).
   (c) Perform a test of these hypotheses at the $\alpha = 0.10$ significance level using equation (11.5.17), assuming the data were obtained from paired samples with $s_D^2 = 81$.
   (d) Test $H_0 : \sigma_2^2/\sigma_1^2 \leqslant 1$ versus $H_a : \sigma_2^2/\sigma_1^2 > 1$ at the $\alpha = 0.05$ level.
   (e) Use Table 7 (Appendix C) to find the power of this test if $\sigma_2^2/\sigma_1^2 = 1.33$.

10. A certain type of component is manufactured by two different companies, and the respective probabilities of a nondefective component are p_1 and p_2. In random samples of 200 components each, 180 from company 1 are nondefective, and 190 from company 2 are nondefective. Test $H_0 : p_1 = p_2$ against $H_a : p_1 \neq p_2$ at significance level $\alpha = 0.05$.

11. Consider a distribution with pdf $f(x; \theta) = \theta x^{\theta - 1}$ if $0 < x < 1$ and zero otherwise.

(a) Based on a random sample of size $n = 1$, find the most powerful test of $H_0 : \theta = 1$ against $H_a : \theta = 2$ with $\alpha = 0.05$.

(b) Compute the power of the test in (a) for the alternative $\theta = 2$.

(c) Derive the most powerful test for the hypotheses of (a) based on a random sample of size n.

12. Suppose that $X \sim \text{POI}(\mu)$.

(a) Derive the most powerful test of $H_0 : \mu = \mu_0$ versus $H_a : \mu = \mu_1$ $(\mu_1 > \mu_0)$ based on an observed value of X.

(b) Rework (a) based on a random sample of size n.

13. Let $X \sim \text{NB}(r, 1/2)$.

(a) Derive the most powerful test of size $\alpha = 0.125$ of $H_0 : r = 1$ against $H_a : r = 2$ based on an observed value of X.

(b) Compute the power of this test for the alternative $r = 2$.

14. Assume that X is a discrete random variable.

(a) Based on an observed value of X, derive the most powerful test of H_0 : $X = \text{GEO}(0.05)$ versus $H_a : X \sim \text{POI}(0.95)$ with $\alpha = 0.0975$.

(b) Find the power of this test under the alternative.

15. Let $X_1, \ldots, X_n$ have joint pdf $f(x_1, \ldots, x_n; \theta)$ and S be a sufficient statistic for θ. Show that a most powerful test of $H_0 : \theta = \theta_0$ versus $H_a : \theta = \theta_1$ can be expressed in terms of S.

16. Consider a random sample of size n from a distribution with pdf $f(x; \theta) = (3x^2/\theta)e^{-x^3/\theta}$ if $0 < x$, and zero otherwise. Derive the form of the critical region for a uniformly most powerful (UMP) test of size α of $H_0 : \theta = \theta_0$ against $H_a : \theta > \theta_0$.

17. Suppose that $X_1, \ldots, X_n$ is a random sample from a normal distribution, $X_i \sim \text{N}(0, \sigma^2)$.

(a) Derive the UMP size α test of $H_0 : \sigma = \sigma_0$ against $H_a : \sigma > \sigma_0$.

(b) Express the power function of this test in terms of a chi-square distribution.

(c) If $n = 20$, $\sigma_0 = 1$, and $\alpha = 0.005$, use Table 5 (Appendix C) to compute the power of the test in (a) when $\sigma = 2$.

18. Consider a random sample of size n from a uniform distribution, $X_i \sim \text{UNIF}(0, \theta)$. Find the UMP test of size α of $H_0 : \theta \geqslant \theta_0$ versus $H_a : \theta < \theta_0$ by first deriving a most powerful test of simple hypotheses and then extending it to composite hypotheses.

19. Let $X_1, \ldots, X_n$ be a random sample from a normal distribution, $X_i \sim \text{N}(\mu, 1)$.

(a) Find a UMP test of $H_0 : \mu = \mu_0$ against $H_a : \mu < \mu_0$.

(b) Find a UMP test of $H_0 : \mu = \mu_0$ against $H_a : \mu > \mu_0$.

(c) Show that there is no UMP test of $H_0 : \mu = \mu_0$ against $H_a : \mu \neq \mu_0$.

20. Suppose that X is a continuous random variable with pdf $f(x; \theta) = 1 - \theta^2(x - 1/2)$ if $0 < x < 1$ and zero otherwise, where $-1 < \theta < 1$. Show that a UMP size α test of $H_0 : \theta = 0$ versus $H_a : \theta \neq 0$, based on an observed value x of X, is to reject H_0 if $x \leqslant \alpha$.

21. Consider a random sample $X_1, \ldots, X_n$ from a discrete distribution with pdf $f(x; \theta) = [\theta/(\theta + 1)]^x/(\theta + 1)$ if $x = 0, 1, \ldots$ where $\theta > 0$. Find a UMP test of $H_0: \theta = \theta_0$ against $H_a: \theta > \theta_0$.

22. Let $X_1, \ldots, X_n$ denote a random sample from a gamma distribution, $X_i \sim \text{GAM}(\theta, \kappa)$.

(a) If κ is known, derive a UMP size α test of $H_0: \theta \leqslant \theta_0$ against $H_a: \theta > \theta_0$.

(b) Express the power function for the test in (a) in terms of a chi-square distribution.

(c) If $n = 4$, $\kappa = 2$, $\theta_0 = 1$, and $\alpha = 0.01$, use Table 5 (Appendix C) to find the power of this test when $\theta = 2$.

(d) If θ is known, derive a UMP size α test of $H_0: \kappa \leqslant \kappa_0$ against $H_a: \kappa > \kappa_0$.

23. Consider a random sample of size n from a Bernoulli distribution, $X_i \sim \text{BIN}(1, p)$.

(a) Derive a UMP test of $H_0: p \leqslant p_0$ versus $H_a: p > p_0$ using Theorem 12.7.1.

(b) Derive the test in (a) using Theorem 12.7.2.

24. Suppose that $X_1, \ldots, X_n$ is a random sample from a Weibull distribution, $X_i \sim \text{WEI}(\theta, 2)$. Derive a UMP test of $H_0: \theta \geqslant \theta_0$ versus $H_a: \theta < \theta_0$ using Theorem 12.7.2.

25. Show that the test of Exercise 20 is unbiased.

26. Consider the hypotheses of Example 12.7.5, and consider a test with critical region of the form $C = \{(x_1, \ldots, x_n) \mid s_0 \leqslant c_1 \text{ or } s_0 \geqslant c_2\}$ where $s_0 = \sum x_i^2/\sigma_0^2$ and where c_1 and c_2 are chosen to provide a test of size α.

(a) Show that the power function of such a test has the form $\pi(\sigma^2) = 1 - H(c_2 \sigma_0^2/\sigma^2; n) + H(c_1 \sigma_0^2/\sigma^2; n)$ where $H(c; n)$ is the CDF of $\chi^2(n)$.

(b) Show that for this test to be unbiased, it is necessary that c_1 and c_2 satisfy the equations $H(c_2; n) - H(c_1; n) = 1 - \alpha$ and $c_2 h(c_2; n) - c_1 h(c_1; n) = 0$ where $h(c; n) = H'(c; n)$. *Hint:* For the test to be unbiased, the minimum of $\pi(\sigma^2)$ must occur at $\sigma^2 = \sigma_0^2$. In Example 12.7.5, $\alpha_1 = H(c_1; n)$ and $\alpha_2 = 1 - H(c_2; n)$.

27. Let $X_1, \ldots, X_n$ be a random sample from an exponential distribution, $X_i \sim \text{EXP}(\theta)$.

(a) Derive the generalized likelihood ratio (GLR) test of $H_0: \theta = \theta_0$ against $H_a: \theta \neq \theta_0$. Determine an approximate critical value for size α using the large-sample chi-square approximation.

(b) Derive the GLR test of $H_0: \theta = \theta_0$ against $H_a: \theta > \theta_0$.

28. Consider independent random samples of size n_1 and n_2 from respective exponential distributions, $X_i \sim \text{EXP}(\theta_1)$ and $Y_j \sim \text{EXP}(\theta_2)$. Derive the GLR test of $H_0: \theta_1 = \theta_2$ versus $H_a: \theta_1 \neq \theta_2$.

29. Let $X_1, \ldots, X_n$ be a random sample from a distribution with pdf $f(x; \theta) = 1/\theta$ if $0 \leqslant x \leqslant \theta$ and zero otherwise. Derive the GLR test of $H_0: \theta = \theta_0$ versus $H_a: \theta \neq \theta_0$.

30. Consider independent random samples of size n_1 and n_2 from respective normal distributions, $X_i \sim N(\mu_1, \sigma_1^2)$ and $Y_j \sim N(\mu_2, \sigma_2^2)$.

(a) Derive the GLR test of $H_0: \sigma_1^2 = \sigma_2^2$ against $H_a: \sigma_1^2 \neq \sigma_2^2$, assuming that μ_1 and μ_2 are known.

(b) Rework (a) assuming that μ_1 and μ_2 are unknown.

(c) Derive the GLR for testing $H_0 : \mu_1 = \mu_2$ and $\sigma_1^2 = \sigma_2^2$ against the alternative $H_0 : \mu_1 \neq \mu_2$ or $\sigma_1^2 \neq \sigma_2^2$.

31. Suppose that X is a continuous random variable with pdf $f(x; \theta) = \theta x^{\theta - 1}$ if $0 < x < 1$, and zero otherwise. Derive the GLR test of $H_0 : \theta = \theta_0$ against $H_a : \theta \neq \theta_0$ based on a random sample of size n. Determine an approximate critical value for a size α test based on a large-sample approximation.

32. To compare the effectiveness of three competing weight-loss systems, 10 dieters were assigned randomly to each system, and the results measured after six months. The following weight losses (in pounds) were reported:

System 1: 4.3, 10.2, 4.4, 23.5, 54.0, 5.7, 10.6, 47.3, 9.9, 37.5

System 2: 10.7, 6.4, 33.5, 54.1, 25.7, 11.6, 17.3, 9.4, 7.5, 5.0

System 3: 51.0, 5.6, 10.3, 47.3, 2.9, 27.5, 14.3, 1.2, 3.4, 13.5

(a) Assuming that the data are normally distributed, use the result of Theorem 12.8.1 to test the hypothesis that all three systems are equally effective in reducing weight.

(b) Use the results of Theorems 12.8.2 and 12.8.3 to test the hypothesis that the variance in weight loss is the same for all three systems.

33. Consider a random sample of size n from a distribution with pdf $f(x; \theta, \mu) = 1/2\theta$ if $|x - \mu| \leqslant \theta$, and zero otherwise. Test $H_0 : \mu = 0$ against $H_a : \mu \neq 0$. Show that the GLR $\lambda = \lambda(x)$ is given by

$$\lambda^{1/n} = (X_{n:n} - X_{1:n})/[2 \max(-X_{1:n}, X_{n:n})]$$

Note that in this case approximately $-2 \ln \lambda \sim \chi^2(2)$ because of equation (12.8.2).

34. Let $X_1, \ldots, X_n$ be a random sample from a continuous distribution.

(a) Show that the GLR for testing $H_0 : X \sim \mathrm{N}(\mu, \sigma^2)$ against $H_a : X \sim \mathrm{EXP}(\theta, \eta)$ is a function of $\hat{\theta}/\hat{\sigma}$. Is the distribution of this statistic free of unknown parameters under H_0?

(b) Show that the GLR for testing $H_0 : X \sim \mathrm{N}(\mu, \sigma^2)$ against $H_a : X \sim \mathrm{DE}(\theta, \eta)$ is a function of $\hat{\theta}/\hat{\sigma}$.

35. Consider a random sample of size n from a two-parameter exponential distribution, $X_i \sim \mathrm{EXP}(\theta, \eta)$, and let $\hat{\eta}$ and $\hat{\theta}$ be the MLEs.

(a) Show that $\hat{\eta}$ and $\hat{\theta}$ are independent. *Hint:* Use the results of Exercise 30 of Chapter 10.

(b) Let $V_1 = 2n(\bar{X} - \eta)/\theta$, $V_2 = 2n(\hat{\eta} - \eta)/\theta$, and $V_3 = 2n\hat{\theta}/\theta$. Show that $V_1 \sim \chi^2(2n)$, $V_2 \sim \chi^2(2)$, and $V_3 \sim \chi^2(2n - 2)$. *Hint:* Note that $V_1 = V_2 + V_3$ and that V_2 and V_3 are independent. Find the MGF of V_3 by the approach used in Theorem 8.3.6.

(c) Show that $(n - 1)(\hat{\eta} - \eta)/\hat{\theta} \sim F(2, 2n - 2)$.

(d) Derive the GLR for a test of $H_0 : \eta = \eta_0$ versus $H_a : \eta > \eta_0$.

(e) Show that the critical region for a size α GLR test is equivalent to $(n - 1)(\hat{\eta} - \eta_0)/\hat{\theta} \geqslant f_{1-\alpha}(2, 2n - 2)$.

36. Suppose that X and Y are independent with $X \sim \text{POI}(\mu_1)$ and $Y \sim \text{POI}(\mu_2)$, and let $S = X + Y$.

(a) Show that the conditional distribution of X given $S = s$ is $\text{BIN}(s, p)$ where $p = \mu_1/(\mu_1 + \mu_2)$.

(b) Use the result of (a) to construct a conditional test of $H_0 : \mu_1 = \mu_2$ versus $H_a : \mu_1 \neq \mu_2$.

(c) Construct a conditional test of $H_0 : \mu_1/\mu_2 = c_0$ versus $H_a : \mu_1/\mu_2 \neq c_0$ for some specified c_0.

37. Consider the hypotheses $H_0 : \sigma = 1$ versus $H_a : \sigma = 3$ when the distribution is normal with $\mu = 0$. If x_i denotes the ith sample value:

(a) Compute the approximate critical values k_1^* and k_2^* for a SPRT with $P[\text{Type I error}] = 0.10$ and $P[\text{Type II error}] = 0.05$.

(b) Derive the SPRT for testing these hypotheses.

(c) Find a sequential test procedure that is stated in terms of the sequence of sums $s_n = \sum_{i=1}^{n} x_i^2$ and is equivalent to the SPRT for testing H_0 against H_a.

(d) Find the approximate expected sample size for the test in (a) if H_0 is true. What is the approximate expected sample size if H_a is true?

(e) Suppose the first 10 values of x_i are: -2.20, 0.50, 2.55, -1.85, -0.45, -1.15, -0.58, 5.65, 0.49, and -1.16. Would the test in (a) terminate before more data is needed?

38. Suppose a population is Poisson distributed with mean μ. Consider a SPRT for testing $H_0 : \mu = 1$ versus $H_a : \mu = 2$.

(a) Express the SPRT in terms of the sequence of sums $s_n = \sum_{i=1}^{n} x_i$.

(b) Find the approximate expected sample size if H_a is true when $\alpha = 0.01$ and $\beta = 0.02$.

39. Gross and Clark (1975, page 105) consider the following relief times (in hours) of 20 patients who received an analgesic:

1.1, 1.4, 1.3, 1.7, 1.9, 1.8, 1.6, 2.2, 1.7, 2.7, 4.1, 1.8, 1.5, 1.2, 1.4, 3.0, 1.7, 2.3, 1.6, 2.0

(a) Assuming that the times were taken sequentially and that relief times are independent and exponentially distributed, $X_i \sim \text{EXP}(\theta)$, use an approximate SPRT to test the hypotheses $H_0 : \theta = 2.0$ versus $H_a : \theta = 4.0$ with $\alpha = 0.10$ and $\beta = 0.05$.

(b) Approximate the expected sample size for the test in (a) when H_0 is true.

(c) Approximate the expected sample size for the test in (a) when H_a is true.

40. Prove the identity (12.8.5). *Hint:* Within the squared terms of the left side, add and subtract $\bar{x}_j$, and then use the binomial expansion.

CHAPTER 13

CONTINGENCY TABLES AND GOODNESS-OF-FIT

13.1

INTRODUCTION

Most of the models discussed to this point have been expressed in terms of a pdf $f(x; \theta)$ that has a known functional form. Moreover, most of the statistical methods discussed in the preceding chapters, such as maximum likelihood estimation, are derived relative to specific models. In many situations, it is not possible to identify precisely which model applies. Thus, general statistical methods for testing how well a given model "fits," relative to a set of data, are desirable.

Another question of interest, which cannot always be answered without the aid of statistical methods, concerns whether random variables are independent. One possible answer to this question involves the notion of contingency tables.

13.2

ONE-SAMPLE BINOMIAL CASE

First let us consider a Bernoulli-trial type of situation with two possible outcomes, A_1 and A_2, with $P(A_1) = p_1$ and $P(A_2) = p_2 = 1 - p_1$. A random sample of n trials is observed, and we let $o_1 = x$ and $o_2 = n - x$ denote the observed number of outcomes of type A_1 and type A_2, respectively. We wish to test $H_0 : p_1 = p_{10}$ against $H_a : p_1 \neq p_{10}$. Under H_0 the expected number of outcomes of each type is $e_1 = np_{10}$ and $e_2 = np_{20} = n(1 - p_{10})$. This situation is illustrated in Table 13.1.

We have discussed an exact small-sample binomial test, and we also have discussed an approximate test based on the normal approximation to the binomial. The approximate test also can be expressed in terms of a chi-square variable, and this form can be generalized for the case of more than two possible types of outcomes and more than one sample.

The square of the approximately normally distributed test statistic will be approximately distributed as $\chi^2(1)$, and it can be expressed as

$$\chi^2 = \frac{(x - np_{10})^2}{np_{10}(1 - p_{10})} = \frac{(x - np_{10})^2}{np_{10}} + \frac{(x - np_{10})^2}{n(1 - p_{10})}$$

$$= \frac{(x - np_{10})^2}{np_{10}} + \frac{[(n - x) - n(1 - p_{10})]^2}{n(1 - p_{10})}$$

$$= \sum_{j=1}^{2} \frac{(o_j - e_j)^2}{e_j} \qquad \textbf{(13.2.1)}$$

An approximate size α test of H_0 is to reject H_0 if $\chi^2 > \chi^2_{1-\alpha}(1)$.

The χ^2 statistic reflects the amount of disagreement between the observed outcomes and the expected outcomes under H_0, and it is an intuitively appealing form. In this form, the differences in both cells are squared, but the differences are linearly dependent because $\sum_{j=1}^{2} (o_j - e_j) = 0$, and the number of degrees of freedom is one less than the number of cells.

TABLE 13.1 **Values of expected and observed outcomes for a binomial experiment**

Possible outcomes	A_1	A_2	Total
Probabilities	p_{10}	p_{20}	1
Expected outcomes	$e_1 = np_{10}$	$e_2 = np_{20}$	n
Observed outcomes	$o_1 = x$	$o_2 = n - x$	n

We should note that the chi-square approximation can be improved in this case by introducing a correction for discontinuity. Thus, a somewhat more accurate result is obtained by using the statistic

$$\chi^2 = \sum_{j=1}^{2} \frac{(|o_j - e_j| - 0.5)^2}{e_j} \tag{13.2.2}$$

The chi-squared test statistic may be generalized in two ways: It can be extended to apply to a multinomial problem with k types of outcomes, and it can be generalized to an r-sample binomial problem. We will consider the r-sample binomial problem first.

13.3

r-SAMPLE BINOMIAL TEST (COMPLETELY SPECIFIED H_0)

Suppose now that $X_i \sim \text{BIN}(n_i, p_i)$ for $i = 1, \ldots, r$, and we wish to test $H_0 : p_i = p_{i0}$, where the p_{i0} are known constants. Now let $o_{i1} = x_i$ and $o_{i2} = n_i - x_i$ denote the observed outcomes in the ith sample, and let $e_{i1} = n_i p_{i0}$ and $e_{i2} = n_i(1 - p_{i0})$ denote the expected outcomes under H_0.

Because a sum of independent chi-square variables is chi-square distributed, we have approximately

$$\chi^2 = \sum_{i=1}^{r} \sum_{j=1}^{2} \frac{(o_{ij} - e_{ij})^2}{e_{ij}} \sim \chi^2(r) \tag{13.3.1}$$

An approximate size α test is to reject H_0 if $\chi^2 > \chi^2_{1-\alpha}(r)$.

Example 13.3.1 A certain characteristic is believed to be present in 20% of the population. Random samples of size 50, 100, and 50 are tested from each of three different races, with the observed outcomes shown in Table 13.2.

The expected numbers of outcomes under $H_0 : p_1 = p_2 = p_3 = 0.20$ then are $e_{11} = 50(0.2) = 10$, $e_{21} = 100(0.2) = 20$, and $e_{31} = 50(0.2) = 10$. The remaining e_{ij} may be obtained by subtraction and are shown in Table 13.3.

TABLE 13.2 Observed outcomes for a three-sample binomial test

	Observed Outcomes		
	Present	Absent	Total
Race 1	20	30	50
Race 2	25	75	100
Race 3	15	35	50

TABLE 13.3 Expected outcomes for a three-sample binomial test

	Present	Absent	Total
Race 1	10	40	50
Race 2	20	80	100
Race 3	10	40	50

We have

$$\chi^2 = \sum_{i=1}^{3} \sum_{j=1}^{2} \frac{(o_{ij} - e_{ij})^2}{e_{ij}} = \frac{(20-10)^2}{10} + \frac{(30-40)^2}{40} + \cdots + \frac{(35-40)^2}{40}$$
$$= 17.18$$

Because $\chi^2_{0.99}(3) = 11.3$, we may reject H_0 at the $\alpha = 0.01$ significance level.

TEST OF COMMON *p*

Perhaps a more common problem is to test whether the p_i are all equal, $H_0: p_1 = p_2 = \cdots = p_r = p$, where the common value p is not specified. We still have the same $r \times 2$ table of observed outcomes, but the value p must be estimated to estimate the expected numbers under H_0. Under H_0 the MLE of p is the pooled estimate

$$\hat{p} = \frac{\sum x_i}{N} = \frac{\sum_{i=1}^{N} o_{i1}}{N}$$

and $\hat{e}_{i1} = n_i \hat{p}$, $\hat{e}_{i2} = n_i(1 - \hat{p})$, where $N = \sum_{i=1}^{r} n_i$.

The test statistic is

$$\chi^2 = \sum_{i=1}^{r} \sum_{j=1}^{2} \frac{(o_{ij} - \hat{e}_{ij})^2}{\hat{e}_{ij}} \tag{13.3.2}$$

The limiting distribution of this statistic can be shown to be chi-squared with $r - 1$ degrees of freedom, so an approximate size α test is to reject H_0 if

$$\chi^2 > \chi^2_{1-\alpha}(r-1) \tag{13.3.3}$$

Quite generally in problems of this nature, one degree of freedom is lost for each unknown parameter estimated. This is quite similar to normal sampling, where

$$\sum_{i=1}^{n} \left(\frac{X_i - \mu}{\sigma}\right)^2 \sim \chi^2(n) \quad \text{and} \quad \sum_{i=1}^{n} \left(\frac{X_i - \bar{X}}{\sigma}\right)^2 \sim \chi^2(n-1)$$

TABLE 13.4 **Table of r-sample binomial observations**

Sample	A_1	A_2	Total
1	o_{11}	o_{12}	n_1
2	o_{21}	o_{22}	n_2
⋮	⋮	⋮	⋮
r	o_{r1}	o_{r2}	n_r
	$N\hat{p}$	$N(1-\hat{p})$	N

We again have a linear relationship, because $\sum (x_i - \bar{x}) = 0$, and the effect is that the latter sum is equivalent to a sum of $n - 1$ independent squared standard normal variables. The degrees of freedom also can be heuristically illustrated by considering the $r \times 2$ table of observed outcomes in Table 13.4. Having a fixed value of the estimate $\hat{p}$ corresponds to having all of the marginal totals in the table fixed. Thus $r - 1$ numbers in the interior of the table can be assigned, and all the remaining numbers then will be determined. In general, the number of degrees of freedom associated with an $r \times c$ table with fixed marginal totals is $(r - 1) \cdot (c - 1)$.

Example 13.3.2 Consider again the data in Example 13.3.1, but suppose we had chosen to test the simple hypothesis that the proportion containing the characteristic is the same in the three races, $H_0 : p_1 = p_2 = p_3 = p$. In this case, $\hat{p} = 60/200 = 0.3$, $\hat{e}_{11} = 50(60/200) = 15$, $\hat{e}_{21} = 100(60/200) = 30$, and the remaining expected numbers may be obtained by subtraction. These expected values are included in Table 13.5 in parentheses. In this case

$$\chi^2 = \sum_{i=1}^{3} \sum_{j=1}^{2} (o_{ij} - \hat{e}_{ij})^2/\hat{e}_{ij} = \frac{(20 - 15)^2}{15} + \cdots + \frac{(35 - 35)^2}{35} = 3.57$$

Because $\chi^2_{0.99}(2) = 9.21$, we cannot reject the hypothesis of common proportions at the $\alpha = 0.01$ level.

TABLE 13.5 **Observed and expected numbers under null hypothesis of equal proportions**

	Present	Absent	Total
Race 1	20 (15)	30 (35)	50
Race 2	25 (30)	75 (70)	100
Race 3	15 (15)	35 (35)	50
Total	60	140	200

Note that reducing the degrees of freedom when a parameter is estimated results in a smaller critical value. This seems reasonable, because estimating p automatically forces some agreement between the $\hat{e}_{ij}$ and o_{ij}, and a smaller computed chi-squared value therefore is considered significant. In this example the smaller critical value still is not exceeded, which suggests that the proportions may all be equal, although we had rejected the simple hypothesis that they were all equal to 0.2 in Example 13.3.1.

In Section 13.2 we mentioned that a correction for discontinuity may enhance the accuracy of the chi-squared approximation for small sample sizes. The chi-squared approximation generally is considered to be sufficiently accurate for practical purposes if all the expected values in each cell are at least 2 and at least 80% of them are 5 or more.

13.4

ONE-SAMPLE MULTINOMIAL

Suppose now that there are c possible types of outcomes, $A_1, A_2, \ldots, A_c$, and in a sample size n let $o_1, \ldots, o_c$ denote the number of observed outcomes of each type. We assume probabilities $P(A_j) = p_j$, $j = 1, \ldots, c$, where $\sum_{j=1}^{c} p_j = 1$, and we wish to test the completely specified hypothesis $H_0: p_j = p_{j0}$, $j = 1, \ldots, c$. Under H_0 the expected values for each type are given by $e_j = np_{j0}$. The chi-square statistic again provides an appealing and convenient test statistic, where approximately

$$\chi^2 = \sum_{j=1}^{c} \frac{(o_j - e_j)^2}{e_j} \sim \chi^2(c-1) \tag{13.4.1}$$

It is possible to show that the limiting distribution of this statistic under H_0 is $\chi^2(c-1)$, and this is consistent with earlier remarks concerning what the appropriate number of degrees of freedom turns out to be in these problems. Equation (13.2.1) illustrated that one degree of freedom was appropriate for the binomial case with $c = 2$. Also, for fixed sample size, $c - 1$ observed values determine the remaining observed value.

Example 13.4.1 A die is rolled 60 times, and we wish to test whether it is a fair die, $H_0: p_j = 1/6$, $i = 1, \ldots, 6$. Under H_0 the expected outcome in each case is $e_j = np_{j0} = 10$, and the results are depicted in Table 13.6.

In this case

$$\chi^2 = \sum_{j=1}^{6} \frac{(o_j - e_j)^2}{e_j} = 6.0 < \chi^2_{0.90}(5) = 9.24$$

so we cannot reject H_0 at the $\alpha = 0.10$ level of significance.

TABLE 13.6 **Observed and expected frequencies for a die-rolling experiment**

	1	2	3	4	5	6	
Observed	8	11	5	12	15	9	60
Expected	10	10	10	10	10	10	60

As suggested in Section 13.3, if the model is not specified completely under H_0, then it is necessary to estimate the expected numbers, $\hat{e}_i$, and the number of degrees of freedom is reduced by one for each parameter estimated. This aspect is discussed further in the later sections.

13.5

r-SAMPLE MULTINOMIAL

We may wish to test whether r samples come from the same multinomial population or that r multinomial populations are the same. Let $A_1, A_2, \ldots, A_c$ denote c possible types of outcomes, and let the probability that an outcome of type A_j will occur for the ith population (or ith sample) be denoted by $p_{j|i}$. Note that $\sum_{j=1}^{c} p_{j|i} = 1$ for each $i = 1, \ldots, r$. Also let o_{ij} denote the observed number of outcomes of type A_j in sample i. For a completely specified $H_0 : p_{j|i} = p_{j|i}^{(0)}$, then $e_{ij} = n_i p_{j|i}^{(0)}$ under H_0, and it is clear that equation (13.3.1) can be extended to this case. Approximately for each i,

$$\chi_i^2 = \sum_{j=1}^{c} \frac{(o_{ij} - e_{ij})^2}{e_{ij}} \sim \chi^2(c - 1)$$

and

$$\chi^2 = \sum_{i=1}^{r} \chi_i^2 = \sum_{i=1}^{r} \sum_{j=1}^{c} (o_{ij} - e_{ij})^2/e_{ij} \sim \chi^2(r(c - 1)) \tag{13.5.1}$$

under H_0.

The more common problem is to test whether the r multinomial populations are the same without specifying the values of the $p_{j|i}$. Thus we consider

$$H_0 : p_{j|1} = p_{j|2} = \cdots = p_{j|r} = p_j \quad \text{for } j = 1, 2, \ldots, c$$

We must estimate $c - 1$ parameters $p_1, \ldots, p_{c-1}$, which also will determine the estimate of p_c because $\sum_{j=1}^{c} p_j = 1$. Under H_0 the MLE of p_j will be the pooled estimate from the pooled sample of $N = \sum_{i=1}^{r} n_i$ items, which gives

$$\hat{p}_j = \frac{\sum_{i=1}^{r} o_{ij}}{N} = \frac{c_j}{N}$$

where c_j is the jth column total, and

$$\hat{e}_{ij} = n_i \hat{p}_j = n_i c_j/N$$

The number of degrees of freedom in this case is $r(c-1)-(c-1) = (r-1)(c-1)$, and approximately

$$\chi^2 = \sum_{i=1}^{r} \sum_{j=1}^{c} (o_{ij} - \hat{e}_{ij})^2/\hat{e}_{ij} \sim \chi^2((r-1)(c-1)) \tag{13.5.2}$$

Example 13.5.1 In Example 13.3.1, suppose that the characteristic of interest may occur at three different levels: absent, moderate, or severe. We are interested in knowing whether the proportion of each level is the same for different races.

The notation is depicted in Table 13.7. We wish to test $H_0: p_{j|1} = p_{j|2} = p_{j|3} = p_j$ for $j = 1, 2, 3$.

The observed outcomes are shown in Table 13.8. The estimated expected numbers under H_0 are given in parentheses, where $\hat{e}_{11} = n_1 \hat{p}_1 = n_1(c_1/N)$

TABLE 13.7 **Conditional probabilities for a three-sample binomial test**

	A_1 Severe	A_2 Moderate	A_3 Absent	
Sample 1 (Race 1)	$p_{1\|1}$	$p_{2\|1}$	$p_{3\|1}$	1
Sample 2 (Race 2)	$p_{1\|2}$	$p_{2\|2}$	$p_{3\|2}$	1
Sample 3 (Race 3)	$p_{1\|3}$	$p_{2\|3}$	$p_{3\|3}$	1
	p_1	p_2	p_3	1

TABLE 13.8 **Observed and expected outcomes**

	Observed Outcomes (Expected Outcomes)			
	Severe	**Moderate**	**Absent**	**Total**
s_1 (Race 1)	10 (6)	10 (9)	30 (35)	50
s_2 (Race 2)	4 (12)	21 (18)	75 (70)	100
s_3 (Race 3)	10 (6)	5 (9)	35 (35)	50
Total	24	36	140	200

$= 50(24)/200 = 6$, $\hat{e}_{12} = n_1(c_2/N) = 9$, $\hat{e}_{21} = n_2(c_1/N) = 12$, $\hat{e}_{22} = n_2(c_2/N) = 18$, and the others may be obtained by subtraction.

The number of degrees of freedom in this case is $(r - 1)(c - 1) = 2(2) = 4$ and

$$\chi^2 = \sum_{i=1}^{3} \sum_{j=1}^{3} \frac{(o_{ij} - \hat{e}_{ij})^2}{\hat{e}_{ij}} = 14.13 > 13.3 = \chi^2_{0.99}(4)$$

so H_0 can be rejected at the $\alpha = 0.01$ level.

The question of whether the characteristic proportions change over samples (in this example, over races) is similar to the question of whether these two factors (characteristic and race) are independent. Note that in the r-sample multinomial case, the row totals (sample sizes) are fixed, and it appears natural to consider the test of common proportions over samples, sometimes referred to as a test of homogeneity. In the example, one could have selected 200 individuals at random and then counted the number that fall in each race category and each characteristic category. In this case, the row totals and column totals are both random. We can look at the conditional test in this case, given the row totals, and analyze the data in the same way as before, but it appears somewhat more natural to look directly at a test of independence in this case. It turns out that the same test statistic is applicable under this interpretation, as discussed in the next section.

13.6

TEST FOR INDEPENDENCE, $r \times c$ CONTINGENCY TABLE

Suppose that one factor with c categories is associated with columns and a second factor with r categories is associated with rows in an $r \times c$ contingency table. Let p_{ij} denote the probability that a sampled item is classified in the ith row category and the jth column category. Let $p_{i\cdot} = \sum_{j=1}^{c} p_{ij}$ denote the marginal probability that an individual is classified in row i, and let $p_{\cdot j} = \sum_{i=1}^{r} p_{ij}$ denote the marginal probability that an individual is classified in the jth column, as illustrated in Table 13.9.

Note that the total joint probabilities in this case add to 1, whereas in Table 13.7 the probabilities under consideration correspond to conditional probabilities, and each row adds to 1.

If the classification of an individual according to one factor is not affected by its classification relative to the other category, then the two factors are independent. That is, they are independent if the joint classification probabilities are the products of the marginal classification probabilities, $p_{ij} = p_{i\cdot}\, p_{\cdot j}$. Thus, to test

TABLE 13.9 **Contingency table of joint and marginal probabilities**

		Columns 1	2	3	
Rows	1	p_{11}	p_{12}	p_{13}	$p_{1\cdot}$
	2	p_{21}	p_{22}	p_{23}	$p_{2\cdot}$
	3	p_{31}	p_{32}	p_{33}	$p_{3\cdot}$
		$p_{\cdot 1}$	$p_{\cdot 2}$	$p_{\cdot 3}$	1

independence we test $H_0: p_{ij} = p_{i\cdot}\,p_{\cdot j}$. Let $n_i = \sum_{j=1}^{c} o_{ij}$ and $c_j = \sum_{i=1}^{r} o_{ij}$ denote the row and column totals as before, although the n_i are not fixed before the sample in this case. Let $N = \sum_{i=1}^{r} n_i$ denote the total number of outcomes. Then $\hat{p}_{i\cdot} = n_i/N$, $\hat{p}_{\cdot j} = c_j/N$, and under H_0 the expected number of outcomes to fall in the (i, j) cell is estimated to be

$$\hat{e}_{ij} = N\hat{p}_{ij} = N\hat{p}_{i\cdot}\,\hat{p}_{\cdot j} = N\left(\frac{n_i}{N}\right)\left(\frac{c_j}{N}\right) = n_i\,c_j/N$$

We note that $\hat{e}_{ij}$ reduces to exactly the same values obtained in the previous problem of testing equal proportions over samples. Thus the chi-squared statistic for measuring the agreement between the observed outcomes o_{ij} and the expected numbers under H_0, $\hat{e}_{ij}$, is computed exactly the same as before. Also, as before, asymptotic results show that approximately

$$\chi^2 = \sum\sum (o_{ij} - \hat{e}_{ij})^2/e_{ij} \sim \chi^2((r-1)(c-1)) \tag{13.6.1}$$

With regard to the number of degrees of freedom, estimating the marginal probabilities $\hat{p}_{i\cdot}$ and $\hat{p}_{\cdot j}$ amounts to fixing the marginal totals, which then leaves $(r-1)(c-1)$ degrees of freedom. This test is also similar to the asymptotic GLR test based on $-2 \ln \lambda$. In that case, the number of degrees of freedom in $\boldsymbol{\Omega}$ is $rc - 1$ because $\sum_{i=1}^{r} \sum_{j=1}^{c} p_{ij} = 1$, and the degrees of freedom for $\boldsymbol{\Omega}_0$ and H_0 is $(r-1) + (c-1)$ because $\sum_{i=1}^{r} p_{i\cdot} = 1$ and $\sum_{j=1}^{c} p_{\cdot j} = 1$; thus the dimension of the parameters specified by H_0 is $(rc-1) - (r-1) - (c-1) = (r-1)(c-1)$. This result also is consistent with the interpretation discussed in Section 13.4. For a completely specified H_0 the number of degrees of freedom is one less than the total number of cells, which in this problem becomes $rc - 1$. If H_0 is not completely specified, then the number of degrees of freedom is reduced by one for each parameter estimated, which in this case is $r - 1 + c - 1$, which again results in $(r - 1)(c - 1)$ degrees of freedom. The formal justification for the approximate distributions and choice of degrees of freedom is based on asymptotic results that are not derived here.

Example 13.6.1 A survey is taken to determine whether there is a relationship between political affiliation and strength of support for space exploration. We randomly select 100 individuals and ask their political affiliation and their support level to obtain the (artificial) data in Table 13.10.

TABLE 13.10 **Contingency table for testing independence of two factors**

	Support						
Affiliation	**Increase**		**Same**		**Decrease**		
Republican	8	(9)	12	(10.5)	10	(10.5)	30
Democrat	10	(12)	17	(14)	13	(14)	40
Independent	12	(9)	6	(10.5)	12	(10.5)	30
	30		35		35		100

Under the hypothesis of independence, $H_0 : p_{ij} = p_{i\cdot}\, p_{\cdot j}$, the expected values are computed and given in parentheses. We have

$$\chi^2 = \sum_i \sum_j (o_{ij} - \hat{e}_{ij})^2/\hat{e}_{ij} = 4.54 < 7.78 = \chi^2_{0.90}(4)$$

thus we do not have sufficient evidence to reject the hypothesis of independence at the $\alpha = 0.10$ level of error. Of course, we would obtain the identical result if we considered a conditional test given fixed row totals, and tested whether the support level probabilities are the same over the political affiliation categories. This is reasonable because, for example, if the "independents" had a higher probability for increased support, then they would have to have a lower probability in some other category, which would represent a dependence between the two factors.

Indeed, if we express the notation in Table 13.10 in terms of the joint probabilities, then $p_{j|i} = p_{ij}/p_{i\cdot}$ represents the conditional probability of being in column j given the ith row, and $p_j = p_{\cdot j}$ is the marginal probability of falling in the jth column classification; thus H_0: $p_{j|i} = p_j$ in the sampling setup in Section 13.5 is equivalent to the test of independence in this section, because $p_{j|i} = p_{ij}/p_{i\cdot} = p_{\cdot j}$ implies $p_{ij} = p_{i\cdot}\, p_{\cdot j}$.

13.7

CHI-SQUARED GOODNESS-OF-FIT TEST

The one-sample multinomial case discussed in Section 13.4 corresponds to testing whether a random sample comes from a completely specified multinomial distribution. This test can be adapted to test that a random sample comes from any completely specified distribution, $H_0 : X \sim F(x)$. Simply divide the sample space into c cells, say $A_1, \ldots, A_c$, and let $p_{j0} = P[X \in A_j]$ where $X \sim F(x)$. Then for a random sample of size n, let o_j denote the number of observations that fall into the jth cell, and under H_0 the expected number in the jth cell is $e_j = np_{j0}$. This is now back in the form of the multinomial problem, and $H_0 : X \sim F(x)$ is rejected at the α significance level if

$$\chi^2 = \sum_{j=1}^{c} (o_j - e_j)^2/e_j > \chi^2_{1-\alpha}(c - 1) \tag{13.7.1}$$

In some cases there may be a natural choice for the cells or the data may be grouped to begin with; otherwise, artificial cells may be chosen. As a general principle, as many cells as possible should be used to increase the number of degrees of freedom, as long as $e_j \geqslant 5$ or so is maintained to ensure that the chi-squared approximation is fairly accurate.

Example 13.7.1 Let X denote the repair time in days required for a certain component in an airplane. We wish to test whether a Poisson model with a mean of three days appears to be a reasonable model for this variable. The repair times for 40 components were recorded, with the results shown in Table 13.11. In some cases the component could be repaired immediately on-site, which is interpreted as zero days.

Under $H_0 : X \sim \text{POI}(3)$, we have $f(x) = e^{-3}3^x/x!$, and the cell probabilities are given by $p_{10} = P[X = 0] = f(0) = e^{-3} = 0.050$, $p_{20} = f(1) = e^{-3}3 = 0.149$, $p_{30} = f(2) = 0.224$, and so on. The expected numbers are then $e_j = np_{j0}$. The

TABLE 13.11 **Observed and expected frequencies for chi-square goodness-of-fit test of Poisson model with mean 3**

Repair Time (Days)	0	1	2	3	4	5	6	> 7
Observed (o_j)	1	3	7	6	10	7	6	0
Probabilities (p_{j0})	0.050	0.149	0.224	0.224	0.168	0.101	0.050	0.034
Expected (e_j)	2.00	5.96	8.96	8.96	6.72	4.04	2.00	1.36
	7.96					7.40		

right-hand tail cells are pooled to achieve an $e_j \geqslant 5$, and the first two cells also are pooled. This leaves $c = 5$ cells and

$$\chi^2 = \frac{(4 - 7.96)^2}{7.96} + \frac{(7 - 8.96)^2}{8.96} + \cdots + \frac{(13 - 7.40)^2}{7.40} = 9.22 > 7.78$$

$$= \chi^2_{0.90}(4)$$

so we can reject H_0 at the $\alpha = 0.10$ significance level.

UNKNOWN PARAMETER CASE

When using goodness-of-fit procedures to help select an appropriate population model, we usually are interested in testing whether some family of distributions seems appropriate, such as $H_0 : X \sim \text{POI}(\mu)$ or $H_0 : X \sim \text{N}(\mu, \sigma^2)$, where the parameter values are unspecified, rather than testing a completely specified hypothesis such as $H_0 : X \sim \text{POI}(3)$. That is, we are not interested in the lack of fit because of the wrong parameter value at this point, but we are interested in whether the general model has an appropriate form and will be a suitable model when appropriate parameter values are used.

Suppose we wish to test $H_0 : X \sim f(x; \theta_1, \ldots, \theta_k)$, where there are k unknown parameters. To compute the χ^2 statistic, the expected numbers under H_0 now must be estimated. If the original data are grouped into cells, then the joint density of the observed values, o_j, is multinomial where the true but unknown $p_{j0} = P[X \in A_j]$ are functions of $\theta_1, \ldots, \theta_k$. If maximum likelihood estimation is used to estimate $\theta_1, \ldots, \theta_k$ (based on the multinomial distribution of grouped data values o_j), then the limiting distribution of the χ^2 statistic is chi-squared with degrees of freedom $c - 1 - k$, where c is the number of cells and k is the number of parameters estimated. That is, approximately,

$$\chi^2 = \sum_{j=1}^{c} \frac{(o_j - \hat{e}_j)^2}{\hat{e}_j} \sim \chi^2(c - 1 - k) \tag{13.7.2}$$

where $\hat{e}_j = n\hat{p}_{j0}$.

In many cases, MLE estimation based on the grouped data multinomial model is not convenient to carry out, and in practice the usual MLEs based on the ungrouped data, or on grouped data approximations of these, most often are used.

If MLEs based on the individual observations are used, then the number of degrees of freedom may be greater than $c - 1 - k$, but the limiting distribution is bounded between chi-square distributions with $c - 1$ and $c - 1 - k$ degrees of freedom. Our policy here will be to use $c - k - 1$ degrees of freedom if k parameters are estimated by any ML procedure. A more conservative approach would be to bound the p-value of the test using $c - 1 - k$ and $c - 1$ degrees of freedom if the MLEs are not based directly on the grouped data (Kendall and Stuart, 1967, page 430).

Example 13.7.2 Consider again the data given in Example 13.7.1, and suppose now that we wish to test $H_0 : X \sim \text{POI}(\mu)$. The usual MLE of μ is the average of the 40 repair times, which in this case is computed as

$$\hat{\mu} = [0(1) + 1(3) + 2(7) + \cdots + 6(6)]/40 = 3.65$$

Under H_0 the estimated probabilities are now $\hat{p}_{j0} = e^{-3.65}(3.65)^j/j!$ and the estimated expected values, $\hat{e}_j = n\hat{p}_{j0}$, are computed using a Poisson distribution with $\mu = 3.65$. Retaining the same five cells used before, the results are as shown in Table 13.12.

TABLE 13.12 **Observed and expected frequencies for chi-square test of Poisson model with estimated mean 3.65**

Repair Times (Days)	(0, 1)	2	3	4	$\geqslant 5$
Observed (o_j)	4	7	6	10	13
Probabilities ($\hat{p}_{j0}$)	0.121	0.173	0.211	0.192	0.303
Expected ($\hat{e}_j$)	4.84	6.92	8.44	7.68	12.12

We have

$$\chi^2 = \frac{(4 - 4.84)^2}{4.84} + \cdots + \frac{(13 - 12.12)^2}{12.12} = 1.62 < 6.25 = \chi^2_{0.90}(3)$$

so a Poisson model appears to be quite reasonable for these data, although the Poisson model with $\mu = 3$ was found not to fit well. The number of degrees of freedom here is 3, because one parameter is estimated.

Note that the question of how to choose the cells is not quite so clear when H_0 is not completely specified. For a completely specified H_0 the best choice is to choose cells so that all the e_j are approximately equal to 5. This makes the e_j large enough to ensure the accuracy of the chi-squared approximation and still gives the largest possible number of degrees of freedom. Of course, with discrete distributions this may not be completely achievable. If H_0 is not completely specified, then the e_j or $\hat{e}_j$ cannot be computed before taking the sample. The usual procedure is to choose some natural or reasonable cell division initially, and then pool adjacent cells after the data are taken to achieve $\hat{e}_j \geqslant 5$. This pooling should not be done in a capricious manner. In some cases the data already are grouped, and this provides an initial cell division. Indeed, one advantage of the chi-squared goodness-of-fit statistic is that it is applicable to grouped data. On the other hand, if the individual observations are available, then some information may be lost by using only grouped data. Some additional goodness-of-fit tests based more directly on the individual observations are mentioned in the next section.

Example 13.7.3 Trumpler and Weaver (1953) provided data collected at the Lick observatory on the radial velocities of 80 bright stars, as shown in Table 13.13.

TABLE 13.13 **Observed and expected frequencies for chi-square goodness-of-fit test with estimates $\hat{\mu} = -20.3$ and $\hat{\sigma} = 12.7$**

Intervals of Velocities	o_j		$\hat{p}_{jo}$		$\hat{e}_j$
(−80, −70)	1		.000		
(−70, −60)	2		.001		
(−60, −50)	2	15	.009	.224	17.92
(−50, −40)	2		.051		
(−40, −30)	8		.163		
(−30, −20)	24		.284		22.72
(−20, −10)	26		.283		22.64
(−10, 0)	11		.154		
(0, 10)	2	15	.046	.209	16.72
(10, 20)	1		.008		
(20, 30)	1		.001		
	80		1.000		80.00

We wish to test for normality, $H_0 : X \sim \mathrm{N}(\mu, \sigma^2)$. We will use the chi-square test with μ and σ estimated by the MLEs based on the grouped data of Table 13.13. If we denote the arbitrary jth cell by $A_j = (a_j, a_{j+1})$ and if $z_j = (a_j - \mu)/\sigma$, then the likelihood equation based on the multinomial data is

$$L = \frac{n!}{o_1! \cdots o_c!} \prod_{j=1}^{c} [\Phi(z_{j+1}) - \Phi(z_j)]^{o_j}$$

The likelihood equations are obtained by equating to zero the partials of the logarithm of L with respect to μ and σ. Specifically, we obtain

$$\sum_{j=1}^{c} o_j[\Phi'(z_{j+1}) - \Phi'(z_j)]/[\Phi(z_{j+1}) - \Phi(z_j)] = 0 \tag{13.7.3}$$

$$\sum_{j=1}^{c} o_j[z_{j+1}\Phi'(z_{j+1}) - z_j\Phi'(z_j)]/[\Phi(z_{j+1}) - \Phi(z_j)] = 0 \tag{13.7.4}$$

Equations (13.7.3) and (13.7.4) must be solved by an iterative numerical method, and for the data of Table 13.13 the estimates of μ and σ are

$\hat{\mu} = -20.3$ and $\hat{\sigma} = 12.7$. The estimated cell probabilities are of the form $\hat{p}_{j0} = \Phi(\hat{z}_{j+1}) - \Phi(\hat{z}_j)$ with $\hat{z}_j = (a_j - \hat{\mu})/\hat{\sigma}$. Each $\hat{p}_{j0}$ must be at least $5/80 = .0625$ to ensure that $\hat{e}_j \geqslant 5$. To satisfy this requirement in the present example, it is necessary to pool the first five cells and, similarly, the last four cells must be pooled. This reduces the number of cells to $c = 4$ with the pooled results shown in Table 13.13. It follows that

$$\chi^2 = \sum_{i=1}^{4} (o_j - \hat{e}_j)^2/\hat{e}_j = 1.22 < 3.84 = \chi^2_{0.95}(1)$$

Thus, the normal model gives a reasonable fit in this example.

It might seem that a simpler method could be based on the following often-used grouped-data estimates:

$$\tilde{\mu} = \frac{\sum m_j o_j}{n} \qquad \tilde{\sigma} = \sqrt{\frac{\sum m_j^2 o_j - n\tilde{\mu}^2}{n-1}}$$

where m_j is the midpoint of the jth interval. However, for this example, these estimates are $\tilde{\mu} = -21$ and $\tilde{\sigma} = 16$. The latter estimate of σ is somewhat larger than the grouped-data MLE. In fact, it is large enough that the chi-square test based on this estimate rejects the normal model, contrary to our earlier conclusion. Another type of simple closed-form estimate for grouped-data will be discussed in Chapter 15.

13.8

OTHER GOODNESS-OF-FIT TESTS

Several goodness-of-fit tests have been developed in terms of the empirical distribution function. The basic principle is to see how closely the observed sample cumulative distribution agrees with the hypothesized theoretical cumulative distribution. Several methods for measuring the closeness of agreement have been proposed.

The EDF tests generally are considered to be more powerful than the chi-squared goodness-of-fit test, because they make more direct use of the individual observations. Of course, then they are not applicable if the data are available only as grouped data.

Let $x_{1:n}, \ldots, x_{n:n}$ denote an ordered random sample of size n. Then the EDF or sample CDF is denoted by $\hat{F}(x)$, and at the ordered values note that

$$\hat{F}(x_{i:n}) = i/n$$

If we wish to test a completely specified hypothesis, $H_0: X \sim F(x)$, then the general approach is to measure how close the agreement is between $F(x_{i:n})$ and

$\hat{F}(x_{i:n})$ for $i = 1, \ldots, n$. Some slight modifications have been found helpful, such as using a half-point correction and comparing the hypothesized $F(x_{i:n})$ values to $(i - 0.5)/n$ rather than to i/n.

Because $U = F(X) \sim \text{UNIF}(0, 1)$, the test of any completely specified H_0 can be expressed equivalently as a test for uniformity, where $U_{i:n} = F(X_{i:n})$ are distributed as ordered uniform variables.

CRAMÉR-VON MISES TEST FOR COMPLETELY SPECIFIED H_0

The Cramér-Von Mises (CVM) test can be modified to apply to Type II censored samples. If the r smallest ordered observations from a sample of size n are available, then the CVM test statistic for testing $H_0 : X \sim F(x)$ is given by

$$\text{CM} = \frac{1}{12n} + \sum_{i=1}^{r} \left(F(x_{i:n}) - \frac{i - 0.5}{n} \right)^2 \tag{13.8.1}$$

The distribution of CM under H_0 is the same as the distribution of

$$\frac{1}{12n} + \sum_{i=1}^{r} \left(U_{i:n} - \frac{i - 0.5}{n} \right)^2$$

where the $U_{i:n}$ are ordered uniform variables. The asymptotic percentage points for CM have been obtained by Pettitt and Stephens (1976). They appear to be sufficiently accurate for practical purposes for samples as small as 10 or so.

An approximate size α test of $H_0 : X \sim F$ is to reject H_0 if $\text{CM} \geqslant \text{CM}_{1-\alpha}$, where the critical values $\text{CM}_{1-\alpha}$ are provided in Table 9 (Appendix C) for several values of α and censoring levels.

Example 13.8.1 We are given that 25 system failures have occurred in a 100 day period, and we wish to test whether the failure times are uniformly distributed, $H_0 : F(x) = x/100$, $0 < x < 100$, where the 25 ordered observations are as follows:

5.2, 13.6, 14.5, 14.6, 20.5, 38.4, 42.0, 44.5, 46.7, 48.5, 50.3, 56.4, 61.7, 62.9, 64.1, 67.1, 71.6, 79.2, 82.6, 83.1, 85.5, 90.8, 92.7, 95.5, 95.6

We have

$$\text{CM} = \frac{1}{12(25)} + \sum_{i=1}^{25} \left(\frac{x_{i:25}}{100} - \frac{i - 0.5}{25} \right)^2 = 0.182$$

Because $\text{CM}_{0.90} = 0.347$, we cannot reject H_0 at the $\alpha = 0.10$ level of significance.

If a Poisson process is observed for a fixed time t, then given the number of occurrences, the successive failure times are conditionally distributed as ordered uniform variables. This suggests that the above data could represent data from a Poisson process (see Chapter 16).

If the above data represent successive failure times from a Poisson process, then the times between failures should be independent exponential variables. The interarrival times are as follows:

5.2, 8.4, 0.9, 0.1, 5.9, 17.9, 3.6, 2.5, 1.2, 1.8, 1.8, 6.1, 5.3, 1.2, 1.2, 3.0, 3.5, 7.6, 3.4, 0.5, 2.4, 5.3, 1.9, 2.8, 0.1

If we wish to use the CM statistic to test these data for exponentiality, we must completely specify H_0. Suppose that we test $H_0 : Y \sim \text{EXP}(5)$. We have

$$\text{CM} = \frac{1}{12(25)} + \sum_{i=1}^{25} \left(1 - e^{-y_{i:n}/5} - \frac{i - 0.5}{25}\right)^2 = 0.173$$

Because $\text{CM}_{0.99} = 0.743$, we see that we cannot reject the hypothesis that the interarrival times follow an EXP(5) distribution at the $\alpha = 0.01$ significance level.

CRAMER-VON MISES TEST, PARAMETERS ESTIMATED

As suggested in the previous section, we often are more interested in testing whether a certain family of distributions is applicable rather than testing a completely specified hypothesis. To test $H_0 : X \sim F(x; \boldsymbol{\theta})$, where $\boldsymbol{\theta}$ is unspecified, we may consider

$$\widehat{\text{CM}} = \frac{1}{12n} + \sum_{i=1}^{n} \left(F(x_{i:n}; \hat{\boldsymbol{\theta}}) - \frac{i - 0.5}{n}\right)^2 \tag{13.8.2}$$

where $\hat{\boldsymbol{\theta}}$ denotes the MLE of $\boldsymbol{\theta}$. In general, the distribution of $\widehat{\text{CM}}$ may or may not depend on unknown parameters; however, we know that if $\boldsymbol{\theta} = (\theta_1, \theta_2)$ are location-scale parameters, then $F(X; \hat{\theta}_1, \hat{\theta}_2)$ and $F(X_{i:n}; \hat{\theta}_1, \hat{\theta}_2)$ are pivotal quantities whose distribution does not depend on the parameters. Thus, at least in the case of location-scale parameters, $\widehat{\text{CM}}$ provides a suitable test statistic whose critical values can be determined. We have the disadvantage in this case that the critical values depend on the form of F being tested, whereas in the original situation the same critical values are applicable for testing any completely specified hypothesis. Some asymptotic and simulated critical values are available in the literature for certain models such as the exponential, normal, and Weibull distributions. Stephens considers slight modifications of the test statistic so that the asymptotic critical values are quite accurate even for small n for complete sample tests of normality and exponentiality. Some of these results, along with the Weibull case, are included in Table 10 (Appendix C). Pettitt and Stephens (1976) and Stephens (1977) provide additional results, including the censored case.

Example 13.8.2 Now we may test whether the interarrival times in the previous example follow any exponential distribution, EXP(θ). In this case, $\hat{\theta} = \bar{y} = 3.7$, and

$$\widehat{\text{CM}} = \frac{1}{12(25)} + \sum_{i=1}^{25} \left(1 - e^{-y_{i:n}/3.7} - \frac{i - 0.5}{25}\right)^2 = 0.051$$

Because $(1 + 0.16/n)\widehat{\text{CM}} = 0.051 < 0.177$, we cannot reject H_0 at the $\alpha = 0.10$ level.

KOLMOGOROV-SMIRNOV OR KUIPER TEST

The Kolmogorov-Smirnov (KS) test statistic is based on the maximum difference between the sample CDF and the hypothesized CDF. To test a completely specified $H_0 : X \sim F(x)$, let

$$D^+ = \max_i \, (i/n - F(x_{i:n})) \tag{13.8.3}$$

$$D^- = \max_i \, (F(x_{i:n}) - (i - 1)/n)) \tag{13.8.4}$$

$$D = \max \, (D^+, D^-) \tag{13.8.5}$$

$$V = D^+ + D^- \tag{13.8.6}$$

Carets will be added if unknown parameters must be estimated. The first three statistics are KS statistics, and V is Kuiper's test statistic. The distributions of these statistics do not depend on F. Also, as in the CVM case, if location-scale parameters are estimated, then the distributions will not depend on the parameters, but they then will depend on the form of F. The KS statistics allow for one-sided alternatives, and they also have been extended to two-sample problems.

Stephens (1974, 1977) has derived asymptotic critical values for these statistics, and he has considered modifications so that these critical values are good for small n. Some of these results are summarized in Table 11 (Appendix C). The Weibull results are provided by Chandra et al. (1981), and they also provide more accurate small sample results for this case, as well as percentage points for D^+ and D^-. These results were developed for the extreme-value distribution for maximums, which is related to the Weibull distribution by a monotonically decreasing transformation; thus the D^+ and D^- critical values are interchanged when applying them directly to the Weibull distribution.

Many other EDF type test statistics, as well as tests devised specifically for a certain model, are available in the literature. Other references in this area include Aho et al. (1983), Dufour and Maag (1978), Koziol (1980), and Bain and Engelhardt (1983).

Example 13.8.3 Let us rework Example 13.8.2 using the Kolmogorov test statistic. A plot of the EDF $\hat{F}(y_{i:n}) = i/n$ and the estimated CDF

$$F(y_{i:n}\,;\,\hat{\theta}) = 1 - e^{-y_{i:n}/3.7}$$

is given in Figure 13.1. The value of D occurs at the fifth order statistic

$$D = D^- = 1 - e^{-1.2/3.7} - \frac{5-1}{25} = 0.117$$

The modified test statistic is

$$(5 + 0.26 + 0.1)(0.117 - 0.2/25) = 0.584 < 0.995$$

so again we cannot reject the hypothesis of exponentiality at the 0.10 significance level.

FIGURE 13.1 Comparison of an empirical CDF with an exponential CDF with estimated mean $\hat{\theta} = 3.7$

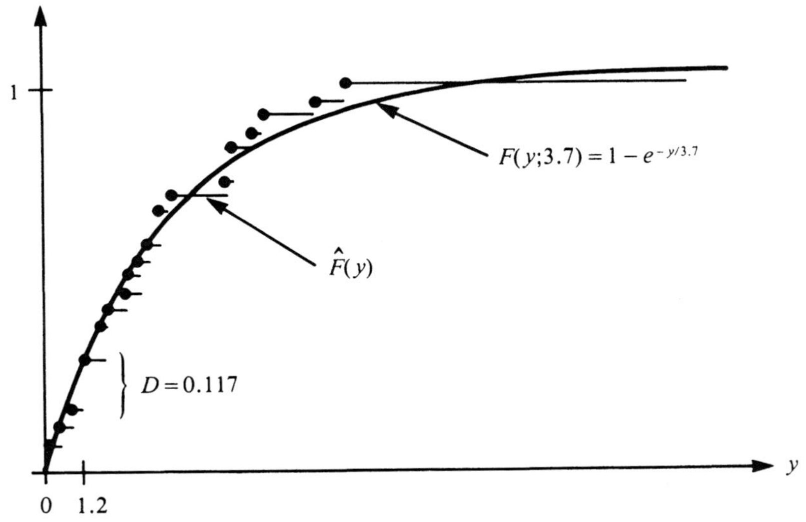

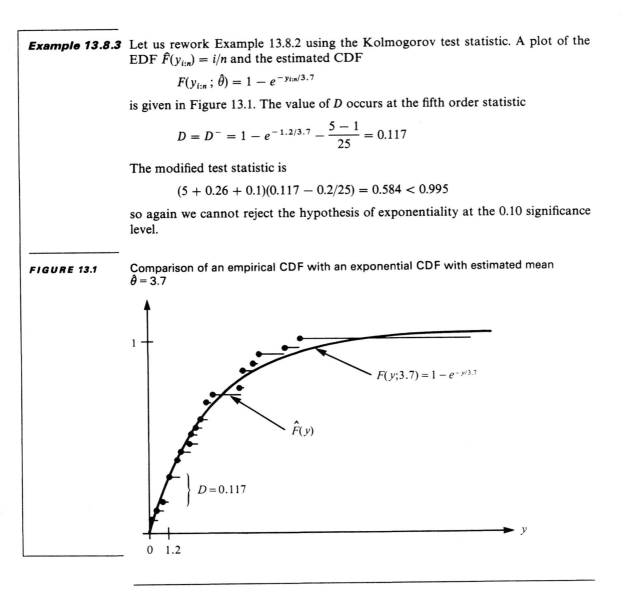

SUMMARY

Our purpose in this chapter was to introduce several tests that are designed either to determine whether a hypothesized distribution provides an adequate model or to determine whether random variables are independent.

Chi-square tests are based on the relative sizes of the differences between observed frequencies and the theoretical frequencies predicted by the model. They have the advantages of being fairly simple and also of having an approximate chi-square distribution.

Tests for independence based on contingency tables use the differences between observed frequencies of joint occurrence and estimates of these frequencies made under the assumption of independence. Other goodness-of-fit tests, such as the CVM and KS tests, are based on the differences between the empirical CDF, computed from a random sample, and the CDF of the hypothesized model. These tests generally are harder to work with numerically, but they tend to have good power relative to many commonly used alternatives.

All of the goodness-of-fit tests considered here are designed primarily for testing completely specified hypotheses, but all can be adapted for testing composite hypotheses by estimating unknown parameters of the model. When this is the case, the power of the test is less than for completely specified hypotheses, and the critical values needed to perform the test are changed. This is taken care of easily for chi-square tests by adjusting the number of degrees of freedom (one degree of freedom is subtracted for each parameter estimated). The situation is not as convenient for the CVM and KS tests, because new tables of critical values are required, and these must be obtained for the specific parametric form that is being tested (normal, Weibull, etc.).

EXERCISES

1. A baseball player is believed to be a .300 hitter. In his first 100 at bats in a season he gets 20 hits. Use equation (13.2.1) to test $H_0 : p = 0.3$ against $H_a : p \neq 0.3$ at the $\alpha = 0.10$ significance level. What would you do if you wanted a one-sided test?

2. You flip a coin 20 times and observe 7 heads. Test whether the coin is unbiased at the $\alpha = 0.10$ significance level. Use equations (13.2.1) and (13.2.2).

3. Consider Example 13.3.1, but suppose that the following data are observed:

	Present	Absent	Total
Race 1	10	40	50
Race 2	50	50	100
Race 3	30	20	50

(a) Test $H_0 : p_1 = 0.25$, $p_2 = 0.50$, $p_3 = 0.50$ at $\alpha = 0.10$.
(b) Test $H_0 : p_1 = p_2 = p_3 = 0.50$ at $\alpha = 0.10$.
(c) Test $H_0 : p_1 = p_2 = p_3$ at $\alpha = 0.10$.

4. A system contains four components that operate independently. Let p_i denote the probability of successful operation of the ith component. Test $H_0 : p_1 = 0.90$, $p_2 = 0.90$,

$p_3 = 0.80$, $p_4 = 0.80$, if in 50 trials the components operated successfully as follows:

Component	1	2	3	4
Successful	40	48	45	40

5. In a certain genetic problem it is believed that brown will occur with probability 1/4, white with probability 1/4, and spotted with probability 1/2.

(a) Test the hypothesis that this model is correct at $\alpha = 0.10$, if the following results were observed in 40 trials:

	Brown	White	Spotted
Observed	5	15	20

(b) Test the hypothesis that the probabilities are 1/9, 4/9, and 4/9, respectively, at $\alpha = 0.10$.

6. A sample of 36 cards are drawn with replacement from a stack of 52 cards with the following results:

spades	hearts	diamonds	clubs
6	8	9	13

Test the hypothesis at $\alpha = 0.05$ that equal numbers of each suit are in the stack of cards.

7. Three cards are drawn from a standard deck of 52 cards, and we are interested in the number of hearts obtained. The possible outcomes are $x_i = 0, 1, 2,$ or 3, and if we assumed the usual sampling-without-replacement scheme, we would hypothesize that these values would occur with probabilities $p_i = f(x_i)$, where $f(x) = \binom{13}{x}\binom{39}{3-x}\Big/\binom{52}{3}$. The following data are available from 100 trials of this experiment. Test $H_0: p_i = f(x_i)$, $i = 1, \ldots, 4$ at the $\alpha = 0.05$ level based on these data:

No. of Hearts	0	1	2	3
Times Occurred	40	45	12	3

Note: Combine the last two cells.

8. A box contains five black marbles and 10 white marbles. Player A and Player B each are asked to draw three marbles from the box and record the number of black marbles obtained. They each do this 100 times, with the following results:

Observed Outcomes

	0	1	2	3
Player A	25	40	25	10
Player B	40	40	15	5

(a) Use equation (13.5.1) to test the hypothesis that Player A drew the marbles without replacement and that Player B drew the marbles with replacement. Let $\alpha = 0.10$.

(b) Similarly, test the hypothesis that A drew with replacement and B drew without replacement.

(c) Use equation (13.5.2) to test the hypothesis that the two multinomial populations are the same at $\alpha = 0.10$.

9. A question was raised as to whether the county distribution of farm tenancy over a given period of time in Audubon County, Iowa, was the same for three different levels of soil fertility. The following results are quoted from Snedecor (1956, page 225):

Soil	Owned	Rented	Mixed	
I	36	67	49	152
II	31	60	49	140
III	58	87	80	225

Test the hypothesis that the multinomial populations are the same for the three different soil fertility levels at $\alpha = 0.10$.

10. Certain airplane component failures may be classified as mechanical, electrical, or otherwise. Two airplane designs are under consideration, and it is desired to test the hypothesis that the type of falure is independent of the airplane design. Test this hypothesis at $\alpha = 0.05$ based on the following data:

	Mech.	Elect.	Other
Design I	50	30	60
Design II	40	30	40

11. A sample of 400 people was asked their degree of support of a balanced budget and their degree of support of public education, with the following results:

	Supported Balanced Budget		
Public Education	Strong	Undecided	Weak
Strong	100	80	20
Undecided	60	50	15
Weak	20	50	5

Test the hypothesis of independence at $\alpha = 0.05$.

12. A sample of 750 people was selected and classified according to income and stature, with the following results:

Stature	Income		
	Poor	Middle	Rich
Thin	100	50	50
Average	50	200	70
Fat	120	60	50

Test the hypothesis that these two factors are independent at $\alpha = 0.10$.

13. A fleet of 50 airplanes was observed for 1000 flying hours, and the number of planes, m_x, that suffered x component failures in that time is recorded below:

x	0	1	2	3	4	$\geqslant 5$
m_x	7	10	9	7	6	11

(a) Test $H_0 : X \sim \text{POI}(2)$ at $\alpha = 0.01$.

(b) Test the hypothesis that X follows a form of the negative binomial distribution given by

$$f_x(x) = \binom{x + k - 1}{x} \frac{(\gamma t)^x}{(\gamma t + 1)^{k+x}} \qquad x = 0, 1, 2, \ldots$$

with $k = 3$ and $\gamma t = 1$. Use $\alpha = 0.10$.

14. Consider the data in Example 4.6.3.

(a) Test $H_0 : X \sim \text{EXP}(100)$ at $\alpha = 0.10$.

(b) Test $H_0 : X \sim \text{EXP}(\theta)$ at $\alpha = 0.10$.

15. The following data concerning the number of thousands of miles traveled between bus motor failures were adapted from Davis (1952):

Miles (1000)	Observed Bus Motor Failures				
	First	Second	Third	Fourth	Fifth
0–20	6	19	27	34	29
20–40	11	13	16	20	27
40–60	16	13	18	15	14
60–80	25	15	13	15	8
80–100	34	15	11	8	5
100–120	46	18	10	3	2
120–140	33	5	4	1	—
140–160	16	2	0	—	—
160–180	2	2	0	—	—
180–200	2	2	2	—	—

(a) Test the hypothesis that the data for the first bus motor failure follow an exponential distribution at $\alpha = 0.05$.

(b) Test the hypothesis that the data for the fifth bus motor failure follow an exponential distribution at $\alpha = 0.10$.

(c) Test the hypothesis that the data for the first bus motor failure follow a normal distribution at $\alpha = 0.10$.

16. The number of areas m_y receiving y flying bomb hits is given as follows

y	0	1	2	3	4	$\geqslant 5$
m_y	229	211	93	35	7	1

Test $H_0 : Y \sim \text{POI}(\mu)$ at $\alpha = 0.05$.

17. In Problem 13, test $H_0 : X \sim \text{POI}(\mu)$ at $\alpha = 0.05$. Assume that $\bar{x} \doteq 3.25$.

18. The lifetimes in minutes of 100 flashlight cells were observed as follows [Davis (1952)]:

Number of Minutes	0–706	706–746	746–786	786–∞
Observed Frequency	13	36	38	13

Test $H_0 : X \sim N(\mu, \sigma^2)$ at $\alpha = 0.10$. Note that $\bar{x} = 746$ and $s = 40$.

19. Consider the weights of 60 major league baseballs given in Exercise 24 of Chapter 4. Test $H_0 : X \sim N(\mu, \sigma^2)$.

20. Consider the data in Example 4.6.3.

(a) Use the CM statistic to test $H_0 : X \sim \text{EXP}(100)$.

(b) Use the CM statistic to test $H_0 : X \sim \text{EXP}(\theta)$.

(c) Use the CM statistic based on the first 20 observations to test $H_0 : X \sim \text{EXP}(100)$.

(d) Use the Kolmogorov-Smirnov statistic to test $H_0 : X \sim \text{EXP}(100)$. Let $\alpha = 0.10$ throughout.

21. Lieblein and Zelen (1956) provide the following data for the endurance, in millions of revolutions, of deep-groove ball bearings:

17.88, 28.92, 33.00, 41.52, 42.12, 45.60, 48.48, 51.84, 51.96, 54.12, 55.56, 67.80, 68.64, 68.64, 68.88, 84.12, 93.12, 98.64, 105.12, 105.84, 127.92, 128.04, 173.40.

Test $H_0 : X \sim \text{WEI}(\theta, \beta)$ at $\alpha = 0.10$. *Note:* $\hat{\beta} = 2.102$ and $\hat{\theta} = 81.88$.

(a) Use a chi-squared test.

(b) Use a CVM test.

(c) Use a KS test.

22. Gross and Clark (1975, page 105) consider the following relief times in hours of 20 patients receiving an analgesic:

1.1, 1.4, 1.3, 1.7, 1.9, 1.8, 1.6, 2.2, 1.7, 2.7, 4.1, 1.8, 1.5, 1.2, 1.4, 3.0, 1.7, 2.3, 1.6, 2.0.

(a) Test $H_0 : X \sim \text{EXP}(\theta)$ at $\alpha = 0.10$. Note that $\bar{x} = 1.90$.

(b) Test $H_0 : X \sim \text{N}(\mu, \sigma^2)$ at $\alpha = 0.10$.

(c) Test $H_0 : X \sim \text{LOG N}(\mu, \sigma^2)$ at $\alpha = 0.10$.

(d) Test $H_0 : X \sim \text{WEI}(\theta, \beta)$ at $\alpha = 0.10$. Note that $\hat{\beta} = 2.79$ and $\hat{\theta} = 2.14$.

CHAPTER 14

NONPARAMETRIC METHODS

14.1

INTRODUCTION

Most of the statistical procedures discussed so far have been developed under the assumption that the population or random variable is distributed according to some specified family of distributions, such as normal, exponential, Weibull, or Poisson. In the previous chapter we considered goodness-of-fit tests that are helpful in deciding what model may be applicable in a given problem. Some types of questions can be answered and some inference procedures can be developed without assuming a specific model, and these results are referred to as nonparametric or distribution-free methods. The advantages of nonparametric methods are that fewer assumptions are required, and in many cases only nominal (categorized) data or ordinal (ranked) data are required, rather than numerical (interval) data. A disadvantage of nonparametric methods is that we usually prefer to have a well-defined model with important parameters such as means and variances included in the model for interpretation purposes. In any event, many important questions can be answered by a nonparametric approach, and some of these results will be given here.

The CVM goodness-of-fit test already discussed is an example of one type of distribution-free result. This type of result depends on the fact that $U = F(X) \sim \text{UNIF}(0, 1)$ for a continuous variable, and distributional results can be obtained for functions of $F(X)$ in terms of the uniform distribution, which hold for all F. For the more classical nonparametric tests, the probability structure is induced by the sampling or randomization procedures used, as in the counting type probability problems considered in Chapter 1.

14.2 ONE-SAMPLE SIGN TEST

Consider a continuous random variable $X \sim F(x)$, and let m denote the median of the distribution. That is, $P[X \leqslant m] = P[X \geqslant m] = 1/2$. We wish to test $H_0 : m = m_0$ against $H_a : m > m_0$. This is a test for location, and it is thought of as analogous to a test for means in a parametric case. Indeed, for a symmetric distribution, the mean and the median are equal.

Now we take a random sample of n observations and let T be the number of x_i's that are less than m_0. That is, we could consider the sign of $(x_i - m_0)$, $i = 1, \ldots, n$, and let

$$T = \text{Number of negative signs} \tag{14.2.1}$$

Note that we do not really need numerical interval scale data here; we need only to be able to rank the responses as less than m_0 or greater than m_0.

Under $H_0 : m = m_0$, we have $P[X \leqslant m_0] = P[X_i - m_0 \leqslant 0] = 1/2$, so the probability of a negative sign is $p_0 = 1/2$. Under the alternative $m_1 > m_0$, we have $p_1 = P[X \leqslant m_0] < P[X \leqslant m_1] = 1/2$. Clearly the statistic T follows a binomial distribution, and when $m = m_0$,

$$T \sim \text{BIN}(n, p_0) \qquad \text{where } p_0 = P[X \leqslant m_0] = 1/2$$

A test of $H_0 : m = m_0$ against $H_a : m > m_0$ based on T is equivalent to the binomial test of $H_0 : p = p_0 = 1/2$ against $H_a : p < 1/2$, where T represents the number of successes, and a success corresponds to a negative sign for $(x_i - m_0)$. That is, for the alternative $m > m_0$, H_0 is rejected if T is small, as described in Theorem 12.4.1 earlier.

Theorem 14.2.1 Let $X \sim F(x)$ and $F(m) = 1/2$. A size α test of $H_0 : m = m_0$ against $H_a : m > m_0$ is to reject H_0 if

$$B(t; n, 1/2) \leqslant \alpha$$

where t = number of negative signs of $(x_i - m_0)$ for $i = 1, \ldots, n$. ■

The other one-sided or two-sided alternatives would be carried out similarly. The usual normal approximation could be used for moderate sample sizes. Also note that the sign test corresponds to a special case of the chi-square goodness-of-fit procedure considered earlier. In that case, c categories were used to reduce the problem to a multinomial problem; here we have just two categories, plus and minus, with each category equally likely under H_0. Of course, the sign test is designed to interpret inferences specifically related to the median of the original population.

The sign test for the median has the advantage that it is valid whatever the distribution F. If the true F happens to be normal, then one may wonder how much is lost by using the sign test compared to using the usual t test for means, which was derived under the normality assumption. One way of comparing two tests is to consider the ratio of the sample sizes required to achieve a given power. To test $H_0 : \theta = \theta_0$, let $n_1(\theta_i)$ be the sample size required to achieve a specified power at θ_i for test one, and $n_2(\theta_i)$ the sample size required to achieve the same power for test two. Then the (Pitman) asymptotic relative efficiency (ARE) of test two compared to test one is given by

$$\text{ARE} = \lim_{\theta_i \to \theta_0} \frac{n_1(\theta_i)}{n_2(\theta_i)} \tag{14.2.2}$$

If the test statistics are expressed in terms of point estimators of the parameter, then in many cases the ARE of the tests corresponds to the ratio of the variances of the corresponding point estimators. Thus there is often a connection between the relative efficiency of a test and the relative efficiency of point estimators as defined earlier. This aspect of the problem will not be developed further here, but it can be shown that under normality, the ARE of the sign test compared with the t test is given by $2/\pi = 0.64$, and it increases to approximately 95% for small n. That is, when normality holds, a t test based on 64 observations would give about the same power as a sign test based on 100 observations. Of course, if the normality assumption is not true, then the t test is not valid. Another restriction is that interval scale data are needed for the t test.

Example 14.2.1 The median income in a certain profession is \$24,500. The contention is that taller men earn higher wages than shorter men, so a random sample of 20 men who are six feet or taller is obtained. Their (ordered) incomes in thousands of dollars are as follows:

10.8, 12.7, 13.9, 18.1, 19.4, 21.3, 23.5, 24.0, 24.6, 25.0,

25.4, 27.7, 30.1, 30.6, 32.3, 33.3, 34.7, 38.8, 40.3, 55.5

To test $H_0 : m = 24{,}500$ against $H_a : m > 24{,}500$, we compute $T = 8$ negative signs. The p-value for this test based on this statistic is $B(8;\ 20,\ 0.5) = 0.2517$. Thus we do not have strong evidence based on this statistic to reject H_0 and support the claim that taller men have higher incomes.

Note that if any of the observed values are exactly m_0, then these values should be discarded and the sample size reduced accordingly. Note also that the sign test would have been unaffected in this example if the workers had been unwilling to give their exact incomes but willing to indicate whether it was less than \$24,500 or more than \$24,500.

14.3

BINOMIAL TEST (TEST ON QUANTILES)

Clearly a test for any quantile (or percentile) can be set up in the same manner as the sign test for medians. We may wish to test the hypothesis that x_0 is the p_0th percentile of a distribution $F(x)$, for some specified value p_0; that is, we wish to test

$$H_0 : x_{p_0} = x_0 \qquad (H_0 : P[X \leqslant x_0] = F(x_0) = p_0)$$

against

$$H_a : x_{p_0} > x_0 \qquad (H_a : F(x_0) < p_0)$$

Let t = number of negative signs of $(x_i - x_0)$ for $i = 1, \ldots, n$. Then when H_0 is true, $T \sim \text{BIN}(n, p_0)$, and this test is equivalent to a binomial test of $H_0 : p = p_0$ against $H_a : p < p_0$, where p_0 is the probability of a negative sign under H_0.

Theorem 14.3.1 Let $X \sim F(x)$ and $F(x_p) = p$. For a specified p_0, a size α test of $H_0 : x_{p_0} = x_0$ against $H_a : x_{p_0} > x_0$ is to reject H_0 if

$$B(t; n, p_0) \leqslant \alpha$$

where t is the number of x_i that are smaller than x_0 in a random sample of size n. ■

The other one-sided and two-sided alternative tests may be carried out in a similar manner, using the binomial test described in Theorem 12.4.2. These tests could be modified to apply to a discrete random variable X if care is taken with the details involved.

Example 14.3.1 In the study of Example 14.2.1, we wish to establish that the 25th percentile for tall men is less than \$24,500. That is, we test $H_0 : x_{0.25} = \$24{,}500$ against $H_a : x_{0.25} < \$24{,}500$. This is equivalent to testing $H_0 : F(24{,}500) = 0.25$ against $H_a : F(24{,}500) > 0.25$. We find from the data that $t = 8$, and the corresponding p value is

$$1 - B(t - 1, n, p_0) = 1 - B(7; 20, 0.25) = 0.102$$

An alternate expression for the test on quantiles can be given in terms of order statistics. The outcome $T = t$ is equivalent to having $x_{t:n} < x_0 < x_{t+1:n}$. We know that confidence intervals for a parameter can be associated with the values of the parameter for which rejection of H_0 would not occur in a test of hypothesis. In developing distribution-free confidence intervals for a quantile, the common practice is to express these directly in terms of the order statistics.

CONFIDENCE INTERVAL FOR A QUANTILE

Consider a continuous random variable $X \sim F(x)$ and let $F(x_p) = p$. Let $Z_i = F(X_{i:n})$. Then

$$g(x_{1:n}, \ldots, x_{n:n}) = n!f(x_{1:n}) \cdots f(x_{n:n}) \tag{14.3.1}$$

$$h(z_1, \ldots, z_n) = n! \qquad 0 < z_1 < \cdots < z_n < 1 \tag{14.3.2}$$

and

$$h_k(z_k) = \frac{n!}{(k-1)!(n-k)!} z_k^{k-1}(1 - z_k)^{n-k} \qquad 0 < z_k < 1 \tag{14.3.3}$$

Now

$$\begin{aligned} P[X_{k:n} \leqslant x_p] &= P[F(X_{k:n}) \leqslant F(x_p)] \\ &= P[Z_k \leqslant p] \\ &= \int_0^p h_k(z_k)\, dz_k \end{aligned} \tag{14.3.4}$$

This integral represents an incomplete beta function, and for integer k it can be expressed in terms of the cumulative binomial distribution, where

$$\begin{aligned} P[X_{k:n} \leqslant x_p] &= \sum_{j=k}^{n} \binom{n}{j} p^j (1-p)^{n-j} \\ &= 1 - B(k-1; n, p) \\ &= \gamma(k, n, p) \end{aligned} \tag{14.3.5}$$

Thus for a given pth percentile, the kth order statistic provides a lower $\gamma(k, n, p)$ level confidence limit for x_p, where k and n can be chosen to achieve a particular desired level. Binomial tables can be used for small n and the normal approximation for larger n.

In a similar fashion,

$$\begin{aligned} P[X_{k:n} \geqslant x_p] &= 1 - P[X_{k:n} < x_p] \\ &= B(k-1; n, p) \end{aligned} \tag{14.3.6}$$

and a desired upper confidence limit can be obtained by proper choice of k and n.

A two-sided confidence interval for a given percentile x_p also can be developed. Note that

$$P[X_{i:n} \leqslant x_p] = P[X_{i:n} \leqslant x_p \leqslant X_{j:n}] + P[X_{j:n} < x_p] \tag{14.3.7}$$

because either $X_{j:n} < x_p$ or $X_{j:n} \geqslant x_p$, so

$$\begin{aligned} P[X_{i:n} \leqslant x_p \leqslant X_{j:n}] &= P[X_{i:n} \leqslant x_p] - P[X_{j:n} < x_p] \\ &= B(j-1; n, p) - B(i-1; n, p) \end{aligned} \tag{14.3.8}$$

Again one would attempt to find combinations of i, j, and n to provide the desired confidence level.

It can be shown that equation (14.3.8) provides a conservative confidence interval if F is discrete.

Example 14.3.2 We now wish to compute a confidence interval for the 25th percentile in the previous example. We note that

$$P[X_{2:20} \leqslant x_{0.25}] = 1 - B(1; 20, 0.25) = 0.9757$$

and

$$P[X_{10:20} \geqslant x_{0.25}] = B(9; 20, 0.25) = 0.9861$$

Thus, $(x_{2:20}, x_{10:20}) = (12.7, 25.0)$ is a two-sided confidence interval for $x_{0.25}$ with confidence coefficient $1 - 0.0243 - 0.0139 = 0.9618$.

For large n, a normal approximation may be used. For example, for an upper limit, in equation (14.3.6) set $B(k-1; n, p) = \Phi(z)$, where $z = (k - 1 + 0.5 - np)/\sqrt{np(1-p)}$. For a specified level $1 - \alpha$, setting $z = z_{1-\alpha}$ gives an approximate expression for k in terms of n,

$$k = 0.5 + np + z_{1-\alpha}\sqrt{np(1-p)}$$

If k is rounded to the nearest integer, then $x_{k:n}$ is the approximate upper $1 - \alpha$ confidence limit for x_p. For the lower limit case, replace $z_{1-\alpha}$ with z_α.

TOLERANCE LIMITS

A function of the sample $L(x)$ is said to be a lower γ probability tolerance limit for proportion p^* if

$$\begin{aligned} P\left[\int_{L(\mathbf{X})}^{\infty} f(x)\,dx \geqslant p^*\right] &= P[1 - F(L(X)) \geqslant p^*] \\ &= P[F(L(X)) \leqslant 1 - p^*] \\ &= P[L(X) \leqslant x_{1-p^*}] = \gamma \end{aligned} \tag{14.3.9}$$

That is, we wish to have an interval that will contain a prescribed proportion p^* of the population. It is not possible to determine such an interval exactly if F is unknown, but the tolerance interval $(L(X), \infty)$ will contain at least a proportion p^* of the population with confidence level γ. The proportion p^* is referred to as the content of the tolerance interval. It is clear that a lower γ probability tolerance limit for proportion p^* is simply a lower γ level confidence limit for the $1 - p^*$ percentile, x_{1-p^*}. Thus $L(X) = X_{k:n}$ is a distribution-free lower γ tolerance limit for proportion p^* if k and n are chosen so that

$$\gamma(k, n, 1 - p^*) = 1 - B(k - 1; n, 1 - p^*) = \gamma \tag{14.3.10}$$

One may also wish to have a two-sided tolerance interval $(L(X), U(X))$ such that

$$P\{F[U(X)] - F[L(X)] \geq p\} = \gamma$$

A two-sided tolerance interval cannot be obtained from a two-sided confidence interval on a percentile, but a two-sided distribution-free tolerance interval can be obtained in the form $(L(X), U(X)) = (X_{i:n}, X_{j:n})$, by the proper choice of i, j, and n. We need to choose i, j, and n to satisfy

$$P[F(X_{j:n}) - F(X_{i:n}) \geq p] = P[Z_j - Z_i \geq p] = \gamma$$

where $f(z_1, \ldots, z_n) = n!;\ 0 < z_1 < \cdots < z_n < 1$.

A few comments will be made before determining the distribution of $Z_j - Z_i$. The content between two consecutive order statistics is known as a coverage, say

$$W_i = F(X_{i:n}) - F(X_{i-1:n}) = Z_i - Z_{i-1} \tag{14.3.11}$$

and $E(W_i) = 1/(n + 1)$. The expected content between two consecutive order statistics is $1/(n + 1)$. It follows that the expected content between any two order statistics $X_{i:n}$ and $X_{j:n}$, $i < j$, is given by

$$E(Z_j - Z_i) = E\left(\sum_{k=i+1}^{j} W_k\right) = \frac{j - i}{n + 1} \tag{14.3.12}$$

That is, the expected content depends only on the difference $j - i$ and does not depend on which i and j are involved. It turns out in general that the density of $Z_j - Z_i$ or the sum of any $j - i$ coverages depends only on $j - i$ and not on i and j separately.

Consider the transformation

$$w_1 = z_1 \qquad w_2 = z_2 - z_1, \ldots \qquad w_n = z_n - z_{n-1}$$

with inverse transformation

$$z_k = \sum_{i=1}^{k} w_i \qquad k = 1, 2, \ldots, n$$

The Jacobian is 1, so

$$f(w_1, \ldots, w_n) = n! \qquad w_i > 0 \qquad \sum_{i=1}^{n} w_i < 1 \tag{14.3.13}$$

This density is symmetric with respect to the w_i. That is, the density of any function of the w_i will not depend on which of the w_i are involved. In particular the density of the variable,

$$U_{j-i} = F(X_{j:n}) - F(X_{i:n}) = Z_j - Z_i = \sum_{k=i+1}^{j} W_k \tag{14.3.14}$$

depends only on the number of coverages summed, $j - i$, and the density is the same as the density of the sum of the first $j - i$ coverages,

$$\sum_{k=1}^{j-i} W_k = Z_{j-i}$$

The marginal density of Z_{j-i} is given by equation (14.3.3) with $k = j - i$, which is a beta density, so

$$U_{j-i} = Z_j - Z_i \sim \text{BETA}(j - i, n - j + i + 1) \tag{14.3.15}$$

Expressed in terms of the binomial CDF,

$$P[Z_j - Z_i > p] = \sum_{k=0}^{j-i-1} \binom{n}{k} p^k (1-p)^{n-k} = B(j - i - 1; n, p) \tag{14.3.16}$$

Thus the interval $(X_{i:n}, X_{j:n})$ provides a two-sided γ probability tolerance interval for proportion p if i and j are chosen to satisfy

$$B(j - i - 1; n, p) = \gamma$$

Theorem 14.3.2 For a continuous random variable $X \sim F(x)$, $L(X) = X_{k:n}$ is a lower γ probability tolerance limit for proportion p^*, where $\gamma = 1 - B(k - 1; n, 1 - p^*)$. Also $X_{k:n}$ is an upper γ probability tolerance limit for proportion p, where $\gamma = B(k - 1; n, p)$. The interval $(X_{i:n}, X_{j:n})$ is a two-sided γ probability tolerance interval for proportion p, where $\gamma = B(j - i - 1; n, p)$. ■

Example 14.3.3 In Example 14.3.2, we see that the lower confidence limit on $x_p = x_{0.25}$, given by $L(X) = X_{2:20}$, also may be interpreted as a $\gamma = 0.9757$ probability tolerance limit for proportion $p^* = 1 - 0.25 = 0.75$. That is, we are 97.57% confident that at least 75% of the incomes of tall men in this profession will exceed $x_{2:20} = 12.7$ thousands of dollars.

If we were interested in a lower tolerance limit for proportion $p^* = 0.90$, then Table 1 (Appendix C) at $1 - p^* = 0.10$ shows

$$P[X_{1:20} \leqslant x_{0.10}] = 1 - B(0; 20, 0.10) = 0.8784$$

Thus, for example, if a 95% tolerance limit is desired for proportion 0.90, a larger sample size is required.

We see that $(L(X), U(X)) = (X_{1:20}, X_{20:20})$ provides a two-sided tolerance interval for proportion 0.80 with probability level $\gamma = B(20 - 1 - 1;\ 20,\ 0.80) = 0.9308$.

For specified γ and p it is of interest to know what sample size is required so that $(X_{1:n}, X_{n:n})$ will provide the desired tolerance interval. Setting

$$B(n - 2;\ n,\ p) = \gamma$$

yields n as the solution to

$$np^{n-1} - (n - 1)p^n = 1 - \gamma$$

For this n, we find $P[F(X_{n:n}) - F(X_{1:n}) \geqslant p] = \gamma$.

14.4

TWO-SAMPLE SIGN TEST

The one-sample sign test is modified easily for use in a paired sample problem, and this approach can be used as an alternative to the paired sample t test when normality cannot be assumed. Assume that n independent pairs of observations (x_i, y_i), $i = 1, \ldots, n$, are available, and let T equal the number of times X_i is less than Y_i. In terms of the differences $X_i - Y_i$, we say that T = number of negative signs of $X_i - Y_i$, $i = 1, \ldots, n$. Again we will assume that X and Y are continuous random variables so that $P[X = Y] = 0$, but if an observed $x_i - y_i = 0$ because of roundoff or other reasons, then that outcome will be discarded and the number of pairs reduced by one. The sign test is sensitive to shifts in location, and it should be useful in detecting differences in means or differences in medians, although strictly speaking it is a test of whether the median of the differences is zero.

Theorem 14.4.1 Suppose that X and Y are continuous random variables and n independent pairs of observation (x_i, y_i) are available. Consider

$$H_0 : P[X < Y] = P[X > Y] = \tfrac{1}{2} \qquad (H_0 : \text{median } (X - Y) = 0)$$

against

$$H_a : P[X < Y] < P[X > Y] \qquad (H_a : \text{median } (X - Y) > 0)$$

Let t be the number of negative signs of $(x_i - y_i)$, $i = 1, \ldots, n$. Then, under H_0,

$$T \sim \text{BIN}(n, 1/2),$$

and a size α test of H_0 against H_a is to reject H_0 if $B(t;\ n,\ 1/2) \leqslant \alpha$. ■

Example 14.4.1 A campaign manager wishes to measure the effectiveness of a certain politician's speech. Eighteen people were selected at random and asked to rate the politician before and after the speech. Of these, 11 had a positive reaction, four had a negative reaction, and three had no reaction. To test H_0 : no effect against H_a : positive effect, we have $t = 4$ negative reactions and $n = 15$, and the p-value for the test is

$$B(4;\, 15,\, 0.5) = 0.0592$$

Thus there is statistical evidence at this error level that the speech was effective for the sampled population.

14.5 WILCOXON PAIRED-SAMPLE SIGNED-RANK TEST

The two-sample sign test makes use only of the signs of the differences $x_i - y_i$. A test would be expected to be more powerful or more efficient if it also makes some use of the magnitude of the differences, which the Wilcoxon signed-rank test does.

Let $d_i = x_i - y_i$ for $i = 1, \ldots, n$ denote the differences of the matched pairs; then rank the differences without regard to sign. That is, rank the $|d_i|$ according to magnitude, but keep track of the signs associated with each one. Now replace the $|d_i|$ with their ranks, and let T be the sum of the ranks of the positive differences. Again this test statistic will be sensitive to differences in location between the two populations. In the sign test the positive signs and negative signs were assumed to be equally likely to occur under H_0. In this case, to determine a critical value for T, we need to assume that the positive signs and negative signs are equally likely to be assigned to the ranks under H_0. The signs will be equally likely if the joint density of x and y is symmetric in the variables; that is, we could consider $H_0 : F(x_i, y_i) = F(y_i, x_i)$. This corresponds to the distribution of the differences being symmetric about zero or $H_0 : F_D(d_i) = 1 - F_D(-d_i)$. Note that the probability of a negative sign is $F_D(0) = 1/2$, and the median is $m_D = 0$. Also, for a symmetric distribution, the mean and the median are the same. Thus, under the symmetry assumptions mentioned, this test may be considered a test for equality of means for the two populations.

In general, the signed-rank test is considered a test of the equality of two populations and has good power against the alternative of a difference in location, but the specific assumption under H_0 is that any sequence of signs is equally likely to be associated with the ranked differences. That is, if the alternative is stated as $H_a : E(X) < E(Y)$, then rejection of H_0 could occur because of some other lack of symmetry. For a one-sided alternative, say $H_a : E(X) < E(Y)$, one would reject H_0 for small values of T, the sum of positive ranks of the differences $d_i = x_i - y_i$. To illustrate how a critical value for T may be computed, consider $n = 8$ pairs; then there are $2^8 = 256$ equally likely possible sequences of pluses and minuses that can be associated with the eight ranks. If we are interested in small values of T, then we will order these 256 possible outcomes by putting the ones associated with the smallest values of T in the critical region. The outcomes associated with small values of T are illustrated in Table 14.1.

TABLE 14.1 **Signs associated with the Wilcoxon paired-sample signed-rank test**

	Ranks								
	1	2	3	4	5	6	7	8	T
Signs	−	−	−	−	−	−	−	−	0
	+	−	−	−	−	−	−	−	1
	−	+	−	−	−	−	−	−	2
	−	−	+	−	−	−	−	−	3
	+	+	−	−	−	−	−	−	3
	+	−	+	−	−	−	−	−	4
	−	−	−	+	−	−	−	−	4

Placing the first five possible outcomes in the critical region corresponds to rejecting H_0 if $T \leqslant 3$, and this gives $\alpha = 5/256 = 0.0195$. Rejecting H_0 if $T \leqslant 4$ results in a significance level of $\alpha = 7/256 = 0.027$, and so on.

Conservative critical values, t_α, are provided in Table 12 (Appendix C) for the usual prescribed α levels for $n \leqslant 20$. The true Type I error may be slightly less than α because of discreteness. A normal approximation is adequate for $n > 20$. The mean and variance of T may be determined as follows.

Without loss of generality, the subscripts of the original differences can be rearranged so that the absolute differences are in ascending order, $|d_1| < \cdots < |d_n|$, in which case the rank of $|d_i|$ is i, and the signed-rank statistic can be written as $T = \sum_{i=1}^{n} iU_i$ where $U_i = 1$ if the difference whose absolute value has rank i is positive, and $U_i = 0$ if it is negative. Under H_0, the variables $U_1, \ldots, U_n$ are independent identically distributed Bernoulli variables, $U_i \sim \text{BIN}(1, 1/2)$. Thus,

$$E(T) = E\left(\sum_{i=1}^{n} iU_i\right) = \sum_{i=1}^{n} iE(U_i) = \frac{1}{2}\sum_{i=1}^{n} i$$

$$= \frac{1}{2}\,\frac{n(n+1)}{2} = \frac{n(n+1)}{4}$$

and

$$\text{Var}(T) = \text{Var}\left(\sum_{i=1}^{n} iU_i\right) = \sum_{i=1}^{n} i^2 \text{Var}(U_i)$$

$$= \frac{1}{4}\sum_{i=1}^{n} i^2 = \frac{n(n+1)(2n+1)}{24}$$

For large n,

$$\frac{T - E(T)}{\sqrt{\text{Var}(T)}} \xrightarrow{d} Z \sim \text{N}(0, 1)$$

Theorem 14.5.1 Let $d_i = x_i - y_i$, $i = 1, \ldots, n$, denote the differences of n independent matched pairs. Rank the d_i without regard to sign, and let

$$T = \text{Sum of ranks associated with positive signed differences}$$

A (conservative) size α test of

$$H_0 : F(x_i, y_i) = F(y_i, x_i) \qquad (H_0 : F_D(d_i) = 1 - F_D(-d_i))$$

against

$$H_a : X \text{ is stochastically smaller than } Y \quad (P[X > a] < P[Y > a], \text{ all } a)$$

is to reject H_0 if $t \leqslant t_\alpha$ (where t_α is given in Table 12). ∎

This test also may be used as a test of the hypothesis that $f_D(d)$ is symmetric with $\mu_D = E(X) - E(Y) = 0$ against the alternative $E(X) < E(Y)$.

For a two-sided alternative, let t^* be the smaller sum of like signed ranks; then a (conservative) size α test is to reject H_0 if $t^* \leqslant t_{\alpha/2}$.

For $n \geqslant 20$, approximately

$$\frac{T - \dfrac{n(n+1)}{4}}{\sqrt{\dfrac{n(n+1)(2n+1)}{24}}} \sim \text{N}(0, 1)$$

Note that the signed-rank test also can be used as a one-sample test for the median of a symmetric population. Consider the null hypothesis H_0 that X is a continuous random variable with a symmetric distribution about the median m_0. Let $d_i = x_i - m_0$ and T = sum of positive signed ranks as above. Then a size α test of H_0 against the alternative $H_a : m < m_0$ is to reject H_0 if $t \leqslant t_\alpha$. For $H_a : m > m_0$, let T = sum of ranks for the negative d_i.

If the differences actually are normally distributed, then a paired-sample t test is applicable for testing

$$H_0 : \mu_D = E(X) - E(Y) = 0 \qquad \text{against} \qquad H_a : E(X) - E(Y) < 0$$

If the Wilcoxon signed-rank test is used in this case, its asymptotic relative efficiency is $3/\pi = 0.955$.

Also note that any observed $d_i = 0$ should be discarded and the sample size reduced. If there is a tie in the value of two or more d_i, then it is common practice to use the average of their ranks for all of the tied differences in the group.

Example 14.5.1 To illustrate the one-sample signed-rank test, consider again Example 14.2.1, where we wish to test the median income $H_0: m = 24.5$ thousand dollars against $H_a: m > 24.5$. If we assume that the distribution of incomes is symmetric, then the signs of $x_i - m_0$ are equally likely to be positive or negative, and we can apply the Wilcoxon paired-sample signed-rank test. The test in this case also will be a test of $H_0: \mu = 24.5$, because the mean and median are the same for symmetric distributions.

Note that if the assumption of symmetry is not valid, then H_0 could be rejected even though $m = m_0$, because lack of symmetry could cause the signs of $x_i - m_0$ not to be equally likely to be positive or negative. Indeed, if the median $m = m_0$ can be assumed known, then the Wilcoxon signed-rank test can be used as a test of symmetry. That is, we really are testing both that the distribution is symmetric and that it has median m_0.

In our example we first determine the ranks of the d_i according to their absolute value $|d_i| = |x_i - m_0|$, as follows:

d_i	−13.7	−11.8	−10.6	−6.4	−5.1	−3.2	−1.0	−0.5	0.1	0.5
rank ($\|d_i\|$)	17	16	15	11	8	6.5	5	2.5	1	2.5

d_i	0.9	3.2	5.6	6.1	7.8	8.8	10.2	14.3	15.8	31.0
rank ($\|d_i\|$)	4	6.5	9	10	12	13	14	18	19	20

For $H_a: m > 24.5$, we reject H_0 for a small sum of negative signed ranks, where for this set of data

$$T = 2.5 + 5 + 6.5 + \cdots = 81$$

From Table 12 (Appendix C) we see that $T = 81$ gives a p value of 0.20, so we cannot reject H_0 at the usual prescribed significance levels. However, this test does give some indication that the hypothesis that the incomes are symmetrically distributed about a median of $24,500 is false. The lack of symmetry may be the greater source for disagreement with H_0 in this example.

Example 14.5.2 To illustrate the paired-sample case, suppose that in the previous example a second sample, $y_1, \ldots, y_{20}$, of 20 men under six feet tall is available, and we wish to see if there is statistical evidence that the median income of tall men is greater than the median income of shorter men, $H_0: x_{0.50} = y_{0.50}$ against $H_a: x_{0.50} > y_{0.50}$. In this two-sample case, if the medians are equal, then there would be little reason to suspect the assumption of symmetry. Note that if the two samples are independent, then the paired sample test still can be applied, but a more powerful independent samples test will be discussed later. If the samples are paired in a meaningful way, then the paired-sample test may be preferable to an independent samples test. For example, the pairs could be of short and tall men who have the same level of education or the same age. Of course, the samples would not be independent in that case.

Consider the following 20 observations, where it is assumed that the first observation was paired with the first (ordered) observation in the first sample, and so on. The differences $d_i = x_i - y_i$ also are recorded.

x_i	10.8	12.7	13.9	18.1	19.4	21.3	23.5	24.0	24.6	25.0
y_i	9.8	13.0	10.7	19.2	18.0	20.1	20.0	21.2	21.3	25.5
d_i	1.0	−0.3	3.2	−1.1	1.4	1.2	3.5	2.8	3.3	−0.5
rank ($\|d_i\|$)	4	1	12	5	8	6	14	10	13	3

x_i	25.4	27.7	30.1	30.6	32.3	33.3	34.7	38.8	40.3	55.5
y_i	25.7	26.4	24.5	27.5	25.0	28.0	37.4	43.8	35.8	60.9
d_i	−0.3	1.3	5.6	3.1	7.3	5.3	−2.7	−5.0	4.5	−5.4
rank ($\|d_i\|$)	2	7	19	11	20	17	9	16	15	18

For the alternative hypothesis as stated, we reject H_0 if the sum of ranks of the negative differences is small. An alternative approach would have been to relabel or to let $d_i = y_i - x_i$; then we would have used the sum of positive signed ranks. Note also that $T^+ + T^- = n(n + 1)/2$, which is useful for computing the smaller sum of like-signed ranks for a two-sided alternative. We have $T = 1 + 5 + 3 + 2 + 9 + 16 + 18 = 54$. Because $t_{0.05} = 60$, according to this set of data we can reject H_0 at the 0.05 level.

The approximate large sample 0.05 critical value for this case is given by

$$t_{0.05} \doteq z_{0.05}\sqrt{\frac{20(21)(41)}{24}} + \frac{20(21)}{4} = 60.9$$

14.6

PAIRED-SAMPLE RANDOMIZATION TEST

In the Wilcoxon signed-rank test, the actual observations were replaced by ranks. The advantage of this is that predetermined critical values can be tabulated as a function only of n and α. It is possible to retain the actual observations and develop a probability structure for testing by assuming that all ordered outcomes of data are equally likely under H_0. For example, if two samples are selected from the same population, then the assignment of which came from population 1 and which from population 2 could be made at random. There would be $N = (n_1 + n_2)!/n_1!n_2!$ equally likely possible assignments to the given set of data. Thus a size $\alpha = k/N$ size test of equality can be obtained by choosing k of these outcomes to be included in our critical region. Of course, we want to pick the k outcomes that are most likely to occur when the alternative hypothesis is true. Thus, we need some test statistic, T, that will identify what order we want to use in putting the possible outcomes into the critical region, and we need to know the critical value for T for a given α. In the signed-rank test, we used the sum of the positive signed ranks, and we were able to tabulate critical values. We now may use the sum of the positive differences as our test statistic, although we cannot determine the proper critical value until we know all the values of the d_i. For example, the following eight differences are observed in a paired sample problem:

$$-20, \ -10, \ -8, \ -7, \ +5, \ -4, \ +2, \ -1$$

There are $2^8 = 256$ possible ways of assigning pluses and minuses to eight numbers, and each outcome is equally likely under the hypothesis that the D_i are symmetrically distributed about $m_D = 0$. We may rank these possible outcomes according to T, the sum of the positive d_i; the first dozen outcomes are shown in Table 14.2

TABLE 14.2 **Differences d_i for the paired-sample randomization test**

								T
−20	−10	−8	−7	−5	−4	−2	−1	0
−20	−10	−8	−7	−5	−4	−2	+1	1
−20	−10	−8	−7	−5	−4	+2	−1	2
−20	−10	−8	−7	−5	−4	+2	+1	3
−20	−10	−8	−7	−5	+4	−2	−1	4
−20	−10	−8	−7	−5	+4	−2	+1	5
−20	−10	−8	−7	+5	−4	−2	−1	5
−20	−10	−8	−7	+5	−4	−2	+1	6
−20	−10	−8	−7	−5	+4	+2	−1	6
−20	−10	−8	−7	−5	+4	+2	+1	7
−20	−10	−8	−7	+5	−4	+2	−1	7
−20	−10	−8	+7	−5	−4	−2	−1	7

Thus to test $H_0: m_D = 0$ (and symmetry) against a one-sided alternative $H_a: m_D < 0$ is to reject H_0 for small T. Given these eight numerical values, a size $\alpha = 12/256 = 0.047$ test is to reject H_0 if $T \leqslant 7$. Thus we can reject H_0 at this α for the data as presented. That is, there were only 12 cases as extreme as the one observed (using the statistic T).

Tests such as the one described based on the actual observations have high efficiency, but the test is much more convenient if the observations are replaced by ranks so that fixed critical values can be tabulated. A normal approximation can be used for larger n, and quite generally the normal theory test procedures can be considered as approximations to the corresponding "exact" randomization tests. For each normal test described, the same test statistic can be used to order the set of possible outcomes produced under the randomization concept that gives equally likely outcomes under H_0. Approximating the distribution of the statistic then returns us to a normal type test. For small n, exact critical values can be computed as described, but these are in general quite inconvenient to determine.

14.7

WILCOXON AND MANN-WHITNEY (WMW) TESTS

We now will consider a nonparametric analog to the t test for independent samples. The Wilcoxon rank sum test is designed to be sensitive to differences in location, but strictly speaking it is a test of the equality of two distributions. To illustrate the case of a one-sided alternative, consider $H_0: F_X = F_Y$ against $H_a: F_X > F_Y$ (X stochastically smaller than Y). Suppose that n_1 observations $x_1, \ldots, x_{n_1}$ and n_2 observations $y_1, \ldots, y_{n_2}$ are available from the two populations. Combine these two samples and then rank the combined samples in ascending order. Under H_0 any arrangement of the x's and y's is equally likely to occur, and there are $N = (n_1 + n_2)!/n_1!n_2!$ possible arrangements. Again we can produce a size $\alpha = k/N$ test by choosing k of the possible arrangements to include in a critical region.

We wish to select arrangements to go into the critical region that are likely to occur when H_0 is true to minimize our Type II error. The Wilcoxon test says to replace the observations with their combined sample ranks, and then reject H_0 if the sum of the ranks of the x's is small. That is, the order of preference for including an arrangement in the critical region is based on $W_x = \sum \text{rank } (x\text{'s})$. For example, if $n_1 = 4$ and $n_2 = 5$, then there are $\binom{9}{4} = 126$ possible arrangements; the ones with the smallest values of W_x are shown in Table 14.3.

A size $\alpha = 7/126 = 0.056$ test is achieved by rejecting H_0 if the observed $w_x \leqslant 13$. Note that $W_x + W_y = (n_1 + n_2)(n_1 + n_2 + 1)/2$.

TABLE 14.3 Arrangements of x's and y's for the Wilcoxon–Mann-Whitney tests

1	2	3	4	5	6	7	8	9	W_x	U_x
x	x	x	x	y	y	y	y	y	10	0
x	x	x	y	x	y	y	y	y	11	1
x	x	y	x	x	y	y	y	y	12	2
x	x	x	y	y	x	y	y	y	12	2
x	y	x	x	x	y	y	y	y	13	3
x	x	y	x	y	x	y	y	y	13	3
x	x	x	y	y	y	x	y	y	13	3

Mann and Whitney suggested using the statistic

$$U_x = \text{Number of times a } y \text{ precedes an } x$$

It turns out that U_x and W_x are equivalent statistics. The minimum value of W_x is $n_1(n_1 + 1)/2$, and this corresponds to $U_x = 0$. If one y precedes one x, then $U_x = 1$ and this increases W_x by 1. Similarly, each time a y precedes an x, this increases W_x by one more so that

$$W_x = \frac{n_1(n_1 + 1)}{2} + U_x$$

Similarly,

$$W_y = \frac{n_2(n_2 + 1)}{2} + U_y$$

where U_y is the number of times an x precedes a y. Note that $U_x + U_y = n_1 n_2$.

For the alternative H_a : (X stochastically larger than Y), we would reject H_0 if W_x is large or if W_y and U_y are small. The seven sequences corresponding to the smallest values of W_y in the example are the seven sequences in Table 14.3 ranked in the reverse order. In this case $W_y = 18$ for the last sequence, for example, and $U_y = 18 - [5(6)/2] = 3 = U_x$ for the original table. Indeed, for a given sequence, U_x computed under the first order of ranking is the same as U_y for the reverse ranking, because the same number of interchanges between the x's and y's occurs whichever direction the ranks are applied to the sequences. Thus, U_x and U_y are identically distributed and the same critical values can be used with either U_x or U_y. Sometimes the subscript will be suppressed and the notation U will be used. The notations u_x and u_y will refer to the observed values of U_x and U_y, respectively, and u_α is the notation for a 100αth percentile of U (where $U \sim U_x \sim U_y$).

Table 13A (Appendix C) gives $P[U_x \leqslant u] = P[U_y \leqslant u]$ for values of $m = \min(n_1, n_2)$ and $n = \max(n_1, n_2)$ less than or equal to 8. Table 13B (Appendix C) gives critical values u_α such that $P[U \leqslant u_\alpha] \leqslant \alpha$ for $9 \leqslant n \leqslant 14$. A normal approximation may be used for larger sample sizes.

Theorem 14.7.1 Let $x_1, \ldots, x_{n_1}$ and $y_1, \ldots, y_{n_2}$ be independent random samples. Then for an observed value of U_x, reject $H_0 : F_X = F_Y$ in favor of $H_a : F_X > F_Y$ (X stochastically smaller than Y) if $P[U \leqslant u_x] \leqslant \alpha$, or if $U_x \leqslant u_\alpha$. Reject H_0 in favor of $H_a : F_X < F_Y$ (X stochastically larger than Y) if $P[U \leqslant u_y] \leqslant \alpha$, or if $U_y \leqslant u_\alpha$. Reject H_0 in favor of a two-sided alternative $H_a : F_X \neq F_Y$ if $P[U \leqslant u_x] \leqslant \alpha/2$ or $P[U \leqslant u_y] \leqslant \alpha/2$. Alternately, reject H_0 against the two-sided alternative if $\min(U_x, U_y) = \min(U_x, n_1 n_2 - U_x) \leqslant u_\alpha$.

A normal approximation for larger sample sizes may be determined as follows. Let

$$Z_{ij} = \begin{cases} 0 & X_i < Y_j \\ 1 & X_i > Y_j \end{cases} \tag{14.7.1}$$

Then

$$U = \sum_{i=1}^{n_1} \sum_{j=1}^{n_2} Z_{ij} \tag{14.7.2}$$

Under H_0,

$$E(Z_{ij}) = 1 \cdot P[Z_{ij} = 1] = \tfrac{1}{2} \tag{14.7.3}$$

and

$$E(U) = \frac{n_1 n_2}{2} \tag{14.7.4}$$

The expected values of products of the Z_{ij} are required to determine the variance of U. For example, if $j \neq k$, then

$$\begin{aligned} E(Z_{ij} Z_{ik}) &= 1 \cdot 1 \cdot P[Z_{ij} = 1, Z_{ik} = 1] \\ &= P[X_i > Y_j ; X_i > Y_k] \\ &= \frac{2}{3!} = \frac{1}{3} \end{aligned} \tag{14.7.5}$$

There are two ways to have a success in the 3! arrangements of X_i, Y_j, and Y_k. It can be shown (see Exercise 26) that

$$\text{Var}(U) = \frac{n_1 n_2 (n_1 + n_2 + 1)}{12} \tag{14.7.6}$$

Thus the normal approximation for the α level critical value is

$$u_\alpha \doteq \frac{n_1 n_2}{2} - z_{1-\alpha} \sqrt{n_1 n_2 (n_1 + n_2 + 1)/12} \tag{14.7.7}$$

It is possible to express the exact distribution of U recursively, but that will not be considered here. If ties occur and their number is not excessive, then it is common practice to assign the average of the ranks to each tied observation. Other adjustments also have been studied for the case of ties. The asymptotic relative efficiency of the WMW test compared to the usual t test under normal assumptions is $3/\pi = 0.955$.

Example 14.7.1 The times to failure of airplane air conditioners for two different airplanes were recorded as follows:

x	23	261	87	7	120	14	62	47	225	71	246	21
y	55	320	56	104	220	239	47	246	176	182	33	

We wish to test $H_0 : F_X = F_Y$ against $H_a : X$ is stochastically smaller than Y. This alternative could be interpreted as $H_a : \mu_X < \mu_Y$, if it is assumed that the distributions are otherwise the same. Associating ranks with the combined samples gives the following results.

x	x	x	x	y	y	x	y	y	x	x	x
7	14	21	23	33	47	47	55	56	62	71	87
1	2	3	4	5	6.5	6.5	8	9	10	11	12

y	x	y	y	y	x	y	x	y	x	y
104	120	176	182	220	225	239	246	246	261	320
13	14	15	16	17	18	19	20.5	20.5	22	23

We have $n_x = 12$, $n_y = 11$, and the sum of the ranks of the x's is

$$W_x = 1 + 2 + 3 + 4 + 6.5 + \cdots = 124$$

and $U_x = W_x - n_x(n_x + 1)/2 = 124 - 78 = 46$.

For the given alternative, we wish to reject H_0 if W_x or U_x is small. From Table 13B (Appendix C), the $\alpha = 0.10$ critical value is $u_{0.10} = 44$, so we cannot reject H_0 at the $\alpha = 0.10$ significance level.

To illustrate the asymptotic normal approximation, $E(U) = 66$, $\text{Var}(U) = 264$, and the approximate p-value for this test is

$$P[U \leqslant 46] \doteq \Phi\left(\frac{46 - 66}{\sqrt{264}}\right) = \Phi(-1.23) = 0.1093$$

14.8

CORRELATION TESTS—TESTS OF INDEPENDENCE

Suppose that we have n pairs of observations (x_i, y_i) from a continuous bivariate distribution function $F(x, y)$ with continuous marginal distributions $F_1(x)$ and $F_2(y)$. We wish to test for independence of X and Y, $H_0 : F(x, y) = F_1(x)F_2(y)$ against, say, the alternative of a positive correlation.

For a given set of observations there are $n!$ possible pairings, which are all equally likely under H_0 that X and Y are independent. For example, we may consider a fixed ordering of the y's; then there are $n!$ permutations of the x's that can be paired with the y's. Let us consider a test that is based on a measure of relationship known as the sample correlation coefficient,

$$r = \frac{\sum x_i y_i - n\bar{x}\bar{y}}{\sqrt{[\sum x_i^2 - n\bar{x}^2][\sum y_i^2 - n\bar{y}^2]}}$$

That is, for a size $\alpha = k/n!$ level test of H_0 against the alternative of a positive correlation, we will compute r for each of the $n!$ possible permutations, and then place the k permutations with the largest values of r in the critical region. If the observed ordering in our sample is one of these permutations, then we reject H_0. Note that $\bar{x}$, $\bar{y}$, s_x, and s_y do not change under permutations of the observations, so we may equivalently consider

$$t = \sum x_i y_i$$

as our test statistic.

Again it becomes too tedious for large n to compute t for all $n!$ permutations to determine the critical value for T. We may use a normal approximation for large n, and for smaller n we may again consider replacing the observations with their ranks so that fixed critical values can be computed and tabulated once and for all.

NORMAL APPROXIMATION

For fixed y_i, let us consider the moments of the x_i relative to the $n!$ equally likely permutations. The notation is somewhat ambiguous, but suppose we let X_1 denote a random variable that takes on the n values $x_1, \ldots, x_n$, each with probability $1/n$. Similarly, the variable $X_i X_j$ will take on the $n(n-1)$ values $x_i x_j$ for $i \neq j$, each with probability $1/(n)(n-1)$, and so on. Now

$$E(X_i) = \sum_{i=1}^{n} x_i \frac{1}{n} = \bar{x}$$

$$\text{Var}(X_i) = E(X_i^2) - \bar{x}^2 = \sum x_i^2 \frac{1}{n} - \bar{x}^2 = (n-1)s_x^2/n$$

and

$$\text{Cov}(X_i, X_j) = E(X_i X_j) - \bar{x}^2 = \frac{1}{n(n-1)} \sum\sum_{i \neq j} x_i x_j - \bar{x}^2 \qquad i \neq j$$

Now

$$E(T) = E(\sum X_i y_i) = \sum y_i E(X_i) = \sum y_i \bar{x} = n\bar{x}\bar{y}$$

and

$$E(r) = 0$$

Because the correlation coefficient is invariant over shifts in location, for convenience we will assume temporarily that the x's and y's are shifted so that $\bar{x} = \bar{y} = 0$; then

$$\begin{aligned}
\text{Var}\left(\sum X_i y_i\right) &= \sum_{i=1}^{n} y_i^2 \,\text{Var}(X_i) + \sum\sum_{i \neq j} y_i y_j \,\text{Cov}(X_i, X_j) \\
&= \sum_{i=1}^{n} y_i^2 (n-1)s_x^2/n + \sum\sum_{i \neq j} y_i y_j \sum\sum_{i \neq j} x_i x_j / n(n-1) \\
&= \frac{(n-1)^2 s_y^2 s_x^2}{n} + \left[\left(\sum y_i\right)^2 - \sum y_i^2\right]\left[\left(\sum x_i\right)^2 - \sum x_i^2\right]/n(n-1) \\
&= \frac{(n-1)^2 s_y^2 s_x^2}{n} + \frac{(n-1)s_y^2 s_x^2}{n} \\
&= (n-1)s_y^2 s_x^2
\end{aligned}$$

Thus

$$\text{Var}(r) = \frac{1}{(n-1)^2 s_x^2 s_y^2} \text{Var}\left(\sum X_i y_i\right) = \frac{1}{n-1}$$

These moments were calculated conditionally, given fixed values of (x_i, y_i), but because the results do not depend on (x_i, y_i), the moments are also true unconditionally.

It can be shown that a good large sample approximation is given by

$$r \sim N\left(0, \frac{1}{n-1}\right)$$

It is interesting that for large n, approximately,

$$E(r^3) = 0 \qquad \text{and} \qquad E(r^4) = \frac{3}{n^2 - 1}$$

These four moments are precisely the first four moments of the exact distribution of r based on random samples from a bivariate normal distribution. Thus a very close approximation for the "permutation" distribution of r, which is quite accurate even for small n, is obtained by using the exact distribution of r under normal theory. We will find in the next chapter that under the hypothesis of independence, the sample correlation coefficient can be transformed into a statistic that is t distributed. In particular,

$$\frac{\sqrt{n-2r}}{\sqrt{1-r^2}} \sim t(n-2)$$

Basically, the preceding results suggest that the test for independence developed under normal theory is very robust in this case, and one does not need to worry much about the validity of the normal assumptions for moderate sample sizes. If one wishes to determine an exact nonparametric test for very small n, say 5 or 10, then it is again convenient to make use of ranks. The rank correlation coefficient also may be useful for testing randomness.

Example 14.8.1 Consider again the paired-sample data given in Example 14.5.2. The correlation coefficient for that set of paired data is $r = 0.96$. It is clear that the pairing was effective in this case and that the samples are highly correlated without performing tests of independence. The approximate t statistic in this case is $t = 0.96\sqrt{18}/\sqrt{1 - 0.96^2} = 14.8$.

Again, it appears safe to use Student's t distribution based on normal theory unless n is very small.

SPEARMAN'S RANK CORRELATION COEFFICIENT

Again consider n pairs of observations; this time, however, the pairs already are ordered according to the y_i. Thus the pairs will be denoted by $(X_i, y_{i:n})$, $i = 1, \ldots, n$, where the $y_{i:n}$ are the fixed ordered y observations, and x_i denotes the x value paired with the ith largest y value. We will replace the observed values with ranks. Let $W_i = \text{rank}(y_{i:n}) = i$, and let $U_i = \text{rank}(x_i)$ denote the rank of the x value that is paired with the ith largest y value. The sample correlation coefficient based on these ranks is referred to as Spearman's rank correlation coefficient, R_s. It may be conveniently expressed in terms of the difference of the ranks,

$$d_i = U_i - i$$

We have

$$\bar{W} = \bar{U} = (n + 1)/2$$

$$\begin{aligned}(n - 1)s_W^2 = (n - 1)s_U^2 &= \sum i^2 - n\bar{U}^2 \\ &= \frac{n(n + 1)(2n + 1)}{6} - \frac{n(n + 1)^2}{4} \\ &= \frac{n(n^2 - 1)}{12}\end{aligned}$$

and

$$\begin{aligned}\sum d_i^2 = \sum (U_i - i)^2 &= \sum U_i^2 - 2\sum iU_i + \sum i^2 \\ &= 2\sum i^2 - 2\sum iU_i \\ &= \frac{n(n + 1)(2n + 1)}{3} - 2\sum iU_i\end{aligned}$$

so

$$R_s = \frac{(n-1)s_{WU}}{(n-1)s_W s_U} = \frac{\sum iU_i - n(n+1)^2/4}{n(n^2-1)/12}$$
$$= 1 - \frac{6\sum d_i^2}{n(n^2-1)}$$

If there is total agreement in the rankings, then each $d_i = 0$ and $R_s = 1$. In the case of perfect disagreement, $d_i = n - i + 1 - i = n - 2i + 1$, and $R_s = -1$ as it should. Of course, one would reject the hypothesis of independence in favor of a positive correlation alternative for large values of R_s. Alternatively, one could compute the p value of the test and reject H_0 if the p value is less than or equal to α. Note also that the distribution of R_s is symmetric, so Table 14 (Appendix C) gives the p values

$$p = P[R_s \leqslant -r] = P[R_s \geqslant r]$$

for possible observed values r or $-r$ of R_s for $n \leqslant 10$. For $n > 10$, approximate p-values or approximate critical values may be obtained using Student's t distribution approximation,

$$T = \frac{\sqrt{n-2}R_s}{\sqrt{1-R_s^2}} \sim t(n-2)$$

Theorem 14.8.1 For an observed value of $R_s = r_s$, a size α test of $H_0 : F(x, y) = F_1(x)F_2(y)$ against H_a : "positive correlation" is to reject H_0 if $p = p[R_s \geqslant r_s] \leqslant \alpha$, or approximately if $t = \sqrt{n-2}r_s/\sqrt{1-r_s^2} \geqslant t_{1-\alpha}(n-2)$. For the alternative H_a : "negative correlation," reject H_0 if $p = P[R_s \leqslant r_s] \leqslant \alpha$, or approximately if $t \leqslant -t_{1-\alpha}(n-2)$. ■

The ARE of R_s compared to R under normal assumptions is $(3/\pi)^2 = 0.91$.

Example 14.8.2 Now we will compute Spearman's rank correlation coefficient for the paired data considered in Examples 14.5.2 and 14.8.1. Replacing the observations with their ranks gives the following results:

Rank (x_i)	1	2	3	4	5	6	7	8	9	10
Rank (y_i)	1	3	2	5	4	7	6	8	9	12
d_i	0	1	−1	1	−1	1	−1	0	0	2

Rank (x_i)	11	12	13	14	15	16	17	18	19	20
Rank (y_i)	13	14	10	15	11	16	18	19	17	20
d_i	2	2	3	1	−4	0	1	1	−2	0

Any convenient procedure for computing the sample correlation coefficient may be applied to the ranks to obtain Spearman's rank correlation coefficient, but because the x_i already were ordered, it is convenient to use the simplified formula based on the d_i, which gives

$$R_s = 1 - \frac{6 \sum d_i^2}{n(n^2 - 1)} = 1 - \frac{6(50)}{20(20^2 - 1)} = 0.96$$

TEST OF RANDOMNESS

Consider n observations $x_1, \ldots, x_n$. The order of this sequence of observations may be determined by some other variable, Y, such as time. We may ask whether these observations represent a random sample or whether there is some sort of trend associated with the order of the observations. A test of randomness against a trend alternative is accomplished by a test of independence of X and Y. The Y variable usually is not a random variable but is a labeling, such as a fixed sequence of times at which the observations are taken. In terms of ranks, the subscripts of the x's are the ranks of the y variable, and Spearman's rank correlation coefficient is computed as described earlier. A test of $H_0: F_1(x) = \cdots = F_n(x)$ against a one-sided alternative of the type $H_a: F_1(x) > F_2(x) \cdots > F_n(x)$, for all x, would be carried out by rejecting H_0 for large values of R_s. This alternative represents an upward trend alternative.

Under normal assumptions, a particular type of trend alternative is one in which the mean of the variable x_i is a linear function of i,

$$X_i \sim \mathrm{N}(\beta_0 + \beta_1 i, \sigma^2) \qquad i = 1, \ldots, n$$

In this framework, a test of $H_0: \beta_1 = 0$ corresponds to a test of randomness. The usual likelihood ratio test for this case is UMP for one-sided alternatives, and it turns out that the ARE of the nonparametric test based on R_s compared to the likelihood ratio test is $(3/\pi)^{1/3} = 0.98$ when the normality assumptions hold.

There are, of course, other types of nonrandomness besides upward or downward trends. For example, there could be a cyclic effect. Various tests have been developed based on runs, and one of these is discussed in the next section.

Example 14.8.3 In Example 14.7.1 the lifetimes between successive repairs of airplane air conditioners were considered as a random sample. If the air conditioners were not restored to like-new conditions, one might suspect a downward trend in the lifetimes. The lifetimes from the first plane and their order of occurrences are shown below:

i	1	2	3	4	5	6	7	8	9	10	11	12
x_i	23	261	87	7	120	14	62	47	225	71	246	21
Rank (x_i)	4	12	8	1	9	2	6	5	10	7	11	3
d_i	3	10	5	−3	4	−4	−1	−3	1	−3	0	−9

We find

$$R_s = 1 - \frac{6(276)}{12(12^2 - 1)} = 0.035$$

Because $R_s > 0$, there is certainly no evidence of a downward trend. If we consider a two-sided alternative, for $\alpha = 0.10$, then we find $c_{0.95} = 0.497$ and there is still no evidence to reject randomness.

14.9 WALD-WOLFOWITZ RUNS TEST

Consider a sequence of observations listed in order of occurrence, which we wish to test for randomness. Suppose that the observations can be reduced to two types, say a and b. Let T be the total number of runs of like elements in the sequence. For example, the following numbers were obtained from a "random number generator" on a computer.

0.1, 0.4, 0.2, 0.8, 0.6, 0.9, 0.3, 0.4, 0.1, 0.2

Let a denote a number less than 0.5 and b denote a number greater than 0.5, which gives the sequence

$$a\ a\ a\ b\ b\ b\ a\ a\ a\ a$$

For this sequence, $T = 3$. A very small value of T suggests nonrandomness, and a very large value of T also may suggest nonrandomness because of a cyclic effect.

In this application the number of a's, say A, is a random variable, but given the number of a's and b's, A and B, there are $N = (A + B)!/A!B!$ equally likely permutations of the $A + B$ elements under H_0. Thus the permutations associated with very small values of T or very large values of T are placed in the critical region. Again, for a specified value of $\alpha = k/N$, it is necessary to know what critical values for T will result in k permutations being included in the critical region. It is possible to work out the probability distribution analytically for the number of runs under H_0.

Given A and B, the conditional probability distribution of the number of runs under H_0 is

$$P[T = r] = \frac{2\binom{A-1}{r/2-1}\binom{B-1}{r/2-1}}{\binom{A+B}{A}}, \quad r \text{ even}$$

$$= \frac{\binom{A-1}{(r-1)/2}\binom{B-1}{(r-3)/2} + \binom{A-1}{(r-3)/2}\binom{B-1}{(r-1)/2}}{\binom{A+B}{A}}, \quad r \text{ odd}$$

For example, for even r there are exactly $r/2$ runs of a's and $r/2$ runs of b's. The sequence may start with either an a sequence or a b sequence, hence the factor 2. Now suppose that the sequence starts with an a. The number of ways of having $r/2$ runs of A a's is the number of ways of putting $r/2 - 1$ slashes into the $A - 1$ spaces between the a's, which is $\binom{A-1}{(r/2)-1}$. Similarly, the number of ways of dividing the B b's into $r/2$ runs is $\binom{B-1}{(r/2)-1}$, which gives $\binom{A-1}{(r/2)-1}\binom{B-1}{(r/2)-1}$ for the total number of ways of having r runs starting with a. The number of runs starting with b would be the same, and this leads to the first equation.

The odd case would be similar, except that if the sequence begins and ends with an a, then there are $(r+1)/2$ runs of a's and $(r-1)/2$ runs of b's. In this case, the number of ways of placing $[(r+1)/2] - 1 = (r-1)/2$ slashes in the $A - 1$ spaces is $\binom{A-1}{(r-1)/2}$, and the number of ways of placing $[(r-1)/2] - 1 = (r-3)/2$ slashes in the $B - 1$ spaces is $\binom{B-1}{(r-3)/2}$.

The total number of ways of having r runs beginning and ending with b is $\binom{B-1}{(r-1)/2}\binom{A-1}{(r-3)/2}$.

In the above example $A = 7$, $B = 3$, $r = 3$, and

$$P[T \leqslant 3] = P[T = 2] + P[T = 3]$$

$$= \frac{2\binom{6}{0}\binom{2}{0}}{\binom{10}{7}} + \frac{\binom{6}{1}\binom{2}{0} + \binom{6}{0}\binom{2}{1}}{\binom{10}{7}}$$

$$= \frac{10}{120} = 0.083$$

Thus for a one-sided alternative associated with small T, one could reject H_0 in this example at the $\alpha = 0.083$ level.

Tabulated critical values for this test are available in the literature (see, for example, Walpole and Myers, 1985, Table A.18), as are large-sample normal approximations.

The runs test is applicable to testing equality of distributions in two sample problems by ranking the combined samples of x's and y's, and then counting the number of runs. The runs test is not as powerful as the Wilcoxon–Mann-Whitney test in this case.

It can be shown that

$$E(T) = \frac{2AB}{A+B} + 1$$

and

$$\operatorname{Var}(T) = \frac{2AB(2AB - A - B)}{(A + B)^2(A + B - 1)}$$

and for A and B greater than 10 or so the normal approximation is adequate, where

$$t_\alpha \doteq E(T) + z_\alpha\sqrt{\operatorname{Var}(T)}$$

Example 14.9.1 We wish to apply the runs test for randomness in Example 14.8.3. The median of the x_i is, say, $(62 + 71)/2 = 66.5$, and we obtain the following sequence of a's and b's:

$$a\ b\ b\ a\ b\ a\ a\ a\ b\ b\ b\ a$$

We have $r = 7$, $A = B = 6$, and $P[T \leqslant 7] = 0.61$, $P(T = 7) = 0.22$, and $P[T \geqslant 7] = 0.61$, so as before we have no evidence at all of nonrandomness.

In this example $E(T) = 7$, $\operatorname{Var}(T) = 2.72$, and the normal approximation with correction for discontinuity gives

$$P[T \leqslant 7.5] \doteq \Phi\left(\frac{7.5 - 7}{\sqrt{2.72}}\right) \doteq 0.62$$

SUMMARY

Our purpose in this chapter was to develop tests of hypotheses, and in some cases confidence intervals, that do not require parametric assumptions about the model. In many cases, only nominal (categorized) data or ordinal (ranked) data are required, rather than numerical (interval) data.

The one-sample sign test can be used to test a hypothesis about the median of a continuous distribution, using binomial tables. In the case of a normal distribution, this would provide an alternative to tests based on parametric assumptions such as the t test. However, the sign test is less powerful than the t test. A similar test can be used to test hypotheses about a percentile of a continuous distribution. Nonparametric confidence intervals also are possible, and this is related to the problem of nonparametric tolerance limits, which also can be derived.

It is also possible, by means of the two-sample sign test, to test for a difference in location of two continuous distributions. However, as one might suspect, if it is applied to test the difference of normal means, then the power is not as high as it would be with a two-sample t test. The power situation is somewhat better for a test based on ranks, such as the Wilcoxon–Mann-Whitney tests.

It is also possible to test nonparametrically for independence. One possibility is to adapt the usual sample correlation coefficient by applying it to the ranks rather than to the values of the variables. This yields the Spearman's rank correlation coefficient.

Another question that arises frequently concerns whether the order of a sequence of observations has occurred at random or whether it was affected by some sort of trend associated with the order. A nonparametric test of correlation, such as the Spearman test, can be used in this situation, but another common choice is the Wald-Wolfowitz runs test. As noted earlier, this test can be used to test equality of two distributions, but it is not as powerful as the Wilcoxon–Mann-Whitney test in this application.

EXERCISES

1. The following 20 observations are obtained from a random number generator.

0.48, 0.10, 0.29, 0.31, 0.86, 0.91, 0.81, 0.92, 0.27, 0.21,

0.31, 0.39, 0.39, 0.47, 0.84, 0.81, 0.97, 0.51, 0.59, 0.70

(a) Test $H_0 : m = 0.5$ against $H_a : m > 0.5$ at $\alpha = 0.10$.
(b) Test $H_0 : m = 0.25$ against $H_a : m > 0.25$ at $\alpha = 0.10$.

2. The median U.S. family income in 1983 was \$24,580.00. The following 20 family incomes were observed in a random sample from a certain city.

23,470, 48,160, 15,350, 13,670, 5,850, 20,130, 25,570,

20,410, 30,700, 19,340, 26,370, 25,630, 18,920, 21,310,

4,910, 24,840, 17,880, 27,620, 21,660, 12,110

For the median city family income, m, test $H_0 : m = 24{,}800$ against $H_a : m < 24{,}800$ at $\alpha = 0.10$.

3. The median number of hours of weekly TV viewing for children ages 6–11 in 1983 was 25 hours. In an honors class of 50 students, 22 students watched TV more than 25 hours per week and 28 students watched TV less than 25 hours per week. For this class, test $H_0 : m = 25$ against $H_a : m < 25$, at $\alpha = 0.05$.

4. For the data in Exercise 2, test the hypothesis that 10% of the families make less than \$16,000 per year against the alternative that the tenth percentile is less than \$16,000.

5. Using the first bus motor failure data in Exercise 15 of Chapter 13, test $H_0 : x_{0.25} = 40{,}000$ miles against $H_a : x_{0.25} > 40{,}000$ miles at $\alpha = 0.01$.

6. Use the data in Exercise 2.
(a) What level lower confidence limit for $x_{0.25}$ can be obtained using $x_{2:n}$?
(b) Obtain an upper confidence limit for $x_{0.25}$.
(c) Obtain an approximate 95% lower confidence limit on the median family income.

7. Consider the data in Exercise 24 of Chapter 4.

(a) Test $H_0 : x_{0.50} = 5.20$ against $H_a : x_{0.50} > 5.20$ at $\alpha = 0.05$. Use the normal approximation

$$B(x;\, n,\, p) \doteq \Phi\left(\frac{x + 0.5 - np}{\sqrt{npq}}\right)$$

(b) Find an approximate 90% two-sided confidence interval on the median weight.

(c) Find an approximate 95% lower confidence limit on the 25th percentile, $x_{0.25}$.

8. Consider the data in Exercise 24 of Chapter 4.

(a) Set a 95% lower tolerance limit for proportion 0.60.

(b) Set a 90% two-sided tolerance interval for proportion 0.60.

9. Repeat Exercise 8 for the data in Example 4.6.3.

10. Consider the data in Exercise 24 of Chapter 4. Determine an interval such that one could expect 94.5% of the weights of such major league baseballs to fall.

11. Ten brand A tires and 10 brand B tires were selected at random, and one brand A tire and one brand B tire were placed on the back wheels of each of 10 cars. The following distances to wearout in thousands of miles were recorded:

	Car									
	1	2	3	4	5	6	7	8	9	10
A	23	20	26	25	48	26	25	24	15	20
B	20	30	16	33	23	24	8	21	13	18

(a) Assume that the differences are normally distributed, and use a paired-sample t test to test $H_0 : \mu_A = \mu_B$ against $H_a : \mu_A > \mu_B$ at $\alpha = 0.10$.

(b) Rework (a) using the two-sample sign test.

12. Suppose that 20 people are selected at random and asked to compare soda drink A against soda drink B. If 15 prefer A over B, then test the hypothesis that soda B is preferable against the alternative that more people prefer brand A at $\alpha = 0.05$.

13. Twelve pairs of twin male lambs were selected; diet plan I was given to one twin and diet plan II to the other twin in each case. The weights at eight months were as follows.

Diet I:	111	102	90	110	108	125	99	121	133	115	90	101
Diet II:	97	90	96	95	110	107	85	104	119	98	97	104

(a) Use the sign test to test the hypothesis that there is no difference in the diets against the alternative that diet I is preferable to diet II at $\alpha = 0.10$.

(b) Repeat (a) using the Wilcoxon paired-sample signed-rank test. Because n is only 12, use the table, but also work using the large-sample normal results for illustration purposes.

14. Siegel (1956, page 85) gives the following data on the number of nonsense syllables remembered under shock and nonshock conditions for 15 subjects:

Subject	1	2	3	4	5	6	7	8	9	10	11	12	13	14	15
Nonshock	5	4	3	5	2	4	2	2	4	4	3	1	5	3	1
Shock	2	2	0	3	3	2	3	1	1	3	4	2	2	4	0

Test H_0 that there is no difference in retention under the two conditions against the alternative that more syllables are remembered under nonshock conditions at $\alpha = 0.10$. Use the Wilcoxon paired-sample signed-rank test.

15. Davis (1952) gives the lifetimes in hours of 40-watt incandescent lamps from a forced-life test on lamps produced in the two indicated weeks:

1–2–47	1067	919	1196	785	1126	936	918	1156	920	948
10–2–47	1105	1243	1204	1203	1310	1262	1234	1104	1303	1185

Test the hypothesis that the manufacturing process is unchanged for the two different periods at $\alpha = 0.10$.

(a) Work out both the small-sample and large-sample tests based on U.

(b) Although these are not paired samples, work out the test based on the Wilcoxon paired-sample test.

16. The following fatigue failure times of ball bearings were obtained from two different testers. Test the hypothesis that there is no difference in testers, $H_0: F(x) = F(y)$, against $H_a: F(x) \neq F(y)$. (Use $\alpha = 0.10$).

Tester 1	140.3	158.0	183.9	132.7	117.8	98.7	164.8	136.6	93.4	116.6
Tester 2	193.0	172.5	173.3	204.7	172.0	152.7	234.9	216.5	422.6	

17. In Exercise 11, test the hypothesis that the brand A and brand B samples are independent at $\alpha = 0.10$.

(a) Use Pearson's r with the approximate t distribution.

(b) Use Spearman's R_s. Compare the small-sample and large-sample approximations in this case.

18. Consider the data in Exercise 13.

(a) Estimate the correlation between the responses (x_i, y_i) on the twin lambs.

(b) Test $H_0: F(x, y) = F_1(x)F_2(y)$ against the alternative of a positive correlation at level $\alpha = 0.05$, using Pearson's r. Compare the asymptotic normal approximation with the approximate t result in this case.

(c) Repeat the test in (b) using R_s.

19. In a pig-judging contest, an official judge and a 4-H Club member each ranked 10 pigs as follows (Dixon and Massey, 1957, page 303):

Judge	9	4	3	7	2	1	5	8	10	6
4-H Member	7	6	4	9	2	3	8	5	10	1

Test the hypothesis of independence against the alternative of a positive correlation at $\alpha = 0.10$.

20. Use R_s to test whether the data in Exercise 1 are random against the alternative of an upward trend at $\alpha = 0.10$.

21. Proschan (1963) gives the times of successive failures of the air-conditioning system of Boeing 720 jet airplanes. The times between failures on Plane 7908 are given below:

413, 14, 58, 37, 100, 65, 9, 169, 447, 184, 36, 201, 118, 34, 31, 18, 18, 67, 57, 62, 7, 22, 34

If the failures occur according to a Poisson process, then the times between failures should be independent exponential variables. Otherwise, wearout or degradation may be occurring, and one might expect a downward trend in the times between failures. Test the hypothesis of randomness against the alternative of a downward trend at $\alpha = 0.10$.

22. The following values represent the times between accidents in a large factory:

8.66, 11.28, 10.43, 10.89, 11.49, 11.44, 15.92, 12.50, 13.86, 13.32

Test the hypothesis of randomness against an upward trend at $\alpha = 0.05$.

23. Use the runs test to test randomness of the numbers in Exercise 1.

24. Use the runs test to work Exercise 16.

25. Suppose that a runs test is based on the number of runs of a's rather than the total number of runs.

(a) Show that the probability of k runs of a's is given by

$$p_k = \frac{\binom{A-1}{k-1}\binom{B+1}{k}}{\binom{A+B}{A}}$$

(b) Rework Exercise 23 by using (a).

26. For the Mann-Whitney statistic U, show that Var (U) is given by equation (14.7.6).

CHAPTER 15

REGRESSION AND LINEAR MODELS

15.1 INTRODUCTION

Random variables that are observed in an experiment often are related to one or more other variables. For example, the yield of a chemical reaction will be affected by variables such as temperature or reaction time. We will consider a statistical method known as **regression analysis** that deals with such problems.

The term *regression* was used by Francis Galton, a nineteenth-century scientist, to describe a phenomenon involving heights of fathers and sons. Specifically, the study considered paired data, $(x_1, y_1), \ldots, (x_n, y_n)$, where x_i and y_i represent, respectively, the heights of the ith father and his son. One result of this work was the derivation of a linear relationship $y = a + bx$ for use in predicting a son's height given the father's. It was observed that if a father was taller than average, then the son tended also to be taller than average, but not by as much as the father. Similarly, sons of fathers who were shorter than average tended to be shorter than average, but not by as much as the father. This effect, which is known as **regression toward the mean**, provides the origin of the term **regression analysis**, although the method is applicable to a wide variety of problems.

15.2

LINEAR REGRESSION

We will consider situations in which the result of an experiment is modeled as a random variable Y_x whose distribution depends on another variable x or vector of variables $\boldsymbol{x} = (x_0, x_1, \ldots, x_p)$. Typically, the distribution of Y_x also will involve one or more unknown parameters. We will consider the situation in which the expectation is a linear function of the parameters

$$E(Y_{\boldsymbol{x}}) = \beta_0 x_0 + \beta_1 x_1 + \cdots + \beta_p x_p \tag{15.2.1}$$

with unknown parameters, $\beta_0, \beta_1, \ldots, \beta_p$. Usually, it also is assumed that the variance does not depend on $\boldsymbol{x}$, $\text{Var}(Y_{\boldsymbol{x}}) = \sigma^2$. Other notations, which are sometimes used for the expectation in equation (15.2.1) are $\mu_{Y|\boldsymbol{x}}$ or $E(Y \mid \boldsymbol{x})$, but these notations will not represent a conditional expectation in the usual sense unless $x_0, x_1, \ldots, x_p$ are values of a set of random variables. Unless otherwise indicated, we will assume that the values $x_0, x_1, \ldots, x_p$ are fixed or measured without error by the experimenter.

A model whose expectation is a linear function of the parameters, such as (15.2.1) will be called a **linear regression model**. This does not require that the model be linear in the x_i's. For example, one might wish to consider such models as $E(Y_{\boldsymbol{x}}) = \beta_0 x_0 + \beta_1 x_1 + \beta_2 x_0 x_1$ or $E(Y_x) = \beta_0 e^x + \beta_1 e^{2x}$, which are both linear in the coefficients but not in the variables. Another important example is the **polynomial regression model** in which the x_i's are integer powers of a common variable x. In particular, for some $p = 1, 2, \ldots,$

$$E(Y_x) = \beta_0 + \beta_1 x + \beta_2 x^2 + \cdots + \beta_p x^p \tag{15.2.2}$$

Some regression models involve functions that are not linear in the parameters, but we will not consider nonlinear regression models here.

Another way to formulate a linear regression model is

$$Y_{\boldsymbol{x}} = \beta_0 x_0 + \beta_1 x_1 + \cdots + \beta_p x_p + \varepsilon_{\boldsymbol{x}} \tag{15.2.3}$$

in which $\varepsilon_{\boldsymbol{x}}$ is interpreted as a random error with $E(\varepsilon_{\boldsymbol{x}}) = 0$ and $\text{Var}(\varepsilon_{\boldsymbol{x}}) = \sigma^2$.

It also is possible to have a constant term by taking the first component in $\boldsymbol{x}$ to be 1. That is, if $\boldsymbol{x} = (1, x_1, \ldots, x_p)$, then $E(Y_{\boldsymbol{x}}) = \beta_0 + \beta_1 x_1 + \cdots + \beta_p x_p$. In the next section we will study the important special case in which $p = 1$ and the model is linear in $x_1 = x$.

15.3

SIMPLE LINEAR REGRESSION

Consider a model of the form

$$Y_x = \beta_0 + \beta_1 x + \varepsilon_x \tag{15.3.1}$$

with $E(\varepsilon_x) = 0$ and $\text{Var}(\varepsilon_x) = \sigma^2$. In this section we will develop the properties of such a model, called the **simple linear model**, under two different sets of assumptions. First we will consider the problem of estimation of the coefficients β_0 and β_1 under the assumption that errors are uncorrelated.

LEAST-SQUARES APPROACH

Suppose $x_1, \ldots, x_n$ are fixed real numbers and that experiments are performed at each of these values, yielding observed values of a set of n uncorrelated random variables of form (15.3.1). For convenience we will denote the subscripts by i rather than x_i. Thus, we will assume that for $i = 1, \ldots, n$,

$$E(Y_i) = \beta_0 + \beta_1 x_i \qquad \text{Var}(Y_i) = \sigma^2 \qquad \text{Cov}(Y_i, Y_j) = 0 \quad i \neq j$$

The resulting data will be represented as pairs $(x_1, y_1), \ldots, (x_n, y_n)$.

Suppose we write the observed value of each Y_i as $y_i = \beta_0 + \beta_1 x_i + e_i$ so that e_i is the difference between what is actually observed on the ith trial and the theoretical value $E(Y_i)$. The ideal situation would be for the pairs (x_i, y_i) to all fall on a straight line, with all the $e_i = 0$, in which case a linear function could be determined algebraically. However, this is not likely because the y_i's are observed values of a set of random variables. The next best thing would be to fit a straight line through the points (x_i, y_i) in such a way as to minimize, in some sense, the resulting observed deviations of the y_i from the fitted line. That is, we choose a line that minimizes some function of the $e_i = y_i - \beta_0 - \beta_1 x_i$. Different criteria for goodness-of-fit lead to different functions of e_i, but we will use a standard approach called the **Principle of Least Squares**, which says to minimize the sum of the squared deviations from the fitted line. That is, we wish to find the values of β_0 and β_1, say $\hat{\beta}_0$ and $\hat{\beta}_1$, that minimize the sum

$$S = \sum_{i=1}^{n} (y_i - \beta_0 - \beta_1 x_i)^2$$

Taking derivatives of S with respect to β_0 and β_1 and setting them equal to zero gives the **least-squares** (LS) estimates $\hat{\beta}_0$ and $\hat{\beta}_1$ as solutions to the equations

$$2 \sum_{i=1}^{n} [y_i - \hat{\beta}_0 - \hat{\beta}_1 x_i](-1) = 0$$

$$2 \sum_{i=1}^{n} [y_i - \hat{\beta}_0 - \hat{\beta}_1 x_i](-x_i) = 0$$

Simultaneous solution gives

$$\hat{\beta}_1 = \frac{\sum x_i y_i - (\sum x_i)(\sum y_i)/n}{\sum x_i^2 - (\sum x_i)^2/n}$$

$$= \frac{\sum (x_i - \bar{x})(y_i - \bar{y})}{\sum (x_i - \bar{x})^2}$$

$$= \frac{\sum (x_i - \bar{x})y_i}{\sum (x_i - \bar{x})^2}$$

and

$$\hat{\beta}_0 = \bar{y} - \hat{\beta}_1 \bar{x}$$

Thus, if one wishes to fit a straight line through a set of points, the equation $y = \hat{\beta}_0 + \hat{\beta}_1 x$ provides a straight line that minimizes the sum of squares of the errors between observed values and the points on the line, say

$$\text{SSE} = \sum_{i=1}^{n} \hat{e}_i^2 = \sum_{i=1}^{n} [y_i - \hat{\beta}_0 - \hat{\beta}_1 x_i]^2$$

The quantities $\hat{e}_i = y_i - \hat{\beta}_0 - \hat{\beta}_1 x_i$ are known as **residuals** and SSE is called the **error sum of squares**. The least-squares principle does not provide a direct estimate of σ^2, but the magnitude of the variance is reflected in the quantity SSE. It can be shown (see Exercise 8) that an unbiased estimate of σ^2 is given by

$$\tilde{\sigma}^2 = \frac{\text{SSE}}{n-2}$$

The notation $\tilde{\sigma}^2$ will be used throughout the chapter for an unbiased estimator of σ^2. The notation $\hat{\sigma}^2$ will represent the MLE, which is derived later in the section. The following convenient form also can be derived (see Exercise 5):

$$\text{SSE} = \sum y_i^2 - \hat{\beta}_0 \sum y_i - \hat{\beta}_1 \sum x_i y_i$$

Also

$$\hat{y} = \hat{\beta}_0 + \hat{\beta}_1 x$$

may be used to predict the value of Y_x, and the same quantity would be used to estimate the expected value of Y_x. That is, an estimate of $E(Y_x) = \beta_0 + \beta_1 x$ is given by

$$\hat{E}(Y_x) = \hat{\beta}_0 + \hat{\beta}_1 x$$

Note also that $\hat{y} = \bar{y} + \hat{\beta}_1(x - \bar{x})$, which reflects the regression adjustment being made to the overall mean, $\bar{y}$.

Other linear combinations of β_0 and β_1 could be estimated in a similar manner. The LS estimators are linear functions of the Y_i's, and it can be shown that among all such linear unbiased estimators, the LS estimators have minimum variance. Thus, the LS estimators often are referred to as **Best Linear Unbiased Estimators** (BLUEs).

Theorem 15.3.1 If $E(Y_i) = \beta_0 + \beta_1 x_i$, $\text{Var}(Y_i) = \sigma^2$ and $\text{Cov}(Y_i, Y_j) = 0$ for $i \neq j$ and $i = 1, \ldots, n$, then the LS estimators have the following properties:

1. $E(\hat{\beta}_1) = \beta_1, \quad \text{Var}(\hat{\beta}_1) = \dfrac{\sigma^2}{\sum_{i=1}^{n} (x_i - \bar{x})^2}$

2. $E(\hat{\beta}_0) = \beta_0, \quad \text{Var}(\hat{\beta}_0) = \dfrac{\sigma^2 \sum_{i=1}^{n} x_i^2}{\left[n \sum_{i=1}^{n} (x_i - \bar{x})^2\right]}$

3. $E(c_1\hat{\beta}_0 + c_2\hat{\beta}_1) = c_1\beta_0 + c_2\beta_1$

4. $c_1\hat{\beta}_0 + c_2\hat{\beta}_1$ is the BLUE of $c_1\beta_0 + c_2\beta_1$

Proof

Part 1:

$$\begin{aligned}
E[\sum (x_i - \bar{x})Y_i] &= \sum (x_i - \bar{x})E(Y_i) \\
&= \sum (x_i - \bar{x})(\beta_0 + \beta_1 x_i) \\
&= \sum (x_i - \bar{x})\beta_0 + \beta_1 \sum (x_i - \bar{x})x_i \\
&= 0 \cdot \beta_0 + \beta_1 \sum (x_i - \bar{x})x_i \\
&= \beta_1 \sum (x_i - \bar{x})(x_i - \bar{x} + \bar{x}) \\
&= \beta_1 \sum (x_i - \bar{x})^2
\end{aligned}$$

it follows that

$$E(\hat{\beta}_1) = \frac{\beta_1 \sum (x_i - \bar{x})^2}{\sum (x_i - \bar{x})^2} = \beta_1$$

Also

$$\begin{aligned}
\text{Var}(\hat{\beta}_1) &= \text{Var}\left[\frac{\sum (x_i - \bar{x})Y_i}{\sum (x_i - \bar{x})^2}\right] \\
&= \frac{1}{[\sum (x_i - \bar{x})^2]^2} \sum \text{Var}[(x_i - \bar{x})Y_i] \\
&= \frac{\sum (x_i - \bar{x})^2\sigma^2}{[\sum (x_i - \bar{x})^2]^2} \\
&= \frac{\sigma^2}{\sum (x_i - \bar{x})^2}
\end{aligned}$$

Part 2:

$$\begin{aligned} E(\hat{\beta}_0) &= E(\bar{Y} - \beta_1 \bar{x}) \\ &= \frac{1}{n} \sum E(Y_i) - \beta_1 \bar{x} \\ &= \frac{1}{n} \sum (\beta_0 + \beta_1 x_i) - \beta_1 \bar{x} \\ &= \beta_0 + \beta_1 \bar{x} - \beta_1 \bar{x} \\ &= \beta_0 \end{aligned}$$

Now $\bar{Y}$ and $\hat{\beta}_1$ are not uncorrelated, but $\hat{\beta}_0$ can be expressed as a linear combination of the uncorrelated Y_i's, and the variance of this linear combination can be derived. If we let $b_i = (x_i - \bar{x})/\sum (x_j - \bar{x})^2$, then $\hat{\beta}_1 = \sum b_i Y_i$ and

$$\begin{aligned} \hat{\beta}_0 &= \frac{1}{n} \sum Y_i - \bar{x} \sum b_i Y_i \\ &= \sum d_i Y_i \end{aligned}$$

where $d_i = \frac{1}{n} - \bar{x} b_i$ and

$$\begin{aligned} \mathrm{Var}(\hat{\beta}_0) &= \sigma^2 \sum d_i^2 \\ &= \frac{\sigma^2 \sum x_i^2}{n \sum (x_i - \bar{x})^2} \end{aligned}$$

after some algebraic simplification.

Part 3 follows from Parts 1–2 and the linearity of expected values.

Part 4:
Any linear function of the Y_i's can be expressed in the form $c_1 \hat{\beta}_0 + c_2 \hat{\beta}_1 + \sum a_i Y_i$ for some set of constants $a_1, \ldots, a_n$. For this to be an unbiased estimator of $c_1 \beta_0 + c_2 \beta_1$ requires that $\sum a_i(\beta_0 + \beta_1 x_i) = 0$ for all β_0 and β_1, because

$$E(c_1 \hat{\beta}_0 + c_2 \hat{\beta}_1 + \sum a_i Y_i) = c_1 \beta_0 + c_2 \beta_1 + \sum a_i(\beta_0 + \beta_1 x_i)$$

But $\sum a_i(\beta_0 + \beta_1 x_i) = 0$ for all β_0 and β_1 implies that $\sum a_i = 0$ and $\sum a_i x_i = 0$. Now,

$$c_1 \hat{\beta}_0 + c_2 \hat{\beta}_1 = \sum (c_1 d_i + c_2 b_i) Y_i$$

and

$$\begin{aligned}\text{Cov}(c_1\hat{\beta}_0 + c_2\hat{\beta}_1, \textstyle\sum a_i Y_i) &= \textstyle\sum a_i(c_1 d_i + c_2 b_i)\sigma^2 \\ &= [c_1 \textstyle\sum a_i d_i + c_2 \sum a_i b_i]\sigma^2 \\ &= \left[c_1 \textstyle\sum a_i\left(\frac{1}{n} - \bar{x}b_i\right) + c_2 \sum a_i b_i\right]\sigma^2 \\ &= \left[\frac{c_1}{n} \textstyle\sum a_i + (c_2 - c_1\bar{x}) \sum a_i b_i\right]\sigma^2 \\ &= 0\end{aligned}$$

The last step follows from the result that $\sum a_i = 0$ and $\sum a_i x_i = 0$ imply $\sum a_i b_i = 0$, which is left as an exercise. Thus

$$\text{Var}(c_1\hat{\beta}_0 + c_2\hat{\beta}_1 + \textstyle\sum a_i Y_i) = \text{Var}(c_1\hat{\beta}_0 + c_2\hat{\beta}_1) + \sum a_i^2 \sigma^2$$

This variance is minimized by taking $\sum a_i^2 = 0$, which requires $a_i = 0$, $i = 1, \ldots, n$. Thus, $c_1\hat{\beta}_0 + c_2\hat{\beta}_1$ is the minimum variance linear-unbiased estimator of $c_1\beta_0 + c_2\beta_1$, and this concludes the proof. ■

Example 15.3.1 In an article about automobile emissions, hydrocarbon emissions (grams per mile) were given by McDonald and Studden (1990) for several values of accumulated mileage (in 1000s of miles). The following paired data was reported on mileage (x) versus hydrocarbons (y).

x: 5.133, 10.124, 15.060, 19.946, 24.899, 29.792, 29.877, 35.011, 39.878, 44.862, 49.795

y: 0.265, 0.287, 0.282, 0.286, 0.310, 0.333, 0.343, 0.335, 0.311, 0.345, 0.319

To compute $\hat{\beta}_0$ and $\hat{\beta}_1$, we note that $n = 11$ and compute $\sum x_i = 304.377$, $\sum x_i^2 = 10461.814$, $\sum y_i = 3.407$, $\sum y_i^2 = 1.063$, and $\sum x_i y_i = 97.506$. Thus, $\bar{x} = 27.671$, $\bar{y} = 0.310$,

$$\hat{\beta}_1 = \frac{(97.506) - (304.377)(3.407)/11}{(10461.814) - (304.377)^2/11} = 0.00158$$

$$\hat{\beta}_0 = 0.310 - (0.00158)(27.671) = 0.266$$

Thus, if it is desired to predict the amount of hydrocarbons after 30,000 miles, we compute $\hat{y} = 0.266 + 0.00158(30) = 0.313$. Furthermore, $\text{SSE} = 1.063 - (0.266)(3.407) - (0.00158)(97.506) = 0.00268$, and $\tilde{\sigma}^2 = 0.00268/9 = 0.000298$. The estimated linear regression function and the plotted data are shown in Figure 15.1.

FIGURE 15.1 Hydrocarbon emissions as a function of accumulated mileage

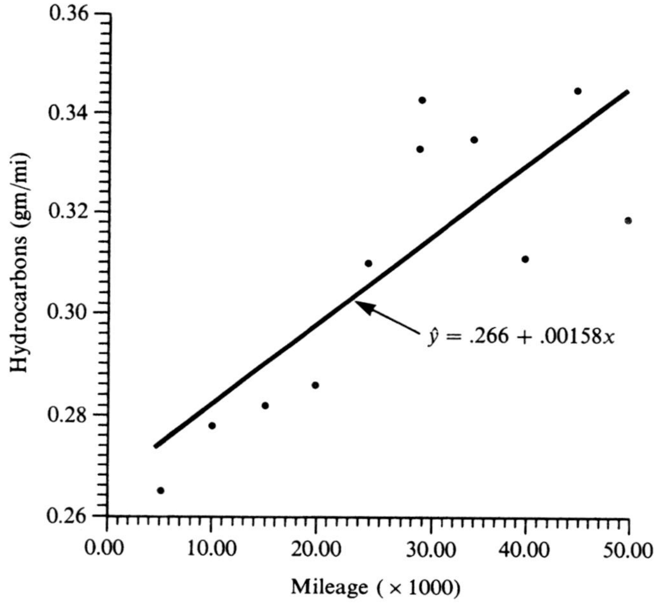

Example 15.3.2 Consider now another problem that was encountered in Example 13.7.3. Recall that the chi-square goodness-of-fit test requires estimates of the unknown parameters μ and σ. At that time it was discovered that the usual grouped sample estimate of σ was somewhat larger than the grouped sample MLE and that this adversely affected the outcome of the test. It also was noted that the grouped sample MLEs are difficult to compute and a simpler method would be desirable. The following method makes use of the least-squares approach and provides estimates that are simple to compute and that appear to be comparable.

Let the data be grouped into c cells and denote the ith cell by $A_i = (a_{i-1}, a_i]$ for the $i = 1, \ldots, c$, and let o_i be the number of observations in A_i. Let $F_i = (o_1 + \cdots + o_i)/n$. Because a_0 is chosen to be less than the smallest observation, the value $F(a_0)$ will be negligible, particularly if n is large. Thus, F_i is an estimate of the CDF value $F(a_i)$, which in the present example is $\Phi((a_i - \mu)/\sigma)$. It follows that approximately

$$\Phi^{-1}(F_i) = -\frac{\mu}{\sigma} + \frac{1}{\sigma} a_i \quad i = 1, \ldots, c - 1$$

which suggests applying the simple linear regression method with $x_i = a_i$ and $y_i = \Phi^{-1}(F_i)$, $\beta_0 = -\mu/\sigma$, and $\beta_1 = 1/\sigma$. The last cell is not used because $F_c = 1$.

For the radial velocity data of Example 13.7.3, the estimates are $\hat{\beta}_0 = 1.606$ and $\hat{\beta}_1 = .0778$, which expressed in terms of μ and σ give estimates $\hat{\mu} = -20.64$ and $\hat{\sigma} = 12.85$. These are remarkably close to the MLEs for grouped data, which are -20.32 and 12.70. Of course, another possibility would be to apply the simple linear method with $x_i = \Phi^{-1}(F_i)$ and $y_i = a_i$. In this case the parameterization is simpler because $\beta_0 = \mu$ and $\beta_1 = \sigma$, although in this situation the y is fixed and the x is variable. However, this approach also gives reasonable estimates in the present example. Specifically, with this modification, the estimates of μ and σ are -20.63 and 12.73, respectively.

This approach is appropriate for location-scale models in general. Specifically, if $F(x) = G\left(\frac{x - \eta}{\theta}\right)$, then approximately, $G^{-1}(F_i) = -\eta/\theta + (1/\theta)a_i$, so the simple linear model can be used with $\beta_0 = -\eta/\theta$, $\beta_1 = 1/\theta$, $y_i = G^{-1}(F_i)$, and $x_i = a_i$, and the resulting estimates of θ and η would be $\hat{\theta} = 1/\hat{\beta}_1$ and $\hat{\eta} = -\hat{\beta}_0/\hat{\beta}_1$.

MAXIMUM LIKELIHOOD APPROACH

Now we will derive the MLEs of β_0, β_1, and σ^2 under the assumption that the random errors are independent and normal, $\varepsilon_x \sim N(0, \sigma^2)$. Assume that $x_1, \ldots, x_n$ are fixed real numbers and that experiments are performed at each of these values, yielding observed values of a set of n independent normal random variables of form (15.3.1). Thus, $Y_1, \ldots, Y_n$ are independent, $Y_i \sim N(\beta_0 + \beta_1 x_i, \sigma^2)$. The resulting data will be represented as pairs $(x_1, y_1), \ldots, (x_n, y_n)$.

Theorem 15.3.2 If $Y_1, \ldots, Y_n$ are independent with $Y_i \sim N(\beta_0 + \beta_1 x_i, \sigma^2)$ for $i = 1, \ldots, n$, then the MLEs are

$$\hat{\beta}_0 = \bar{y} - \hat{\beta}_1 \bar{x}$$

$$\hat{\beta}_1 = \frac{\sum_{i=1}^{n} (x_i - \bar{x}) y_i}{\sum_{i=1}^{n} (x_i - \bar{x})^2}$$

$$\hat{\sigma}^2 = \frac{1}{n} \sum_{i=1}^{n} (y_i - \hat{\beta}_0 - \hat{\beta}_1 x_i)^2$$

Proof

The likelihood function is

$$L = L(\beta_0, \beta_1, \sigma^2) = \prod_{i=1}^{n} \frac{1}{\sqrt{2\pi\sigma^2}} \exp\left[-\frac{1}{2\sigma^2}(y_i - \beta_0 - \beta_1 x_i)^2\right] \quad \textbf{(15.3.2)}$$

Thus, the log-likelihood is

$$\ln L = -\frac{n}{2}\ln(2\pi\sigma^2) - \frac{1}{2\sigma^2}\sum_{i=1}^{n}(y_i - \beta_0 - \beta_1 x_i)^2$$

If we set the partials to zero with respect to β_0, β_1, and σ^2, then we have the ML equations

$$n\hat{\beta}_0 + \left(\sum_{i=1}^{n} x_i\right)\hat{\beta}_1 = \sum_{i=1}^{n} y_i \quad \textbf{(15.3.3)}$$

$$\left(\sum_{i=1}^{n} x_i\right)\hat{\beta}_0 + \left(\sum_{i=1}^{n} x_i^2\right)\hat{\beta}_1 = \sum_{i=1}^{n} x_i y_i \quad \textbf{(15.3.4)}$$

$$n\hat{\sigma}^2 = \sum_{i=1}^{n}(y_i - \hat{\beta}_0 - \hat{\beta}_1 x_i)^2 \quad \textbf{(15.3.5)}$$

The MLEs of β_0 and β_1 are obtained by solving equations (15.3.3) and (15.3.4), which are linear equations in $\hat{\beta}_0$ and $\hat{\beta}_1$. ■

Notice that the MLEs of β_0 and β_1 are identical in form to the BLUEs, which were derived under a much less restrictive set of assumptions. However, it is possible to establish some useful properties under the present assumptions.

Theorem 15.3.3 If $Y_1, \ldots, Y_n$ are independent with $Y_i \sim N(\beta_0 + \beta_1 x_i, \sigma^2)$ and

$$S_1 = \sum_{i=1}^{n} Y_i, \; S_2 = \sum_{i=1}^{n} x_i Y_i, \quad \text{and} \quad S_3 = \sum_{i=1}^{n} Y_i^2, \quad \text{then}$$

1. The statistics S_1, S_2, and S_3 are jointly complete and sufficient for β_0, β_1, and σ^2.

2. If σ^2 is fixed, then S_1 and S_2 are jointly complete and sufficient for β_0 and β_1.

Proof

Part 1:
The joint pdf of $Y_1, \ldots, Y_n$ given by equation (15.3.2), can be written as

$$f(y_1, \ldots, y_n) = (2\pi\sigma^2)^{-n/2} \exp\left[-\frac{1}{2\sigma^2}\sum_{i=1}^{n} y_i^2 + \frac{\beta_0}{\sigma^2}\sum_{i=1}^{n} y_i + \frac{\beta_1}{\sigma^2}\sum_{i=1}^{n} x_i y_i - \frac{1}{2\sigma^2}\sum_{i=1}^{n}(\beta_0 + \beta_1 x_i)^2\right]$$

$$= C(\boldsymbol{\theta})h(y_1, \ldots, y_n) \exp\left[q_1(\boldsymbol{\theta})\sum_{i=1}^{n} y_i + q_2(\boldsymbol{\theta})\sum_{i=1}^{n} x_i y_i + q_3(\boldsymbol{\theta})\sum_{i=1}^{n} y_i^2\right]$$

with

$$\boldsymbol{\theta} = (\beta_0, \beta_1, \sigma^2), \quad C(\boldsymbol{\theta}) = (2\pi\sigma^2)^{-n/2} \exp\left[-\sum_{i=1}^{n}(\beta_0 + \beta_1 x_i)^2/(2\sigma^2)\right]$$

$h(y_1, \ldots, y_n) = 1$, $q_1(\boldsymbol{\theta}) = \beta_0/\sigma^2$, $q_2(\boldsymbol{\theta}) = \beta_1/\sigma^2$, and $q_3(\boldsymbol{\theta}) = -1/(2\sigma^2)$. This is the multivariate REC form of Chapter 10.

Part 2 follows by rewriting the pdf as

$$f(y_1, \ldots, y_n) = C(\boldsymbol{\theta})h(y_1, \ldots, y_n) \exp\left[q_1(\boldsymbol{\theta})\sum_{i=1}^{n} y_i + q_2(\boldsymbol{\theta})\sum_{i=1}^{n} x_i y_i\right]$$

with the notation now defined as $\boldsymbol{\theta} = (\beta_0, \beta_1)$, $q_1(\boldsymbol{\theta}) = \beta_0/\sigma^2$, and $q_2(\boldsymbol{\theta}) = \beta_1/\sigma^2$ and

$$h(y_1, \ldots, y_n) = \exp\left[-\sum_{i=1}^{n} y_i^2/(2\sigma^2)\right]$$

Notice that the MLEs $\hat{\beta}_0$, $\hat{\beta}_1$, and $\hat{\sigma}^2$ are jointly complete and sufficient for β_0, β_1, and σ^2 because they can be expressed as functions of $\sum Y_i$, $\sum x_i Y_i$, and $\sum Y_i^2$. Similarly, if σ^2 is fixed then $\hat{\beta}_0$ and $\hat{\beta}_1$ are functions of $\sum Y_i$ and $\sum x_i Y_i$ and thus they are jointly complete and sufficient for β_0 and β_1. ■

We note at this point that some aspects of the analysis are simplified if we consider a related problem in which the x_i's are centered about $\bar{x}$. Specifically, if we let $x_i^* = x_i - \bar{x}$ and $\beta_c = \beta_0 + \beta_1\bar{x}$, then the variables of Theorem 15.3.2 can be represented as $Y_i = \beta_c + \beta_1(x_i - \bar{x}) + \varepsilon_i = \beta_c + \beta_1 x_i^* + \varepsilon_i$ where $\sum x_i^* = 0$. In this representation, $\hat{\beta}_c$ has the form

$$\hat{\beta}_c = \bar{y}$$

and

$$\hat{\sigma}^2 = \frac{1}{n} \sum_{i=1}^{n} [y_i - \bar{y} - \hat{\beta}_1(x_i - \bar{x})]^2$$

It also is easily verified that $\hat{\beta}_c$, $\hat{\beta}_1$, and $\hat{\sigma}^2$ are jointly complete and sufficient for β_c, β_1, and σ^2, and if σ^2 is fixed, then $\hat{\beta}_c$ and $\hat{\beta}_1$ are jointly complete and sufficient for β_c and β_1.

An interesting property of the centered version that we will verify shortly is that the MLEs of β_c, β_1, and σ^2 are independent. This property will be useful in proving the following distributional properties of the MLEs of β_0, β_1, and σ^2.

Theorem 15.3.4 If $Y_i = \beta_0 + \beta_1 x_i + \varepsilon_i$ with independent errors $\varepsilon_i \sim N(0, \sigma^2)$, then the MLEs of β_0 and β_1 have a bivariate normal distribution with $E(\hat{\beta}_0) = \beta_0$, $E(\hat{\beta}_1) = \beta_1$, and

$$\text{Var}(\hat{\beta}_0) = \frac{\sigma^2 \sum x_i^2}{n \sum (x_i - \bar{x})^2}$$

$$\text{Var}(\hat{\beta}_1) = \frac{\sigma^2}{\sum (x_i - \bar{x})^2}$$

$$\text{Cov}(\hat{\beta}_0, \hat{\beta}_1) = -\frac{\bar{x}\sigma^2}{\sum (x_i - \bar{x})^2}$$

Furthermore, $(\hat{\beta}_0, \hat{\beta}_1)$ is independent of $\hat{\sigma}^2$, and $n\hat{\sigma}^2/\sigma^2 \sim \chi^2(n-2)$.

Proof

The proof is somewhat simpler for the centered problem, because in this case the estimators of the coefficients are uncorrelated and all three MLEs will be independent. Our approach will be to prove the result for the centered case and then extend it to the general case.

We define a set of $n + 2$ statistics, $W_i = Y_i - \hat{\beta}_c - \hat{\beta}_1(x_i - \bar{x})$ if $i = 1, \ldots, n$, $U_1 = \hat{\beta}_c$, and $U_2 = \hat{\beta}_1$. Notice that each variable is a linear combination in the independent normal random variables $Y_1, \ldots, Y_n$. Specifically,

$$U_1 = \sum_{j=1}^{n} (1/n)Y_j \qquad U_2 = \sum_{j=1}^{n} b_j Y_j \qquad W_i = \sum_{j=1}^{n} c_{ij} Y_j$$

with coefficients $b_j = (x_j - \bar{x}) \Big/ \sum_{i=1}^{n} (x_i - \bar{x})^2$, $c_{ii} = 1 - 1/n - (x_i - \bar{x})b_i$, and $c_{ij} = -1/n - (x_i - \bar{x})b_j$ if $j \neq i$. It is easily verified that $\sum_{j=1}^{n} b_j = 0$ and $\sum_{j=1}^{n} c_{ij} = 0$ for each $i = 1, \ldots, n$. Using these identities first we will derive the joint MGF of the U_i's and then derive the joint MGF of the W_i's.

Let $\boldsymbol{U} = (U_1, U_2)$ and $\boldsymbol{t} = (t_1, t_2)$, and consider

$$
\begin{aligned}
M_{\boldsymbol{U}}(\boldsymbol{t}) &= E[\exp(t_1 U_1 + t_2 U_2)] \\
&= E\left[\exp\left(t_1 \sum_{j=1}^{n} (1/n) Y_j + t_2 \sum_{j=1}^{n} b_j Y_j\right)\right] \\
&= E\left[\exp\left(\sum_{j=1}^{n} a_j Y_j\right)\right] \qquad \text{(15.3.6)}
\end{aligned}
$$

with $a_j = (t_1/n + t_2 b_j)$. It can be verified that $\sum_{j=1}^{n} a_j = t_1$, $\sum_{j=1}^{n} (x_j - \bar{x}) a_j = t_2$, and $\sum_{j=1}^{n} a_j^2 = t_1^2/n + t_2^2 \Big/ \sum_{i=1}^{n} (x_i - \bar{x})^2$ (see Exercise 12).

Because each Y_j is normal, its MGF, evaluated at a_j, has the form

$$
M_{Y_j}(a_j) = \exp\left[(\beta_c + \beta_1(x_j - \bar{x}))a_j + \tfrac{1}{2}\sigma^2 a_j^2\right]
$$

Thus, from equation (15.3.6) we have that

$$
\begin{aligned}
M_{\boldsymbol{U}}(\boldsymbol{t}) &= \prod_{i=1}^{n} M_{Y_j}(a_j) \\
&= \prod_{j=1}^{n} \exp\left[(\beta_c + \beta_1(x_j - \bar{x}))a_j + \tfrac{1}{2}\sigma^2 a_j^2\right] \\
&= \exp\left[\sum_{j=1}^{n} (\beta_c + \beta_1(x_j - \bar{x}))a_j + \tfrac{1}{2}\sigma^2 \sum_{i=1}^{n} a_j^2\right] \\
&= \exp\left[\beta_c \sum_{j=1}^{n} a_j + \beta_1 \sum_{j=1}^{n} (x_j - \bar{x})a_j + \tfrac{1}{2}\sigma^2 \sum_{j=1}^{n} a_j^2\right] \\
&= \exp\left[\beta_c t_1 + \beta_1 t_2 + \tfrac{1}{2}\sigma^2 t_1^2/n + \tfrac{1}{2}\sigma^2 t_2^2 \Big/ \sum_{i=1}^{n} (x_i - \bar{x})^2\right] \\
&= \exp\left[\beta_c t_1 + \tfrac{1}{2}(\sigma^2/n) t_1^2\right] \exp\left[\beta_1 t_2 + \tfrac{1}{2}\sigma^2 t_2^2 \Big/ \sum_{i=1}^{n} (x_i - \bar{x})^2\right]
\end{aligned}
$$

The first factor, a function of t_1 only, is the MGF of $\mathrm{N}(\beta_c, \sigma^2/n)$, and the second factor, a function of t_2 only, is the MGF of $\mathrm{N}(\beta_1, \sigma^2/\sum (x_i - \bar{x})^2)$. Thus, $\hat{\beta}_c$ and $\hat{\beta}_1$ are independent with $\hat{\beta}_c \sim \mathrm{N}(\beta_c, \sigma^2/n)$ and

$$
\hat{\beta}_1 \sim \mathrm{N}(\beta_1, \sigma^2/\textstyle\sum (x_i - \bar{x})^2)
$$

We know that for fixed σ^2, $\hat{\beta}_c$ and $\hat{\beta}_1$ are complete and sufficient for β_c and β_1. We also know from Theorem 10.4.7 that any other statistic whose distribution does not depend on β_c and β_1 must be independent of $\hat{\beta}_c$ and $\hat{\beta}_1$. Furthermore, if the other statistic is free of σ^2, then it is independent of $\hat{\beta}_c$ and $\hat{\beta}_1$ even when σ^2 is not fixed.

The next part consists of showing that the joint distribution of the W_i's does not depend on β_c or β_1. If this can be established, then it follows from Theorem 10.4.7 that the W_i's are independent of the U_i's.

Let $\boldsymbol{W} = (W_1, \ldots, W_n)$ and $\boldsymbol{t} = (t_1, \ldots, t_n)$, and consider

$$M_{\boldsymbol{W}}(\boldsymbol{t}) = E\left[\exp\left(\sum_{i=1}^{n} t_i W_i\right)\right]$$
$$= E\left[\exp\left(\sum_{i=1}^{n} t_i \sum_{j=1}^{n} c_{ij} Y_j\right)\right]$$
$$= E\left[\exp\left(\sum_{j=1}^{n} \sum_{i=1}^{n} t_i c_{ij} Y_j\right)\right]$$
$$= E\left[\exp\left(\sum_{j=1}^{n} d_j Y_j\right)\right] \tag{15.3.7}$$

with $d_j = \sum_{i=1}^{n} t_i c_{ij}$. The following identities are useful in deriving the MGF (see Exercise 12):

$$\sum_{j=1}^{n} d_j = 0$$

$$\sum_{j=1}^{n} (x_j - \bar{x}) d_j = 0$$

$$\sum_{j=1}^{n} d_j^2 = \sum_{j=1}^{n} \left(\sum_{i=1}^{n} c_{ij} t_i\right)^2$$

The rest of the derivation is similar to that of the MGF of U_1 and U_2 except that a_j is replaced with d_j for all $j = 1, \ldots, n$. We note from (15.3.7) that

$$M_{\boldsymbol{W}}(\boldsymbol{t}) = \prod_{i=1}^{n} M_{Y_j}(d_j)$$
$$= \exp\left[\beta_c \sum_{j=1}^{n} d_j + \beta_1 \sum_{j=1}^{n} (x_j - \bar{x}) d_j + \tfrac{1}{2}\sigma^2 \sum_{j=1}^{n} d_j^2\right]$$
$$= \exp\left[\tfrac{1}{2}\sigma^2 \sum_{j=1}^{n} \left(\sum_{i=1}^{n} c_{ij} t_i\right)^2\right]$$

This last function depends on neither β_c nor β_1, which as noted earlier means that $\boldsymbol{W} = (W_1, \ldots, W_n)$ is independent of $\hat{\beta}_c$ and $\hat{\beta}_1$.

The rest of the proof relies on the following identity (see Exercise 13):

$$\sum_{i=1}^{n} [Y_i - \beta_c - \beta_1(x_i - \bar{x})]^2 = n\hat{\sigma}^2 + n(\hat{\beta}_c - \beta_c)^2 + \sum_{i=1}^{n} (x_i - \bar{x})^2(\hat{\beta}_1 - \beta_1)^2 \tag{15.3.8}$$

We defined random variables $Z_i = [Y_i - \beta_c - \beta_1(x_i - \bar{x})]/\sigma$, for $i = 1, \ldots, n$, $Z_{n+1} = \sqrt{n}(\hat{\beta}_c - \beta_c)/\sigma$ and $Z_{n+2} = \sqrt{\sum_{i=1}^{n} (x_i - \bar{x})^2}(\hat{\beta}_1 - \beta_1)/\sigma$, which are all standard normal. From equation (15.3.8) note that

$$\sum_{i=1}^{n} Z_i^2 = \frac{n\hat{\sigma}^2}{\sigma^2} + Z_{n+1}^2 + Z_{n+2}^2 \tag{15.3.9}$$

If we define $V_1 = \sum_{i=1}^{n} Z_i^2$, $V_2 = n\hat{\sigma}^2/\sigma^2$, and $V_3 = Z_{n+1}^2 + Z_{n+2}^2$, then $V_1 \sim \chi^2(n)$ and $V_3 \sim \chi^2(2)$. Furthermore, because $V_2 = n\hat{\sigma}^2/\sigma^2$ is a function of $W_1, \ldots, W_n$, we know that V_2 and V_3 are independent. From equation (15.3.9), it follows that $V_1 = V_2 + V_3$, and the MGF of V_1 factors as follows:

$$M_{V_1}(t) = M_{V_2}(t)M_{V_3}(t)$$
$$(1 - 2t)^{-n/2} = M_{V_2}(t)(1 - 2t)^{-1}$$

Thus, $M_{V_2}(t) = (1 - 2t)^{-(n-2)/2}$, from which it follows that $V_2 \sim \chi^2(n - 2)$. This proves the theorem for the centered problem. The extension to the general case can be accomplished by using the fact that $\hat{\beta}_0$ is a linear function of $\hat{\beta}_c$ and $\hat{\beta}_1$, $\hat{\beta}_0 = \hat{\beta}_c - \hat{\beta}_1\bar{x}$. Using this relationship, it is straightforward to derive the joint MGF of $\hat{\beta}_0$ and $\hat{\beta}_1$ based on the MGF of $U = (\hat{\beta}_c, \hat{\beta}_1)$ (see Exercise 14). This concludes the proof. ■

According to Theorem 15.3.4, $E(\hat{\beta}_0) = \beta_0$, $E(\hat{\beta}_1) = \beta_1$, and $E(n\hat{\sigma}^2/\sigma^2) = n - 2$, from which it follows that $\hat{\beta}_0$, $\hat{\beta}_1$, and $\tilde{\sigma}^2 = \sum [Y_i - \hat{\beta}_0 - \hat{\beta}_1 x_i]^2/(n - 2)$ are unbiased estimators. We also know from Theorem 15.3.2 that these estimators are complete and sufficient, yielding the following theorem.

Theorem 15.3.5 If $Y_1, \ldots, Y_n$ are independent, $Y_i \sim N(\beta_0 + \beta_1 x_i, \sigma^2)$, then $\hat{\beta}_0$, $\hat{\beta}_1$, and $\tilde{\sigma}^2$ are UMVUEs of β_0, β_1, and σ^2. ■

It also is possible to derive confidence intervals for the parameters based on the above results.

Theorem 15.3.6 If $Y_1, \ldots, Y_n$ are independent, $Y_i \sim N(\beta_0 + \beta_1 x_i, \sigma^2)$, then $\gamma \times 100\%$ confidence intervals for β_0, β_1, and σ^2 are given respectively by

1. $$\left(\hat{\beta}_0 - t_{(1+\gamma)/2}\,\tilde{\sigma}\sqrt{\frac{\sum x_i^2}{n\sum (x_i - \bar{x})^2}},\ \hat{\beta}_0 + t_{(1+\gamma)/2}\,\tilde{\sigma}\sqrt{\frac{\sum x_i^2}{n\sum (x_i - \bar{x})^2}}\right)$$
2. $$\left(\hat{\beta}_1 - \frac{t_{(1+\gamma)/2}\,\tilde{\sigma}}{\sqrt{\sum (x_i - \bar{x})^2}},\ \hat{\beta}_1 + \frac{t_{(1+\gamma)/2}\,\tilde{\sigma}}{\sqrt{\sum (x_i - \bar{x})^2}}\right)$$
3. $$\left(\frac{(n-2)\tilde{\sigma}^2}{\chi^2_{(1+\gamma)/2}},\ \frac{(n-2)\tilde{\sigma}^2}{\chi^2_{(1-\gamma)/2}}\right)$$

where the respective t and χ^2 percentiles have $n - 2$ degrees of freedom.

Proof

It follows from Theorem 15.3.4 that

$$Z_0 = \frac{(\hat{\beta}_0 - \beta_0)}{\sigma\sqrt{[\sum x_i^2/[n\sum (x_i - \bar{x})^2]]}} \sim N(0, 1)$$

$$Z_1 = \frac{\sqrt{\sum (x_i - \bar{x})^2}(\hat{\beta}_1 - \beta_1)}{\sigma} \sim N(0, 1)$$

$$V = \frac{(n-2)\tilde{\sigma}^2}{\sigma^2} \sim \chi^2(n-2)$$

Furthermore, each Z_0 and Z_1 is independent of V. Thus,

$$T_0 = \frac{Z_0}{\sqrt{\tilde{\sigma}^2/\sigma^2}} = \frac{(\hat{\beta}_0 - \beta_0)}{\tilde{\sigma}\sqrt{[\sum x_i^2/[n\sum (x_i - \bar{x})^2]]}} \sim t(n-2)$$

and

$$T_1 = \frac{Z_1}{\sqrt{\tilde{\sigma}^2/\sigma^2}} = \frac{\sqrt{\sum (x_i - \bar{x})^2}(\hat{\beta}_1 - \beta_1)}{\tilde{\sigma}} \sim t(n-2)$$

The confidence intervals are derived from the pivotal quantities T_0, T_1, and V. ■

It also is possible to derive tests of hypotheses based on these pivotal quantities.

Theorem 15.3.7 Assume that $Y_1, \ldots Y_n$ are independent, $Y_i \sim N(\beta_0 + \beta_1 x_i, \sigma^2)$, and denote by t_0, t_1, and v computed values of T_0, T_1, and V with β_0, β_1, and σ^2 replaced by β_{00}, β_{10}, and σ_0^2, respectively.

1. A size α test of $H_0 : \beta_0 = \beta_{00}$ versus $H_a : \beta_0 \neq \beta_{00}$ is to reject H_0 if

$$|t_0| \geq t_{1-\alpha/2}(n-2).$$

2. A size α test of $H_0 : \beta_1 = \beta_{10}$ versus $H_a : \beta_1 \neq \beta_{10}$ is to reject H_0 if

$$|t_1| \geqslant t_{1-\alpha/2}(n-2)$$

3. A size α test of $H_0 : \sigma^2 = \sigma_0^2$ versus $H_a : \sigma^2 \neq \sigma_0^2$ is to reject H_0 if

$$v \geqslant \chi^2_{1-\alpha/2}(n-2) \text{ or } v \leqslant \chi^2_{\alpha/2}(n-2)$$ ■

One-sided tests can be obtained in a similar manner, but we will not state them here (see Exercise 16).

Example 15.3.3 Consider the auto emission data of Exercise 15.3.1. Recall that $\sum x_i = 304.377$ and $\sum x_i^2 = 10461.814$ from which we obtain $\sum (x_i - \bar{x})^2 = 10461.814 - 11(27.671)^2 = 2039.287$. We also have that $\hat{\beta}_0 = 0.266$, $\hat{\beta}_1 = 0.00158$, and $\tilde{\sigma}^2 = 0.00030$ so that $\tilde{\sigma} = 0.017$.

Note that $t_{.975}(9) = 2.262$, $\chi^2_{.025}(9) = 2.70$, and $\chi^2_{.975}(9) = 19.02$. If we apply Theorem 15.3.6, then 95% confidence limits for β_0 are given by $0.266 \pm (2.262)(0.017)\sqrt{(10461.814)/[(11)(2039.287)]}$ or 0.266 ± 0.038, and a 95% confidence interval is (0.228, 0.304). Similarly, 95% confidence limits for β_1 are given by $0.00158 \pm (2.262)(0.017)/\sqrt{2039.287}$ or 0.00158 ± 0.00085, and a 95% confidence interval is (0.0007, 0.0024). The 95% confidence limits for σ^2 are $9(0.00030)/19.02 = 0.00014$ and $9(0.00030)/2.70 = 0.00100$, and thus 95% confidence interval for σ^2 and σ are (0.00014, 0.00100) and (0.012, 0.032), respectively.

15.4

GENERAL LINEAR MODEL

Many of the results derived for the simple linear model can be extended to the general linear case. It is not possible to develop the general model conveniently without introducing matrix notation. A few basic results will be stated in matrix notation for the purpose of illustration, but the topic will not be developed fully here. We will denote the *transpose* of an arbitrary matrix A by A'. That is, if $A = \{a_{ij}\}$, then $A' = \{a_{ji}\}$. Furthermore, if A is a square nonsingular matrix, then we denote its *inverse* by A^{-1}. We also will make no distinction between a $1 \times k$ matrix and a k-dimensional row vector. Similarly, a $k \times 1$ matrix will be regarded the same as a k-dimensional column vector. Thus, if $\boldsymbol{c}$ represents a k-dimensional column vector, then its transpose $\boldsymbol{c}'$ will represent the corresponding row vector.

Consider the linear regression model (15.2.3) and assume that a response y_i is observed at the values $x_{i0}, x_{i1}, \ldots, x_{ip}$, $i = 1, \ldots, n$ with $n \geqslant p + 1$. That is, assume that

$$E(Y_i) = \sum_{j=0}^{p} \beta_j x_{ij} \qquad \text{Var}(Y_i) = \sigma^2 \qquad \text{Cov}(Y_i, Y_j) = 0 \quad i \neq j$$

We will denote by V a matrix such that the ijth element is the covariance of the variables Y_i and Y_j, $V = \{\text{Cov}(Y_i, Y_j)\}$. The matrix V is called the **covariance matrix** of $Y_1, \ldots, Y_n$. We will define the expected value of a vector of random variables to be the corresponding vector of expected values. For example, if $W = (W_1, \ldots, W_k)$ is a row vector whose components are random variables, then $E(W) = (E(W_1), \ldots E(W_k))$, and similarly for column vectors of random variables. It is possible to reformulate the model in terms of matrices as follows:

$$E(Y) = X\boldsymbol{\beta} \qquad V = \sigma^2 I \tag{15.4.1}$$

where I is the $n \times n$ identity matrix, and Y, $\boldsymbol{\beta}$, and X are

$$Y = \begin{pmatrix} Y_1 \\ \vdots \\ Y_n \end{pmatrix} \qquad \boldsymbol{\beta} = \begin{pmatrix} \beta_0 \\ \vdots \\ \beta_p \end{pmatrix} \qquad X = \begin{pmatrix} x_{10} & \cdots & x_{1p} \\ \vdots & & \vdots \\ x_{n0} & \cdots & x_{np} \end{pmatrix}$$

LEAST-SQUARES APPROACH

The least-squares estimates are the values $\beta_j = \hat{\beta}_j$ that minimize the quantity

$$S = \sum_{i=1}^{n} \left[y_i - \sum_{j=0}^{p} \beta_j x_{ij} \right]^2 = (Y - X\boldsymbol{\beta})'(Y - X\boldsymbol{\beta}) \tag{15.4.2}$$

The approach used with the simple linear model generalizes readily. In other words, if we set the partials of S with respect to the β_j's to zero and solve the resulting system of equations, then we obtain the LS estimates. Specifically, we solve

$$\frac{\partial}{\partial \beta_k} S = \sum_{i=1}^{n} 2 \left[y_i - \sum_{j=0}^{p} \beta_j x_{ij} \right] (-x_{ik}) = 0 \qquad k = 0, 1, \ldots, p$$

This system of equations is linear in the β_j's, and it is conveniently expressed in terms of the matrix equation

$$X'Y = X'X\boldsymbol{\beta} \tag{15.4.3}$$

If the matrix $X'X$ is nonsingular, then there exists a unique solution of the form

$$\hat{\boldsymbol{\beta}} = (X'X)^{-1}X'Y \tag{15.4.4}$$

Unless indicated otherwise, we will assume, that $X'X$ is nonsingular. Of course, a more basic assumption would be that X has *full rank.*

The estimators $\hat{\beta}_j$ are linear functions of $Y_1, \ldots Y_n$, and it can be shown that they are unbiased. Specifically, it follows from matrix (15.4.1) and equation (15.4.4) that

$$\begin{aligned} E(\hat{\boldsymbol{\beta}}) &= (\boldsymbol{X}'\boldsymbol{X})^{-1}\boldsymbol{X}'E(\boldsymbol{Y}) \\ &= (\boldsymbol{X}'\boldsymbol{X})^{-1}\boldsymbol{X}'\boldsymbol{X}\boldsymbol{\beta} \\ &= \boldsymbol{\beta} \end{aligned} \tag{15.4.5}$$

As in the case of the simple linear model, the LS estimates of the β_j's for the general model are referred to as the BLUEs.

It also can be shown that the respective variances and covariances of the BLUEs are the elements of the matrix

$$\boldsymbol{C} = \{\text{Cov}(\hat{\beta}_i, \hat{\beta}_j)\} = \sigma^2(\boldsymbol{X}'\boldsymbol{X})^{-1} \tag{15.4.6}$$

and that the BLUE of any linear combination of the β_j's, say $\boldsymbol{r}'\beta = \sum_{j=0}^{p} r_j\beta_j$ is given by $\boldsymbol{r}'\hat{\boldsymbol{\beta}}$ (see Scheffe, 1959).

Theorem 15.4.1 **Gauss-Markov** If $E(\boldsymbol{Y}) = \boldsymbol{X}\boldsymbol{\beta}$ and $\boldsymbol{V} = \{\text{Cov}(Y_i, Y_j)\} = \sigma^2\boldsymbol{I}$, then $\boldsymbol{r}'\hat{\boldsymbol{\beta}}$ is the BLUE of $\boldsymbol{r}'\boldsymbol{\beta}$, where $\hat{\boldsymbol{\beta}} = (\boldsymbol{X}'\boldsymbol{X})^{-1}\boldsymbol{X}'\boldsymbol{Y}$. ■

This theorem can be generalized to the case in which

$$\boldsymbol{V} = \{\text{Cov}(Y_i, Y_j)\} = \sigma^2\boldsymbol{A}$$

where $\boldsymbol{A}$ is a known matrix. That is, the Y_i's may be correlated and have unequal variances as long as $\boldsymbol{A}$ is known. It turns out that the BLUEs in this case are obtained by minimizing the weighted sum of squares $(\boldsymbol{Y} - \boldsymbol{X}\boldsymbol{\beta})'\boldsymbol{A}^{-1}(\boldsymbol{Y} - \boldsymbol{X}\boldsymbol{\beta})$. Note that σ^2 also may be a function of the unknown β_j's, say $\sigma^2 = c(\boldsymbol{\beta})$.

Theorem 15.4.2 **Generalized Gauss-Markov** Let $E(\boldsymbol{Y}) = \boldsymbol{X}\boldsymbol{\beta}$ and $\boldsymbol{V} = \{\text{Cov}(Y_i, Y_j)\} = c(\boldsymbol{\beta})\boldsymbol{A}$, where $\boldsymbol{A}$ is a matrix of known constants. The generalized least-squares estimates of $\boldsymbol{\beta}$ are the values that minimize $\boldsymbol{S} = (\boldsymbol{Y} - \boldsymbol{X}\boldsymbol{\beta})'\boldsymbol{A}^{-1}(\boldsymbol{Y} - \boldsymbol{X}\boldsymbol{\beta})$, and they are given by $\hat{\boldsymbol{\beta}} = (\boldsymbol{X}'\boldsymbol{A}^{-1}\boldsymbol{X})^{-1}\boldsymbol{X}'\boldsymbol{A}^{-1}\boldsymbol{Y}$. Also, $\boldsymbol{r}'\hat{\boldsymbol{\beta}}$ is the BLUE of $\boldsymbol{r}'\boldsymbol{\beta}$. ■

Note that all of these results have been developed in terms of the means, variances, and covariances and that no other distributional assumptions were made.

LINEAR FUNCTIONS OF ORDER STATISTICS

One interesting application of Theorem 15.4.2, which arises in a slightly different context, is that of finding estimators of location and scale parameters based on linear functions of the order statistics. Consider a pdf of the form

$$f(w; \beta_0, \beta_1) = \frac{1}{\beta_1} g\left(\frac{w - \beta_0}{\beta_1}\right)$$

where g is a function independent of β_0 and β_1 and let

$$Z = \frac{W - \beta_0}{\beta_1} \sim g(z)$$

For an ordered random sample of size n, let

$$\boldsymbol{Y} = \begin{pmatrix} Y_1 \\ \vdots \\ Y_n \end{pmatrix} = \begin{pmatrix} W_{1:n} \\ \vdots \\ W_{n:n} \end{pmatrix} \qquad \boldsymbol{\beta} = \begin{pmatrix} \beta_0 \\ \beta_1 \end{pmatrix}$$

It follows that

$$\begin{aligned} E(Y_i) = E(W_{i:n}) &= \beta_0 + \beta_1 E\left[\frac{W_{i:n} - \beta_0}{\beta_1}\right] \\ &= \beta_0 + \beta_1 E(Z_{i:n}) \\ &= \beta_0 + \beta_1 k_i \end{aligned}$$

That is, in this case let

$$\boldsymbol{X} = \begin{pmatrix} 1 & k_1 \\ \vdots & \vdots \\ 1 & k_n \end{pmatrix}$$

Also,

$$v_{ij} = \text{Cov}(W_{i:n}, W_{j:n}) = \beta_1^2 \text{ Cov}(Z_{i:n}, Z_{j:n}) = \beta_1^2 a_{ij}$$

and

$$\boldsymbol{V} = \beta_1^2 \boldsymbol{A}$$

It then follows that the components of

$$\hat{\boldsymbol{\beta}} = \begin{pmatrix} \hat{\beta}_0 \\ \hat{\beta}_1 \end{pmatrix} = (\boldsymbol{X}'\boldsymbol{A}^{-1}\boldsymbol{X})^{-1}\boldsymbol{X}'\boldsymbol{A}^{-1}\boldsymbol{Y}$$

are unbiased estimators of β_0 and β_1 that have minimum variance among all linear unbiased functions of the order statistics. The main drawback of this method is that the constants k_i and a_{ij} often are not convenient to compute. In some cases, asymptotic approximations of the constants have been useful.

It is interesting to note that if $\boldsymbol{A}$ is not used, then the ordinary LS estimates $\hat{\boldsymbol{\beta}} = (\boldsymbol{X}'\boldsymbol{X})^{-1}\boldsymbol{X}'\boldsymbol{Y}$ still are unbiased, although they will not be the BLUEs in this case.

MAXIMUM LIKELIHOOD APPROACH

In this section we will develop the MLEs and related properties of the general linear model under the assumption that the random errors from different experiments are independent and normal, $\varepsilon_x \sim N(0, \sigma^2)$.

Suppose $Y_1, \ldots, Y_n$ are independent normally distributed random variables, and they are the elements of a vector Y that satisfies model (15.4.1).

Theorem 15.4.3 If $Y_1, \ldots, Y_n$ are independent with $Y_i \sim N\left(\sum_{j=0}^{p} \beta_j x_{ij}, \sigma^2\right)$, for $i = 1, \ldots, n$, then the MLEs of $\beta_0, \ldots, \beta_p$ and σ^2 are given by

$$\hat{\boldsymbol{\beta}} = (X'X)^{-1}X'Y \tag{15.4.7}$$

$$\hat{\sigma}^2 = \frac{(Y - X\hat{\boldsymbol{\beta}})'(Y - X\hat{\boldsymbol{\beta}})}{n} \tag{15.4.8}$$

Proof

The likelihood function is

$$L = L(\beta_0, \ldots, \beta_p, \sigma^2) = \prod_{i=1}^{n} \frac{1}{\sqrt{2\pi\sigma^2}} \exp\left[-\frac{1}{2\sigma^2}\left(y_i - \sum_{j=0}^{p} \beta_j x_{ij}\right)^2\right] \tag{15.4.9}$$

Thus, the log-likelihood is

$$\begin{aligned} \ln L &= -\frac{n}{2} \ln (2\pi\sigma^2) - \frac{1}{2\sigma^2} \sum_{i=1}^{n} \left(y_i - \sum_{j=0}^{p} \beta_j x_{ij}\right)^2 \\ &= -\frac{n}{2} \ln (2\pi\sigma^2) - \frac{1}{2\sigma^2} (Y - X\boldsymbol{\beta})'(Y - X\boldsymbol{\beta}) \\ &= -\frac{n}{2} \ln (2\pi\sigma^2) - \frac{1}{2\sigma^2} S \end{aligned} \tag{15.4.10}$$

where S is the quantity (15.4.2). Clearly the values of $\beta_0, \ldots, \beta_p$ that maximize function (15.4.9) are the same ones that minimize S. Of course, this means that the MLEs of the β_j's under the present assumptions are the same as the BLUEs that were derived earlier. The MLE of σ^2 is obtained by setting the partial derivative (15.4.10) with respect to σ^2 to zero and replacing the β_j's with the $\hat{\beta}_j$'s. Another convenient form for the MLE of σ^2 (see Exercise 25) is

$$\hat{\sigma}^2 = \frac{Y'(Y - X\hat{\boldsymbol{\beta}})}{n} \tag{15.4.11}$$

■

As in the case of the simple linear model, the minimum value of S reflects the amount of variation of the data about the estimated regression function, and this defines the error sum of squares for the general linear regression model, denoted by $\text{SSE} = (\boldsymbol{Y} - \boldsymbol{X}\hat{\boldsymbol{\beta}})'(\boldsymbol{Y} - \boldsymbol{X}\hat{\boldsymbol{\beta}})$.

Many of the results that were derived for the MLEs in the case of the simple linear model have counterparts for the general linear model. However, to state some of these results, it is necessary to introduce a multivariate generalization of the normal distribution.

MULTIVARIATE NORMAL DISTRIBUTION

In Chapter 5, a bivariate generalization of the normal distribution was presented. Specifically, a pair of continuous random variables X_1 and X_2 are said to be bivariate normal, $\text{BVN}(\mu_1, \mu_2, \sigma_1^2, \sigma_2^2, \rho)$, if they have a joint pdf of the form

$$f(x_1, x_2) = \frac{1}{2\pi\sigma_1\sigma_2\sqrt{1-\rho^2}} \exp\left(-\frac{1}{2}Q\right) \tag{15.4.12}$$

with

$$Q = \frac{1}{(1-\rho^2)}\left[\left(\frac{x_1-\mu_1}{\sigma_1}\right)^2 - 2\rho\left(\frac{x_1-\mu_1}{\sigma_1}\right)\left(\frac{x_2-\mu_2}{\sigma_2}\right) + \left(\frac{x_2-\mu_2}{\sigma_2}\right)^2\right]$$

This pdf can be expressed more conveniently using matrix notation. In particular, we define the following vectors and matrices:

$$\boldsymbol{x} = \begin{pmatrix} x_1 \\ x_2 \end{pmatrix} \qquad \boldsymbol{\mu} = \begin{pmatrix} \mu_1 \\ \mu_2 \end{pmatrix} \qquad \boldsymbol{V} = \{\text{Cov}(X_i, X_j)\}$$

It can be shown that $Q = (\boldsymbol{x} - \boldsymbol{\mu})'\boldsymbol{V}^{-1}(\boldsymbol{x} - \boldsymbol{\mu})$ and that the determinant of $\boldsymbol{V}$ is $|\boldsymbol{V}| = \sigma_1^2\sigma_2^2(1-\rho^2)$. Notice that we are assuming that $\boldsymbol{V}$ is nonsingular. We will restrict our attention to this case throughout this discussion. This provides a way to generalize the normal distribution to a k-dimensional version.

Definition 15.4.1

A set of continuous random variables $X_1, \ldots, X_k$ are said to have a **multivariate normal** or **k-variate normal** distribution if the joint pdf has the form

$$f(x_1, \ldots, x_k) = \frac{1}{\sqrt{(2\pi)^k |\boldsymbol{V}|}} \exp\left[-\frac{1}{2}(\boldsymbol{x} - \boldsymbol{\mu})'\boldsymbol{V}^{-1}(\boldsymbol{x} - \boldsymbol{\mu})\right] \tag{15.4.13}$$

with $\boldsymbol{x}' = (x_1, \ldots, x_k)$, $\boldsymbol{\mu}' = (\mu_1, \ldots, \mu_k)$, and $\boldsymbol{V} = \{\text{Cov}(X_i, X_j)\}$, and where $\mu_i = E(X_i)$ and $\boldsymbol{V}$ is a $k \times k$ nonsingular covariance matrix.

Notice that for a set of multivariate normal random variables $X_1, \ldots, X_k$, the distribution is determined completely by the mean vector $\boldsymbol{\mu}$ and covariance matrix $\boldsymbol{V}$. An important property of multivariate normal random variables is that their marginals are normal, $X_i \sim \text{N}(\mu_i, \sigma_i^2)$ (see Exercise 27).

Another quantity that was encountered in this development is the variable $Q = (\boldsymbol{x} - \boldsymbol{\mu})'\boldsymbol{V}^{-1}(\boldsymbol{x} - \boldsymbol{\mu})$. This is a special case of a function called a quadratic form. More generally, a **quadratic form** in the k variables $x_1, \ldots, x_k$ is a function of the form

$$Q = \sum_{i=1}^{k} \sum_{j=0}^{k} a_{ij} x_i x_j \tag{15.4.14}$$

where the a_{ij}'s are constants. It often is more convenient to express Q in matrix notation. Specifically, if $\boldsymbol{A} = \{a_{ij}\}$ and $\boldsymbol{x}' = (x_1, \ldots, x_k)$, then $Q = \boldsymbol{x}'\boldsymbol{A}\boldsymbol{x}$. Strictly speaking, $Q = (\boldsymbol{x} - \boldsymbol{\mu})'\boldsymbol{V}^{-1}(\boldsymbol{x} - \boldsymbol{\mu})$ is a quadratic form in the differences $x_1 - \mu_1, \ldots, x_k - \mu_k$ rather than $x_1, \ldots, x_k$. An example of a quadratic form in the x_i's would be $Q = \boldsymbol{x}'\boldsymbol{I}\boldsymbol{x} = \sum x_i^2$. Some quadratic forms in the y_i's that have been encountered in this section are $\tilde{\sigma}^2$ and $\hat{\sigma}^2$.

PROPERTIES OF THE ESTIMATORS

Most of the properties of the MLEs for the simple linear model can be extended using the approach of Section 15.3, but the details are more complicated for the higher-dimensional problem. We will state some of the properties of the MLEs in the following theorems.

Theorem 15.4.4 Under the assumptions of Theorem 15.4.3 the following properties hold:

1. The MLEs $\hat{\beta}_0, \ldots, \hat{\beta}_p$ and $\hat{\sigma}^2$ are jointly complete and sufficient.
2. $\hat{\boldsymbol{\beta}}$ has a multivariate normal distribution with mean vector $\boldsymbol{\beta}$ and covariance matrix $\sigma^2(\boldsymbol{X}'\boldsymbol{X})^{-1}$.
3. $n\hat{\sigma}^2/\sigma^2 \sim \chi^2(n - p - 1)$.
4. $\hat{\boldsymbol{\beta}}$ and $\hat{\sigma}^2$ are independent.
5. Each $\hat{\beta}_j$ is the UMVUE of β_j.
6. $\tilde{\sigma}^2 = (\boldsymbol{Y} - \boldsymbol{X}\hat{\beta})'(\boldsymbol{Y} - \boldsymbol{X}\hat{\beta})/(n - p - 1)$ is the UMVUE of σ^2. ■

It also is possible to derive confidence intervals for the parameters. In the following theorem, let $\boldsymbol{A} = \{a_{ij}\} = (\boldsymbol{X}'\boldsymbol{X})^{-1}$ so that $\boldsymbol{C} = \{\text{Cov}(\hat{\beta}_i, \hat{\beta}_j)\} = \sigma^2\boldsymbol{A}$.

Theorem 15.4.5 Under the assumptions of Theorem 15.4.3, the following intervals are $\gamma \times 100\%$ confidence intervals for the β_j's and σ^2:

1. $(\hat{\beta}_j - t_{(1+\gamma)/2}\sqrt{\tilde{\sigma}^2 a_{jj}}, \hat{\beta}_j + t_{(1+\gamma)/2}\sqrt{\tilde{\sigma}^2 a_{jj}})$
2. $\left(\dfrac{(n-p-1)\tilde{\sigma}^2}{\chi^2_{(1+\gamma)/2}}, \dfrac{(n-p-1)\tilde{\sigma}^2}{\chi^2_{(1-\gamma)/2}}\right)$

where the respective t and χ^2 percentiles have $n - p - 1$ degrees of freedom.

Proof

It follows from Theorem 15.4.4 that $\hat{\beta}_j \sim N(\beta_j, \sigma^2 a_{jj})$, so that

$$Z = \frac{(\hat{\beta}_j - \beta_j)}{\sqrt{\sigma^2 a_{jj}}} \sim N(0, 1)$$

$$V = \frac{(n-p-1)\tilde{\sigma}^2}{\sigma^2} \sim \chi^2(n-p-1)$$

Furthermore, because Z and V are independent,

$$T = \frac{Z}{\sqrt{\tilde{\sigma}^2/\sigma^2}} = \frac{(\hat{\beta}_j - \beta_j)}{\sqrt{\tilde{\sigma}^2 a_{jj}}} \sim t(n-p-1)$$

The confidence intervals follow from these results. ■

It also is possible to derive tests of hypotheses based on these pivotal quantities.

Theorem 15.4.6 Assume the conditions Theorem 15.4.3 and denote by t and v computed values of T and V with $\beta_j = \beta_{j0}$ and $\sigma^2 = \sigma_0^2$.

1. A size α test of $H_0 : \beta_j = \beta_{j0}$ versus $H_a : \beta_j \neq \beta_{j0}$ is to reject H_0 if
$$|t| \geqslant t_{1-\alpha/2}(n-p-1)$$
2. A size α test of $H_0 : \beta_j = \beta_{j0}$ versus $H_a : \beta_j > \beta_{j0}$ is to reject H_0 if
$$t \geqslant t_{1-\alpha}(n-p-1)$$
3. A size α test of $H_0 : \sigma^2 = \sigma_0^2$ versus $H_a : \sigma^2 \neq \sigma_0^2$ is to reject H_0 if
$$v \geqslant \chi^2_{1-\alpha/2}(n-p-1) \text{ or } v \leqslant \chi^2_{\alpha/2}(n-p-1)$$
4. A size α test of $H_0 : \sigma^2 = \sigma_0^2$ versus $H_a : \sigma^2 > \sigma_0^2$ is to reject H_0 if
$$v \geqslant \chi^2_{1-\alpha}(n-p-1)$$ ■

Lower one-sided tests can be obtained in a similar manner, but we will not state them here.

Example 15.4.1 Consider the auto emission data of Example 15.3.1. Recall from that example that we carried out the analysis under the assumption of a simple linear model. Although there is no theoretical reason for assuming that the relationship between mileage and hydrocarbon emissions is nonlinear in the variables, the plotted data in Figure 15.1 suggests such a possibility, particularly after approximately 40,000 miles. An obvious extension of the original analysis would be to consider a second-degree polynomial model. That is, each measurement y_i is an observation on a random variable $Y_i = \beta_0 + \beta_1 x_i + \beta_2 x_i^2 + \varepsilon_i$, with independent normal errors, $\varepsilon_i \sim N(0, \sigma^2)$. This can be formulated in terms of the general linear model (15.4.1) using the matrices

$$\boldsymbol{\beta} = \begin{pmatrix} \beta_0 \\ \beta_1 \\ \beta_2 \end{pmatrix} \quad X = \begin{pmatrix} 1 & x_1 & x_1^2 \\ \vdots & \vdots & \vdots \\ 1 & x_n & x_n^2 \end{pmatrix}$$

with $n = 11$. Recall from equation (15.4.4) that if the matrix $X'X$ is nonsingular, then the LS estimates, which also are the ML estimates, are the components of the vector $\hat{\boldsymbol{\beta}} = (X'X)^{-1}X'Y$. Although the recommended procedure is to use a statistical software package such as Minitab or SAS, it is possible to give explicit formulas for the estimates. In particular,

$$\hat{\beta}_1 = \frac{s_{1y}s_{22} - s_{2y}s_{12}}{s_{11}s_{22} - s_{12}^2} \qquad \hat{\beta}_2 = \frac{s_{2y}s_{11} - s_{1y}s_{12}}{s_{11}s_{22} - s_{12}^2} \qquad \hat{\beta}_0 = \bar{y} - \hat{\beta}_1\bar{x} - \hat{\beta}_2\overline{x^2}$$

with

$$\overline{x^2} = \frac{\sum x_i^2}{n}, \quad s_{1y} = \sum x_i y_i - n\bar{x}\bar{y}, \quad s_{2y} = \sum x_i^2 y_i - n(\bar{x})^2\bar{y},$$

$$s_{11} = \sum x_i^2 - n(\bar{x})^2, \quad s_{12} = \sum x_i^3 - \bar{x}\overline{x^2}, \quad \text{and} \quad s_{22} = \sum x_i^4 - n(\overline{x^2})^2.$$

The LS estimates of the regression coefficients are $\hat{\beta}_0 = 0.2347$, $\hat{\beta}_1 = 0.0046$, and $\hat{\beta}_2 = -0.000055$, yielding the regression function $\hat{y} = 0.2347 + 0.0046x - 0.000055x^2$. A graph of $\hat{y}$ is provided in Figure 15.2 along with a plot of the data. For comparison, the graph of the linear regression function obtained in Example 15.3.1 also is shown as a dashed line.

An obvious question would be whether the second-degree term is necessary. One approach to answering this would be to test whether β_2 is significantly different from zero. According to Theorem 15.4.6, a size $\alpha = .05$ test of $H_0 : \beta_2 = 0$ versus $H_1 : \beta_2 \neq 0$ is to reject H_0 if $|t| \geqslant 2.306$. Because in this example $t = -2.235$, β_2 does not differ significantly from zero, at least at the .05 level. It should be noted, however, that the test would reject at the .10 level.

FIGURE 15.2 Hydrocarbon emissions as a function of accumulated mileage (second-degree polynomial fit)

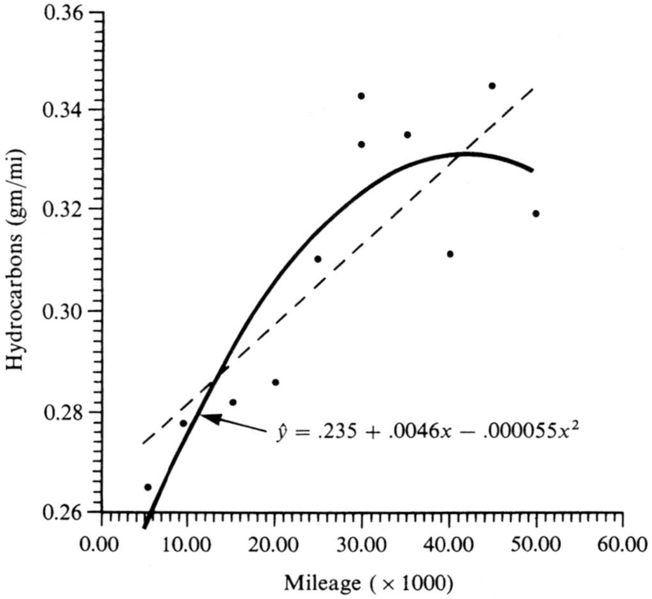

Another question of interest involves joint tests and confidence regions for two or more of the coefficients.

JOINT TESTS AND CONFIDENCE CONTOURS

Part 1 of Theorem 15.4.6 provides a means for testing individual regression coefficients with the other parameters of the model regarded as unknown nuisance parameters. It also is possible to develop methods for simultaneously testing two or more of the regression coefficients.

Example 15.4.2 Although the plots of the hydrocarbon data in Figures 15.1 and 15.2 strongly suggest that some sort of regression model is appropriate, it would be desirable in general to test whether terms beyond the constant term β_0 really are needed. Such a test can be constructed using the approach of the generalized likelihood ratio (GLR) test of Chapter 12. In the auto emissions example, suppose it is desired to test jointly the hypothesis $H_0 : \beta_1 = \beta_2 = 0$ versus the alternative that at least one of these coefficients is nonzero. The parameters β_0 and σ^2 would be unknown. Let Ω be the set of all quadruples $(\beta_0, \beta_1, \beta_2, \sigma^2)$ with $-\infty < \beta_j < \infty$ and $\sigma^2 > 0$. Furthermore, let Ω_0 be the subset $(\beta_0, 0, 0, \sigma^2)$. Under the assumption of independent errors, $\varepsilon_i \sim N(0, \sigma^2)$, and over the subspace Ω_0 the MLEs of β_0 and σ^2 are $\hat{\beta}_{00} = \bar{y}$ and $\hat{\sigma}_0^2 = \sum (y_i - \bar{y})^2/n$, while over the unrestricted space Ω the joint MLEs are the components of $\hat{\boldsymbol{\beta}} = (X'X)^{-1}X'Y$ and $\hat{\sigma}^2 = \sum (y_i - \hat{y}_i)^2/n$ with $\hat{y}_i = \hat{\beta}_0 + \hat{\beta}_1 x_i + \hat{\beta}_2 x_i^2$. Consequently, the GLR is

$$\lambda(y) = \frac{f(y; \hat{\beta}_{00}, 0, 0, \hat{\sigma}_0^2)}{f(y; \hat{\beta}_0, \hat{\beta}_1, \hat{\beta}_2, \hat{\sigma}^2)} = \frac{[(2\pi)\hat{\sigma}_0^2]^{-n/2} \exp\left[-(1/2\hat{\sigma}_0^2) \sum (y_i - \bar{y})^2\right]}{[(2\pi)\hat{\sigma}^2]^{-n/2} \exp\left[-(1/2\hat{\sigma}^2) \sum (y_i - \hat{y}_i)^2\right]}$$

$$= \left(\frac{\hat{\sigma}_0^2}{\hat{\sigma}^2}\right)^{-n/2}$$

Thus, the GLR test would reject $H_0 : \beta_1 = \beta_2 = 0$ if $\lambda(y) \leqslant k$, where k is chosen to provide a size α test. A simple way to proceed would be to use the approximate test given by equation (12.8.3). The test is based on the statistic $-2 \ln \lambda(y) = n \ln (\hat{\sigma}_0^2/\hat{\sigma}^2) = 16.67$. Because $r = 2$ parameters are being tested, the approximate critical value is $\chi^2_{.95}(2) = 5.99$, and the test rejects H_0. In fact, the test is highly significant because the p-value is .0002. We will see shortly that an exact test can be constructed, but first we will consider joint confidence regions.

Consider the quantity $S = S(\boldsymbol{\beta}) = (Y - X\boldsymbol{\beta})'(Y - X\boldsymbol{\beta})$, which is the sum of squares that we minimized to obtain the LS and ML estimates of the regression coefficients. In other words, the minimum value of S is $S(\hat{\boldsymbol{\beta}})$. Specifically, it can be shown (see Exercise 29) that

$$S(\boldsymbol{\beta}) = S(\hat{\boldsymbol{\beta}}) + (\hat{\boldsymbol{\beta}} - \boldsymbol{\beta})'(X'X)(\hat{\boldsymbol{\beta}} - \boldsymbol{\beta}) \tag{15.4.15}$$

With the assumptions of Theorem 15.4.4 we have that $S(\boldsymbol{\beta})/\sigma^2 \sim \chi^2(n)$ and $S(\hat{\boldsymbol{\beta}})/\sigma^2 = n\hat{\sigma}^2/\sigma^2 \sim \chi^2(n - p - 1)$. Furthermore, from Part 4 of Theorem 15.4.4 we know that $\hat{\boldsymbol{\beta}}$ and $\hat{\sigma}^2$ are independent; with equation (15.4.15), this implies that $S(\boldsymbol{\beta}) - S(\hat{\boldsymbol{\beta}}) = (\hat{\boldsymbol{\beta}} - \boldsymbol{\beta})'(X'X)(\hat{\boldsymbol{\beta}} - \boldsymbol{\beta})$ and $S(\hat{\boldsymbol{\beta}}) = n\hat{\sigma}^2$ are independent. Based on the rationale of Theorem 15.3.4, it follows that $[S(\boldsymbol{\beta}) - S(\hat{\boldsymbol{\beta}})]/\sigma^2 \sim \chi^2(n - (n - p - 1)) = \chi^2(p + 1)$. Thus, we can derive an F-variable

$$\frac{[S(\boldsymbol{\beta}) - S(\hat{\boldsymbol{\beta}})]/(p + 1)}{S(\hat{\boldsymbol{\beta}})/(n - p - 1)} \sim F(p + 1, n - p - 1) \tag{15.4.16}$$

It follows that a $\gamma \times 100\%$ confidence region for $\beta_0, \beta_1, \ldots, \beta_p$ is defined by the inequality

$$S(\boldsymbol{\beta}) \leqslant S(\hat{\boldsymbol{\beta}})\left[1 + \frac{p + 1}{n - p - 1} f_\gamma(p + 1, n - p - 1)\right] \tag{15.4.17}$$

The boundary of such a region is known as a **confidence contour**. Of course, another way to express such a confidence region is in terms of the quantity $Q = (\hat{\boldsymbol{\beta}} - \boldsymbol{\beta})'(\boldsymbol{X}'\boldsymbol{X})(\hat{\boldsymbol{\beta}} - \boldsymbol{\beta})$, which is a quadratic form in the differences $\hat{\beta}_j - \beta_j$. In particular, we have

$$(\hat{\boldsymbol{\beta}} - \boldsymbol{\beta})'(\boldsymbol{X}'\boldsymbol{X})(\hat{\boldsymbol{\beta}} - \boldsymbol{\beta}) \leqslant (p + 1)\tilde{\sigma}^2 f_\gamma(p + 1, n - p - 1) \tag{15.4.18}$$

For the important special case of the simple linear model, the corresponding confidence contour is an ordinary two-dimensional ellipse with center at $(\hat{\beta}_0, \hat{\beta}_1)$.

Example 15.4.3 Consider the auto emission data of Example 15.3.1. To obtain a confidence contour by quantity (15.4.18), it is necessary to find $\boldsymbol{X}'\boldsymbol{X}$. For the simple linear model we have $\boldsymbol{X}' = \begin{pmatrix} 1 & \cdots & 1 \\ x_1 & \cdots & x_n \end{pmatrix}$, and consequently

$$\boldsymbol{X}'\boldsymbol{X} = \begin{pmatrix} n & \sum x_i \\ \sum x_i & \sum x_i^2 \end{pmatrix} = \begin{pmatrix} n & n\bar{x} \\ n\bar{x} & n\overline{x^2} \end{pmatrix}$$

FIGURE 15.3 95% confidence contour for (β_0, β_1)

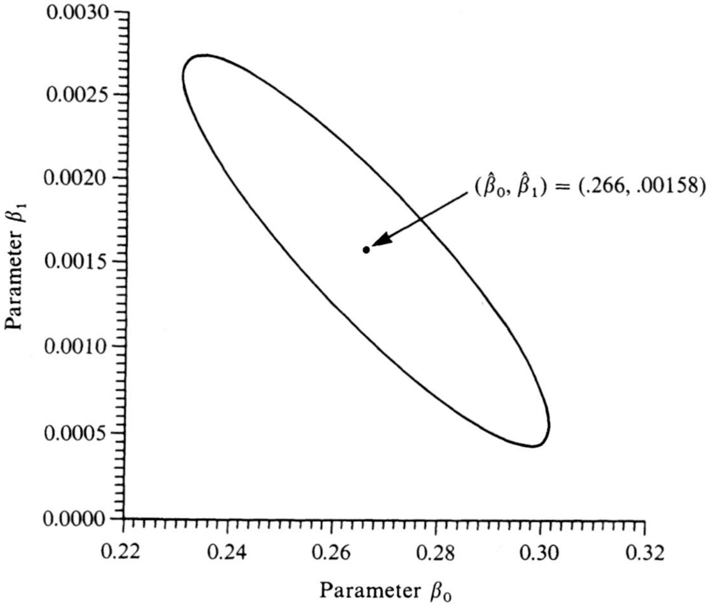

From this we obtain $Q = n(\hat{\beta}_0 - \beta_0)^2 + 2n\bar{x}(\hat{\beta}_0 - \beta_0)(\hat{\beta}_1 - \beta_1) + n\overline{x^2}(\hat{\beta}_1 - \beta_1)^2$, verifying the earlier comment that the confidence contour is an ellipse with center at $(\hat{\beta}_0, \hat{\beta}_1)$. For this example, $n = 11$, $\bar{x} = 27.67$, and $\overline{x^2} = 951.07$. Using the percentile $f_{.95}(2, 9)$, the graph of a 95% confidence contour for this data is shown in Figure 15.3. Of course, the same confidence region would be obtained using inequality (15.4.17).

The above results also suggest how to construct a joint test of hypotheses about all of the β_j's. For example, a size α test of the hypothesis $H_0 : \beta_0 = \beta_1 = \cdots = \beta_p = 0$ versus the alternative that at least one of the β_j's is nonzero would be to reject if

$$\frac{[S(\mathbf{0}) - S(\hat{\boldsymbol{\beta}})]/(p + 1)}{S(\hat{\boldsymbol{\beta}})/(n - p - 1)} > f_{1-\alpha}(p + 1, n - p - 1) \tag{15.4.19}$$

where $\mathbf{0}$ is the $(p + 1)$-dimensional column vector with all zero components. This can be generalized to provide a test that the $\hat{\beta}_j$'s in some subset are all zero.

Consider a null hypothesis of the form $H_0 : \beta_m = \cdots = \beta_p = 0$ where $0 \leqslant m \leqslant p$. This is a generalization the test of Example 15.4.2, and in fact the GLR approach extends immediately to the case of the general linear model with $p + 1$ coefficients. The parameters $\beta_0, \ldots, \beta_{m-1}$ and σ^2 would be unknown. Let Ω be the set of all $(p + 2)$-tuples $(\beta_0, \beta_1, \ldots, \beta_p, \sigma^2)$ with $-\infty < \beta_j < \infty$ and $\sigma^2 > 0$, and let Ω_0 be the subset of Ω such that $\beta_j = 0$ for $j = m, \ldots, p$. We assume independent errors, $\varepsilon_i \sim \mathrm{N}(0, \sigma^2)$. The MLEs over the subspace Ω_0 are the components of $\hat{\boldsymbol{\beta}}_0 = (X'_0 X_0)^{-1} X_0' Y$, where X_0 is the $n \times m$ matrix consisting of the first m columns of X, and $\hat{\sigma}_0^2 = S(\hat{\boldsymbol{\beta}}_0)/n = \sum_{i=1}^{n} (y_i - \hat{y}_{0i})^2/n$ with $\hat{y}_{0i} = \sum_{j=0}^{m-1} \hat{\beta}_{0j} x_{ij}$. On the other hand, over the unrestricted space Ω the joint MLEs are the usual $\hat{\boldsymbol{\beta}} = (X'X)^{-1}X'Y$ and $\hat{\sigma}^2 = S(\hat{\beta})/n = \sum_{i=1}^{n} (y_i - \hat{y}_i)^2/n$ with $\hat{y}_i = \sum_{j=0}^{p} \hat{\beta}_j x_{ij}$. The GLR derivation of Example 15.4.2 extends easily to yield a GLR of the form $\lambda(y) = (\hat{\sigma}_0^2/\hat{\sigma}^2)^{-n/2}$. Thus, the GLR test would reject $H_0 : \beta_m = \cdots = \beta_p = 0$ if $\lambda(y) \leqslant k$ where k is chosen to provide a size α test. As in the example, we could employ the chi-square approximation of $-2 \ln \lambda(y)$, but an exact test is possible. Note that

$$\frac{n - p - 1}{p - m + 1} [\lambda(y)^{-2/n} - 1] = \frac{n - p - 1}{p - m + 1} \left[\frac{\hat{\sigma}_0^2}{\hat{\sigma}^2} - 1\right]$$
$$= \frac{[S(\hat{\boldsymbol{\beta}}_0) - S(\hat{\boldsymbol{\beta}})]/(p - m + 1)}{S(\hat{\boldsymbol{\beta}})/(n - p - 1)} \tag{15.4.20}$$

which is a decreasing function of $\lambda(y)$. Furthermore, under the null hypothesis $H_0 : \beta_m = \cdots = \beta_p = 0$, it can be shown that the ratio on the right of equation (15.4.20) is distributed as $F(p - m + 1, n - p - 1)$. Consequently, a size α test, which is equivalent to the GLR test, would reject H_0 if

$$\frac{[S(\hat{\boldsymbol{\beta}}_0) - S(\hat{\boldsymbol{\beta}})]/(p - m + 1)}{S(\hat{\boldsymbol{\beta}})/(n - p - 1)} > f_{1-\alpha}(p - m + 1, n - p - 1) \tag{15.4.21}$$

For additional information on these distributional results, see Scheffe (1959). We summarize the above remarks in the following theorem.

Theorem 15.4.7 Assume the conditions of Theorem 15.4.4 and let $S(\boldsymbol{\beta}) = (Y - X\boldsymbol{\beta})'(Y - X\boldsymbol{\beta})$.

1. A $\gamma \times 100\%$ confidence region for $\beta_0, \beta_1, \ldots, \beta_p$ is given by the set of solutions to the inequality

$$S(\boldsymbol{\beta}) \leqslant S(\hat{\boldsymbol{\beta}})\left[1 + \frac{p+1}{n-p-1} f_\gamma(p + 1, n - p - 1)\right]$$

2. A size α test of $H_0 : \beta_m = \cdots = \beta_p = 0$, where $0 \leqslant m \leqslant p$, would reject H_0 if

$$\frac{[S(\hat{\boldsymbol{\beta}}_0) - S(\hat{\boldsymbol{\beta}})]/(p - m + 1)}{S(\hat{\boldsymbol{\beta}})/(n - p - 1)} > f_{1-\alpha}(p - m + 1, n - p - 1)$$

where $\hat{\beta} = (X'X)^{-1}X'Y$ and $\hat{\sigma}^2 = S(\hat{\boldsymbol{\beta}})/n$ are the MLEs over the full parameter space $\boldsymbol{\Omega} = \{\beta_0, \beta_1, \ldots, \beta_p, \sigma^2)$: $-\infty < \beta_j < \infty, \sigma^2 > 0\}$, and $\hat{\boldsymbol{\beta}}_0$ and $\hat{\sigma}_0^2$ are the MLEs over the subset Ω_0 of Ω such that $\beta_j = 0$ for $j = m, \ldots, p$. Furthermore, $\hat{\boldsymbol{\beta}}_0 = (X_0' X_0)^{-1} X_0' Y$, where X_0 is the $n \times m$ matrix consisting of the first m columns of X, $\hat{\sigma}_0^2 = S(\hat{\boldsymbol{\beta}}_0)/n$, and the resulting test is equivalent to the GLR test. ■

Example 15.4.4 In Example 15.4.2 we considered a GLR test using the auto emissions data. There we used the chi-square approximation of $-2 \ln \lambda(y)$ to carry out the test. Now we can perform an exact test using the results of Theorem 15.4.7. Recall that it was desired to test the hypothesis $H_0 : \beta_1 = \beta_2 = 0$ versus the alternative that at least one of these coefficients is nonzero. Over the subspace Ω_0 there is only one undetermined coefficient β_0 with MLE $\hat{\beta}_0 = \bar{y}$ over Ω_0, and in this case $S(\hat{\beta}_0) = \sum (y_i - \bar{y})^2 = 0.00796$, while over the unrestricted space Ω the MLE is $\hat{\boldsymbol{\beta}} = (X'X)^{-1}X'Y$ and $S(\hat{\boldsymbol{\beta}}) = 0.00175$. Because $n = 11$, $p = 2$, and $m = 1$ in this example, the value of the F statistic is $\dfrac{[0.00796 - 0.00175]/2}{0.00175/8} = 14.194$, which would reject H_0 at the $\alpha = .05$ level because $f_{.95}(2, 8) = 4.46$. As in the test of Example 15.4.2, evidence in favor of using a regression model is overwhelming because the p-value is .0023. It is interesting to note that both procedures led to a rejection, and in both cases the result was highly significant.

15.5

ANALYSIS OF BIVARIATE DATA

To this point, we have assumed that the variable x can be fixed or measured without error by the experimenter. However, there are situations in which both X and Y are random variables. Assume $(X_1, Y_1), \ldots, (X_n, Y_n)$ is a random sample from a bivariate population with pdf $f(x, y)$. That is, each pair has the same joint pdf, and there is independence between pairs, but the variables within a pair may be dependent.

Our purpose in this section is to show how the results about the simple linear model can be used in the analysis of data from bivariate distributions and to develop methods for testing the hypothesis that X and Y are independent. We first define an estimator of the correlation coefficient ρ.

Definition 15.5.1

If $(X_1, Y_1), \ldots, (X_n, Y_n)$ is a random sample from a bivariate population, then the **sample correlation coefficient** is

$$R = \frac{\sum_{i=1}^{n} (X_i - \bar{X})(Y_i - \bar{Y})}{\sqrt{\sum_{i=1}^{n} (X_i - \bar{X})^2 \sum_{i=1}^{n} (Y_i - \bar{Y})^2}} \tag{15.5.1}$$

The corresponding quantity computed from paired data $(x_1, y_1), \ldots, (x_n, y_n)$, denoted by r, is an estimate of ρ.

We will consider the important special case in which the paired data are observations of a random sample from a bivariate normal population, $(X, Y) \sim \text{BVN}(\mu_1, \mu_2, \sigma_1^2, \sigma_2^2, \rho)$. We know from Theorem 5.4.8 that the conditional distribution of Y given $X = x$ is normal,

$$Y \mid x \sim \text{N}\left(\mu_2 + \rho \frac{\sigma_2}{\sigma_1}(x - \mu_1), \sigma_2^2(1 - \rho^2)\right)$$

This can be related to a simple linear model because $E(Y \mid x) = \beta_0 + \beta_1 x$ with $\beta_0 = \mu_2 - \rho \dfrac{\sigma_2}{\sigma_1} \mu_1$, $\beta_1 = \rho \dfrac{\sigma_2}{\sigma_1}$, and $\text{Var}(Y \mid x) = \sigma_2^2(1 - \rho^2)$. Thus, we have the following theorem

Theorem 15.5.1 Consider a random sample from a bivariate normal population with means μ_1 and μ_2, variances σ_1^2 and σ_2^2, and correlation coefficient ρ, and let $\boldsymbol{X} = (X_1, \ldots, X_n)$ and $\boldsymbol{x} = (x_1, \ldots, x_n)$. Then, conditional on $\boldsymbol{X} = \boldsymbol{x}$, the variables $Y_1, \ldots, Y_n$ are distributed as independent variables $Y_i \sim \text{N}(\beta_0 + \beta_1 x_i, \sigma^2)$ with $\beta_0 = \mu_2 - \rho \dfrac{\sigma_2}{\sigma_1} \mu_1$, $\beta_1 = \rho \dfrac{\sigma_2}{\sigma_1}$, and $\sigma^2 = \sigma_2^2(1 - \rho^2)$. ■

If it is desired to test the hypothesis $H_0 : \rho = 0$, the fact that $\beta_1 = \rho \dfrac{\sigma_2}{\sigma_1}$ suggests using the test of $H_0 : \beta_1 = 0$ in Part 2 of Theorem 15.3.7. That is, reject H_0 at level α if $|t_1| \geqslant t_{1-\alpha/2}(n - 2)$. Of course, the resulting test is a conditional test, but a conditional test of size α also gives an unconditional test of size α.

The following derivation shows that an equivalent test can be stated in terms of the sample correlation coefficient r. Specifically, if

$$t = \sqrt{\sum (x_i - \bar{x})^2}\hat{\beta}_1/\tilde{\sigma} \tag{15.5.2}$$

then under $H_0 : \rho = 0$ and conditional on $\boldsymbol{X} = \boldsymbol{x}$, t is the observed value of a random variable T that is distributed the same as T_1. In other words, under $H_0 : \rho = 0$ and, conditional on the observed x_i's, the variable $T \sim t(n - 2)$. But the MLE of β_1 for the simple linear model is

$$\hat{\beta}_1 = \frac{\sum (x_1 - \bar{x})y_i}{\sum (x_i - \bar{x})^2}$$

$$= \frac{\sum (x_i - \bar{x})(y_i - \bar{y})}{\sum (x_i - \bar{x})^2}$$

$$= \frac{\sum (x_i - \bar{x})(y_i - \bar{y})}{\sqrt{\sum (x_i - \bar{x})^2 \sum (y_i - \bar{y})^2}} \sqrt{\frac{\sum (y_i - \bar{y})^2}{\sum (x_i - \bar{x})^2}}$$

$$= r \sqrt{\frac{\sum (y_i - \bar{y})^2}{\sum (x_i - \bar{x})^2}} \tag{15.5.3}$$

where r is the estimate (15.5.1). Recall that for the centered regression problem, $\hat{\beta}_c = \bar{y}$, and thus

$$\begin{aligned}(n-2)\tilde{\sigma}^2 &= \sum [y_i - \hat{\beta}_0 - \hat{\beta}_1 x_i]^2 \\ &= \sum [y_i - \bar{y} - \hat{\beta}_1(x_i - \bar{x})]^2 \\ &= \sum (y_i - \bar{y})^2 - \hat{\beta}_1^2 \sum (x_i - \bar{x})^2 \\ &= \sum (y_i - \bar{y})^2 \left[1 - \hat{\beta}_1^2 \frac{\sum (x_i - \bar{x})^2}{\sum (y_i - \bar{y})^2}\right]\end{aligned}$$

It follows that

$$\tilde{\sigma}^2 = \frac{\sum (y_i - \bar{y})^2[1 - r^2]}{n-2} \qquad (15.5.4)$$

The last expression is based on substituting the right side of equation (15.5.3) for $\hat{\beta}_1$ and some simplification. By substituting equations (15.5.3) and (15.5.4) into (15.5.2), we obtain

$$\begin{aligned}t &= \frac{\sqrt{\sum (x_i - \bar{x})^2}\, r \sqrt{\sum (y_i - \bar{y})^2 / \sum (x_i - \bar{x})^2}}{\sqrt{\sum (y_i - \bar{y})^2 [1 - r^2]/(n-2)}} \\ &= \frac{\sqrt{n-2}\, r}{\sqrt{1 - r^2}} \qquad (15.5.5)\end{aligned}$$

It follows that a size α test of $H_0 : \rho = 0$ versus $H_a : \rho \neq 0$ is to reject H_0 if $|t| \geqslant t_{1-\alpha/2}(n-2)$ where $t = \sqrt{n-2}\,r/\sqrt{1-r^2}$. This provides a convenient test for independence of X and Y because bivariate normal random variables are independent if and only if they are uncorrelated. These results are summarized in the following theorem.

Theorem 15.5.2 Assume that $(X_1, Y_1), \ldots, (X_n, Y_n)$ is a random sample from a bivariate normal distribution and let $t = \sqrt{n-2}\,r/\sqrt{1-r^2}$.

1. A size α test of the hypothesis $H_0 : \rho = 0$ versus $H_a : \rho \neq 0$ is to reject if

$$|t| \geqslant t_{1-\alpha/2}(n-2).$$

2. A size α test of the hypothesis $H_0 : \rho \leqslant 0$ versus $H_a : \rho > 0$ is to reject if

$$t \geqslant t_{1-\alpha}(n-2).$$

3. A size α test of the hypothesis $H_0 : \rho \geqslant 0$ versus $H_a : \rho < 0$ is to reject if

$$t \leqslant -t_{1-\alpha}(n-2).$$

■

It is interesting to note that for the above conditional distribution of T to hold, it is only necessary to assume that the $2n$ variables $X_1, \ldots, X_n$ and $Y_1, \ldots, Y_n$ are independent and that each $Y_i \sim N(\mu, \sigma^2)$ for some μ and σ^2. Thus, the test described above is appropriate for testing independence of X and Y, even if the distribution of X is arbitrary. This is summarized by the following theorem.

Theorem 15.5.3 Assume that $(X_1, Y_1), \ldots, (X_n, Y_n)$ is a random sample from a bivariate population with pdf $f(x, y)$. If $Y_i \sim N(\mu, \sigma^2)$ for each $i = 1, \ldots, n$, then a size α test of the null hypothesis that the variables X_i and Y_i are independent is to reject if $|t| \geqslant t_{1-\alpha/2}(n-2)$ where $t = \sqrt{n-2}\,r/\sqrt{1-r^2}$. ■

The test defined in Theorem 15.5.2 does not extend readily to testing other values of ρ except $\rho = 0$. However, tests for nonzero values of ρ can be based on the following theorem, which is stated without proof.

Theorem 15.5.4 Assume that $(X_1, Y_1), \ldots, (X_n, Y_n)$ is a random sample from a bivariate normal distribution, $BVN(\mu_1, \mu_2, \sigma_1^2, \sigma_2^2, \rho)$, and define

$$V = \frac{1}{2}\ln\left(\frac{1+R}{1-R}\right) \qquad m = \frac{1}{2}\ln\left(\frac{1+\rho}{1-\rho}\right)$$

Then $Z = \sqrt{n-3}(V - m) \xrightarrow{d} N(0, 1)$ as $n \to \infty$. ■

That is, for large n, V is approximately normal with mean m and variance $1/(n-3)$. This provides an obvious approach for testing hypotheses about ρ, and such tests are given in the following theorem.

Theorem 15.5.5 If $(X_1, Y_1), \ldots, (X_n, Y_n)$ is a random sample from a bivariate normal distribution, $BVN(\mu_1, \mu_2, \sigma_1^2, \sigma_2^2, \rho)$, and $z_0 = \sqrt{n-3}(v - m_0)$ with $v = (1/2)\ln[(1+r)/(1-r)]$ and $m_0 = (1/2)\ln[(1+\rho_0)/(1-\rho_0)]$, then

1. An approximate size α test of $H_0: \rho = \rho_0$ versus $H_a: \rho \neq \rho_0$ is to reject H_0 if $|z_0| \geqslant z_{1-\alpha/2}$.
2. An approximate size α test of $H_0: \rho \leqslant \rho_0$ versus $H_a: \rho > \rho_0$ is to reject H_0 if $z_0 \geqslant z_{1-\alpha}$.
3. An approximate size α test of $H_0: \rho \geqslant \rho_0$ versus $H_a: \rho < \rho_0$ is to reject H_0 if $z_0 \leqslant -z_{1-\alpha}$. ■

It also is possible to construct confidence intervals for ρ based on this approximation. For example, the approximate normal variable Z can be used to derive a confidence interval for m of the form

$$(m_1, m_2) = (v - z_{1-\alpha/2}/\sqrt{n-3},\ v + z_{1-\alpha/2}/\sqrt{n-3})$$

Limits for a confidence interval for ρ are obtained by solving equations $m_i = 1/2 \ln [(1 + \rho_i)/(1 - \rho_i)]$ for $i = 1, 2$. The resulting confidence interval is of the form (ρ_1, ρ_2) where $\rho_i = [\exp(2m_i) - 1]/[\exp(2m_i) + 1]$ for $i = 1, 2$.

Example 15.5.1 Consider the auto emissions data of Example 15.3.1. Although, strictly speaking, the variable x is not random in this example, we will use it to illustrate the computation of r and a test of hypotheses. Recall that $n = 11$, $\sum x_i = 304.377$, $\sum x_i^2 = 10461.814$, $\sum y_i = 3.4075$, $\sum y_i^2 = 1.063$, $\sum x_i y_i = 97.506$, $\bar{x} = 27.671$, and $\bar{y} = 0.310$. Thus, $\sum (x_i - \bar{x})^2 = \sum x_i^2 - n\bar{x}^2 = 2039.287$, $\sum (y_i - \bar{y})^2 = \sum y_i^2 - n\bar{y}^2 = 0.0059$, and $\sum (x_i - \bar{x})(y_i - \bar{y}) = \sum x_i y_i - n\bar{x}\bar{y} = 3.1479$. It follows that $r = 3.1479/\sqrt{(2039.287)(0.0059)} = 0.908$. A test of $H_0 : \rho = 0$ versus $H_a : \rho \neq 0$ based on $t = \sqrt{9}\,(0.908)/\sqrt{1 - (0.908)^2} = 6.502$, which, of course, would reject at any practical level of significance, indicating a near linear relationship among the variables.

In Chapter 12, a paired-sample t test was discussed for testing the difference of means for a bivariate normal distribution with the variances and the correlation coefficient unknown nuisance parameters. We now consider the problem of a simultaneous test of equality of the means and variances of a bivariate normal population with unknown correlation coefficient. This test was suggested by Bradley and Blackwood (1989). Suppose X and Y are bivariate normal, $(X, Y) \sim \text{BVN}(\mu_1, \mu_2, \sigma_1^2, \sigma_2^2, \rho)$. It follows from the results of Chapter 5 that the sum $S = X + Y$ and difference $D = X - Y$ also are bivariate normal with means $\mu_S = \mu_1 + \mu_2$ and $\mu_D = \mu_1 - \mu_2$, variances $\sigma_S^2 = \sigma_1^2 + \sigma_2^2 + 2\rho$ and $\sigma_D^2 = \sigma_1^2 + \sigma_2^2 - 2\rho$ and covariance

$$\begin{aligned}\text{Cov}(S, D) &= \text{Cov}(X + Y, X - Y)\\ &= \text{Var}(X) + \text{Cov}(X, Y) - \text{Cov}(X, Y) - \text{Var}(Y)\\ &= \sigma_1^2 - \sigma_2^2\end{aligned}$$

Thus, the correlation coefficient of S and D is

$$\rho_{SD} = \frac{\sigma_1^2 - \sigma_2^2}{\sigma_S \sigma_D}$$

As a consequence of Theorem 5.4.8, we know that the conditional distribution of D given $S = s$ is normal,

$$D \mid s \sim \text{N}\left(\mu_D + \rho_{SD}\frac{\sigma_D}{\sigma_S}(s - \mu_S),\ \sigma_D^2(1 - \rho_{SD}^2)\right)$$

Notice that if we let

$$\beta_0 = \mu_D - \rho_{SD}\frac{\sigma_D}{\sigma_S}\mu_S = (\mu_1 - \mu_2) - \frac{\sigma_1^2 - \sigma_2^2}{\sigma_S^2}(\mu_1 + \mu_2)$$

and

$$\beta_1 = \frac{\sigma_1^2 - \sigma_2^2}{\sigma_S^2}$$

then, conditional on $S = s$, D is normal with mean $E(D \mid s) = \beta_0 + \beta_1 s$ and variance $\text{Var}(D \mid s) = \sigma_D^2(1 - \rho_{SD}^2)$. Thus, conditionally we have a simple linear regression model. Because $\mu_1 = \mu_2$ and $\sigma_1^2 = \sigma_2^2$ if and only if $\beta_0 = \beta_1 = 0$, the joint null hypothesis $H_0 : \mu_1 = \mu_2, \sigma_1^2 = \sigma_2^2$ is equivalent to the joint null hypothesis $H_0 : \beta_0 = \beta_1 = 0$, which can be tested easily with the results of Theorem 15.4.7. In particular, if $s_1, \ldots, s_n$ and $d_1, \ldots, d_n$ are the sums and differences of n pairs (x_i, y_i) based on a random sample from a bivariate normal population, then a size α test of H_0 would reject if

$$\frac{[\sum d_i^2 - \text{SSE}]/2}{\text{SSE}/(n-2)} = \frac{[S(\mathbf{0}) - S(\hat{\boldsymbol{\beta}})]/2}{S(\hat{\boldsymbol{\beta}})/(n-2)} > f_{1-\alpha}(2, n-2)$$

with $\hat{\boldsymbol{\beta}} = (X'X)^{-1}X'Y$, $X' = \begin{pmatrix} 1 & \cdots & 1 \\ s_1 & \cdots & s_n \end{pmatrix}$, and $Y' = (d_1, \ldots, d_n)$.

It also is possible to test the equality of variances because $\sigma_1^2 = \sigma_2^2$ if and only if $\beta_1 = 0$. Theorem 15.4.7 yields a test of $H_0 : \sigma_1^2 = \sigma_2^2$ versus the alternative $H_a : \sigma_1^2 \neq \sigma_2^2$, namely, we reject H_0 if

$$\frac{\sum (d_i - \bar{d})^2 - \text{SSE}}{\text{SSE}/(n-2)} = \frac{S(\hat{\beta}_0) - S(\hat{\boldsymbol{\beta}})}{S(\hat{\boldsymbol{\beta}})/(n-2)} > f_{1-\alpha}(1, n-2)$$

with $\hat{\beta}_0 = \bar{d}$ over the parameter subspace with $\beta_1 = 0$ and $\hat{\boldsymbol{\beta}}$ the unrestricted ML solution of the previous test.

SUMMARY

Many problems in statistics involve modeling the relationship between a variable Y, which is observed in an experiment, and one or more variables x, which the experimenter assumes can be controlled or measured without error. We have considered the approach of linear regression analysis, which assumes that Y can be represented as a function that is linear in the coefficients, plus an error term whose expectation is zero. It was shown that estimates with minimum variance among the class on linear unbiased estimates could be obtained with the mild assumption that the errors are uncorrelated with equal variances. With the additional assumption that the errors are independent normal, it also was possible to obtain confidence limits and tests of hypotheses about the parameters of the model. We have only scratched the surface of the general problem of regression analysis. For additional reading, the book by Draper and Smith (1981) is recommended.

EXERCISES

1. In a study of the effect of thermal pollution on fish, the proportion of a certain variety of sunfish surviving a fixed level of thermal pollution was determined by Matis and Wehrly (1979) for various exposure times. The following paired data were reported on scaled time (x) versus proportion surviving (y).

x: 0.10, 0.15, 0.20, 0.25, 0.30, 0.35, 0.40, 0.45, 0.50, 0.55

y: 1.00, 0.95, 0.95, 0.90, 0.85, 0.70, 0.65, 0.60, 0.55, 0.40

(a) Plot the paired data as points in an x–y coordinate system as in Figure 15.1.
(b) Assuming a simple linear model, compute the LS estimates of β_0 and β_1.
(c) Estimate $E(Y_x) = \beta_0 + \beta_1 x$ if the exposure time is $x = 0.325$ units. Include the graph of $\hat{y} = \hat{\beta}_0 + \hat{\beta}_1 x$ with the plotted data from (a).
(d) Compute the SSE and give an unbiased estimate of $\sigma^2 = \mathrm{Var}(Y_x)$.

2. Mullet (1977) considers the goals scored per game by the teams in the National Hockey League. The average number of goals scored per game at home and away by each team in the 1973–74 season was:

	At Home	Away
Boston	4.95	4.00
Montreal	4.10	3.41
N.Y. Rangers	4.26	3.44
Toronto	3.69	3.33
Buffalo	3.64	2.56
Detroit	4.36	2.18
Vancouver	3.08	2.67
N.Y. Islanders	2.46	2.21
Philadelphia	3.90	3.10
Chicago	3.64	3.33
Los Angeles	3.36	2.62
Atlanta	3.10	2.38
Pittsburgh	3.18	3.03
St. Louis	3.08	2.21
Minnesota	3.69	2.33
California	2.87	2.13

(a) Plot the paired data in an x–y coordinate system with x the average number of goals at home and y the average number of goals away.
(b) Assuming a simple linear model, compute the LS estimates of β_0 and β_1.
(c) Predict the average number of away goals per game scored by a team that scored four goals per game at home.
(d) Include the graph of the estimated regression function with the plotted data from (a).
(e) Compute an unbiased estimate of σ^2.

3. Rework Exercise 1 after centering the x_i's about $\bar{x}$. That is, first replace each x_i with $x_i - \bar{x}$.

4. Rework Exercise 2 after centering the x_i's about $\bar{x}$.

5. Show that the error sum of squares can be written as

$$\text{SSE} = \sum y_i^2 - \hat{\beta}_0 \sum y_i - \hat{\beta}_1 \sum x_i y_i$$

6. Show each of the following:

(a) Residuals can be written as $\hat{e}_i = y_i - \bar{y} - \hat{\beta}_1(x_i - \bar{x})$.
(b) $\text{SSE} = \sum (y_i - \bar{y})^2 - \hat{\beta}_1^2 \sum (x_i - \bar{x})^2$.

7. For the constants $a_1, \ldots, a_n$ in part 4 of Theorem 15.3.1, show that

$$\sum a_i = 0 \text{ and } \sum a_i x_i = 0 \text{ imply } \sum a_i b_i = 0.$$

8. Assume $Y_1, \ldots, Y_n$ are uncorrelated random variables with $E(Y_i) = \beta_0 + \beta_1 x_i$ and $\text{Var}(Y_i) = \sigma^2$, and that $\hat{\beta}_0$ and $\hat{\beta}_1$ are the LS estimates of β_0 and β_1. Show each of the following:

(a) $E(\hat{\beta}_0^2) = \beta_0^2 + \dfrac{\sigma^2}{n}[1 + n\bar{x}^2/\sum (x_i - \bar{x})^2]$.

Hint: Use general properties of variance and the results of Theorem 15.3.1.
(b) $E(\hat{\beta}_1^2) = \beta_1^2 + \sigma^2/\sum (x_i - \bar{x})^2$.
(c) $E[\sum (Y_i - \bar{Y})^2] = (n - 1)\sigma^2 + \beta_1^2 \sum (x_i - \bar{x})^2$.
Hint: Use an argument similar to the proof of Theorem 8.2.2.
(d) $\tilde{\sigma}^2 = \sum [Y_i - \hat{\beta}_0 - \hat{\beta}_1 x_i]^2/(n - 2)$ is an unbiased estimate of σ^2.
Hint: Use (b) and (c) together with Exercise 6.

9. Consider the bus motor failure data in Exercise 15 of Chapter 13.

(a) Assume that the mileages for the first set of data are normally distributed with mean μ and variance σ^2. Apply the method described in Example 15.3.2, with $x_i = a_i$ and $y_i = \Phi^{-1}(F_i)$ to estimate μ and σ^2.
(b) For the fifth set of bus motor failure data, assume that the mileages have a two-parameter exponential distribution with location parameter η and scale parameter θ. Apply the method described in Example 15.3.2 with $x_i = a_i$ and $y_i = G^{-1}(F_i)$.

10. Rework (a) from Exercise 9 but use the method in which $x_i = \Phi^{-1}(F_i)$ and $y_i = a_i$.

11. Verify that the MLEs $\hat{\beta}_c$, $\hat{\beta}_1$, and $\hat{\sigma}^2$ for the centered regression model of Section 3 are jointly complete and sufficient for β_c, β_1, and σ^2.

12. Verify the identities that follow equations (15.3.6) and (15.3.7).
Hint: Note that $\sum (x_i - \bar{x}) = 0$ and $\sum (x_j - \bar{x})b_j = 1$.

13. Derive equation (15.3.8). *Hint:* Add and subtract $\hat{\beta}_c + \hat{\beta}_1(x_i - \bar{x})$ within the squared brackets, then use the binomial expansion. The term involving $(\hat{\beta}_c - \beta_c)(\hat{\beta}_1 - \beta_1)$ is zero because $\sum (x_i - \bar{x}) = 0$.

14. Derive the joint MGF of $\hat{\beta}_0$ and $\hat{\beta}_1$ under the assumptions of Theorem 15.3.4. *Hint:* Use the fact that $\hat{\beta}_c$ and $\hat{\beta}_1$ are independent normally distributed random variables and that $\hat{\beta}_0 = \hat{\beta}_c - \hat{\beta}_1 \bar{x}$. Recall that the MGF of bivariate normal random variables is given in Example 5.5.2.

15. For the thermal pollution data of Exercise 1, assume that errors are independent and normally distributed, $\varepsilon_i \sim N(0, \sigma^2)$.

(a) Compute 95% confidence limits for β_0.

(b) Compute 95% confidence limits for β_1.

(c) Compute 95% confidence limits for σ^2.

(d) Perform a size $\alpha = .01$ test of $H_0 : \beta_0 = 1$ versus $H_a : \beta_0 \neq 1$.

(e) Perform a size $\alpha = .10$ test of $H_0 : \beta_1 = -1.5$ versus $H_a : \beta_1 \neq -1.5$.

(f) Perform a size $\alpha = .10$ test of $H_0 : \sigma = .05$ versus $H_a : \sigma \neq .05$.

16. Under the assumptions of Theorem 15.3.7,

(a) Derive both upper and lower one-sided tests of size α for β_0.

(b) Redo (a) for β_1.

(c) Redo (a) for σ^2.

17. Let $Y_1, \ldots, Y_n$ be independent where $Y_i \sim N(\beta x_i, \sigma^2)$ with both β and σ^2 unknown.

(a) If $y_1, \ldots, y_n$ are observed, derive the MLEs $\hat{\beta}$ and $\hat{\sigma}^2$ based on the pairs $(x_1, y_1), \ldots, (x_n, y_n)$.

(b) Show that the estimator $\hat{\beta}$ is normally distributed. What are $E(\hat{\beta})$ and $\text{Var}(\hat{\beta})$?

(c) Show that the estimators $\hat{\beta}$ and $\hat{\sigma}^2$ are independent.

(d) Find an unbiased estimator $\tilde{\sigma}^2$ of σ^2 and constants, say c and v, such that $c\tilde{\sigma}^2 \sim \chi^2(v)$.

(e) Find a pivotal quantity for β.

(f) Derive a $(1 - \alpha)100\%$ confidence interval for β.

(g) Derive a $(1 - \alpha)100\%$ confidence interval for σ^2.

18. In a test to determine the static stiffness of major league baseballs, each of six balls was subjected to a different amount of force x (in pounds), and the resulting displacement y (in inches) was measured. The data are given as follows:

x_i:	10	20	30	40	50	60
y_i:	.045	.071	.070	.112	.120	.131

Assuming the regression model of Exercise 17,

(a) Compute the MLEs $\hat{\beta}$ and $\hat{\sigma}^2$.

(b) Compute an unbiased estimate of σ^2.

(c) Compute a 95% confidence interval for β.

(d) Compute a 90% confidence interval for σ^2.

19. Assume independent $Y_i \sim EXP(\beta x_i)$; $i = 1, \ldots, n$.

(a) Find the LS estimator of β.

(b) Find the MLE of β.

20. Apply the regression model of Exercise 19 to the baseball data of Exercise 18 and obtain the MLE of β.

21. Assume that $Y_1, \ldots, Y_n$ are independent Poisson-distributed random variables, $Y_i \sim \text{POI}(\lambda x_i)$.

(a) Find the LS estimator of λ.

(b) Find the ML estimator of λ.

(c) Are both the LS and ML estimators unbiased for λ?

(d) Find the variances of both estimators.

22. Assume that $Y_1, \ldots, Y_n$ are uncorrelated random variables with means $E(Y_i) = \beta_0 + \beta_1 x_i$ and let $w_1, \ldots, w_n$ be known positive constants.

(a) Derive the solutions $\beta_0 = \hat{\beta}_0$ and $\beta_1 = \hat{\beta}_1$ that minimize the sum $S = \sum w_i[y_i - \beta_0 - \beta_1 x_i]^2$. The solutions $\hat{\beta}_0$ and $\hat{\beta}_1$ are called **weighted least squares estimates.**

(b) Show that if $\text{Var}(Y_i) = \sigma^2/w_i$ for each $i = 1, \ldots, n$, and some unknown $\sigma^2 > 0$, then $\hat{\beta}_0$ and $\hat{\beta}_1$ are the BLUEs of β_0 and β_1.
Hint: Use Theorem 15.4.2.

23. Let $X_1, \ldots, X_n$ be a random sample from $\text{EXP}(\theta, \eta)$, and let $Z_i = (X_i - \eta)/\theta$.

(a) Show that $E(Z_{i:n}) = \sum_{k=1}^{i} \dfrac{1}{n-k+1}$.

(b) Show that $a_{ij} = \text{Cov}(Z_{i:n}, Z_{j:n}) = \sum_{k=1}^{m} \dfrac{1}{(n-k+1)^2}$, where $m = \min(i, j)$.

(c) Show that $(A^{-1})_{ii} = (n-i+1)^2 + (n-i)^2$, $i = 1, \ldots, n$; $(A^{-1})_{i,i+1} = (A^{-1})_{i+1,i} = -(n-i)^2$, $i = 1, \ldots, n-1$; and all other elements equal zero.

(d) Show that the BLUEs, based on order statistics, are $\hat{\theta} = n(\bar{X} - X_{1:n})/(n-1)$ and $\hat{\eta} = X_{1:n} - (\bar{X} - X_{1:n})/(n-1)$.

(e) Compare the estimators in (d) to the MLEs.

24. Assume that the data of Example 4.6.3 are the observed values of order statistics for a random sample from $\text{EXP}(\theta, \eta)$. Use the results of Exercise 23 to compute the BLUEs of θ and η.

25. Under the assumptions of Theorem 15.4.3, verify equation (15.4.11).
Hint: Note that $(\mathbf{Y} - \mathbf{X}\hat{\boldsymbol{\beta}})'(\mathbf{Y} - \mathbf{X}\hat{\boldsymbol{\beta}}) = \mathbf{Y}'(\mathbf{Y} - \mathbf{X}\hat{\boldsymbol{\beta}}) + \hat{\boldsymbol{\beta}}'[\mathbf{X}'\mathbf{Y} - (\mathbf{X}'\mathbf{X})\hat{\boldsymbol{\beta}}]$ and make use of equation (15.4.7).

26. Assume that $(X_1, X_2) \sim \text{BVN}(\mu_1, \mu_2, \sigma_1^2, \sigma_2^2, \rho)$. Recall the joint MGF $M(t_1, t_2)$ is given in Example 5.5.2. Show that if $\boldsymbol{\mu}' = (\mu_1, \mu_2)$, $\mathbf{t}' = (t_1, t_2)$ and $V = \{\text{Cov}(X_i, X_j)\}$, that the joint MGF can be written $M(t_1, t_2) = \exp[\mathbf{t}'\boldsymbol{\mu} + (1/2)\mathbf{t}'V\mathbf{t}]$.

27. Using advanced matrix theory, it can be shown that the joint MGF of a vector of k-variate normal random variables, $\mathbf{X}' = (X_1, \ldots, X_k)$, has the form given in Exercise 26, namely $M(t_1, \ldots, t_k) = \exp[\mathbf{t}'\boldsymbol{\mu} + (1/2)\mathbf{t}'V\mathbf{t}]$ with $\boldsymbol{\mu}' = (\mu_1, \ldots, \mu_k)$ and $\mathbf{t}' = (t_1, \ldots, t_k)$. Assuming this result, show the following:

(a) The marginal distribution of each component is normal, $X_i \sim N(\mu_i, \sigma_i^2)$.

(b) If $\boldsymbol{a}' = (a_1, \ldots, a_k)$ and $\boldsymbol{b}' = (b_1, \ldots, b_k)$ and $U = \boldsymbol{a}'\boldsymbol{X}$ and $W = \boldsymbol{b}'\boldsymbol{X}$, then U and W have a bivariate normal distribution.

(c) U and W are independent if and only if $\boldsymbol{a}'\boldsymbol{V}\boldsymbol{b} = 0$.

28. Using only the first eight pairs of hydrocarbon emission data of Example 15.3.1:

(a) Compute the LS estimates of β_0, β_1, and β_2 for the second-degree polynomial model; also compute the unbiased estimate $\tilde{\sigma}^2$.

(b) Express the regression function $\hat{y}$ based on the smaller set of data and sketch its graph. Compare this to the function $\hat{y}$, which was based on all $n = 11$ pairs. Does it make sense in either case to predict hydrocarbon emissions past the range of the data, say for $x = 60{,}000$ miles?

(c) Assuming the conditions of Theorem 15.4.3, compute a 95% confidence interval for β_0 based on the estimate from (a).

(d) Repeat (c) for β_1.

(e) Repeat (c) for β_2.

(f) Compute a 95% confidence interval for σ^2.

29. Verify equation (15.4.15).
Hint: In the formula for $S(\beta)$, add and subtract $\boldsymbol{X}\hat{\boldsymbol{\beta}}$ in the expression $\boldsymbol{Y} - \hat{\boldsymbol{\beta}}\boldsymbol{X}$ and then simplify.

30. Sketch the confidence contour for the coefficients β_0 and β_1 in the simple linear model using the baseball data of Exercise 18. Use $\gamma = .95$.

31. Test the hypothesis $H_0: \beta_1 = \beta_2 = 0$ concerning the coefficients of the second-degree polynomial model in Exercise 28. Use $\alpha = 0.005$.

32. Compute the sample correlation coefficient for the hockey goal data of Exercise 2, with x_i = average goals at home, and y_i = average goals away for the ith team.

33. Under the assumptions of Theorem 15.5.2 and using the data of Exercise 2, perform a size $\alpha = .10$ test of $H_0: \rho = 0$ versus $H_a: \rho \neq 0$. *Note:* This assumes that the pairs (X_i, Y_i) are identically distributed from one team to the next, which is questionable, but we will assume this for the sake of the problem.

34. Using the hockey goal data, and assuming bivariate normality, construct 95% confidence limits for ρ.

35. For the hockey goal data, assuming bivariate normality, test each of the following hypotheses at level $\alpha = .05$:

(a) $H_0: \mu_1 = \mu_2$ and $\sigma_1^2 = \sigma_2^2$ versus $H_a: \mu_1 \neq \mu_2$ or $\sigma_1^2 \neq \sigma_2^2$.

(b) $H_0: \sigma_1^2 = \sigma_2^2$ versus $H_a: \sigma_1^2 \neq \sigma_2^2$.

CHAPTER 16

RELIABILITY AND SURVIVAL DISTRIBUTIONS

16.1 INTRODUCTION

Many important statistical applications occur in the area of reliability and life testing. If the random variable X represents the lifetime or time to failure of a unit, then X will assume only nonnegative values. Thus, distributions such as the Weibull, gamma, exponential, and lognormal distributions are of particular interest in this area. The Weibull distribution is a rather flexible two-parameter model, and it has become the most important model in this area. One possible theoretical justification for this in certain cases is that it is a limiting extreme-value distribution.

One aspect of life testing that is not so common in other areas is that of censored sampling. If a random sample of n items are placed on life test, then the first observed failure time is automatically the smallest order statistic, $x_{1:n}$. Similarly, the second recorded failure time is $x_{2:n}$, and so on. If the experiment is terminated after the first r ordered observations are obtained, then this is referred to as Type II censored sampling on the right. If for some reason the first s

ordered observations are not available, then this is referred to as Type II censored sampling on the left. If the experiment is terminated after a fixed time, x_0, then this is known as Type I censored sampling, or sometimes truncated sampling. If all n ordered observations are obtained, then this is called complete sampling.

Because the observations are naturally ordered, all the information is not lost for the censored items. It is known that items censored on the right have survived at least until time x_0. Also, a great savings in time may result from censoring. If 100 light bulbs are placed on life test, then the first 50 may fail in one year, whereas it may take 20 years for the last one to fail. Similarly, if 50 light bulbs are placed on test, then it may take 10, 15, or 20 years for all 50 to fail, yet the first 50 failure times from a sample size 100 obtained in one year may contain as much information in some cases as the 50 failure times from a complete sample of size 50. The expected length of experiment required to obtain the first r ordered observations from a sample of size n is $E(X_{r:n})$. These values can be compared for different values of r and n for different distributions.

If a complete random sample is available, then statistical techniques can be expressed in terms of either the random sample or the associated order statistics. However, if a censored sample is used, then the statistical techniques and distributional results must be developed in terms of the order statistics.

16.2

RELIABILITY CONCEPTS

If a random variable X represents the lifetime or time to failure of a unit, then the **reliability** of the unit at time t is defined to be

$$R(t) = P[X > t] = 1 - F_X(t) \tag{16.2.1}$$

The same function, with the notation $S(x) = 1 - F_X(x)$, is called the **survivor function** in biomedical applications.

Properties of a distribution that we previously studied, such as the mean and variance, remain important in the reliability area, but an additional property that is quite useful is the **hazard function** (HF) or **failure-rate function**. The hazard function, $h(x)$, for a pdf is defined to be

$$h(x) = \frac{f(x)}{1 - F(x)} = \frac{-R'(x)}{R(x)} = \frac{-d[\log R(x)]}{dx} \tag{16.2.2}$$

The HF may be interpreted as the instantaneous failure rate, or the conditional density of failure at time x, given that the unit has survived until time x,

$$\begin{aligned} f(x \mid X \geqslant x) &= F'(x \mid X \geqslant x) \\ &= \lim_{\Delta x \to 0} \frac{F(x + \Delta x \mid X \geqslant x) - F(x \mid X \geqslant x)}{\Delta x} \\ &= \lim_{\Delta x \to 0} \frac{P[x \leqslant X \leqslant x + \Delta x \mid X \geqslant x]}{\Delta x} \\ &= \lim_{\Delta x \to 0} \frac{P[x \leqslant X \leqslant x + \Delta x, X \geqslant x]}{\Delta x\, P[X \geqslant x]} \\ &= \lim_{\Delta x \to 0} \frac{P[x \leqslant X \leqslant x + \Delta x]}{\Delta x\, [1 - F(x)]} \\ &= \frac{f(x)}{1 - F(x)} \\ &= h(x) \end{aligned} \tag{16.2.3}$$

An increasing HF at time x indicates that the unit is more likely to fail in the next increment of time $(x, x + \Delta x)$ than it would be in an earlier interval of the same length. That is, the unit is wearing out or deteriorating with age. Similarly, a decreasing HF means that the unit is improving with age. A constant hazard function occurs for the exponential distribution, and it reflects the no-memory property of that distribution mentioned earlier.

If $X \sim \text{EXP}(\theta)$,

$$h(x) = \frac{f(x)}{1 - F(x)} = \frac{\left(\dfrac{1}{\theta}\right) e^{-x/\theta}}{e^{-x/\theta}} = \frac{1}{\theta}$$

In this case the failure rate is the reciprocal of the mean time to failure, and it does not depend on the age of the unit. This assumption may be reasonable for certain types of electrical components, but it would tend not to be true for mechanical components. However, the no-wearout assumption may be reasonable over some restricted time span. The exponential distribution has been an important model in the life-testing area, partly because of its simplicity. The Weibull distribution is a generalization of the exponential distribution, and it is much more flexible.

If $X \sim \text{WEI}(\theta, \beta)$, then

$$\begin{aligned} h(x) &= \frac{\beta \theta^{-\beta} x^{\beta - 1} e^{-(x/\theta)^\beta}}{e^{-(x/\theta)^\beta}} \\ &= \frac{\beta}{\theta} \left(\frac{x}{\theta}\right)^{\beta - 1} \end{aligned}$$

FIGURE 16.1 Weibull HFs

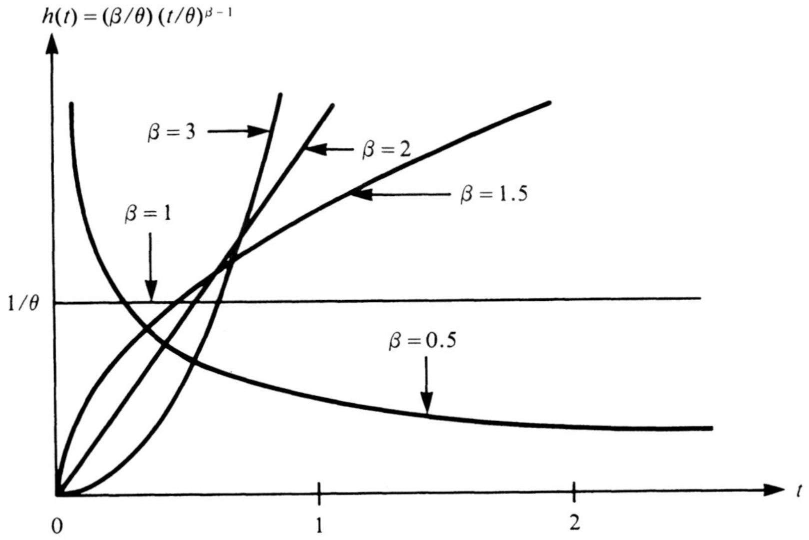

This reduces to the exponential case for $\beta = 1$. For $\beta > 1$, the Weibull HF is an increasing function of x; and for $\beta < 1$, the HF is a decreasing function of x. The Weibull HF is illustrated in Figure 16.1.

The gamma distribution also is an important model in life testing. It is not easy to express its HF, but if $X \sim \text{GAM}(\theta, k)$, then the HF is increasing for $k > 1$ and decreasing for $k < 1$. For $k > 1$ the HF approaches $1/\theta$ asymptotically from below, while for $k < 1$, the HF approaches $1/\theta$ asymptotically from above. This is substantially different from the Weibull distribution, where the HF approaches ∞ or 0 in these cases. The HF for a lognormal distribution is hump-shaped. Although the pdf's in these cases appear quite similar, they clearly have somewhat different characteristics as life-testing distributions; the HF is a very meaningful property for distinguishing between these densities. Indeed, specifying a HF completely determines the CDF and vice versa.

Theorem 16.2.1 For any HF $h(x)$, the associated CDF is determined by the relationship

$$F(x) = 1 - \exp\left[-\int_0^x h(t)\,dt\right] \tag{16.2.4}$$

or

$$f(x) = h(x)\exp\left[-\int_0^x h(t)\,dt\right] \tag{16.2.5}$$

Proof

This result follows because

$$h(x) = \frac{-d[\log R(x)]}{dx}$$

and

$$\int_0^x h(t)\,dt = \int_0^x -d \log R(t) = -\log R(x)$$

which gives

$$R(x) = 1 - F(x) = \exp\left[-\int_0^x h(t)\,dt\right] \tag{16.2.6}$$

■

Note that a function must satisfy certain properties to be a HF.

Theorem 16.2.2 A function $h(x)$ is a HF if and only if it satisfies the following properties:

$$h(x) \geqslant 0, \text{ for all } x \tag{16.2.7}$$

$$\int_0^\infty h(x)\,dx = \infty \tag{16.2.8}$$

Proof

The properties are necessary because

$$\frac{f(x)}{1 - F(x)} \geqslant 0$$

and

$$\int_0^\infty h(x)\,dx = \int_0^\infty -d[\log R(x)] = -\log R(x)\,|_0^\infty = \infty$$

The properties are sufficient because the resulting $F(x)$ will be a valid CDF; that is, in terms of $h(x)$,

$$F(-\infty) = F(0) = 1 - \exp\left[-\int_0^0 h(t)\,dt\right] = 0$$

$$F(\infty) = 1 - \exp\left[-\int_0^\infty h(t)\,dt\right] = 1$$

and $F(x)$ is an increasing function of x because $\int_0^x h(t)\,dt$ is an increasing function of x.

■

One typical life-testing form of HF is a U shaped or bathtub shape. For example, a unit may have a fairly high failure rate when it is first put into operation, because of the presence of manufacturing defects. If the unit survives the early period, then a nearly constant HF may apply for some period, where the causes of failure occur "at random." Later on, the failure rate may begin to increase as wearout or old age becomes a factor. In life sciences, such early failures correspond to the "infant mortality" effect.

Unfortunately, none of the common standard distributions will accommodate a U-shaped HF. Of course, following Theorem 16.2.1, an $F(x)$ can be derived that has a specified U-shaped HF. Quite often it is possible to consider the analysis after some "burn-in" period has taken place, and then the more common distributions are suitable. The exponential distribution is used extensively in this area because of its simplicity. Although its applicability is somewhat limited by its constant HF, it may often be useful over a limited time span as suggested above, and it is convenient for illustrating many of the concepts and techniques applicable to life testing. Also the homogeneous Poisson process assumes that the times between occurrences of failures are independent exponential variables, as we shall see in Section 16.5.

The preceding discussion refers to the time to failure of a nonrepairable system or the time to first failure of a repairable system. The times between failures of a repairable system often would be related to more general stochastic processes. Note that the HF of a density should not be confused with the failure rate or failure intensity of a stochastic process, although there is a connection between the two for the case of a nonhomogeneous Poisson process. In that case the HF of the time to first failure is also the failure intensity of the process, although its interpretation is somewhat different depending on whether you are concerned with the continuing process or only the first failure.

PARALLEL AND SERIES SYSTEMS

Redundancies may be introduced into a system to increase its reliability. If X_i, $i = 1, \ldots, m$, denotes the lifetimes of m independent components connected in **parallel**, then the lifetime of the system, Y, is the maximum of the individual components,

$$Y = \max(X_i) = X_{m:m}$$

and

$$Y \sim F_Y(y) = \prod_{i=1}^{m} F_{X_i}(y)$$

The distributions of maximums in general are not very convenient to work with, but the reliability of the system for a fixed time at least can be expressed. The reliability of the parallel system at time t, $R_Y(t)$, in terms of the reliabilities of the components, $R_i(t)$, is given by

$$R_Y(t) = 1 - \prod_{i=1}^{m} [1 - R_i(t)] \tag{16.2.9}$$

If P_i is the probability that the ith component functions properly, then the probability that the parallel system of independent components functions properly is

$$P = 1 - \prod_{i=1}^{m} (1 - P_i) \tag{16.2.10}$$

For example, if $X_i \sim \text{EXP}(\theta_i)$, then the reliability of the system at time t is

$$R_Y(t) = 1 - \prod_{i=1}^{m} (1 - e^{-t/\theta_i}) \tag{16.2.11}$$

If the components all have a common mean, $\theta_i = \theta$, then it can be shown that the mean time to failure of the system is

$$E(Y) = E(X_{m:m}) = \left(1 + \frac{1}{2} + \cdots + \frac{1}{m}\right)\theta \tag{16.2.12}$$

Thus the mean time to failure increases as each additional parallel component is added; however, the relative gain decreases as m increases.

To illustrate the effect on the HF from using a parallel system, consider a parallel system of two independent identically distributed components, $X_i \sim F_X(x)$. The HF for the system $h_Y(y)$, in terms of the HF, $h_X(x)$, of each component is

$$\begin{aligned} h_Y(y) &= \frac{f_Y(y)}{1 - F_Y(y)} \\ &= \frac{2 f_X(y) F_X(y)}{1 - [F_X(y)]^2} \\ &= \left[\frac{2F_X(y)}{1 + F_X(y)}\right] h_X(y) \end{aligned} \tag{16.2.13}$$

For positive continuous variables, the term in brackets goes from 0 to 1 as y goes from 0 to ∞. The failure rate of the parallel system is always less than the failure rate of the individual components, but it approaches the failure rate of an individual component as $y \to \infty$.

The HF of a system connected in **series** is somewhat easier to express. If X_i denotes the failure times of m independent components connected in series, then the failure time of the system, W, is the minimum of those of the individual components,

$$W = X_{1:m}$$

In this case

$$\begin{aligned} F_W(w) &= P[W \leqslant w] \\ &= 1 - P[W > w] \\ &= 1 - P[\text{all } X_i > w] \\ &= 1 - \prod_{i=1}^{m} [1 - F_{X_i}(w)] \end{aligned} \tag{16.2.14}$$

In terms of the reliability of the system at time w,

$$R_W(w) = \prod_{i=1}^{m} R_i(w) \tag{16.2.15}$$

If one is simply interested in the proper functioning of some sort, then for the series system

$$P = \prod_{i=1}^{m} P_i \tag{16.2.16}$$

where P_i is the probability of proper functioning of each component. Clearly the reliability of the system decreases as more components are added.

In terms of HFs, for the ith component,

$$R_i(x) = \exp\left[-\int_0^x h_i(z)\, dz \right]$$

$$\begin{aligned} R_W(w) = 1 - F_W(w) &= \prod_{i=1}^{m} R_i(w) \\ &= \exp\left[-\int_0^x \sum_{i=1}^{m} h_i(z)\, dz \right] \\ &= \exp\left[-\int_0^x h_W(z)\, dz \right] \end{aligned} \tag{16.2.17}$$

thus

$$h_W(w) = \sum_{i=1}^{m} h_i(w) \tag{16.2.18}$$

That is, the HF of the system is the sum of the HFs of the individual components.

If $X_i \sim \text{EXP}(\theta_i)$, then

$$h_W(w) = \sum_{i=1}^{m} h_i(w) = \sum_{i=1}^{m} \frac{1}{\theta_i} = c \tag{16.2.19}$$

Because $h_W(w)$ is constant, this implies that

$$W \sim \text{EXP}(1/c)$$

If $\theta_i = \theta$, then

$$E(W) = \theta/m \tag{16.2.20}$$

16.3

EXPONENTIAL DISTRIBUTION

COMPLETE SAMPLES

If $X \sim \text{EXP}(\theta)$, then we know that $\bar{X}$ is the MLE and the UMVUE of θ, the mean of the random variable. The MLE of the reliability at time t is

$$\hat{R}(t) = e^{-t/\hat{\theta}} = e^{-t/\bar{x}} \tag{16.3.1}$$

The MLE of $R(t)$ is not unbiased, and it can be verified that the UMVUE is given by

$$\tilde{R}(t) = \begin{cases} [1 - t/(n\bar{x})]^{n-1} & n\bar{x} > t \\ 0 & n\bar{x} < t \end{cases} \tag{16.3.2}$$

The MLE does have smaller MSE except when $t/(n\bar{x})$ is relatively large or close to zero.

Tests of hypotheses about θ, or monotonic functions of θ such as the reliability or the HF, are carried out easily based on the property that

$$\frac{2n\bar{X}}{\theta} \sim \chi^2(2n)$$

We already have seen that a one-sided confidence interval on a percentile is also a one-sided tolerance interval. There also is a close connection between tolerance limits and confidence limits on reliability. For a lower (γ, p) tolerance limit, $L(\boldsymbol{X}; \gamma, p)$, the content p is fixed and the lower limit is a random variable. For a lower confidence limit on reliability, $R_L(t, \gamma)$, the lower limit t is fixed and the proportion R_L is a random variable. However, for a given set of sample data, if one determines the value p^* that results in

$$L(\boldsymbol{x}; \gamma, p^*) = t$$

then

$$p^* = R_L(t, \gamma)$$

If p^* is a random variable as defined above, and if $R(t) = p$ and $t = x_{1-p}$, then $p \geq p^*$ if and only if $L(\boldsymbol{x}; \gamma, p) \leq L(\boldsymbol{x}; \gamma, p^*) = t$, because increasing the content decreases the lower limit. Thus

$$\begin{aligned} P[R(t) \geqslant p^*] &= P[p \geqslant p^*] \\ &= P[L(X; \gamma, p) \leqslant L(X; \gamma, p^*)] \\ &= P[L(X; \gamma, p) \leqslant t] \\ &= P[L(X; \gamma, p) \leqslant x_{1-p}] \\ &= \gamma \end{aligned} \quad \textbf{(16.3.3)}$$

For the exponential distribution

$$x_{1-p} = -\theta \ln p$$

and a γ probability tolerance limit for proportion p is

$$L(\boldsymbol{x}, \gamma, p) = -\theta_L \ln p$$

where

$$\theta_L = \frac{2n\bar{x}}{\chi^2_\gamma(2n)} \quad \textbf{(16.3.4)}$$

is the lower γ level confidence limit for θ.

To obtain a lower confidence limit on the reliability at time t, we may set $t = L(\boldsymbol{x}, \gamma, p^*) = -\theta_L \ln p^*$, and

$$R_L(t) = p^* = e^{-t/\theta_L} = \exp\,[-t\chi^2_\gamma(2n)/2n\bar{x}] \quad \textbf{(16.3.5)}$$

is a lower γ level confidence limit for $R(t)$. Of course, this could be obtained directly in this case because

$$R(t) = e^{-t/\theta}$$

which is a monotonically increasing function of θ, so

$$\begin{aligned} \gamma &= P[\theta_L \leqslant \theta] \\ &= P[e^{-t/\theta_L} \leqslant e^{-t/\theta}] \\ &= P[R_L(t) \leqslant R(t)] \end{aligned}$$

as before.

Example 16.3.1 Consider the following 30 failure times in flying hours of airplane air conditioners (Proschan, 1963):

23, 261, 87, 7, 120, 14, 62, 47, 3, 95, 225, 71, 246, 21, 42, 20, 5, 12, 120, 11, 14, 71, 11, 14, 11, 16, 90, 1, 16, 52

The MLE of θ is $\hat{\theta} = \bar{x} = 59.6$. The MLE of the HF is $\hat{h}(x) = 1/\hat{\theta} = 0.017$, and the MLE of the reliability at time $t = 20$ hours is $\hat{R}(20) = e^{-20/59.6} = 0.715$. A lower

0.95 confidence limit for θ is

$$\theta_L = \frac{2n\bar{x}}{\chi^2_{0.95}(2n)} = \frac{60(59.6)}{79.08} = 45.22$$

A 95% lower confidence limit for the reliability at $t = 20$ hours is

$$R_L = e^{-t/\theta_L} = e^{-20/45.22} = 0.643$$

We are 95% confident that 64.3% of the air conditioners will last at least 20 hours. If we are interested in 90% reliability, then we may set $p = 0.90$ and determine the lower tolerance limit $L(\mathbf{X})$ such that

$$P[L(\mathbf{X}) \geqslant 0.90] = 0.95$$

We have

$$L(\mathbf{X}; 0.95, 0.90) = \theta_L(-\ln 0.90) = 45.22(0.105) = 4.75$$

We are 95% confident that 90% of the air conditioners will last at least 4.75 hours.

Many of the results for complete samples can be extended to censored samples. We will make use of Theorems 6.5.4 and 6.5.5, with the notation $X_{i:n}$ representing the ith order statistic for a random sample of size n.

TYPE II CENSORED SAMPLING

The special properties of the exponential distribution make it a particularly convenient model for analyzing censored data. The joint density function of the r smallest order statistics from a sample of size n is given by

$$g(x_{1:n}, \ldots, x_{r:n}; \theta) = \frac{n!}{(n-r)!\,\theta^r} \exp\left\{-\left[\sum_{i=1}^{r} x_{i:n} + (n-r)x_{r:n}\right]\Big/\theta\right\}$$

$$0 < x_{1:n} < \cdots < x_{r:n} < \infty \qquad \textbf{(16.3.6)}$$

A useful property of the exponential distribution is that differences of consecutive order statistics are distributed as independent exponential variables.

Theorem 16.3.1 Let $Y_i = X_i/\theta \sim \text{EXP}(1)$, $i = 1, \ldots, n$, be n independent exponential variables, and let

$$W_1 = Y_{1:n} \qquad W_2 = Y_{2:n} - Y_{1:n}, \ldots \qquad W_n = Y_{n:n} - Y_{n-1:n}$$

then

1. $W_1, \ldots, W_n$ are independent.
2. $W_i \sim \text{EXP}(1/(n-i+1))$.

3. The kth order statistic may be expressed as a linear combination of independent exponential variables, $Y_{k:n} = \sum_{j=1}^{k} Z_j/(n - j + 1)$, where $Z_j \sim \text{EXP}(1)$.
4. $E(Y_{k:n}) = \sum_{j=1}^{k} (n - j + 1)^{-1}$.

Proof

$$f(y_{1:n}, \ldots, y_{n:n}) = n! \exp\left[-\sum_{i=1}^{n} y_{i:n}\right]$$

$$= n! \exp\left[-\sum_{i=1}^{n} (n - i + 1)(y_{i:n} - y_{i-1:n})\right]$$

where $y_{0:n} = 0$. Consider the joint transformation

$$w_i = y_{i:n} - y_{i-1:n} \qquad i = 1, \ldots, n$$

with inverse transformation

$$y_{1:n} = w_1 \qquad y_{2:n} = w_1 + w_2, \ldots, \qquad y_{n:n} = w_1 + \cdots + w_n$$

with Jacobian $J = 1$. This gives

$$f(w_1, \ldots, w_n) = n! \exp\left[-\sum_{i=1}^{n} (n - i + 1)w_i\right]$$

$$= \prod_{i=j}^{n} (n - i + 1)e^{-(n-i+1)w_i} \qquad 0 < w_i < \infty$$

Thus we recognize that the w_i are independent exponential variables as stated in parts 1 and 2. Also, from the above transformation it follows that we may express $Y_{k:n}$ as

$$Y_{k:n} = W_1 + \cdots + W_k \tag{16.3.7}$$

but $Z_i \sim \text{EXP}(1)$ implies $Z_i/(n - i + 1) \sim \text{EXP}(1/(n - i + 1)$, so part 3 follows, and part 4 follows immediately from part 3. ■

Let us now consider the problem of estimation and hypothesis testing for the Type II censored sampling case. The MLE of θ based on the joint density, $g(x_{1:n}, \ldots, x_{r:n}; \theta)$, of the first r order statistics is easily seen to be

$$\hat{\theta} = \frac{\sum_{i=1}^{r} x_{i:n} + (n - r)x_{r:n}}{r} \tag{16.3.8}$$

The properties of $\hat{\theta}$ for the censored sampling case are amazingly similar to those for the complete sample case.

Theorem 16.3.2 Let $\hat{\theta}$ denote the MLE based on the first r order statistics from a sample of size n from EXP(θ). Then

1. $\hat{\theta}$ is a complete, sufficient statistic for θ.
2. $\hat{\theta}$ is the UMVUE of θ.
3. $2r\hat{\theta}/\theta \sim \chi^2(2r)$.

Proof

Because $g(x_{1:n}, \ldots, x_{r:n}; \theta)$ is a member of the (multivariate) exponential class, it follows that $\hat{\theta}$ is a complete, sufficient statistic for θ. By uniqueness, $\hat{\theta}$ will be the UMVUE of θ if it is unbiased. The unbiasedness of $\hat{\theta}$ will follow easily from part 3. To verify part 3, note that $\hat{\theta}$ may be rearranged to obtain

$$\frac{2r\hat{\theta}}{\theta} = \frac{2\sum_{i=1}^{r}(n-i+1)(X_{i:n} - X_{i-1:n})}{\theta}$$

$$= 2\sum_{i=1}^{r}(n-i+1)W_i$$

$$= 2\sum_{i=1}^{r} Z_i$$

where the W_i are as defined in Theorem 16.3.1, and the $Z_i \sim \text{EXP}(1)$ are independent. Thus part 3 follows. ∎

Recall in the complete sample case that

$$\frac{2n\hat{\theta}}{\theta} \sim \chi^2(2n) \tag{16.3.9}$$

This is a very unusual situation in which the censored sampling results essentially are identical to the complete sample results if n is replaced by r. Indeed, all of the confidence intervals and tests described earlier also apply to the Type II censored sampling case by replacing n by r.

The disadvantage of censored sampling is the extra cost involved with sampling the additional $n - r$ items and placing them on test. The principal advantage is that it may take much less time for the first r failures of n items to occur than for all r items in a random sample of size r. Yet the efficiency and precision involved in the two cases are exactly the same. The relative expected experiment time in the two cases may be expressed as

$$\text{REET} = \frac{EX_{r:n}}{EX_{r:r}} = \frac{\sum_{i=1}^{r} (n - i + 1)^{-1}}{\sum_{i=1}^{r} (r - i + 1)^{-1}} \tag{16.3.10}$$

A few values of REET are given below for $r = 10$ to illustrate the substantial savings in time that may be realized:

n:	11	12	13	15	20	30
REET:	0.69	0.55	0.46	0.35	0.23	0.14

A reasonable approach is to choose r and n to minimize the expected value of some cost function involving the cost of the units and the cost related to the length of the experiment.

Example 16.3.2 Consider again the data in Example 16.3.1. As mentioned earlier, in many life-testing cases, the data will occur naturally ordered. This was perhaps not the case in the air conditioner data, but let us consider an analysis based on the 20 smallest ordered observations for illustration purposes. These are

1, 3, 5, 7, 11, 11, 11, 12, 14, 14, 14, 16, 16, 20, 21, 23, 42,
47, 52, 62

The MLE and UMVUE of θ is $\hat{\theta} = 51.1$ based on these 20 observations. The MLE for reliability at time $t = 20$ is now $\hat{R}(20) = 0.676$. A lower 95% confidence limit for θ is

$$\theta_L = \frac{2r\hat{\theta}}{\chi^2_{0.95}(2r)} = \frac{40(51.1)}{55.76} = 36.66$$

A 95% lower confidence limit for reliability at $t = 20$ hours is

$$R_L = e^{-t/\theta_L} = e^{-20/36.66} = 0.58$$

The lower (0.95, 0.90) tolerance limit becomes

$$L(X; 0.95, 0.90) = \theta_L(-\ln 0.90) = 36.66(0.105) = 3.85$$

There are, of course, some differences between these numbers and the complete sample numbers because of random variation, but the censored values have the same accuracy as if they had been based on a complete sample of size 20.

TYPE I CENSORED SAMPLING

Statistical analyses based on Type I censored sampling generally are more complicated than for the Type II case. In this case the length of the experiment, say t, is fixed but the number of values observed in time t is a random variable. The number of items failing before time t follows a binomial distribution

$$R \sim \text{BIN}(n, p)$$

where $p = F_X(t) = 1 - e^{-t/\theta}$, when sampling from an exponential distribution.

Now the distribution of an exponential variable X given that $X \leqslant t$ is a truncated exponential distribution

$$F_t(x) = P[X \leqslant x \mid X \leqslant t] = \frac{P[X \leqslant x]}{P[X \leqslant t]} \qquad x \leqslant t$$

$$= \begin{cases} \dfrac{1 - e^{-x/\theta}}{1 - e^{-t/\theta}} & x \leqslant t \\ 1 & x > t \end{cases} \tag{16.3.11}$$

Also

$$f_t(x) = \frac{f_x(x)}{F_X(t)} = \frac{e^{-x/\theta}}{\theta(1 - e^{-t/\theta})} \qquad 0 < x < t \tag{16.3.12}$$

Thus, given $R = r$, the conditional density of the first r failure times is equivalent to the joint density of an ordered random sample of size r from a truncated exponential distribution,

$$g(x_{1:n}, \ldots, x_{r:n} \mid R = r) = r! \prod_{i=1}^{r} f_t(x_{i:n})$$

$$= r! \prod_{i=1}^{r} \left[\frac{f_x(x_{i:n})}{F_X(t)}\right]$$

$$= \frac{r! \exp\left[-\sum_{i=1}^{r} x_{i:n}/\theta\right]}{\theta^r(1 - e^{-t/\theta})^r} \tag{16.3.13}$$

The joint density of obtaining $R = r$ ordered observations at the values $x_{1:n}, \ldots,$ $x_{r:n}$ before time t, may be expressed as

$$g(x_{1:n}, \ldots, x_{r:n}) = g(x_{1:n}, \ldots, x_{r:n} \mid r) b(r; n, p)$$

$$= \frac{n!}{(n - r)!\,\theta^r} \exp\{-[\textstyle\sum x_{i:n} + (n - r)t]/\theta\} \tag{16.3.14}$$

If $0 < x_{1:n} < \cdots < x_{r:n} < t$ for $r = 1, 2, \ldots, n$, and $P[R = 0] = (1 - p)^n = e^{-nt/\theta}$ otherwise.

Note that this joint density has exactly the same form as in the Type II case with $x_{r:n}$ replaced by t. In this case, the MLE of θ is

$$\hat{\theta} = \frac{\sum_{i=1}^{r} x_{i:n} + (n - r)t}{r} \tag{16.3.15}$$

It is interesting to note that in both cases $\hat{\theta}$ is of the form T/r, where T represents the total surviving time of the n units on test until the termination of the experiment.

In this case $\hat{\theta}$ and r or $\sum_{i=1}^{r} x_{i:n}$ and r are joint sufficient statistics for θ. Generally it is not clear how to develop optimal statistical procedures for a single parameter based on two sufficient statistics. Reasonably good inference procedures for θ can be based on R alone, even though this makes use of only the number of failures and not their values. In this case $R \sim \text{BIN}(n, p)$, where p is a monotonic function of θ, so the usual binomial procedures can be adapted to apply to θ. Additional results are given by Bain and Engelhardt (1991).

TYPE I CENSORED SAMPLING (WITH REPLACEMENT)

As we have seen, the manner in which the sampling is carried out affects the probability structure and statistical analysis. It may be of interest to consider the effect of sampling with replacement in these cases.

Suppose that test equipment is available for testing n units simultaneously. It may make more economical use of the test equipment to replace failed items immediately after failure and continue testing. As the experiment continues, the failure times are measured from the start of the experiment, whether the failed item was an original or a replacement item. These failure times will be naturally ordered, but they are not the usual order statistics, so we still denote these successive failure times by s_i, $i = 1, 2, 3, \ldots$. Note that the number of observations may exeed n in this case.

A physical example of how such data may occur is as follows. Consider a system of n identical components in series, in which a component is replaced each time it fails. Each time a component fails, the system fails, so the system failure times would correspond to the type of observations described above. The special properties of the exponential distribution make the mathematics tractable for this type of sampling. In particular, this situation can be related to the Poisson process for the Type I censoring case. (See Section 16.5 for details.)

Suppose that the successive failure times from n positions are recorded until a fixed time t. If the times between failures are independent exponential variables with mean θ, then the failure times for each position represent the occurrences of a Poisson process with $\lambda = 1/\theta$. Now it follows from the properties of Poisson processes that if n independent Poisson processes with intensity parameters $1/\theta$

are occurring simultaneously, then the combined occurrences may be considered to come from a single Poisson process with intensity parameter $v = n/\theta$. Thus, for example, the number of failures in time t from the total experiment follows a Poisson distribution with parameter $vt = nt/\theta$, $R \sim \text{POI}(nt/\theta)$, so

$$f_R(r) = \frac{e^{-nt/\theta}(nt/\theta)^r}{r!} \qquad r = 0, 1, \ldots \tag{16.3.16}$$

If we let $T_1 = S_1$ and $T_i = S_i - S_{i-1}$, $i = 2, 3, \ldots$ denote the interarrival times of occurrences from the n superimposed Poisson processes, then the joint density of $T_1, \ldots, T_r$ is

$$f(t_1, \ldots, t_r) = v^r \exp\left[-v \sum_{i=1}^{r} t_i\right]$$

and transforming gives the joint density of the first r successive failures,

$$\begin{aligned} f(s_1, \ldots, s_r) &= v^r \exp\left\{-v\left[\sum_{i=2}^{r} (s_i - s_{i-1}) + s_1\right]\right\} \\ &= v^r e^{-vs_r} \end{aligned} \tag{16.3.17}$$

where $v = n/\theta$. Now for Type I censoring at t, the likelihood function on $0 < s_1 < \cdots < s_r < t$ for $r = 1, 2, 3, \ldots$ is given by

$$\begin{aligned} f(s_1, \ldots, s_r, r) &= f(r \mid s_1, \ldots, s_r) f(s_1, \ldots, s_r) \\ &= P[T_{r+1} > t - s_r] f(s_1, \ldots, s_r) \\ &= e^{-v(t - s_r)} v^r e^{-vs_r} \\ &= v^r e^{-vt} \\ &= \left(\frac{n}{\theta}\right)^r e^{-nt/\theta} \end{aligned}$$

$$0 < s_1 < \cdots < s_r < t \qquad r = 1, 2, \ldots \tag{16.3.18}$$

Also

$$P[R = 0] = P[S_1 > t] = e^{-nt/\theta} \qquad r = 0$$

It is interesting to note at this point that given $R = r$, the successive failure times are distributed as ordered uniform variables.

Theorem 16.3.3 Let events occur according to a Poisson process with intensity v, and let $S_1, S_2, \ldots$ denote the successive times of occurrence. Given that r events occurred in the interval $(0, t)$, then conditionally $S_1, \ldots, S_r$ are distributed as ordered observations from the uniform distribution on $(0, t)$.

Proof

We have

$$f(s_1, \ldots, s_r \mid r) = \frac{f(s_1, \ldots, s_r)}{f_r(r)}$$

$$= \frac{v^r e^{-vt}}{e^{-vt}(vt)^r/r!}$$

$$= \frac{r!}{t^r}, \quad 0 < s_1 < \cdots < s_r < t \tag{16.3.19}$$

This is the joint density of the order statistics of a sample of size r from the uniform density $f(x) = 1/t$; $0 < x < t$, and zero otherwise. ■

Returning to the likelihood function, the MLE of θ is seen easily to be

$$\hat{\theta} = \frac{nt}{r} \quad \text{if } r > 0 \tag{16.3.20}$$

in this case. The MLE is again in the form T/r where $T = nt$ represents the total test time accrued by the items in the experiment before termination.

It is interesting that with replacement the statistic R now becomes a single sufficient statistic. Furthermore, $R \sim \text{POI}(nt/\theta)$, so that previously developed techniques for the Poisson distribution can be readily applied here to obtain tests or confidence intervals on θ. For example, a lower $1 - \alpha$ level confidence limit for θ based on r is

$$\theta_L = 2nt/\chi^2_{1-\alpha}(2r + 2) \tag{16.3.21}$$

and an upper $1 - \alpha$ level confidence limit for θ is

$$\theta_U = 2nt/\chi^2_{\alpha}(2r) \tag{16.3.22}$$

These again may be slightly conservative because of the discreteness of the Poisson distribution.

Example 16.3.3 Consider a chain with 20 links, with the failure time of each link distributed as EXP(θ). The chain is placed in service; each time a link breaks, it is replaced by a new one and the failure time is recorded. The experiment is conducted for 100 hours, and the following 25 successive failure times are recorded:

5.2, 13.6, 14.5, 14.6, 20.5, 38.4, 42.0, 44.5, 46.7, 48.5, 50.3, 56.4, 61.7, 62.9, 64.1, 67.1, 71.6, 79.2, 82.6, 83.1, 85.5, 90.8, 92.7, 95.5, 95.6

The MLE of θ is $\hat{\theta} = 20(100)/25 = 80$. Suppose that we wish to test $H_0 : \theta \geqslant 100$ against $H_a : \theta < 100$ at the $\alpha = 0.05$ significance level. We have

$$\frac{2nt}{\theta_0} = \frac{2(20)(100)}{100} = 40 \nless \chi^2_{0.05}(50) = 34.8$$

so H_0 cannot be rejected at the 0.05 level. A lower 95% confidence limit for θ is

$$\theta_L = \frac{2(20)(100)}{\chi^2_{0.95}(52)} = 57.3$$

Note that this set of data actually was generated from an exponential distribution with $\theta = 100$.

TYPE II CENSORED SAMPLING (WITH REPLACEMENT)

Again suppose that n units are being tested with replacement, but that the experiment continues until r failures have occurred. As before, the ordered successive failure times, $s_1, \ldots, s_r$, are measured from the beginning of the experiment without regard to whether they were original or replacement units when they fail.

If the failure time of an individual unit follows the exponential distribution, EXP(θ), then the superimposed failures from the n positions may be considered to be the occurrences of a Poisson process with intensity parameter $v = n/\theta$, as discussed earlier for the time-truncated case. Thus, the interarrival times $y_1 = s_1$, $y_i = s_i - s_{i-1}$, $i = 2, \ldots, r$, are independent exponential variables with mean θ/n,

$$f(y_1, \ldots, y_r) = \left(\frac{n}{\theta}\right)^r \exp\left[-n \sum_{i=1}^{r} y_i/\theta\right]$$

Transforming back to the s_i, $\sum_{i=1}^{r} y_i = s_r$, with Jacobian $J = 1$, and

$$f(s_1, \ldots, s_r) = \left(\frac{n}{\theta}\right)^r e^{-ns_r/\theta} \qquad 0 < s_1 < \cdots < s_r < \infty \tag{16.3.23}$$

The MLE in this case is

$$\hat{\theta} = \frac{ns_r}{r} \tag{16.3.24}$$

where ns_r is again the accrued survival time of all units involved in the test until the experiment ends. The likelihood function is a member of the multivariate exponential class, and s_r is a complete, sufficient statistic for θ.

As noted above, in terms of the independent exponential variables Y_i,

$$S_r = \sum_{i=1}^{r} Y_i$$

so

$$\frac{2r\hat{\theta}}{\theta} = \frac{2nS_r}{\theta} = \frac{2\sum_{i=1}^{r} Y_i}{\theta/n} \sim \chi^2(2r) \tag{16.3.25}$$

This is again a somewhat astonishing result, because exactly the same result is obtained for the Type II censored sampling without replacement, as well as for the complete sample case with $n = r$. It follows that all of the statistical analyses available for the complete sample case may be applied directly to the Type II censored sampling case with replacement by using the above $\hat{\theta}$ and replacing n by r in the previous formulas.

It is clear that identical results may be achieved by placing r units on test and conducting the experiment with replacement until r units fail, or beginning the experiment with n units on test and conducting the experiment with replacement until r units fail. The expected experiment time in the latter case is

$$E(S_{r,n}) = \frac{\theta}{2n} E(\chi^2_{2r}) = \frac{2r\theta}{2n} = \frac{r\theta}{n} \tag{16.3.26}$$

so the relative expected experiment time of the latter case to the first case is

$$\text{REET} = \frac{E(S_{r,n})}{E(S_{r,r})} = \frac{r}{n} \tag{16.3.27}$$

Thus substantial savings in time may be achieved by beginning with additional units. The value in saving time must be weighed against the cost of testing additional units to decide on the appropriate censoring fraction to use.

The relative experiment time gained by using replacement also can be determined by comparing the expected time required to obtain n failures from n units with replacement, $E(S_{n,n})$, with the time required to obtain n failures without replacement, $E(X_{n:n})$.

Example 16.3.4 Consider again the data given in Example 16.3.3, but suppose that Type II censored sampling with $r = 20$ was used. Based on the first 20 observations, the MLE of θ is

$$\hat{\theta} = \frac{ns_r}{r} = \frac{20(83.1)}{20} = 83.1$$

To test $H_0 : \theta \geqslant 100$ against $H_a : \theta < 100$ in this case, we consider

$$\frac{2r\hat{\theta}}{\theta_0} = \frac{2(20)(83.1)}{100} = 33.2 > \chi^2_{0.05}(40) = 26.5$$

so again H_0 cannot be rejected at the 0.05 level. Similarly, tests for reliability, tolerance limits, and so on can be carried out for this case.

16.4

WEIBULL DISTRIBUTION

We have seen that the exponential distribution is an important life-testing model that is very simple to analyze statistically. However, it is somewhat restrictive in that it is applicable only where a constant HF is reasonably appropriate. The Weibull distribution represents a generalization of the exponential distribution that is considerably more flexible, because it allows for either an increasing or a decreasing HF. The Weibull distribution has been shown empirically to provide a good model for a great many types of variables. Also recall that the Weibull distribution is one of the three limiting extreme-value distributions. This may provide some theoretical justification for its use in certain cases. For example, the strength of a long chain (or the failure time of a system in series) is equal to that of the weakest link or component, and the limiting distribution of the minimum is a Weibull distribution in many cases. Similarly, the breaking strength of a ceramic would be that of its weakest flaw.

Also recall that if $X \sim \text{WEI}(\theta, \beta)$, then $Y = \ln X \sim \text{EV}(1/\beta, \ln \theta)$. Thus, any statistical results developed for the Weibull distribution also can be applied easily to the Type I extreme-value model and vice versa. Indeed, in the Type I extreme-value notation, the parameters are location-scale parameters, so it often is more convenient to develop techniques in the extreme-value notation first. If $Y \sim \text{EV}(\delta, \xi)$, then

$$F_Y(y) = 1 - \exp\left[-\exp\left(\frac{y - \xi}{\delta}\right)\right] \qquad -\infty < y < \infty$$

$$-\infty < \xi < \infty \qquad \delta > 0$$

and $X = e^Y \sim \text{WEI}(\theta, \beta)$, where $\xi = \ln \theta$ and $\delta = 1/\beta$. For example, if $x_1, \ldots, x_n$ represents a sample of size n from a Weibull distribution, then letting $y_i = \ln x_i$, $i = 1, \ldots, n$, produces a sample from an extreme-value distribution. A test for ξ may be developed based on the y_i, and then this test could be restated in terms of $\theta = \exp(\xi)$.

MAXIMUM LIKELIHOOD ESTIMATION

Let $X \sim \text{WEI}(\theta, \beta)$. Then the likelihood function for the first r ordered observations from a random sample of size n under Type II censored sampling is given by

$$\begin{aligned} g(x_{1:n}, \ldots, x_{r:n}) &= \frac{n!}{(n-r)!}\left[\prod_{i=1}^{r} f_X(x_{i:n})\right][1 - F_X(x_{r:n})]^{n-r} \\ &= \frac{n!}{(n-r)!}\left(\frac{\beta}{\theta}\right)^r \prod_{i=1}^{r}\left(\frac{x_{i:n}}{\theta}\right)^{\beta-1} \\ &\quad \times \exp\left\{-\left[\sum_{i=1}^{r}\left(\frac{x_{i:n}}{\theta}\right)^{\beta} + (n-r)\left(\frac{x_{r:n}}{\theta}\right)^{\beta}\right]\right\} \end{aligned} \qquad \textbf{(16.4.1)}$$

Setting the partial derivatives with respect to θ and β equal to zero gives the MLEs $\hat{\theta}$ and $\hat{\beta}$ as solutions to the equations

$$\frac{\sum_{i=1}^{r} x_{i:n}^{\hat{\beta}} \ln x_{i:n} + (n-r)x_{r:n}^{\hat{\beta}} \ln x_{r:n}}{\sum_{i=1}^{r} x_{i:n}^{\hat{\beta}} + (n-r)x_{r:n}^{\hat{\beta}}} - \frac{1}{\hat{\beta}} = \frac{1}{r}\sum_{i=1}^{r} \ln x_{i:n} \tag{16.4.2}$$

and

$$\hat{\theta} = \left[\frac{\sum_{i=1}^{r} x_{i:n}^{\hat{\beta}} + (n-r)x_{r:n}^{\hat{\beta}}}{r}\right]^{1/\hat{\beta}} \tag{16.4.3}$$

For the special case of complete samples, where $n = r$, the equations reduce to

$$\frac{\sum_{i=1}^{n} x_i^{\hat{\beta}} \ln x_i}{\sum_{i=1}^{n} x_i^{\hat{\beta}}} - \frac{1}{\hat{\beta}} = \frac{1}{n}\sum_{i=1}^{n} \ln x_i \tag{16.4.4}$$

and

$$\hat{\theta} = \left[\frac{\sum_{i=1}^{n} x_i^{\hat{\beta}}}{n}\right]^{1/\hat{\beta}} \tag{16.4.5}$$

In either case, the first equation cannot be solved in closed form. However, it has been shown that the MLEs are unique solutions of these equations. The Newton-Raphson procedure for solving an equation $g(\hat{\beta}) = 0$ is to determine successive approximations $\hat{\beta}_j$, where $\hat{\beta}_{j+1} = \hat{\beta}_j - g(\hat{\beta}_j)/g'(\hat{\beta}_j)$. Many other techniques also are available with a computer.

Note that the MLEs for ξ and δ in the extreme-value notation are simply

$$\hat{\xi} = \ln \hat{\theta} \qquad \hat{\delta} = 1/\hat{\beta} \tag{16.4.6}$$

It may initially seem unclear how to develop inference procedures based on the MLEs in this case. If the estimators cannot be expressed explicitly, then how can their distributions be determined? Two key factors are involved in determining distributional results in this case. These are the recognition of pivotal quantities and the ability to determine their distributions by Monte Carlo simulation.

It follows from Theorem 11.3.2 for the extreme-value model with location-scale parameters that

$$\frac{\hat{\xi} - \xi}{\delta} \qquad \frac{\hat{\xi} - \xi}{\hat{\delta}} \quad \text{and} \quad \frac{\hat{\delta}}{\delta} \tag{16.4.7}$$

are pivotal quantities with distributions that do not depend on any unknown parameters. Thus, in the Weibull notation, it follows that

$$\left(\frac{\hat{\theta}}{\theta}\right)^{\beta} \quad \left(\frac{\hat{\theta}}{\theta}\right)^{\hat{\beta}} \quad \text{and} \quad \frac{\hat{\beta}}{\beta} \tag{16.4.8}$$

are also pivotal quantities. For the Weibull case, the reliability at time t is given by

$$R = R(t) = e^{-(t/\theta)^{\beta}}$$

A pivotal quantity for R is not available, but the distribution of $\hat{R}$ depends only on R and not on t, θ, and β individually. This result is true in general for location-scale models (or related distributions such as the Weibull), but it is shown directly in this case, because

$$\begin{aligned} -\ln \hat{R} &= \left(\frac{t}{\hat{\theta}}\right)^{\hat{\beta}} \\ &= \left[\frac{(t/\theta)^{\beta}}{(\hat{\theta}/\theta)^{\beta}}\right]^{\hat{\beta}/\beta} \\ &= \left[\frac{-\ln R}{(\hat{\theta}/\theta)^{\beta}}\right]^{\hat{\beta}/\beta} \end{aligned} \tag{16.4.9}$$

which is a function only of R and the previous pivotal quantities. This result makes it possible to test hypotheses about R or set confidence intervals on R, if the distribution of $\hat{R}$ can be obtained for various R values. Recognition of these pivotal quantity properties makes it quite feasible to determine percentiles for the necessary distributions by Monte Carlo simulation.

For example, we may desire to know the percentile, q_{γ}, such that

$$P[\hat{\beta}/\beta \leqslant q_{\gamma}] = \gamma$$

for some sample size n. Let us generate, say, 1000 random samples of size n from a standard Weibull distribution, WEI(1, 1), and compute the MLE of β, say $\hat{\beta}_{11}$. In particular, we could determine the number, $\tilde{q}_{\gamma}$, for which $100\gamma\%$ of the calculated values of $\hat{\beta}_{11}$ were smaller than $\tilde{q}_{\gamma}$. Approximately, then,

$$P[\hat{\beta}_{11} \leqslant \tilde{q}_{\gamma}] = \gamma$$

This approximation can be improved by increasing the number of simulated samples within the limits of the random number generator. Now, because the distribution of $\hat{\beta}/\beta$ does not depend on the values of the unknown parameters, the distribution of $\hat{\beta}/\beta$ is the same as the distribution of $\hat{\beta}_{11}/1$; thus, approximately,

$$P[\hat{\beta}/\beta \leqslant \tilde{q}_{\gamma}] = \gamma$$

For example, within simulation error, $\hat{\beta}/\tilde{q}_{\gamma}$ is a lower $100\gamma\%$ confidence limit for β.

Tables of percentage points for the quantities $\sqrt{n}(\hat{\beta}/\beta - 1)$ and $\sqrt{n}\hat{\beta} \ln(\hat{\theta}/\theta)$, and other tables for determining tolerance limits and confidence limits on reliability, are provided by Bain and Engelhardt (1991) for both complete and censored sampling cases.

ASYMPTOTIC RESULTS

Convergence is somewhat slow in the Weibull case, but for reasonably large n the asymptotic normality properties of the MLEs become useful.

As $n \to \infty$ and $r/n \to p$, the following properties hold asymptotically:

$$\sqrt{r}(\hat{\beta} - \beta)/\beta \sim N(0, a_{22}) \tag{16.4.10}$$

$$\sqrt{r}\hat{\beta} \ln(\hat{\theta}/\theta) \sim N(0, a_{11}) \tag{16.4.11}$$

$$\sqrt{r}[\hat{R}(t) - R(t)] \sim N(0, V_R) \tag{16.4.12}$$

where $a_{11} = r\beta^2 \operatorname{Var}(\hat{\theta}/\theta)$, $a_{22} = r \operatorname{Var}(\hat{\beta}/\beta)$, and $a_{12} = r \operatorname{Cov}(\hat{\theta}/\theta, \hat{\beta}/\beta)$ are the asymptotic variances and covariances, and

$$V_R = R^2\{a_{11}(\ln R)^2 - 2a_{12} \ln(-\ln R) + a_{22}[\ln(-\ln R)]^2\} \tag{16.4.13}$$

The a_{ij} are included in Table 15 (Appendix C) for censoring levels $p = 0.1, 0.2, \ldots, 1.0$. See also Harter (1969) for $c_{ij} = (n/r)a_{ij}$.

Similar results hold for the extreme-value case where, asymptotically,

$$\sqrt{r}(\hat{\delta} - \delta)/\delta \sim N(0, a_{22}) \tag{16.4.14}$$

$$\sqrt{r}(\hat{\xi} - \xi)/\hat{\delta} \sim N(0, a_{11}) \tag{16.4.15}$$

with $a_{11} = r \operatorname{Var}(\hat{\xi}/\delta)$, $a_{22} = r \operatorname{Var}(\hat{\delta}/\delta)$, and $r \operatorname{Cov}(\hat{\xi}/\delta, \hat{\delta}/\delta) = -a_{12}$.

For the Weibull distribution, it appears that convergence to normality occurs faster for an alternate pivotal quantity of the form

$$W(d) = \frac{\hat{\xi} - \xi}{\hat{\delta}} - d\,\frac{\delta}{\hat{\delta}} \tag{16.4.16}$$

Confidence limits on ξ, $R(t)$, or percentiles, based on $W(d)$, will be equivalent to those based directly on the earlier pivotal quantities. Johns and Lieberman (1966) consider confidence limits using $W(d)$ based on simpler estimators, and Jones et al. (1984) develop limits essentially based on $W(d)$ [see also Bain and Engelhardt (1986)]. Let $w_\gamma(d)$ be the γ percentage point such that

$$P[W(d) \leqslant w_\gamma(d)] = \gamma$$

then the asymptotic normal approximation for $w_\gamma(d)$ is

$$w_\gamma(d) = -d + \sigma_w z_\gamma \tag{16.4.17}$$

where $r\sigma_w^2 = a_{11} + d^2 a_{22} - 2da_{12} = A(d)$.

We now consider first approximate confidence limits on reliability. Note that in the extreme-value notation, the reliability at time y is related to the Weibull reliability as

$$R_Y(y) = P[Y \geqslant y] = P[\ln X \geqslant y] = P[X \geqslant e^y] = R_X(e^y) \tag{16.4.18}$$

and

$$R_X(x) = R_Y(\ln x) \tag{16.4.19}$$

Now for a fixed time x, let $y = \ln x$, and a lower γ confidence limit for $R_X(x)$ is given by

$$L(R_X(x)) = \exp\left[-\exp\left(-\hat{d} + z_\gamma \sigma_w\right)\right] \tag{16.4.20}$$

where $\hat{d} = -\ln[-\ln \hat{R}_X(x)] = (\hat{\xi} - y)/\hat{\delta}$ and $\sigma_w = \sigma_w(\hat{d}) = \sqrt{A(\hat{d})/r}$. This follows because

$$\begin{aligned}
P[L(\hat{R}_X(x)) \leqslant R_X(x)] &= P\{\exp[-\exp(-\hat{d} + z_\gamma \sigma_w)] \leqslant \exp[-(x/\theta)^\beta]\} \\
&= P\{-\hat{d} + z_\gamma \sigma_w \geqslant \beta \ln(x/\theta)]\} \\
&= P\left[\frac{-\xi + \ln x}{\delta} \leqslant -\hat{d} + z_\gamma \sigma_w\right] \\
&= P\left[\frac{\hat{\xi} - \xi}{\delta} - \frac{\hat{\xi} - \ln x}{\hat{\delta}} \frac{\hat{\delta}}{\delta} \leqslant w_\gamma(\hat{d})\right] \\
&= P[W(\hat{d}) \leqslant w_\gamma(\hat{d})] \\
&= E_{\hat{d}} P[W(\hat{d}) \leqslant w_\gamma(\hat{d}) \mid \hat{d}] \\
&= E_{\hat{d}}(\gamma) \\
&= \gamma
\end{aligned}$$

Approximate confidence limits for percentiles can be similarly determined. The 100α percentile for the Weibull distribution is given by

$$x_\alpha = \theta[-\ln(1 - \alpha)]^{1/\beta} = \exp(y_\alpha) \tag{16.4.21}$$

where

$$y_\alpha = \xi + \delta \ln[-\ln(1 - \alpha)] = \xi + \delta\lambda_\alpha \tag{16.4.22}$$

is the α percentile for the extreme-value distribution. For the special case of $\alpha = 1 - 1/e$, $\lambda_\alpha = 0$ and these percentiles reduce to $y_\alpha = \xi$ and $x_\alpha = \theta$. Note that a lower γ level confidence limit for x_α, say $L(x_\alpha; \gamma)$, is also a γ-probability tolerance limit for proportion $1 - \alpha$ for the Weibull distribution. For the extreme-value distribution

$$L(y_\alpha; \gamma) = \ln L(x_\alpha; \gamma) \tag{16.4.23}$$

In terms of the pivotal quantity

$$Q = \frac{\hat{y}_\alpha - y_\alpha}{\hat{\delta}} \tag{16.4.24}$$

$$L(y_\alpha\,;\,\gamma) = \hat{y}_\alpha - q_\gamma \hat{\delta} \tag{16.4.25}$$

where $P[Q \leqslant q_\gamma] = \gamma$. The Monte Carlo value q_γ may be determined from Bain and Engelhardt (1991). In terms of $W(d)$,

$$\begin{aligned} P[Q \leqslant q_\gamma] &= P\left[\frac{\hat{\xi} - \xi}{\delta} - (q_\gamma - \lambda_\alpha)\frac{\hat{\delta}}{\delta} \leqslant \lambda_\alpha\right] \\ &= P[W(d) \leqslant \lambda_\alpha] \\ &= \gamma \end{aligned}$$

which gives $w_\gamma(d) = \lambda_\alpha$, where $d = q_\gamma - \lambda_\alpha$. That is, $q_\gamma = d + \lambda_\alpha$ where d is the solution to $w_\gamma(d) = \lambda_\alpha$. Using the asymptotic normal approximation for $w_\gamma(d)$, we may solve d to obtain

$$d \doteq \frac{-a_{12} z_\gamma^2 - r\lambda_\alpha + z_\gamma[(a_{12}^2 - a_{11}a_{22})z_\gamma^2 + ra_{11} + 2ra_{12}\lambda_\alpha + ra_{22}\lambda_\alpha^2]^{1/2}}{r - a_{22} z_\gamma^2} \tag{16.4.26}$$

Then

$$L(y_\alpha\,;\,\gamma) \doteq \hat{y}_\alpha - (d + \lambda_\alpha)\hat{\delta} = \hat{\xi} - d\hat{\delta} \tag{16.4.27}$$

and

$$\begin{aligned} L(x_\alpha\,;\,\gamma) &\doteq \exp\,[L(y_\alpha\,;\,\gamma)] \\ &= \hat{\theta}\, e^{-d/\hat{\beta}} \end{aligned} \tag{16.4.28}$$

Lower confidence limits for ξ and θ are obtained by letting $\lambda_\alpha = 0$ in computing d.

INFERENCES ON δ OR β

A chi-square approximation often is useful for positive variables. An approximate distribution for a variable U with two correct moments is achieved by considering

$$cU \sim \chi^2(\nu) \tag{16.4.29}$$

where c and ν are chosen to satisfy $cE(U) = \nu$ and $c^2\ \text{Var}(U) = 2\nu$. Following along these lines, Bain and Engelhardt (1986) propose the simple approximate distributional result given by

$$cr(\hat{\delta}/\delta)^{1+p^2} \sim \chi^2(c(r-1)) \tag{16.4.30}$$

where $p = r/n$, $c = 2/[(1 + p^2)^2 a_{22}]$, and a_{22} is the asymptotic variance of $\sqrt{r}\hat{\delta}/\delta$ as $n \to \infty$ and $r/n \to p$. Values of a_{22} are given in Table 15 (Appendix C), and values of c also are included for convenience. The constant c makes this approximation become correct asymptotically.

It is clear that inferences on β or δ can be carried out easily based on the above approximation. For example, a lower γ level confidence limit for δ is given by

$$\delta_L = \hat{\delta}[\chi_\gamma^2(c(r-1))/cr)]^{-1/(1+p^2)} \tag{16.4.31}$$

An upper limit is obtained by replacing γ with $1 - \gamma$.

Example 16.4.1 The 20 smallest ordered observations from a simulated random sample of size 40 from a Weibull distribution with $\theta = 100$ and $\beta = 2$ are given by Harter (1969) as follows:

5, 10, 17, 32, 32, 33, 34, 36, 54, 55, 55, 58, 58, 61, 64, 65, 65, 66, 67, 68

It is possible to observe how the statistical results relate to the known model for this data. Also, Monte Carlo tables happen to be available for this particular sample size and censoring level, so the approximate results can be compared to the results that would be obtained from using the Monte Carlo tables in this case.

For this set of data, $r = 20$, $n = 40$, $p = 0.5$, $\hat{\beta} = 2.09$, and $\hat{\theta} = 83.8$. In the extreme-value notation, $\hat{\xi} = \ln \hat{\theta} = 4.43$ and $\hat{\delta} = 1/\hat{\beta} = 0.478$. These also are the values that would be obtained if one directly computed the maximum likelihood estimators of ξ and δ in an extreme-value distribution based on the natural logarithms of the above data, $y_i = \ln x_i$.

Now, an upper γ level confidence limit β_U such that $P[\beta < \beta_U] = \gamma$ is given by

$$\beta_U = \hat{\beta}h_1(\gamma, p, r) = \hat{\beta}[\chi_\gamma^2(c(r-1))/cr)]^{1/(1+p^2)} \tag{16.4.32}$$

Similarly, a lower γ level confidence limit is given by

$$\beta_L = \hat{\beta}h_1(1-\gamma, p, r) = \hat{\beta}[\chi_{1-\gamma}^2(c(r-1))/cr]^{1/(1+p^2)} \tag{16.4.33}$$

For $\gamma = 0.95$, based on the above data, $c = 1.49$ from Table 15 (Appendix C) and

$$\beta_L = 2.09[\chi_{0.05}^2(28.3)/29.8]^{1/1.25} = 1.34$$

and

$$\beta_U = 2.09[\chi_{0.95}^2(28.3)/29.8]^{1/1.25} = 2.73$$

If the tables in Bain and Engelhardt (1991) based on Monte Carlo simulation are used, one obtains $\beta_L = 1.34$ and $\beta_U = 2.72$ in this case.

Note that in the extreme-value notation, $\delta_L = 1/\beta_U$ and $\delta_U = 1/\beta_L$.

We now will illustrate the lower confidence limit for reliability at time t, $R_X(t)$. The true reliability at time $t = 32.46$ is 0.90 for a Weibull distribution with $\theta = 100$ and $\beta = 2$. Thus, let us compute a lower confidence limit for $R_X(32.46)$ based on the above data. The lower confidence limit for $R_X(t)$ is given by

$$L(R_X(t); \gamma) = \exp[-\exp(-\hat{d} + \sigma_w z_\gamma)] \tag{16.4.34}$$

where $\hat{d} = -\ln[-\ln \hat{R}_X(t)]$, $A = a_{11} + \hat{d}^2 a_{22} - 2\hat{d}a_{12}$, and the a_{ij} are given in Table 15. We have

$$\hat{R}_X(32.46) = \exp[-(32.46/83.8)^{2.09}] = 0.871$$

$$\hat{d} = -\ln[-\ln(0.871)] = 1.98$$

$$r\sigma_w^2 = A = 1.25512 + (1.98)^2(0.85809) - 2(1.98)(0.46788) = 2.766$$

and for, say, $\gamma = 0.90$, we have $z_\gamma = 1.282$ and

$$L(R_X(32.46);\, 0.90) = \exp[-\exp(-1.98 + \sqrt{2.766}(1.282)/\sqrt{20})] = 0.801$$

Again, direct use of the Monte Carlo tables gives nearly the same result, 0.797.

Considering the reliability at time $t = 32.46$ in the Weibull distribution is comparable to considering the reliability at $y_0 = \ln(32.46) = 3.48$ in the analogous extreme-value model. Thus, for example, a 90% lower confidence limit for $R_Y(3.48)$ is also

$$L(R_Y(3.48);\, 0.90) = L(R_X(e^{3.48});\, 0.90) = 0.801$$

We now will illustrate a tolerance limit or confidence limit on a percentile. The 100α percentile, x_α, for the Weibull distribution and y_α for the extreme-value distribution are given in equations (16.4.21) and (16.4.22). If, for example, $\alpha = 0.10$, then

$$\hat{x}_\alpha = \hat{\theta}[-\ln(1 - 0.10)]^{1/\hat{\beta}} = 28.6$$

and

$$\hat{y}_\alpha = \ln(\hat{x}_\alpha) = 3.35$$

A lower γ level tolerance limit for proportion $1 - \alpha$ for the extreme-value distribution is, from equation (16.4.27),

$$L(y_\alpha;\, \gamma) = \hat{y}_\alpha - (d + \lambda_\alpha)\hat{\delta} = \hat{\xi} - d\hat{\delta} \tag{16.4.35}$$

and for the Weibull distribution

$$L(x_\alpha;\, \gamma) = \exp[L(y_\alpha;\, \gamma)] \tag{16.4.36}$$

where $\lambda_\alpha = \ln[-\ln(1 - \alpha)]$ and d is given by equation (16.4.26).

We have $\lambda_{0.10} = \ln[-\ln(1 - 0.10)] = -2.25$, and if we choose $\gamma = 0.90$ in our example, then $z_{0.90} = 1.282$, and

$$\begin{aligned} d = \{&-0.46788(1.282)^2 - 20(-2.25) + 1.282[((0.46788)^2 \\ &-(1.25512)(0.85809))(1.282)^2 + 20(1.25512) \\ &+2(20)(0.46788)(-2.25) \\ &+20(0.85809)(-2.25)^2]^{1/2}\}/[20 - 0.85809(1.282)^2] \\ = {}& 2.95 \end{aligned}$$

$$L(y_{0.10};\, 0.90) = 3.35 - (2.90 - 2.25)(0.478) = 3.02$$

and

$$L(x_{0.10}\,;\,0.90) = \exp\,(3.04) = 20.4$$

Again, direct use of the Monte Carlo tables gives almost the same result, 20.3.

A lower γ level confidence limit for the parameter ξ or θ is given by

$$L(\xi;\,\gamma) = \hat{\xi} - d\hat{\delta} \qquad \textbf{(16.4.37)}$$

or

$$L(\theta;\,\gamma) = \hat{\theta}\;e^{-d/\hat{\beta}} \qquad \textbf{(16.4.38)}$$

where $\alpha = 1 - 1/e$ and $\lambda_\alpha = 0$.

Note that upper γ level confidence limits are given by replacing γ with $1 - \gamma$:

$$U(\theta;\,\gamma) = L(\theta;\,1 - \gamma) \qquad \textbf{(16.4.39)}$$

Let us find a two-sided 90% confidence limit for θ. We must compute d with $z_{0.95} = 1.645$ and $z_{0.05} = -1.645$. This simply changes the sign of the term in brackets; thus the two values of d are given by

$$d = \frac{-0.46788(1.645)^2 \pm 1.645[((0.46788)^2 - (1.25512)(0.85809))(1.645)^2 + 20(1.25512)]^{1/2}}{20 - (0.85809)(1.645)^2}$$

$$= 0.372 \quad \text{or} \quad -0.515$$

This gives

$$L(\theta,\,0.95) = 83.8e^{-0.372/2.09} = 70.1$$

and

$$U(\theta,\,0.95) = 83.8e^{+0.515/2.09} = 107.2$$

Thus a two-sided 90% confidence interval for θ is (70.1, 107.2), and a two-sided 90% confidence interval for ξ is

$$(\ln 70.1,\,\ln 107.2) = (4.25,\,4.67)$$

The corresponding interval from the Monte Carlo tables is found to be (4.27, 4.71).

SIMPLE ESTIMATORS

Computation of the MLEs is relatively simple if a computer is available; however, it sometimes is more convenient to have simpler closed-form estimators available. A small set of sufficient statistics does not exist for the Weibull distribution, so it is not completely clear how to proceed with this model. We know that the MLEs are asymptotically efficient, and they are good estimators for small n except for their difficulty of computation. The MLEs also are somewhat biased for small n and heavy censoring.

Simpler unbiased estimators have been developed for the location-scale parameters of the extreme-value distribution (see Engelhardt and Bain, 1977b), and these can be applied to the Weibull parameters. These estimators are very similar to the MLEs, and for the most part can be used interchangeably with them. If an adjustment is made for the bias of the MLEs, then these two methods are essentially equivalent, particularly for the censored sampling case. The simple estimators still require some tabulated constants for their computation.

Let $x_{1:n}, \ldots, x_{r:n}$ denote the r smallest observations from a sample of size n from a Weibull distribution, and let $y_i = \ln x_{i:n}$ denote the corresponding ordered extreme-value observations. The simple estimators then are computed as follows:

1. Complete sample case, $r = n$:

$$\tilde{\delta} = 1/\tilde{\beta} = \left[-\sum_{i=1}^{s} y_i + \frac{s}{n-s} \sum_{i=s+1}^{n} y_i \right] \Big/ nk_n \tag{16.4.40}$$

$$\tilde{\xi} = \ln \tilde{\theta} = \bar{y} + \gamma \tilde{\delta} \tag{16.4.41}$$

where $s = [0.84n] =$ largest integer $\leqslant 0.84n$, $\bar{y}$ is the mean, and $\gamma = 0.5772$. Some values of k_n are provided in Table 16 (Appendix C).

2. Censored samples, $r < n$:

$$\tilde{\delta} = 1/\tilde{\beta} = \left[(r-1)y_r - \sum_{i=1}^{r-1} y_i \right] \Big/ nk_{r,n} \tag{16.4.42}$$

$$\tilde{\xi} = \ln \tilde{\theta} = y_r - c_{r,n} \tilde{\delta} \tag{16.4.43}$$

Quadratic approximations for computing $k_{r,n}$ and $c_{r,n}$ are given by

$$k_{r,n} \doteq k_0 + k_1/n + k_2/n^2 \tag{16.4.44}$$

$$c_{r,n} \doteq E(Y_r - \xi)/\delta = c_0 + c_1/n + c_2/n^2 \tag{16.4.45}$$

where the coefficients are tabulated in Table 17 (Appendix C). These constants make $\tilde{\delta}$ and $\tilde{\xi}$ unbiased estimators of δ and ξ. The values k_0 and c_0 are the asymptotic values as $n \to \infty$ and $r/n \to p$.

If one wishes to substitute simple estimators for the MLEs, then slightly improved results are obtained by using the following modified simple estimators:

$$\delta^* = \tilde{\delta}/[1 + \text{Var}(\tilde{\delta}/\delta)] = h\tilde{\delta}/(h+2) \tag{16.4.46}$$

$$\xi^* = \tilde{\xi} - \text{Cov}(\tilde{\xi}/\delta, \tilde{\delta}/\delta)\delta^* \tag{16.4.47}$$

where

$$\text{Cov}(\tilde{\xi}/\delta, \tilde{\delta}/\delta) = d_{r,n} - c_{r,n} 2n/h$$

$$d_{r,n} = n \, \text{Cov}(Y_r/\delta, \tilde{\delta}/\delta) \doteq d_0 + d_1/n + d_2/n^2$$

$$h/n = 2/n \, \text{Var}(\tilde{\delta}/\delta) \doteq a_0 + a_1/n + a_2/n^2$$

and the coefficients are included in Table 17 (Appendix C).

Similarly, approximately debiased MLEs are given by

$$\hat{\beta}_u = \frac{(h-2)\hat{\beta}}{(h+2)} \qquad \hat{\delta}_u = \frac{(h+2)\hat{\delta}}{h}$$

$$\hat{\xi}_u = \hat{\xi} + \hat{\delta} \, \text{Cov}(\tilde{\xi}/\delta, \tilde{\delta}/\delta) \tag{16.4.48}$$

Again, we have the approximations $\hat{\delta}_u \approx \tilde{\delta}$, $\hat{\xi}_u \approx \tilde{\xi}$, $\hat{\delta} \approx \delta^*$, and $\hat{\xi} \approx \xi^*$.

16.5 REPAIRABLE SYSTEMS

Much of the theory of reliability deals with nonrepairable systems or devices, and it emphasizes the study of lifetime models. It is important to distinguish between models for repairable and nonrepairable systems. A nonrepairable system can fail only once, and a lifetime model such as the Weibull distribution provides the distribution of the time at which such a system fails. This was the situation in the earlier sections of this chapter. On the other hand, a repairable system can be repaired and placed back in service. Thus, a model for repairable systems must allow for a whole sequence of repeated failures.

One such model is the homogeneous Poisson process or HPP which was introduced in Chapter 3. In this section we will consider additional properties of the HPP and discuss some more general processes that are capable of reflecting changes in the reliability of the system as it ages.

HOMOGENEOUS POISSON PROCESS

We denote by $X(t)$ the number of occurrences (failures) in the time interval $[0, t]$. It was found in Theorem 3.2.4 that under the following conditions $X(t)$ is an HPP:

1. $X(0) = 0$.
2. $P[X(t+h) - X(t) = n \mid X(s) = m] = P[X(t+h) - X(t) = n]$ for all $0 \leqslant s \leqslant t$ and $0 < h$.
3. $P[X(t+\Delta t) - X(t) = 1] = \lambda \Delta t + o(\Delta t)$ for some constant $\lambda > 0$.
4. $P[X(t+\Delta t) - X(t) \geqslant 2] = o(\Delta t)$.

In other words, if conditions 1 through 4 are satisfied, then

$$P[X(t) = n] = e^{-\lambda t}(\lambda t)^n/n!$$

for all $n = 0, 1, \ldots$, and some $\lambda > 0$.

Thus, $X(t) \sim \text{POI}(\lambda t)$, where $\mu = E[X(t)] = \lambda t$. The proportionality constant λ reflects the **rate of occurrence** or **intensity** of the Poisson process. Because λ is assumed constant over t, and the increments are independent, it turns out that one does not need to be concerned about the location of the interval under question, and the model $X \sim \text{POI}(\mu)$ is applicable for any interval of length t, $[s, s + t]$, with $\mu = \lambda t$. The constant λ is the rate of occurrence per unit length, and the interval is t units long. This also is consistent with Theorem 3.2.4. In particular, the interval $[0, t]$ can be represented as a union of n disjoint subintervals, each of length t_i. If $Y = \sum_{i=1}^{n} X_i$, where X_i is the number of occurrences in the ith subinterval, then $\mu_Y = \sum \mu_i = \lambda \sum t_i$, and Y represents a Poisson variable with intensity rate λ relative to the interval of length $\sum t_i$. That is, one can choose any interval, but the variable remains Poisson with the appropriate mean.

Example 16.5.1 Let X denote the number of alpha particles emitted from a bar of polonium in one second, and assume that the rate of emission is $\lambda = 0.5$ per second. Thus $\mu_X = 0.5(1) = 0.5$, and the Poisson model for this variable would be

$$f_X(x) = e^{-0.5}(0.5)^x/x! \qquad x = 0, 1, \ldots$$

For example,

$$P[X = 1] = f_x(1) \doteq 0.3$$

Let Y denote the number of emissions in an eight-second interval. One may consider $Y = \sum_{i=1}^{8} X_i$ with $\mu_Y = \sum_{i=1}^{8} (0.5) = 4$, or one may consider the mean of Y as $\lambda t = 0.5(8) = 4$. In any case

$$\begin{aligned} f_Y(y) &= e^{-\lambda t}(\lambda t)^y/y! \\ &= e^{-4}4^y/y! \qquad y = 0, 1, \ldots \end{aligned}$$

In practice one may wish to estimate the value of μ_Y from data. A frequency histogram also would be useful to help evaluate whether the Poisson model provides an appropriate distribution of probability. Rutherford and Geiger (1910) observed the number of emissions in 2608 intervals of 7.5 seconds each, with the results shown in Table 16.1. Note that y denotes the number of emissions in a 7.5-second interval in this example.

The table indicates that no emissions were observed in 57 intervals, 1 emission was observed in 203 of the intervals, and so on. If we let Y denote the number of emissions in a 7.5-second period, and if we assume a Poisson model $Y \sim \text{POI}(\mu)$,

TABLE 16.1 **Observed number of alpha particle emissions in 2608 intervals of 7.5 seconds each**

No. of Particles Emitted, y	No. of Intervals with y Emissions, m_y	Estimated Expected Nos., e_i
0	57	54.40
1	203	210.52
2	383	407.36
3	525	525.50
4	532	508.42
5	408	393.52
6	273	253.82
7	139	140.32
8	45	67.88
9	27	29.19
10	10	11.30
11	4	3.97
12	2	1.28
⩾13	0	0.52
	2608	

then it would be reasonable to estimate μ with the sample mean, $\bar{y}$. In this example the data are grouped, so

$$\sum_{i=1}^{2608} y_i = \sum_{y=0}^{\infty} ym_y = 0(57) + 1(203) + 2(383) + \cdots = 10094$$

and $\bar{y} = 3.870$.

Using the fitted model $Y \sim \text{POI}(3.870)$, in 2608 observed intervals one would expect $(2608)P[Y = 0] = (2608)e^{-3.870}(3.870)^0/0! = 54.4$ intervals with no emissions, $(2608)P[Y = 1] = 210.52$ intervals with 1 emission, and so on. These computed expected numbers are included in Table 16.1. The computed expected numbers appear to agree quite closely with the observed numbers, and the suggested Poisson model seems appropriate for this problem. More formal statistical tests for the goodness-of-fit of a model can be performed using the results of Section 13.7. If we combine the cases where $y \geqslant 11$, the chi-square value is $\chi^2 = 12.97$ and $\nu = 12 - 1 - 1 = 10$. Because $\chi^2_{0.90}(10) = 15.99$, we cannot reject the Poisson model at the $\alpha = 0.10$ level.

EXPONENTIAL WAITING TIMES

With any Poisson process there is an associated sequence of continuous waiting times for successive occurrences.

Theorem 16.5.1 If events are occurring according to an HPP with intensity parameter λ, then the waiting time until the first occurrence, T_1, follows an exponential distribution, $T_1 \sim \text{EXP}(\theta)$ with $\theta = 1/\lambda$. Furthermore, the waiting times between consecutive occurrences are independent exponential variables with the same mean time between occurrences, $1/\lambda$.

Proof

The CDF of T_1 at time t is given by

$$F_1(t) = P[T_1 \leqslant t] = 1 - P[T_1 > t]$$

Now $T_1 > t$ if and only if no events occur in the interval $[0, t]$, that is, if $X(t) = 0$. Hence,

$$F_1(t) = 1 - P[X(t) = 0] = 1 - P_0(t) = 1 - e^{-\lambda t}$$

which is an exponential CDF with mean $1/\lambda$.

The proof of the second part is beyond the scope of this book (see Parzen, 1962, p. 135). ■

We see that the mean time to failure, θ, is inversely related to the failure intensity λ. The HPP assumptions are rather restrictive, but at least a very tractable and easily analyzed model is realized.

Theorem 16.5.2 If T_k denotes the waiting time until the kth occurrence in an HPP, then $T_k \sim \text{GAM}(1/\lambda, k)$.

Proof

The CDF of T_k at time t is given by

$$\begin{aligned} F_k(t) &= 1 - P[T_k > t] \\ &= 1 - P[k - 1 \text{ or fewer occurrences in } [0, t]] \\ &= 1 - \sum_{i=0}^{k-1} P_i(t) \\ &= 1 - \sum_{i=0}^{k-1} (\lambda t)^i e^{-\lambda t}/i! \end{aligned}$$

which is the CDF of a gamma variable with parameters k and $1/\lambda$. ■

This result also is consistent with the second part of Theorem 16.5.1; if we assume independent $Y_i \sim \text{EXP}(1/\lambda)$, then

$$T_k = \sum_{i=1}^{k} Y_i \sim \text{GAM}(1/\lambda, k)$$

It was observed earlier that the exponential distribution has a no-memory property. This is related to the assumption of a constant failure intensity λ, which implies that no wearout is occurring. It sometimes is said that the exponential distribution is applicable when the failures occur "at random" and are not affected by aging. We know that the "at random" terminology is related to the uniform distribution, but it is used in this framework for the following reason. If $T_1, T_2, \ldots$ denote the successive times of occurrence of a Poisson process measured from time 0, then given that n events have occurred in the interval $[0, t]$, the successive occurrence times $T_1, \ldots, T_n$ are conditionally distributed as ordered observations from a uniform distribution on $[0, t]$.

Example 16.5.2 Proschan (1963) gives the times of successive failures of the air conditioning system of each member of a fleet of Boeing 720 jet airplanes. The hours of flying time, y_i, between 30 failures on plane 7912 are listed below.

23, 261, 87, 7, 120, 14, 62, 47, 225, 71, 246, 21, 42, 20, 5, 12, 120,

11, 3, 14, 71, 11, 14, 11, 16, 90, 1, 16, 52, 95

If we assume that the failures follow an HPP with intensity λ, then this set of data represents a random sample of size 30 from EXP($1/\lambda$). Using the sample mean to estimate the population mean gives $\hat{\theta} = \bar{y} = 59.6 = 1/\hat{\lambda}$, and $\hat{\lambda} = 1/59.6 = 0.0168$.

Let X denote the number of failures for a 200-hour interval from this process. Then X follows a Poisson distribution with $\mu = \lambda t = 200\lambda$. If we wish to estimate λ using the Poisson count data, we first consider the successive failure times of the observed data, given by

23, 284, 371, 378, 498, 512, 574, 621, 846, 917, 1163, 1184,

1226, 1246, 1251, 1263, 1383, 1394, 1397, 1411, 1482, 1493, 1507,

1518, 1534, 1624, 1625, 1641, 1693, 1788

Considering the first eight consecutive intervals of length 200 hours, the numbers of observed failures per interval are 1, 3, 3, 1, 2, 2, 7, 6. Thus, these eight values represent a sample of size eight from POI(200λ). Estimating the mean of the Poisson variable from these count data gives $\hat{\mu} = \bar{x} = 3.125 = 200\hat{\lambda}$, and $\hat{\lambda} = 0.0156$. The two estimates obtained for λ are quite consistent, although, of course, they are not identical.

NONHOMOGENEOUS POISSON PROCESS

The Poisson process is an important model for the failure times of a repairable system. In this terminology, the HPP assumptions imply that the time to first failure is a random variable that follows the exponential distribution, and also that the time between failures is an independent exponential variable. The assumption of a constant failure intensity parameter λ suggests that the system is being maintained and not wearing out or degrading. If the system is wearing out, then the model should be generalized to allow λ to be an increasing function of t. More generally, we might want to allow the intensity to be an arbitrary non-negative function of t.

We can model this if, in part 3 of Theorem 3.2.4, we replace the constant λ with a function of t, denoted by $\lambda(t)$. A similar derivation yields another type of Poisson process, known as a **nonhomogeneous Poisson process** (NHPP).

If $X(t)$ denotes the number of occurrences in a specified interval $[0, t]$ for a NHPP, then it can be shown that

$$X(t) \sim \text{POI}(\mu(t))$$

where

$$\mu(t) = \int_0^t \lambda(s)\, ds$$

The CDF for the time to first occurrence, T_1, now becomes

$$F_1(t) = 1 - \exp\left[-\mu(t)\right]$$

An important choice for a nonhomogeneous intensity function is

$$\lambda(t) = (\beta/\theta)(t/\theta)^{\beta - 1}$$

which gives

$$\mu(t) = (t/\theta)^{\beta}$$

In this case the time to first occurrence follows a Weibull distribution, WEI(θ, β). This intensity parameter is an increasing function of t if $\beta > 1$ and a decreasing function of t if $\beta < 1$. The $\beta < 1$ case might apply to a developmental situation, in which the system is being improved over time. Note that the times between consecutive failures are not independent Weibull variables in this case.

COMPOUND POISSON PROCESS

It was noted earlier that one characteristic of the Poisson distribution is that the mean and variance have the same value. In some cases this property may not be valid, and a more flexible model is required. One type of generalization is to consider mixtures of distributions. For example, if a fraction p, called type 1, of the population follows POI(μ_1), and the fraction $1 - p$, called type 2, follows POI(μ_2), then the CDF for the population distribution is given by

$$\begin{aligned} F(x) &= P[X \leqslant x \mid \text{type 1}]P[\text{type 1}] + P[X \leqslant x \mid \text{type 2}]P[\text{type 2}] \\ &= F_1(x; \mu_1)p + F_2(x; \mu_2)(1 - p) \end{aligned}$$

The pdf would be a similar mixture of the two separate pdf's.

More generally, if the population is a mixture of k types, with fraction p_i following the pdf $f_i(x)$, then

$$f(x) = \sum_{i=1}^{k} p_i f_i(x) \qquad \sum_{i=1}^{k} p_i = 1$$

If the $f_i(x)$ are all Poisson pdf's, but with differing means, then

$$f(x) = \sum_{i=1}^{k} p_i e^{-\mu_i}\mu_i^x/x! \qquad \sum_{i=1}^{k} p_i = 1 \tag{16.5.1}$$

For example, suppose that a fleet of k types of airplanes is considered, with fraction p_i of type i. These also could be the same type of airplane used under k different conditions. Assume that the number of air conditioner failures in a specified interval $[0, t]$ from an airplane of type i follows POI(μ_i). Now, if an airplane is selected at random from this fleet of airplanes, then the number of air conditioner failures in time $[0, t]$ for that airplane is a random variable that follows the mixed Poisson distribution, given by equation (16.5.1).

This situation is equivalent to assuming that, given μ_i, the conditional density of the number of failures, $f_{X|\mu_i}(x)$, is POI(μ_i), and that μ is a random variable that takes on the value μ_i with probability p_i. In this example the μ_i are fixed values, and the effect of drawing an airplane at random is to produce a random variable μ distributed over these values.

Now we consider, at least conceptually, a large fleet of airplanes in which, for any given airplane, the number of air conditioner failures in $[0, t]$ follows a POI(μ) with $\mu = \lambda t$; however, λ may be different from plane to plane. In particular, for this conceptually large population, we assume that λ is a continuous random variable that follows a gamma distribution, GAM(γ, κ). That is,

$$f_\lambda(\lambda) = \frac{\lambda^{\kappa-1}e^{-\lambda/\gamma}}{\Gamma(\kappa)\gamma^\kappa} \qquad 0 < \lambda < \infty$$

and

$$f_{X|\lambda}(x) = e^{-\lambda t}(\lambda t)^x/x! \qquad x = 0, 1, \ldots$$

Note that in the context of a Bayesian analysis of a Poisson model, the density $f_\lambda(\lambda)$ corresponds to a prior density for the parameter, and the mathematics involved here is essentially equivalent to that involved in the associated Bayesian development. The differences between the two problems depend on the philosophy for introducing a density function for the parameter, and the interpretation of the results. In a Bayesian analysis the parameter may have been considered fixed, but the prior density reflects a degree of belief about the value of the parameter or some previous information about the value of the parameter.

The marginal density for the number of failures in time $[0, t]$ for an airplane selected at random is given in this case by

$$\begin{aligned} f_X(x) &= \int_0^\infty f(x, \lambda)\, d\lambda \\ &= \int_0^\infty f_{X|\lambda}(x) f_\lambda(\lambda)\, d\lambda \\ &= \int_0^\infty \frac{e^{-\lambda t}(\lambda t)^x}{x!} \frac{\lambda^{\kappa-1} e^{-\lambda/\gamma}}{\Gamma(\kappa)\gamma^\kappa}\, d\lambda \\ &= \binom{x+\kappa-1}{x} \frac{(\gamma t)^x}{(\gamma t + 1)^{\kappa + x}} \qquad x = 0, 1, \ldots \end{aligned}$$

where $0 < \kappa < \infty$ and $0 < \gamma < \infty$. This is a form of negative binomial distribution, with $p = 1/(1 + \gamma t)$, and it is referred to as a **compound Poisson distribution** with a gamma-compounding density.

Thus, the negative binomial distribution represents a generalization of the Poisson distribution, and it converges to the Poisson distribution when the gamma prior density becomes degenerate at a constant. The negative binomial model is used frequently as an alternative to the Poisson model in analyzing count data, particularly when the variance and the mean cannot be assumed equal.

The mean and variance for the negative binomial variable in the above notation are given by

$$E(X) = \kappa\gamma t = tE(\lambda)$$

$$\mathrm{Var}(X) = \kappa\gamma t(\gamma t + 1) = tE(\lambda) + t^2 \mathrm{Var}(\lambda)$$

We see that $\mathrm{Var}(X) \geqslant E(X)$, and the Poisson case holds as $\mathrm{Var}(\lambda) = \kappa\gamma^2 \to 0$. Of course, other compound Poisson models can be obtained by considering compounding densities other than the gamma density; however, the gamma density is a very flexible two-parameter density, and it is mathematically convenient. The unknown parameters in this case are κ and γ, and techniques developed for the negative binomial model may be used to estimate these parameters based on observed values of x.

If one follows through the Bayes' Rule, then an expression for the conditional density of λ given x may be obtained:

$$\begin{aligned} f_{\lambda|x}(\lambda) &= \frac{f(x, \lambda)}{f_X(x)} \\ &= \frac{f_{X|\lambda}(x) f_\lambda(\lambda)}{f_X(x)} \end{aligned}$$

Simplification shows that

$$\lambda \mid x \sim \mathrm{GAM}(\gamma/(\gamma t + 1), x + \kappa)$$

COMPOUND EXPONENTIAL DISTRIBUTION

If air conditioner failure for a given airplane occurs according to a Poisson process with failure intensity λ, then the time to first failure, T, follows an exponential distribution. If we again assume that the intensity parameter varies from airplane to airplane according to a gamma distribution, then the time to first failure for an airplane selected at random follows a compound exponential distribution.

$$
\begin{aligned}
f_T(t) &= \int_0^\infty f_{T,\lambda}(t, \lambda)\, d\lambda \\
&= \int_0^\infty f_{T|\lambda}(t) f_\lambda(\lambda)\, d\lambda \\
&= \int_0^\infty \frac{\lambda e^{-\lambda t} \lambda^{\kappa-1} e^{-\lambda/\gamma}}{\Gamma(\kappa)\gamma^\kappa}\, d\lambda \\
&= \kappa\gamma(\gamma t + 1)^{-(\kappa+1)} \qquad 0 < t < \infty
\end{aligned}
$$

This is a form of the Pareto distribution, $T \sim \text{PAR}(1/\gamma, \kappa)$.

Example 16.5.3 Proschan (1963) gave air conditioner failure data for several airplanes. Ten of these airplanes had at least 1000 flying hours. For these 10 planes the numbers of failure, x, in 1000 hours are recorded below:

Airplane:	7908	7909	7910	7911	7912	7913	7914	7915	8044	8045
x:	8	16	9	6	10	13	16	4	9	12

For this set of data, $s^2 = 15.79$ and $\bar{x} = 10.30$. These results suggest that the mean and variance of X may not be equal and that the compound Poisson (negative binomial) model may be preferable to the Poisson model in this case; however, additional distributional results are needed to indicate whether this magnitude of difference between s^2 and $\bar{x}$ reflects a true difference, or whether it could result from random variation. It has been shown in the literature for this case that approximately

$$\frac{(n-1)S^2}{\bar{X}} \sim \chi^2(n-1)$$

when X does follow a Poisson model. In our problem $(n-1)s^2/\bar{x} = 9(15.79)/10.30 = 13.80$. Now $P[\chi^2(9) \geqslant 13.8] \doteq 0.13$; thus, the observed ratio is larger than would be likely. However, there is an approximate 13% chance of getting such a result when the Poisson model is valid.

Greenwood and Yule (1920) studied the number of accidents during a five-week period for 647 women in a shell factory. It turned out that a Poisson process appeared reasonable for each individual, but the intensity varied from individual to individual. That is, some workers were more accident-prone than others. They found that a gamma distribution provided a good compounding distribution for the accident intensities over workers, and that the negative binomial model provided a good model for the number of accidents of a worker selected at random.

SUMMARY

Our purpose in this chapter was to introduce some basic concepts of reliability and to develop the mathematical aspects of the statistical analyses of some common life-testing models.

Various characterizations of models in reliability theory can be given. The most basic is the reliability function (or survivor function) that corresponds to the probability of failure after time t, for each positive t. The hazard function (or failure-rate function) provides another way to characterize a reliability model. The hazard function gives a means of interpreting the model in terms of aging or wearout. If the hazard function is constant, then the model is exponential. An increasing hazard function is generally interpreted as reflecting aging or wearout of the unit under test.

The gamma and Weibull distributions are two different models that include the exponential model, but also allow, by proper choice of the shape parameter, for an increasing hazard function. These models also admit the possibility of a decreasing hazard function, although this is less common in reliability applications.

Most of the statistical analyses for parametric life-testing models have been developed for the exponential and Weibull models. The exponential model is generally easier to analyze because of the simplicity of the functional form and some special mathematical properties that hold for the exponential model. However, the Weibull model is more flexible, and thus it provides a more realistic model in many applications, particularly those involving wearout or aging. Although the Weibull distribution is not a location-scale model, it is related by means of a log transformation to the extreme-value model. This makes the derivation of confidence intervals and tests of hypotheses possible, because pivotal quantities can be constructed from the MLEs.

EXERCISES

1. Consider a random sample of size 25 from $f(x) = (1 + x)^{-2}$, $0 < x < \infty$.

(a) Give the likelihood function for the first 10 ordered observations.

(b) Give the likelihood function for the data censored at time $x = 9$.

(c) What is the probability of getting zero observations by $x = 9$?

(d) What is the expected number of observations by time $x = 9$?

(e) What sample size would be needed so that the expected number of observations by $x = 9$ would be 20?

(f) Approximately what sample size would be needed to be 90% sure of observing 40 or more observations by $x = 9$?

(g) Given that $R = 20$ observations occurred by $x = 9$, what is the joint conditional density of these 20 observations?

2. Rework Exercise 1, assuming $X \sim \text{EXP}(4)$.

3. Suppose that $f(x) = 1/\theta$, $0 < x < \theta$. Find the MLE of θ based on the first r ordered observations from a sample of size n.

4. Suppose that $X \sim \text{PAR}(\theta, k)$.

(a) Determine the hazard function.

(b) Express the $1 - p$ percentile, x_{1-p}.

5. Can $h(x) = e^{-x}$ be a hazard function?

6. Find the pdf associated with the hazard function $h(x) = e^x$.

7. A component in a repairable system has a mean time to failure of 100 hours. Five spare components are available.

(a) What is the expected operation time to be obtained?

(b) If $T_i \sim \text{EXP}(100)$ for each component and the five spares, what is the probability that the system will still be in operation after 300 hours?

(c) How many spares are needed to have a system reliability of 0.95 at 300 hours?

8. The six identical components considered in Exercise 7(b) are connected as a parallel system.

(a) What is the mean time to failure of this parallel system?

(b) How many of these components would be needed in a parallel system to achieve a mean time to failure of 300 hours?

(c) What is the reliability of this parallel system at 200 hours?

9. The six components considered in Exercise 7(b) now are connected in series.

(a) What is the mean time to failure of this series system?

(b) What is the reliability of the system at 10 hours?

(c) What mean time to failure would be required for each component for the series system to have a reliability of 0.90 at 20 hours?

(d) Give the hazard function for the series system.

10. Rework Exercise 9, assuming that the $T_i \sim \text{PAR}(400, 5)$, $i = 1, \ldots, 6$.

11. Rework Exercise 8(c), assuming that $T_i \sim \text{PAR}(400, 5)$.

12. The failure time of a certain electronic component follows an exponential distribution, $X \sim \text{EXP}(\theta)$. In a random sample of size $n = 25$, one observes $\bar{x} = 75$ days.

(a) Compute the MLE of the reliability at time 8 days, $R(8)$.

(b) Compute an unbiased estimate of $R(8)$.

(c) Compute a 95% lower confidence limit for $R(8)$.

(d) Compute a 95% lower tolerance limit for proportion 0.90.

13. Grubbs (1971) gives the following mileages for the failure times of 19 personnel carriers:

162, 200, 271, 302, 393, 508, 539, 629, 706, 777, 884, 1008,

1101, 1182, 1463, 1603, 1984, 2355, 2880

Assume that these observations follow a two-parameter exponential distribution with known threshold parameter, $\eta = 100$; that is, $X \sim \text{EXP}(\theta, 100)$.

(a) Give the distribution of $2n(\bar{X} - 100)/\theta$.

(b) Compute a lower 90% confidence limit for θ.

(c) Compute a 95% lower tolerance limit for proportion 0.80.

14. Rework Exercise 13, assuming that only the first 15 failure times for the 19 carriers were recorded.

15. Wilk et al. (1962) give the first 31 failure times (in weeks) from an accelerated life test of 34 transistors as follows:

3, 4, 5, 6, 6, 7, 8, 8, 9, 9, 9, 10, 10, 11, 11, 11, 13,

13, 13, 13, 13, 17, 17, 19, 19, 25, 29, 33, 42, 42, 52

It may be that a threshold parameter is needed in this problem, but for illustration purposes suppose that $X \sim \text{EXP}(\theta)$.

(a) Estimate θ.

(b) Compute a 90% lower confidence limit for θ.

(c) Compute a 90% lower tolerance limit for proportion 0.95.

(d) Compute a 50% lower tolerance limit for proportion 0.95.

(e) Estimate $x_{0.05}$.

16. Suppose in Exercise 15 that the experiment on the 34 transistors had been terminated after 50 weeks.

(a) Estimate $\hat{\theta}$.

(b) Set a lower 0.90 confidence limit on θ.

(c) Compute a lower (0.90, 0.95) tolerance limit.

17. One hundred light bulbs are placed on test, and the experiment is continued for one year. As light bulbs fail they are replaced with new bulbs, and at the end of one year a total of 85 bulbs have failed. Assume $\text{EXP}(\theta)$.

(a) Estimate θ.

(b) Test $H_0 : \theta \geqslant 1.5$ years against $H_a : \theta < 1.5$ at $\alpha = 0.05$.

(c) Compute a 90% two-sided confidence interval for θ.

(d) If bulbs are guaranteed for six months, estimate what percentage of the bulbs will have to be replaced.

(e) What warranty period should be offered if one wishes to be 90% confident that at least 95% of the bulbs will survive the warranty period?

18. Consider the ball bearing data of Exercise 21 in Chapter 13. If we assume that this set of data is a complete sample from a Weibull distribution, then the MLEs are $\hat{\beta} = 2.102$ and $\hat{\theta} = 81.88$.

(a) Use equation (16.4.33) to compute an approximate lower 0.95 level confidence limit for β.

(b) Use equation (16.4.34) to compute an approximate lower 0.90 level confidence limit for $R_X(75)$.

(c) Compute the MLE of the 10th percentile, $x_{0.10}$.

(d) Use equation (16.4.36) to compute a lower 0.95 level tolerance limit for proportion 0.90.

(e) Use equation (16.4.38) to compute an approximate lower 0.90 level confidence limit for θ.

19. Consider the censored Weibull data of Example 16.4.1.

(a) Compute the simple estimates $\tilde{\delta}$ and $\tilde{\beta} = 1/\tilde{\delta}$, using equation (16.4.42).

(b) Compute the simple estimates $\tilde{\xi}$ and $\tilde{\theta} = \exp(\tilde{\xi})$, using equation (16.4.43).

20. Compute the simple estimates of δ, β, ξ, and θ, using equations (16.4.40) and (16.4.41), with the ball bearing data of Exercise 21 in Chapter 13.

21. Let $X \sim \text{POI}(\mu)$. Show that

$$f(x-1;\mu) < f(x;\mu) \qquad \text{for } x < \mu$$

and

$$f(x-1;\mu) > f(x;\mu) \qquad \text{for } x > \mu$$

22. Verify equation (16.2.9).

23. Let X denote the number of people seeking a haircut during a one-hour period, and suppose that $X \sim \text{POI}(4)$. If a barber will service three people in an hour:

(a) What is the probability that all customers arriving can be serviced?

(b) What is the probability that all but one potential customer can be serviced?

(c) How many people must the barber be able to service in an hour to be 90% likely to service everyone who arrives?

(d) What is the expected number of customers arriving per hour? Per 8-hour day?

(e) What is the expected number of customers serviced per hour?

(f) If two barbers are available, what is the expected number of customers serviced per hour?

24. Assume that the number of emissions of particles from a radioactive source is a homogeneous Poisson process with intensity $\lambda = 2$ per second.

(a) What is the probability of 0 emissions in 1 second?

(b) What is the probability of 0 emissions in 10 seconds?
(c) What is the probability of 3 emissions in 1 second?
(d) What is the probability of 30 emissions in 10 seconds?
(e) What is the probability of 20 emissions or less in 10 seconds?

25. In World War II, London was divided into $n = 576$ small areas of 1/4 square kilometer each. The number of areas, m_y, receiving y flying bomb hits is given by Clarke (1946), and these are listed below.

y:	0	1	2	3	4	$\geqslant 5$
m_y:	229	211	93	35	7	1

The total number of hits was 537. Although clustering might be expected in this case, the Poisson model was found to provide a good fit to this data. Assume that $Y \sim \text{POI}(\mu)$.

(a) Estimate μ from the data.
(b) Under the Poisson assumption, compute the estimated expected number of areas receiving y hits, $\hat{e}_y = nf(y, \hat{\mu})$, for each y, and compare these values to the observed values m_y.
(c) What is the estimated probability of an area receiving more than one hit?

26. Mullet (1977) suggests that the goals scored per game by the teams in the National Hockey League follow independent Poisson variables. The average numbers of goals scored per game at home and away by each team in the 1973–74 season are given in Exercise 2 of Chapter 15.
Assume a Poisson model with these means.

(a) What is the probability that Boston scores more than three goals in any away game?
(b) What is the probability that Boston scores more than six goals in two away games?
(c) What is the most likely number of goals scored by Boston in one away game?
(d) If the first eight teams play at home against the other eight teams, what is the distribution of S, the total number of home goals scored? What is the distribution of T, the total number of home and away goals scored in the eight games?
(e) What is the distribution of the total number of goals scored by Boston in a 78-game season?
(f) If Boston plays Atlanta in Atlanta, what is the probability that Boston wins? That is $P[X < Y]$, where Y represents the number of Boston goals and X represents the number of Atlanta goals.

27. The probability of a typographical error on a page is 0.005. Using a Poisson approximation:

(a) What is the expected number of errors in a 500-page book?
(b) What is the probability of having five or fewer errors in a 500-page book?
(c) What size sample (of pages) is needed to be 90% sure of finding at least one error?

28. A certain mutation occurs in one out of 1000 offspring. How many offspring must be examined to be 20% sure of observing at least one mutation?

29. Suppose that $X \sim \text{BIN}(20, 0.1)$.

(a) Compute $P[X \leqslant 5]$.
(b) Approximate $P[X \leqslant 5]$ with a Poisson distribution.
(c) Approximate $P[X \leqslant 5]$ with a normal distribution.

30. In Exercise 23, what is the probability that the barber can finish a 10-minute coffee break before the first customer shows up? What is the probability that a person arriving after 30 minutes will not get serviced? What is the mean waiting time until the first customer arrives?

31. In Exercise 24, what is the probability of at least one emission within 0.5 seconds? What is the probability that the time until the third emission is less than 0.5 seconds? What is the mean time until the third emission?

32. Suppose that the breakdowns of a repairable system occur according to a nonhomogeneous Poisson process with failure intensity $\lambda(t) = 2t/9$, where time is measured in days.

(a) What is the mean number of breakdowns in one week?
(b) What is the probability of five or fewer breakdowns in a week?
(c) What is the probability that the first breakdown will occur in less than one day?
(d) What is the average time to the first breakdown?
(e) If 10 independent systems were in operation, what would be the mean time to the first breakdown from any of the 10 systems?

33. Let $X_i \sim f_i(x_i)$ with $E(X_i) = \mu_i$ and $\text{Var}(X_i) = \sigma_i^2$. Find the mean and variance of the mixed density

$$f(x) = \sum_{i=1}^{k} p_i f_i(x)$$

34. In Exercise 26, Boston has two home games and one away game.

(a) If one of these games is selected at random, what is the probability that the number of goals scored in it will be less than or equal to 4?
(b) What is the expected number of goals scored in the game?

35. Assume that the 16 NHL teams given in Exercise 26 represent a random sample from a conceptually large population of teams, and that the number of goals scored per home game for any team selected at random follows a Poisson model for fixed μ, $X \mid \mu \sim \text{POI}(\mu)$, where $\mu \sim \text{GAM}(\gamma, 2)$.

(a) Estimate γ by using the average of the 16 at-home values to estimate the mean of the gamma distribution.
(b) What is the marginal pdf of the number of at-home goals scored by a team selected at random?
(c) Find $P[X \leqslant 4]$ using k and γ values from (a).
(d) Estimate $E(X)$ and $\text{Var}(X)$.

36. For the airplane air conditioner time-to-failure example in Section 15.5, suppose that $\lambda \sim GAM(0.0005, 20)$.

(a) What is the probability of no failure in 100 hours for a plane selected at random?

(b) What is the mean time to first failure of an airplane selected at random?

(c) What is the probability that the time to first failure is less than 100 hours?

APPENDIX A

REVIEW OF SETS

The study of probability models requires a familiarity with some of the basic notions of set theory.

A **set** is a collection of distinct objects. Other terms that sometimes are used instead of set or collection are *family* and *class*. Sets usually are designated by capital letters, A, B, C, ..., or in some instances with subscripted letters A_1, A_2, A_3, ... In describing which objects are contained in a set A, two methods are available:

1. The objects can be listed. For example, $A = \{1, 2, 3\}$ is the set consisting of the integers 1, 2, and 3.
2. A verbal description can be used. For example, the set A above consists of "the first three positive integers." A more formal way is to write $A = \{x \mid x \text{ is an integer and } 1 \leqslant x \leqslant 3\}$. More generally, if $p(x)$ is a statement about the object x, then $\{x \mid p(x)\}$ consists of all objects x such that $p(x)$ is a true statement. Thus, if $A = \{x \mid p(x)\}$, then a is in A if and only if $p(a)$ is true. This also can be related to the listing method if $p(x)$ is the statement $x = a_1$ or $x = a_2$ or ..., or $x = a_n$ when $A = \{a_1, a_2, \ldots, a_n\}$.

The individual objects in a set A are called **elements**. Other terms that sometimes are used instead of element are *member* and *point*. In the context of probability, the objects usually are called **outcomes**. When a is an element of A we write $a \in A$, and otherwise $a \notin A$. For example, $3 \in \{1, 2, 3\}$, but $4 \notin \{1, 2, 3\}$.

In most problems we can restrict attention to a specific set of elements and no others. The **universal set**, which we will denote by S, is the set of all elements under consideration. In probability applications, such a set usually is called the **sample space**, and it consists of all outcomes of some experiment that is to be performed.

Another special set, called the **empty set** or **null set**, is denoted by $\varnothing$. It is the set that contains no elements. For example $\{x \mid x \text{ is an integer and } x^2 = 2\} = \varnothing$, because the solutions $x = \pm\sqrt{2}$ are not integers.

In some cases all of the elements in a set A also are contained in another set B. If this is the case, then we say that A is a **subset** of B, denoted $A \subset B$. For example, if $A = \{1, 2, 3\}$ and $B = \{1, 2, 3, 4\}$, then $A \subset B$. It is always the case that $\varnothing \subset A \subset S$, for any set A under consideration.

There are standard ways to combine two or more sets into a new set:

1. The **intersection** of two sets A and B, denoted by $A \cap B$, is

$$A \cap B = \{x \mid x \in A \text{ and } x \in B\}$$

For example, if $A = \{1, 2, 3\}$ and $B = \{2, 3, 4\}$, then $A \cap B = \{2, 3\}$.

2. The **union** of two sets A and B, denoted by $A \cup B$, is

$$A \cup B = \{x \mid x \in A \text{ or } x \in B\}$$

For example, if A and B are the sets given in part 1, then $A \cup B = \{1, 2, 3, 4\}$.

3. The **complement** of a set A, denoted by A' or $\bar{A}$, is

$$A' = \bar{A} = \{x \mid x \in S \text{ and } x \notin A\}$$

For example, if A is the set given in part 1 and $S = \{1, 2, 3, 4, 5\}$, then $A' = \{4, 5\}$.

4. The **difference** of A and B is $A - B = A \cap B'$.

Sometimes it is convenient to use a graphical device known as a Venn diagram. Such diagrams for intersection, union, and complement are given in Figure A.1. The points inside the rectangles are associated with S, and the points inside the circles are associated with the sets A and B. The shaded regions correspond to the intersection, union, and complement respectively.

In some cases, two sets A and B have no elements in common. This can be expressed by writing $A \cap B = \varnothing$, and saying that A and B are **disjoint**. In probability applications we say that A and B are **mutually exclusive** in this case. The Venn diagram of disjoint sets corresponds to nonoverlapping circles, as shown in Figure A.2.

FIGURE A.1

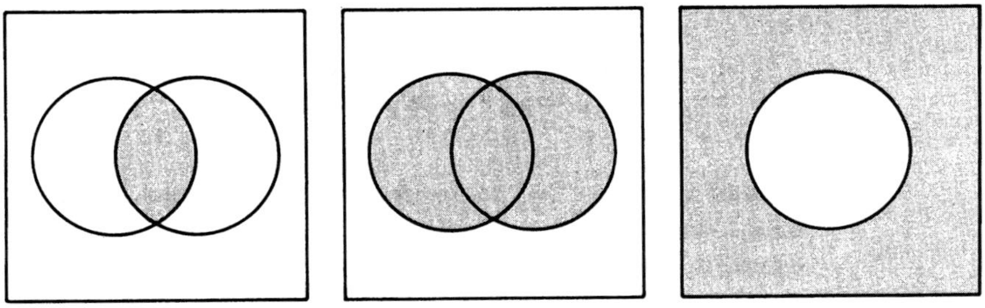

$A \cap B$ $\qquad$ $A \cup B$ $\qquad$ A'

FIGURE A.2

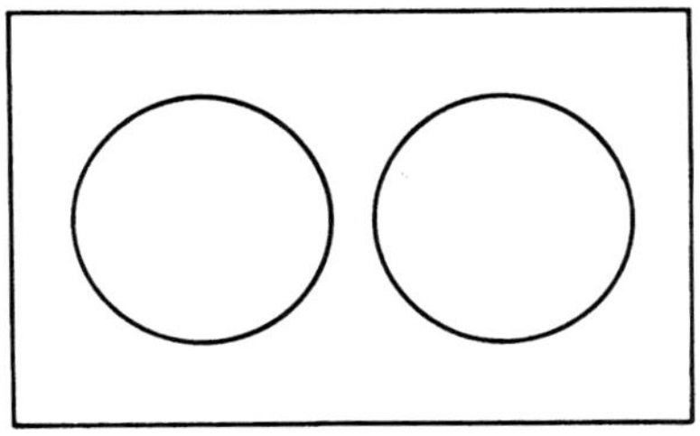

$A \cap B = \phi$

The notions of intersection and union can be extended to more than two sets. We can define the intersection and union of three sets A, B, and C to be respectively

$$A \cap B \cap C = \{x \mid x \in A \text{ and } x \in B \text{ and } x \in C\}$$

and

$$A \cup B \cup C = \{x \mid x \in A \text{ or } x \in B \text{ or } x \in C\}$$

Another way to accomplish this would be to use parentheses along with the definitions of intersection and union of two sets. For example, $A \cap B \cap C = A \cap (B \cap C) = (A \cap B) \cap C$. To avoid ambiguity, it would be desirable to establish that the way the sets are grouped with parentheses does not make a difference. This and several other properties of "set algebra" are stated in the following theorem.

Theorem A.1 For any subsets A, B, and C of S, the following equations are true:

1. $A \cup (B \cup C) = (A \cup B) \cup C$ and $A \cap (B \cap C) = (A \cap B) \cap C$.
2. $A \cup B = B \cup A$ and $A \cap B = B \cap A$.

3. $A \cup (B \cap C) = (A \cup B) \cap (A \cup C)$ and
   $A \cap (B \cup C) = (A \cap B) \cup (A \cap C)$.
4. $A \cup \emptyset = A$ and $A \cap S = A$.
5. $A \cup A' = S$ and $A \cap A' = \emptyset$. ■

Each assertion can be verified easily by Venn diagrams; however, a more formal way would involve showing that the set on either side of the inequality is included in the set on the other side. Equations 1, 2, and 3 are referred to as associative, commutative, and distributive laws, respectively.

Other useful identities are given in the following theorem.

Theorem A.2 For any subsets A and B of S, the following equations are true:

1. $(A')' = A$.
2. $\emptyset' = S$ and $S' = \emptyset$.
3. $A \cup A = A$ and $A \cap A = A$.
4. $A \cup S = S$ and $A \cap \emptyset = \emptyset$.
5. $A \cup (A \cap B) = \mathrm{A}$ and $A \cap (A \cup B) = A$.
6. $(A \cup B)' = A' \cap B'$ and $(A \cap B)' = A' \cup B'$. ■

The identities given in part 6, known as De Morgan's laws, are particularly useful in many probability applications.

A third theorem gives identities that are useful when one set is a subset of another.

Theorem A.3 The following statements about sets A and B are equivalent:

1. $A \subset B$.
2. $A \cap B = A$.
3. $A \cup B = B$. ■

Notice that property 4 of Theorem A.1 and properties 3, 4, and 5 of Theorem A.2 can be viewed as corollaries of Theorem A.3, because $\emptyset \subset A \subset S$, $A \subset A$, $A \cap B$ is a subset of both A and B, A and B are both subsets of $A \cup B$, and $A \cap B \subset A \cup B$.

The notions of intersection and union are extended easily to more than three sets, but it is more convenient in this case to use subscripted set notation $A_1, A_2, \ldots, A_n$.

1. The intersection of $A_1, A_2, \ldots, A_n$ is defined as

$$A_1 \cap A_2 \cap \cdots \cap A_n = \{x \mid x \in A_i \text{ for all } i = 1, 2, \ldots, n\}$$

2. The union of $A_1, A_2, \ldots, A_n$ is defined as

$$A_1 \cup A_2 \cup \cdots \cup A_n = \{x \mid x \in A_i \text{ for at least one } i = 1, 2, \ldots, n\}$$

More concise notations for these expressions are, respectively, $\bigcap_{i=1}^{n} A_i$ and $\bigcup_{i=1}^{n} A_i$, and the terms **finite intersection** and **finite union**, respectively usually are applied to them.

There are counterparts, in the case of n sets, to many of the properties in Theorems A.1, A.2, and A.3, but they generally are harder to state. One property that is very useful in the area of probability is a generalization of the distributive law.

Theorem A.4 If $A_1, A_2, \ldots, A_n$ and B are subsets of S, then the following equations are true:

1. $B \cap (A_1 \cup A_2 \cup \cdots \cup A_n) = (B \cap A_1) \cup (B \cap A_2) \cup \cdots \cup (B \cap A_n)$.
2. $B \cup (A_1 \cap A_2 \cap \cdots \cap A_n) = (B \cup A_1) \cap (B \cup A_2) \cap \cdots \cap (B \cup A_n)$.

■

Property 1 is the most frequently used of the two statements, because it provides a way to partition a set B into subsets. In particular, suppose that $A_1, A_2, \ldots, A_n$ are pairwise disjoint sets ($A_i \cap A_j = \varnothing$ if $i \neq j$), which also are **exhaustive** in the sense that $A_1 \cup A_2 \cup \cdots \cup A_n = S$. It can be established from the preceding theorems that

$$B = (B \cap A_1) \cup (B \cap A_2) \cup \cdots \cup (B \cap A_n)$$

which partitions B into disjoint sets $B \cap A_1, B \cap A_2, \ldots, B \cap A_n$. This partitioning also is seen easily by means of an appropriate Venn diagram, such as that in Figure A.3.

FIGURE A.3

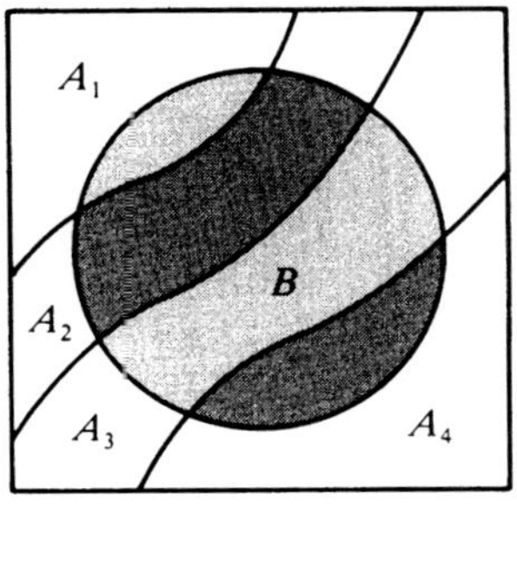

In the development of probability it often is necessary to consider higher dimensional or vector quantities. The notion of Cartesian products is useful in this regard.

If A and B are two sets, then the Cartesian product of A and B, denoted by $A \times B$, is defined to be the following set of ordered pairs:

$$A \times B = \{(x, y) \mid x \in A \text{ and } y \in B\}$$

For example, if A and B are the closed intervals $A = [1, 3] = \{x \mid x \text{ is real and } 1 \leqslant x \leqslant 3\}$ and $B = [1, 2] = \{y \mid y \text{ is real and } 1 \leqslant y \leqslant 2\}$, then $A \times B$ can be represented as a rectangle in the x–y plane, as shown in Figure A.4.

Notice that if we associate A and B with the corresponding Cartesian product sets $A^* = A \times (-\infty, \infty)$ and $B^* = (-\infty, \infty) \times B$, then the Cartesian product set $A \times B$ is identical to the intersection $A^* \cap B^*$. This correspondence is useful in certain probability problems in which an experiment consists of performing two successive steps, such as tossing a coin twice or drawing two cards from a deck.

Some problems also require higher-dimensional Cartesian product sets. If A_1, $A_2, \ldots, A_n$ are sets, then the n-fold Cartesian product consists of the following set of n-tuples:

$$A_1 \times A_2 \times \cdots \times A_n = \{(x_1, x_2, \ldots, x_n) \mid x_i \in A_i \text{ for all } i = 1, 2, \ldots, n\}$$

The question of how many elements are in a set often is of considerable importance in probability applications. A set A is said to be **finite** if its elements correspond in a one-to-one manner to the elements in a set of integers of the form $\{1, 2, \ldots, n\}$ for some positive integer n. It is said to be **countably infinite** if its elements correspond in a one-to-one manner to the elements in the set of all positive integers $\{1, 2, \ldots\}$. For example, the set of all positive *even* integers

FIGURE A.4

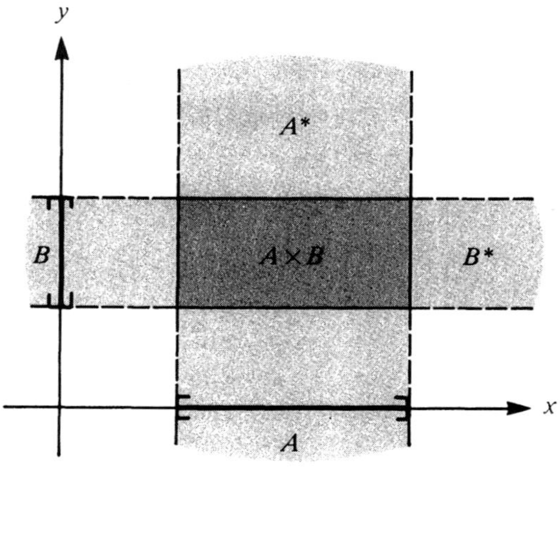

$\{2, 4, 6, \ldots\}$ is countably infinite, because each positive integer has the form $2i$ for some positive integer i. This establishes a one-to-one correspondence, $i \leftrightarrow 2i$. Although the correspondence is harder to describe, the set of all integers also can be put into one-to-one correspondence with the set of positive integers, and hence is countably infinite. A set is said to be **countable** if it is either finite or countably infinite.

The notions of intersection and union are extended easily to a countably infinite collection (or an infinite sequence) of sets.

1. The intersection of $A_1, A_2, \ldots$ is defined as

$$A_1 \cap A_2 \cap \cdots = \{x \mid x \in A_i \text{ for all } i = 1, 2, \ldots\}$$

2. The union of $A_1, A_2, \ldots$ is defined as

$$A_1 \cup A_2 \cup \cdots = \{x \mid x \in A_i \text{ for at least one } i = 1, 2, \ldots\}$$

More concise notations for these expressions are, respectively, $\bigcap_{i=1}^{\infty} A_i$ and $\bigcup_{i=1}^{\infty} A_i$.

As an example, let $A_1 = \{1\}$, and let $A_i = \{1, 2, \ldots, i\}$ for $i = 2, 3, \ldots$. Then $\bigcap_{i=1}^{\infty} A_i = \{1\}$ and $\bigcup_{i=1}^{\infty} A_i = \{1, 2, \ldots\}$. These are called, respectively, countably infinite intersections and countably infinite unions. An intersection (or union) is called a **countable intersection** (or **union**) if it is either a finite intersection (or union) or a countably infinite intersection (or union).

Infinite sets that are not countably infinite are difficult to characterize in general. However, the only ones that will be of interest in this development are intervals, Cartesian products of intervals, and finite or countably infinite unions and intersections of such sets.

APPENDIX B

SPECIAL DISTRIBUTIONS

TABLE B.1

Special discrete distributions

Name of Distribution	Notation and Parameters	Discrete pdf $f(x)$	Mean	Variance	MGF $M_X(t)$
Binomial	$X \sim \text{BIN}(n, p)$ $0 < p < 1$ $q = 1 - p$	$\binom{n}{x} p^x q^{n-x}$ $x = 0, 1, \ldots, n$	np	npq	$(pe^t + q)^n$
Bernoulli	$X \sim \text{BIN}(1, p)$ $0 < p < 1$ $q = 1 - p$	$p^x q^{1-x}$ $x = 0, 1$	p	pq	$pe^t + q$
Negative Binomial	$X \sim \text{NB}(r, p)$ $0 < p < 1$ $r = 1, 2, \ldots$	$\binom{x-1}{r-1} p^r q^{x-r}$ $x = r, r+1, \ldots$	r/p	rq/p^2	$\left(\frac{pe^t}{1 - qe^t}\right)^r$
Geometric	$X \sim \text{GEO}(p)$ $0 < p < 1$ $q = 1 - p$	pq^{x-1} $x = 1, 2, \ldots$	$1/p$	q/p^2	$\frac{pe^t}{1 - qe^t}$
Hypergeometric	$X \sim \text{HYP}(n, M, N)$ $n = 1, 2, \ldots, N$ $M = 0, 1, \ldots, N$	$\binom{M}{x}\binom{N-M}{n-x} \Big/ \binom{N}{n}$ $x = 0, 1, \ldots, n$	nM/N	$n\frac{M}{N}\left(1 - \frac{M}{N}\right)\frac{N-n}{N-1}$	*
Poisson	$X \sim \text{POI}(\mu)$ $0 < \mu$	$\frac{e^{-\mu}\mu^x}{x!}$ $x = 0, 1, \ldots$	μ	μ	$e^{\mu(e^t - 1)}$
Discrete Uniform	$X \sim \text{DU}(N)$ $N = 1, 2, \ldots$	$1/N$ $x = 1, 2, \ldots, N$	$\frac{N+1}{2}$	$\frac{N^2 - 1}{12}$	$\frac{1}{N}\frac{e^t - e^{(N+1)t}}{1 - e^t}$

* Not tractable.

TABLE B.2

Special continuous distributions

Name of Distribution	Notation and Parameters	Continuous pdf $f(x)$	Mean	Variance	MGF $M_X(t)$
Uniform	$X \sim \text{UNIF}(a, b)$ $a < b$	$\frac{1}{b-a}$ $a < x < b$	$\frac{a+b}{2}$	$\frac{(b-a)^2}{12}$	$\frac{e^{bt}-e^{at}}{(b-a)t}$
Normal	$X \sim N(\mu, \sigma^2)$ $0 < \sigma^2$	$\frac{1}{\sqrt{2\pi}\sigma} e^{-[(x-\mu)/\sigma]^2/2}$	μ	σ^2	$e^{\mu t + \sigma^2 t^2/2}$
Gamma	$X \sim \text{GAM}(\theta, \kappa)$ $0 < \theta$ $0 < \kappa$	$\frac{1}{\theta^\kappa \Gamma(\kappa)} x^{\kappa-1} e^{-x/\theta}$ $0 < x$	$\kappa\theta$	$\kappa\theta^2$	$\left(\frac{1}{1-\theta t}\right)^\kappa$
Exponential	$X \sim \text{EXP}(\theta)$ $0 < \theta$	$\frac{1}{\theta} e^{-x/\theta}$ $0 < x$	θ	θ^2	$\frac{1}{1-\theta t}$
Two-Parameter Exponential	$X \sim \text{EXP}(\theta, \eta)$ $0 < \theta$	$\frac{1}{\theta} e^{-(x-\eta)/\theta}$ $\eta < x$	$\eta + \theta$	θ^2	$\frac{e^{\eta t}}{1-\theta t}$
Double-Exponential	$X \sim \text{DE}(\theta, \eta)$ $0 < \theta$	$\frac{1}{2\theta} e^{-\lvert x-\eta \rvert/\theta}$	η	$2\theta^2$	$\frac{e^{\eta t}}{1-\theta^2 t^2}$
Weibull	$X \sim \text{WEI}(\theta, \beta)$ $0 < \theta$ $0 < \beta$ $0 < \beta$	$\frac{\beta}{\theta^\beta} x^{\beta-1} e^{-(x/\theta)^\beta}$ $0 < x$	$\theta\Gamma\left(1+\frac{1}{\beta}\right)$	$\theta^2\left[\Gamma\left(1+\frac{2}{\beta}\right) - \Gamma^2\left(1+\frac{1}{\beta}\right)\right]$	*
Extreme value	$X \sim \text{EV}(\theta, \eta)$ $0 < \theta$	$\frac{1}{\theta} \exp\{[(x-\eta)/\theta]$ $-\exp[(x-\eta)/\theta]\}$	$\eta - \gamma\theta$ $\gamma \doteq 0.5772$ (Euler's constant)	$\frac{\pi^2\theta^2}{6}$	$e^{\eta t}\Gamma(1+\theta t)$
Cauchy	$X \sim \text{CAU}(\theta, \eta)$ $0 < \theta$	$\frac{1}{\theta\pi\{1+[(x-\eta)/\theta]^2\}}$	**	**	**
Pareto	$X \sim \text{PAR}(\theta, \kappa)$ $0 < \theta$ $0 < \kappa$	$\frac{\kappa}{\theta(1+x/\theta)^{\kappa+1}}$ $0 < x$	$\frac{\theta}{\kappa-1}$ $1 < \kappa$	$\frac{\theta^2\kappa}{(\kappa-2)(\kappa-1)^2}$ $2 < \kappa$	**

TABLE B.2 (continued)

Name of Distribution	Notation and Parameters	Continuous pdf $f(x)$	Mean	Variance	MGF $M_X(t)$
Lognormal	$X \sim \text{LOGN}(\mu, \sigma^2)$ $0 < \sigma^2$	$\frac{1}{\sqrt{2\pi}x\sigma} e^{-[(\ln x - \mu)/\sigma]^2/2}$	$e^{\mu + \sigma^2/2}$	$e^{2\mu+\sigma^2}(e^{\sigma^2} - 1)$	*
Logistic	$X \sim \text{LOG}(\theta, \eta)$ $0 < \theta$	$\frac{1}{\theta} \frac{\exp[(x-\eta)/\theta]}{\{1 + \exp[(x-\eta)/\theta]\}^2}$	η	$\frac{\pi^2}{3}\theta^2$	$e^{\eta t}\pi\theta t \csc(\pi\theta t)$
Chi-Square	$X \sim \chi^2(\nu)$ $\nu = 1, 2, \ldots$	$\frac{1}{2^{\nu/2}\Gamma(\nu/2)} x^{\nu/2-1} e^{-x/2}$ $0 < x$	ν	2ν	$\left(\frac{1}{1-2t}\right)^{\nu/2}$
Student's t	$X \sim t(\nu)$ $\nu = 1, 2, \ldots$	$\frac{\Gamma\left(\frac{\nu+1}{2}\right)}{\Gamma\left(\frac{\nu}{2}\right)} \frac{1}{\sqrt{\nu\pi}} \left(1 + \frac{x^2}{\nu}\right)^{-(\nu+1)/2}$	0 $1 < \nu$	$\frac{\nu}{\nu-2}$ $2 < \nu$	**
Snedecor's F	$X \sim F(\nu_1, \nu_2)$ $\nu_1 = 1, 2, \ldots$ $\nu_2 = 1, 2, \ldots$	$\frac{\Gamma\left(\frac{\nu_1+\nu_2}{2}\right)}{\Gamma\left(\frac{\nu_1}{2}\right)\Gamma\left(\frac{\nu_2}{2}\right)} \left(\frac{\nu_1}{\nu_2}\right)^{\nu_1/2} x^{(\nu_1/2)-1}$ $\times \left(1 + \frac{\nu_1}{\nu_2}x\right)^{-(\nu_1+\nu_2)/2}$	$\frac{\nu_2}{\nu_2 - 2}$ $2 < \nu_2$	$\frac{2\nu_2^2(\nu_1 + \nu_2 - 2)}{\nu_1(\nu_2-2)^2(\nu_2-4)}$ $4 < \nu_2$	**
Beta	$X \sim \text{BETA}(a, b)$ $0 < a$ $0 < b$	$\frac{\Gamma(a+b)}{\Gamma(a)\Gamma(b)} x^{a-1}(1-x)^{b-1}$ $0 < x < 1$	$\frac{a}{a+b}$	$\frac{ab}{(a+b+1)(a+b)^2}$	*

* Not tractable.
** Does not exist.

APPENDIX C

TABLES OF DISTRIBUTIONS

The authors thank the organizations mentioned below for granting permission to use certain materials in constructing some of the tables in Appendix C:

Portions of Table 8 are excerpted from Table A-12C of Dixon & Massey, *Introduction to Statistical Analysis*, 2nd ed., McGraw-Hill Book Co., New York.

Portions of Table 9 are excerpted from Table 1 of "Modified Cramer-Von Mises Statistics for Censored Data," *Biometrika*, **63**, 1976, with the permission of the Biometrika Trustees.

Portions of Tables 10 and 11 are excerpted from Table 1 of "EDF Statistics for Goodness-of-Fit and Some Comparisons," *Journal of the American Statistical Association*, **69**, 1974.

Portions of Tables 10 and 11 are excerpted from Table 1 of "Goodness-of-Fit for the Extreme-Value Distribution," *Biometrika*, **64**, 1977, with the permission of the Biometrika Trustees.

TABLE 1

Binomial Cumulative Distribution Function

$$B(x;n,p)=\sum_{k=0}^{x}\binom{n}{k}p^k(1-p)^{n-k}$$

n	x	0.05	0.10	0.15	0.20	0.25	0.30	0.35	0.40	0.45	0.50	0.55	0.60	0.65	0.70	0.75	0.80	0.85	0.90	0.95
4	0	0.8145	0.6561	0.5220	0.4096	0.3164	0.2401	0.1785	0.1296	0.0915	0.0625	0.0410	0.0256	0.0150	0.0081	0.0039	0.0016	0.0005	0.0001	0.0000
	1	0.9860	0.9477	0.8905	0.8192	0.7383	0.6517	0.5630	0.4752	0.3910	0.3125	0.2415	0.1792	0.1265	0.0837	0.0508	0.0272	0.0120	0.0037	0.0005
	2	0.9995	0.9963	0.9880	0.9728	0.9492	0.9163	0.8735	0.8208	0.7585	0.6875	0.6090	0.5248	0.4370	0.3483	0.2617	0.1808	0.1095	0.0523	0.0140
	3	1.0000	0.9999	0.9995	0.9984	0.9961	0.9919	0.9850	0.9744	0.9590	0.9375	0.9085	0.8704	0.8215	0.7599	0.6836	0.5904	0.4780	0.3439	0.1855
5	0	0.7738	0.5905	0.4437	0.3277	0.2373	0.1681	0.1160	0.0778	0.0503	0.0312	0.0185	0.0102	0.0053	0.0024	0.0010	0.0003	0.0001	0.0000	0.0000
	1	0.9774	0.9185	0.8352	0.7373	0.6328	0.5282	0.4284	0.3370	0.2562	0.1875	0.1312	0.0870	0.0540	0.0308	0.0156	0.0067	0.0022	0.0005	0.0000
	2	0.9988	0.9914	0.9734	0.9421	0.8965	0.8369	0.7648	0.6826	0.5931	0.5000	0.4069	0.3174	0.2352	0.1631	0.1035	0.0579	0.0266	0.0086	0.0012
	3	1.0000	0.9995	0.9978	0.9933	0.9844	0.9692	0.9460	0.9130	0.8688	0.8125	0.7438	0.6630	0.5716	0.4718	0.3672	0.2627	0.1648	0.0815	0.0226
	4	1.0000	1.0000	0.9999	0.9997	0.9990	0.9976	0.9947	0.9898	0.9815	0.9688	0.9497	0.9222	0.8840	0.8319	0.7627	0.6723	0.5563	0.4095	0.2262
6	0	0.7351	0.5314	0.3771	0.2621	0.1780	0.1176	0.0754	0.0467	0.0277	0.0156	0.0083	0.0041	0.0018	0.0007	0.0002	0.0001	0.0000	0.0000	0.0000
	1	0.9672	0.8857	0.7765	0.6554	0.5339	0.4202	0.3191	0.2333	0.1636	0.1094	0.0692	0.0410	0.0223	0.0109	0.0046	0.0016	0.0004	0.0001	0.0000
	2	0.9978	0.9842	0.9527	0.0911	0.8306	0.7443	0.6471	0.5443	0.4415	0.3438	0.2553	0.1792	0.1174	0.0705	0.0376	0.0170	0.0059	0.0013	0.0001
	3	0.9999	0.9987	0.9941	0.9830	0.9624	0.9295	0.8826	0.8208	0.7447	0.6562	0.5585	0.4557	0.3529	0.2557	0.1694	0.0989	0.0473	0.0158	0.0022
	4	1.0000	0.9999	0.9996	0.9984	0.9954	0.9891	0.9777	0.9590	0.9308	0.8906	0.8364	0.7667	0.6809	0.5798	0.4661	0.3446	0.2235	0.1143	0.0328
	5	1.0000	1.0000	1.0000	0.9999	0.9998	0.9993	0.9982	0.9959	0.9917	0.9844	0.9723	0.9533	0.9246	0.8824	0.8220	0.7379	0.6229	0.4686	0.2649
7	0	0.6983	0.4783	0.3206	0.2097	0.1335	0.0824	0.0490	0.0280	0.0152	0.0078	0.0037	0.0016	0.0006	0.0002	0.0001	0.0000	0.0000	0.0000	0.0000
	1	0.9556	0.8503	0.7166	0.5767	0.4449	0.3294	0.2338	0.1586	0.1024	0.0625	0.0357	0.0188	0.0090	0.0038	0.0013	0.0004	0.0001	0.0000	0.0000
	2	0.9962	0.9743	0.9262	0.8520	0.7564	0.6471	0.5323	0.4199	0.3164	0.2266	0.1529	0.0963	0.0556	0.0288	0.0129	0.0047	0.0012	0.0002	0.0000
	3	0.9998	0.9973	0.9879	0.9667	0.9294	0.8740	0.8002	0.7102	0.6083	0.5000	0.3917	0.2898	0.1998	0.1260	0.0706	0.0333	0.0121	0.0027	0.0002
	4	1.0000	0.9998	0.9988	0.9953	0.9871	0.9712	0.9444	0.9037	0.8471	0.7734	0.6836	0.5801	0.4677	0.3529	0.2436	0.1480	0.0738	0.0257	0.0038
	5	1.0000	1.0000	0.9999	0.9996	0.9987	0.9962	0.9910	0.9812	0.9643	0.9375	0.8976	0.8414	0.7662	0.6706	0.5551	0.4233	0.2834	0.1497	0.0444
	6	1.0000	1.0000	1.0000	1.0000	0.9999	0.9998	0.9994	0.9984	0.9963	0.9922	0.9848	0.9720	0.9510	0.9176	0.8665	0.7903	0.6794	0.5217	0.3017

TABLE 1 (continued)

		p																		
n	x	0.05	0.10	0.15	0.20	0.25	0.30	0.35	0.40	0.45	0.50	0.55	0.60	0.65	0.70	0.75	0.80	0.85	0.90	0.95
8	0	0.6634	0.4305	0.2725	0.1678	0.1001	0.0576	0.0319	0.0168	0.0084	0.0039	0.0017	0.0007	0.0002	0.0001	0.0000	0.0000	0.0000	0.0000	0.0000
	1	0.9428	0.8131	0.6572	0.5033	0.3671	0.2553	0.1691	0.1064	0.0632	0.0352	0.0181	0.0085	0.0036	0.0013	0.0004	0.0001	0.0000	0.0000	0.0000
	2	0.9942	0.9619	0.8948	0.7969	0.6785	0.5518	0.4278	0.3154	0.2201	0.1445	0.0885	0.0498	0.0253	0.0113	0.0042	0.0012	0.0002	0.0000	0.0000
	3	0.9996	0.9950	0.9786	0.9437	0.8862	0.8059	0.7064	0.5941	0.4770	0.3633	0.2604	0.1737	0.1061	0.0580	0.0273	0.0104	0.0029	0.0004	0.0000
	4	1.0000	0.9996	0.9971	0.9896	0.9727	0.9420	0.8939	0.8263	0.7396	0.6367	0.5230	0.4059	0.2936	0.1941	0.1138	0.0563	0.0214	0.0050	0.0004
	5	1.0000	1.0000	0.9998	0.9988	0.9958	0.9887	0.9747	0.9502	0.9115	0.8555	0.7799	0.6846	0.5722	0.4482	0.3215	0.2031	0.1052	0.0381	0.0058
	6	1.0000	1.0000	1.0000	0.9999	0.9996	0.9987	0.9964	0.9915	0.9819	0.9648	0.9368	0.8936	0.8309	0.7447	0.6329	0.4967	0.3428	0.1869	0.0572
	7	1.0000	1.0000	1.0000	1.0000	1.0000	0.9999	0.9998	0.9993	0.9983	0.9961	0.9916	0.9832	0.9681	0.9424	0.8999	0.8322	0.7275	0.5695	0.3366
9	0	0.6302	0.3874	0.2316	0.1342	0.0751	0.0404	0.0207	0.0101	0.0046	0.0020	0.0008	0.0003	0.0001	0.0000	0.0000	0.0000	0.0000	0.0000	0.0000
	1	0.9288	0.7748	0.5995	0.4362	0.3003	0.1960	0.1211	0.0705	0.0385	0.0195	0.0091	0.0038	0.0014	0.0004	0.0001	0.0000	0.0000	0.0000	0.0000
	2	0.9916	0.9470	0.8591	0.7382	0.6007	0.4628	0.3373	0.2318	0.1495	0.0898	0.0498	0.0250	0.0112	0.0043	0.0013	0.0003	0.0000	0.0000	0.0000
	3	0.9994	0.9917	0.9661	0.9144	0.8343	0.7297	0.6089	0.4826	0.3614	0.2539	0.1658	0.0994	0.0536	0.0253	0.0100	0.0031	0.0006	0.0001	0.0000
	4	1.0000	0.9991	0.9944	0.9804	0.9511	0.9012	0.8283	0.7334	0.6214	0.5000	0.3786	0.2666	0.1717	0.0988	0.0489	0.0196	0.0056	0.0009	0.0000
	5	1.0000	0.9999	0.9994	0.9969	0.9900	0.9747	0.9464	0.9006	0.8342	0.7461	0.6386	0.5174	0.3911	0.2703	0.1657	0.0856	0.0339	0.0083	0.0006
	6	1.0000	1.0000	1.0000	0.9997	0.9987	0.9957	0.9888	0.9750	0.9502	0.9102	0.8505	0.7682	0.6627	0.5372	0.3993	0.2618	0.1409	0.0530	0.0084
	7	1.0000	1.0000	1.0000	1.0000	0.9999	0.9996	0.9986	0.9962	0.9909	0.9805	0.9615	0.9295	0.8789	0.8040	0.6997	0.5638	0.4005	0.2252	0.0712
	8	1.0000	1.0000	1.0000	1.0000	1.0000	1.0000	0.9999	0.9997	0.9992	0.9980	0.9954	0.9899	0.9793	0.9596	0.9249	0.8658	0.7684	0.6126	0.3698
10	0	0.5987	0.3487	0.1969	0.1074	0.0563	0.0282	0.0135	0.0060	0.0025	0.0010	0.0003	0.0001	0.0000	0.0000	0.0000	0.0000	0.0000	0.0000	0.0000
	1	0.9139	0.7361	0.5443	0.3758	0.2440	0.1493	0.0860	0.0464	0.0233	0.0107	0.0045	0.0017	0.0005	0.0001	0.0000	0.0000	0.0000	0.0000	0.0000
	2	0.9885	0.9298	0.8202	0.6778	0.5256	0.3828	0.2616	0.1673	0.0996	0.0547	0.0274	0.0123	0.0048	0.0016	0.0004	0.0001	0.0000	0.0000	0.0000
	3	0.9990	0.9872	0.9500	0.8791	0.7759	0.6496	0.5138	0.3823	0.2660	0.1719	0.1020	0.0548	0.0260	0.0106	0.0035	0.0009	0.0001	0.0000	0.0000
	4	0.9999	0.9984	0.9901	0.9672	0.9219	0.8497	0.7515	0.6331	0.5044	0.3770	0.2616	0.1662	0.0949	0.0473	0.0197	0.0064	0.0014	0.0001	0.0000
	5	1.0000	0.9999	0.9986	0.9936	0.9803	0.9527	0.9051	0.8338	0.7384	0.6230	0.4956	0.3669	0.2485	0.1503	0.0781	0.0328	0.0099	0.0016	0.0001
	6	1.0000	1.0000	0.9999	0.9991	0.9965	0.9894	0.9740	0.9452	0.8980	0.8281	0.7340	0.6177	0.4862	0.3504	0.2241	0.1209	0.0500	0.0128	0.0010
	7	1.0000	1.0000	1.0000	0.9999	0.9996	0.9984	0.9952	0.9877	0.9726	0.9453	0.9004	0.8327	0.7384	0.6172	0.4474	0.3222	0.1798	0.0702	0.0115
	8	1.0000	1.0000	1.0000	1.0000	1.0000	0.9999	0.9995	0.9983	0.9955	0.9893	0.9767	0.9536	0.9140	0.8507	0.7560	0.6242	0.4557	0.2639	0.0861
	9	1.0000	1.0000	1.0000	1.0000	1.0000	1.0000	1.0000	0.9999	0.9997	0.9990	0.9975	0.9940	0.9865	0.9718	0.9437	0.8926	0.8031	0.6513	0.4013

15	0	0.4633	0.2059	0.0874	0.0352	0.0134	0.0047	0.0016	0.0005	0.0001	0.0000	0.0000	0.0000	0.0000	0.0000	0.0000	0.0000	0.0000	0.0000	0.0000
	1	0.8290	0.5490	0.3186	0.1671	0.0802	0.0353	0.0142	0.0052	0.0017	0.0005	0.0001	0.0000	0.0000	0.0000	0.0000	0.0000	0.0000	0.0000	0.0000
	2	0.9638	0.8159	0.6042	0.3980	0.2361	0.1268	0.0617	0.0271	0.0107	0.0037	0.0011	0.0003	0.0001	0.0000	0.0000	0.0000	0.0000	0.0000	0.0000
	3	0.9945	0.9444	0.8227	0.6482	0.4613	0.2969	0.1727	0.0905	0.0424	0.0176	0.0063	0.0019	0.0005	0.0001	0.0000	0.0000	0.0000	0.0000	0.0000
	4	0.9994	0.9873	0.9383	0.8358	0.6865	0.5155	0.3519	0.2173	0.1204	0.0592	0.0255	0.0093	0.0028	0.0007	0.0001	0.0000	0.0000	0.0000	0.0000
	5	0.9999	0.9978	0.9832	0.9389	0.8516	0.7216	0.5643	0.4032	0.2608	0.1509	0.0769	0.0338	0.0124	0.0037	0.0008	0.0001	0.0000	0.0000	0.0000
	6	1.0000	0.9997	0.9964	0.9819	0.9434	0.8689	0.7548	0.6098	0.4522	0.3036	0.1818	0.0950	0.0422	0.0152	0.0042	0.0008	0.0001	0.0000	0.0000
	7	1.0000	1.0000	0.9994	0.9958	0.9827	0.9500	0.8868	0.7869	0.6535	0.5000	0.3465	0.2131	0.1132	0.0500	0.0173	0.0042	0.0006	0.0000	0.0000
	8	1.0000	1.0000	0.9999	0.9992	0.9958	0.9848	0.9578	0.9050	0.8182	0.6964	0.5478	0.3902	0.2452	0.1311	0.0566	0.0181	0.0036	0.0003	0.0000
	9	1.0000	1.0000	1.0000	0.9999	0.9992	0.9963	0.9876	0.9662	0.9231	0.8491	0.7392	0.5968	0.4357	0.2784	0.1484	0.0611	0.0168	0.0022	0.0001
	10	1.0000	1.0000	1.0000	1.0000	0.9999	0.9993	0.9972	0.9907	0.9745	0.9408	0.8796	0.7827	0.6481	0.4845	0.3135	0.1642	0.0617	0.0127	0.0006
	11	1.0000	1.0000	1.0000	1.0000	1.0000	0.9999	0.9995	0.9981	0.9937	0.9824	0.9576	0.9095	0.8273	0.7031	0.5387	0.3518	0.1773	0.0556	0.0055
	12	1.0000	1.0000	1.0000	1.0000	1.0000	1.0000	0.9999	0.9997	0.9989	0.9963	0.9893	0.9729	0.9383	0.8732	0.7639	0.6020	0.3958	0.1841	0.0362
	13	1.0000	1.0000	1.0000	1.0000	1.0000	1.0000	1.0000	1.0000	0.9999	0.9995	0.9983	0.9948	0.9858	0.9647	0.9198	0.8329	0.6814	0.4510	0.1710
	14	1.0000	1.0000	1.0000	1.0000	1.0000	1.0000	1.0000	1.0000	1.0000	1.0000	0.9999	0.9995	0.9984	0.9953	0.9866	0.9648	0.9126	0.7941	0.5367
20	0	0.3585	0.1216	0.0388	0.0115	0.0032	0.0008	0.0002	0.0000	0.0000	0.0000	0.0000	0.0000	0.0000	0.0000	0.0000	0.0000	0.0000	0.0000	0.0000
	1	0.7358	0.3917	0.1756	0.0692	0.0243	0.0076	0.0021	0.0005	0.0001	0.0000	0.0000	0.0000	0.0000	0.0000	0.0000	0.0000	0.0000	0.0000	0.0000
	2	0.9245	0.6769	0.4049	0.2061	0.0913	0.0355	0.0121	0.0036	0.0009	0.0002	0.0000	0.0000	0.0000	0.0000	0.0000	0.0000	0.0000	0.0000	0.0000
	3	0.9841	0.8670	0.6477	0.4114	0.2252	0.1071	0.0444	0.0160	0.0049	0.0013	0.0003	0.0000	0.0000	0.0000	0.0000	0.0000	0.0000	0.0000	0.0000
	4	0.9974	0.9568	0.8298	0.6296	0.4148	0.2375	0.1182	0.0510	0.0189	0.0059	0.0015	0.0003	0.0000	0.0000	0.0000	0.0000	0.0000	0.0000	0.0000
	5	0.9997	0.9887	0.9327	0.8042	0.6172	0.4164	0.2454	0.1256	0.0553	0.0207	0.0064	0.0016	0.0003	0.0000	0.0000	0.0000	0.0000	0.0000	0.0000
	6	1.0000	0.9976	0.9781	0.9133	0.7858	0.6080	0.4166	0.2500	0.1299	0.0577	0.0214	0.0065	0.0015	0.0003	0.0000	0.0000	0.0000	0.0000	0.0000
	7	1.0000	0.9996	0.9941	0.9679	0.8982	0.7723	0.6010	0.4159	0.2520	0.1316	0.0580	0.0210	0.0060	0.0013	0.0002	0.0000	0.0000	0.0000	0.0000
	8	1.0000	0.9999	0.9987	0.9900	0.9591	0.8867	0.7624	0.5956	0.4143	0.2517	0.1308	0.0565	0.0196	0.0051	0.0009	0.0001	0.0000	0.0000	0.0000
	9	1.0000	1.0000	0.9998	0.9974	0.9861	0.9520	0.8782	0.7553	0.5914	0.4119	0.2493	0.1275	0.0532	0.0171	0.0039	0.0006	0.0000	0.0000	0.0000
	10	1.0000	1.0000	1.0000	0.9994	0.9961	0.9829	0.9468	0.8725	0.7507	0.5881	0.4086	0.2447	0.1218	0.0480	0.0139	0.0026	0.0002	0.0000	0.0000
	11	1.0000	1.0000	1.0000	0.9999	0.9991	0.9949	0.9804	0.9435	0.8692	0.7483	0.5857	0.4044	0.2376	0.1133	0.0409	0.0100	0.0013	0.0001	0.0000
	12	1.0000	1.0000	1.0000	1.0000	0.9998	0.9987	0.9940	0.9790	0.9420	0.8684	0.7480	0.5841	0.3990	0.2277	0.1018	0.0321	0.0059	0.0004	0.0000
	13	1.0000	1.0000	1.0000	1.0000	1.0000	0.9997	0.9985	0.9935	0.9786	0.9423	0.8701	0.7500	0.5834	0.3920	0.2142	0.0867	0.0219	0.0024	0.0000
	14	1.0000	1.0000	1.0000	1.0000	1.0000	1.0000	0.9997	0.9984	0.9936	0.9793	0.9447	0.8744	0.7546	0.5836	0.3828	0.1958	0.0673	0.0113	0.0003
	15	1.0000	1.0000	1.0000	1.0000	1.0000	1.0000	1.0000	0.9997	0.9985	0.9941	0.9811	0.9490	0.8818	0.7625	0.5852	0.3704	0.1702	0.0432	0.0026
	16	1.0000	1.0000	1.0000	1.0000	1.0000	1.0000	1.0000	1.0000	0.9997	0.9987	0.9951	0.9840	0.9556	0.8929	0.7748	0.5886	0.3523	0.1330	0.0159
	17	1.0000	1.0000	1.0000	1.0000	1.0000	1.0000	1.0000	1.0000	1.0000	0.9998	0.9991	0.9964	0.9879	0.9645	0.9087	0.7939	0.5951	0.3231	0.0755
	18	1.0000	1.0000	1.0000	1.0000	1.0000	1.0000	1.0000	1.0000	1.0000	1.0000	0.9999	0.9995	0.9979	0.9924	0.9757	0.9308	0.8244	0.6083	0.2642
	19	1.0000	1.0000	1.0000	1.0000	1.0000	1.0000	1.0000	1.0000	1.0000	1.0000	1.0000	1.0000	0.9998	0.9992	0.9968	0.9885	0.9612	0.8784	0.6415
	20	1.0000	1.0000	1.0000	1.0000	1.0000	1.0000	1.0000	1.0000	1.0000	1.0000	1.0000	1.0000	1.0000	1.0000	1.0000	1.0000	1.0000	1.0000	1.0000

TABLE 2

Poisson cumulative distribution function

$$F(x; \mu) = \sum_{k=0}^{x} e^{-\mu}\mu^k/k!$$

	μ									
x	0.1	0.2	0.3	0.4	0.5	0.6	0.7	0.8	0.9	1.0
0	0.9048	0.8187	0.7408	0.6730	0.6065	0.5488	0.4966	0.4493	0.4066	0.3679
1	0.9953	0.9825	0.9631	0.9384	0.9098	0.8781	0.8442	0.8088	0.7725	0.7358
2	0.9998	0.9989	0.9964	0.9921	0.9856	0.9769	0.9659	0.9526	0.9371	0.9197
3	1.0000	0.9999	0.9997	0.9992	0.9982	0.9966	0.9942	0.9909	0.9865	0.9810
4		1.0000	1.0000	0.9999	0.9998	0.9996	0.9992	0.9986	0.9977	0.9963
5				1.0000	1.0000	1.0000	0.9999	0.9998	0.9997	0.9994
6							1.0000	1.0000	1.0000	0.9999

	μ									
x	2.0	3.0	4.0	5.0	6.0	7.0	8.0	9.0	10.0	15.0
0	0.1353	0.0498	0.0183	0.0067	0.0025	0.0009	0.0003	0.0001	0.0000	
1	0.4060	0.1991	0.0916	0.0404	0.0174	0.0073	0.0030	0.0012	0.0005	
2	0.6767	0.4232	0.2381	0.1247	0.0620	0.0296	0.0138	0.0062	0.0028	0.0000
3	0.8571	0.6472	0.4335	0.2650	0.1512	0.0818	0.0424	0.0212	0.0103	0.0002
4	0.9473	0.8153	0.6288	0.4405	0.2851	0.1730	0.0996	0.0550	0.0293	0.0009
5	0.9834	0.9161	0.7851	0.6160	0.4457	0.3007	0.1912	0.1157	0.0671	0.0028
6	0.9955	0.9665	0.8893	0.7622	0.6063	0.4497	0.3134	0.2068	0.1301	0.0076
7	0.9989	0.9881	0.9489	0.8666	0.7440	0.5987	0.4530	0.3239	0.2202	0.0180
8	0.9998	0.9962	0.9786	0.9319	0.8472	0.7291	0.5925	0.4557	0.3328	0.0374
9	1.0000	0.9989	0.9919	0.9682	0.9161	0.8305	0.7166	0.5874	0.4579	0.0699
10		0.9997	0.9972	0.9863	0.9574	0.9015	0.8159	0.7060	0.5830	0.1185
11		0.9999	0.9991	0.9945	0.9799	0.9466	0.8881	0.8030	0.6968	0.1848
12		1.0000	0.9997	0.9980	0.9912	0.9730	0.9362	0.8758	0.7916	0.2676
13			0.9999	0.9993	0.9964	0.9872	0.9658	0.9261	0.8645	0.3632
14			1.0000	0.9998	0.9986	0.9943	0.9827	0.9585	0.9165	0.4657
15				0.9999	0.9995	0.9976	0.9918	0.9780	0.9513	0.5681
16				1.0000	0.9998	0.9990	0.9963	0.9889	0.9730	0.6641
17					0.9999	0.9996	0.9984	0.9947	0.9857	0.7489
18					1.0000	0.9999	0.9994	0.9976	0.9928	0.8195
19						1.0000	0.9997	0.9989	0.9965	0.8752
20							0.9999	0.9996	0.9984	0.9170
21							1.0000	0.9998	0.9993	0.9469
22								0.9999	0.9997	0.9673
23								1.0000	0.9999	0.9805
24									1.0000	0.9888
25										0.9938
26										0.9967
27										0.9983
28										0.9991
29										0.9996
30										0.9998
31										0.9999
32										1.0000

TABLE 3

Standard normal cumulative distribution function $\Phi(z)$ and $100 \times \gamma$th percentiles z_γ

$$\Phi(z) = \int_{-\infty}^{z} \frac{1}{\sqrt{2\pi}} e^{-t^2/2}\, dt$$

z	0.00	0.01	0.02	0.03	0.04	0.05	0.06	0.07	0.08	0.09
0.0	0.5000	0.5040	0.5080	0.5120	0.5160	0.5199	0.5239	0.5279	0.5319	0.5359
0.1	0.5398	0.5438	0.5478	0.5517	0.5557	0.5596	0.5636	0.5675	0.5714	0.5753
0.2	0.5793	0.5832	0.5871	0.5910	0.5948	0.5987	0.6026	0.6064	0.6103	0.6141
0.3	0.6179	0.6217	0.6255	0.6293	0.6331	0.6368	0.6406	0.6443	0.6480	0.6517
0.4	0.6554	0.6591	0.6628	0.6664	0.6700	0.6736	0.6772	0.6808	0.6844	0.6879
0.5	0.6915	0.6950	0.6985	0.7019	0.7054	0.7088	0.7123	0.7157	0.7190	0.7224
0.6	0.7257	0.7291	0.7324	0.7357	0.7389	0.7422	0.7454	0.7486	0.7517	0.7549
0.7	0.7580	0.7611	0.7642	0.7673	0.7704	0.7734	0.7764	0.7794	0.7823	0.7852
0.8	0.7881	0.7910	0.7939	0.7967	0.7995	0.8023	0.8051	0.8078	0.8106	0.8133
0.9	0.8159	0.8186	0.8212	0.8238	0.8264	0.8289	0.8314	0.8340	0.8365	0.8389
1.0	0.8413	0.8438	0.8461	0.8485	0.8508	0.8531	0.8554	0.8577	0.8599	0.8621
1.1	0.8643	0.8665	0.8686	0.8708	0.8729	0.8749	0.8770	0.8790	0.8810	0.8830
1.2	0.8849	0.8869	0.8888	0.8907	0.8925	0.8944	0.8962	0.8980	0.8997	0.9015
1.3	0.9032	0.9049	0.9066	0.9082	0.9099	0.9115	0.9131	0.9147	0.9162	0.9177
1.4	0.9192	0.9207	0.9222	0.9236	0.9251	0.9265	0.9279	0.9292	0.9306	0.9319
1.5	0.9332	0.9345	0.9357	0.9370	0.9382	0.9394	0.9406	0.9418	0.9429	0.9441
1.6	0.9452	0.9463	0.9474	0.9484	0.9495	0.9505	0.9515	0.9525	0.9535	0.9545
1.7	0.9554	0.9564	0.9573	0.9582	0.9591	0.9599	0.9608	0.9616	0.9625	0.9633
1.8	0.9641	0.9649	0.9656	0.9664	0.9671	0.9678	0.9686	0.9693	0.9699	0.9706
1.9	0.9713	0.9719	0.9726	0.9732	0.9738	0.9744	0.9750	0.9756	0.9761	0.9767
2.0	0.9772	0.9778	0.9783	0.9788	0.9793	0.9798	0.9803	0.9808	0.9812	0.9817
2.1	0.9821	0.9826	0.9830	0.9834	0.9838	0.9842	0.9846	0.9850	0.9854	0.9857
2.2	0.9861	0.9864	0.9868	0.9871	0.9875	0.9878	0.9881	0.9884	0.9887	0.9890
2.3	0.9893	0.9896	0.9898	0.9901	0.9904	0.9906	0.9909	0.9911	0.9913	0.9916
2.4	0.9918	0.9920	0.9922	0.9925	0.9927	0.9929	0.9931	0.9932	0.9934	0.9936
2.5	0.9938	0.9940	0.9941	0.9943	0.9945	0.9946	0.9948	0.9949	0.9951	0.9952
2.6	0.9953	0.9955	0.9956	0.9957	0.9959	0.9960	0.9961	0.9962	0.9963	0.9964
2.7	0.9965	0.9966	0.9967	0.9968	0.9969	0.9970	0.9971	0.9972	0.9973	0.9974
2.8	0.9974	0.9975	0.9976	0.9977	0.9977	0.9978	0.9979	0.9979	0.9980	0.9981
2.9	0.9981	0.9982	0.9982	0.9983	0.9984	0.9984	0.9985	0.9985	0.9986	0.9986
3.0	0.9987	0.9987	0.9987	0.9988	0.9988	0.9989	0.9989	0.9989	0.9990	0.9990
3.1	0.9990	0.9991	0.9991	0.9991	0.9992	0.9992	0.9992	0.9992	0.9993	0.9993
3.2	0.9993	0.9993	0.9994	0.9994	0.9994	0.9994	0.9994	0.9995	0.9995	0.9995
3.3	0.9995	0.9995	0.9995	0.9996	0.9996	0.9996	0.9996	0.9996	0.9996	0.9997
3.4	0.9997	0.9997	0.9997	0.9997	0.9997	0.9997	0.9997	0.9997	0.9997	0.9998

γ	0.90	0.95	0.975	0.99	0.995	0.999	0.9995	0.99995	0.999995
z_γ	1.282	1.645	1.960	2.326	2.576	3.090	3.291	3.891	4.417

TABLE 4

100 × γth Percentiles $\chi^2_\gamma(\nu)$ of the chi-square distribution with ν degrees of freedom

$$\gamma = \int_0^{\chi^2_\gamma(\nu)} h(y;\, \nu)\, dy$$

ν	0.005	0.010	0.025	0.050	0.100	0.250	0.500	0.750	0.900	0.950	0.975	0.990	0.995	0.999
1					0.02	0.10	0.45	1.32	2.71	3.84	5.02	6.63	7.88	10.83
2	0.01	0.02	0.05	0.10	0.21	0.58	1.39	2.77	4.61	5.99	7.38	9.21	10.60	13.82
3	0.07	0.11	0.22	0.35	0.58	1.21	2.37	4.11	6.25	7.81	9.35	11.34	12.84	16.27
4	0.21	0.30	0.48	0.71	1.06	1.92	3.36	5.39	7.78	9.49	11.14	13.28	14.86	18.47
5	0.41	0.55	0.83	1.15	1.61	2.67	4.35	6.63	9.24	11.07	12.83	15.09	16.75	20.52
6	0.68	0.87	1.24	1.64	2.20	3.45	5.35	7.84	10.64	12.59	14.45	16.81	18.55	22.46
7	0.99	1.24	1.69	2.17	2.83	4.25	6.35	9.04	12.02	14.07	16.01	18.48	20.28	24.32
8	1.34	1.65	2.18	2.73	3.49	5.07	7.34	10.22	13.36	15.51	17.53	20.09	21.96	26.12
9	1.73	2.09	2.70	3.33	4.17	5.90	8.34	11.39	14.68	16.92	19.02	21.67	23.59	27.88
10	2.16	2.56	3.25	3.94	4.87	6.74	9.34	12.55	15.99	18.31	20.48	23.21	25.19	29.59
11	2.60	3.05	3.82	4.57	5.58	7.58	10.34	13.70	17.28	19.68	21.92	24.72	26.76	31.26
12	3.07	3.57	4.40	5.23	6.30	8.44	11.34	14.85	18.55	21.03	23.34	26.22	28.30	32.91
13	3.57	4.11	5.01	5.89	7.04	9.30	12.34	15.98	19.81	22.36	24.74	27.69	29.82	34.53
14	4.07	4.66	5.63	6.57	7.79	10.17	13.34	17.12	21.06	23.68	26.12	29.14	31.32	36.12
15	4.60	5.23	6.26	7.26	8.55	11.04	14.34	18.25	22.31	25.00	27.49	30.58	32.80	37.70

16	5.14	5.81	6.91	7.96	9.31	11.91	15.34	19.37	23.54	26.30	28.85	32.00	34.27	39.25
17	5.70	6.41	7.56	8.67	10.09	12.79	16.34	20.49	24.77	27.59	30.19	33.41	35.73	40.79
18	6.26	7.01	8.23	9.39	10.86	13.68	17.34	21.60	25.99	28.87	31.53	34.81	37.16	42.31
19	6.84	7.63	8.91	10.12	11.65	14.56	18.34	22.72	27.20	30.14	32.85	36.19	35.58	43.82
20	7.43	8.26	9.59	10.85	12.44	15.45	19.34	23.83	28.41	31.41	34.17	37.57	40.00	45.32
21	8.03	8.90	10.28	11.59	13.24	16.34	20.34	24.93	29.62	32.67	35.48	38.93	41.40	46.80
22	8.64	9.54	10.98	12.34	14.04	17.24	21.34	26.04	30.81	33.92	36.78	40.29	42.80	48.27
23	9.26	10.20	11.69	13.09	14.85	18.14	22.34	27.14	32.01	35.17	38.08	41.64	44.18	49.73
24	9.89	10.86	12.40	13.85	15.66	19.04	23.34	28.24	33.20	36.42	39.36	42.98	45.56	51.18
25	10.52	11.52	13.12	14.61	16.47	19.94	24.34	29.34	34.38	37.65	40.65	44.31	46.93	52.62
30	13.79	14.95	16.79	18.49	20.60	24.48	29.34	34.80	40.26	43.77	46.98	50.89	53.67	59.70
40	20.71	22.16	24.43	26.51	29.05	33.66	39.34	45.62	51.80	55.76	59.34	63.69	66.77	73.40
50	27.99	29.71	32.36	34.76	37.69	42.94	49.33	56.33	63.17	67.50	71.42	76.15	79.49	86.66
60	35.53	37.48	40.48	43.19	46.46	52.29	59.33	66.98	74.40	79.08	83.30	88.38	91.95	99.61
70	43.28	45.44	48.76	51.74	55.33	61.70	69.33	77.58	85.53	90.53	95.02	100.42	104.22	112.32
80	51.17	53.54	57.15	60.39	64.28	71.14	79.33	88.13	96.58	101.88	106.63	112.33	116.32	124.84
90	59.20	61.75	65.65	69.13	73.29	80.62	89.33	98.64	107.56	113.14	118.14	124.12	128.30	137.21
100	67.33	70.06	74.22	77.93	82.36	90.13	99.33	109.14	118.50	124.34	129.56	135.81	140.17	149.45

For large ν, $\chi^2_\gamma(\nu) \doteq \nu[1 - (2/9\nu) + z_\gamma \sqrt{(2/9\nu)}]^3$.

TABLE 5 **Cumulative distribution function $H(c; \nu)$ of the chi-square distribution with ν degrees of freedom**

$$H(c;\, \nu) = \int_0^c h(y;\, \nu)\, dy$$

	ν											
c	1	2	3	4	5	6	7	8	9	10	11	12
0.001	0.025											
0.002	0.036	0.001										
0.004	0.050	0.002										
0.006	0.062	0.003										
0.01	0.080	0.005										
0.02	0.112	0.010	0.001									
0.06	0.194	0.030	0.004									
0.10	0.248	0.049	0.008	0.001								
0.20	0.345	0.095	0.022	0.005	0.001							
0.60	0.561	0.259	0.104	0.037	0.012	0.004	0.001					
1.0	0.683	0.393	0.199	0.090	0.037	0.014	0.002	0.001				
1.4	0.763	0.503	0.294	0.156	0.076	0.034	0.014	0.006	0.002	0.001		
1.8	0.820	0.593	0.385	0.229	0.124	0.063	0.030	0.013	0.006	0.002	0.001	
2.2	0.862	0.667	0.468	0.301	0.179	0.100	0.052	0.026	0.012	0.005	0.002	0.001
2.6	0.893	0.727	0.543	0.373	0.239	0.143	0.081	0.043	0.022	0.011	0.005	0.002
3.0	0.917	0.777	0.608	0.442	0.300	0.191	0.115	0.066	0.036	0.019	0.009	0.004
3.4	0.935	0.817	0.666	0.507	0.361	0.243	0.154	0.093	0.054	0.030	0.016	0.008
3.8	0.949	0.850	0.716	0.566	0.421	0.296	0.198	0.125	0.076	0.044	0.025	0.013
4.2	0.960	0.878	0.759	0.620	0.479	0.350	0.244	0.161	0.102	0.062	0.036	0.020
4.6	0.968	0.900	0.796	0.669	0.533	0.404	0.291	0.201	0.132	0.084	0.051	0.030
5.0	0.975	0.918	0.828	0.713	0.584	0.456	0.340	0.242	0.166	0.109	0.069	0.042
5.4	0.980	0.933	0.855	0.751	0.631	0.506	0.389	0.286	0.202	0.137	0.090	0.057
5.8	0.984	0.945	0.878	0.785	0.674	0.554	0.437	0.330	0.240	0.168	0.114	0.074
6.2	0.987	0.955	0.898	0.815	0.713	0.599	0.483	0.375	0.280	0.202	0.140	0.094
6.6	0.990	0.963	0.914	0.841	0.748	0.641	0.528	0.420	0.321	0.237	0.170	0.117
7.0	0.992	0.970	0.928	0.864	0.779	0.679	0.571	0.463	0.362	0.275	0.201	0.142
7.4	0.994	0.975	0.940	0.884	0.807	0.715	0.612	0.506	0.404	0.313	0.234	0.170
7.8	0.995	0.980	0.950	0.901	0.832	0.747	0.649	0.547	0.446	0.352	0.269	0.199
8.2	0.996	0.983	0.958	0.915	0.854	0.776	0.685	0.586	0.486	0.391	0.305	0.231
8.6	0.997	0.986	0.965	0.928	0.874	0.803	0.717	0.623	0.525	0.430	0.341	0.263
9.0	0.997	0.989	0.971	0.939	0.891	0.826	0.747	0.658	0.563	0.468	0.378	0.297
10.0	0.998	0.993	0.981	0.960	0.925	0.875	0.811	0.735	0.650	0.560	0.470	0.384
11.0	0.999	0.996	0.988	0.973	0.949	0.912	0.861	0.798	0.724	0.642	0.557	0.471
12.0	0.999	0.998	0.993	0.983	0.965	0.938	0.899	0.849	0.787	0.715	0.636	0.554
13.0		0.998	0.995	0.989	0.977	0.957	0.928	0.888	0.837	0.776	0.707	0.631
14.0		0.999	0.997	0.993	0.984	0.970	0.949	0.918	0.878	0.827	0.767	0.699
15.0		0.999	0.998	0.995	0.990	0.980	0.964	0.941	0 909	0.868	0.818	0.759
16.0			0.999	0.997	0.993	0.986	0.975	0.958	0.933	0.900	0.859	0.809
17.0			0.999	0.998	0.996	0.991	0.983	0.970	0.951	0.926	0.892	0.850
18.0				0.999	0.997	0.994	0.988	0.979	0.965	0.945	0.918	0.884
19.0				0.999	0.998	0.996	0.992	0.985	0.975	0.960	0.939	0.911
20.0				0.999	0.999	0.997	0.994	0.990	0.982	0.971	0.955	0.933
25.0							0.999	0.998	0.997	0.995	0.991	0.985
30.0										0.999	0.998	0.997

***TABLE 5* (*continued*)**

	v											
c	13	14	15	16	17	18	19	20	21	22	23	24
2.6	0.001											
3.0	0.002	0.001										
3.4	0.004	0.002	0.001									
3.8	0.007	0.003	0.002	0.001								
4.2	0.011	0.006	0.003	0.001	0.001							
4.6	0.017	0.009	0.005	0.003	0.001	0.001						
5.0	0.025	0.014	0.008	0.004	0.002	0.001	0.001					
5.4	0.035	0.021	0.012	0.007	0.004	0.002	0.001					
5.8	0.047	0.029	0.017	0.010	0.006	0.003	0.002	0.001				
6.2	0.061	0.039	0.024	0.014	0.008	0.005	0.003	0.001	0.001			
6.6	0.078	0.051	0.032	0.020	0.012	0.007	0.004	0.002	0.001	0.001		
7.0	0.098	0.065	0.042	0.027	0.016	0.010	0.006	0.003	0.002	0.001	0.001	
7.4	0.120	0.082	0.054	0.035	0.022	0.014	0.008	0.005	0.003	0.002	0.001	
7.8	0.144	0.101	0.068	0.045	0.029	0.019	0.011	0.007	0.004	0.002	0.001	0.001
8.2	0.170	0.121	0.084	0.057	0.038	0.024	0.015	0.011	0.006	0.003	0.002	0.001
8.6	0.197	0.144	0.103	0.071	0.048	0.032	0.020	0.013	0.008	0.005	0.003	0.002
9.0	0.227	0.169	0.122	0.087	0.060	0.040	0.027	0.017	0.011	0.007	0.004	0.002
10.0	0.306	0.238	0.180	0.133	0.096	0.068	0.047	0.032	0.021	0.014	0.009	0.005
11.0	0.389	0.314	0.247	0.191	0.143	0.106	0.076	0.054	0.037	0.025	0.017	0.011
12.0	0.472	0.394	0.321	0.256	0.200	0.153	0.114	0.084	0.060	0.043	0.030	0.020
13.0	0.552	0.473	0.398	0.327	0.264	0.208	0.161	0.123	0.091	0.067	0.048	0.034
14.0	0.626	0.550	0.474	0.401	0.333	0.271	0.216	0.169	0.130	0.099	0.073	0.053
15.0	0.693	0.622	0.549	0.475	0.405	0.338	0.277	0.244	0.177	0.138	0.105	0.079
16.0	0.751	0.687	0.618	0.547	0.476	0.407	0.343	0.283	0.230	0.184	0.145	0.112
17.0	0.801	0.744	0.681	0.614	0.546	0.477	0.410	0.347	0.289	0.237	0.191	0.151
18.0	0.842	0.793	0.737	0.676	0.611	0.544	0.478	0.413	0.351	0.294	0.243	0.197
19.0	0.877	0.835	0.786	0.731	0.671	0.608	0.543	0.478	0.415	0.355	0.299	0.248
20.0	0.905	0.870	0.828	0.780	0.726	0.667	0.605	0.542	0.479	0.417	0.358	0.303
21.0	0.927	0.898	0.863	0.821	0.774	0.721	0.663	0.603	0.541	0.479	0.419	0.361
22.0	0.945	0.921	0.892	0.857	0.815	0.768	0.716	0.659	0.600	0.540	0.480	0.421
23.0	0.958	0.940	0.916	0.886	0.851	0.809	0.763	0.711	0.656	0.598	0.539	0.480
24.0	0.969	0.954	0.935	0.910	0.881	0.845	0.804	0.758	0.707	0.653	0.596	0.538
25.0	0.977	0.965	0.950	0.930	0.905	0.875	0.839	0.799	0.753	0.703	0.650	0.594
26.0	0.983	0.974	0.962	0.946	0.926	0.900	0.870	0.834	0.794	0.748	0.699	0.647
27.0	0.988	0.981	0.971	0.959	0.942	0.921	0.895	0.865	0.829	0.789	0.744	0.696
28.0	0.991	0.986	0.978	0.968	0.955	0.938	0.917	0.891	0.860	0.824	0.784	0.740
29.0	0.993	0.990	0.984	0.976	0.965	0.952	0.934	0.912	0.886	0.855	0.820	0.780
30.0	0.995	0.992	0.988	0.982	0.974	0.963	0.948	0.930	0.908	0.882	0.851	0.815
35.0	0.999	0.999	0.998	0.996	0.993	0.991	0.986	0.980	0.972	0.961	0.948	0.932
40.0				0.999	0.999	0.998	0.997	0.995	0.993	0.989	0.985	0.979
50.0										0.999	0.999	0.999

For large v, $H(c;\, v) \doteq \Phi(z)$; $z = [(c/v)^{1/3} - 1 + 2/9v]/(2/9v)^{1/2}$

TABLE 6

100 × γth Percentiles $t_\gamma(\nu)$ of Student's t distribution with ν degrees of freedom

$$\gamma = \int_{-\infty}^{t_\gamma(\nu)} f(t;\ \nu)\ dt$$

ν	γ 0.60	0.70	0.80	0.90	0.95	0.975	0.99	0.995	0.9995
1	0.325	0.727	1.376	3.078	6.314	12.706	31.821	63.657	636.619
2	0.289	0.617	1.061	1.886	2.920	4.303	6.965	9.925	31.598
3	0.277	0.584	0.978	1.638	2.353	3.182	4.541	5.841	12.924
4	0.271	0.569	0.941	1.533	2.132	2.776	3.747	4.604	8.610
5	0.267	0.559	0.920	1.476	2.015	2.571	3.365	4.032	6.869
6	0.265	0.553	0.906	1.440	1.943	2.447	3.143	3.707	5.959
7	0.263	0.549	0.896	1.415	1.895	2.365	2.998	3.499	5.408
8	0.262	0.546	0.889	1.397	1.860	2.306	2.896	3.355	5.041
9	0.261	0.543	0.883	1.383	1.833	2.262	2.821	3.250	4.781
10	0.260	0.542	0.879	1.372	1.812	2.228	2.764	3.169	4.587
11	0.260	0.540	0.876	1.363	1.796	2.201	2.718	3.106	4.437
12	0.259	0.539	0.873	1.356	1.782	2.179	2.681	3.055	4.318
13	0.259	0.538	0.870	1.350	1.771	2.160	2.650	3.012	4.221
14	0.258	0.537	0.868	1.345	1.761	2.145	2.624	2.977	4.140
15	0.258	0.536	0.866	1.341	1.753	2.131	2.602	2.947	4.073
16	0.258	0.535	0.865	1.337	1.746	2.120	2.583	2.921	4.015
17	0.257	0.534	0.863	1.333	1.740	2.110	2.567	2.898	3.965
18	0.257	0.534	0.862	1.330	1.734	2.101	2.552	2.878	3.922
19	0.257	0.533	0.861	1.328	1.729	2.093	2.539	2.861	3.883
20	0.257	0.533	0.860	1.325	1.725	2.086	2.528	2.845	3.850
21	0.257	0.532	0.859	1.323	1.721	2.080	2.518	2.831	3.819
22	0.256	0.532	0.858	1.321	1.717	2.074	2.508	2.819	3.792
23	0.256	0.532	0.858	1.319	1.714	2.069	2.500	2.807	3.767
24	0.256	0.531	0.857	1.318	1.711	2.064	2.492	2.797	3.745
25	0.256	0.531	0.856	1.316	1.708	2.060	2.485	2.787	3.725
26	0.256	0.531	0.856	1.315	1.706	2.056	2.479	2.779	3.707
27	0.256	0.531	0.855	1.314	1.703	2.052	2.473	2.771	3.690
28	0.256	0.530	0.855	1.313	1.701	2.048	2.467	2.763	3.674
29	0.256	0.530	0.854	1.311	1.699	2.045	2.462	2.756	3.659
30	0.256	0.530	0.854	1.310	1.697	2.042	2.457	2.750	3.646
40	0.255	0.529	0.851	1.303	1.684	2.021	2.423	2.704	3.551
60	0.254	0.527	0.848	1.296	1.671	2.000	2.390	2.660	3.460
120	0.254	0.526	0.845	1.289	1.658	1.980	2.358	2.617	3.373
∞	0.253	0.524	0.842	1.282	1.645	1.960	2.326	2.576	3.291

TABLE 7

100 × γth Percentiles $f_\gamma(\nu_1, \nu_2)$ of Snedecor's Distribution with ν_1 and ν_2 degrees of freedom

$$\gamma = \int_0^{f_\gamma(\nu_1, \nu_2)} g(x; \nu_1, \nu_2)\, dx$$

γ	ν_2 \ ν_1	1	2	3	4	5	6	7	8	9	10	12	15	20	30	60	120	∞
0.90		39.9	49.5	53.6	55.8	57.2	58.2	58.9	59.4	59.9	60.2	60.7	61.2	61.7	62.3	62.8	63.1	63.3
0.95		161	200	216	225	230	234	237	239	241	242	244	246	248	250	252	253	254
0.975	1	648	800	864	900	922	937	948	957	963	969	977	985	993	1,000	1,010	1,010	1,020
0.99		4,050	5,000	5,400	5,620	5,760	5,860	5,930	5,980	6,020	6,060	6,110	6,160	6,210	6,260	6,310	6,340	6,370
0.995		16,200	20,000	21,600	22,500	23,100	23,400	23,700	23,900	24,100	24,200	24,400	24,600	24,800	25,000	25,200	25,400	25,500
0.90		8.53	9.00	9.16	9.24	9.29	9.33	9.35	9.37	9.38	9.39	9.41	9.42	9.44	9.46	9.47	9.48	9.49
0.95		18.5	19.0	19.2	19.2	19.3	19.3	19.4	19.4	19.4	19.4	19.4	19.4	19.5	19.5	19.5	19.5	19.5
0.975	2	38.5	39.0	39.2	39.2	39.3	39.3	39.4	39.4	39.4	39.4	39.4	39.4	39.4	39.5	39.5	39.5	39.5
0.99		98.5	99.0	99.2	99.2	99.3	99.3	99.4	99.4	99.4	99.4	99.4	99.4	99.4	99.5	99.5	99.5	99.5
0.995		199	199	199	199	199	199	199	199	199	199	199	199	199	199	199	199	199
0.90		5.54	5.46	5.39	5.34	5.31	5.28	5.27	5.25	5.24	5.23	5.22	5.20	5.18	5.17	5.15	5.14	5.13
0.95		10.1	9.55	9.28	9.12	9.01	8.94	8.89	8.85	8.81	8.79	8.74	8.70	8.66	8.62	8.57	8.55	8.53
0.975	3	17.4	16.0	15.4	15.1	14.9	14.7	14.6	14.5	14.5	14.4	14.3	14.3	14.2	14.1	14.0	13.9	13.9
0.99		34.1	30.8	29.5	28.7	28.2	27.9	27.7	27.5	27.3	27.2	27.1	26.9	26.7	26.5	26.3	26.2	26.1
0.995		55.6	49.8	47.5	46.2	45.4	44.8	44.4	44.1	43.9	43.7	43.4	43.1	42.8	42.5	42.1	42.0	41.8
0.90		4.54	4.32	4.19	4.11	4.05	4.01	3.98	3.95	3.93	3.92	3.90	3.87	3.84	3.82	3.79	3.78	3.76
0.95		7.71	6.94	6.59	6.39	6.26	6.16	6.09	6.04	6.00	5.96	5.91	5.86	5.80	5.75	5.69	5.66	5.63
0.975	4	12.2	10.6	9.98	9.60	9.36	9.20	9.07	8.98	8.90	8.84	8.75	8.66	8.56	8.46	8.36	8.31	8.26
0.99		21.2	18.0	16.7	16.0	15.5	15.2	15.0	14.8	14.7	14.5	14.4	14.2	14.0	13.8	13.7	13.6	13.5
0.995		31.3	26.3	24.3	23.2	22.5	22.0	21.6	21.4	21.1	21.0	20.7	20.4	20.2	19.9	19.6	19.5	19.3
0.90		4.06	3.78	3.62	3.52	3.45	3.40	3.37	3.34	3.32	3.30	3.27	3.24	3.21	3.17	3.14	3.12	3.11
0.95		6.61	5.79	5.41	5.19	5.05	4.95	4.88	4.82	4.77	4.74	4.68	4.62	4.56	4.50	4.43	4.40	4.37
0.975	5	10.0	8.43	7.76	7.39	7.15	6.98	6.85	6.76	6.68	6.62	6.52	6.43	6.33	6.23	6.12	6.07	6.02
0.99		16.3	13.3	12.1	11.4	11.0	10.7	10.5	10.3	10.2	10.1	9.89	9.72	9.55	9.38	9.20	9.11	9.02
0.995		22.8	18.3	16.5	15.6	14.9	14.5	14.2	14.0	13.8	13.6	13.4	13.1	12.9	12.7	12.4	12.3	12.1
0.90		3.78	3.46	3.29	3.18	3.11	3.05	3.01	2.98	2.96	2.94	2.90	2.87	2.84	2.80	2.76	2.74	2.72
0.95		5.99	5.14	4.76	4.53	4.39	4.28	4.21	4.15	4.10	4.06	4.00	3.94	3.87	3.81	3.74	3.70	3.67
0.975	6	8.81	7.26	6.60	6.23	5.99	5.82	5.70	5.60	5.52	5.46	5.37	5.27	5.17	5.07	4.96	4.90	4.85
0.99		13.7	10.9	9.78	9.15	8.75	8.47	8.26	8.10	7.98	7.87	7.72	7.56	7.40	7.23	7.06	6.97	6.88
0.995		18.6	14.5	12.9	12.0	11.5	11.1	10.8	10.6	10.4	10.2	10.0	9.81	9.59	9.36	9.12	9.00	8.88

TABLE 7 (continued)

γ	ν_2	ν_1 = 1	2	3	4	5	6	7	8	9	10	12	15	20	30	60	120	∞
0.90		3.59	3.26	3.07	2.96	2.88	2.83	2.78	2.75	2.72	2.70	2.67	2.63	2.59	2.56	2.51	2.49	2.47
0.95		5.59	4.74	4.35	4.12	3.97	3.87	3.79	3.73	3.68	3.64	3.57	3.51	3.44	3.38	3.30	3.27	3.23
0.975	7	8.07	6.54	5.89	5.52	5.29	5.12	4.99	4.90	4.82	4.76	4.67	4.57	4.47	4.36	4.25	4.20	4.14
0.99		12.2	9.55	8.45	7.85	7.46	7.19	6.99	6.84	6.72	6.62	6.47	6.31	6.16	5.99	5.82	5.74	5.65
0.995		16.2	12.4	10.9	10.1	9.52	9.16	8.89	8.68	8.51	8.38	8.18	7.97	7.75	7.53	7.31	7.19	7.08
0.90		3.46	3.11	2.92	2.81	2.73	2.67	2.62	2.59	2.56	2.54	2.50	2.46	2.42	2.38	2.34	2.31	2.29
0.95		5.32	4.46	4.07	3.84	3.69	3.58	3.50	3.44	3.39	3.35	3.28	3.22	3.15	3.08	3.01	2.97	2.93
0.975	8	7.57	6.06	5.42	5.05	4.82	4.65	4.53	4.43	4.36	4.30	4.20	4.10	4.00	3.89	3.78	3.73	3.67
0.99		11.3	8.65	7.59	7.01	6.63	6.37	6.18	6.03	5.91	5.81	5.67	5.52	5.36	5.20	5.03	4.95	4.86
0.995		14.7	11.0	9.60	8.81	8.30	7.95	7.69	7.50	7.34	7.21	7.01	6.81	6.61	6.40	6.18	6.06	5.95
0.90		3.36	3.01	2.81	2.69	2.61	2.55	2.51	2.47	2.44	2.42	2.38	2.34	2.30	2.25	2.21	2.18	2.16
0.95		5.12	4.26	3.86	3.63	3.48	3.37	3.29	3.23	3.18	3.14	3.07	3.01	2.94	2.86	2.79	2.75	2.71
0.975	9	7.21	5.71	5.08	4.72	4.48	4.32	4.20	4.10	4.03	3.96	3.87	3.77	3.67	3.56	3.45	3.39	3.33
0.99		10.6	8.02	6.99	6.42	6.06	5.80	5.61	5.47	5.35	5.26	5.11	4.96	4.81	4.65	4.48	4.40	4.31
0.995		13.6	10.1	8.72	7.96	7.47	7.13	6.88	6.69	6.54	6.42	6.23	6.03	5.83	5.62	5.41	5.30	5.19
0.90		3.29	2.92	2.73	2.61	2.52	2.46	2.41	2.38	2.35	2.32	2.28	2.24	2.20	2.15	2.11	2.08	2.06
0.95		4.96	4.10	3.71	3.48	3.33	3.22	3.14	3.07	3.02	2.98	2.91	2.84	2.77	2.70	2.62	2.58	2.54
0.975	10	6.94	5.46	4.83	4.47	4.24	4.07	3.95	3.85	3.78	3.72	3.62	3.52	3.42	3.31	3.20	3.14	3.08
0.99		10.0	7.56	6.55	5.99	5.64	5.39	5.20	5.06	4.94	4.85	4.71	4.56	4.41	4.25	4.08	4.00	3.91
0.995		12.8	9.43	8.08	7.34	6.87	6.54	6.30	6.12	5.97	5.85	5.66	5.47	5.27	5.07	4.86	4.75	4.64
0.90		3.18	2.81	2.61	2.48	2.39	2.33	2.28	2.24	2.21	2.19	2.15	2.10	2.06	2.01	1.96	1.93	1.90
0.95		4.75	3.89	3.49	3.26	3.11	3.00	2.91	2.85	2.80	2.75	2.69	2.62	2.54	2.47	2.38	2.34	2.30
0.975	12	6.55	5.10	4.47	4.12	3.89	3.73	3.61	3.51	3.44	3.37	3.28	3.18	3.07	2.96	2.85	2.79	2.72
0.99		9.33	6.93	5.95	5.41	5.06	4.82	4.64	4.50	4.39	4.30	4.16	4.01	3.86	3.70	3.54	3.45	3.36
0.995		11.8	8.51	7.23	6.52	6.07	5.76	5.52	5.35	5.20	5.09	4.91	4.72	4.53	4.33	4.12	4.01	3.90
0.90		3.07	2.70	2.49	2.36	2.27	2.21	2.16	2.12	2.09	2.06	2.02	1.97	1.92	1.87	1.82	1.79	1.76
0.95		4.54	3.68	3.29	3.06	2.90	2.79	2.71	2.64	2.59	2.54	2.48	2.40	2.33	2.25	2.16	2.11	2.07
0.975	15	6.20	4.77	4.15	3.80	3.58	3.41	3.29	3.20	3.12	3.06	2.96	2.86	2.76	2.64	2.52	2.46	2.40
0.99		8.68	6.36	5.42	4.89	4.56	4.32	4.14	4.00	3.89	3.80	3.67	3.52	3.37	3.21	3.05	2.96	2.87
0.995		10.8	7.70	6.48	5.80	5.37	5.07	4.85	4.67	4.54	4.42	4.25	4.07	3.88	3.69	3.48	3.37	3.26

0.90		2.97	2.59	2.38	2.25	2.16	2.09	2.04	2.00	1.96	1.94	1.89	1.84	1.79	1.74	1.68	1.64	1.61
0.95		4.35	3.49	3.10	2.87	2.71	2.60	2.51	2.45	2.39	2.35	2.28	2.20	2.12	2.04	1.95	1.90	1.84
0.975	20	5.87	4.46	3.86	3.51	3.29	3.13	3.01	2.91	2.84	2.77	2.68	2.57	2.46	2.35	2.22	2.16	2.09
0.99		8.10	5.85	4.94	4.43	4.10	3.87	3.70	3.56	3.46	3.37	3.23	3.09	2.94	2.78	2.61	2.52	2.42
0.995		9.94	6.99	5.82	5.17	4.76	4.47	4.26	4.09	3.96	3.85	3.68	3.50	3.32	3.12	2.92	2.81	2.69
0.90		2.88	2.49	2.28	2.14	2.05	1.98	1.93	1.88	1.85	1.82	1.77	1.72	1.67	1.61	1.54	1.50	1.46
0.95		4.17	3.32	2.92	2.69	2.53	2.42	2.33	2.27	2.21	2.16	2.09	2.01	1.93	1.84	1.74	1.68	1.62
0.975	30	5.57	4.18	3.59	3.25	3.03	2.87	2.75	2.65	2.57	2.51	2.41	2.31	2.20	2.07	1.94	1.87	1.79
0.99		7.56	5.39	4.51	4.02	3.70	3.47	3.30	3.17	3.07	2.98	2.84	2.70	2.55	2.39	2.21	2.11	2.01
0.995		9.18	6.35	5.24	4.62	4.23	3.95	3.74	3.58	3.45	3.34	3.18	3.01	2.82	2.63	2.42	2.30	2.18
0.90		2.79	2.39	2.18	2.04	1.95	1.87	1.82	1.77	1.74	1.71	1.66	1.60	1.54	1.48	1.40	1.35	1.29
0.95		4.00	3.15	2.76	2.53	2.37	2.25	2.17	2.10	2.04	1.99	1.92	1.84	1.75	1.65	1.53	1.47	1.39
0.975	60	5.29	3.93	3.34	3.01	2.79	2.63	2.51	2.41	2.33	2.27	2.17	2.06	1.94	1.82	1.67	1.58	1.48
0.99		7.08	4.98	4.13	3.65	3.34	3.12	2.95	2.82	2.72	2.63	2.50	2.53	2.20	2.03	1.84	1.73	1.60
0.995		8.49	5.80	4.73	4.14	3.76	3.49	3.29	3.13	3.01	2.90	2.74	2.57	2.39	2.19	1.96	1.83	1.69
0.90		2.75	2.35	2.13	1.99	1.90	1.82	1.77	1.72	1.68	1.65	1.60	1.54	1.48	1.41	1.32	1.26	1.19
0.95		3.92	3.07	2.68	2.45	2.29	2.18	2.09	2.02	1.96	1.91	1.83	1.75	1.66	1.55	1.43	1.35	1.25
0.975	120	5.15	3.80	3.23	2.89	2.67	2.52	2.39	2.30	2.22	2.16	2.05	1.94	1.82	1.69	1.53	1.43	1.31
0.99		6.85	4.79	3.95	3.48	3.17	2.96	2.79	2.66	2.56	2.47	2.34	2.19	2.03	1.86	1.66	1.53	1.38
0.995		8.18	5.54	4.50	3.92	3.55	3.28	3.09	2.93	2.81	2.71	2.54	2.37	2.19	1.98	1.75	1.61	1.43
0.90		2.71	2.30	2.08	1.94	1.85	1.77	1.72	1.67	1.63	1.60	1.55	1.49	1.42	1.34	1.24	1.17	1.00
0.95		3.84	3.00	2.60	2.37	2.21	2.10	2.01	1.94	1.88	1.83	1.75	1.67	1.57	1.46	1.32	1.22	1.00
0.975	∞	5.02	3.69	3.12	2.79	2.57	2.41	2.29	2.19	2.11	2.05	1.94	1.83	1.71	1.57	1.39	1.27	1.00
0.99		6.63	4.61	3.78	3.32	3.02	2.80	2.64	2.51	2.41	2.32	2.18	2.04	1.88	1.70	1.47	1.32	1.00
0.995		7.88	5.30	4.28	3.72	3.35	3.09	2.90	2.74	2.62	2.52	2.36	2.19	2.00	1.79	1.53	1.36	1.00

$f_{1-\gamma}(\nu_1, \nu_2) = 1/f_\gamma(\nu_2, \nu_1)$

TABLE 8

Sample size for *t* test

Sample size n to achieve power $1-\beta$ for $d = |\mu - \mu_0|/\sigma$ in one-sample case and $n = n_1 = n_2$ for $d' = |\mu_1 - \mu_2|/\sigma$ in two-sample case, for one sided test at significance level α. These are approximate n for two-sided test at significance level 2α.

α (2α)	d	One-Sample Test							Two-Sample Test							d'
		$1-\beta$														
		0.50	0.60	0.70	0.80	0.90	0.95	0.99	0.50	0.60	0.70	0.80	0.90	0.95	0.99	
0.005 (0.01)	0.1	669	805	966	1173	1493	1785	2403	1327	1602	1922	2337	2977	3567	4806	0.1
	0.2	169	204	244	296	377	450	605	333	403	484	588	749	894	1206	0.2
	0.4	42	53	64	77	97	115	154	87	101	124	150	189	226	304	0.4
	0.6	22	26	31	36	45	53	71	37	44	55	65	85	100	138	0.6
	0.8	14	16	19	22	27	32	41	23	27	32	39	49	56	85	0.8
	1.0	10	12	13	14	19	22	28	15	18	21	26	32	38	49	1.0
	1.2	8	9	10	12	14	16	21	11	13	16	18	23	27	36	1.2
	1.4	7	8	9	10	12	13	16	9	10	12	14	18	20	27	1.4
	1.6	6	7	8	8	10	11	13	7	9	10	11	14	16	21	1.6
	1.8	6	6	7	8	9	9	11	6	7	8	9	11	13	17	1.8
	2.0	4	6	6	7	8	8	10	6	6	7	8	10	11	14	2.0
	3.0	3	4	5	5	6	6	7	4	4	5	5	6	6	8	3.0
0.0125 (0.025)	0.1	506	625	768	954	1245	1514	2090	1004	1244	1529	1901	2482	3020	4180	0.1
	0.2	127	158	194	240	312	380	524	255	316	388	482	629	766	1057	0.2
	0.4	34	42	52	63	81	98	135	64	81	97	120	157	191	265	0.4
	0.6	17	19	24	30	37	45	61	30	37	44	54	71	84	117	0.6
	0.8	11	13	15	18	23	27	36	18	21	26	32	40	49	69	0.8
	1.0	8	9	11	13	16	18	24	12	14	17	21	27	32	44	1.0
	1.2	7	7	8	10	12	14	18	9	11	13	15	19	23	31	1.2
	1.4	6	6	7	8	10	11	14	7	8	10	12	15	17	23	1.4
	1.6	5	6	6	7	8	9	11	6	7	8	9	12	14	18	1.6
	1.8		5	6	6	7	8	10	5	6	7	8	10	11	15	1.8
	2.0			5	6	6	7	9	4	5	6	7	8	9	12	2.0
	3.0							6		3	4	4	5	5	7	3.0
0.025 (0.05)	0.1	386	492	619	788	1054	1302	1840	771	982	1237	1574	2106	2603	3680	0.1
	0.2	98	124	156	201	265	327	459	193	245	310	395	527	650	922	0.2
	0.4	26	33	41	52	68	85	117	49	62	80	100	133	164	231	0.4
	0.6	13	16	20	24	32	39	53	23	29	36	45	60	77	104	0.6
	0.8	9	10	12	14	18	23	31	14	17	21	26	34	42	59	0.8
	1.0	6	7	9	10	13	16	21	9	11	14	17	22	28	38	1.0
	1.2	6	6	7	8	10	12	15	7	8	10	12	16	19	27	1.2
	1.4	5	5	6	7	8	9	12	5	7	8	10	12	15	20	1.4
	1.6			5	6	7	8	10	5	5	6	8	10	12	16	1.6
	1.8				5	6	7	8	4	4	5	6	8	10	13	1.8
	2.0					5	6	7	4	4	5	6	7	8	11	2.0
	3.0											4	4	5	6	3.0
0.05 (0.10)	0.1	272	362	473	620	858	1084	1580	543	722	943	1235	1715	2166	3160	0.1
	0.2	69	92	119	156	215	272	396	138	182	237	312	430	543	793	0.2
	0.4	19	24	31	41	55	70	101	35	46	59	79	109	137	199	0.4
	0.6	9	12	15	19	26	32	46	16	21	28	35	48	62	89	0.6
	0.8	6	8	9	12	15	19	27	10	12	16	20	28	35	50	0.8
	1.0	5	6	7	8	11	13	18	7	8	11	13	18	23	33	1.0
	1.2		5	5	6	8	10	13	5	6	8	10	13	16	23	1.2
	1.4				5	6	8	10	4	5	6	7	10	12	17	1.4
	1.6					6	6	8	4	4	5	6	8	10	14	1.6
	1.8					5	6	7		4	4	5	7	8	11	1.8
	2.0						5	6			4	4	6	7	9	2.0
	3.0												3	4	5	3.0

TABLE 9 Upper critical values $CM_{1-\alpha}$ for completely specified H_0

n	r/n	α 0.10	0.05	0.025	0.01
∞	0.5	0.189	0.258	0.330	0.427
	0.6	0.241	0.327	0.417	0.539
	0.7	0.286	0.386	0.491	0.633
	0.8	0.321	0.430	0.544	0.700
	0.9	0.341	0.455	0.573	0.735
	1.0	0.347	0.461	0.581	0.743

TABLE 10 Critical values for CVM test of $H_0: X \sim F(x)$ with parameters estimated

H_0	Statistic	$1-\alpha$ 0.90	0.95	0.975	0.99
EXP(θ)	$(1+0.16/n)\widehat{CM}$	0.177	0.224	0.273	0.337
WEI(θ, β)	$(1+0.2/\sqrt{n})\widehat{CM}$	0.102	0.124	0.146	0.175
N(μ, σ^2)	$(1+0.5/n)\widehat{CM}$	0.104	0.126	0.148	0.178

TABLE 11 Critical values for KS and Kuiper statistics

H_0	Statistic	$1-\alpha$ 0.90	0.95	0.975	0.99
$F(x)$	$(\sqrt{n}+0.12+0.11/\sqrt{n})D$	1.224	1.358	1.480	1.628
EXP(θ)	$(\sqrt{n}+0.26+0.5/\sqrt{n})(\hat{D}-0.2/n)$	0.995	1.094	1.184	1.298
N(μ, σ^2)	$(\sqrt{n}-0.1+0.85/\sqrt{n})\hat{D}$	0.819	0.895	0.955	1.035
WEI(θ, β)	$\sqrt{n}\hat{D}$	0.803	0.874	0.939	1.007
$F(x)$	$(\sqrt{n}+0.155+0.24/\sqrt{n})V$	1.620	1.747	1.862	2.001
EXP(θ)	$(\sqrt{n}+0.24+0.35/\sqrt{n})(\hat{V}-0.2/n)$	1.527	1.655	1.774	1.910
N(μ, σ^2)	$(\sqrt{n}+0.05+0.82/\sqrt{n})\hat{V}$	1.386	1.489	1.585	1.693
WEI(θ, β)	$\sqrt{n}\hat{V}$	1.372	1.477	1.557	1.671

TABLE 12 **Critical values t_α for the Wilcoxon signed-rank test, $P[T \leqslant t_\alpha] \leqslant \alpha$**

	α						
n	0.005	0.01	0.025	0.05	0.10	0.20	0.50
4					0	2	4
5				0	2	3	7
6			0	2	3	5	10
7		0	2	3	5	8	13
8	0	1	3	5	8	11	17
9	1	3	5	8	10	14	22
10	3	5	8	10	14	18	27
11	5	7	10	13	17	22	32
12	7	9	13	17	21	27	38
13	9	12	17	21	26	32	45
14	12	15	21	25	31	38	52
15	15	19	25	30	36	44	59
16	19	23	29	35	42	50	67
17	23	27	34	41	48	57	76
18	27	32	40	47	55	65	85
19	32	37	46	53	62	73	94
20	37	43	52	60	69	81	104

TABLE 13A Values of $P[U \leqslant u]$ for the Mann-Whitney statistic with $m = \min(n_1, n_2)$, $n = \max(n_1, n_2)$, $2 \leqslant m \leqslant n \leqslant 8$

u	m: 2 n: 3	4	5	6	7	8	3 3	4	5	6	7	8
0	0.100	0.067	0.047	0.036	0.028	0.022	0.050	0.028	0.018	0.012	0.008	0.006
1	0.200	0.133	0.095	0.071	0.056	0.044	0.100	0.057	0.036	0.024	0.017	0.012
2	0.400	0.267	0.190	0.143	0.111	0.089	0.200	0.114	0.071	0.048	0.033	0.024
3	0.600	0.400	0.286	0.214	0.167	0.133	0.350	0.200	0.125	0.083	0.058	0.042
4		0.600	0.429	0.321	0.250	0.200	0.500	0.314	0.196	0.131	0.092	0.067
5			0.571	0.429	0.333	0.267	0.650	0.429	0.286	0.190	0.133	0.097

u	m: 4 n: 4	5	6	7	8	5 5	6	7	8	6 6	7	8
1	0.014	0.008	0.005	0.003	0.002	0.004	0.002	0.001	0.001	0.001	0.001	0.000
1	0.029	0.016	0.010	0.006	0.004	0.008	0.004	0.003	0.002	0.002	0.001	0.001
2	0.057	0.032	0.019	0.012	0.008	0.016	0.009	0.005	0.003	0.004	0.002	0.001
3	0.100	0.056	0.033	0.021	0.014	0.028	0.015	0.009	0.005	0.008	0.004	0.002
4	0.171	0.095	0.057	0.036	0.024	0.048	0.026	0.015	0.009	0.013	0.007	0.004
5	0.243	0.143	0.086	0.055	0.036	0.075	0.041	0.024	0.015	0.021	0.011	0.006
6	0.343	0.206	0.129	0.082	0.055	0.111	0.063	0.037	0.023	0.032	0.017	0.010
7	0.443	0.278	0.176	0.115	0.077	0.155	0.089	0.053	0.033	0.047	0.026	0.015
8	0.557	0.365	0.238	0.158	0.107	0.210	0.123	0.074	0.047	0.066	0.037	0.021
9		0.452	0.305	0.206	0.141	0.274	0.165	0.101	0.064	0.090	0.051	0.030
10		0.548	0.381	0.264	0.184	0.345	0.214	0.134	0.085	0.120	0.069	0.041
11			0.457	0.324	0.230	0.421	0.268	0.172	0.111	0.155	0.090	0.054
12			0.545	0.394	0.285	0.500	0.331	0.216	0.142	0.197	0.117	0.071
13				0.464	0.341	0.579	0.396	0.265	0.177	0.242	0.147	0.091
14				0.538	0.404		0.465	0.319	0.217	0.294	0.183	0.114

u	m: 7 n: 7	8	8 8
0	0.000	0.000	0.000
1	0.001	0.000	0.000
2	0.001	0.001	0.000
3	0.002	0.001	0.001
4	0.003	0.002	0.001
5	0.006	0.003	0.001
6	0.009	0.005	0.002
7	0.013	0.007	0.003
8	0.019	0.010	0.005
9	0.027	0.014	0.007
10	0.036	0.020	0.010

u	m: 7 n: 7	8	8 8
11	0.049	0.027	0.014
12	0.064	0.036	0.019
13	0.082	0.047	0.025
14	0.104	0.060	0.032
15	0.130	0.076	0.041
16	0.159	0.095	0.052
17	0.191	0.116	0.065
18	0.228	0.140	0.080
19	0.267	0.168	0.097
20	0.310	0.198	0.117
21	0.355	0.232	0.139

TABLE 13B **Critical values u_α such that $P[U \leqslant u_\alpha] \leqslant \alpha$ for the Mann-Whitney statistic with $m = \min(n_1, n_2)$, $n = \max(n_1, n_2)$, $2 \leqslant m \leqslant n \leqslant 20$**

α	$m = 2$, $n = 9$	10	11	12	13	14	$m = 3$, 9	10	11	12	13	14
0.01					0	0	1	1	1	2	2	2
0.025	0	0	0	1	1	1	2	3	3	4	4	5
0.05	1	1	1	2	2	3	4	4	5	5	6	7
0.10	2	3	3	4	4	4	5	6	7	8	9	10
	$m = 4$						$m = 5$					
0.01	3	3	4	5	5	6	5	6	7	8	9	10
0.025	4	5	6	7	8	9	7	8	9	11	12	13
0.05	6	7	8	9	10	11	9	11	12	13	15	16
0.10	9	10	11	12	13	15	12	13	15	17	18	20
	$m = 6$						$m = 7$					
0.01	7	8	9	11	12	13	9	11	12	14	16	17
0.025	10	11	13	14	16	17	12	14	16	18	20	22
0.05	12	14	16	17	19	21	15	17	19	21	24	26
0.10	15	17	19	21	23	25	18	21	23	26	28	31
	$m = 8$						$m = 9$					
0.01	11	13	15	17	20	22	14	16	18	21	23	26
0.025	15	17	19	22	24	26	17	20	23	26	28	31
0.05	18	20	23	26	28	31	21	24	27	30	33	36
0.10	22	24	27	30	33	36	25	28	31	35	38	41

α	$m = 10$, $n = 10$	11	12	13	14	$m = 11$, 11	12	13	14	$m = 12$, 12	13	14
0.01	19	22	24	27	30	25	28	31	34	31	35	38
0.025	23	26	29	33	36	30	33	37	40	37	41	45
0.05	27	31	34	37	41	34	38	42	46	42	47	51
0.10	32	36	39	43	47	40	44	48	52	49	53	58

α	$m = 13$, $n = 13$	14	$m = 14$, $n = 14$	$m = 15$, $n = 15$	$m = 20$, $n = 20$
0.01	39	43	47	56	114
0.025	45	50	55	64	127
0.05	51	56	61	72	138
0.10	58	63	69	80	151

$$u_\alpha \doteq \frac{n_1 n_2}{2} - z_{1-\alpha}\sqrt{n_1 n_2(n_1 + n_2 + 1)/12}$$

TABLE 14

p Values for Spearman's rank correlation coefficient

$$P[R_s \leqslant -r] = P[R_s \geqslant r] = p$$

$n = 3$

r	p
1.000	0.167
0.500	0.500

$n = 4$

r	p
1.000	0.042
0.800	0.167
0.600	0.208
0.400	0.375
0.200	0.458
0.000	0.542

$n = 5$

r	p
1.000	0.008
0.900	0.042
0.800	0.067
0.700	0.117
0.600	0.175
0.500	0.225
0.400	0.258
0.300	0.342
0.200	0.392
0.100	0.475
0.000	0.525

$n = 6$

r	p
1.000	0.001
0.943	0.008
0.886	0.017
0.829	0.029
0.771	0.051
0.714	0.068
0.657	0.088
0.600	0.121
0.543	0.149
0.486	0.178
0.429	0.210
0.371	0.249
0.314	0.282
0.257	0.329
0.200	0.357
0.143	0.401
0.086	0.460
0.029	0.500

$n = 7$

r	p
1.000	0.000
0.964	0.001
0.929	0.003
0.893	0.006
0.857	0.012
0.821	0.017
0.786	0.024
0.750	0.033
0.714	0.044
0.679	0.055
0.643	0.069
0.607	0.083
0.571	0.100
0.536	0.118
0.500	0.133
0.464	0.151
0.429	0.177
0.393	0.198
0.357	0.222
0.321	0.249
0.286	0.278
0.250	0.297
0.214	0.331
0.179	0.357
0.143	0.391
0.107	0.420
0.071	0.453
0.036	0.482
0.000	0.518

$n = 8$

r	p
1.000	0.000
0.976	0.000
0.952	0.001
0.929	0.001
0.905	0.002
0.881	0.004
0.857	0.005
0.833	0.008
0.810	0.001
0.786	0.014
0.762	0.018
0.738	0.023
0.714	0.029
0.690	0.035
0.667	0.042
0.643	0.048
0.619	0.057
0.595	0.066
0.571	0.076
0.548	0.085
0.524	0.098
0.500	0.108
0.476	0.122
0.452	0.134
0.429	0.150
0.405	0.163
0.381	0.180
0.357	0.195
0.333	0.214
0.310	0.231
0.286	0.250
0.262	0.268
0.238	0.291
0.214	0.310
0.190	0.332
0.167	0.352
0.143	0.376
0.119	0.397
0.095	0.420
0.071	0.441
0.048	0.467
0.024	0.488
0.000	0.512

$n = 9$

r	p
1.000	0.000
0.983	0.000
0.967	0.000
0.950	0.000
0.933	0.000
0.917	0.001
0.900	0.001
0.883	0.002
0.867	0.002
0.850	0.003
0.833	0.004
0.817	0.005
0.800	0.007
0.783	0.009
0.767	0.011
0.750	0.013
0.733	0.016
0.717	0.018
0.700	0.022
0.683	0.025
0.667	0.029
0.650	0.033
0.633	0.038
0.617	0.043
0.600	0.048
0.583	0.054
0.567	0.060
0.550	0.066
0.533	0.074
0.517	0.081
0.500	0.089
0.483	0.097
0.467	0.106
0.450	0.115
0.433	0.125
0.417	0.135
0.400	0.146
0.383	0.156
0.367	0.168
0.350	0.179
0.333	0.193
0.317	0.205
0.300	0.218
0.283	0.231
0.267	0.247
0.250	0.260
0.233	0.276
0.217	0.290
0.200	0.307
0.183	0.322
0.167	0.339
0.150	0.354
0.133	0.372
0.117	0.388
0.100	0.405
0.083	0.422
0.067	0.440
0.050	0.456
0.033	0.474
0.017	0.491
0.000	0.509

$n = 10$

r	p
1.000	0.000
0.988	0.000
0.976	0.000
0.964	0.000
0.952	0.000
0.939	0.000
0.927	0.000
0.915	0.000
0.903	0.000
0.891	0.001
0.879	0.001
0.867	0.001
0.855	0.001
0.842	0.002
0.830	0.002
0.818	0.003
0.806	0.004
0.794	0.004
0.782	0.005
0.770	0.007
0.758	0.008
0.745	0.009
0.733	0.010
0.721	0.012
0.709	0.013
0.697	0.015
0.685	0.017
0.673	0.019
0.661	0.022
0.648	0.025
0.636	0.027
0.624	0.030
0.612	0.033
0.600	0.037
0.588	0.040
0.576	0.044
0.564	0.048
0.552	0.052
0.539	0.057
0.527	0.062
0.515	0.067
0.503	0.072
0.491	0.077
0.479	0.083
0.467	0.089
0.455	0.096
0.442	0.102
0.430	0.109
0.418	0.116
0.406	0.124
0.394	0.132
0.382	0.139
0.370	0.148
0.358	0.156
0.345	0.165
0.333	0.174
0.321	0.184
0.309	0.193
0.297	0.203
0.285	0.214
0.273	0.224
0.261	0.235
0.248	0.246
0.236	0.257
0.224	0.268
0.212	0.280
0.200	0.292
0.188	0.304
0.176	0.316
0.164	0.328
0.152	0.341
0.139	0.354
0.127	0.367
0.115	0.379
0.103	0.393
0.091	0.406
0.079	0.419
0.067	0.433
0.055	0.446
0.042	0.459
0.030	0.473
0.018	0.486
0.006	0.500

TABLE 15

Asymptotic variances and covariances a_{11}, a_{22}, and a_{12} of $\sqrt{r}\hat{\delta}/\delta$ and $\sqrt{r}\hat{\xi}/\delta$ for MLEs $\hat{\xi}$ and $\hat{\delta}$, and $c = 2/[(1+p^2)^2 a_{22}]$

	$p = r/n$										
	1.0	0.9	0.8	0.7	0.6	0.5	0.4	0.3	0.2	0.1	0
c	0.82	0.88	1.0	1.15	1.31	1.49	1.67	1.83	1.95	2.01	2.00
a_{11}	1.10867	1.03652	1.00209	1.01308	1.08718	1.25512	1.57321	2.15713	3.29575	6.05171	
a_{22}	0.60793	0.69034	0.74255	0.78571	0.82367	0.85809	0.88990	0.91965	0.94775	0.97447	
a_{12}	−0.25702	−0.15877	−0.03943	0.10138	0.26796	0.46788	0.71421	1.03158	1.47506	2.21872	

TABLE 16

Values of k_n, $n\,\mathrm{Var}(\tilde{\delta}/\delta)$, $n\,\mathrm{Cov}(\tilde{\xi}/\delta, \tilde{\delta}/\delta)$, and $n\,\mathrm{Var}(\tilde{\xi}/\delta)$ for simple estimators $\tilde{\delta}$ and $\tilde{\xi}$

	n												
	2	3	4	5	6	7	8	9	10	15	30	60	∞
k_n	0.69	0.98	1.15	1.27	1.35	1.18	1.25	1.31	1.36	1.40	1.50	1.53	1.57
$n\,\mathrm{Var}(\tilde{\delta}/\delta)$	1.42	1.04	0.92	0.86	0.83	0.80	0.77	0.75	0.74	0.71	0.68	0.66	0.65
$n\,\mathrm{Cov}(\tilde{\xi}/\delta, \tilde{\delta}/\delta)$	0.13	−0.06	−0.12	−0.14	−0.14	−0.19	−0.19	−0.20	−0.20	−0.21	−0.22	−0.23	−0.23
$n\,\mathrm{Var}(\tilde{\xi}/\delta)$	1.32	1.23	1.21	1.20	1.20	1.16	1.16	1.17	1.17	1.16	1.17	1.16	1.16

TABLE 17

Coefficients for the quadratic approximations of $k_{r,n} \doteq k_0 + k_1/n + k_2/n^2$, $c_{r,n} \doteq c_0 + c_1/n + c_2/n^2$, $d_{r,n} \doteq d_0 + d_1/n + d_2/n^2$, $h/n \doteq a_0 + a_1/n + a_2/n^2$

	r/n								
	0.1	0.2	0.3	0.4	0.5	0.6	0.7	0.8	0.9
k_0	0.10265	0.21129	0.32723	0.45234	0.58937	0.74274	0.92026	1.1382	1.4436
k_1	−1.0271	−1.0622	−1.1060	−1.1634	−1.2415	−1.1340	−1.5313	−1.8567	−2.6929
k_2	0.000	0.030	0.054	0.089	0.145	0.242	0.433	0.906	2.796
c_0	−2.2504	−1.4999	−1.0309	−0.6717	−0.3665	−0.0874	0.1856	0.4759	0.8340
c_1	−5.5743	−3.070	−2.2859	−1.9301	−1.7619	−1.7114	−1.7727	−2.0110	−2.7773
c_2	−7.201	−1.886	−0.767	−0.335	−0.091	−0.111	−0.369	−0.891	−2.825
d_0	0.25973	0.27113	0.28480	0.30160	0.32305	0.35188	0.39384	0.46402	0.62397
d_1	−0.1259	−0.1436	−0.1681	−0.2026	−0.2537	−0.3365	−0.4887	−0.8394	−2.1509
d_2	0.044	0.046	0.067	0.102	0.162	0.280	0.550	1.383	5.934
a_0	0.2052	0.4218	0.6514	0.8959	1.1577	1.4391	1.7416	2.0598	2.3394
a_1	−2.052	−2.111	−2.175	−2.244	−2.314	−2.376	−2.390	−2.205	−0.856
a_2	0.000	0.008	0.002	−0.106	−0.064	−0.188	−0.526	−1.682	−7.928

ANSWERS TO SELECTED EXERCISES

CHAPTER 1

1 **a.** $S = \{r, g, b\}$

b. $\{r\}, \{g\}, \{b\}, \{r, g\}, \{r, b\}, \{g, b\}, S, \varnothing$ **c.** $\{b, g\}$ **d.** $\varnothing$

2 **a.** $S = \left\{\begin{matrix}(r, r), (r, b), (r, g)\\(b, r), (b, b), (b, g)\\(g, r), (g, b), (g, g)\end{matrix}\right\}$

b. 9

c. $C_1 = \{(r, r), (r, b), (r, g)\} = \{(r, r)\} \cup \{(r, b)\} \cup \{(r, g)\}$
$C_2 = \{(r, r), (r, b), (r, g), (b, r), (g, r)\} = \{(r, r)\} \cup \cdots \cup \{(g, r)\}$
$C_1 \cap C_2 = C_1, \quad C_1' \cap C_2 = \{(b, r), (g, r)\} = \{b, r)\} \cup \{g, r)\}$

3 **a.** (O, O), (O, A), (O, B), (O, AB), (A, O), (A, A), (A, B), (A, AB), (B, O), (B, A), (B, B), (B, AB), (AB, O), (AB, A), (AB, B), (AB, AB).

b. (O, O), (O, A), (O, B), (O, AB), (A, A), (B, B), (AB, AB), (A, AB), (B, AB).

c. (O, O), (A, A), (B, B), (AB, AB).

4 $S = \{r, br, gr, bbr, ggr, bgr, gbr, \ldots\}$
$\{x \mid x = r \text{ or } x = c_1c_2 \cdots c_kr\}$ where $c_i = b$ or g

5 **a.** $S = \{0. 1, 2, \ldots\}$ **b.** $S = [0, \infty) = \{t \mid t \geqslant 0\}$

6 $S = [0, 1] = \{x \mid 0 \leqslant x \leqslant 1\}$

7 $S = [0, \infty) = \{t \mid t \geqslant 0\}$

8 **a.** Yes. $p_1 + p_2 + p_3 = 1,\ p_i \geqslant 0$
b. No. $p_1 + p_2 + p_3 + p_4 > 1$

9 **a.** 1/9 **b.** 1/3 **c.** 5/9 **d.** 1/3 **e.** 2/9 **f.** 5/9

10 **a.** 9/16 **b.** 1/4 **c.** 1/8

13 **a.** 1/3, 1/3, 1/3 **b.** 1/4, 1/4, 1/2 **c.** 6/11, 3/11, 2/11

14 **a.** 1/4 **b.** 15/16 **c.** 3/8 **d.** 5/16

15 $S = \{(t, t), (t, a), (t, c), (a, t), (a, a), (a, c), (c, t), (c, a), (c, c)\}$, 2/3

19 **a.** 2/3 **b.** 23/30 **c.** 7/30 **d.** 9/10

20 **a.** 7/8 **b.** 1/8

22 **a.** 0.8 **b.** 0.3 **c.** 0.2

23 **a.** 0.2 **b.** 0.2 **c.** 0.3 **d.** 0.5

24 319/420

25 **a.** 3/5 **b.** 1/2 **c.** 3/4 **d.** 3/10 **f.** 3/5 **g.** 1/2

26 **a.** 3/5 **b.** 3/5 **c.** 3/5 **d.** 9/25 **f.** 3/5 **g.** 3/5

27 **a.** 5/14 **b.** 15/28 **c.** 25/28 **d.** 3/28

28 **a.** 1/56 **b.** 15/56 **c.** 5/28 **d.** 3/8

29 1/3

30 **a.** 12/51 **b.** 2/15

31 **a.** 1/5 **b.** 34/105

32 **a.** 0.66 **b.** 0.1212

33 **a.** 1/13 **b.** 53/715 **c.** 5/53

34 **a.** 5/12 **b.** 4/5

35 **a.** 29/1000 **b.** 10/29

36 **a.** 43/80 **b.** 28/43

37 **a.** 0.2 **b.** 1/3

39 $(1 - p)^3$

40 $1 - p^3$

41 0.97412

42 27/50

43 **a.** 25/64 **b.** 15/32 **c.** 55/64 **d.** 9/64

46 **a.** 47/60 **b.** 3/5

47 No. $P(A_1 \cap A_2 \cap A_3) = 0 \neq 1/8 = P(A_1)P(A_2)P(A_3)$

48 **a.** 26! **b.** 7,893,600 **c.** 11,881,376

49 **a.** 6,760,000 **b.** 3,407,040 **c.** 1,514,240

50 72

51 40

52 ***a.*** 0.504 ***b.*** 0.496

53 8

54 ***a.*** $\binom{30}{5}$ ***b.*** $\binom{17}{3}\binom{13}{2}$ ***c.*** $\binom{1}{1}\binom{16}{2}\binom{13}{2}$

55 ***a.*** $\binom{49}{11}$ ***b.*** $\binom{24}{5}\binom{25}{6}\Big/\binom{49}{11}$

57 ***a.*** 8 ***b.*** 32 ***c.*** 2^{2n-1}

58 ***a.*** $\binom{7}{3}$ ***b.*** 1/7 ***c.*** 7! ***d.*** 576

59 24,360

60 10! − (2)(9)!

61 ***a.*** 365^n ***b.*** ${}_{365}P_n$ ***c.*** $\dfrac{{}_{365}P_n}{365^n}$ ***d.*** 0.5073

62 ***a.*** 27,720 ***b.*** 27,720

63 (26!)/(9!)(11!)(6!)

64 Same as 63

65 ***a.*** 60 ***b.*** 13 ***c.*** 170

66 (60!)/(15!)(20!)(25!)

67 ***a.*** 3^9 ***b.*** $(9!)/(3!)^3$

68 ***a.*** 11,550 ***b.*** (12!)(3!)/14!

69 ***a.*** 126 ***b.*** 0.0397 ***c.*** 3024

71 ***a.*** 0.9722 ***b.*** 0.6475

72 ***a.*** No ***b.*** Yes ***c.*** Yes. *B*, *D*, and *C*, *D*

CHAPTER 2

1 ***a.***

y	2	3	4	5	6	7	8
$f_Y(y)$	1/16	1/8	3/16	1/4	3/16	1/8	1/16

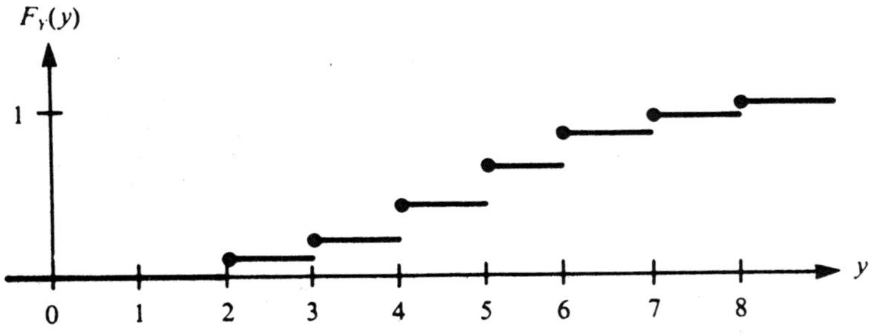

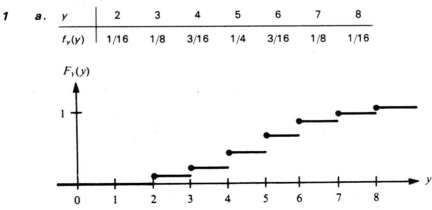

c.

w	0	1	4	9
$f_W(w)$	1/4	3/8	1/4	1/8

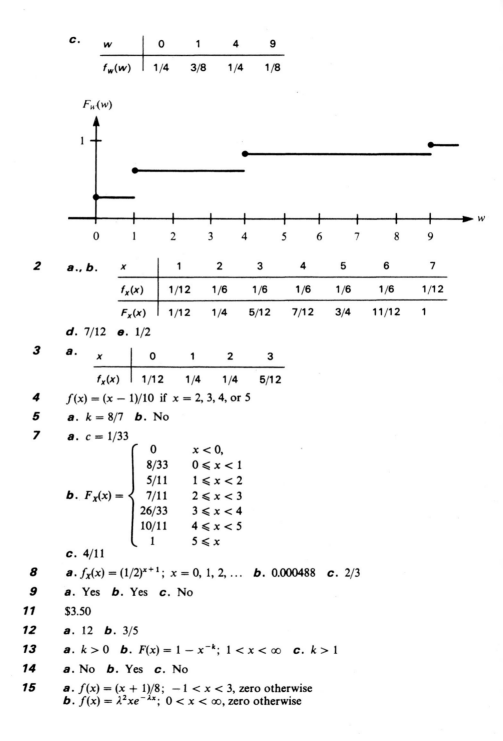

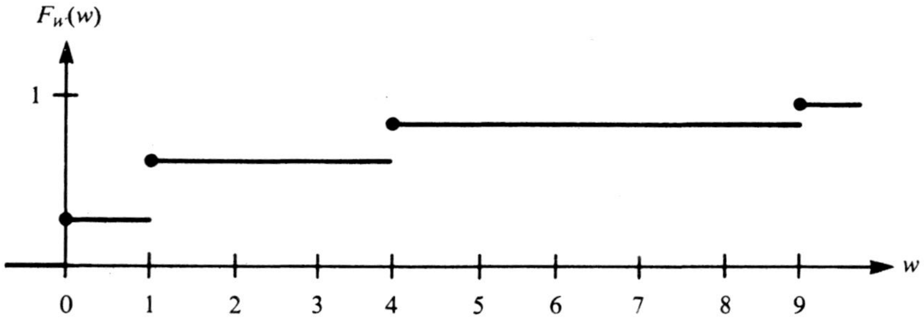

2 **a., b.**

x	1	2	3	4	5	6	7
$f_X(x)$	1/12	1/6	1/6	1/6	1/6	1/6	1/12
$F_X(x)$	1/12	1/4	5/12	7/12	3/4	11/12	1

d. 7/12 **e.** 1/2

3 **a.**

x	0	1	2	3
$f_X(x)$	1/12	1/4	1/4	5/12

4 $f(x) = (x - 1)/10$ if $x = 2, 3, 4,$ or 5

5 **a.** $k = 8/7$ **b.** No

7 **a.** $c = 1/33$

b. $$F_X(x) = \begin{cases} 0 & x < 0, \\ 8/33 & 0 \leqslant x < 1 \\ 5/11 & 1 \leqslant x < 2 \\ 7/11 & 2 \leqslant x < 3 \\ 26/33 & 3 \leqslant x < 4 \\ 10/11 & 4 \leqslant x < 5 \\ 1 & 5 \leqslant x \end{cases}$$

c. 4/11

8 **a.** $f_X(x) = (1/2)^{x+1}$; $x = 0, 1, 2, \ldots$ **b.** 0.000488 **c.** 2/3

9 **a.** Yes **b.** Yes **c.** No

11 \$3.50

12 **a.** 12 **b.** 3/5

13 **a.** $k > 0$ **b.** $F(x) = 1 - x^{-k}$; $1 < x < \infty$ **c.** $k > 1$

14 **a.** No **b.** Yes **c.** No

15 **a.** $f(x) = (x + 1)/8$; $-1 < x < 3$, zero otherwise
b. $f(x) = \lambda^2 x e^{-\lambda x}$; $0 < x < \infty$, zero otherwise

17 **a.** **b.**

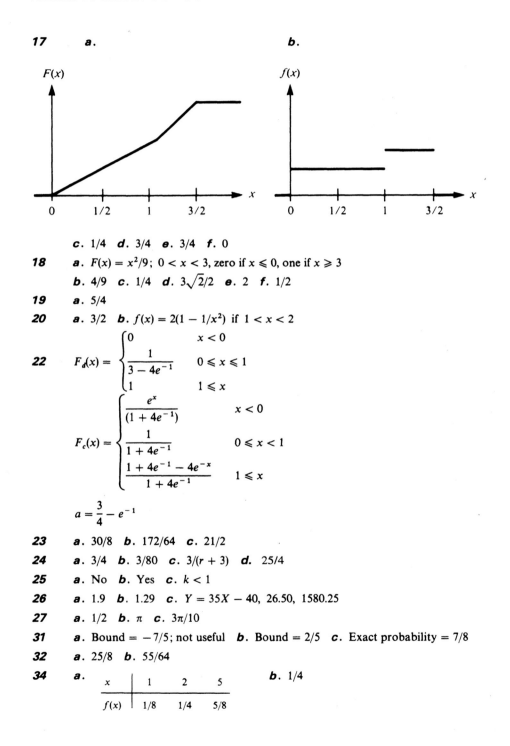

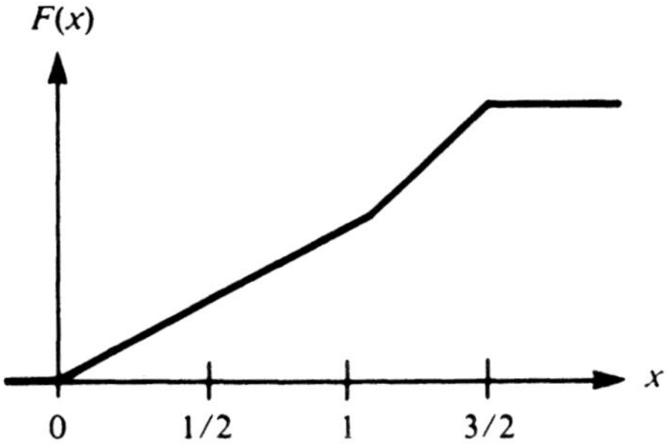

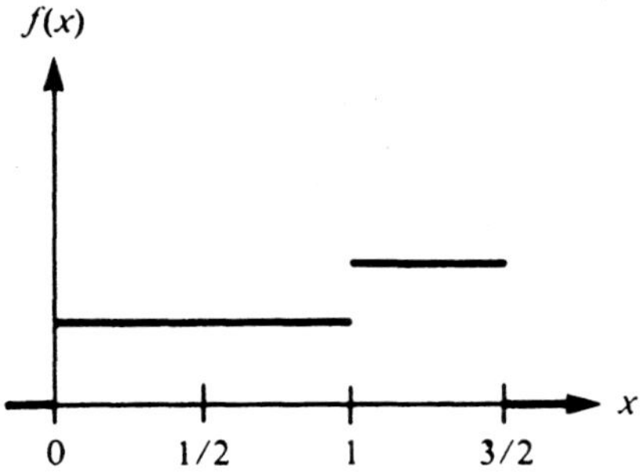

c. 1/4 **d.** 3/4 **e.** 3/4 **f.** 0

18 **a.** $F(x) = x^2/9$; $0 < x < 3$, zero if $x \leqslant 0$, one if $x \geqslant 3$

b. 4/9 **c.** 1/4 **d.** $3\sqrt{2}/2$ **e.** 2 **f.** 1/2

19 **a.** 5/4

20 **a.** 3/2 **b.** $f(x) = 2(1 - 1/x^2)$ if $1 < x < 2$

22 $$F_d(x) = \begin{cases} 0 & x < 0 \\ \dfrac{1}{3 - 4e^{-1}} & 0 \leqslant x \leqslant 1 \\ 1 & 1 \leqslant x \end{cases}$$

$$F_c(x) = \begin{cases} \dfrac{e^x}{(1 + 4e^{-1})} & x < 0 \\ \dfrac{1}{1 + 4e^{-1}} & 0 \leqslant x < 1 \\ \dfrac{1 + 4e^{-1} - 4e^{-x}}{1 + 4e^{-1}} & 1 \leqslant x \end{cases}$$

$$a = \frac{3}{4} - e^{-1}$$

23 **a.** 30/8 **b.** 172/64 **c.** 21/2

24 **a.** 3/4 **b.** 3/80 **c.** $3/(r + 3)$ **d.** 25/4

25 **a.** No **b.** Yes **c.** $k < 1$

26 **a.** 1.9 **b.** 1.29 **c.** $Y = 35X - 40$, 26.50, 1580.25

27 **a.** 1/2 **b.** π **c.** $3\pi/10$

31 **a.** Bound $= -7/5$; not useful **b.** Bound $= 2/5$ **c.** Exact probability $= 7/8$

32 **a.** 25/8 **b.** 55/64

34 **a.**

x	1	2	5
$f(x)$	1/8	1/4	5/8

b. 1/4

36 **a.** $M_X(t) = e^{-2t}/(1-t)$ **b.** $-1, 2$

37 6, 13

38 **a.** $Y = 35S - 10c$ **b.** 3

CHAPTER 3

1 0.008

2 **a.** 0.000977 **b.** 0.0439

3 **a.** 0.0439 **b.** 0.1209

4 **a.** $1 - 4q^3 + 3q^4$ **b.** $1 - q^2$

c. Two engines safer if $q > 1/3$; four engines safer if $q < 1/3$; equally safe if $q = 1/3$

5 **a.** $P(\text{he wins}) = 0.5177 > 0.5$ (good bet)

b. $P(\text{he wins}) = 0.4914$

c. $P(\text{at least one 6 in six rolls}) = 0.6651$
$P(\text{at least two 6's in twelve rolls}) = 0.6187$

6 **a.** 0.2182 **b.** 0.6292 **c.** $E(X) = 4$, $\text{Var}(X) = 3.2$

8 **a.** 0.2007 **b.** 0.2151

9 **a.** 3/10 **b.** 2/3

10 **a.** 0.03993 **b.** 0.03993 **c.** 0.9583 **d.** 0.0417

11 **a.** 0.1517 **b.** 0.1759

12 **a.** 0.0465 **b.** 0.0465 **c.** 0.9494 **d.** 0.0506

13 **a.** 0.09 **b.** $0.1(0.9)^{x-1}$ **c.** 0.729 **d.** 10

14 **a.** 1/4 **b.** 15/16

15 **a.** 0.01722 **b.** 0.999 **c.** 30

16 **a.** $\binom{x-1}{3}(0.6)^4(0.4)^{x-4}$; $x = 4, 5, 6, 7$ **b.** 0.2765 **c.** $p = 1/2$ **d.** $x = 6$

17 **a.** 0.00729 **b.** 0.99144 **c.** 0.271

20 **a.** 0.0335 **b.** 0.8385

21 **a.** 0.090 **b.** 0.220 **c.** 0.217

22 0.0242

23 **a.** 0.1234 **b.** 0.1247

24 0.677

25 0.1008

26 **a.** 0.8009 **b.** 0.1493

28 **a.** 3/5 **b.** 3/5 **c.** $1 - 1/(10k^2)$

29 **a.** $f_Y(y) = (41 - 2y)/400$ if $y = 1, 2, \ldots, 20$

b. 19/8000 **c.** $E(X) = 21/2$, $\text{Var}(X) = 133/4$

32 **a.** $F(x) = (x - 50)/25,\ 50 < x < 75$, zero if $z \leqslant 50$, one if $x \geqslant 75$
b. 2/5 **c.** 125/2 **d.** 625/12

33 1/3

34 **a.** UNIF(0, 5) **b.** 2/5

36 **a.** 24 **b.** $3\sqrt{\pi}/4$

37 **a.** 0.353 **b.** 0.214 **c.** 20 days

38 **a.** 0.122 **b.** 0.0028

39 0.2636

40 **a.** 0.861 **b.** 0.333 **c.** Same as *a* (no memory) **d.** 10,000

41 $m_0 = 1$

42 0.5768

43 $\theta^k\Gamma(1 + k/\beta)$

44 **a.** $\theta/(\kappa - 1)$ **b.** $2\theta^2/[(\kappa - 1)(\kappa - 2)]$

45 $E(X) = 50$, $\mathrm{Var}(X) = 10{,}000$

46 **a.** 0.362 **b.** 0.967 **c.** $300\sqrt{\pi}$ **d.** $3(100)^2(32 - 3\pi)$

47 **a.** 0.0183 **c.** $E(X) = 5\sqrt{\pi}$, $\mathrm{Var}(X) = 100(1 - \pi/4)$

48 **a.** $x_p = \theta[(1 - p)^{-1/\kappa} - 1]$
b. $m = 10(\sqrt{2} - 1)$

49 **a.** 0.846 **b.** 0.0377 **c.** 20 days

51 **a.** 0.937 **b.** 0.688 **c.** 0.341 **d.** 0.20 **e.** 0.38 **f.** 1.96

52 **a.** 0.5 **b.** 0.227 **c.** 0.290 **d.** 3.822 **e.** 0.658

53 **a.** 0.816 **b.** 5.88

54 **a.** 0.8413 **b.** 0.9104 **c.** 0.8413 **d.** 16.58

55 **a.** 0.6366 **b.** 324.36

56 **a.** 12 **b.** 25

57 **a.** $E(Y) = 350p - 100$ **b.** $p > 2/7$ **c.** No

58 **c.** 1002.857 **d.** 1001.8

59 **d.** 6 **e.** 6

CHAPTER 4

1 **a.**

y \ z	−3	−2	−1	0	1	2	3
2	0	0	0	1/16	0	0	0
3	0	0	1/16	0	1/16	0	0
4	0	1/16	0	1/16	0	1/16	0
5	1/16	0	1/16	0	1/16	0	1/16
6	0	1/16	0	1/16	0	1/16	0
7	0	0	1/16	0	1/16	0	0
8	0	0	0	1/16	0	0	0

b.

z \ w	0	1	4	9
−3	0	0	0	1/16
−2	0	0	1/8	0
−1	0	3/16	0	0
0	1/4	0	0	0
1	0	3/16	0	0
2	0	0	1/8	0
3	0	0	0	1/16

2.

y \ x	1	2	3	4	5	6	7
1	1/12	1/12	0	0	0	0	0
2	0	1/12	1/12	0	0	0	0
3	0	0	1/12	1/12	0	0	0
4	0	0	0	1/12	1/12	0	0
5	0	0	0	0	1/12	1/12	0
6	0	0	0	0	0	1/12	1/12

3 **a.** 0.0399 **b.** Same as (a) **c.** 0.000609 **d.** 0.0793 **e.** 0.0153

f. $\binom{4}{x}\binom{48}{5-x}\Big/\binom{52}{5}$; $x = 0, 1, 2, 3, 4$

g. 0.9583 **h.** 0.0417 **i.** 0.00005541

j. $\binom{4}{x}\binom{4}{y}\binom{4}{z}\binom{40}{5-x-y-z}\Big/\binom{52}{5}$; $0 \leqslant x,\ 0 \leqslant y,\ 0 \leqslant z,\ x + y + z \leqslant 5$

4 **a.** 0.0331 **b.** 0.0666

5 **a.** 0.0465 **b.** Same as **a** **c.** 0.00089 **d.** 0.0922 **e.** 0.0191

f. $\binom{5}{x}(1/13)^x(12/13)^{5-x}$; $x = 0, 1, 2, 3, 4, 5$

6 **a.** 0.000382 **b.** 0.00344 **c.** 0.00153

d. $\dfrac{12!}{x_1!x_3!x_5!(12 - x_1 - x_3 - x_5)!}(1/6)^{x_1+x_3+x_5}(1/2)^{12-x_1-x_3-x_5}$

7 1/18

8 **a.** e^{-4} **b.** Both POI(2) **c.** Yes

9 **a.**

x_1	1	2	3
$f_1(x_1)$	1/4	14/45	79/180

x_2	1	2	3
$f_2(x_2)$	5/36	19/36	1/3

b. No **c.** 101/180 **d.** 25/36

10 **a.** $f(x, y) = \dfrac{\binom{13}{x}\binom{26}{y}\binom{13}{2-x-y}}{\binom{52}{2}}$; $x \geqslant 0,\ y \geqslant 0,\ x + y \leqslant 2$

c. $f_1(x) = \binom{13}{x}\binom{39}{2-x}\Big/\binom{52}{2}$; $f_2(y) = \binom{26}{y}\binom{26}{2-y}\Big/\binom{52}{2}$

d. No **e.** 0.668

f. $P[Y = y \mid X = 1] = \dfrac{f(1, y)}{f_1(1)} = \dfrac{\binom{26}{y}\binom{13}{1-y}}{\binom{39}{1}}$; $y = 0, 1$

g. $P[Y = y \mid X = x] = \dfrac{\binom{26}{y}\binom{13}{2-x-y}}{\binom{39}{2-x}}$; $0 \leqslant y \leqslant 2 - x$

h. $P[X + Y \leqslant z] = \begin{cases} 0 & \text{if } z < 0 \\ 0.059 & \text{if } 0 \leqslant z < 1 \\ 0.441 & \text{if } 1 \leqslant z < 2 \\ 1 & \text{if } 2 \leqslant z \end{cases}$

11 **a.** $f(x, y) = \dfrac{2!}{x!y!(2-x-y)!}(1/4)^x(1/2)^y(1/4)^{2-x-y}$

c. $f_1(x) = \binom{2}{x}(1/4)^x(3/4)^{2-x}$; $f_2(y) = \binom{2}{y}(1/4)$

d. No **e.** 2/3

f. $P[Y = y \mid X = 1] = \dfrac{f(1, y)}{f_1(y)} = (2/3)^y(1/3)^{1-y}$; $y = 0, 1$

g. $P[Y = y \mid X = x] = \binom{2-x}{y}(2/3)^y(1/3)^{2-x-y}$; $0 \leqslant y \leqslant 2 - x$

12 No

14 **a.** Both EXP(1)

b. $F(x, y) = (1 - e^{-x})(1 - e^{-y})$ if $x > 0$ and $y > 0$, 0 otherwise

c. e^{-2} **d.** 1/2 **e.** $3e^{-2}$ **f.** Yes

15 **a.** $f(x_1, x_2) = 2(1 + x_1)^{-2}(1 + x_2)^{-3}$ if $x_1 > 0$ and $x_2 > 0$ **b.** 1/3

16 1/3

17 **a.** $f_1(x_1) = 3(1 - x_1)^2$; $0 < x_1 < 1$

b. $f_2(x_2) = 6x_2(1 - x_2)$; $0 < x_2 < 1$

c. $f_{1,2}(x_1, x_2) = 6(1 - x_2)$; $0 < x_1 < x_2 < 1$

18 **a.** $f(x, y) = x + y$; $0 < x < 1$ and $0 < y < 1$

b. 1/8 **c.** 1/2 **d.** 1/24 **e.** 19/24

f. $P[X + Y \leqslant z] = \begin{cases} z^3/3 & \text{if } 0 < z < 1 \\ z^2 - z^3/3 - 1/3 & \text{if } 1 \leqslant z < 2 \\ 1 & \text{if } 2 \leqslant z \end{cases}$

19 **a.** $k = 2$

b. $f_1(x) = (1 + 3x)(1 - x)$ if $0 < x < 1$; $f_2(y) = 3y^2$ if $0 < y < 1$

c. $$F(x, y) = \begin{cases} 0 & \text{if } x < 0 \text{ or } y < 0 \\ xy^2 + x^2y - x^3 & \text{if } 0 \leqslant x \leqslant y \leqslant 1 \\ y^3 & \text{if } x > y \text{ and } 0 < y < 1 \\ x + x^2 - x^3 & \text{if } y > 1 \text{ and } 0 < x < 1 \\ 1 & \text{if } y > 1 \text{ and } x > 1 \end{cases}$$

d. $f(y \mid x) = \dfrac{2(x + y)}{(1 + 3x)(1 - x)}$ if $x \leqslant y \leqslant 1$

e. $f(x \mid y) = \dfrac{2(x + y)}{3y^2}$ if $0 \leqslant x \leqslant y$

21 **a.** $f_1(x) = (2/3)(x + 1)$ if $0 < x < 1$

b. $f_2(y) = 1$ if $0 < y < 1$

c. $f(y \mid x) = 1$ if $0 < y < 1,\ 0 < x < 1$

d. 4/9 **e.** 73/162 **f.** Yes

22 **a.** $f(x_1, x_2, \ldots, x_n) = 3^n(x_1 x_2 \cdots x_n)^2$ if all $0 < x_i < 1$

b. 1/8 **c.** $(1/8)^n$

23 **a.** $f(x_1, x_2, \ldots, x_n) = 2^n(x_1 x_2 \cdots x_n)\exp(-x_1^2 - x_2^2 - \cdots - x_n^2)$ if all $0 < x_i$

b. $1 - e^{-1/4}$ **c.** $(1 - e^{-1/4})^n$

26 **a.** $$f(x_1, x_3) = \begin{cases} 6(x_3 - x_1) & \text{if } 0 < x_1 < x_3 < 1 \\ 0 & \text{otherwise} \end{cases}$$

c. $$f(x_2 \mid x_1, x_3) = \begin{cases} 1/(x_3 - x_1) & \text{if } 0 < x_1 < x_2 < x_3 < 1 \\ 0 & \text{otherwise} \end{cases}$$

e. $$f(x_1, x_2 \mid x_3) = \begin{cases} 2/x_3^2 & \text{if } 0 < x_1 < x_2 < x_3 < 1 \\ 0 & \text{otherwise} \end{cases}$$

27 **a.**

$x_2 \backslash x_1$	1	2	3
1	0.04	0.12	0.04
2	0.12	0.36	0.12
3	0.04	0.12	0.04

b.

$x_2 \backslash x_1$	1	2	3
1	0.04	0.16	0.20
2	0.16	0.64	0.80
3	0.20	0.80	1.00

c. 0.72

28 **a.** Yes $f(x, y) = g(x)h(y)$ over $(0, 1) \times (0, 1)$ **b.** 1/6

29 **a.** No $f(x, y) = 0 < f_1(x)f_2(y)$ if $0 < 1 - x < y < 1$

b. 7/27 **c.** $f_1(x) = 12x(1 - x)^2$ if $0 < x < 1$

30 **a.** $f_1(x) = 30x^2(1-x)^2$ if $0 < x < 1$
b. $f(y \mid x) = 2y/(1-x)^2$ if $0 < y < 1 - x$ **c.** 0.96

31 **a.** 5/48 **b.** $f_2(x_2) = \dfrac{3x_2^2}{4}$ if $0 < x_2 < 1$
c. $f(x_1 \mid x_2) = \dfrac{2}{3}\dfrac{x_1 + x_2}{x_2^2}$ if $0 < x_1 < x_2 < 1$

CHAPTER 5

1 **a.** 15 **b.** 63
2 **a.** 720 **b.** 23.04
3 **a.** 2/15 **b.** −2/75 **c.** −2/3 **d.** −2/5
e. −2/75 **f.** −2/15 **g.** 3/25
4 **a.** 4/3 **b.** 12/5 **c.** 16/5 **d.** 0
5 **a.** 7/12 **b.** 7/6 **c.** 1/3 **d.** −1/24 **e.** $E(Y \mid x) = (3x+2)/(6x+3)$; $0 < x < 1$
7 **a.** 7 **b.** 116 **c.** 16 **d.** 20
11 **a.** $f_1(x) = 6x(1-x)$; $0 < x < 1$
b. $f_2(y) = 3y^2$; $0 < y < 1$
c. 1/40 **d.** $\sqrt{3}/3$ **e.** $f(y \mid x) = 1/(1-x)$; $x < y < 1$ **f.** $(x+1)/2$
12 **a.** 1/6 **b.** 1/3 **c.** 1/9 **d.** 0 **e.** 1/3
13 **a.** $f(y \mid x) = \dfrac{1}{2x}$ if $0 < y < 2x$ **b.** $E(Y \mid x) = x$ **c.** 1/2
15 **a.** 12/5 **b.** 6/25
17 **a.** 1 **b.** 2
18 115/88
19 **a.** 1.44 **b.** 1.0944
20 **a.** 3/2 **b.** 25/12 **c.** 6
21 $M(t_1, t_2) = 2/[(1-t_2)(2-t_1-t_2)]$; $t_2 < 1$, $t_1 + t_2 < 2$
22 $M(t_1, t_2) = [1/(t_2 - 1) - 1/(t_1 + t_2 - 1)]/t_1$; $t_2 < 1$, $t_1 + t_2 < 1$

CHAPTER 6

1 **a.** $f_Y(y) = 1$; $0 < y < 1$
b. $f_W(w) = 4(\ln w)^3/w$; $1 < w < e$
c. $f_Z(z) = 4e^{4z}$; $-\infty < z < 0$
d. $f_U(u) = 0.5(1 + 12u)u^{-1/2}$; $0 < u < 0.25$

2 **a.** $f_Y(y) = 4y^3;\quad 0 < y < 1$

b. $f_W(w) = 1/w;\quad e^{-1} < w < 1$

c. $f_Z(z) = 1/(1 - z);\quad 0 < z < 1 - e^{-1}$

d. $f_U(u) = 2(1 - 4u)^{-1/2};\quad 0 < u < 1/4$

3 **a.** $X = 2\pi R;\quad f_X(x) = 6x(2\pi - x)/(2\pi)^3;\quad 0 < x < 2\pi$

b. $Y = \pi R^2;\quad f_Y(y) = 3(\pi^{1/2} - y^{1/2})/\pi^{3/2};\quad 0 < y < \pi$

6 $y = x^4$ or $y = 1 - x^4$

7 **a.** $y = -\ln(1 - u)$

b. $w = x$ if $B(x - 1; 3, 1/2) < w \leqslant B(x; 3, 1/2);\; x = 0, 1, 2, 3$

10 **a.** $f_Y(y) = \exp(-y);\; y > 0$

b. $F_W(w) = 1/2$ if $0 \leqslant w < 1$, one if $1 \leqslant w$, zero otherwise

11 $Y \sim \text{BIN}(n, 1 - p)$

12 $f_Y(y) = \binom{y + r - 1}{r - 1} p^r(1 - p)^y;\; y = 0, 1, 2, \ldots$

13 $f_Y(y) = y^{1/2}/24$ if $0 < y < 8;\; y^{1/2}/48$ if $4 \leqslant y < 16$

14 **a.** $F_W(w) = 1 - (1 + 2w)e^{-2w};\; w > 0$

b. $f_{U,V}(u, v) = 4(v/u^2)\exp(-2v(1 + 1/u));\; u > 0, v > 0$

c. $f_U(u) = 1/(1 + u)^2\;\; u > 0$

15 $Y \sim \text{POI}(2\lambda)$

16 **a.** $f_{U,V}(u, v) = 1/(u^2 v);\; 1 \leqslant v < u$

b. $f_U(u) = (\ln u)/u^2;\; 1 \leqslant u$

17 **a.** $f_Y(y) = y\exp(-y^2/2);\; y > 0$

b. $f_W(w) = w^{-1/2}(1 + w)^{-1}/\pi;\; w > 0$

18 **a.** $f_{S,T}(s, t) = e^{t-s};\; 0 < t < s/2$

b. $f_T(t) = e^{-t};\; 0 < t$

c. $f_S(s) = e^{-s}(e^{s/2} - 1);\; s > 0$

22 0.8508

23 $M_Y(t) = \left(\dfrac{pe^t}{1 - qe^t}\right)^k;\quad Y \sim \text{NB}(k, p)$

24 **a.** $M_Y(t) = (1 - 2t)^{-10}$ **b.** $Y \sim \text{GAM}(2, 10)$

25 **a.** $X_1 \sim \text{POI}(10)$ **b.** $W \sim \text{POI}(15)$

27 **a.** $\text{LOGN}(\sum \mu_i, \sum \sigma_i^2)$ **b.** $\text{LOGN}(\sum a_i\mu_i, \sum a_i^2\sigma_i^2)$

c. $\text{LOGN}(\mu_1 - \mu_2, \sigma_1^2 + \sigma_2^2)$

29 **a.** $g(y_1, \cdots, y_n) = n!(y_1 \cdots y_n)^{-2};\; 1 < y_1 < \cdots < y_n$

b. $g_1(y_1) = n/y_1^{n+1};\quad y_1 > 1$

c. $g_n(y_n) = n(y_n - 1)^{n-1}/y_n^{n+1};\quad 1 < y_n$

e. $g_r(y_r) = \dfrac{(2r-1)!(y_r - 1)^{r-1}}{[(r-1)!]^2 y_r^{2r}}; \quad y_r > 1$

31 **a**. $g_1(y_1) = ne^{-ny}; \quad 0 < y_1$

b. $g_n(y_n) = ne^{-y_n}(1 - e^{-y_n})^{n-1}; \quad 0 < y_n$

c. $f_R(r) = (n-1)e^{-r}(1 - e^{-r})^{n-2}; \quad 0 < r$

d. $g(y_1, \ldots, y_r) = \dfrac{n!}{(n-r)!} \exp\left(-\sum_{i=1}^{r} y_i - (n-r)y_r\right); \quad 0 < y_1 < \cdots < y_r$

32 **a**. $g_1(y_1) = 5e^{-5y_1}; \quad 0 < y_1$

b. $g_5(y_5) = 5e^{-y_5}(1 - e^{-y_5})^4; \quad 0 < y_5$

c. $g_3(y_3) = 30e^{-3y_3}(1 - e^{-y_3})^2; \quad 0 < y_3$

d. $g_1(y_1) = (1/\theta)e^{-y_1/\theta}; \quad \theta = 1/(1/\theta_1 + \cdots + 1/\theta_n); \quad y_1 > 0$

33 **a**. $G_1(y_1) = 1 - (1-p)^{ny_1}; \quad y_1 = 1, 2, 3, \ldots$

b. $G_k(y_k) = \sum_{i=k}^{n} \binom{n}{i} [1 - (1-p)^{y_k}]^i (1-p)^{(n-i)y_k}; \quad y_k = 1, 2, \ldots$

c. $G_n(y_n) = [1 - (1-p)^{y_n}]^n; \quad y_n = 1, 2, \ldots$

d. $P[Y_1 \leqslant 1] = G_1(1) = 1 - (1-p)^n$

35 **a**. $M_Y(t) = (1 - \theta^2 t^2)^{-1}$ **b**. $Y \sim \text{DE}(\theta, 0)$

CHAPTER 7

1 **a**. $G_n(y) = \begin{cases} 0 & \text{if } y < 1 \\ 1 - (1/y)^n & \text{if } y \geqslant 1 \end{cases}$

b. Degenerate at $y = 1$

c. $G(y) = \begin{cases} 0 & \text{if } y \leqslant 1 \\ 1 - 1/y & \text{if } y > 1 \end{cases}$

2 **a**. No **b**. Yes, $G(y) = \exp(-\exp(-y))$

3 **a**. Degenerate at $y = 1$ **b**. No limiting distribution

c. $G(y) = \begin{cases} 0 & \text{if } y \leqslant 0 \\ \exp(-1/y^2) & \text{if } y > 0 \end{cases}$

5 N(0, 1)

7 **a**. $a \doteq 0.733$, $b \doteq 1.193$ **b**. $a \doteq 0.634$, $b \doteq 1.032$

8 0.8508

9 **a**. 0.1587 **b**. 0.1359

11 **a**. 0.9394 **b**. 11.655

12 **a**. 90 **b**. 122 **c**. 92

15 **a**. 0.4364

d. $G(y) = \begin{cases} 0 & \text{if} \quad y \leqslant 0 \\ y^{\theta} & \text{if } 0 < y < 1 \\ 1 & \text{if} \quad y \geqslant 1 \end{cases}$

e. $N(B/2^{1/\theta}, [B/(\theta 2^{1/\theta})]^2/n)$ **f.** $B/2^{1/\theta}$ **g.** WEI(1, θ)

17 $m = \mu$, $c = \sigma\sqrt{\pi/2}$

19 $G(y) = \begin{cases} 0 & \text{if } y \leqslant 0 \\ e^{-1/y} & \text{if } y > 0 \end{cases}$

23 $N(0, c^2)$, $N(0, 1/c^2)$

24 **a.** 0.2206 **b.** 0.6044

25 **a.** Type I (for maximums), $a_n = 1$, $b_n = \ln n$

Type I (for minimums), $a_n = 1$, $b_n = -\ln n$

d. Type II (for maximums), $a_n = \theta n^{1/\kappa}(e^{1/\kappa} - 1)$, $b_n = 0$

Type III (for minimums), $a_n = \theta/(\kappa n)$, $b_n = 0$

28 Type II (for maximums), $a_n = \tan[\pi(1/2 - 1/n)]$, $b_n = 0$

CHAPTER 8

1 0.987

2 **a.** 0.39 **b.** 0.0043

3 **a.** $2U/n - 5(W - U^2/n)/(n - 1)$ **b.** W/n **c.** $\bar{Y}$

5 **a.** GAM(100, 10) **b.** 0.95 **c.** 12 spares

7 approx. 0.90

8 No

10 0.685

11 **a.** 0.95 **b.** 0.95

12 approx. 0.95

13 **a.** 0.9772 **b.** 0.921 **c.** 0.921 **d.** 0.988 **e.** 0.95

15 **a.** $N(0, 2\sigma^2)$ **b.** $N(3\mu, 5\sigma^2)$ **c.** $t(k - 1)$ **d.** $\chi^2(1)$ **e.** $t(k - 1)$

f. $\chi^2(2)$ **g.** unknown **h.** $t(1)$ **i.** $F(1, 1)$ **j.** Cauchy **k.** unknown

l. $t(k)$ **m.** $\chi^2(n + k - 1)$ **n.** $N\left(\frac{\mu}{\sigma^2}, \frac{1}{n\sigma^2} + \frac{1}{k}\right)$ **o.** $\chi^2(1)$ **p.** $F(n - 1, k - 1)$

16 **a.** 0.6898 **b.** 0.05 **c.** 0.75

17 **a.** 0.85 **b.** 27.14 **c.** 0.90 **d.** 0.19 **e.** 0.256 **f.** 2.50

g. 0.045 **h.** 0.975

18 **a.** 0.144 **b.** 0.95 **c.** 0.05 **d.** 0.90 **e.** 0.592

23 0.9929

25 **b.** 0.924 **c.** 13

27 $\chi^2_{0.95}(10) = 18.31$; approx. = 18.29 $\chi^2_{0.05}(10) = 3.94$; approx. = 3.93

CHAPTER 9

1 **a.** $\bar{X}/(1-\bar{X})$ **b.** $1/(\bar{X}-1)$ **c.** $2/\bar{X}$

3 **a.** $-n/\sum \ln X_i$ **b.** $-1+n/\sum \ln X_i$ **c.** $2/\bar{X}$

5 $X_{1:n}$

7 **a.** $\bar{X}$ **b.** $\bar{X}^2-\bar{X}$ **c.** $(1-1/\bar{X})^k$

11 **a.** $\sum [x_i/(x_i+\theta)] = n/3$ **b.** $\hat{\theta} = 1.404$

15 **a.** $c = n/(n-1)$ **b.** $[n^2/(n-1)]\hat{p}(1-\hat{p})$

c. $\sum_{i=1}^{N} X_i/(nN)$, $\sum_{i=1}^{N} [n^2/(n-1)](X_i/n)(1-X_i/n)/N$

19 $c = (n+1)/[2(n-1)]$

21 **a.** $p(1-p)/n$ **b.** $p(1-p)(1-2p)^2/n$ **c.** $\bar{X}$

23 **a.** Yes **b.** Yes

27 **a.**

Estimator:	$\hat{\theta}_1$	$\hat{\theta}_2$	$\hat{\theta}_3$	$\hat{\theta}_4$
Risk:	1/2	5/8	$\frac{4+\theta^2}{9}$	$\frac{2+\theta^2}{9}$

b. Max Risk: 1/2 5/8 5/9 1/3 ($\hat{\theta}_4$ is minimax)

c. Bayes Risk: 1/2 5/8 13/27 7/27

29 **a.** $\dfrac{\sum x_i + 1}{n+2}$ **b.** $\dfrac{\sum x_i + 1}{n+2}\left(1-\dfrac{\sum x_i + 2}{n+3}\right)$ **c.** $1/[6(n+2)]$

33 **a.** μ/n **b.** $\mu e^{-2\mu}/n$ **c.** $\bar{X}$ **d.** $e^{-\bar{X}}$ **e.** No **f.** Yes

35 $N(p, p(1-p)/n)$

37 **a.** ARE $= 1/2$ **b.** ARE $= 2/\pi \doteq 0.64$

44 **b.** $(n+1)/(\beta+\sum x_i)$ **c.** $(\beta+\sum x_i)/n$

d. $\dfrac{\chi^2_{0.50}(2n+2)}{2(\beta+\sum x_i)}$ **e.** $\dfrac{2(\beta+\sum x_i)}{\chi^2_{0.50}(2n+2)}$

CHAPTER 10

1 $s!\bigg/\left(n^s \prod_{i=1}^{n} x_i!\right)$ if s $= \sum x_i$, zero otherwise

3 $\Gamma(n/2)/(\pi^{n/2}s^{n/2-1})$ if s $= \sum x_i^2$, zero otherwise

5 **a.** $\left(\prod_{i=1}^{n} x_i\right)\Gamma(2n)/s^{2n-1}$ if s $= \sum x_i$, zero otherwise

b. $f(x_1, \ldots, x_n; \theta) = \theta^{-2n}\exp(-\sum x_i/\theta)(\prod x_i)$; all $x_i > 0$

7 $S = \sum_{i=1}^{n} X_i$

9 **a.** $S = \sum_{i=1}^{n} X_i^2$ **b.** Only when $n = 1$

13 $S_1 = \prod_{i=1}^{n} X_i,\ S_2 = \prod_{i=1}^{n} (1 - X_i)$

15 $\hat{p} = n/s$. Same as usual MLE

17 **c.** $X_{1:n} - 1/n - \ln(1 - p)$

21 **a.** $[\sum X_i - (\sum X_i)^2/n]/(n - 1)$ **b.** $[(\sum X_i)^2 - X_1]/[n(n - 1)]$

23 **a.** $\bar{X} + 3(1.645)$ **b.** $\Phi([\sqrt{n}(c - \bar{X})]/[3\sqrt{n - 1}])$

25 **a.** $(-1/n)(\sum \ln X_i)$ **b.** $(1 - n)/(\sum \ln X_i)$

27 **a.** $S^2 = [\sum X_i^2 - (\sum X_i)^2/n]/(n - 1)$

b. $\bar{X} + 1.645cS$ with $c = \Gamma((n - 1)/2)/[\sqrt{2}\Gamma(n/2)]$

29 **a.** Use regular exponential class **b.** $\bar{X}$ **c.** $\prod_{i=1}^{n} X_i = \tilde{X}^n$

31 **a.** $n/\sum \ln(1 + X_i)$ **b.** $\sum \ln(1 + X_i)$ **c.** $1/(n\theta^2)$ **d.** $(1/n) \sum \ln(1 + X_i)$

e. $N(\theta, \theta^2/n)$ for $\tilde{\theta}$, $N(\theta^{-1}, \theta^{-2}/n)$ for $1/\tilde{\theta}$ **f.** $(n - 1)/\sum \ln(1 + X_i)$

33 **a.** $(1/r)\left[\sum_{i=1}^{r} X_{i:n} + (n - r)X_{r:n}\right]$ **b.** $\sum_{i=1}^{r} X_{i:n} + (n - r)X_{r:n}$

CHAPTER 11

1 **a.** (18.1, 20.5) **b.** lower: 18.3, upper: 20.3 **c.** $n \doteq 25$ for length = 2

d. (17.9, 20.7) **e.** (4.68, 33.39)

3 **a.** 14.40 **b.** $\exp(-t/14.40)$

5 **a.** $Q \sim EXP(1/n)$

b. $(x_{1:n} + (1/n)\ln(\alpha/2),\ x_{1:n} + (1/n)\ln(1 - \alpha/2))$ with $\alpha = 1 - \gamma$

c. (161.842, 161.997)

9 (0.81, 2.88)

11 (0.04, 0.21)

13 **a.** $(2n\bar{x}/\chi^2_{1-\alpha/2}(2n\kappa),\ 2n\bar{x}/\chi^2_{\alpha/2}(2n\kappa))$ **b.** (5.62, 15.94)

17 **a.** $p_L = 1 - \gamma^{1/x}$ **b.** 0.0209

21 (1.35, 2.10)

23 **a.** $(\chi^2_{\alpha/2}(2n + 2)/[2(n\bar{x} + \beta)],\ \chi^2_{1-\alpha/2}(2n + 2)/[2(n\bar{x} + \beta)])$ **c.** (0.119, 0.326)

29 **c.** $(\hat{\kappa} - z_{1-\alpha/2}\,\hat{\kappa}/\sqrt{n},\ \ \hat{\kappa} + z_{1-\alpha/2}\,\hat{\kappa}/\sqrt{n})$ **d.** (0.64, 2.73)

CHAPTER 12

1 **a.** $a = 19.59$, $b = 20.41$

b. For $A: \beta \doteq 1$ For $B: \beta = 0.0091$

c. For $A: \beta = 0.0091$ For $B: \beta \doteq 1$ **d.** $\alpha = 0.10$ **e.** $\beta \doteq 0.0091$

3 **a.** $z_0 = -2.236 > -2.326$ Can't reject H_0 **b.** $\beta = 0.1515$ **c.** $n = 24$

d. $t_0 = -1.12 > -2.539$ Can't reject H_0

e. $v_0 = 33.78 < 36.19$ Can't reject H_0 **f.** $n \doteq 50$

5 $n = 32$ from Table 8. Appendix C ($d = 0.8$, $1 - \beta = 0.95$)
$t_{0.995}(31) = 2.7454$ (by interpolation in Table 6)
$C = \{t_0 \mid t_0 \leqslant -2.7454 \text{ or } t_0 \geqslant 2.7454\}$

7 **a.** $z = -1.79 < -1.28$ Reject H_0

b. $B(6; 20, 0.5) = 0.0577 < 0.10$ Reject H_0

c. $\pi(0.2) = 0.9133$ **d.** p value $= 0.0577$

9 **a.** $-2 < -1.746$ Reject H_0

b. $-2 < -1.75$ Reject H_0

c. $-2 < -1.86$ Reject H_0

d. $0.29 < 0.8$ Can't reject H_0

11 **a.** Reject H_0 if $x \geqslant 0.95$ **b.** 0.0975

c. Reject H_0 if $\prod_{i=1}^{n} x_i \geqslant c$ where $P\left[\prod_{i=1}^{n} X_i \geqslant c \mid \theta = 1\right] = 0.05$

13 **a.** Reject H_0 if $x \geqslant 4$ **b.** power $= 0.5$

17 **a.** Reject H_0 if $\sum x_i^2/\sigma_0^2 \geqslant \chi^2_{1-\alpha}(n)$

b. $\pi(\sigma) = 1 - H[(\sigma_0^2/\sigma^2)\chi^2_{1-\alpha}(n); n]$ **c.** 0.968

19 **a.** Reject H_0 if $\bar{x} \leqslant \mu_0 + z_\alpha/\sqrt{n}$

b. Reject H_0 if $\bar{x} \geqslant \mu_0 + z_{1-\alpha}/\sqrt{n}$

21 Reject H_0 if $\sum x_i \geqslant c$ where $P[\sum X_i \geqslant c \mid \theta = \theta_0] = \alpha$

23 Reject H_0 if $t = \sum x_i \geqslant k_0$ where k_0 is the smallest k such that
$B(k - 1; n, p_0) \geqslant 1 - \alpha$

27 **a.** Reject H_0 if $-2n[\ln(\bar{x}/\theta_0) - \bar{x}/\theta_0 + 1] \geqslant \chi^2_{1-\alpha}(1)$

b. Reject H_0 if $2n\bar{x}/\theta_0 \geqslant \chi^2_{1-\alpha}(2n)$

29 Reject H_0 if $x_{n:n} \leqslant \theta_0 \alpha^{1/n}$

31 Reject H_0 if $-2n[\ln(\theta_0/\hat{\theta}) + (\theta_0/\hat{\theta} - 1)] \geqslant \chi^2_{1-\alpha}(1)$

37 **a.** $k_0^* = 0.1053$, $k_1^* = 18$

c. Reject H_0 if $\sum x_i^2 \geqslant 2.47n + 5.07$ Accept H_0 if $\sum x_i^2 \leqslant 2.47n - 6.50$

d. $E(N \mid \sigma = 1) \doteq 4$, $E(N \mid \sigma = 3) \doteq 1$ **e.** Yes. Rejects with $n = 4$

39 **a.** Accepts H_0 with $n = 14$ **b.** 12 **c.** 6

CHAPTER 13

1 $\chi^2 = 4.76$ (without continuity correction)
$\chi^2 = 4.30$ (with continuity correction)
$4.30 > 2.71$ Reject H_0
For a one-sided test use binomial test

3 **a.** $2.67 < 6.25$ Can't reject H_0

b. $20 > 6.25$ Reject H_0

c. $18.18 > 4.61$ Reject H_0

5 **a.** $5 > 4.61$ Reject H_0

b. $0.78 < 4.61$ Can't reject H_0

7 $0.090 < 5.99$ Can't reject H_0

9 $1.54 < 7.78$ Can't reject H_0

11 $17.93 > 9.49$ Reject H_0

13 **a.** $16.504 > 13.28$ Reject H_0

b. $0.24 < 9.24$ Can't reject H_0

15 **a.** $\hat{\mu} = 96.70$ $235.36 > 15.51$ Reject H_0

b. $\hat{\mu} = 35.65$ $4.42 < 6.26$ Can't reject H_0

c. $\hat{\mu} = 96.70$ $\hat{\sigma} = 37.46$ $10.31 < 12.02$ Can't reject H_0

17 $18.05 > 9.49$

21 **b.** $\text{CM} = 0.058$ $(1 + 0.2/\sqrt{23})(0.058) = 0.060 < 0.102$ Can't reject H_0

c. $D = 0.151$ $\sqrt{23}(0.151) = 0.724 < 0.803$ Can't reject H_0

CHAPTER 14

1 **a.** $t = 10$; $B(10; 20, 0.5) = 0.588 > 0.05$ Can't reject H_0

b. $t = 2$; $B(2; 30, 0.25) = 0.091 < 0.10$ Reject H_0

3 Based on large-sample binomial test, $z = 0.85 < 1.96$ Can't reject H_0

5 Based on large-sample binomial test, $\Phi(-5.14) < 0.01$ Reject H_0

7 **a.** $t = 22$; $B(22; 60, 0.5) \doteq \Phi(-1.94) = 0.0262 < 0.05$ Reject H_0

b. $(5.22, 5.25)$ $(i = 24, j = 37)$ **c.** 5.13 $(k = 10)$

9 **a.** 35.39 $(k = 11)$

b. $(22.40, 112.26)$ based on $i = 9$ and $j = 30$; other intervals are possible provided $j - i = 21$

11 **a.** $1.349 < 1.383$ Can't reject H_0

b. $t = 2$; $B(2; 10, 0.5) = 0.055 < 0.10$ Reject H_0

13 **a.** $t = 4$; $B(4; 12, 0.5) = 0.1938 > 0.10$ Can't reject H_0

b. $t = 10 < 21$ Reject H_0

Normal approx. $z = -2.27 < -1.28$ Reject H_0

15 **a.** $U_x = 5 < 27$ Reject H_0

Normal approx. $U_{0.05} \doteq 28.2$

b. t = sum of ranks of positive differences = 3

Two-sided test would reject at $\alpha = 0.01$ $(\alpha/2 = 0.005)$

17 **a.** $r = 0.165$; $t = 0.472 < 1.860$ ($\alpha/2 = 0.05$)
Can't reject H_0 : independence
b. $R_s = 0.215$; p value between 0.280 and 0.268;
$t = 0.623 < 1.860$ Can't reject H_0 : independence

21 $R_s = -0.259$; $t = 1.229 > -1.383$
Suggests moderate decreasing trend, but can't reject at $\alpha = 0.10$

23 $t = 4$ runs; $P(T \leqslant 4 \mid H_0 \text{ true}) = 0.001$
p-value $\leqslant 0.002$ (for two-sided test)

CHAPTER 15

1 **b.** $\hat{\beta}_0 = 1.182, \hat{\beta}_1 = -1.315$ **c.** 0.76 **d.** SSE = 0.0184, $\tilde{\sigma}^2 = 0.0023$

2 **b.** $\hat{\beta}_0 = 0.628, \hat{\beta}_1 = 0.608$ **c.** $\hat{y} = 3.06$ **e.** $\tilde{\sigma}^2 = 0.194$

3 **b.** $\hat{\beta}_0 = 1.182, \hat{\beta}_c = 0.755$

9 **a.** $\hat{\mu} = 93.90, \hat{\sigma} = 37.13$ **b.** $\hat{\eta} = 13.14, \hat{\theta} = 24.73$

10 $\hat{\mu} = 93.96, \hat{\sigma} = 36.77$

18 **a.** $\hat{\beta} = 0.00245, \hat{\sigma}^2 = 0.00024$ **b.** $\tilde{\sigma}^2 = 0.00028$
c. (0.0020, 0.0029) **d.** (0.00013, 0.00123)

20 $\hat{\beta} = 0.00296$

31 $f = 22.17 > 12.4$, Reject H_0

32 $r = 0.672$

33 $t = 3.391 > 1.761$, Reject H_0

35 **a.** $f = 18.97 > 3.75$ Reject H_0 **b.** $f = 0.252 < 4.60$ Can't reject H_0

CHAPTER 16

1 **c.** $(1/10)^{25}$ **d.** 22.5 **e.** 23 (rounded up) **f.** 47 (by normal approx.)

3 $\hat{\theta} = (n/r)x_{r:n}$

5 No (convergent integral)

7 **a.** 600 hours **b.** 0.916 **c.** $n = 7$

9 **a.** $100/6 = 16.67$ **b.** 0.5488 **c.** 1139 hours **d.** 6/100

11 0.5714

13 **a.** $\chi^2(38)$ **b.** 688.63 **c.** 142.52

17 **a.** 1.177 **b.** $133.33 < 140.85$ Reject H_0
c. (0.982, 1.420) **d.** 0.346 **e.** 0.05 year or 2.6 weeks

19 **a.** $\tilde{\delta} = 0.482, \tilde{\beta} = 2.07$ **b.** $\tilde{\xi} = 4.42, \tilde{\theta} = 83.1$

23 **a.** 0.4335 **b.** 0.6288 **c.** $k = 6$; $P[X \leqslant 6] = 0.8893$ **d.** 4, 32
e. $Y = \min(X, 3)$; $E(Y) = 2.652$
f. $W = X_1 + X_2 \sim \text{POI}(8)$; $Z = \min(W, 6)$; $E(Z) = 5.650$

25 **a.** $\bar{x} \doteq [1(211) + 2(93) + 3(35) + 4(7)]/575 = 0.92$ (neglecting $y \geqslant 5$)
Another approach: If $p = P[Y = 0] = e^{-\mu}$, then $\hat{p} = 229/576 = 0.398$ and $\hat{\mu} = -\ln(0.398) = 0.92$, which is the MLE based on binomial data for p.
b.

m_y:	229	211	93	35	7
$\hat{e}_y$:	229.5	211.2	97.1	29.8	6.9

c. 0.602

27 **a.** 2.5 **b.** 0.958 **c.** 461

29 **a.** 0.9887 **b.** 0.9834 **c.** 0.9955

31 0.6321, 0.0803, 3/2

35 **a.** $\hat{\gamma} = 3.583/2 = 1.7925$

b. $f_X(x) = \binom{x+1}{x}\gamma^x/(1 + \gamma)^{2+x}$; $x = 0, 1, \ldots$ **c.** 0.6959

d. $2\hat{\gamma} = 3.583$, $2\hat{\gamma}(1 + \hat{\gamma}) = 10.011$

REFERENCES

Aho, M.; Bain, L.J.; and Englehardt, M. 1983. "Goodness-of-Fit Tests for the Weibull Distribution with Unknown Parameters and Censored Sampling." *J. Statist. Comput. Simul.* **18**, p. 59.

Antle, C.E., and Bain, L.J. 1969. "A Property of Maximum Likelihood Estimators of Location and Scale Parameters." *Siam Review.* **11**, p. 251.

Bain, L.J., and Engelhardt, M. 1991. *Statistical Analysis of Reliability and Life-Testing Models.* 2nd ed. New York: Marcel Dekker.

Bain, L.J., and Engelhardt, M. 1983. "A Review of Model Selection Procedures Relevant to the Weibull Distribution." *Commun. Statist.-Theor. Meth.* **12**(5), p. 589.

Bain, L.J., and Engelhardt, M. 1986. "Approximate Distributional Results Based on the Maximum Likelihood Estimators for the Weibull Distribution." *J. Qual. Tech.* p. 18.

Barr, D.R., and Zehna, P.W. 1983. *Probability: Modeling Uncertainty.* Reading, Mass.: Addison-Wesley.

Basu, D. 1955. "On Statistics Independent of a Complete Sufficient Statistic." *Sankya.* **15**, p. 377.

Bickel, P.J., and Doksum, K.A. 1977. *Mathematical Statistics: Basic Ideas and Selected Topics.* San Francisco: Holden-Day.

Blackwell, D. 1947. "Conditional Expectation and Unbiased Sequential Estimation." *Ann. Math. Statist.* **18**, p. 105.

Bradley, E.L., and Blackwood, L.G. 1989. "Comparing Paired Data: A Simultaneous Test for Means and Variances." *Amer. Statist.* **43**, p. 234.

Brownlee, K.A. 1960. *Statistical Theory and Methodology in Science and Engineering.* New York: John Wiley & Sons.

Chandra, M.; Singpurwalla, N.D.; and Stephens, M.A. 1981. "Kolmogorov Statistics for Tests of Fit for the Extreme-Value and Weibull Distributions." *J. Amer. Statist. Assoc.* **76**, p. 729.

Clark, R.D. 1946. "An Application of the Poisson Distribution." *J. Instit. Actuar.* **72**.

Cramér, H. 1946. *Mathematical Methods of Statistics.* Princeton, N.J.: Princeton University Press.

Davis, D.J. 1952. "An Analysis of Some Failure Data." *J. Amer. Statist. Assoc.* **47**, p. 113.

DeGroot, M.H. 1970. *Optimal Statistical Decisions.* New York: McGraw-Hill.

Dixon, W.J., and Massey, F.J. 1957. *Introduction to Statistical Analysis.* New York: McGraw-Hill.

Draper, N., and Smith, H. 1981. *Applied Regression Analysis* (2nd ed.). New York: John Wiley & Sons.

Dufour, R., and Maag, U.R. 1978. "Distribution Results for Modified Kolmogorov-Smirnov Statistics for Truncated or Censored Samples." *Technometrics.* **20**, p. 29.

Eastman, J., and Bain, L.J. 1973. "A Property of Maximum Likelihood Estimators in the Presence of Location-Scale Nuisance Parameters." *Commun. Statist.* **2**(1), p. 23.

Engelhardt, M., and Bain, L.J. 1976. "Tolerance Limits and Confidence Limits on Reliability for the Two-Parameter Exponential Distribution." *Technometrics.* **18**, p. 37.

Engelhardt, M., and Bain, L.J. 1977a. "Uniformly Most Powerful Tests on the Scale Parameter of a Gamma Distribution with a Nuisance Shape Parameter." *Technometrics.* **19**, p. 77.

Engelhardt, M., and Bain, L.J. 1977b. "Simplified Statistical Procedures for the Weibull or Extreme-Value Distribution." *Technometrics.* **19**, p. 323.

Ferguson, T.S. 1967. *Mathematical Statistics: a Decision Theoretic Approach.* New York: Academic Press.

Fisher, R.A., and Tippett, L.M.C. 1928. "Limiting Forms of the Frequency Distribution of the Largest or Smallest Member of a Sample." *Proceedings of the Cambridge Philosophical Society.* **24**, p. 180.

Ghosh, B.K. 1970. *Sequential Tests of Statistical Hypotheses.* Reading, Mass.: Addison-Wesley.

Greenwood, J.A., and Durand, D. 1960. "Aids for Fitting the Gamma Distribution by Maximum Likelihood." *Technometrics*, **2**, p. 55.

Greenwood, M., and Yule, G.U. 1920. "An Inquiry into the Nature of Frequency Distributions Representative of Multiple Happenings." *J. Royal Statist. Soc.* **83**, p. 255.

Gross, A.J., and Clark, V.A. 1975. *Survival Distributions: Reliability Applications in the Biomedical Sciences.* New York: John Wiley & Sons.

Grubbs, F.E. 1971. "Fiducial Bounds on Reliability for the Two-Parameter Negative Exponential Distribution." *Technometrics.* **13**, p. 873.

Harter, H.L. 1964. *New Tables of the Incomplete Gamma Function Ratio and of Percentage Points of the Chi-Square and Beta Distributions.* Washington, D.C.: U.S. Government Printing Office.

Harter, H.L. 1969. *Order Statistics and Their Use in Testing and Estimation. Vol. 2: Estimates Based on Order Statistics from Various Populations.* Washington, D.C.: U.S. Government Printing Office.

Hogg, R.V., and Craig, A.T. 1978. *Introduction to Mathematical Statistics*, 4th ed. New York: Macmillan.

Johns, M.V., and Lieberman, G.J. 1966. "An Exact Asymptotically Efficient Confidence Bound for Reliability in the Case of the Weibull Distribution." *Technometrics.* **8**, p. 135.

Jones, R.A.; Scholz, F.W.; Ossiander, M.; and Shorack, G.R. 1985. "Tolerance Bounds for Log Gamma Regression Models." *Technometrics.* **27**, p. 109.

Kendall, M.G., and Stuart, A. 1967. *The Advanced Theory of Statistics*, 2nd ed., Vol. 2. New York: Hafner.

Kitagawa, T. 1952. *Tables of Poisson Distribution.* Tokyo: Baifukan.

Koziol, J.A. 1980. "Percentage Points of the Asymptotic Distributions of One and Two Sample Kuiper Statistics for Truncated or Censored Data." *Technometrics.* **22**, p. 437.

Lehmann, E.L. 1959. *Testing Statistical Hypotheses.* New York: John Wiley & Sons.

Lehmann, E.L., and Scheffe, H. 1955. "Completeness, Similar Regions and Unbiased Estimates." *Sankya.* **10**, p. 305.

Lieberman, G.J., and Owen, D.B. 1961. *Tables of the Hypergeometric Distribution.* Stanford, Ca.: Stanford University Press.

Lieblein, J., and Zelen, M. 1956. "Statistical Investigation of the Fatigue Life of Deep-Groove Ball Bearings." *J. Res. Nat. Bur. Stand.* **47**, p. 273.

Mann, H.B., and Whitney, D.R. 1947. "On a Test Whether One of Two Random Variables is Stochastically Larger than the Other." *Ann. Math. Statist.* **18**, p. 50.

Matis, J.H., and Wehrly, T.E. 1979. "Stochastic Models of Compartmental Systems." *Biometrics.* **35**, p. 199.

McDonald, G.C., and Studden, W.J. 1990. "Design Aspects of Regression-Based Ratio Estimation." *Technometrics.* **32**, p. 417.

Molina, E.C. 1942. *Poisson's Exponential Binomial Limit.* New York: D. Van Nostrand.

Mood, A.M.; Graybill, F.A., and Boes, D.C. 1974. *Introduction to the Theory of Statistics.* New York: McGraw-Hill.

Mullet, G.M. 1977. "Simeon Poisson and the National Hockey League." *The American Statistician.* **31**, p. 8.

National Bureau of Standards. 1950. *Tables of the Binomial Probability Distribution, Applied Mathematics, Series 6.* Washington, D.C.: U.S. Printing Office.

Parzen, E. 1962. *Stochastic Processes.* San Francisco: Holden-Day.

Patel, J.K., and Read, C.B. 1982. *Handbook of the Normal Distribution.* New York: Marcel Dekker.

Pearson, E.S., and Hartley, H.O. 1958. *Biometrika Tables for Statisticians,* Vol. 1, 2nd ed. London: Cambridge University Press.

Pearson, E.S., and Hartley, H.O. 1966. *Biometrika Tables for Statisticians,* Vol. 1, 3rd ed. London: Cambridge University Press.

Pearson, K. (Ed.) 1934. *Tables of the Incomplete Beta Function.* London: The Biometrika Office, University College.

Pearson, K. 1951. *Tables of the Incomplete Gamma Function.* Cambridge: Cambridge University Press.

Pettitt, A.N., and Stephans, M.A. 1976. "Modified Cramer-Von Mises Statistics for Censored Data." *Biometrika.* **63**, p. 291.

Proschan, F. 1963. "Theoretical Explanation of Observed Decreasing Failure Rate." *Technometrics.* **5**, p. 375.

Rao, C.R. 1949. "Sufficient Statistics and Minimum Variance Unbiased Estimates." *Proc. Camb. Phil. Soc.* **45**, p. 213.

Romig, H.G. 1953. *50–100 Binomial Tables.* New York: John Wiley & Sons.

Rutherford, E., and Geiger, H. 1910. "The Probability Variations in the Distribution of a Particle." *Philosophical Magazine.* **20**, p. 698.

Scheffé, H. 1959. *The Analysis of Variance.* New York: John Wiley & Sons.

Snedecor, G.W. 1959. *Statistical Methods.* Ames, Iowa: Iowa State University Press.

Stephens, M.A. 1974. "EDF Statistics for Goodness-of-Fit and Some Comparisons." *J. Amer. Statist. Assoc.* **69**, p. 730.

Stephens, M.A. 1977. "Goodness-of-Fit for the Extreme-Value Distribution." *Biometrika.* **64**, p. 583.

Thoman, D.R.; Bain, L.J.; and Antle, C.E. 1969. "Inferences on the Parameters of the Weibull Distribution." *Technometrics.* **11**, p. 445.

Trumpler, R.J., and Weaver, H.F. 1953. *Statistical Astronomy.* Berkeley: University of California Press.

Wald, A., and Wolfowitz, J. 1940. "On a Test Whether Two Samples are From the Sample Population." *Ann. Math. Statist.* **11**, p. 147.

Walpole, R.E., and Myers, R.H. 1985. *Probability and Statistics for Engineers and Scientists.* New York: Macmillan.

Wang, Y. 1971. "Probabilities of the Type I errors of the Welch tests . . . ." *J. Amer. Statist. Assoc.* **66**, p. 605.

Wasan, M.T. 1970. *Parametric Estimation.* New York: McGraw-Hill.

Weibull, W. 1939. "A Statistical Theory of the Strength of Materials." *Ing. Velenskaps Akad. Handl.* **151**, p. 1.

Welch, B. 1949. "Further Notes on Mrs. Aspin's Tables." *Biometrika.* **36**, p. 243.

Wilcoxon, F. 1945. "Individual Comparisons by Ranking Methods." *Biometrics.* **1**, p. 80.

Wilcoxon, F. 1947. "Probability Tables for Individual Comparisons by Ranking Methods." *Biometrics.* **3**, p. 119.

Williamson, E., and Bretherton, M.K. 1963. *Tables of the Negative Binomial Probability Distribution.* New York: John Wiley & Sons.

Zellner, A. 1971. *An Introduction to Bayesian Inference in Econometrics.* New York: John Wiley & Sons.

INDEX

Special Continuous Distributions

Notation and Parameters	Continuous pdf $f(x)$	Mean	Variance	MGF $M_X(t)$
Uniform				
$X \sim \text{UNIF}(a, b)$	$\frac{1}{b-a}$	$\frac{a+b}{2}$	$\frac{(b-a)^2}{12}$	$\frac{e^{bt} - e^{at}}{(b-a)t}$
$a < b$	$a < x < b$			
Normal				
$X \sim \text{N}(\mu, \sigma^2)$	$\frac{1}{\sqrt{2\pi}\,\sigma} e^{-[(x-\mu)/\sigma]^2/2}$	μ	σ^2	$e^{\mu t + \sigma^2 t^2/2}$
$0 < \sigma^2$				
Gamma				
$X \sim \text{GAM}(\theta, \kappa)$	$\frac{1}{\theta^\kappa \Gamma(\kappa)} x^{\kappa - 1} e^{-x/\theta}$	$\kappa\theta$	$\kappa\theta^2$	$\left(\frac{1}{1-\theta t}\right)^\kappa$
$0 < \theta$ $0 < \kappa$	$0 < x$			
Exponential				
$X \sim \text{EXP}(\theta)$	$\frac{1}{\theta} e^{-x/\theta}$	θ	θ^2	$\frac{1}{1-\theta t}$
$0 < \theta$	$0 < x$			
Two-Parameter Exponential				
$X \sim \text{EXP}(\theta, \eta)$	$\frac{1}{\theta} e^{-(x-\eta)/\theta}$	$\eta + \theta$	θ^2	$\frac{e^{\eta t}}{1-\theta t}$
$0 < \theta$	$\eta < x$			
Double Exponential				
$X \sim \text{DE}(\theta, \eta)$	$\frac{1}{2\theta} e^{-\lvert x-\eta \rvert/\theta}$	η	$2\theta^2$	$\frac{e^{\eta t}}{1-\theta^2 t^2}$
$0 < \theta$				

Special Discrete Distributions

Notation and Parameters	Discrete pdf $f(x)$	Mean	Variance	MGF $M_X(t)$
Binomial				
$X \sim \text{BIN}(n, p)$ $0 < p < 1$ $q = 1 - p$	$\binom{n}{x} p^x q^{n-x}$ $x = 0, 1, \ldots, n$	np	npq	$(pe^t + q)^n$
Bernoulli				
$X \sim \text{BIN}(1, p)$ $0 < p < 1$ $q = 1 - p$	$p^x q^{1-x}$ $x = 0, 1$	p	pq	$pe^t + q$
Negative Binomial				
$X \sim \text{NB}(r, p)$ $0 < p < 1$ $r = 1, 2, \ldots$	$\binom{x-1}{r-1} p^r q^{x-r}$ $x = r, r+1, \ldots$	r/p	rq/p^2	$\left(\frac{pe^t}{1 - qe^t}\right)^r$
Geometric				
$X \sim \text{GEO}(p)$ $0 < p < 1$ $q = 1 - p$	pq^{x-1} $x = 1, 2, \ldots$	$1/p$	q/p^2	$\frac{pe^t}{1 - qe^t}$
Hypergeometric				
$X \sim \text{HYP}(n, M, N)$ $n = 1, 2, \ldots, N$ $M = 0, 1, \ldots, N$	$\binom{M}{x}\binom{N-M}{n-x} \Big/ \binom{N}{n}$ $x = 0, 1, \ldots, n$	nM/N	$n \frac{M}{N}\left(1 - \frac{M}{N}\right)\frac{N-n}{N-1}$	*
Poisson				
$X \sim \text{POI}(\mu)$ $0 < \mu$	$\frac{e^{-\mu}\mu^x}{x!}$ $x = 0, 1, \ldots$	μ	μ	$e^{\mu(e^t - 1)}$
Discrete Uniform				
$X \sim \text{DU}(N)$ $N = 1, 2, \ldots$	$1/N$ $x = 1, 2, \ldots, N$	$\frac{N+1}{2}$	$\frac{N^2 - 1}{12}$	$\frac{1}{N}\,\frac{e^t - e^{(N+1)t}}{1 - e^t}$

*Not tractable.